SECOND INTERNA

ADVANCES IN UNDERGROUND PIPELINE ENGINEERING

Sponsored by the
ASCE Pipeline Division
ASCE Seattle Section
ASCE Technical Council on Lifeline Earthquake Engineering

Endorsed by the
North American Society for Trenchless Technology
American Concrete Pressure Pipe Association
National Clay Pipe Institute
Ductile Iron Pipe Research Association
Steel Plate Fabricators Association

Bellevue, Washington
June 25-28, 1995

Edited by Jey K. Jeyapalan and Menaka Jeyapalan

Published by the
American Society of Civil Engineers
345 East 47th Street
New York, New York 10017-2398

ABSTRACT

The first International Conference on Advances in Underground Pipeline Engineering was held in 1985. Some 50 authors and over 300 delegates from 20 countries took part in the presentations and discussions on what the state-ofthe-art of pipeline technology was at that time. A decade later, the second international conference reviews the progress we have made, if any, on materials, design, corrosion control, coatings, linings, seismic design, hydraulics, special design considerations, trenchless methods, and rehabilitation of aging pipelines and manholes, and field performance. Two plenary sessions discussed progress around the globe in many areas and on adequacy of consensus standards. The papers were presented by authors from U.S.A., Canada, U.K., Germany, Sweden, Saudi Arabia, Japan, France, and Australia. They covered such topics as: 1) Special design considerations for pipes made of a variety of materials such as plastic, concrete, and steel; 2) trenchless technology for pipeline rehabilitation; and 3) protective coatings and linings. Shifts in types of pipe materials used in different applications over the past decade and projection of future technological needs in this field are given throughout the proceedings.

Library of Congress Cataloging-in-Publication Data

Advances in underground pipeline engineering: proceedings of the second international conference, Bellevue, Washington, June 25-28, 1995 / sponsored by the ASCE Pipeline Division, ASCE Seattle Section, ASCE Technical Council on Lifeline Earthquake Engineering; endorsed by the North American society for Trenchless Technology...[et al]; edited by Jey K. Jeyapalan and Menaka Jeyapalan.
    p.   cm.
    Includes index.
    ISBN 0-7844-0093-8
    1. Underground pipelines—Congresses. I. Jeyapalan, Jey K. II. Jeyapalan, Menaka. III. American Society of Civil Engineers. Pipeline Division. IV. American Society of Civil Engineers. Seattle Section. V. American Society of Civil Engineers. Technical Council on Lifeline Earthquake Engineering.
TJ930.A32 1995   95-19043
621.8'672—dc20   CIP

The Society is not responsible for any statements made or opinions expressed in its publications.

Photocopies. Authorization to photocopy material for internal or personal use under circumstances not falling within the fair use provisions of the Copyright Act is granted by ASCE to libraries and other users registered with the Copyright Clearance Center (CCC) Transactional Reporting Service, provided that the base fee of $2.00 per article plus $.25 per page copied is paid directly to CCC, 222 Rosewood, Drive, Danvers, MA 01923. The identification for ASCE Books is 0-7844-0093-8/95 $2.00 + $.25. Requests for special permission or bulk copying should be addressed to Permissions & Copyright Dept., ASCE.

Copyright © 1995 by the American Society of Civil Engineers,
All Rights Reserved.
Library of Congress Catalog Card No: 95-19043
ISBN 0-7844-0093-8
Manufactured in the United States of America.

FOREWORD

The Conference Committee is pleased to present these proceedings on behalf of the Pipeline Division, the Seattle Section, and the Technical Council on Lifeline Earthquake Engineering of the American Society of Civil Engineers for the Second International Conference on Advances in Underground Pipeline Engineering, 1995.

The field of underground pipeline engineering has come a long way in some areas since our last conference on this subject in 1985. A decade has past and it's time to bring together researchers, educators, practitioners, manufacturers, and those from the construction industry to discuss the advances which have taken place in this field. The papers in this program discuss up and coming technology in areas such as trenchless technology, design considerations, methods for structural analysis, plastic pipe, welded steel pipe, ductile iron pipe, trenchless clay pipe, concrete pressure pipe, and rehabilitation, to name a few.

Each of the papers included in the proceedings has received two positive peer reviews. All papers are eligible for discussion in the ASCE Journal of Transportation Engineering and are also eligible for the ASCE Awards.

The international participation at this conference compliments the industry. The information you hold in your hands is great and new, so please read and be both interested and enlightened.

On behalf of the committee,

Jey K. Jeyapalan, Ph.D., P.E.
Co-Editor

Menaka Jeyapalan
Co-Editor

ORGANIZING COMMITTEE

Dr. Jey K. Jeyapalan, P.E., Chair
Mr. Joseph Castronovo, P.E. Vice Chair
Mr. James Clark, P.E., Vice Chair
Mr. William Quinn, P.E., Oil and Gas Pipe
Mr. Mark Pickell, P.E., Public Agencies
Mr. Richard Mueller, P.E., Concrete Pipe
Mr. Richard Bonds, P.E., Ductile Iron Pipe
Mr. Bruce Vanderploeg, P.E., Welded Steel Pipe
Mr. Edward Sikora, Vitrified Clay Pipe
Mr. Lynn Osborne, P.E., Pipeline Rehabilitation
Mr. Bob Meinzer, P.E., Exhibits
Mr. Larry Petroff, P.E., HDPE Pipe
Mr. Ron Bishop, P.E., PVC Pipe
Mr. William Shook, P.E., Manhole Rehabilitation
Professor A.P.S. Selvadurai, P.Eng., Canada
Professor Dietrich Stein, Germany
Professor Lars-Eric Jansen, Sweden
Professor Marcel Gerbault, France
Mr. Bala K. Balasubramaniam, C. Eng, Australasia
Mr. Abdullah Al-Shaikh, Saudi Arabia
Professor Jun Tohda, Japan
Dr. C.D.F. Rogers, United Kingdom
Dr. James B. Thompson, ASCE Seattle Section
Mr. Don Ballentine, ASCE Technical Council on Lifeline Earthquake Eng.
Ms. Shiela Menaker, ASCE Book Production
Ms. Andi Simon, ASCE Conference Liasion
Ms. Menaka Jeyapalan, Conference Secretary

ACKNOWLEDGEMENTS

In order to include the best papers in this conference, the help of others was required in reviewing the papers sent to us. Members of the steering committee, authors who submitted manuscripts, and respected senior members in the pipeline industry helped in this task. We would like to thank the following people:

Anderson, Kenneth	Anderson Seal Company
Anspach, James	So-Deep
Argent, Mike	Permalok Corporation
Ballantyne, Donald	Dames & Moore
Bardakjian, Henry	Ameron
Barsoom, Joseph	City & County of Denver
Beieler, Roger	CH2M Hill
Bennett, Bruce	Stowe Engineering
Bonds, Richard	Ductile Iron Pipe Research Assoc.
Bradish, Bryan	Newport News Waterworks
Brunner, Phyllis	Woodward-Clyde Consultants
Butler, Thomas	Howard County
Card, Robert	Thompson Steel Pipe
Carpenter, Ralph	American Cast Iron Pipe
Castronovo, Joseph	CH2M Hill
Clark, James	Engineering Consultant
Conner, Randall	American Cast Iron Pipe
Craft, Gerald	U.S. Pipe & Foundry
Cronin, Roger	Greeley & Hansen
Damak, Samir	Corrosion Consultant
Dechant, Dennis	Northwest Pipe & Casing Comp.
Dodge, Christopher	John Carollo Engineers
Donnelly, Gerald	STV Group
Felio, Guy	NRC Canada
Fowles, Deon	Ductile Iron Pipe Research Assoc.
Fuerst, Richard	U.S. Bureau of Reclamation
Galeziewski, Tom	HDR Engineering
Gamble, William	University of Illinois
Garcia, Jorge	City of Las Cruces
Gemperline, Mark	U.S. Bureau of Reclamation
Ghaboussi, Jamshid	University of Illinois
Gill, Mohammad	City of Detroit
Gnanapragasam, N.	Seattle University
Gnanapragasam, E.	Argon National Laboratory
Granning, Eric	Northwest Pipe & Casing Comp.
Guastella, David	Tucker, Young, Jackson, Tull, Inc.
Habibian, Ahmad	American Society of Civil Engineers
Hall, Sylvia	Ameron
Hartinger, Christina	John Carollo Engineers

Hausmann, Del	Engineering Consultant
Hermanson, Glenn	Montgomery Watson
Hook, David	Santa Clara Valley
Horn, L. Gregg	Ductile Iron Pipe Research Assoc.
Horton, A.M.	U.S. Pipe & Foundry Co.
Howard, Amster	Engineering Consultant
Jaramillo, Peggy	Federal Energy Regulatory Commission
Jurgens, John	Gelco, Inc.
Kaneshiro, Jon	Parsons Engineering Science
Klein, Steve	Woodward Clyde Consultants
Kramer, Steve	Jason Consultants
Kuraoka, Senro	NRC Canada
Little, Bryan	Hancor, Inc.
Martinez, Hector	Sweetwater Authority
Mueller, Richard	Gifford Hill American
Nafaji, Mohammad	Missouri Western State College
Noonan, James	Anticorrosion Consultant
Osborn, Lynn	Insituform Technologies
Paul, Stanley	University of Illinois
Peabody, Michael	U.S. Bureau of Reclamation
Petroff, Larry	Plexco-Spirolite
Pickell, Mark	City of Tulsa
Poiriez, James	Killam Assoc.
Price, Ted	Sonex Ltd.
Prosser, David	American Concrete Pressure Pipe Assoc.
Rajani, Balvant	NRC Canada
Randolph, Randy	U.S. Bureau of Reclamation
Read, Don	U.S. Bureau of Reclamation
Rhodes, George	U.S. Pipe & Foundry
Robinson, William	Intermountain Corrosion Service Inc.
Saleira, Wesley	Monenco-Agra Inc.
Schiff, Mel	Engineering Consultant
Schrock, B. Jay	JSC International Engineering
Selvadurai, A.P.S.	McGill University
Sikora, Edward	National Clay Pipe Institute
Smith, Mark	WGP Engineering
Smith, David	West Consultants, Inc.
Spruch, Arthur	Sea Consultants
Stephens, Malcolm	Engineering Consultant
Struzziery, John	Sea Consultants
Taylor, Craig	John L. Wallace & Assoc.
Teal, Martin	West Consultants, Inc.
Thiyagaram, Mike	TAMS Consultants
Thompson, D.B.	Texas Tech University
Timmermann, David	Black & Veatch
Tucker, Michael	Ductile Iron Pipe Research Assoc.
Tupac, George	G.J. Tupac & Assoc.
Turnipseed, Steven	Chevron Research & Technology
Turpin, James	West Consultants, Inc.

Vanderploeg, Bruce	Northwest Pipe & Casing Comp.
Walsh, Terry	Greeley & Hansen
Watkins, Reynold	Utah State University
Zhan, Caizhao	NRC Canada
Zoumaras, Dave	City of San Diego

CONTENTS

PLENARY SESSION I
Chairman: Jey K. Jeyapalan

Diagnosis and Assessment of Damaged Sewers Concerning Their Structural Capacity
Dietrich Stein 1
Behavior of Buried Large Thin Wall Flexible Pipe—Field Test and Numerical Analysis Considered With Stage of Construction of Buried Flexible Pipe
T. Kawabata and Y. Mohri 13

PLENARY SESSION II
Chairman: James Clark

Underground Pipeline Materials, Design, and Construction: What did we learn during 1985–95? Where do we go from here?
Jey K. Jeyapalan, Wesley S. Saleira, Abdullah Al-Shaikh, Bala K. Balasubramaniam, and John Jurgens 25
A Soil-Structure Interactive Model
Marcel Gerbault 42

Session 1A
SPECIAL DESIGN CONSIDERATIONS I
Chairman: Phyllis Brunner

Supporting Spacing of Buried and Above-Ground Piping for Stress and Vibration
Phil Kormann and Joe Zhou 54
Avoiding Pipeline Failures at Waterway Crossings—Part 1: Scour Assessment Methods
Brian Doeing, Jeffrey Bradley, and Samuel Carreon 65
Avoiding Pipeline Failures at Waterway Crossings—Part 2: Scour Countermeasures
Martin Teal, Jeffrey Bradley, and Samuel Carreon 77
Criterion for Predicting Failure of Corroded Linepipe
Bin Fu and Mike Kirkwood 89

Session 2A
PLASTIC PIPE I
Chairman: Bruce Bennett

Earth Pressure on Pipelines in Centrifuged Models
L. Li and Jun Tohda 102

CONTENTS

Large-scale Model Test of Leachate Pipes in Landfills Under Heavy Load
Helmut Zanzinger and Erwin Gartung 114
Inspecting Buried Plastic Pipe Using a Rotating Sonic Caliper
Ted Price 126
Field Performance of PVC Water Mains Buried in Different Backfills
Balvant Rajani and Senro Kuraoka 138

Session 3A
CONCRETE PIPE
Chairman: **Joseph Castronovo**

Improved Joint System for Concrete Pipe
Kenneth Anderson 150
When Should PCCP With Interpace Class IV Wire Be Replaced?
Bryan Bradish, Roger Cronin, and Richard Lewis 156
Cathodic Protection Requirements of Prestressed Concrete Cylinder Pipe
Sylvia Hall and Ivan Mathew 168
Cathodic Protection Retrofit of a 35-Year-Old Concrete Pressure Pipeline
Stephen Turnipseed and Richard Mueller 183

Session 1B
SPECIAL DESIGN CONSIDERATIONS II
Chairman: **Deon Fowles**

Impact Factors for Estimating Vehicle Live Loads
Mark Gemperline and Thomas Siller 194
Earthquake Loss Estimation Techniques for Pipelines
Donald Ballantyne 205
Advances in Transmission Pipelines for Desalted Water in Saudi Arabia
Abdullah Al-Shaikh, Mohammed Al-Amry, and Saleh Al-Najdi 217
Lanslides, Liquefaction, and Artesians—Pipeline Crossing of the Snoqualimie River Valley
Roger Beieler, Karen Dawson, and Stephanie Murphy 226

Session 2B
PLASTIC PIPE II
Chairman: **Abdullah Al-Shaikh**

Stress Relaxation in Constantly Deflected PE Pipes
Lars-Eric Janson 238
Buried Plastic Pipe-Performance Versus Prediction
C.D.F. Rogers, P.R. Fleming, M.W.J. Loeppky, and E. Farapher 248
Installation Technique and Field Performance of HDPE, Profile Pipe
Larry Petroff 260

CONTENTS   xiii

Time-Deflection Field Test of 120-cm Steel, Fiberglass and Pretensioned
Concrete Pipe
   Amster Howard 272

Session 3B
DUCTILE IRON PIPE
Chairman: **Bill Quinn**

Developments in Ductile Iron Pipe
   Randall Conner 285
Pressure Class Ductile Iron Pipe and its Design
   L. Gregg Horn 298
The Development and Installation of Ductile Iron Microtunneling Pipe
   Ralph Carpenter and Richard Croxton 310
Polyethylene Encasement and the 1993 Revision of ANSI/AWWA C105/A21.5 Standard
   Richard Bonds 322

Session 1C
SPECIAL DESIGN CONSIDERATIONS III
Chairman: **Donald Ballentine**

Reexamining Seismic Risk Assessment for Buried Pipelines
   Douglas G. Honegger 334
Influence of Tertiary Creep on the Uplift Behavior of a Pipe Embedded in a Frozen Soil
   J. Hu and A.P.S. Selvadurai 345
Ductile Iron Pipe in Earthquake/Seismic Activity
   Michael S. Tucker 359
An Economic Approach to Sewer Pipe Selection
   Edwin Lamb 370

Session 2C
WELDED STEEL PIPE
Chairman: **Bruce Vanderploeg**

Protective Coatings—Current Practices
   William Robinson 375
Specify the Right Steel for Your Steel Water Pipe
   George Tupac 383
The Strain Equation—Incidence on Buried Steel Pipe Design
   R. Prevost 394
The Effects of High-Strength Steel in the Design of Steel Water Pipe
   Robert Card and Dennis Dechant 406

CONTENTS

Session 3C
DESIGN METHODS FOR HYDRAULICS AND BEDDING
Chairman: **Ed Sikora**

Newton Meets Darcy and Colebrook in a Programmable Calculator
Jorge Garcia 413
Hydraulic Transient Analysis in a Large Water Transmission System
Awni Qaquish, David Guastella, James Dillingham, and Donald Chase .. 425
Sleeve Valves for a Wastewater Application
David Timmermann, Arne Sandvik, and Hans Torabi 437
Trench Widths for Buried Pipes
Reynold Watkins 445

Session 1D
TRENCHLESS TECHNOLOGY I
Chairman: **William Shook**

Mill Woods Sanitary Storage Tunnel
Ken Chua and John Kelly 456
Point Loma Tunnel Outfall
Richard Trembath, F. Stuart Seymore, and Thomas Willoughby 468
Choice of Pipes for Microtunnels
Y.G. Diab 480
Construction of the Big No-Dig
John Struzziery and Arthur Spruch 489

Session 2D
TECHNOLOGY FOR STRUCTURAL ANALYSES
Chairman: **Patrick Selvadurai**

Computer Aided Pipeline Design: A Step-by-Step Description of the Process
Jorge Garcia 501
Transfer Matrix Technic and Pipe Structural Analysis
Marcel Gerbault 512
PCCP Design Concepts Made Simple
Richard Mueller 524
Structural Performance Criteria for Fitness-for-Service Evaluations of Underground Natural Gas Pipelines
Wen-Shou Tseng and Chih-Hung Lee 536

Session 3D
REHABILITATION I
Chairman: **James B. Thompson**

State-of-the-Art Review: Trenchless Pipeline Rehabilitation Systems
M. Najafi and V.K. Varma 548

CONTENTS xv

Condition Assessment & Rehabilitation Program For Large Diameter Sanitary Sewers in Phoenix, Arizona
Thomas Galeziewski, Samuel Edmondson, and Robert Webb 560
Denver's Experience in Trenchless Technology
Joseph Barsoom 572
The Value of Internal Manhole Inspections
James McGregor 589

Session 1E
TRENCHLESS TECHNOLOGY II
Chairman: **Larry Petroff**

Microtunneling Design Considerations
Steve Klein and Randy Essex 603
From Conception to Completion: Watershed 22 Trunk Sewer Upgrade
Curtis Swanson, Glenn Hermanson, and Charles Joyce 615
Microtunneling Forces: The Pipe's Perspective
Robert Lys, Jr. and Thomas Garrett 627
Trenchless Replacement & Corrosion Protection of Deteriorated Manholes
William Shook 635

Session 2E
FIELD PERFORMANCE I
Chairman: **Lars-Eric Jansen**

Pipe Performance During The 1993 Flood
Gary Moore and Charles Nance 640
Thermal Performance of Trench Backfills for Buried Watermains
Caizhao Zhan, Laurel Goodrich, and Balvant Rajani 650
Tacoma's Second Supply Project—Challenges and Solutions
Roger Beieler 662
Sewer Trench Subsidence Due to Severe Flooding
Mohammed Islam, Quazi Hashmi, and Steven Helfrich 673

Session 3E
REHABILITATION II
Chairman: **Mark Pickell**

Repair Study for the Lafayette Aqueduct No. 1
Christopher Dodge and Christina Hartinger 685
New Steel Pipe Joining System For Trenchless Construction & Rehabilitation
Michael Argent, David Pecknold, and Rami Hajali 697
Rehabilitation of Masonry Combined Sewers in the City of St. Louis
Marie Collins and Carl Ted Stude 709
Managed Operation and Repair of a Deteriorating Large Diameter Pipeline
Kenneth L. Cramblitt, Thomas J. Lawson and Adrian T. Ciolko 721

Session 1F
COATINGS AND LININGS
Chairman: Ron Bishop

Quality Enhancements of Cement-Mortar Coatings
  Henry Bardakjian 734
Special Protective Coatings and Linings For Ductile Iron Pipe
  A.M. Horton 745
Cost Effective Corrosion Mitigation
  Gerald Craft 757
New Bonded Tape Coating Systems and Cathodic Protection applied to Non-Steel Water Pipelines: Quality Through Proper Design Specifications
  James Noonan, and Bryan Bradish 765

Session 2F
FIELD PERFORMANCE II
Chairman: C.D.F. Rogers

CDF—A New Bedding System for Clay Pipe
  Edward Sikora, John Butler, and Robert Lys 775
Deformation of HDPE Pipes due to Ground Saturation
  Jun Tohda, L. Li, T. Hamada, J. Hinobayashi, and M. Inuki 786
Open-Cut Sewer Crossing of Railroad Completed in Just 48 Hours
  Jeff Garvey and Afshin Oskoui 798
Restrained Joint Systems
  George Rhodes 806

Session 3F
REHABILITATION III
Chairman: Lynn Osborne

Subsurface Utility Engineering: Upgrading the Quality of Utility Information
  James Anspach 813
Pipe Rehabilitation Using a Slipline Pipe
  Ernest Hanna and Allan Scarpine 825
Wailupe Reconstructed Trunk Sewer-ASCE Hawaiian Chapter Project of the Year 1993
  John Jurgens 837
Pipeline Integrity Assessment and Rehabilitation
  Peter Brooks and Mark Smith 845

Subject Index 1223
Author Index 1229

Diagnosis and assessment of damaged sewers
concerning their structural capacity

Prof. Dr.-Ing. Dietrich Stein[1],
Dipl.-Ing. Susanne Kentgens[2], Dipl.-Ing. Andreas Bornmann[2]

Abstract

To manage the extensive renovation program for the sewer system in Germany it is necessary to fix a scientifically proved strategy of maintenance. To set up a priority lists it is essential to use an assessment schema, which allows the objective assessment of the current state and the possible endangering potential of damaged sewers on hydraulic, environmental and structural aspects. The presented paper offers a scientific proved suggestion for the structural assessment of damages in non-accessible sewers.

1. Introduction

The present redevelopment programme for reconstructing or improving the sewer systems in the Federal Republic of Germany has such an enormous scope that, for technical and financial reasons, it will have to take place over a longer period of time, depending on the size of the sewage network, as well as on the type and extent of the damage which has been caused. In the case of larger systems, redevelopment will even be a permanent task, and for this reason maintenance strategies will have to be drawn up in which priorities are determined for planning and implementing measurements for maintenance, inspection and redevelopment.

In order to realise this, the ATV M 143 Part 1 [M143] lays down that there must be a "scientifically grounded base for establishing criteria for a uniform description and objective assessment of the actual situation and of the potential hazards caused by damaged sewers".
The aspect first referred to in the requirements laid down in 1986 has already been realised with ATV M 143 Part 2 [M143]. The second aspect is to be realised with the BMFT joint project "Water hazards caused by leaking sewers - recording and evaluating". The final result is an assessment schema [Bütow94] which permits an

---
[1] Professor of Ruhr-University Bochum, AG Leitungsbau & Leitungsinstandhaltung, Universitätsstrasse 150, 44780 Bochum, Germany,
[2] Members of Prof. Stein´s scientific staff at the Ruhr-University

assessment of the actual constructional, hydraulic engineering and environmentally-relevant condition of each sewage network and of any potential hazards which may result.

The object of part project A3, which will be introduced in the following pages, is an assessment of constructional damage to sewers with regard to its effects on stability.

The examinations covered constructional damage which was determined with the assistance of optical internal inspection devices and which is listed and described in ATV M 143 Part 1 and Part 2 [M143, Stein92]:

- leaks
- discharge obstructions
- positional deviations
- mechanical wear
- corrosion
- deformation
- cracks
- broken pipe
- collapse.

The explanations apply to pipes with a non-accessible nominal width of < DN 800, which is found in approx. 80 % of local authority sewer networks [Keding87]. An exception to this is the assessment of corrosion which covered all concrete pipes in accordance with DIN 4032, i.e. to nominal width DN 1400. Constructional damage was rated in classes 0 (immediate measures required) to 4 (damage without effects).

Because of the greater potential danger in cases of failures of the load bearing capacity, accessible sewers require special considerations which could not be taken into account when the assessment schema [Bütow94] based on the optical interior inspection was being prepared.

2   Leaks

Leaks are present either if water can be seen emerging from or entering the system, or if the conditions of a test for watertightness are not fulfilled.

Leaks only affect the stability if infiltrating ground water or discharging waste water lead to changes in the pipeline bedding conditions up to and including cavity formation as a result of erosion or suffusion.

Fig. 1 shows the behaviour of coarse clays and sand in the area of a leak. Where ground water infiltrates, or external water flows in, e.g. from leaking water supply pipes, finer particles of soil are washed out (Fig. 1a) until a stable filter is formed above the leak (Fig. 1b). Depending on the grain distribution and the compactness it is possible, e.g., to bridge leaks in this way, even if the leak dimensions are considerably greater than the maximum grain size. This filter can be destroyed by overloading/flooding, rapid changes in water levels, high-pressure cleaning or leak tests with water (Fig. 1c), so that until a new equilibrium has been set, particles of soil can penetrate which are practically the same size as the leak opening [Stein92,

Jones84].

The cavities created by internal erosion can be made so much larger in the course of time that subsidence can occur or even downfalls reaching to the surface. Both represent danger not only to buildings but also to traffic. As each visual type of damage does not necessarily change during this process, indirect optical interior inspections do not usually supply any information on these reactions and cavity location is therefore necessary to record the extent of the damage, and as a basis for redevelopment measures.

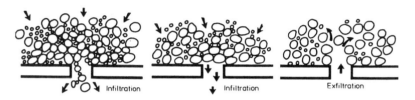

Fig 1: Behaviour of coarse clays and sand with leaks [Stein92]

Cavities have more than likely been formed where there is a leak with visible material yield. For reasons of stability, a leak in combination with entry of material must always be dealt with by immediate measures. Leaks with water infiltration or visible soil can also lead to changes to the bedding through soil being illuviated, e.g. after changes in the operating conditions, and for reasons of stability should therefore be allocated to hazard class 1 (priority damage elimination).

3 Discharge obstructions

Discharge obstructions consist of objects or matter which are found in the pipe cross-section, or project into it or cross it, in such a way that the free cross-sectional area required for the correct discharge of the waste water is no longer fully available [Stein92]. They usually have no direct effect on stability.

4 Positional deviations

In the case of sewers, positional deviations are differentiated according to [M143] in

- vertical (e.g. displacement),
- horizontal and
- longitudinal deviations.

Positional deviations do not usually have an a priori effect on stability until they are caused by bedding changes. In combination with leaks, the explanations in section 2 are valid here as well.

5 Mechanical wear

Mechanical wear is understood as wear in the area of wet internal pipe surface and can be measured by means of loss of mass (wear); in extreme cases this leads to the destruction of the pipe.

The reduction in load bearing capacity through mechanical wear can be treated analogously to internal corrosion in the wet cross-section (cf. Section 6).

6   Corrosion

6.1   General

Corrosion in pipes can occur internally and externally. **External corrosion** is usually caused by soil or ground water aggressiveness, or aggressive substances which have passed into the soil or the ground water and cannot usually be determined by means of an optical internal inspection. This applies as well to microstructural changes resulting from corrosion which, if at all present, is also not usually detectable.

Internal corrosion occurs with aggressive waste water in the wet areas or as a result of biogenous sulphuric acids in the gas-filled space [Bielecki87]. Other causes of corrosion can be non-compliance with limiting values laid down in standards and guidelines, and the absence of, or incorrectly applied or damaged, corrosion protection.

Corrosion leads to material being worn down and thus to a reduction in the pipe wall thickness and, as a consequence of this, a change in stability.

In the framework of the research project the following examinations were carried out for an assessment of the effect of internal corrosion on the stability of sewers made of concrete pipes:

1. Theoretical examinations on the bearing behaviour of circular concrete pipes with and without bases (DN 100 - DN 1400) in accordance with DIN 4032 with differently distributed and marked corrosion on the internal surface.
2. Laboratory experiments with concrete pipes DN 300 KFW with artificially generated, differently distributed and marked internal corrosion to determine the bearing capacity and to verify the theoretical examinations.
3. Three-edge load bearing tests in accordance with DIN 4032 on concrete pipes from drain levels with internal corrosion for comparison with the examinations described above under 1 and 2.

6.2   Theoretical examinations

It has already been mentioned that material wear takes place in dependence of the type of corrosion and that this results in a pipe with an altered wall thickness. The following wall thickness reductions were examined (Fig. 2):
a)   reduction of wall thickness in the gas-filled space
b)   reduction of wall thickness at the crown
c)   reduction of wall thickness in the wet cross-section with in the main complete filling
d)   reduction of wall thickness in the wet cross-section

with in the main partial filling.

The reduction in the wall thickness in the area affected leads to a corresponding reduction in the moment of inertia, which generates a displacement of the bending moment. In accordance with ATV A 127 [A127], internal force coefficients were determined for axial and shear forces and bending moments for corrosion wear between 0 - 60 %. Taking the residual wall thickness into account, the internal force coefficients of ATV A 127 can also be used for corrosion wear distributed evenly over the circumference, and for this reason this type of wear was not taken into consideration.

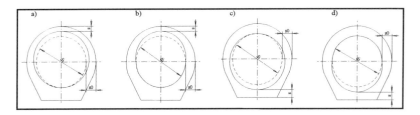

Fig. 2: Overview of the forms of corrosion wear in the examinations of circular concrete pipes with bases.

The basis of the calculation was the dimensions and minimum wall thicknesses of unreinforced pipes with and without bases in accordance with DIN 4032 as amended 1981 [DIN4032].

The internal forces and tensions respectively of the corroded concrete pipes can be calculated analogously to ATV A 127 for any load cases with the coefficients determined in accordance with [A127] for various stratifications and bearing angles.

Evaluation of the three following load combinations took place in the research project:

Case 1	Case 2	Case 3
Total vertical load qv Lateral pressure qh = 0 Deadweight Water filling	Total vertical load qv Lateral pressure qh = 0.3 x qv Deadweight Water filling	Total vertical load qv Lateral pressure qh = 0.5 x qv Deadweight Water filling

These load combinations cover a great range of common loads on pipes. An individual examination must be carried out in each case to see whether, when calculating qv, the load assumptions in ATV 127 (e.g. silo theory) for newly laid pipes also apply for older pipes, i.e. pipes which have been in the ground longer. The actuall safety coefficient of $\gamma = 2.2$ [A127] was used here for failure through fracture for safety class A.

The internal force coefficients for the ring bending tensile stress and the shear stress have been evaluated in 96 bearing capacity diagrams, an example of which is shown in Fig. 3, in dependence on the pipe form and corrosion wear. Fig. 3 shows the maximum absorbable load qv in dependence on the residual wall thickness for a calculated ring bending tensile stress permitted σ RBZ = 6.0 N/mm^2 and a shear stress $\tau = 0.7$ N/mm^2. In contrast to the present calculation standard the shear force was also taken into account, as failure of the shear force cannot be excluded where the wall thickness is reduced.

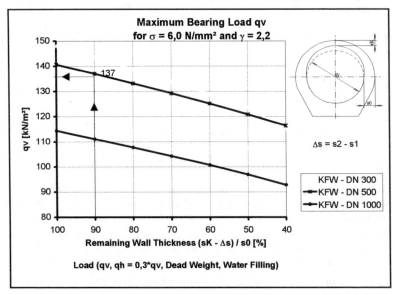

Fig. 3 Example of a bearing capacity diagram for corroded concrete pipes, form KFW (with base = F and increased wall thickness = W), in conformance with DIN 4032, with corrosion in the gas-filled space

In the bearing capacity diagrams, the maximum vertically absorbable load for a given residual wall thickness can be seen in dependence on the nominal width, the type of pipe and the wear, and compared with the existing load. The residual wall thickness is shown for a range between 100 % and 40 % of the original wall thickness. It does not appear practical to treat wear rates above this in this way, as in this case, e.g., the pointed end of jointed socket pipes no longer exists in the corroded area.

6.3 Laboratory experiments

In order to verify the altered static supporting behaviour which was the subject of the theoretical examinations, 18 new concrete pipes, DN 300 KFW, shortened to 50 cm, were subjected to artificially generated corrosion through acids. In two parallel test series concrete pipes were filled with 7 % sulphuric acid and 2 % hydrochloric acid respectively in the gas-filled space and in the bottom area. The choice of these two types of acid was to enable the simulation not only of dissolving but also of driving corrosion [Stein92, Knoblauch92]. The corrosion products were removed regularly with high-pressure cleaning. In spite of acid treatment lasting 10 months, the maximum wear rate was only 20 %. After this period, most of the residual wall thicknesses in the sample pipes were between 84 % and 92 %. These were determined between the aggregate grains projecting from the matrix ("in the valley").

6.4 Summary

Corrosion in waste water conduits made of concrete pipes leads in the first place to a reduction of the pipe wall thickness and thus to a reduction of the load bearing capacity.

The aim of the above-mentioned theoretical and laboratory examinations was to quantify these reductions.

The present research findings show that, from the aspect of statics, corroded concrete pipes do not represent a priori damage which requires immediate elimination. With the corresponding wall thicknesses and boundary conditions it is certainly possible that a pipe with a residual wall thickness of 40 % is still load-bearing with a sufficient degree of safety.

However, for an exact assessment it is necessary above and beyond the indirect optical internal inspection to gain additional information on the consistency of the concrete, the pipe's external shape (with/without base), the residual wall thickness and the wear.

With the help of the special bearing capacity diagrams for all circular concrete pipes complying with DIN 4032 [DIN4032], this information can be used to determine the maximum absorbable load for the corroded pipes and this can then be compared with the given load.

An assessment of the pipes then results in the following cases only:

1. Stable (hazard class 4, no construction measures required to improve the load bearing capacity).
2. Unstable (hazard class 0, immediate measures required)

In the first case it is important to gain information on the further development of the damage. To do this, a shortened inspection interval is required with determination of the wear by means, for example, of calibration measuring. If this procedure shows that there is no further damage, e.g. because the constitution of the waste water has altered, or the preconditions for the formation of "biogenous sulphuric acid" are no longer given, the level can still be assigned to hazard class 4, if its stability has been verified. Of course, any leaks, for example, must be eliminated.

The internal force coefficients and the bearing capacity diagrams which have been developed have enabled the creation of an extensive rated value system for all concrete pipes DN 100 - DN 1400 complying with DIN 4032 which can be worked into all classification models and is easy to use.

7 Deformation

Deformation of flexible pipes can be of significance for an assessment of their bearing capacity and can also be assessed as damage depending on the type and size. Vertical deformations of ≤ 4 % [DIN4033] directly after installation and ≤ 6 % during the subsequent service life can in general be regarded as safe and, with regard to the bearing behaviour of flexural pipes, can even be regarded as desirable [A127]. If there are greater deformations, their influence on the bearing capacity of the pipe-ground system depends on various factors which must omitted here because of the shortage of space.

8 Cracks

8.1 General

The damage represented by "**cracks**" occurs mainly in rigid pipes, whereby three main types of crack which may form the preliminary stages for pipe fractures and finally for the collapse of the sewage pipe [M143] must be differentiated:

– longitudinal cracks
– circumferential cracks
– multiple cracks starting from a point (in many cases with formation of fragments).

The type of crack and the cause of cracking are closely connected, whereby the shape of a crack, its dimensions and its course allow conclusions to be drawn regarding the causes. It is possible here that on the one hand there is *a single* cause for several cracks at various places, and on the other hand that *a single* crack may have *several* causes.

Longitudinal cracks represent an a priori danger to stability and for this reason were examined thoroughly in the research project (cf. section 8.2).

Circumferential cracks are created where the permitted longitudinal bending tensile strength of the pipe, the longitudinal tensile stress or the pipe's shear strength are exceeded. These cracks run mainly over the whole circumference of the pipe but they do not represent an immediate danger to stability, as long as there are no changes to the bedding, as described below.

Along with longitudinal and circumferential cracks, which display a relatively clear cracking course, there are also **cracks** in sewers which **start from a single point** and are star-shaped, or which have an irregular course. In both cases, fragment formation takes place in nearly every case in which single pieces of pipe wall are completely surrounded by cracks. This damage cannot be statically calculated and so should be treated in this case as immediate measures..

8.2 Longitudinal cracks

8.2.1 General

Of the main types of crack referred to above, longitudinal cracks occur most frequently in flexurally rigid pipes and are generated when the ring bending tensile strength is exceeded, e.g. in the case of line displacement as a result of construction defects. They can also be created by positional displacements caused by changes to the bedding as a result of leaks or ground movements, as well as through incorrect pipe connections. They usually occur in the top, bottom and sides (called quarter chord points) and stretch along the overall length of the pipe [WRC90].

The longitudinally cracked pipe carries as a ground-supported four-bar ring, which reacts sensitively to changes in the bedding with deformations (Fig. 4). The initial deformation is essentially dependent on the layer thickness of the bedding and can be considerable in the case of poor installation.

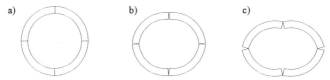

Fig. 4: Deformation course of a longitudinally cracked pipe to collapse of pipe
a) appearance of longitudinal crack
b) initial deformation through reduction of lateral bedding reaction pressure as a result of ground erosion
c) heavy deformation as a result of loss of side bedding.

Changes in the bedding can be aggravated by heavily fluctuating operating conditions (frequent overloads or flooding), by high-pressure cleaning or water-tightness tests (cf. Fig. 2).

Because of the many possible interrelationships, the point at which total failure (collapse) will occur cannot be easily forecast.

The following programme was implemented in the research project for an improved evaluation of the danger to the stability of pipes cracked lengthwise:

1. Theoretical examination of the following problem fields:
    1.1 Bearing behaviour of rigid, non-reinforced pipes with longitudinal cracks in the quarter chords, taking the bedding into consideration.
    1.2 Effects of bedding changes caused, e.g., by illuviation in the pipeline.

These examinations were carried out with the help of the finite elements method (FEM), variing the following parameters in the calculation.

2. Laboratory experiments on concrete pipes with circular cross-sections with and

without bases under simulation of in situ conditions to verify the theoretical examinations listed under 1 above.

3. Construction site experiments on newly laid and old levels made of concrete and vitrified clay pipes with longitudinal cracks to verify the boundary conditions determined in 1 and 2 above.

The laboratory and construction site experiments will not be examined further here. The targets set in these experiments were achieved.

Table 1: Variation quantities for the calculation with FEM

Pipe	Soil	Load
Nominal width	Design of support	Backfill Height
Material (vitrified clay/concrete)	Design of bedding	Traffic Loads
Wall thickness	Compactness of bedding	Ground water
	Trench effect	

8.2 Results

When longitudinal cracks occur in the quarter chords of rigid, non-reinforced pipes, a ground-supported four-bar ring is formed as a static system. Inward deformation of the crown is accompanied by external deformation of the springer. This springer deformation activates the bedding reaction pressure of the ground. The sinking of the crown initiates a load displacement from the area above the pipe to the areas next to the pipe.

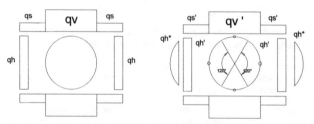

Fig. 5: Loads on the uncracked rigid pipe (left figure) and the ground-supported four-bar ring (right figure).

This load displacement to areas next to the pipe has the effect of increasing the horizontal soil pressure, and supports the possibility of achieving equilibrium through activating the bedding reaction pressure, in that the deformed system consisting of pipe quarter sections and soil can then remain in position (Fig. 5).

This shows that the stability of pipes cracked lengthwise is not a priori endangered.

DAMAGED SEWERS ASSESSMENT 11

The deformation of the circumferential cross-section plays an important part in the evaluation of the longitudinal cracks.

It can be seen from the extensive calculations which were carried out that the stability of concrete and vitrified clay pipes cracked lengthwise is endangered with
- deformations > 5 % of the diameter and/or
- crack widths > 10 % of the wall thickness.

In all other cases in which the crack edges are not displaced, i.e. the joint mechanism is retained, and no bedding changes (ground illuviation) are to be expected, grading can be made in status classes 1 - 4.

The determination of status classes 1 - 4 required special static examinations, for which the following additional data is required:
- crack width
- deformation
- wall thickness.

The calculation model required for this is available.

Until damage elimination measures have been carried out, safety precautions should be taken during maintenance work, in particular with high-pressure cleaning, which may have to be done without completely in the areas affected until the damage has been eliminated. Inspection intervals must be reduced in all cases, in order to obtain an overview of the further development of the damage.

9 Broken pipe and collapse

"Broken pipe" and "collapses" are failure types which require immediate measures.

10 Summary

In order to realise the redevelopment programme for local authority sewer networks, maintenance strategies must be drawn up in which priorities are determined for planning and implementing maintenance, inspection and redevelopment measures. To do this, an evaluation schema is required which permits an objective evaluation of the actual state of the conduits and the possible danger potential caused by damaged conduits from constructional, hydraulic engineering and environmental aspects. In the present article, a scientifically grounded proposal is made for the evaluation of constructional damage to non-accessible sewers and their hazard potential with regard to their stability. Particular attention was paid to the interaction "pipe-ground" and to the damage type "cracks" and "corrosion". The examinations showed that damage which up to now has been regarded as critical does not necessarily have to be classified in the category for immediate measures, but can be classified in status class 4 (damage without effect) on the basis of the findings gained in the examinations and depending on the local situation. If this economically important advantage is to be used, quantitative inspections must be carried out complementary to optical internal inspections, e.g. to record deformations in cracked rigid pipes, deformations of flexible pipes, wall thicknesses, crack widths and cavities in the pipeline zone.

11 References

ATV A 127: Richtlinie für die statische Berechnung von Entwässerungskanälen und -leitungen.. Abwassertechnische Vereinigung e. V., Ausgabe Dezember 1988.

Bielecki, R.; Schremmer, H.: Biogene Schwefelsäure-Korrosion in teilgefüllten Abwasserkanälen. Sonderdruck aus Heft 94/1987 der Mitteilungen des Leichtweiß-Instituts für Wasserbau der Technischen Universität Braunschweig.

Bütow, E; Lühr, H.P.: BMFT-Verbundprojekt "Wassergefährdung durch undichte Kanäle - Erfassung und Bewertung" Teilprojekt WA 9039/2 "Entwicklung von Kriterien zur Erfassung und Beurteilung der Schäden in Kanälen und Erarbeitung eines Modells zur Abschätzung des von defekten Kanälen ausgehenden Gefährdungspotentials", BMFT-Statusseminar, Oktober 1994 in Hamburg

DIN 4032: Betonrohre und Formstücke, Maße - Technische Lieferbedingungen. Ausgabe Januar 1981.

DIN 4033: DIN 4033: Entwässerungskanäle und -leitungen; Richtlinien für die Ausführung (11.79).

Jones, G.M.A.: The Structural Deterioration of Sewers. International Conference on the Planning, Construction, Maintenance & Operation of Sewerage Systems. Reading (England), 12.-14. Sept. 1984. Paper C1, S93-108

Keding, M., Stein, D., Witte, H.: Ergebnisse einer Umfrage zur Erfassung des Istzustandes der Kanalisation in der Bundesrepublik Deutschland. Korrespondenz Abwasser (1987) H. 4, S 118 ff

Knoblauch, H.; Schneider, U.: Bauchemie. Werner-Verlag GmbH,Düsseldorf, 3. Auflage, 1992.

ATV M 143: Inspektion, Sanierung und Erneuerung von Entwässerungskanälen und -leitungen. Teil 1: Grundlagen, Ausgabe Dezember 1989; Teil 2: Optische Inspektion, Ausgabe Juni 1991. Abwassertechnische Vereinigung e. V.

Stein, D.; Niederehe ,W.: Instandhaltung von Kanalisationen. 2., überarbeitete und erweiterte Auflage, Berlin 1992.

Trott, J.J., Stevens, J.B.: The load-carrying capacity of cracked rigid pipes - a preliminary study. TRRL Supplementary Report 534. Transport and Road Research Laboratory, Crowthorne (England) 1980

Water Research Centre: Sewerage Rehabilitation Manual, WRC, Swindon 1990

Behavior of buried large thin wall flexible pipe
— Field Test and Numerical Analysis considered with
Stage of Construction of Buried Flexible Pipe

T.Kawabata [1]   and   Y.Mohri [2]

Abstract

The behavior of buried flexible pipe is influenced to a considerable extent by the properties of the backfill material, ring stiffness of the pipe and the way of construction. In particular, deformation of pipe with low stiffness is mainly determined by the stage of backfill and compaction.

This paper discusses the behavior of buried low stiffness pipe that is obtained by field test of FRP pipe (ϕ 1500mm, ring stiffness factor; $E_p I_p / D_2^3 = 2.56 \text{kN/m}^2$) and finite-element analyses which evaluate compaction process. Natural ground in the test field is KANTO loam and crushed gravel (0-25mm) is used for backfill material.

I . Introduction

The behavior of a buried pipeline or any other underground structure is significantly influenced by the surrounding ground, the construction method employed, and various properties of the backfilling material used. For a flexible pipe, in particular, those conditions should be given careful consideration in the construction plan since they determine in large measure the degree of deformation and settlement of the pipe in the ground. Concerning research on the behavior of buried pipelines, attempts have been made by many researchers to predict the safety of pipelines by an experimental, theoretical, or analytical approach.

In Japan, the application of flexible pipes having a low stiffness will increase in the future because of their good

---

[1] Chief Researcher, Kubota Corporation, 3-1-3, Nihombashi Muromachi , Chuo-ku Tokyo 103, Japan,
[2] Chief Researcher, National Research Institute of Agricultural Engineering, 2-1-2, Kannondai, Tsukuba Ibaragi 305, Japan

workability, economy, etc. Such flexible pipes having a low stiffness tend to be easily deformed by an external force applied thereto during backfilling or compaction. Nevertheless, there have been very few reports dealing with the effect of the backfilling and compacting processes on flexible pipe.

This paper discusses the behavior of a low stiffness pipe buried in the ground based on the results of a large-scale field test using a large-diameter and low stiffness pipe, and on the results of a numerical analysis taking into account the construction process.

II. Field Test
1. Ground Condition and Work Method

The natural ground at the test site consists mainly of layers of KANTO loam and sandy soil. The groundwater level is approximately 4.0 m below the surface of the ground and is relatively stable. Longitudinal and cross sections of the pipeline tested are schematically shown in Fig. 1. The pipeline made up of a total of three pipes (4.5 m, 9.0 m, and 4.5 m) connected end to end. The overburden was 1.8 m.

A foundation bed 0.3 m in thickness was laid in the ditch and the pipes were set in place on the foundation bed. After that, the ditch was backfilled. Each time the backfilling material was spread in a layer of about 0.3 m thick, the backfill was compacted by an 80 kg tamper which was moved back and forth twice. The work procedure is shown in Table 1, and the detailed cross section is shown in Fig. 2.

2. Pipe and Backfilling Material

The pipes used in the field test were FRP pipes having an inside diameter (D) of 1,500 mm and a wall thickness (t) of 15.5 mm ($t/D = 1.03\%$). The elastic modulus (E_p) and ring stiffness ($E_p I_p / D_2^3$) of the pipe were 28.1GN/m^2 and 2.56 kN/m^2 respectively. (I_p: the moment of inertia of area per unit length of the cross section, D_2: the pipe diameter for wall-thickness center)

As shown in Fig. 2, crushed gravel (up to 25 mm in size) was used to backfill the foundation bed and the space around the pipe, and the excavated soil was used to backfill the space from a level 0.3 m above the pipe crown to the ground level. Their physical properties are shown in Table 2.

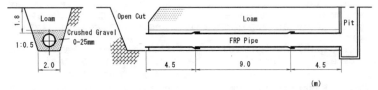

Fig.1 Longitudinal and cross section profile

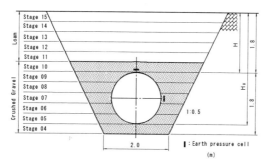

Fig.2 Cross section detail

Table 1 Stage of construction and analyses

Stage No.	Stage of construction	H_B (m)	H (m)
1	Natural ground		
2	Excavation (G.L.-1.80m)		
3	Excavation (G.L.-3.60m)		
4	Backfill & compaction of bed	0.00	
5	Backfill & compaction of 1st layer	0.30	
6	Backfill & compaction of 2nd layer	0.60	
7	Backfill & compaction of 3rd layer	0.90	
8	Backfill & compaction of 4th layer	1.20	
9	Backfill & compaction of 5th layer	1.50	0.00
11	Backfill & compaction of 7th layer	2.10	0.60
13	Backfill & compaction of 9th layer	2.70	1.20
15	Completion of backfill (G.L.)	3.30	1.80

Table 2 Physical properties of KANTO loam and crushed gravel

	KANTO Loam	Crushed gravel (0-25mm)
G_s	2.957	2.853
$\omega_n(\%)$	103.1	3.4
U_c	—	76.7
U_c'	—	11.4
$\omega_L(\%)$	133.2	N.P.
$\omega_P(\%)$	51.7	N.P.
I_P	81.5	N.P.
$JUSCS^*$	MH	G-F

* Japanese unified soil classification system

3. Items Measured

Of the three pipes that made up the pipeline, only the 9.0m long pipe at the center was subjected to various measurements. To measure the pipe behavior, displacement gauges were installed across the diameter of the pipe at the center and the pipe deformation in each of the backfilling stages was measured. In addition, a total of 12 biaxial strain gauges were attached at equal intervals (30°) to the pipe inner circumference near the displacement gauges to measure the circumferential and axial strains occurring in the pipe. Furthermore, strain-type earth pressure cells were installed at the pipe crown and pipe side to measure the vertical earth pressure near the pipe crown and the horizontal earth pressure at the spring line.

III. Nonlinear Analysis Taking Work Process into Account
1. Evaluation of Compaction Process

In the work of backfilling for a pipeline, compacting the backfill sufficiently is important from the viewpoint of restraining the pipe deformation. It is also an important item of work supervision. In the case of the low stiffness pipe used in the experiment, in particular, the effect of compaction is considered so large that the pattern of pipe deformation can be decided in some measure in the early stages of backfilling.

In order to measure the behavior of backfill in the compaction process by a static numerical analysis, it is necessary to evaluate the compacting energy of the compacting machine in terms of load applied to the backfill. In recent years, studies on the effect of soil compaction by a dynamic load (vibration, impact, etc.) and on methods of evaluation of the effect of compaction by a static load have been actively conducted for earth-fill dams and other large structures. Nevertheless, much less attention has been paid to the study of small and impact-type compacting machines, such as tampers.

In this paper, we discussed the compaction effect of an impact given by the 80 kg tamper that was used to compact the backfill in the field test on the basis of the result of measurement of tamper acceleration, and assume it as the load applied to the compaction surface. In concrete, the evaluated load is applied to the compaction surface in the first step of calculation and then removed in the next step of calculation. Namely, we consider the effect of compaction in terms of residual strain in the backfill caused by the loading and unloading.

(1) Compaction test using tamper

It is said that there is generally a hyperbolic relationship between compaction energy (or number of times of compaction) and compacted soil density ρ_t. This means that when a soil has been compacted to a certain degree, additional compaction produces very little effect in terms of soil density. To confirm this, we conducted a compaction test using 80kg tamper. A mound 3.0 m in width and 5.0 m in length was built using the same crushed gravel as used in the field test, and the relationship between number of compaction N_c and compacted layer density ρ_t was studied by changing the tamper traversing speed v and the number of tamper traversing. The test result is shown in Fig. 3. It can be seen that in the case of the crushed gravel used in the test, the effect of compaction after the number of compaction N_c exceeds 11 or so is very little. As shown in Equation (1), the number of compaction N_c was calculated from the measured tamper traversing speed v and number of blows per unit time (frequency: f).

$$N_c = n \cdot f \cdot \frac{A_T}{v \cdot B_b} \quad \quad \quad \quad \quad (1)$$

where, N_c: Number of compaction (blows)
n: Number of tamper traversing
v: Tamper traversing speed (m/s)
f: Number of blows per unit time (blows/s)
B_b: Width of striking plate (= 0.30 m)
A_T: Area of striking plate (= 0.26 · B_b = 0.078 m^2)

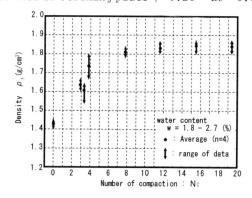

Fig. 3 Relation between number of compaction and density

(2) Evaluation based on acceleration measurement

The operating principle of the tamper is as follows. A heavy weight is caused to spring up by the explosive power of an internal combustion engine, then let fall onto the surface to be compacted. Since the compaction is effected by the impact the weight produces as it falls, we installed an accelerometer to the bottom plate of a tamper and measured the acceleration of vibration. The results of measurement are shown in Fig. 4 (a), (b) and Table 3. Fig. 4 (a) shows the acceleration of vibration in the initial stage of compaction, and Fig. 4 (b) shows the one in the final stage of compaction. It can be seen that the acceleration increases as the compaction of the backfill progresses.

As already mentioned, a dynamic load cannot easily be converted to a static load theoretically. In the present study, therefore, we calculated the static pressure σ_v, applied to the surface of compaction, by the following equation (2) using the measured average acceleration shown in Table 3. The calculation results are shown in Table 4.

$$\sigma_V = \frac{P_V}{A_T} = \frac{m \cdot \alpha}{A_T} \quad \cdots \cdots \cdots \cdots \cdots \cdots \cdots \cdots \cdots \cdots \cdots \quad (2)$$

where, σ_V: Pressure of tamper (kPa)
P_V: Load of tamper (kN)
m: Mass of tamper (kg)
α: Average acceleration (G)

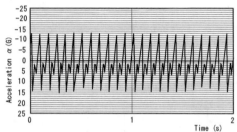
(a) Initial compaction stage : loose

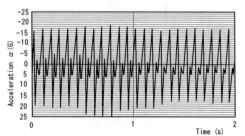
(b) Final compaction stage : dense
Fig. 4 Acceleration of vibration of tamper

Table 3 Characteristic of vibration of tamper

Characteristic of vibration	(a) Initial stage	(b) Final stage	for analyses
Frequency f (blow/s)	12	12	12
Acceleration α (G)	13 ~ 15	17 ~ 25	17

Table 4 Valuation of static load for tamper

Estimation for static load	Measuring of vibration
Mass of tamper m	80 (kg)
Acceleration α	17 (G)
Load for 1 blow P_V	13.60 (kN)
Pressure for 1 blow σ_V *	174 (kPa)

* $\sigma_V = P_V / A_T$

2. Analytical Models and Analytical Methods

In relation to the field test described in the preceding section, we conducted a nonlinear finite-element(F.E.) analysis of elasticity taking into account the work process, including ground excavation and backfilling. Examples of F.E. models used in the major stages of analysis are shown in Fig. 5(a) through (f). The analysis assuming a plane strain model was carried out along the work process (excavation, bed spreading and compaction, pipe laying, and backfill and compaction), with the stress condition of the natural ground before excavation assumed as the initial condition, as shown in Table 1.

The compaction process was expressed as the loading and unloading σ_v by the number of effective blow. As is evident from the measurement results obtained by Fukuoka et al., the effect of compaction by the tamper extends not more than approximately 0.3 m from the surface of compaction, hence the layers below that level are not compacted. In view of this, during compaction of each layer, a boundary condition for restraining vertical action was set at the bottom of the layer being compacted so that the compacting load would not be transmitted to the lower layers.

In the discussion that follows, the analysis conducted using the load evaluated from the measured tamper acceleration shall be called "Analysis ①".

In addition, other two analysis were carried out. One is an ordinary bottom-up sequential analysis leaving the compaction process out of consideration, which is hereinafter called "Analysis ②". The other one is general F.E. analysis at the completion of backfilling (Stage-15), which is hereinafter called "Analysis ③"

A rectangular quadratic-order(8-nodes) element was used for each of the natural ground, pipe, and backfill. For the discontinuous planes of the pipe and backfill, a 6-nodes joint element was used. The pipe was circumferentially divided into 12 elements. To evaluate the nonlinear characteristic of the backfill, the variable moduli model of Nelson was used. The parameters of variable moduli model of KANTO loam and crushed gravel calculated from the results of a triaxial compression test and isotropic consolidation test are shown in Table 5. It was assumed that the natural ground of the test site consisted entirely of KANTO loam.

IV. Results of Field Test and F.E. Ana. and Considerations
1. Pipe Deflection

The deflection of pipe obtained from the field test and F.E. analyses are shown in Fig. 6. The vertical deflection obtained by the "Analysis ③" is indicated by ★, and the horizontal deflection is indicated by ☆.

From the field test results, it can be seen that the pipe was vertically deflected by about 8 mm by backfilling to the spring line, and further deflected vertically by about 14 mm by backfilling to the pipe crown. After that, backfilling

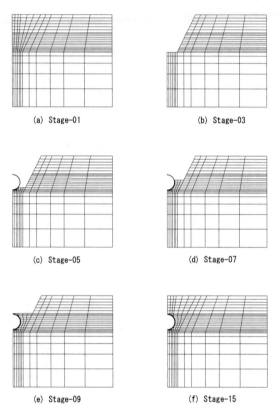

Fig. 5 (a) ～ (f) F.E. models used in analyses

Table 5 Parameters of KANTO loam and crushed gravel for non-linear F.E. analyses

parameter	KANTO loam	crushed gravel	parameter	KANTO loam	crushed gravel
G_0 (MN/m^2)	2.028	1.262	K_0 (MN/m^2)	3.436	2.200
G_{0u} (MN/m^2)	5.631	1.696	K_{0u} (MN/m^2)	1.533	5.449
G_{0r} (MN/m^2)	6.785	4.088	K_{0r} (MN/m^2)	3.628	6.977
$\overline{\gamma_1}$	-126.2	-25.73	K_1 (MN/m^2)	-33.669	-31.967
$\overline{\gamma_{1u}}$	122.30	87.34	K_{1u}	132.47	288.58
$\overline{\gamma_{1r}}$	-422.27	-50.69	K_{1r}	19.78	74.56
γ_1	49.79	16.66	K_2 (GN/m^2)	26.005	19.396
γ_{1u}	27.15	5.85			
γ_{1r}	166.61	43.70			

the space above the level of pipe crown caused the pipe to enter horizontal deflection (flattening) mode, and the amount of pipe vertical deflection began to decrease slightly. At the time when the backfilling was completed, however, the pipe remained vertically deflected by 12 mm. Thus, the implication of the results of the field test is that the deformation behavior of a buried low stiffness pipe is significantly influenced (i.e., the pipe is vertically deflected) by the process of work, especially the process of backfilling to the pipe crown, but that it remains almost unaffected by the backfill from the level of pipe crown upward and shows stability on a long-term basis.

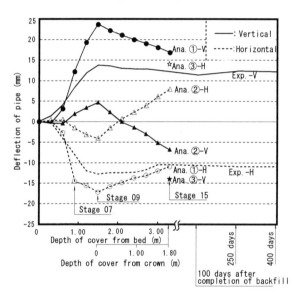

Fig. 6 Change of deflection of pipe

In the analysis taking the work process into account, compaction at the side of the pipe caused the pipe to be vertically deflected. Thus, the analysis expresses the actual pipe deformation behavior qualitatively. However, in the process of backfilling to the pipe crown, "Analysis ① "gave a pipe vertical deflection about 70% larger than the measured deflection, and "Analysis ② " gave a pipe vertical deflection about 70% smaller than the measured deflection. In both analyses, the recovery from vertical deflection (the amount of horizontal deflection) by backfilling of the space above the pipe crown was greater than the measured value.

Thus, it can be seen that by using the technique that expresses the compaction by tamper vibration as cyclic loading, it is possible to roughly estimate the pipe deformation caused by the compacting effect of backfill.

According to the "Analysis ③" conducted at the completion of backfilling, the vertical deflection of pipe was -13.8 mm, while horizontal deflection was 13.9 mm. With this method, therefore, it is impossible to predict the behavior of a buried pipe. The implication is that consideration must be given to the process of backfilling and the effect of compaction.

2. Earth Pressure Acting upon Pipe

Fig. 7 shows the time-serial changes in measured values of the earth pressure acting vertically upon the pipe crown and the earth pressure acting horizontally to the pipe at the spring line. It can be seen that the vertical earth pressure upon the pipe crown is greater than the horizontal earth pressure to the pipe at the spring line. At the end of backfilling, the earth pressure acting upon the pipe was about 1.25 times that of the overburden pressure (Wv = 24.8 kPa), indicating a concentration of the earth pressure onto the pipe crown.

The magnitude and distribution of the earth pressure that acts upon a pipe depend upon the interaction of pipe and backfill deformations. Generally speaking, the distribution of vertical earth pressure acting on a flexible pipe which deflects horizontally has a concave shape, whereas in the case of a high stiffness pipe, or a pipe buried under a condition under which it shows high stiffness relative to the backfill, the earth pressure concentrates on the pipe crown. The pipe used in the present study is a low stiffness pipe. The reason why the crown of this pipe was subjected to a large vertical earth pressure is probably that the apparent relative stiffness of the pipe increased because the pipe deflection due to the backfill above the pipe crown did not increase so much.

Fig. 8 shows the distributions of earth pressure around the pipe obtained by "Analysis ①" and "Analysis ②" at two stages: backfilling up to the pipe crown (Stage-09) and completion of backfilling (Stage-15). The right half of Fig.8 shows the results obtained by "Analysis ①". It can be seen that a large horizontal earth pressure exceeding 80 kPa occurred in the pipe during compaction of the backfill at the spring line (Stage-07) and the layer of backfill right on that (Stage-08), and that the earth pressure distributions as a whole are discontinuous due to a residual earth pressure caused by the compaction.

The vertical earth pressure acting upon the pipe crown was approximately 50 kPa. Like the measured vertical earth pressure, it was higher than the overburden pressure Wv. This is considered to indicate a condition in which a pipe has a length longer than its diameter and is hardly deflected laterally (i.e., the condition of an apparently rigid pipe).

The left half of Fig.8 shows the results obtained by "Analysis ②". The earth pressure distributions are smooth, and the vertical earth pressure acting upon the pipe crown and the horizontal earth pressure acting to the pipe at the spring line at the end of backfilling are both approximately 35 kPa.

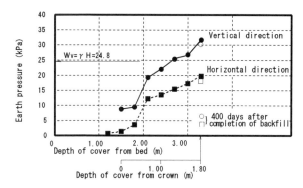

Fig. 7 Changes of vertical and horizontal earth pressure on pipe (Experiment)

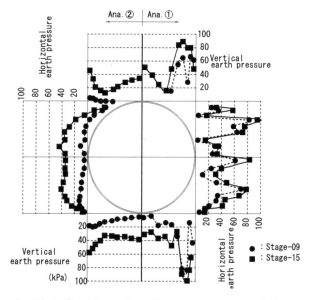

Fig. 8 Distribution of earth pressure around the pipe

V. Conclusion

The authors conducted a field test using a 1,500mm diameter low stiffness pipe and finite-element analyses taking the process of work into consideration to study the behavior of a buried low stiffness pipe.

Concerning the amount of pipe deflection, it was confirmed from the field test that the pipe was vertically deflected by 14 mm during backfilling of the space up to the pipe crown, then slightly deflected horizontally by the backfilling of a 1.8 m thick overburden, and finally became stable with a vertical deflection of 12.5 mm on a permanent basis. It was found that by an analysis taking into the work process into account, it is possible to express the behavior of a buried low stiffness pipe qualitatively. Namely, it was confirmed that compacting the backfill causes the pipe to be deformed vertically and its effect on the strain and earth pressure which act on the pipe resides until the completion of backfilling, thereby determines the final condition of the pipe.

The effects of compaction and backfilling process mentioned above have been left out of consideration in conventional design methods. Nevertheless, it was found that they are very important factors, especially when a thin wall and low stiffness pipe is involved, and that the behavior of a buried pipe cannot be evaluated by an "Analysis ③ "conducted at the completion of the backfilling process.

It was found that the vertical earth pressure acting upon the pipe crown at the end of backfilling was about 1.25 times of the overburden pressure obtained by the prism formula ($Wv = \gamma H$) and that the earth pressure concentrated on the pipe crown.

Nevertheless, the earth pressure and the change in earth pressure distribution on pipe caused by compaction are little known. Therefore, not only the value of static compaction load but also the distribution of earth pressure in the layers of backfill should be studied further.

Acknowledgments: The authors express thanks to Mr. Yujiro Tsurumaru for his help in the field test. We also thank Mr. Motoki Ikeyama, Mr. Masayuki Itashiki and Mr. Ken-ichiro Sadajima for their help in the same field test.

Reference
1) Fukuoka, M. et al., (1987) : Control of soil backfill around pipes, the 22nd Japan national conference on SMFE, pp.1609-1610 (in Japanese)
2) Nelson, I. and Baron, M.L. (1971) : Application of Variable Moduli Models to Soil Behavior, International Journal of Solid Structures, Vol.7, pp.399-417
3) Agricultural Structure Improvement Bureau of Ministry of Agriculture, Forestry and Fisheries (1977) : Planning and Design Standard for Land Improvement Project, Pipeline (in Japanese)

UNDERGROUND PIPELINE MATERIALS, DESIGN, AND CONSTRUCTION: WHAT HAVE WE LEARNED DURING 1985-1995? WHERE DO WE GO FROM HERE?

By

Jey K. Jeyapalan, Ph.D., P.E.[1], Wesley E. Saleira, Ph.D., P.E.[1],
Abdullah Al-Shaikh, M.S.[2],
Bala. K. Balasubramaniam, M.S., C.Eng, FICE[3] and John Jurgens [4]

Abstract

The first international conference on advances in underground pipeline engineering was held in 1985. Some 50 authors and over 300 delegates from 20 countries took part in the presentations and discussions on what the state-of-the-art of pipeline technology was at that time. Now, a decade later, do we know how to design and build pipelines any better? This paper reviews the technologies used for the assessment of geotechnical, environmental, physical, chemical, and hydraulic conditions of new pipeline sites, materials, construction methods, causes of failures, and rehabilitation of aging pipelines and manholes. A detailed description of the progress made and the setbacks encountered in various pipeline materials, design techniques, consensus standards, construction methods, and rehabilitation are given. A case is made for developing a national or even an international standard which will provide sound engineering guidance on all pipe materials and a common design methodology where the design equations would essentially stay the same while, material properties would differ from one pipe material to another. As a matter of fact, the same standard could govern gravity and pressure applications, and pipes installed by open cut and trenchless. Even, the same standard for water, sewage, storm, gas, oil, where the factors of safety and level of risk we engineers are willing to take on behalf of our clients could be varied depending on the type of service and the fluid conveyed. Projection of future technological needs in this field conclude this paper.

---

[1] American Ventures, Inc. 2320 85th Place NE, Bellevue, WA, USA 98004; e-mail jeyjey@aol.com; [2] Saline Water Conversion Corporation, P.O. Box 5968, Riyadh, Saudi Arabia; [3] Trenchless Technology International, 33 Railway Parade, Suite 2, Eastwood, NSW, Australia 2122; [4] Gelco Services, 20606 84th Avenue S, Kent, WA, USA 98035; email nodig@aol.com

INTRODUCTION

Safety concerns and space limitations in large cities have forced electrical transmission and distribution lines, television cables, telephone lines and other utilities to go underground in addition to the conventional water, sanitary sewers, storm sewers, and steam lines. The increasing use of underground transit systems to ease traffic congestion problems further crowds the space below grade in many cities and will in other larger cities in the not too distant future. Disruption due to underground pipe failures and leakages cause frequent maintenance repairs which in turn cost more to cities. The need to maintain or increase existing pipeline capacities for transmission, distribution, and disposal requirements has resulted in an increased number of new pipeline projects as well as increased inspection, maintenance, and rehabilitation programs. Pipe products have established their own niche markets based on their past performances while failures are still too common more with some materials than others. Although some revisions have been attempted in standards and codes to reduce pipeline failures, very little progress has been made in this area. Engineers, planners, pipe manufacturers, and contractors need to update themselves on the recent advances and developments in pipeline technology on a continuous basis to remain competent and competitive.

PIPELINE PROJECTS

Pipelines transporting water, storm run-off, process wastes, petroleum, natural gas, slurries, solids, sewage, steam, and other fluids either by gravity or under pressure, continue to be a major part of the lifeline infrastructure of modern cities. Pipeline projects both new and upgrading of existing pipelines will continue to grow to keep up with the economic growth. Even though personal consumption levels of water and other utilities have been levelling off, new consumer needs require additional underground conveyance systems. In addition, infrastructure of most cities are nearing the end of their design life, have already been operating at capacities greater than designed for, and require extensive assessment of their current and future serviceability. Rehabilitation of sanitary sewers using "trenchless" methods has grown in popularity in recent years. When many new pipeline infrastructure components were constructed, inadequate planning was done in regard to the need for future expansion. To compound the problem, many new state, federal, and local regulations have made open excavation virtually impossible in busy streets, close to waterways, and environmentally sensitive areas. For example, worker safety regulations have become much stricter over the years. Federal and state governments' attempts to protect wetlands also have been more vigilant than ever before. In addition, the cities are forced to keep the losses from closing of traffic and detours from open-cut excavation for pipe installation to minimum. These limits have forced most cities, counties, states, and private entities to turn to doing pipeline repairs and renovation primarily by trenchless techniques.

PIPELINE MATERIALS

Materials technology has evolved the most among all aspects of pipeline engineering in the past 10 years. Many new materials and products appeared in the market, and an equal number of products and trade names disappeared. In metals, structural plates for large span corrugated structures were introduced for highway drainage structures. For pressure lines, welded steel has captured most of a market once controlled by prestressed concrete pipe and has moved to increasing use of higher yield strengths particularly for projects where internal pressure governs the wall thickness. The ductile iron industry is increasing pipe sizes up to 1700 mm in North America and 3400mm in Japan, while decreasing wall thickness, a trend which was long overdue.

Concrete pipe industry simply offered nothing new during the past decade, just like the decades before; same ineffective A-Z sacrificial wall thickness method for sulfide corrosion although, all other pipe materials have abandoned this out-dated design philosophy several decades ago. Most efforts by the concrete industry to provide newer bedding systems and fancy computer software to design the pipe better never were put into practice by public agencies, consulting firms, and even pipe suppliers themselves. Because of these factors, the concrete pipe industry has lost an incredible size of its market niche to plastics, metals, and composites. In 1977 only 15 % of the pipe market was done in plastics, while in year 2000, over 50 % of the pipe would be made of plastics. Clay pipe is being used primarily in the trenchless market in smaller sizes, with the exception of a few cities which use clay for open-cut projects as well.

The plastic pipe industry has brought numerous developments during the past decade. Resins such as PVC, MDPE, ABS, and HDPE entered the industry in smooth walls in smaller sizes and later by including fillers, cores, profiles, and corrugations, the industry has been able to make good competitive pipe for sizes up to 3100 mm. The composites also have gained momentum in the past decade.

In summary, performance of each type of pipe invariably dictated its own demise or success. Conservative designs would save a pipe only temporarily in the market place. The nature of the market continued to steer itself to exclude both bad products, and overly conservative good products. Most specifiers have used lessons learned from both past failures and inadequate performances to improve their pipe specifications to give their clients better service.

CORROSION CONTROL, COATINGS, AND LININGS

If one were to calculate the cost/benefit ratios based on pipe stiffness and pipe strength, steel is by far the most economical material available in the market for all types of application. However, monitoring and maintenance against corrosion requires some engineering thought and planning. While the concrete pipe industry has made little progress in dealing with corrosion problems during the past decade, steel and ductile pipe suppliers have developed many viable coatings and linings made of epoxy, coal tar enamel, coal tar epoxy, coal tar urethane, urethane, tape, and extruded polyolefins. Cathodic protection systems also have seen extensive use over the past decade on pipeline projects. The most common lining for water used is still cement mortar with other types used under special circumstances. Due to stricter environmental regulations, many owners are hoping to have the perfect coating and lining systems in the future.

PIPE-SOIL SYSTEM DESIGN

This topic received considerable attention during the past 10 years. Even the basic definition of what makes a pipe rigid or flexible has to compare relative stiffness of the pipe to that of its embedment-native soil system. Thus, the design has to be of pipe-soil system and not each of these components alone. Rigid pipe designs are still done using 70 year old indirect methods dating back to Marston's time in North America while most Asian, Middle Eastern, and European countries have switched to direct design methods where the critical stresses and strains in the pipe wall are calculated and compared with pipe material capacities. Even in the selection of bedding factors for rigid pipe indirect designs, most countries around the world have been using higher values and fewer bedding systems, while North American concrete pipe industry is attempting to use more types of highly impractical standard bedding systems on one hand and using 70 year old values on the other. In summary, the rigid pipe industry unfortunately spent most of its efforts trying to prevent newer pipe materials penetrating the market, while doing nothing worthwhile to improve its own offering to the public agencies. For example, the age-old debate over "should there be a minimum pipe stiffness for flexible pipe?" could be put to rest when one examines a recent very successful project by the Metropolitan Water Districts of Southern California involving a 3700 mm(144 inch) steel pipe of mere 13 mm(0.5 inch) plate thickness in soil embedment of 20 MN/m^2(3,000 psi). In this project, the stiffness contribution from the pipe ring is mere 0.5 % while, the soil envelope interacting with good native ground provided 99.5 % of the stiffness for the pipe-soil system to keep deflection under 2 %. Indeed, the pipe was stulled for the higher temporary stiffness it needed during transport, handling, backfilling, and compaction. Given this remarkable accomplishment, similar pipeline projects could be built of other pipe materials when proper care is afforded for design, construction, quality control, and inspection.

The flexible pipe industry has displaced the rigid pipe industry in a major sector of the pipeline market in the past 10 years. Given this tremendous progress, one would tend to think that clear engineering guidelines would be available by now for the most important design parameter characterizing the soil stiffness for the design of flexible pipe-soil systems. To the contrary, this modulus of soil reaction, E', is still neither defined nor obtained using rational engineering principles by most design engineers.

SELECTION OF E' FOR OPEN-CUT PIPE-SOIL SYSTEM

The E' is not a fundamental geotechnical engineering property of the soil. This property cannot be measured either in the laboratory or in the field. This is an empirical soil-pipe system parameter which could be obtained only from back-calculating by knowing the values of other parameters in the modified Iowa equation. An experienced soil-pipe interaction design engineer would expect the pipe-soil stiffness ratio to have an effect on the value one uses for E' in design. It is interesting to note that the range of E' used for stiff ductile iron pipe ranges from 1 MN/m^2(150 psi) to 5 MN/m^2(700 psi), while much softer plastic pipe is designed with values in the range of 7 MN/m^2(1000 psi) to 20 MN/m^2(3,000 psi).

Flexible pipe designs rely on deflection predictions based on the modified Spangler's equation and the Bureau of Reclamation's E' values. These E' values are very unreliable and a designer could come up with grossly misleading design calculations and specifications for pipe wall, bedding, and trench widths. The value of E' chosen will affect the overall design of the pipeline project and the cost of construction in a major way. Thus, it is important to recognize that the E' is controlled by the following factors:

o depth of soil over the pipe
o size of pipe
o stiffness of the pipe relative to the soil
o trench width
o location of the water table
o native soil type, compaction density, modulus
o bedding soil type, compaction density, modulus

The authors have developed a well-reasoned design procedure where the above factors are carefully considered in the selection of E' values.

SELECTION OF E' FOR TRENCHLESS

Trenchless pipe experiences soil loadings quite different from pipe used in open-cut construction. Other than the broad guidance, where a lower bound of 3.5 MN/m^2(500 psi) and an upper bound of 7.0 MN/m^2(1,000 psi), there is nothing in the literature in the past decade to guide the design engineer on how to select E' for trenchless pipe. The factors which influence the choice of E' for open-cut pipe will determine the E' for trenchless pipe. In addition, the method of installation and the over-cut ratio also will affect the value of E' chosen. Use of good quality finite element analyses of the trenchless pipe-soil system would be the only manner in which a design engineer could establish reasonable E' values for trenchless projects for all flexible pipe materials. In this method, the soil surrounding the jacked pipe would be modelled as a nonlinear stress-dependent stress-strain material. The construction sequence where, the soil is excavated at the face of the tunnel, followed by the insertion of the pipe with a jacking force would also be accurately represented in a step-by-step sequential analysis. The pipe would be modelled as a series of beam elements with bending and axial resistance in both in-line and perpendicular directions.

In summary, the design engineer needs one universal design methodology and a common national standard for designing all pipe materials where checks are made for internal pressure, external load-induced bending strain/stress, external load-induced deflection, stress/strain from combined loading, handling stiffness and strength, and buckling. Finite element analyses could provide more economical designs and alternate tools to verify designs done by conventional design procedures. It is also essential to recognize that the design process cannot be separated completely from field performance and inspection. In all projects, detailed design calculations need to be done to cope with longitudinal stresses and strains caused by uneven soil support, uneven settlement, varying geological conditions, poisson's effects, seismic loads, and thermal loads. Three dimensional analyses are very vital for specials, bends, and critical sections of the pipeline system.

PIPELINE CONSTRUCTION

Pipeline construction methods have remained the same in the past decade. However, some contractors and designers are now beginning to realize the importance of the pipe-soil system instead of designing the pipe and laying of the pipe as two separate items. Claims by contractors based on change of site conditions continued to be the most frequent method of draining more resources from public agencies. Using the first 30 day period to test how vigilant owners and their agents are in ensuring that the contractor performs in accordance with the project specifications also continues to be the trick of the trade among the constructors of pipeline projects. Many contractors are preferring to use beddings with either no compaction or some amount of cement or other admixtures to avoid conflicts with site inspectors and to increase daily production rates. Groundwater control with the use of the most appropriate dewatering method chosen as a function of geotechnical conditions of the site need to be recognized for all pipeline projects. Use of most suitable joints also is an important aspect of pipeline construction. Acceptance testing of ground surface settlement/heave, pipe deflection, zero leakage at the pipe plant and in the field are becoming most commonly used tools to monitor the quality of workmanship of the pipe supplier and the constructor even in the third world countries. Unfortunately, yielding to strong pipe lobbies, most public agencies in North America are still unable to enforce many quality control/assurance programs such as zero leakage.

The most major advance in pipeline technology in the past decade has been in the area of trenchless methods for new installations and for renovation of aging pipelines. The following sections provide an overview of this progress.

TRENCHLESS TECHNOLOGY FOR NEW PIPELINES

Although many terms are used in the industry to subdivide trenchless technology for new pipe construction, all such methods fit into just three major groups as described below.

Pipe Ramming

Pipe ramming is the least automated, cheapest, and quickest form of pipe installation, but it suffers from lack of steering capability. Only certain pipe materials can be handled by this method and the drift is usually about 2 % or more of the drive length. Hard inclusions cause major difficulties with this method. Pipe sizes up to 1500 mm and drive lengths up to 100 m can be done with these methods. One-third of all projects are done with these methods in North America and Europe.

Directional Drilling

In directional drilling methods, a pilot bore is made and the product pipe is pulled right behind the backreaming operation on the return trip from the exit point to the starting point. This method cannot be used for minimum grade gravity sewer systems. The drift control is within inches using electromagnetic tracking systems. Cobbles and boulders cause serious problems and may prevent finishing of the line. Usually, fluid additions reduce friction on the outer wall of the pipe, assist in cutting of the bore hole, and prevent the borehole from collapsing. Most projects use either steel or HDPE pipe materials for renovation or new line installation but copper pipes and cables also can be installed by this technology. The capability is subdivided into mini, midi, and maxi where pipe sizes up to 900 mm for a distance of up to 1,500 m to a depth of 30 m with pull forces up to 250 tons could be handled for gas, sewage, water, and other utilities.

Microtunneling

Microtunneling and pipe jacking technology is an art of accurately installing non-man entry sized pipelines without disrupting the daily routine of urban life. Remotely controlled tunneling systems in pipes smaller than 900 mm are normally referred to as microtunneling. However lately this method is used for pipes as large as 3600 mm for drive lengths of over 1000 m in most soil conditions. The tunnel excavation is done in controlled conditions in which the tunnel face and the ground water pressure are constantly balanced by counter exerting an equivalent force by the tunnel shield. Either auger or slurry soil removal systems are employed. The precision is within 25 mm on line and grade.

The advantages of this technology are many folds. Primary advantage is that the technology offers an efficient, effective, and faster method of installing underground pipelines without affecting social, commercial, and indutrial activities of the people and is environmentally friendly. Its cost does not increase in direct proportion to the depth of the sewer, thus giving an opportunity to the system designer to put deeper pipes and to do away with pumping stations. It can handle wide variety of soil conditions, it has greater precision than traditional dig and replace methods, it is completely remote controlled, safer, least disruption to surface structures and traffic, and can be used for contaminated soils. The primary disadvantages are this method are uneconomical with limited length of installation, mixed soils cause problems, there is poor access to remove any unexpected obstruction and attend to any breakdown of either the exacavation head or the pipe.

Microtunneling was first introduced in Japan in 1975 and Singapore adopted this technology in 1982. Due to wide national acceptance of this technology in Singapore, most authorities impose stringent requirements, if a contractor were to offer open-cut method as an option. More than any other country in the world, Singapore has provided the very conditions needed for microtunneling to be method of choice for most pipe construction. Singapore has installed more than 100 km of sewers in the past decade using the technology. Thus, Singapore's exposure to the technology and its accumulated know-how had enabled many neighboring countries to adopt the technology to install sewers in their cities. Thailand, Taiwan, Indonesia, Malaysia, Philippines, Hongkong, China, and Korea have been following Singapore's success in recent years. Germany and France used the technology in 1982 and UK in 1983. U.S.A. in 1984 started in microtunneling and now the U.S. provides a considerable market in the world for such projects, while German and Japanese companies supply most equipment. Middle Eastern countries such as U.A.E., Saudi Arabia, and Kuwait have used some microtunneling to install sewers. Australia used the technology in 1987. Since then Australian made microtunneling system is used to install sewers in self supporting dry ground. Currently several microtunneling projects in Sydney and Melbourne are under way using the engineering know-how brought in from Singapore by Trenchless Technology International.

In summary, 38 countries have used microtunneling during the past 10 years for installing sewers and water pipelines in wide variety of ground conditions for pipe sizes up to 3600 mm with record drive lengths in excess of 1000 m.

INSPECTION OF AGING PIPELINES AND MANHOLES

The condition assessment of aging pipeline systems, materials, and methods used for pipeline and manhole rehabilitation have seen numerous advances in the past 10 years. Many well known technologies for investigation have been either supplemented or replaced with newer methods. Closed circuit televising, smoke testing, dye-water testing, infrared thermography, radar and sonic devices, groundwater monitoring, ground surface observations, flow monitoring, corrosion evaluation are all technologies in use to some degree or another in pipeline assessment and most of these tools have seen significant improvement over the years. Many new pipe materials and installation methods for pipeline system rehabilitation have entered the market place while several old ones are becoming obsolete. In a broad sense, sliplining, cured in place lining, deformed lining, fold and form lining, segmented lining, grouting, fill and drain repairing, multi-function robotic repairing are all methods using many new materials year after year.

Closed circuit television (CCTV) inspection is an excellent example. Ten years ago the standard of the industry was black and white equipment. The camera was pulled through the pipeline manhole to manhole. Today the standard is for the picture to be in color, with the ability to look directly into any lateral to determine if the pipe is actively serving a building or whether it is a source for infiltration. Pulling the camera through the system is still the most common method for inspection but the trend is for cameras to be mounted upon a tractor unit, which allows for inspection up to a specific location in the pipeline system and retracking back to the original manhole through which access was made possible.

Preliminary System Evaluation

The main objectives in performing the preliminary system evaluation are to identify the local areas with the most problems and to prepare scope for subsequent detailed investigation. The major sources of information include the following:

o as built pipeline maps
o operation and maintenance records
o geological, geotechnical, climatological records
o topographical maps
o city and municipal planning records
o treatment plant records
o system monitoring records
o historical sewage flow records
o water usage records
o population trends and user surveys
o industrial surveys
o corrosion records

Rehabilitation of pipeline systems using "trenchless" methods has grown in popularity in recent years. With the advancement of television cameras communities can determine if a segment of pipe needs to be repaired or replaced, or whether the total pipeline system needs to be addressed. A condition assessment should be performed prior to making any such repairs, unless the work is of an emergency nature. In recent years, infiltration and inflow in pipeline systems nationwide have caused major problems. Sanitary sewer overflow problems also have come to the front of the national attention. If infiltration and inflow problems are not addressed properly, the pipeline system will become short of their capacity much earlier than their intended design life and the cost of treating sewage would become quite high. Thus, an accurate assessment of the infiltration and inflow problem need to be done very early on in the evaluation of the pipeline system.

Assessing the Structural Condition

The structural condition evaluation involves the following steps:
o visual inspection and recording
o delamination sounding

o estimate of wall thickness loss
o causes of wall thickness loss
o estimate of changes in pipe material properties
o inspection of soil conditions
o record of structural loadings and anticipated changes
o groundwater level and seasonal variations
o changes in pipe wall geometry
o design calculations to determine the remaining structural capacity

Evaluation of structural capacity is the most difficult task among all phases of pipeline system evaluation and it is prudent that the team involved in investigation allocate adequate resources and time for this part of the work. Geotechnical characteristics of sites for sewer systems could be documented with more modern tools such as infrared thermography, seismic measurements, ground penetrating radars, continuous soil sampling and strength evaluation with electric and piezo-cone soundings, and pressuremeter testing. Use of seismic wave propagation techniques across the pipe wall thickness at selected stations and smart pigs which could travel down the pipeline system to determine the wall thickness still left in an aging system are all becoming more reliable for routine use in structural condition evaluations.

Assessing the Hydraulic Condition

Some form of controlled tests are necessary on an aging pipeline system to establish the drop in flow characteristics and to be able to estimate the available hydraulic capacity. It is most common for the Manning's coefficient used for flow roughness calculations to increase with age in a pipeline system and this would normally be worst for concrete pipes due to significant uneven loss of wall thickness from sulfide corrosion. The increase in roughness coefficient is due to laterals protruding into the main waterway of the pipes, walls deflecting excessively into the pipes, structural members such as brickwork moving into the pipes, and joints becoming badly fitted with time, all contributing to a substantial reduction in the flow capacity of the pipeline system.

Considerations in Selection of Suitable Technology

The factors to consider in selecting the most suitable technology for pipeline rehabilitation are as follows:

o design life
o sizes of pipe and manholes
o lengths of the reaches and access
o type of fluid carried
o soil and groundwater conditions
o past track record of the technology
o availability of qualified contractors
o structural condition of the aging system
o hydraulic capacity needs

UNDERGROUND PIPELINE DESIGN

- o sizes and number of laterals
- o depth of the sewer system
- o traffic patterns
- o environmental concerns

Some descriptions of various technologies we have used in the past decade, their potential, strengths and weaknesses are given below.

Cleaning

In many pipeline systems and their components, proper cleaning alone could return most of the lost capacity back to the line. Lack of sufficient flow capacity in old pipelines have generated many new pipeline construction projects. In this regard, jet cleaning is most commonly used to renovate pipeline systems. A CCTV video is prepared before any cleaning is commenced as a baseline measurement of the current condition of the system and after jet cleaning, a second video is prepared to determine the level of rejuvenation done to the pipeline systems. If a lining method using a paint mixed into an epoxy hardener would enhance the pipeline system capacity, this is always undertaken following the jet cleaning operations. It is important to recognize that the skill level and past experience of the operators employed by the city would determine the degree of success of cleaning operations. Although chemical cleaning is being tried with some good results for water distribution systems, use of such technology for sewer system cleaning would have to ensure that the chemicals utilized do not interfere with the treatment functions of the sewage treatment plants. The best tool for verifying the effectiveness of cleaning efforts is CCTV.

Root Control and Removal

Chemical root control is commonly used to kill root growth into the sewer system and to inhibit regrowth without causing significant damage to trees and plants, the ambient environment and the wastewater treatment process. It is necessary to recognize that the tools used in root control could also cause damage to the walls of the pipes, if proper care is not taken. The usual ingredient for killing roots is a special herbicide that kills roots at low concentrations. Common materials include sodium methyldithiocarbonate, diclobenil and other. Although some attempts were made to include formation of copper sulfate solution insitu by providing copper wires to react with other chemicals around pipes with root intrusion problems, due to the contamination caused, this approach is no longer accepted by the U.S.EPA.

TRENCHLESS AND LESS-TRENCH METHODS

There are numerous technologies which have been in use for some time for rehabilitation of sanitary sewers and manholes. These methods could be broadly classified into two types namely, "trenchless" and "less-trench". In the trenchless methods, the process needs to be applied from one existing manhole to the next one with the laterals needing reconnection using robotic cutters. It is most common not to have any open-cut excavation at all in or around the job site. Examples of such methods are Cured in Place Pipes, Fold and Formed Pipes, Pipes by Directional Drilling, Robotic Repairs, and Fill & Drain Technologies. In the less-trench methods, some excavation is always required. These are used to introduce the new pipe into the system and for reactivating the laterals. Some processes require pits, auger holes, while others need sloping trenches leading into the old pipe. Few examples are Sliplinings, Swaged & Rolled Down Pipes, Spiral Wound Pipes, Segemented Linings, Pipe Bursting, Microtunneling, Pipe Ramming, and Manhole Rehabilitation Technologies.

Coatings

Reinforced shotcrete and cast-in-place concrete are possible coatings only when corrosion is not a problem. In the presence of sulfide corrosion, one would not use either of these two coatings. Other engineering materials and resins also could be used with or without fibrous reinforcements for coatings.

Point Repairs

Most of the spot repair systems currently available involve the application of chemical grout to fill cracks, a cured-in-place sleeve or mat, or an epoxy-based resin. Differences lie in the method of application, with some processes using robotic devices, a winched-into-place sleeve, or a combination of methods. The latest advances in point repair include performance liner, link-pipe, amkrete, econoliner, fibers in gunite or shotcrete and chemical grouts. In these methods either a preformed segment of a new pipe is inserted into the right location where the existing sewer system needs renovation and the new pipe is expanded or cured to form a new structural and/or hydraulic liner to provide the point repair. The most effective point repair is to use a grout either cementitious or chemical using an inflatable packer guided by a close circuit television camera to the location of the sewer system needing the repair.

Chemical Grouting

Although the earliest form of pipe rehabilitation technology involved some form of chemical grouting, this method is still seeing many new developments. Use of many new chemical and cementitious grouts with varying properties, set times, and functions are emerging in the market place. The most common type include epoxy, gels, sodium silicate, acrylamide grout, acrylic grout, acrylate grout, urethane grout, and urethane foam. Cementitious grouts continue to see several enhancements such as the use of centrifugally acting spraying machines which travel along the pipeline and provide new coatings and linings completely unmanned, use of admixtures, and fibrous reinforcements to increase tensile and corrosion characteristics of such grouts. The internal grouting is usually applied using a remotely inflatable packer guided by CCTV, the external grouting process involves a detailed study of soil conditions to establish which grout would flow into the subsurface around the pipe effectively. Chemical grouting has been successfully used for over 30 years. The success rates of grouting was based upon the level of competence of the contractor and the ability of the engineer to design and specify the project out in a manner which would guarantee good results. Chemical grouts, which use a packer to first air test a joint to test its water tightness. If the joint fails the air test then the joint is grouted using a grout with a viscosity of less than 10 cps. This process allows for a pipeline system installed in the 1940's to be brought up to 1990 standards. Chemical grouting as a form of maintenance for pipelines is increasing in popularity. Many managers of aging systems are seizing upon this technology to ensure additional longevity of their pipes. The process must be repeated after a period of time. Epoxy placement is performed with robotic tools, again with the aid of CCTV. Some of these processes utilize routing bits to allow the epoxy to bond to a clean surface.

Cured in Place Pipes

Cured in Place Pipe (CIPP) systems enable sewer pipelines to be repaired from within by insertion of a lining material through existing manholes. The liner is composed of a fabric reconstruction tube which is impregnated with a thermosetting resin that hardens into a structurally sound jointless pipe when exposed to hot circulating water or steam. Once cured the pipe is allowed to gradually cool to prevent thermoshock and then the laterals are reconnected. The rehabilitation liner not only serves to repair the deteriorated structure of the existing pipe, but reduces infiltration of unwanted ground water. The CIPP process was introduced in the United States in 1977 and millions of feet of CIPP have since been installed. Until recently, competition was almost non-existent from other trenchless rehabilitation products and the primary competitors with CIPP were open-cut construction and sliplining. In the mid-80's several new CIPP products were introduced. The fact that multiple options exist for owners to consider for pipe renovation has its strengths and weaknesses. For example, on one hand there is more competition among many technologies, while on the other, many of the new products in the market place do not have research results comparable to those collected by the original technology provider.

Cured in Place Pipes are either winched into place or inverted in place using air or water pressure. The curing process is done using either steam or hot water. All thermoset resins commonly used for RPM and FRP pipes are used for Cured in Place Pipes. Insitu cutters guided by CCTV provide means for opening the laterals. The liner pipe is designed either to perform just the hydraulic function or both hydraulic and structural functions depending on whether the existing pipe has structural capacity intact or not. The most significant advantage of Cured in Place Pipe is that it could be used for old pipes of any size and shape. New innovations include using reinforced felt to handle internal pressures of substantially high values.

Fold and Formed Pipes

Fold and Form Pipe (FFP) allows for pipelines to be repaired through existing manholes as the CIPP process. The FFP system uses a thermoplastic material which have been deformed from a circular shape, i.e., folded, to result in a smaller cross-section that can be easily fed into an existing sewer. These products utilize either extruded polyvinyl chloride (PVC) or high density polyethylene (HDPE) pipe that is flattened and folded longitudinally. The plastic pipe is fed from a spool into an existing pipe where hot water or steam is applied until the liner reaches its temperature for rounding. After rounding the materials are allowed to cool and then the laterals are reactivated. As is the case with CIPP process additional testing is needed to determine the longevity of all these products.

Softer PVC cell class resins and HDPE folded pipe require minimal heating to form the pipe back to its initial shape. For folded pipes, it is common to use a rounding device and apply the technology effectively when the old pipe is not surrounded by excess amounts of groundwater. In the presence of groundwater the heat input from the source would not be sufficient to keep up with the loss of heat into the ambient ground, unless a heat containment tube is utilized. It is important to recognize that the fold and formed pipes using HDPE resins could never come close to those pipes using PVC resins in the pipe stiffness property particularly to meet long term buckling capacity requirements. The HDPE fold and formed pipe has a mere 10 % of equally thick PVC fold formed pipe when it comes to long term buckling strength. In all applications of pipeline rehabilitation, the long term buckling strength would govern the choice of the pipe material. Thus, HDPE in essence would never be the material of choice in comparison to those pipes made of PVC resins. It is important to design the fold and formed pipes to withstand external groundwater pressures and for soil and traffic loads if the old pipe has deteriorated far enough not to provide appreciable support to the new pipe. The testing of fold and formed pipes for strength and stiffness should be done preferably on fold and formed pipe and not on the pipe made at the plant before it is folded.

Robotic Repairs

Intelligent robots which could brush away the dirt in the pipeline systems, jet clean the pipe walls, cut down the root growth back to the pipe wall, fill holes with proper grouts, mill away damaged and badly fitted pipes and laterals, perform point structural or hydraulic repairs, provide continuous video record of the internal condition of the pipeline systems are becoming more and more popular.

Fill and Drain Methods

In this method, two chemical solutions which react with one another when they are brought into contact are filled and pumped out one after another into the pipeline system to form a third material which provide structural repair to all components of the system in a monolithic fashion. Although the same method and materials could be used for any existing pipe material and for all components of the pipeline systems, there are some drawbacks in this method of repair such as clogging of the main waterway in the smaller sized pipeline systems and when these products come into contact with roots.

Sliplinings

In this process in its most recent form, a new pipe is inserted either by pulling or pushing in continuous length or in short discrete lengths. It is very common to have the annulus grouted with proper care once the slipliner is inserted, if no compression fit is present. Without the grout the slipliner will not be able to withstand most of the external water pressure and other buckling loads. The most preferred material for sliplining is HDPE due to its superior characteristics in corrosion resistance, abrasion, impact strength, and strain tolerance. The design checks should lead to the selection of proper wall thickness for the liner pipe. In addition to ensuring that the hydraulic capacity does not diminish excessively, the structural strength and stiffness and the liner pipe and its composite action with the existing old pipe needs to be evaluated. The liner pipe should be able to withstand grouting pressures during construction in addition to the pull or push force in the axial direction. Also, the capacity of the liner-existing pipe composite to withstand the soil loads, live loads, and groundwater loads need to be checked. In many situations, use of simple ASTM or AWWA equations for structural checks could lead to either under-design or over-design of the liner pipe. Detailed structural analysis calculations using tools such as the finite element method is always a more cost-effective and reliable means for evaluating such composite structures.

Continuous Pipes

A nose cone is attached to a butt-fusion welded HDPE for continuous pulling or pushing into a sewer requiring rehabilitation. It is common to clean the line before sliplining operations start.

Discrete Pipes

Short pipes using either gasketed or mechanical joints are inserted insitu using hydraulic jacks. Smooth wall HDPE, profiled walled HDPE, profiles walled PVC, FRP, RPM, Ductile Iron, Steel, Clayware are all used quite effectively in discrete pipe form to slipline pipeline systems. The design checks for discrete pipes should include the evaluation of the joints for adequate structural strength and stiffness.

Swaged Pipes

Because HDPE is a soft resin, this permits swaging or rolling down of this pipe to be able to insert into smaller ID old pipe. Use of swaging die, gradually sized down rollers, etc. are used to insert HDPE pipes into old pipe. The design of swaged pipe would be similar to that of slipliner pipe.

Spiral Wound Pipes

Spiral wound pipes are so weak structurally, even with proper grouting behind them, this technology could never be a match to Cured in Place Processes for pipe repair. In the spiral wound process, the seam is made using the ribs located at the edges of the extruded strip and the pipe is wound into an insertable form at the bottom of the manhole in small sizes. In larger sizes, either segmented spiral pipe or continuous spiral pipe is unwound inside the broken pipe while grouting the annulus for strength and stiffness. In essence the spiral pipe acts only as a formwork to make the grout liner and whatever grouting one does is the main structural help this process can give to the old pipe. Some insitu testing of the effectiveness of the grouting is very essential.

Segmented Linings

Segmented linings are suitable for large sized pipes of various shapes. These are installed through either manholes or special access shafts built during the repair contract. The sewer has to be dry and bypassing is commonly employed in this method. The common materials are fiberglass reinforced cement, fiberglass reinforced plastic, reinforced plastic mortar, PVC, HDPE, and steel.

Pipe Bursting

Pipe bursting falls as a matter of fact within the generic category of sliplining when upsizing of a system is required. Pipe bursting is generally not used if congestion underground is a question or if the existing pipeline is not of a brittle nature. In smaller sized pipes, pipe bursting becomes a viable tool where, a hydraulically activated cutting head breaks the old pipe and pushes into the native ground making way for the new pipe to take its place up to almost twice its size. This method however, has major noise and vibration problems, somewhat uneconomical if too many laterals have to be reconnected.

Manhole Rehabilitation

Chemical grouting, epoxy linings, waterproofings, sealings, coatings, structural linings, cementitious grouting are the primary methods available for manholes.

FUTURE OF THE INDUSTRY

Plastics and in particular made of certain recycled resins will see wider usage in pipeline projects. Larger sized plastic pipe can compete better and offer a viable alternative construction material to concrete pipe. Corrosion resistance should be one of the key factors considered in the selection of pipeline materials. Many countries are already using a common standard for designing and specifying all pipe materials. These countries are also writing zero leakage specifications demanding better quality materials and service from suppliers, contractors and design engineers. Most design calculations in the future will be done using fully integrated technologies. Due to environmental concerns and other societal factors, increasing number of pipeline projects would be constructed using trenchless methods in the coming years. Among these, microtunneling and pipe jacking would play a major role. The engineering of pipeline projects by microtunneling methods should be designed using procedures which represent actual field loading conditions in a realistic manner. The pipe should be analysed for both circumferential and longitudinal behavior under construction and service load conditions. Enough care should be exercised in the collection and use of geotechnical data from past records and subsurface explorations in making key decisions on pipeline projects. A textbook selection of E' from sources such as AWWA, ASCE, ASTM, and the Bureau of Reclamation should be avoided on all projects. Estimation and use of E' should be based on field and laboratory testing of native ground conditions and embedment materials, proper evaluation of all factors controlling pipe-soil interaction, and good engineering judgement. The pipeline design should be carried out by experienced engineers who are fully familiar with geotechnical and structural design principles of pipelines and related soil-structure interaction concepts. Field monitoring is very essential as a measure of checking on the quality of the contractor's workmanship and for collecting useful fullscale data to guide future microtunneling project designs. The technologies used for the assessment of current conditions of various components of the sewer system need further improvements. Taking videos of the inside of the pipe, manholes, and laterals, situation analysis for the structural condition of the pipes, chemical and hydraulic condition studies are all which could be done in a single pass of the inspection equipment from one end of the line to the end of the site in question. The industry will continue to use chemical grouting as a versatile and effective method for repairing pipe systems. More monolithic repair technologies such as the fill and drain methods will prove to be very convenient for public agencies to apply for all components of the pipeline systems, particularly when they have pipes made of many different materials. Robots which could think for themselves and select the most appropriate repair methods among its full array of technologies using fuzzy logic systems would make pipeline system evaluation and renovation become highly automated. In this regard, the human bias causing the public agency to use any technology rather than the most effective one could be eliminated.

Acknowledgements

The following companies and groups provided useful data to make this paper possible: Gelco Services, Insituform Technologies, Insituform of Mid America, Pipe Rehab International, Avanti, DeNeef Chemicals, Nupipe, Uliner, AMliner, Airrigation, Tobys, Performance Liner, Chevron Plexco-Spirolite, APM Permaform, Permalok, Lamson Sessions, LinkPipe, Amkrete, Sika, Iseki, Herrenkenecht, Soltau, Akkerman, Flowmole, Vermeer, Straightline, Clearline, TT Technologies, British Gas, Paltem, Hobas, Sarplast, Powermole, Sanipor, Strong Seal, Owens Corning Fiberglass, North American Society for Trenchless Technology, American Society of Civil Engineers, Water Environment Foundation, Trenchless Technology magazine, American Water Works Association, and the U.S. Environmental Protection Agency. Ms. Menaka Jeyapalan offered valuable editorial assistance on the manuscript.

A SOIL-STRUCTURE INTERACTIVE MODEL
Marcel GERBAULT[1]

Abstract

In structural analysis of buried pipes, the distribution of the soil reaction pressures is generally supposed of a given shape, like for instance in the Spangler's model, where this shape is assumed to be parabolic and the deflection elliptic. For flexible pipes that is not always the case. An other model is proposed, where the soil inter-action is simulated by distributed springs placed around the pipe. When the rigidity of the springs is supposed constant, the analysis allows to get results only by way of hand-calculation. When this rigidity varies, simulating pipe installation conditions, and when an elasto-plastic constitutive law, simulating soil behaviour, is given to these springs, an analytic process, using transfer-matrix technic, is proposed. This model allows for 2nd. order effects, taking into account buckling, or 're-rounding' in case of pressure pipe.

1- Introduction

Often, in design of flexible buried pipes, the distribution of the soil reaction pressure is supposed to have a given shape, like for instance a parabolic shape in Spangler's model. These models correspond to an elliptical ovalization. In some circumstances, other types of deflections were observed, like 'squarring shape' or those called 'wild deformations'. Of course these models are not able to predict such deformations because they are only applicable to elliptical deflection.

An other issue of a model, like Spangler's, is the the fact that ovalization is linear versus load. It cannot predict amplification of pipe response with load increase, leading to instability like buckling.

The use of such models and formula for determining an important parameter, the soil elasticity modulus E_s, from measurements, using back calculation, will inevitably lead to misuses and errors. A simple proof is to try to determine the ring stiffness S of a pipe installed in a given soil (E_s given) leading to a given ovalization (for instance 7%). When E_s is large enough, negative S value is found, which is obviously impossible.

----

[1]Prof. E.N.P.C. ; SADE, 28 rue de La Baume, F-75008-PARIS, FRANCE

In this paper, an other model of calculation of buried pipes is proposed. It deals only with ring effects, such as ovalization, and ignores longitudinal effects, such as bending moments due to differential settlements. The pipe is assumed to have a linear elastic behaviour and the surrounding soil is modelized by distributed springs perpendicular to pipe wall. In a first step, in order to allow to do only formal hand-calculation, the rigidity of these springs is supposed constant. It is this model which was used to determine the critical buckling pressure, for which no stability is possible, also known as 'Luscher's formula'. In a second step, more refined assumptions are considered, as well on pressure distribution as on the soil constitutive law, which is elasto-plastic. This last method uses the matrix-transfer technique : the cross section of pipe is divided into a given number of segments, called elements. Each element in contact with soil rests on uniformly distributed springs, orthogonal to the element, representing the soil. This method may be applied to cross section shapes other than circular, for instance ovoid, whose initial geometrical shape has to be described. It supposes the use of a computer.

2- Description of the model

In order to describe simply the method, the cross section of the pipe is supposed circular and the loading symmetrical about the vertical axis, thus only half of the pipe is modelled. A unit length of pipe is considered, of wall-thickness e and mean radius R. For composite or non homogeneous material, the pipe may be characterized by R and its ring stiffness S.

For an homogeneous pipe material : $S = \dfrac{EI}{D^3}$ where E is the elasticity Young's modulus, $D = 2R$ and I is the flexural inertia, $I = \dfrac{e^3}{12\left(1-v^2\right)}$ where v is Poisson's ratio for the material which allows to take into account the shell effect, instead of juxtaposition of simple rings.

The rigidity of the distributed springs representing the soil interaction is β : that means that for a given radial displacement δv, the reaction pressure is $\delta p = -\beta \delta v$. The dimensions of β are [F.L^{-3}]. Its value can be related to the soil elasticity modulus E_s (see for instance Ref [1]). $\beta = \dfrac{E_s}{R\left(1-v_s^2\right)}$ and v_s is Poisson's ratio for the soil. In the 1st part of this paper β is constant. In the 2nd part, the elasticity of these springs may vary around the pipe and is limited by a lower bound, corresponding to the active pressure, and an upper bound, corresponding to the passive pressure of soil. These two limits are determined by the Mohr-Coulomb criterion, defined by the data of the soil cohesion c and its internal friction angle φ.

When necessary, an out-of-roundness e_0 is introduced ; this may be useful when verifying a flexible pipe against buckling problems. e_0 represents the departure from circularity, for instance due to manufacturing tolerances, before any load application (fig. 1).

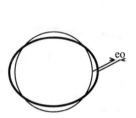
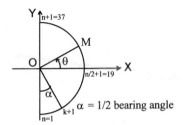

Fig. 1 : out-of-roundness   Fig. 2 : half pipe

3- First part : E_s is constant

In order to simplify the formulation of the equations, E_s as well as vertical and horizontal earth pressure, resp. p_v and p_h, are supposed constant (fig. 3). An hydrostatic pressure p_w may also be applied, orthogonal to pipe wall, with a positive sign when it is an external pressure (in presence of a water table), or negative for an internal pressure.

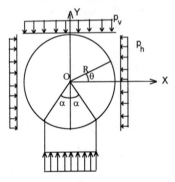

The description of the process which gives the differential equation governing this model was published in [1]. The effects of 2nd order were taken into account. It consists in writing the equilibrium equations for the displaced position, and not the initial one, which is circular. The main results are given here under.

At first, the discussion of the solution shows a critical value of the 'spherical' component of pressure $\bar{p} = p_w + \dfrac{p_v + p_h}{2}$

Fig. 3 : pressure distribution

The critical, or buckling, pressure is : $p_{cr} = (n_0^2 - 1)\dfrac{EI}{R^3} + \dfrac{\beta R}{n_0^2 - 1}$ (1)

with n_0 integer ≥ 2, which minimizes (1). n_0 is the number of 'waves' of the buckling shape.

When β is replaced by $\dfrac{E_s}{R(1 - v_s^2)}$ and when n_0 is great enough $(n_0 \geq 4)$, we get

$p_{cr} \cong \sqrt{\dfrac{32 EI\, E_s}{(1 - v_s^2) D^3}}$, expression very similar to Luscher's one.

In order to make sensible the role played by the parameter $\bar{p}$, fig. 4 shows the decomposition of pressures given in fig. 3 into two components : the spherical one $\bar{p}$ and the deviatoric one $\dfrac{p_v - p_h}{2}$.

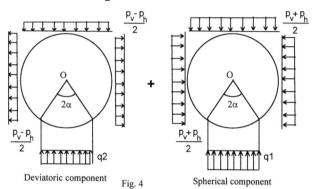

Deviatoric component Fig. 4 Spherical component

The deviatoric component gives the elliptical ovalization to the pipe, while the spherical one, similar to an hydrostatic pressure $\bar{p}$, increases this deflection when $\bar{p} > 0$, or decreases it when $\bar{p} < 0$. That is the reason why for flexible pipes 2nd. order effects have to be taken into account. By the way, with this model, increasing in $\bar{p}$ induces a beginning of buckling.

Let $p_h = k_2 \, p_v$, $s = \dfrac{E_s}{8S(1 - \nu_s^2)}$, $\lambda = \dfrac{\bar{p}}{8S}$, A = amplification factor $\dfrac{1}{1 - \dfrac{\bar{p}}{p_{cr}}}$

The other results are :

Bending moment : $M = p_v \dfrac{D^2}{4} \dfrac{K_\alpha - \dfrac{k_2}{4}}{1 + \dfrac{s}{9} - \dfrac{\lambda}{3}} + (n_0^2 - 1)(A - 1) 8 S e_o \dfrac{D}{2}$ (2)

Ovalization : $\dfrac{\delta D}{D} = OV_1 + OV_2 = p_v \dfrac{k_\alpha - \dfrac{k_2}{12}}{8S + \dfrac{E_s}{9(1 - \nu_s^2)} - \dfrac{\bar{p}}{3}} + 2(A - 1)\dfrac{e_o}{D}$ (3)

Maximum flexural strain : $\varepsilon = \dfrac{K_\alpha - \dfrac{k_2}{4}}{k_\alpha - \dfrac{k_2}{12}} \dfrac{e}{D} OV_1 + (n_0^2 - 1)\dfrac{e}{D} OV_2$ (4)

K_α and k_α are coefficients given in table 1.

2α	0°	60°	90°	120°	150°	180°
K_α Top Spring Bottom	0.2994 -0.3067 0.5872	0.2863 -0.2933 0.3772	0.2736 -0.2794 0.3140	0.2615 -0.2651 0.2754	0.2530 -0.2541 0.2558	0.2500 -0.2500 0.2500
k_α	0.1161	0.1053	0.0966	0.0893	0.0848	0.0833

Table 1

Remarks

a) Amplification factor A becomes a reduction factor when the internal pressure is large enough (re-rounding effect).

b) For rigid pipes s, λ and (A-1) are negligible. Equ. 2 gives the classical result on bending moment.

c) When 1st order only is considered, equ. 3 is reduced to $\dfrac{\delta D}{D} = p_v \dfrac{k_\alpha - \dfrac{k_2}{12}}{8S + \dfrac{E_s}{9(1-v_s^2)}}$

very similar to Spangler's equation. Differences are :

- initial horizontal soil pressure is taken into account $(k_2 \neq 0)$;

- no lag factor ; the change in deflection with time is included in S and E_s using their long term values, including creep ;

- the coefficient of E_s is $\cong 0.126$ (as $v_s \cong 0.3$) which is exactly twice 0.063 the figure in Spangler's equation. This is due to the fact that in Spangler's model the soil springs are horizontal only, while in our model springs are isotropic, which is equivalent to the superposition of horizontal and vertical springs of the same rigidity.

This explains that no reduction factor on vertical soil pressure (i.e. Marston's effect for flexible pipes) has to be applied, because it is already included in the pipe response, as soil-structure interaction is integrated in this model. The Marston's reduction has not to be accounted twice. On the other hand, vertical soil load majoration on rigid pipe, due to silo effect or differential settlements, has to be considered, because it is not part of this model and has to be taken into account as 'boundaries conditions'.

Limitations of validity of this model come from the basic assumptions and the way the results were drawn up :

- initial horizontal soil pressure p_h and hydrostatic pressure p_w constant - no gradient considered -.

- β or E_s constant. As a matter of fact, soil has not a linear behaviour but has lower and upper bounds due to plasticity, and E_s may vary all around the pipe depending on installation conditions.
- The distributed springs are supposed bilateral ; they allow for compressive and tensile reaction pressure. But the 'initial' soil pressures p_v and p_h give a compressive pressure distribution and reaction pressures induce only variation in compression.
- If the pipe is placed too close from surface (H, height of earth cover upon pipe, is such that H < 2D) then E_s (or springs rigidity) should be reduced.
- Ovalization $\dfrac{\delta D}{D}$ has to be limited to 15% and $\bar{p} < p_{cr}/2$

The 3 main parameters 2α, E_s and k_2 depend on the nature of the soil in contact with the pipe and in its vicinity, its degree of compaction. They are closely bound to installation method and the way trench wall protections, if any, are removed.

4 - Second part : E_s is variable

4.1 - General

In this part, more refined hypothesis are made. Gradient of soil and hydrostatic pressure are taken into account. The rigidity of the distributed springs may vary around pipe. Distinction is made between 4 areas : bedding, spring and top of the pipe, and the area between bedding and spring sides, where compaction is difficult to achieve and where possible voids may exist. An elasto-plastic behaviour of soil is considered. Hand calculation is no more possible. Transfer-matrix technic is used in a computer program.

In order to take into account the 2nd order effects, i.e. to write the equilibrium equations in the displaced configuration, loads are applied with a given number of increments. For each increment the state of the soil in contact with each element is considered with respect to plasticity conditions and the matrix of rigidity is reassessed.

The pipe is assumed to be supported vertically on a given bearing angle 2α. Its shape is described in general axes (O,X,Y) ; the origin O is in the centerline.

e_0 being the initial out-of-roundness, (see fig. 1 and 2) the polar radius OM is :

$\rho = R + e_0 \cos 2\theta$, and the coordinates of M are : $\begin{cases} X = (R + e_o \cos 2\theta)\cos\theta \\ Y = (R + e_o \cos 2\theta)\sin\theta \end{cases}$

The half-pipe section is then discretised in n elements and n+1 nodes (for instance n=36). The elements are rectilinear, so the section is a polygonal line. All the elements are rigidly fixed in sequential way one to the other.

The node #1 is at the bottom ($\theta = -\pi/2$), node #19 corresponds to $\theta = 0$, and node #37 to the top ($\theta = \pi/2$). The half bearing arc α is discretised in k elements, such that the corresponding angle is just ≤ 5°. The following arc, going from node #k+1 to #19, is divided into 18-k equal elements. Then the arc between node #19 and the top #37 is divided into 18 equal elements corresponding to an angle of 5°.

4.2 - Loading

Loads applied to the pipe have various origins : vertical pressure due to the weight of earth, pressure due to surface loads (traffic), lateral earth pressure, external hydrostatic pressure (in case of a water table), internal hydrostatic pressure (pressure pipe), and pipe self weight.

4.2.1 - Formulation in general axes

4.2.1.1 *Soil pressure*

Let $\varpi, \text{resp.}\, \varpi_d$, the mean unit weight of dry, resp. saturated, soil ; H = height of earth cover upon the upper generatrix ; H_W = depth of the water table, if any.

Vertical soil pressure

In node #I, of coordinates X(I), Y(I), the vertical pressure is equal to the effect of the weight of soil, multiplied if necessary by a "*concentration factor C*" :

$$\sigma_v(I) = \varpi H_w + \varpi_d \left(H - H_w + R - Y(I) \right)$$

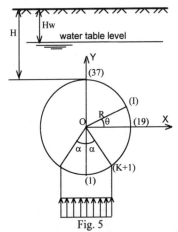

Fig. 5

This pressure is applied on the half upper circle of the pipe. Account is taken of the gradient of pressure. On the bearing arc of projected length $2R\sin\alpha$, the pressure σ_v balances the resultant vertical pressure. For the load due to the weight of soil, we get :

$$\sigma_v = \frac{\varpi H_w + \varpi_d \left(H - H_w + R\left(1 - \frac{\pi}{4}\right)\right)}{\sin\alpha}$$

On the part of circumference between nodes K+1 and 19 no vertical pressure is initially applied.

Of course account is taken of the value of H_w. The concentration factor C represents the effects of boundary conditions, i.e. what are the conditions at a certain distance from pipe, like "silo effects" on trench walls, or differential settlement effects in embankment. So σ_v becomes $C\sigma_v$.

Horizontal soil pressure

The horizontal soil pressure is equal to $\sigma_h = k_2 \sigma_v$, where k_2 is a coefficient depending on pipe installation conditions. For an actual soil, it is necessary that σ_h be consistent with the limit-states of the soil :

$$\sigma_v \tan^2\left(\frac{\pi}{4} - \frac{\varphi}{2}\right) - 2c\tan\left(\frac{\pi}{4} - \frac{\varphi}{2}\right) \leq k_2 \sigma_v \leq \sigma_v \tan^2\left(\frac{\pi}{4} + \frac{\varphi}{2}\right) + 2c\tan\left(\frac{\pi}{4} + \frac{\varphi}{2}\right)$$

σ_h, applied on the circumference, is varying as σ_v and then takes into account the gradient of pressure, as well as the concentration coefficient C.

4.2.1.2 *Pressure due to surface loads*

Vertical pressure σ_v is calculated at the level of the upper generatrix by an adequate formula, such as Boussinesq's or other, and is applied uniformly to the half upper circle. The reaction vertical pressure on bearing arc is $\sigma_v / \sin\alpha$.

The concomitant horizontal pressure is equal to $\sigma_h = k_2 \sigma_v$, or to an other pre-calculated value. This pressure is added to the horizontal soil pressure, and the sum must comply with the plasticity conditions.

4.2.1.3 *External hydrostatic pressure*

This pressure is orthogonal to the elements. If ϖ_w is the unit weight of water, the pressure, acting in the direction of local axis y, is : $\sigma_{w_e} = \varpi_w \left(H - H_w + R - Y(I) \right)$
with the condition   if $\sigma_{w_e} \leq 0$ then $\sigma_{w_e} = 0$.

4.2.1.4 *Internal hydrostatic pressure*

In the case of a pressure pipe, the pressure p_o, at center axis in O, is given. The pressure on the pipe wall, acting in the opposite direction of local axis y, is :

$$\sigma_{w_i} = p_o - \varpi_w Y(I)$$

- In case of internal vacuum $p_o < 0$ *and* $\sigma_{w_i} = -p_o$
- If the pipe is empty $p_o = 0$ *and* $\sigma_{w_i} = 0$.
- If the pipe is full of water without pressure $p_o = \varpi_w R$ and $\sigma_{w_i} = \varpi_w \left(R - Y(I) \right)$.

4.2.2 - Formulation in local axes

Each element has its own local axes (P_i,x,y), with the same orientation as the global axes (O,X,Y). x-axis is the element $P_i P_{i+1}$; y-axis is orthogonal to the element in the internal direction (fig. 6).

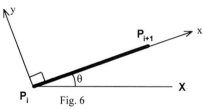

Fig. 6

When using transfer-matrix procedures it is simpler to express the loads in local axes. This is easily done for internal and external water pressure.

For soil pressure the stress tensor σ_p is supposed to have eigen axes parallel to global axes (O,X,Y).

The rotation matrix **R** transforms global axes to local axes attached to the current element, θ being the oriented angle between OX and the element $P_i x$. These matrices are :

$$\sigma_p = \begin{bmatrix} \sigma_h & 0 \\ 0 & \sigma_v \end{bmatrix} \qquad \mathbf{R} = \begin{bmatrix} \cos\theta & \sin\theta \\ -\sin\theta & \cos\theta \end{bmatrix}$$

In local axes, $\mathbf{R}^T$ being the transposed matrix of $\mathbf{R}$, $\mathbf{R}^T = \mathbf{R}^{-1}$, the stress matrix is σ_t, - σ_t, σ_l and τ, being respectively the transverse, longitudinal and shear stress :

$$\sigma = \mathbf{R}^T \sigma_p \mathbf{R} = \begin{bmatrix} \sigma_l & \tau \\ \tau & \sigma_t \end{bmatrix} \quad \begin{cases} \sigma_l = \sigma_h \cos^2\theta + \sigma_v \sin^2\theta \\ \sigma_t = \sigma_h \sin^2\theta + \sigma_v \cos^2\theta \\ \tau = (\sigma_v - \sigma_h)\sin\theta\cos\theta \end{cases}$$

For the pipe self weight, if p_{vol} is the pipe unit weight and e the wall thickness, the equivalent normal pressure is $p_{vol} \, e \cos\theta$ and longitudinal pressure $p_{vol} \, e \sin\theta$.

4.3 - Soil constitutive law

All around the pipe, the soil is represented according to WINKLER's model, i.e. distributed independent springs, orthogonal to pipe wall. Before application of loads the soil is supposed to be in elastic state (*CODE = 2*). When load or the first increment of load is applied, each element moves. Δy_i designates the transverse local displacement of node I. The soil develops then a reaction pressure Δp_i :
$\Delta p_i = -\beta \Delta y_i$ where β is the reaction modulus. Δp_i is orthogonal to pipe wall.

The soil normal stress changes from σ_t to $\sigma_t + \Delta p_i$; σ_l and τ are assumed unchanged during the increment.

However the variation Δp_i is limited by the consideration of limit-states in soil : upper or lower bound pressure, or passive and active pressures.

Expressed in local axes, the soil limit stresses, given in § 2.1.1, are :

$$\sigma_{\lim} = \frac{\sigma_l(1+\sin^2\varphi) + 2c\sin\varphi\cos\varphi + 2\varepsilon\sqrt{(\sigma_l \sin\varphi + c\cos\varphi)^2 - \tau^2 \cos^2\varphi}}{\cos^2\varphi}$$

with $\varepsilon = +1$ for passive pressure, and $\varepsilon = -1$ for active pressure.

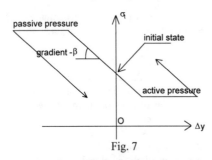

Fig. 7

If Δp_i goes beyond the limit values, Δp_i is adjusted to fit the relevant limit with *CODE=1* for active pressure and *CODE = 3* for passive pressure.

In case of cohesive soil, if it is found that
$\sigma_t + \Delta p_i < 0$ then $\Delta p_i = -\sigma_t$
and *CODE = 0* (no contact soil-structure).

This process is done for all nodes I of the upper-half circle $(19 \leq I \leq 37)$. For the bearing nodes $(1 \leq I < K+1)$, the bearing arc acts like a foundation and the soil ultimate pressure is much higher than the passive one; an elastic state for the soil is imposed : *CODE=2*. For the lower nodes, not within the bearing arc $(K+1 \leq I < 19)$, two different assumptions can be made, accordingly to the quality of compaction beneath the horizontal diametral plane of the pipe : either interaction exists, then the elements are on springs ; either it does not exist, then the corresponding elements are free and *CODE = 0* is imposed.

For following increments, when the sign of Δy changes, if the soil had reached a limit state, then it is assumed that the "*return path*" has an elastic behaviour up to reach the other limit state, etc...(see fig. 7).

The code number given to the soil at each node permits the calculation of the transfer matrix of the relevant element according to the appropriate assumption : element resting on elastic soil if *CODE=2* , or free element if *CODE≠2*. In this last case, the corresponding Δp_i term to reach the limit state, is considered as a load for the element. Δp_i may be equal to 0 if the limit state of the soil does not change during the load increment.

4.4 - Transfer matrix

At node #I a state-vector is defined equal to : $\mathcal{E}_i^T = [u_i, v_i, \omega_i, M_i, N_i, T_i, 1]$ where u_i and v_i are the local coordinates of the displacement of node I, ω_i its rotation, M_i the bending moment oriented as ω_i, N_i the axial force (positive for a compression) and T_i the shear force. The last component 1 is only here to introduce a 7th. column reserved to loads in the transfer matrix. A transfer matrix **T** allows to pass from state-vector $\mathcal{E}_i$ at node I to state-vector $\mathcal{E}_{i+1}$ at node I+1, by the equation :

$$\mathcal{E}_{i+1} = \mathbf{T}_{i,i+1} \mathcal{E}_i$$

The matrix $\mathbf{T}_{i,i+1}$ is a 7x7 matrix. The transfer matrices of elements for which 2nd order is neglected are well known (see Ref [2] and [3]) ; they are given in relevant text books. For buried flexible pipe design analysis, it is important to take into account the 2nd order effects, in order to deal with the amplification of deformations (pipe void or under vacuum) or with the stiffening due to internal pressure (pressure pipe). These matrices were established, see Ref [4].

4.5 - Calculation process

For a given increment of load, the transfer-matrix (T.M.) from node #1, called A, to node #n+1 (n+1=37), called B, is calculated by simple matrix multiplications on the left side. For this purpose the T.M. have to be expressed in the same axes system, the general one, by the relationship : $\tilde{\mathbf{T}}_{i,i+1} = \mathbf{R}\, \mathbf{T}_{i,i+1}\, \mathbf{R}^{-1}$

R is the rotation matrix of angle θ, changing with each element, similar to the matrix mentioned in § 2.2, but of dimension 7x7.

Let : $\mathbf{T}_{A,B} = \prod_{i=1}^{n} \tilde{\mathbf{T}}_{i,i+1}$ we have : $\mathcal{E}_B = \mathbf{T}_{A,B}\, \mathcal{E}_A$ (i)

Due to symmetry, the limit conditions in A and B are imposed :
$u_A = \omega_A = T_A = 0$ and $u_B = \omega_B = T_B = 0$.

The unknowns are : v_A, M_A, N_A in A and v_B, M_B, N_B in B. The 6 algebraic equations corresponding to matricial equation (i) permit solution of this problem. In a first step, the 3 equations giving u_B, ω_B and T_B are equated to zero ; they give v_A, M_A and N_A by solving this 3x3 linear system. The 3 other equations permit calculation of v_B, M_B and N_B.

At this stage, it is possible to calculate the displacement and the internal forces (i.e. ε_I) at each node I by successive multiplications of T.M. from A to node #I (I=2, 37). Knowing the displacement of node I, the corresponding reaction pressure Δp_i is determined, with a test on the figure related to lower and upper bounds of soil stress state ; the relevant code number being given to the node.

The calculation is then redone, assigning to each element the adequate T.M. function of the code number of its nodes. An element is considered to rest on an elastic soil if its two nodes have CODE=2. If it is not the case the T.M. corresponds to free element, and the part of Δp_i is limited in order to reach the relevant plasticity condition.

So several iterations are done until convergence has been reached for a given accuracy.

When, for a given load increment, a solution has been found (successful convergence), the elements have a new position. Then T.M. are re-calculated, as well as the rotation matrices. The rigidity of the system is reassessed, taking into account new axial forces and new position. Equilibrium equations are again written with the following load increment.

In this way elasto-plasticity of soil, geometrical structural non linearity, as well as normal forces effects, have been taken into account.

Several sequential runs of calculation are made, each of them using a certain number of load increments, limited to ten. The two first runs are related to Serviceability Limit States (SLS), without load factors. The 1st one deals with longstanding or permanent loads, using long term E-modulus with respect to creep. The 2nd one deals with short term loads, using the instantaneous E-modulus. Cumulative results are also given, as well as the soil stresses distribution in global axes. The two other runs are related to Ultimate Limit State (ULS) with adequate load factors for long term and short term loads.

During the calculation process, instability or divergence may appear. Special warnings are given, but in any case intermediate results, during iterations, have to be reviewed in order to ensure the validity of the final solution (e.g. not too large displacements).

As the calculation process consists only of matrix multiplications and resolution of a 3x3 linear system, it is very quickly performed.

Remarks

1- The main hypothesis of this method is the WINKLER assumption of normal distributed springs. If a more accurate modelisation of this aspect is needed, the use of a Finite Elements Method is then recommended, allowing for material and geometrical non-linearities.

2- Soil E_s-modulus, introduced as data, may vary from one area of pipe to another (e.g. from top to bottom with special care to haunch area), in order to represent, as far as possible, the soil state all around the pipe due to installation conditions. Options are left to the user of the programme in the data related to soil E-moduli. A first figure $E_{S,1}$ is given for the bedding angle area. A second one $E_{S,2}$ is given for the spring area, and a third one $E_{S,3}$ for the top area. This third value may be evaluated by default by use of the formula $E_{S,3} = E_{S,2} \lambda$ with $\lambda = \text{Inf}\left[1 ; \left(\frac{H}{4R}\right)^2\right]$. Between these two area a suggested variation of E_S is given by $E_S(\theta) = E_{S,2} \cos^2 \theta + E_{S,3} \sin^2 \theta$

Another option is available for the haunch area, between node #k+1 and node #$\frac{n}{2}+1$ (=19), related to the installation conditions : either E_S is equal to zero (poor or no compaction), either $E_S = E_{S,2}$ when compaction is achieved in this area.

3- If soil conditions or loads are not symmetrical about the vertical plane, the full ring has to be considered, and the problem to solve is given by the equation $\mathcal{E}_A = T_{A,A} \mathcal{E}_A$. It corresponds to a linear system of 6 equations, leading to the determination of the 6 unknown parameters : $u_A, v_A, \omega_A, M_A, N_A, T_A$. It is the only change in the calculation process.

4- As mentioned at the beginning, this method of calculation permits analysis of other cross sections than a circular one, provided the initial geometry of the section is given. For instance an ovoid duct can be easily calculated.

5 - Conclusion

Test results, for which pipe and soil characteristics were known (ref. [1] & [5]), have been compared with calculation results, using the 1st and the 2nd formulations of this proposed model. These comparisons were satisfactory, but they are too long to be reported here.

Bibliography

[1] M. GERBAULT : *Calcul de canalisations circulaires semi-rigides.* Annales de l'I.T.B.T.P. N° 439, Novembre 1985.

[2] E.C. PESTEL and F.A. LECKIE : *Matrix Methods in Elastomechanics.* (1963) McGRAW-HILL.

[3] J. COURBON : *Cours de théorie des structures, ch. 1 - Méthode des matrices-transfert.* (1970) Ecole Nationale des Ponts-et-Chaussées.

[4] M. GERBAULT : CEN TC 164/165/JWG 1/TG 1 DOC N 41, *Transfer matrix technic in structural pipe design* (October 1993) - non published -.

[5] J. MOLIN : *Flexible pipes buried in clay, section 6*, International Conference on Underground Plastic Pipe, (April 1981) New Orleans, LA.

SUPPORT SPACING OF BURIED AND ABOVE-GROUND PIPING

Phil Kormann, P.Eng.[1]   Z. Joe Zhou, Ph.D.

ABSTRACT

High pressure gas piping requires supports to protect termination points, prevent excessive vibration, provide alignment, and to limit settlement. The appropriate selection of support locations and piping spans before performing a piping stress and vibration analysis can save considerable time by reducing the number of trial-and-error cycles.

NOVA has developed several hand calculation methods and spreadsheet based equations to aid in the preliminary design. The equations are based on working stress principles. Detailed stress and vibration analysis showed that the spacing calculated by these equations are conservative and need not be modified later.

The paper discusses the derivation of the spacing equations for buried and above-ground high pressure gas piping. It also presents a calculation method to determine allowable loads on saddle supports.

Currently, research is ongoing at NOVA to determine the maximum support load a pipe can sustain. The behaviour of the pipe is modeled into the post-buckling range. A summary of the results to-date will conclude the presentation.

1.0   INTRODUCTION

The design of the piping-support system begins with the layout of the supports based on a spacing equation. The spacing information, together with piping isometric and soil stiffness values, is then entered into a finite element stress analysis model for calculation of piping stresses, support loads, etc..

---

[1] Stress Analysts, NOVA Gas Transmission Ltd., P.O. Box 2535, 801 - 7th Street S.W.,Calgary, Alberta, CANADA T2P 2N6

For the selection of an appropriate spacing formula, the Canadian code (CAN/CSA Z-184-M92, 1992) does not give explicit guidance, except to require that "supports shall be designed to support the pipe without causing excessive local stresses in the pipe ...". In the following, several equations are developed to assist the engineer in meeting this requirement.

2.0 BELOW-GROUND SUPPORTS

Below-ground supports are installed at facilities such as compressor or meter stations to allow for accurate alignment of piping, to control settlement, and to facilitate construction independent of inclement weather. They are designed for downward static loads; any vibration is usually dampened by the soil surrounding the piping. In cases where buried piping vibrates noticeably to a person standing close by, it is recommended to treat the source rather than to fixate the piping. The below-ground supports are usually designed as blocks of concrete with a thin layer of soft material (e.g. wood, plastic) protecting the pipe coating from the rough concrete surface. The size of the blocks depends on static load imposed by the piping and on the allowable soil pressure.

2.1 BELOW-GROUND SUPPORT SPACING EQUATION

In general, pipe supports can be designed to withstand any practical vertical load. Therefore, the piping is the weaker component, and its structural capacity governs the maximum allowable support load. The following spacing equation has been developed based on the principles of working stress and the ASME Boiler and Pressure Vessel (BPV) Code:

$$L = 20,000 \frac{SD}{H+16t}(D/t)^{-1.75} \tag{1}$$

in which
L = support spacing [mm]
D = pipe OD [mm]
H = depth of cover [mm]
t = pipe wall thickness [mm]
S = specified minimum yield strength, SMYS [MPa]

2.2 DERIVATION OF SUPPORT SPACING EQUATION

Equation (1) limits σ_{circ}, the total circumferential stress in the piping at the support location. The components of σ_{circ} are
- hoop stress, σ_h, due to internal pressure
- local stresses, σ_2 and σ'_2, due to the support reaction,

2.2.1 Hoop Stress

The hoop stress due to internal pressure is easily calculated as the product of pressure times diameter, divided by twice the wall thickness. In the case of compressor and meter stations, the code (CAN/CSA Z-184-M92, 1992) limits the hoop stress to 50% of SMYS.

2.2.2 Local Stresses

The calculation of the local stresses due to support reaction is based on an elastic solution shown in Fig. 1, which can be found in the literature (Roark and Young, 1975). For a cylindrical shell subject to line load over a very short length in the longitudinal direction, two stress components at the loading point are given as

$$\sigma_2 = -0.13 B P R^{\frac{3}{4}} b^{\frac{-3}{2}} t^{\frac{-5}{4}} \tag{2a}$$

$$\sigma'_2 = -B^{-1} P R^{\frac{1}{4}} b^{\frac{-1}{2}} t^{\frac{-7}{4}} \tag{2b}$$

in which

σ_2 = circumferential membrane stress (positive in tension)
σ'_2 = circumferential bending stress (positive in tension on outside of pipe wall)
R = pipe radius = OD/2
t = pipe wall thickness
P = applied load or reaction force
$2b$ = width of applied load; the length of the support, usually 1.5 OD
$B = [12(1-v^2)]^{1/8} = 1.348273$ for $v = 0.3$

Assuming the backfill density of $\gamma_{backfill}$ = 15 kN/m³, γ_{steel} = 77 kN/m³, and including the self weight of the pipe but ignoring its contents, the applied load P is

$$\begin{aligned} P &= \gamma_{backfill} DLH + \gamma_{steel} \pi DtL \\ &\approx 15 DLH \times 10^{-6} + 240 DtL \times 10^{-6} \\ &= 15 DL(H + 16t) \times 10^{-6} \end{aligned} \tag{3}$$

in which

H = depth of cover [mm]
D = pipe OD [mm]
L = support spacing [mm]
P = load or reaction force [N]

Substituting Eq.(3) and R, b, B into Eq.(2a) and (2b), they become

$$\sigma_2 = -2.41 \times 10^{-6} L(H + 16t) D^{\frac{1}{4}} t^{\frac{-5}{4}} \tag{4a}$$

$$\sigma'_2 = -10.803 \times 10^{-6} L(H+16t) D^{\frac{3}{4}} t^{\frac{-7}{4}} \tag{4b}$$

The total circumferential stress is then calculated as

$$\sigma_{circ} = \sigma_h - \sigma'_2 + \sigma_2 \tag{5}$$
$$= 0.5 SMYS + 10.803 \times 10^{-6} L(H+16t) D^{\frac{3}{4}} t^{\frac{-7}{4}} (1 - 0.223 \sqrt{t/D})$$

For practical ratios of D/t, the last term in Equation (5), σ_2, is comparatively small and can be neglected. Eq. (5) becomes then,

$$\sigma_{circ} = 0.5 SMYS + 10.803 \times 10^{-6} LHD^{\frac{3}{4}} t^{\frac{-7}{4}} \tag{6}$$

In the case of primary membrane stress (P_m) combined with primary bending stress (P_b), the ASME Code (Boiler and Pressure Vessels Code, Section VIII, Division 2) sets the design stress, $S_{allowable}$, to

$$S_{allowable} = 1.5 S_m \tag{7}$$

where S_m is the allowable stress for the general membrane stress P_m. In the case of compressor and meter station piping (CAN/CSA Z184-M92, 1992)

$$S_m = 0.8 \times 0.625 \times SMYS = 0.5 SMYS \tag{8}$$

The limit for the total circumferential stress is therefore (Eq. (7) and (8))

$$\sigma_{circ} = S_{allowable} = 1.5 \times 0.5 SMYS = 0.75 SMYS \tag{9}$$

Eq. (6) and (9) can be combined and solved for L to yield (note that $S_{allowable}$ can be tensile or compressive)

$$L = 23142 \frac{SD}{H+16t} (D/t)^{-1.75} \tag{10}$$

Considering the neglected weight of the pipe contents, and to simplify the calculation, an even number of 20000 is used to replace 23142. Eq.(10) then becomes Eq.(1) which is repeated here for convenience

$$L = 20,000 \frac{SD}{H+16t} (D/t)^{-1.75}$$

3.0 ABOVE-GROUND SUPPORTS

Above-ground supports, compared to buried ones, have two additional functions: to protect termination points such as compressors, vessels etc., and to avoid gross excitation of the piping. The protection of termination points requires to restrict lateral and sometimes axial movement of the piping as predicted by a suitable piping flexibility computer program; the appropriate design of the supports are the task of the civil engineer and need not be discussed here. To avoid gross excitation of the piping, the designer can mismatch its natural frequencies to possible excitation frequencies. In NOVA's experience, gross excitation will not occur if the natural frequencies of the piping are above 30 Hertz, the starting point of a range where excitation frequencies do not contain enough energy to cause large vibrations even in the case of resonance.

Natural frequencies of the piping are readily controlled by limiting its free span. As a conservative rule, NOVA uses the equation

$$L = \sqrt{D}/3 \qquad (11)$$

in which
  L = free span between supports [m]
  D = pipe OD [mm]

3.1 SPACING EQUATION FOR ABOVE-GROUND SUPPORTS

As outlined above, Eq.(11) is derived in the following from the design condition of the fundamental natural frequency being not less than 30 Hz.

The fundamental frequency may be estimated from the following model. The pipeline is modeled as a beam supported at two ends. Considering a typical continuous span of the pipeline, the end condition of the beam is assumed to be fixed at both ends. The pipe is assumed to be empty (the mass of the contents is neglected). The fundamental frequency is found (Paz, 1985) as

$$\omega^2 = (\frac{4.73}{L})^4 \frac{EI}{\rho A} \qquad (12)$$

in which
  ω = the fundamental circular frequency
  L = spacing of supports
  E = Young's modulus of pipe material (200000 MPa)
  I = Second moment of area of the pipe cross-section
  ρ = density of pipe steel (7800 kg/m^3)
  A = area of the pipe cross-section

The second moment of area and area of the pipe cross-section are defined as

$$I = \frac{\pi}{8} D^3 t \qquad (13)$$

$$A = \pi Dt \tag{14}$$

Substituting Eqs. (13) and (14) into Eq. (12), the fundamental frequency may be obtained as

$$f = \frac{\omega}{2\pi} = 6.37 \times 10^6 \frac{D}{L^2} \tag{15}$$

Limiting f to 30 Hertz, and solving for L yields

$$L = \sqrt{D} \times 10^3 / 2.17 \tag{16}$$

If the constant in Eq. (16) is conservatively chosen as 3, and D,L are expressed in units of millimeter and meter respectively, Eq. (16) becomes the easy-to-remember Eq.(11) with a safety factor of approximately 1.5:

$$L = \sqrt{D}/3$$

3.2   OTHER SPACING EQUATIONS

3.2.1   Spacing Based On Tresca's Yield Criterion

A frequently used formula for estimating free spans is

$$L = 0.123\sqrt{SD} \times 10^3 \tag{17}$$

in which
   L = support spacing [mm]
   D = pipe OD [mm]
   t = pipe wall thickness [mm]
   S = specified minimum yield strength, SMYS [MPa]

This equation, which limits the primary membrane and bending stresses in the pipe, assumes zero axial restraint and disregards any local stresses. In its derivation, longitudinal and hoop stresses are combined to meet Tresca's yield criterion

$$\sigma_{Tresca} = \sigma_{max.principal} - \sigma_{min.principal} = \sigma_{hoop} - \sigma_{longit.} \tag{18}$$

As outlined in Sect. 2.2.1, hoop stress in the case of compressor and meter stations is limited by the code to 50% of SMYS. Longitudinal stresses can be expressed as the sum of bending stress due to weight of pipe (weight of gas is negligible) and membrane stress due to endcap pressure

$$\sigma_{longit.} = \sigma_{bending} + \sigma_{endcap} = \frac{M}{I}D/2 + 0.5\sigma_{hoop} \quad (19)$$

The bending moment M for continuously supported beams reaches a maximum of

$$M = 0.107qL^2 \quad (20)$$

under the uniformly distributed load

$$q = \gamma_{steel}\pi Dt = 77\pi Dt \times 10^6 = 242Dt \times 10^6 \quad (21)$$

Using Eqs.(21) and (13), Eq.(19) becomes

$$\sigma_{longit.} = 0.136\frac{[242Dt \times 10^6]L^2}{D^2 t} + 0.5 \times 50\% S \quad (22)$$

Following the same argument that lead to Eq.(9), the allowable stress in Tresca's yield criterion will be limited to 1.5 times 50% SMYS, and Eq.(18) becomes

$$0.75S = 0.5S - \left[0.25S - 33 L^2/D \times 10^6\right] \quad (23)$$

and solving for the support spacing, L, yields Eq.(17), repeated here for convenience

$$L = \sqrt{\frac{SD}{66}} \times 10^3 = 0.123\sqrt{SD} \times 10^3$$

3.2.2 Spacing Based On Longitudinal Stress Limitation

The Canadian code (CAN/CSA-Z184-M92,1992) limits longitudinal stresses as defined in Eq.(19) to 50% of SMYS in compressor and meter stations. This clause is probably based on a B31.3 (ASME Standard B31.3, 1989) rule which intended to limit creep at high temperature. Modifying Eq.(23) to account for longitudinal stresses only and the reduced allowable stress produce

$$0.5S = \left[33 L^2/D \times 10^6 + 0.25S\right] \quad (24)$$

and solving for L yields

$$L = \frac{0.123}{2}\sqrt{SD} \times 10^3 \quad (25)$$

which is half the spacing of Eq.(17).

3.2.3 Spacing Based On Local Stress

For the local stress at supports, the same consideration as discussed in Sect. 2.2.2 apply, except that the overburden load is omitted for the above-ground piping. By replacing the first term in Eq.(3):

$$\gamma_{backfill} DLH \Rightarrow 0, \tag{26a}$$

Eq.(1) becomes

$$L = 20000 \frac{SD}{16t} (D/t)^{-1.75} \tag{26}$$

The notations in the above equation have the same definition as in Eq. (3).

3.3 COMPARISON OF SPACING EQUATIONS

Spacing values based on Eqs. (1), (26), (17) and (25) for typical pipe sections are listed in Table 1 below.

Pipe Spec.	D [mm]	t [mm]	S (MPa)	Spacing Eq. (11) [m]	Spacing Eq. (26) [m]	Spacing Eq. (17) [m]	Spacing Eq. (25) [m]
1	610	18.3	290	8.2	26.1	51.7	25.8
2	610	11.0	483	8.2	29.7	62.9	31.4
3	914	19.2	414	10.1	28.5	75.6	37.8
4	914	16.5	483	10.1	29.7	81.7	40.8

Table 1 Spacing for Above Ground Piping

The comparison shows that neglecting local stresses or vibration can lead to large pipe spans.

3.3 SPACING EQUATION FOR SADDLE SUPPORTS

The spacing of saddle supports is determined based on local stresses induced by the saddle supports. The local stresses both in the longitudinal and circumferential directions, are estimated (Roark and Young, 1975) as

$$S_{max} = k \frac{P}{t^2} ln(\frac{R}{t}) \tag{27}$$

in which
 P = the total saddle reaction force
 R = the pipe radius
 t = the pipe thickness
and k is coefficient given by

$$k = 0.02 - 0.00012(\beta - 90) \qquad (28)$$

where β is total angle subtended by the arc of contact between pipe and saddle in degree. The load P is defined in Eq.(3) for buried pipelines. Equation (3) may be used for above ground piping if the first term of the equation is replaced according to term (26a).

The equation (27) is established on experimental studies (Hartenberg, 1941, Wilson and Olson, 1941). The stress is mainly due to circumferential bending occurring at points about 15 degree above the saddle tips. Therefore, it could be either in tension or compression. The angle β should be normally not less than 120 degree. The applicability of Eq. (27) for large β (say 260 degree and up) is questionable because the coefficient k becomes negative.

The stress predicted by Eq. (27) is combined with other stresses and checked by a yield criterion, either Von-Mises or Tresca. The principal stresses required by a yield criterion are

$$S_x = S_{max} + S_T + S_{Poisson} \qquad (29a)$$

$$S_\theta = S_{max} + S_h \qquad (29b)$$

$$S_r = 0 \qquad (29c)$$

in which S_x, S_θ and S_r are the stresses components in the longitudinal, circumferential and radial directions, respectively, and S_T, $S_{Poisson}$ and S_h are the longitudinal stresses due to temperature differential and pressure, Poisson effect, and the hoop stress due to the internal pressure. The Von-Mises effective stress is then

$$S_{vm} = \sqrt{\tfrac{1}{2}((S_x - S_\theta)^2 + S_x^2 + S_\theta^2)} = \sqrt{S_x^2 + S_\theta^2 - S_x S_\theta} \qquad (30)$$

and the Tresca effective stress is

$$S_{Tr} = \left| \text{larger of } S_x \text{ and } S_\theta \right| \qquad (31)$$

The spacing of saddle supports cannot be solved explicitly but can easily be found by trial and error if the above equations are incorporated into a spreadsheet. The effective stresses from Eqs. (30) and (31) are limited to SMYS.

4.0   CURRENT PIPE-SUPPORT INTERACTION RESEARCH

4.1   MOTIVATION

Equation (1) returns small spacing values for increasing depth of cover, H. For example, pipe specimen 2 would require support spacing of L=4.4m for H=1m, which is small but still reasonable, and L=1.65m for H=3m, which is impractical for discrete support. Clearly, the working stress principles need to be abandoned in favour of a limit states approach.

4.2   DESCRIPTION OF PROJECT

NOVA and the National Science and Engineering Research Council (NSERC) are sponsors of a research project to investigate the behaviour of pipe during pipe-support interaction. The research includes both the analytical and experimental studies. In the analytical study, non-linear finite element analysis method is employed to calculate stresses, strains and deformations of a pipe that is pushed onto a rigid support. Contact elements simulate the change in bearing surface as the pipe flattens and increases contact with the support. The results of the finite element analysis will be verified by full-scale experiments.

4.3   PRELIMINARY RESULTS

The following conclusions may be drawn based on the results of the finite element analysis (Zhou and Kormann, 1994, 1995a and 1995b):

1. The prediction equations for local stress in Eq. (2a) and (2b) underestimate the maximum local stress in the unpressurized pipe. However, the differences are not too large to invalidate their application in unpressurized pipes.

2. In the elastic range, the local stresses concentrate in the contact areas. The maximum negative moment is located at the bottom of the pipe, and the maximum positive moment at about 50° to 60° measured from the bottom. The local stress distributions are not uniform over the support length and show substantial concentration at the edge of the support. Among the six stress components, the circumferential and longitudinal stresses are the major ones with the circumferential stress being dominant.

3. The elastic-plastic behaviour redistributes the local stresses, which increases the load-carrying capacity. Plastic strain and deformation concentrate at the maximum positive and negative moment areas. Good ductility of the material is required to utilize the extra load carrying capacity. It is essential to include nonlinear effects in order to correctly predict the deformation and strain.

4. The contact condition between the pipe and sleeper is constantly changing as the cross-sectional deformation increases. The contact area is generally increasing and the location moves away from the bottom of pipe. This change toward to a more effective contact condition is the major contribution to the continuing increase of load carrying capacity beyond the elastic range.

5. The load carrying capacity of the pipe is approximately proportional to the out-of-roundness where elastic-plastic behavior of pipe material and large displacement have only limited effect. No limit load has been observed before the pipe deformation become clearly excessive. The out-of-roundness is approximately proportional to the displacement.

6. The pressure stiffening effect has significant influences on the pipe response. It reduces the maximum local stress, the rate at which the local stress increases with load, the plastic strain and the cross-sectional deformation. It is necessary to properly include the pressure stiffening effect in order to predict the local stress of pressurized pipes.

These conclusions indicate that the support spacing may be increased substantially without compromising the integrity of the pipe.

REFERENCES

ASME Standard B31.3 (1993), "ASME Code for Pressure Piping, B31", United Engineering Center, 345 East 47th Street, New York, New York 10017

ASME Boiler and Pressure Vessels Code (1992), "ASME Boiler and Pressure Code, Section VIII, Pressure Vessels, Division 2 : Alternative Rules", United Engineering Center, 345 East 47 Street, New York, N.Y. 10017

CAN/CSA-Z184-M92 (1992), "CAN/CSA-Z184-M92 Gas Pipeline Systems", 178 Rexdale Boulevard, Rexdale (Toronto), Ontario, Canada M9W 1R3

Hartenberg, R.S., (1941). "The Strength and Stiffness of Thin Cylindrical Shells on Saddle Supports," Ph.D. Thesis, University of Wisconsin.

Paz, M. (1985), "Structural Dynamics, 2nd Ed", Van Nostrand Reinhold Company Inc., 135 West 50th Street, New York, New York 10020

Roark, R.J. and Young, W.C., 1975, Formulas for Stress and Strain, Fifth Edition, McGraw-Hill Book Company

Wilson, W.M. and E.D. Olson, (1941). "Tests on Cylindrical Shells," University of Illinois, Engineering Experimental Station, Bulletin No. 331, September 23, p. 129.

Zhou, Z. J. (1994), "Study of Pipe-Support Interaction" , Internal Technical Report, Nova Gas Transimission Ltd., P. O. Box 2535, Station M, Clagary, Alberta, Canada T2P 2N6

Zhou, Z. J. and P. Kormann (1995a), "Nonlinear Analysis of Sleeper-supported Piping", The 1995 ASME/JSME Pressure Vessels and Piping Conference with the Material and NDE Divisions, Hilton Hawaiian Village, Honolulu, Hawaii, July 23-27

Zhou, Z. J. and Kormann, P. (1995b), "Behavior of Sleeper-Supported High Pressure Piping", 14th International Conference on Offshore Mechanics and Arctic Engineering, ASME, SAS Falconer Center, Copenhagen, Denmark, June 18-22

AVOIDING PIPELINE FAILURES AT WATERWAY CROSSINGS
PART 1: SCOUR ASSESSMENT METHODS

By Brian J. Doeing[1], M. ASCE, Jeffrey B. Bradley[2], F. ASCE,
and Samuel Carreon[3], M. ASCE

Abstract

Pipeline crossings at streams, rivers, and washes are at risk due to erosion in the river environment. Buried pipelines may be exposed in the channel bottom by scour of bed materials or in the floodplain through migration of the channel banks. Pipelines carried across streams on bridges are at risk from scour at bridge piers and bridge abutments. This paper, being Part 1 of a two-part series, describes scour assessment methods that have been used by the authors to evaluate the erosion potential at existing or proposed pipeline installations in the Southwest United States. In Part 2, presented in a separate paper, countermeasures that have been used to protect pipelines from scour are described, such as drop structures, riprap stone protection, and other types of erosion control measures.

Introduction

The Nation's attention turned to Texas after severe flooding on the San Jacinto River exposed and ruptured two major pipelines buried 3 feet beneath the river bed on October 20, 1994. The two lines carried nearly one-sixth of U.S. daily gasoline supplies. The breach caused gasoline prices to rise on the futures market,

---

[1]Vice President, WEST Consultants, Inc., 2111 Palomar Airport Road, Suite 180, Carlsbad, CA 92009-1419, Phone (619) 431-8113, FAX (619) 431-8220

[2]Chief Executive Officer, WEST Consultants, Inc., 2101 4th Avenue, Suite 1050, Seattle, WA 98121-2357, Phone (206) 441-4212 FAX (206) 441-4431

[3]Consulting Engineer, Transmission Operations Engineering, El Paso Natural Gas Company, P.O. Box 1492, El Paso, TX 79978, Phone (915) 541-3013 FAX (915) 541-5947

interrupted supplies throughout Northeastern states, and forced the Houston ship channel to close down for several days (Fedarko 1994). Almost two years earlier, in January 1993, severe flooding in the State of Arizona exposed and damaged a number of natural gas pipelines. The failure of one pipeline crossing the Gila River near U.S. Highway 80 threatened nearly one-half of the natural gas supply to Southern California. These incidents underscore the importance of protecting pipelines from scour. In the interest of public safety and economics, there is a need to assess the appropriate burial depth and protection measures for both existing and proposed pipelines.

Although the determination of scour potential in rivers is by no means an exact science, it requires the detailed analysis of hydraulics, hydrology, and sediment transport, and an examination of the geomorphology of the stream system. The experience of a qualified river engineering professional cannot be overemphasized in this process. Yaremko and Cooper (1983) discussed the prediction of scour, assessment of bank stability, and the use of river training structures in the design of buried river and floodplain crossings in Northern pipeline projects. In addition, Veldman (1983) presented lessons learned in scour and bank migration in gravel bed rivers in Arctic pipeline river crossings.

This paper presents two techniques used by the authors to assess scour potential at pipeline crossings in the Southwestern United States: 1) a reconnaissance level screening method, and 2) a detailed scour potential study. A reconnaissance level screening method was developed to quickly assess erosion potential when faced with a large number of crossing locations. For crossings identified as having significant erosion potential, detailed scour and sediment transport studies have been used to compute the scour depth and burial requirements.

Reconnaissance Level Screening

The El Paso Natural Gas Company (EPNG), an interstate natural gas transmission company, has an estimated 11,000 stream crossings in its pipeline network throughout the Southwest. When the company planned an expansion of pipeline system in New Mexico and Northern Arizona, a method was developed to quickly evaluate the scour potential for 715 stream crossings. The purpose of this initial screening process was to quickly determine the number of crossing sites that needed more detailed analysis. The procedure entailed the determination of a critical drainage basin size that would produce significant scour at the crossing. In addition, United States Geological Survey (USGS) quadrangle maps and aerial photographs were examined for stream features that would indicate past or potential channel movement or incision. The "threshold" drainage area that would cause the streambed to erode to a critical scour depth and endanger the pipeline during the 100-year flood was determined with regime-type equations and empirical data. A summary of the overall procedure will be presented here and full details can be found in Williams, et al. (1992).

EPNG has traditionally buried pipelines at a depth of 5 feet below the stream bottom at stream crossings where no indications of significant scour are evident. Using a conservative depth of overburden of 2 feet to control buoyancy, an allowable local scour depth of 3 feet was selected as the critical value. Lacey's regime scour equation (Pemberton and Lara 1984) was used to establish the relationship between the local scour and the mean depth of flow:

$$d_s = Z d_m \qquad (1)$$

where:

d_s = scour depth below streambed, feet
d_m = mean water depth, feet
Z = factor depending on stream characteristics as follows:

Stream Condition	Z
Straight reach	0.25
Moderate bend	0.50
Severe bend	0.75
Right angle bend	1.00
Vertical bank or wall	1.25

From equation (1) for a moderate bend with Z equal to 0.50, a mean water depth d_m equal to 6 feet is expected to produce a local scour d_s of 3 feet. Next, Lacey's regime relationship between mean water depth and discharge was used to determine the discharge associated with the mean depth:

$$d_m = 0.47 \left(\frac{Q}{f}\right)^{1/3} \qquad (2)$$

where:

Q = discharge, cubic feet per second (cfs)
f = Lacey's silt factor = $1.76(D_m)^{1/2}$
D_m = D_{50} (mean bed sediment size), mm

Using a conservative representative D_{50} of 0.3 mm, based on available geotechnical data and references, Lacey's silt factor is:

$$f = 1.76(0.3)^{1/2} = 0.96 \qquad (3)$$

Solving for Q in equation (2):

$$Q = 0.96 \left(\frac{6}{0.47}\right)^3 = 1997 \ cfs \quad (use \ 2000 \ cfs) \qquad (4)$$

A comparison was made to historical scour depths within the Southwest and this method was judged to be reasonable for use in this level of analysis. Using USGS regional regression relationships between discharge, basin area, mean elevation, and mean precipitation for Arizona and New Mexico, a threshold drainage basin area was determined that would produce a 100-year discharge of 2000 cfs. For stream crossings with a drainage area greater than the threshold area, the 100-year discharge has a potential to scour to at least a depth of 3 feet, the allowable scour depth. Crossings with drainage areas less than the threshold value were preliminarily eliminated from further analysis. The threshold drainage areas for the regions considered ranged from 2 to 70 square miles in Arizona and 5 square miles in New Mexico.

Because scour at stream crossings is influenced by the stream's planform characteristics, each crossing was examined using USGS quadrangle maps and stereo aerial photographs for the following features:

1. Meander pattern and historic downstream migration trends
2. Stream angle of attack at the crossing
3. Possible lateral stream migration within the floodplain limits
4. Evidence of headcuts upstream and downstream of the crossing
5. Location of abandoned streams (no obvious outlet, but the 100-year discharge could create one that crosses the pipeline)

The presence of any of the above features was noted and evaluated for its impact on scour potential. For some crossings, a field inspection was conducted. This, in combination with the drainage area criteria, allowed each crossing to be identified as having low, moderate, or high scour potential. Using this procedure, the number of stream crossings with potential scour problems was reduced from 715 to 113. It should be noted that this analysis was based on regime-type relationships and very general procedures for the purpose of identifying potential scour problems at the pipeline crossings. The method neglects long term bed elevation changes and the results are not sufficient for design purposes.

Detailed Scour Analysis

To obtain design pipeline burial depths or information for the design of grade control structures or other streambed protection features, a detailed scour analysis is recommended. Analytical methods may be applied to determine the potential scour depth for the more simple and relatively small crossings. The study of more complex and larger stream crossings should include a sediment transport analysis with the use of a computer model to determine the average bed elevation change during the design flood (typically the 100-year flood). The model can also be used to evaluate the impact of upstream impoundments (reservoirs and debris basins) as well as sand/gravel extractions upstream of the crossings. The computer model can predict scour at stream crossing constrictions which the regime-type relations cannot predict. An example of the utility and application of this type of detailed analysis for pipeline crossing design and construction was given by Carreon and Doeing (1994).

General and Local Scour Analysis

There are essentially two kinds of scour - general (degradation) and local. General scour is associated with general streambed lowering over a significant stream reach length. This occurs under conditions such as depletion of upstream sediment sources (e.g., scour downstream of reservoirs) and changes in stream gradient. Local scour is associated with specific features of the channel such as bed forms, channel alignment, and obstructions to flow. Examples of local scour include the deepening incision of the low flow channel, constriction or pier scour at bridges or between bridge abutments, flow concentrations at severe stream bends, and impinging flows at stream confluences.

The general scour portion of the total scour requires information on the cross sectional geometry of the stream upstream and downstream of the pipeline crossing, hydraulic parameters (depth, velocity, flow width, etc.) of each discharge of the design hydrograph and long term hydrograph, the sediment gradation of the streambed, location of geological controls (bedrock, hardpan, etc.), volume and composition of the sediment entering the reach for each discharge, and the flow and duration of each portion of the design hydrograph. This type of analysis requires the use of a numerical model such as the U.S. Army Corps of Engineers' sediment transport model *HEC-6, Scour and Deposition in Rivers and Reservoirs (HEC-6* 1993).

Description of HEC-6 Model

HEC-6 is a one-dimensional, movable boundary, open channel flow model designed to simulate streambed profile changes over fairly long time periods. Since its initial nationwide distribution by the Hydrologic Engineering Center (HEC) of the Corps of Engineers in 1973 and again in 1977, 1987, 1991, and 1993, it has been the most widely used one-dimensional sediment transport model in the United States, and particularly with the Corps of Engineers.

In general terms, the model first calculates the hydraulic parameters such as flow depth, water velocity, and effective flow width for each cross section. It then computes the sediment transport potential using the hydraulics of the main channel. Sediment contribution at the upstream end of the reach being modeled is simulated by the use of a sediment vs. discharge relation. This load is compared to the sediment transport potential of the cross section. If the inflowing load is larger than the transport potential, the difference is deposited in the cross section. If the inflowing load is less than the transport potential, it is picked up (scoured) from the bed, taking into account the availability of material in the bed (e.g., bedrock, armoring, etc.). The sediment load leaving the cross section then becomes the inflowing load to the next downstream cross section. This continues until the most downstream cross section is simulated. For the next discharge in the hydrograph, the hydraulics are again computed using the new cross sectional geometry formed by the previous discharge. The cycle is repeated until the entire hydrograph is simulated. Further details of the model are presented in the HEC-6 User's Manual (*HEC-6* 1993) and MacArthur et al. (1990).

Field and Laboratory Investigations

Hydraulic and sedimentation engineers need to visit the pipeline crossing site to make a visual assessment of the channel conditions and to determine the need for field data collection. Channel geometry input for HEC-6 may be obtained from field surveys or from detailed topography maps. If field surveys are to be taken, the surveys obtained generally cover a minimum distance from 3 to 5 times the active channel width downstream and 5 to 10 times the active channel width upstream of the crossing location. Locations are also determined for samples of the channel bed materials to obtain representative grain size distributions. About two-thirds of the sediment samples are generally obtained upstream of the crossing and one-third downstream. The sediment samples are analyzed for grain size distribution (sieve analysis) down to approximately 0.0625 mm.

Development of Geometry and Hydraulics

Cross sectional geometry of the channel is required for input into the sediment transport model. Prior to the development of the HEC-6 model, a hydraulic analysis of the river segment containing the pipeline crossing can be completed with computer program *HEC-2 Water Surface Profiles* (*HEC-2* 1991). The HEC-2 program conducts a "fixed bed" analysis, which allows one to check the geometry in the model and the reasonableness of the hydraulic results. The program also provides a rating curve of water elevation versus discharge at the downstream boundary. The information obtained during the field reconnaissance is incorporated into the hydraulic model. Stations for the left and right channel banks are set for each cross section and characteristic reach lengths are determined for the channel and overbanks. Manning's "n" value roughness coefficients are input for each cross section and areas judged to be ineffective in conveying flow are coded within the model.

A sediment transport model can then be developed in HEC-6 using the input data from HEC-2 with slight adjustments. Before conducting the "movable bed" analysis, the HEC-6 program may be executed in a "fixed bed" mode in order to compare the results with the previous HEC-2 run. In general, a small variation in flow depth for all discharges and at all cross sections is sought between the two models for the "fixed bed" run.

Streambed Gradation and Limits of Erodible Bed

Representative stream bed material size gradations within the study reach are a requirement for sediment transport modeling. The laboratory results of particle size distribution obtained from the streambed samples are reviewed and representative gradations are input to the HEC-6 model. At cross sections that do not have sample gradations available, the upstream and downstream cross section gradations are linearly interpolated to produce a representative gradation. This interpolation is performed automatically in HEC-6. If geotechnical borings are taken along the pipeline alignment, information from the boring logs is analyzed to determine the depth of bedrock or strongly cemented (flow resistant) soils. Information obtained from seismic refraction studies conducted along the alignment can also be helpful in determining the maximum depth of potential scour.

From the field reconnaissance and plots of the channel cross sections, the lateral limits of scour are determined and input to the HEC-6 model. The model assumes that erosion is uniform between these limits but deposition can occur outside these limits but only within the wetted portions of the channel. In general, the limits of scour are within what is termed the "active bed" and are often located just within the main channel limits. Once the maximum scour is obtained from the model, the judgement of the sedimentation engineer is used to determine the extent of possible migration of the main channel within the floodplain. This process helps to determine the lateral extent of possible scour.

Inflowing Sediment Load

In many cases, no information is available on the sediment entering the stream crossing sites from upstream sources. To develop a sediment rating curve (the relation between water discharge and sediment discharge by grain size), one technique is to assume that the upstream cross sections are in quasi-equilibrium.

This means that the sediment gradation of the streambed determines the sediment loads passing the cross sections and the amount of sediment entering the reach is generally equal to the amount of sediment exiting the reach.

For a range of flows, the inflowing sediment for the HEC-6 model can be assumed and the amount of sediment passing the upstream cross sections noted. The sediment load passing these cross sections is then used as the inflowing sediment load. This iterative process can be continued until the inflowing sediment load is consistent with the sediment load passing the upstream cross sections and no significant scour or deposition is occurring in these cross sections (consistent with the equilibrium assumption). The inflowing sediment loads from the last iteration can then be used as a basis for the scour analysis.

Development of Streamflow Hydrograph

In order to use HEC-6 for sedimentation analysis, it is necessary to obtain a flood hydrograph containing the values of flow discharge over a specified period of time. The water discharge hydrograph is approximated by a sequence of steady flow discharges in discrete time intervals for input into HEC-6. Using a utility computer program called the Sediment Weighted Histogram Generator (SWHG), an output file of the representative 100-year flood histogram can be created in HEC-6 input format (Williams and Bradley 1990). The SWHG program reduces the number of computations in the HEC-6 program by grouping mean daily flows into periods of longer duration while maintaining the nonlinear relationship between water discharge and sediment discharge.

The methods used to develop the 100-year design discharge hydrographs vary depending on the location of the pipeline crossing, whether the stream is regulated by dams or other controls, and the availability of hydrologic data. The 100-year peak flow and discharge hydrograph at some locations may be found in Federal Emergency Management Agency (FEMA) flood insurance studies or other Federal, State, or County agency reports. On larger streams, historical streamflow gage data can be used for a long term simulation and to determine the 100-year design flood discharge from discharge-frequency relationships. The shape of the design hydrograph can be approximated by taking gage data from a large flow on the stream in recent history and adjusting the values proportionally to obtain the peak flow for the design event. On smaller streams, the design peak flow may be obtained from USGS regional equations. For example, the USGS methods for determining the magnitude and frequency of floods in the Southwestern United States applies to unregulated streams that drain basins less than about 200 square miles (Thomas, et al. 1994).

One method to obtain a flood hydrograph for an ungaged basin is to develop a synthetic unit hydrograph for the basin and apply a rainfall pattern for the design storm to the unit hydrograph. Although a variety of techniques are available to develop unit hydrographs, knowledge of how and where the methods were developed helps one determine which is most suitable to use. For example, the U.S. Bureau of Reclamation (USBR) procedure, as described in *Design of Small Dams* (*Design* 1987) is intended for use in the 11 western United States (Montana, Wyoming, Colorado, New Mexico, Idaho, Utah, Nevada, Arizona, Washington, Oregon, and California). In practice, the method uses regional drainage basin parameters with either a dimensionless unit hydrograph or a dimensionless summation unit hydrograph to derive the basin unit hydrograph.

Results of Detailed Scour Analysis

From the results of the HEC-6 model, a maximum average scour is obtained along the pipeline alignment. It must be understood that the results of the simulation are representative of the general scour and that the actual local scour must be added to obtain the total scour. Determination of local scour is theoretically beyond the capabilities of HEC-6 or any one-dimensional sediment transport model; however, analytical methods can be used to approximate it.

The local scour component is dependant on conditions unique to each site. It may include the thalweg formation, which is an allowance for incisement of the channel due to the concentration of flow. In sand bed channels, it may include the formation of bed forms in the channel such as dunes and antidunes. Other local scour components may include bend scour for flow at bends in the channel and pier scour if the pipeline alignment is close to a bridge in the stream. For some of the local scour components, the magnitude of scour may be a function of the mean depth at the design discharge. For others, such as pier scour, flow velocity and empirical coefficients are used in determining the value.

Lacey's relation (discussed previously) has been used to determine the local scour due to thalweg formation and bend scour. Additional information on flow at bends may be obtained from the U.S. Army Corps of Engineers publication "Hydraulic Design of Flood Control Channels", *EM 1110-2-1601* ("Hydraulic" 1991). Bed forms are flow-induced and are dependant on the flow regime. Bed form scour depends on the type of bed form predicted. For example, Yalin (1964) said that the dune height should be about 1/6 of the average flow depth. Nordin (1965) said that dune height could get as high as 1/3 of the flow depth. We used a scour due to dunes of 1/2 the dune height. An accepted relationship used to determine pier scour is the Colorado State University Equation. The complete procedure is presented by the Federal Highway Administration in *Hydraulic Engineering Circular No. 18* ("Evaluating" 1993).

The scour analysis would not be complete without a lateral scour analysis, which may be based upon inspection of aerial photographs, field reconnaissance, and other site specific information, in order to determine the maximum lateral extent of pipe burial depth for scour. Meander belt width and activity are taken into consideration in the assessment of the lateral migration of the alluvial channel. A summary of the scour results and the expected lateral limits for a number of pipeline crossings recently analyzed in Arizona are provided in Table 1. The crossings represent a variety of channel configurations, from confined to unconfined width, and with stream material size classes ranging from fine sand to cobbles.

TABLE 1. RESULTS OF SCOUR ANALYSES

Location of Pipeline Crossing	100-Year Design Discharge (cfs)	General Scour (ft)	Local Scour (ft)	Total Vertical Scour (ft)	Lateral Scour (ft)
Mariposa Wash near Nogales	5,550	3.5	2.2	5.7	170
Agua Fria River near Black Canyon City	55,000	0.8	11.6	12.4	875
Russell Gulch near Miami	12,700	0.3	6.0	6.3	2,700
Six-Shooter Canyon near Globe	13,400	2.6	7.3	9.9	265
Goodwin Wash near Geronimo	36,000	11.4	14.6	26.0	800
Unnamed wash near Dutch Flat	3,230	2.0	3.8	5.8	225
Gila River at Gillespie Dam	235,000	13.7	12.9	26.6	3,900
Gila River at Coolidge	130,000	6.3	8.4	14.7	6,700
Agua Fria River at New River	95,000	2.0	7.0	9.0	4,300
Gila River at Duncan	37,900	6.2	9.7	15.9	1,500
Aravaipa Creek near Mammoth	28,200	0.6	4.8	5.4	2,950
San Pedro River at Mammoth	46,800	5.4	9.4	14.8	2,200

Summary

Two methods to assess scour potential at pipeline crossings in the Southwestern United States have been presented: 1) a reconnaissance level screening method, and 2) a detailed scour potential study. The reconnaissance level screening method was developed to quickly assess erosion potential when faced with a large number of crossing locations. For crossings identified as having significant erosion potential, detailed scour and sediment transport studies have been used to compute the scour depth and pipeline burial requirements. This detailed analysis of hydraulic and erosional factors can help optimize cost savings in either future maintenance or in present crossing design.

References

Carreon, S., and Doeing, B. (1994). "Slurry Trenching for Waterway Pipeline Installation." *Proceedings, ASCE Pipeline Division, Hydraulics of Pipelines*, Phoenix, AZ.

Design of Small Dams. (1987). 3d. ed., U.S. Bureau of Reclamation. Denver, Colorado.

"Evaluating Scour at Bridges." (1993). *Hydraulic Engineering Circular No. 18: FHWA-IP90-017, 2nd ed.* Office of Research and Development, Federal Highway Administration (FHWA), Washington, D.C.

Fedarko, K. (1994). "Flood, Flames, and Fear," *Time Magazine,* October 28, 1994.

HEC-2, Water Surface Profiles, User's Manual. (1991). U. S. Army Corps of Engineers, Hydrologic Engineering Center, Davis, CA.

HEC-6, Scour and Deposition in Rivers and Reservoirs, User's Manual. (1993). U. S. Army Corps of Engineers, Hydrologic Engineering Center, Davis, CA.

"Hydraulic Design of Flood Control Channels," *EM 1110-2-1601.* (1991). U.S. Army Corps of Engineers, CECW-EH-D, Washington, D.C.

MacArthur, R. C., Williams, D. T., and Thomas, W. A. (1990). "Status and New Capabilities of Computer Program *HEC-6, Scour and Deposition in Rivers and Reservoirs".* *Proceedings, ASCE Hydraulic Engineering Conference*, San Diego, CA.

Nordin, C., and Algert, J.A. (1965). "A Discussion of 'Geometric Properties of Sand Waves,'" *Proceedings of the American Society of Civil Engineers*, Vol. 91, No. HY5.

Pemberton, E. L., and Lara, J. M. (1984). *Computing Degradation and Local Sour, Technical Guideline for Bureau of Reclamation,* U.S. Bureau of Reclamation, Denver, Colorado.

Thomas, B. E., Hjalmarson, H. W., and Waltemayer, S. D. (1994). *Methods for Estimating Magnitude and Frequency of Floods in the Southwestern United States,* U.S. Geological Survey Open-File Report 93-419, Tucson, Arizona.

Veldman, W. M. (1983). "Arctic Pipeline River Crossings - Design Trends and Lessons Learned." *Proceedings, ASCE Pipeline Division, Pipelines in Adverse Environments II,* San Diego, CA.

Williams, D.T. and Bradley, J.B. (1990). "The Sediment Weighted Histogram Generator and Estimation of Sediment Transport Trends." *Proceedings, ASCE Hydraulic Engineering Conference,* San Diego, CA.

Williams, D.T., Carreon, S., and Bradley, J.B. (1992). "Evaluation of Erosion Potential at Pipeline Crossings." *Proceedings, ASCE Water Forum 1992,* Baltimore, MD.

Yalin, M.S., 1964 "Geometric Properties of Sand Waves," *Proceedings of the American Society of Civil Engineers,* Vol. 90, No. HY5.

Yaremko, E. K., and Cooper, R. H. (1983). "Influence of Northern Pipelines on River Crossing Design." *Proceedings, ASCE Pipeline Division, Pipelines in Adverse Environments II,* San Diego, CA.

AVOIDING PIPELINE FAILURES AT WATERWAY CROSSINGS
PART 2: SCOUR COUNTERMEASURES

By Martin J. Teal[1], M. ASCE, Samuel Carreon[2], M. ASCE,
and Jeffrey B. Bradley[3], F. ASCE

Abstract

Pipeline crossings at streams, rivers, and washes are at risk due to erosion in the river environment. Buried pipelines may be exposed in the channel bottom by scour of bed materials or in the floodplain through migration of the channel banks. Pipelines carried across streams on bridges are at risk from scour at bridge piers and bridge abutments. This paper, being Part 2 of a two-part series, describes measures that can be taken to prevent erosion at pipeline crossings, emphasizing the authors' experience in the Southwestern United States (Part 1 of the series described scour assessment methods). Brief reviews of countermeasures to control lateral erosion (channel migration) and vertical erosion (scour or degradation) are followed by a design example for a pipeline crossing threatened by scour.

Introduction

Once erosion potential has been identified near an existing or future pipeline crossing using methods such as those presented in Part 1 of this paper, countermeasures to deal with the erosion can be studied. The most cost effective solution may be to bury the pipeline to a depth below the expected vertical scour along a length that encompasses any expected lateral scour. However, in many

---

[1]Project Engineer, WEST Consultants, Inc., 2111 Palomar Airport Road, Suite 180, Carlsbad, CA 92009-1419

[2]Consulting Engineer, Transmission Operations Engineering, El Paso Natural Gas Company, P.O. Box 1492, El Paso, TX 79978

[3]Chief Executive Officer, WEST Consultants, Inc., 2101 4th Avenue, Suite 1050, Seattle, WA 98121-2357

cases, using structural alternatives to slow or halt erosion can result in substantial cost savings in comparison to simple burial. These structural alternatives are the focus of this paper. The first section of the paper deals with control of lateral erosion (channel migration) while the second section considers control of vertical erosion (scour or degradation). Scour at bridge structures is not addressed in this paper. Bridge scour evaluation guidelines may be found in Federal Highway Administration Hydraulic Engineering Circular No. 18 (Richardson et al., 1993). The third section of the paper presents an example of steps to be considered in design of an erosion control structure.

Control of Lateral Erosion

Where lateral movement of a channel would endanger pipelines in the overbanks or floodplains of a stream (often buried at shallower depths than the main channel crossing), channel bank protection can slow or halt the lateral movement. Bank protection can be classified into two categories: structural measures and vegetative measures. Because the emphasis of this paper is on experiences in the arid Southwest, vegetative methods will not be discussed. However, in areas where enough water is available, vegetative methods are often given consideration (e.g., Henderson and Shields, 1984). Some of the most commonly used structural bank protection methods are described below.

Riprap - Defined as loose, angularly shaped stone, riprap is often the most cost-effective solution for streambank protection projects and is widely used throughout the world. Riprap, as long as it is properly sized to resist the forces imposed by moving water, is often an excellent solution to bank erosion problems. This is because the blanket of stones is flexible, allowing the stones to shift if there is subsidence. Since riprap is porous, it can prevent buildup of pore water pressure in the soil, which can cause more rigid countermeasures (such as concrete lining) to fail. Local failures in the riprap layer can be easily repaired by the addition of more stone. In design of riprap protection, it is extremely important to provide protection at the toe of the installation, down to the maximum predicted scour depth. One way to accomplish this without excavating to the full scour depth is to place a continuous mound of riprap at the toe of the blanket. When the channel scours below the depth of the mound, the rock will fall (launch) down the sideslope, protecting it. Riprap is sometimes used with other methods of bank protection (described subsequently) to provide toe protection by launching. Riprap should also be keyed into the bank at the ends of the installation to prevent outflanking of the protective works. Consideration must be given to filter requirements to prevent leaching of the smaller base material from underneath the riprap layer. In many cases either a synthetic filter fabric or sand/gravel filter can be provided to prevent such leaching.

Several manuals and methods are available for design of riprap, with two of the most popular being the Federal Highway Administration Hydraulic Engineering Circular No. 11 (HEC-11), "Design of Riprap Revetment" (Brown and Clyde, 1989), and the U.S. Army Corps of Engineers (COE) Engineering Manual "Hydraulic Design of Flood Control Channels" (U.S. Army, 1994).

Although usually placed in a layer on the banks to be protected, riprap may also be placed in trenches (windrows) set back from the channel so that the rock is essentially hidden until the channel migrates. If channel migration exposes a windrow, the riprap will launch, protecting the streambank slope (Figure 1).

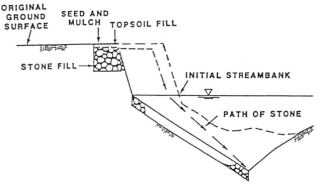

Figure 1. Windrow Revetment

Other Self-adjusting Armor - Concrete blocks or concrete filled sacks are sometimes used in place of rock for bank protection. If the blocks or sacks are loosely placed, and are not connected (e.g., by cables), the installation will function nearly identically to rock riprap.

Rigid or Semi-rigid Armor - Rigid revetments include concrete pavement, concrete-grouted riprap, concrete-filled fabric mat, and soil cement. Relief of pore water pressure is a concern that must be considered with these revetments. Articulating block mats are often considered as semi-rigid. As with riprap, these installations must be protected against degradation and outflanking.

Concrete-grouted riprap (riprap with all or part of the interstices filled with portland cement mortar) allows the use of smaller rocks, a lesser layer thickness, and more latitude in rock gradation than does dumped riprap. However, the grouted riprap is subject to the same types of failure as other impermeable, rigid revetments, and has little or no tensile strength. Design guidance may be found in HEC-11 (Brown and Clyde, 1989) and Engineer Technical Letter No. 1110-2-334 (U.S. Army, 1992).

Concrete-filled fabric mat, sometimes known by the patented product name Fabriform, consists of porous, pre-assembled nylon fabric forms which are placed on the surface to be protected and then filled with high-strength mortar by injection. These mats are attractive for construction in remote regions and underwater installation as all construction is performed in place. However, the mats only allow limited differential movement before failure.

Soil cement can be used for a wide range of stability problems and has been used with success in the Southwest. As soil cement is impervious, it is not recommended for areas where pore water pressure in the underlying soil could cause failure. Also, because a soil cement blanket is relatively brittle, very little if any traffic (vehicular, pedestrian or livestock) can be sustained without cracking the thin protective veneer. In climates with large temperature variations, the blanket can break up during freeze-thaw cycles. The material may be placed in 10-12 inch lifts and on slopes up to 1 horizontal to 1 vertical for stair-stepped soil cement. Procedures for constructing soil-cement protection by a stair step method can be found in Portland Cement Association literature (PCA, 1984a and 1984b). For many applications, soil cement can be more aesthetically pleasing than other types of revetment (its color is the same as the native material used). Soil cement protection can also be low cost and relatively easy to construct with on-site materials.

Gabions - If there are not suitable sizes of rock available for riprap, or if the sizes required are too expensive, wire baskets filled with stones (gabions) often become an option for bank protection. It is important to note that the wire baskets are subject to clipping if coarse material (gravel sizes and larger) is transported by flows in the channel. Installation of gabions can be more labor-intensive than other forms of bank protection. Gabions are also vulnerable to vandalism and are not as flexible as riprap. Gabion manufacturers provide guidelines and technical advice on their particular products. A filter blanket or synthetic filter fabric is used where required to prevent leaching of base material and undermining of the baskets.

Permeable/Impermeable dikes - These are also sometimes called groins (groynes) or spurs. These structures, which can be permeable or impermeable, are constructed with rock (riprap or gabions) or soil cement. They project into a stream at or nearly perpendicular to the flow thereby redirecting and/or retarding the flow. Brown (1985), Kehe (1984) and Copeland (1983) provide design guidance on spur length, spacing, and orientation.

Retards - Rigid retards include structures of concrete, steel or timber (such as cribs filled with rock or debris) and resemble the dikes just described. Examples of flexible retards which allow the retard structure itself to displace downward as scour occurs include Kellner jetties (also called jacks) and cut trees. Use of this technique requires secure connections between retard units and between the retard structure and the landward anchors. Allowance must be made in the retard height such that the downward displacement due to scour will not leave the upper bank exposed to significant erosion during high flows. A good performance record has resulted in extensive use of jetty fields on streams throughout the Southwest. Several agencies such as the Corps of Engineers, the Bureau of Reclamation, the Soil Conservation Service, state highway departments and railroad companies have used this method of streambank protection. In a number of cases, Kellner jetties have proved successful where more rigid protection methods have failed. These fields seem to perform best on streams with high bed material sediment load. Disadvantages of the method include: no immediate improvement to bank stability or surface erosion, interference with access to stream, and an appearance that is not aesthetically pleasing.

Vanes - Submerged vanes can reduce or eliminate the helical motion of flow in and near channel bends, reducing the velocity and scour along the outer bank (Odgaard and Mosconi, 1987). In the United States only a few installations exist, mostly in the Midwest.

Control of Vertical Erosion

Grade control structures are usually employed to reduce the slope of a stream or arroyo and prevent headcuts from migrating upstream. These structures stabilize the upstream area, allowing pipelines to cross the stream at a much shallower burial depth than without the structure. Drop structures (grade control structures with vertical or near-vertical drops) can be constructed of concrete, soil-cement, gabions, sheet piles, timber cribs, or even corrugated metal pipe (for small streams). Sloping grade control structures can be constructed of riprap, gabions, or other armoring types described above for bank protection measures. Many agencies such as the Corps of Engineers (U.S. Army, 1984), the Soil Conservation Service (SCS, 1984) and the Bureau of Reclamation (USBR, 1977) provide guidance for design of grade control structures. These structures must be designed considering the stability of the structure and the depth of the scour hole downstream of the structure.

The depth of the scour hole will be due to the erosion produced by the water jet plunging over the structure. Some type of stilling basin or energy dissipator is usually provided in the design of these structures to dissipate the energy and reduce erosion potential downstream of the structure. Energy dissipation can be accomplished using rigid concrete aprons, baffles, or a plunge pool. Sometimes (especially for smaller structures) the formation of a scour hole (plunge pool) is included in the design. Turbulent flow downstream of the structures can also cause lateral instability of banks and subsequent widening and possible outflanking of the grade control structures.

Allowance of self-forming plunge pools in grade control structure design can be cost effective, especially in remote areas. Care must be taken to allow for the maximum expected scour so that the structure itself is not undermined. Several methods of estimating scour hole depths in plunge pools downstream of drop or grade control structures are given in "Computing Degradation and Local Scour" (Pemberton and Lara, 1984). One of the most widely used equations for estimating scour hole depths for a vertical drop is the Veronese equation:

$$d_s = KH_T^{0.225} q^{0.54} - d_m \qquad (1)$$

where:
- d_s = Maximum depth of scour below stream bed, ft (m)
- K = 1.32 for English units, 1.90 for SI units
- H_T = Head from upstream water surface to tailwater level, ft (m)
- q = Design discharge per unit width, ft^3/s/ft (m^3/s/m)
- d_m = Downstream mean water depth, ft (m)

Laursen and Flick (1983) also proposed relations for scour at both vertical drops and sloping (4 horizontal to 1 vertical) sills based on laboratory experiments. Because large variations in calculated values of maximum scour depths are obtained depending on the equation used, it is always a good idea to calculate scour depths based on more than one relation and use engineering judgement as to the final value to be used for design.

Example of Grade Control Structure Design Steps

The following example is taken from the design of an actual pipeline crossing in Arizona where several structural scour countermeasures were studied. As an alternative to lowering four pipelines at a stream crossing, two grade control concepts were considered. One concept involved one or more vertical drop structures. The other involved sloped grade stabilization structures. Either alternative would prevent erosion of the stream bed at the pipe crossings for flows up to the 100-year discharge of 3,700 cfs.

Equilibrium Slope Concept

The concept of dynamic equilibrium is expressed by the Lane relation:

$$QS \sim Q_s D_{50} \qquad (2)$$

where
- Q = water discharge
- S = energy slope
- Q_s = bed sediment discharge
- D_{50} = median bed material sediment size

This relation indicates that alluvial channels tend toward a state of equilibrium in which the dominant discharge and slope are in balance with the sediment transport capacity and bed material size. The ultimate slope to which the channel will tend is called the equilibrium slope.

The equilibrium slope concept was used in design of the grade control structure alternatives downstream of the pipelines. The channel invert elevation at the structure was fixed at the structure's crest height. The crest height was set at such an elevation that, projecting the channel invert upstream at the equilibrium slope, the pipelines would remain buried with three feet of cover.

Crest Length and Lateral Limits

The crest length of the sill for either of the two structure types was selected to:
a) provide a hydraulically efficient section to pass the design discharge
b) prevent concentration of flow beyond what historical evidence indicated
c) limit velocities at the sill, which affect scour depth and bank protection requirements

A crest length L of 80 feet was chosen. Then the unit discharge q (equal to Q/L) is then 3,700/80 = 46.25 cfs/ft.

Lateral limits of the structures were chosen to prevent channel migration evidenced by abandoned channels in the area. It was assumed that an earthen berm already in place would not provide protection during the 100-year event. Therefore, to maintain flows in the existing channel where the sill would be located, the structures were designed to extend to the limits of the floodplain determined from analysis of the 100-year discharge. For either of the grade control options, it was considered that the extension of the structure could be accomplished either with an earthen berm built to Federal standards, or with a slurry wall/concrete cap combination.

For the berm option, embankment and foundation stability requirements would have to be met. Further engineering analyses would have to be conducted that would, at a minimum, address the following concerns: embankment geometry and length of seepage path at critical locations, embankment and foundation materials, embankment compaction, penetrations and other design factors affecting seepage (such as drainage layers), and other design factors affecting embankment and foundation stability.

Vertical Drop Structure

The vertical structure would essentially be a wall across the wash, with a trapezoidal notch in it to handle all flows up to the design (100-year) event. A transverse section of the single-drop design is shown in Figure 2. A double-drop (two-structure) design was also evaluated, but because the scour depths downstream of the structures did not vary greatly from the scour produced from the single-drop concept, the idea was not pursued further.

Scour Depths - Scour depths produced from the water plunging over the wall were calculated using the Veronese Equation (1) with q = 46.25 cfs/ft, H_T = 6 ft (obtained from the HEC-2 hydraulic model [U.S. Army, 1990]) and d_m assumed to

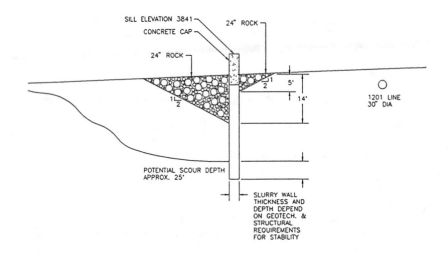

Figure 2. Single Drop Structure Configuration

be zero (no tailwater assumed). Inserting these values into equation (1) yields a scour depth of 16 ft. Scour was also predicted using an equation proposed by Laursen and Flick (1983):

$$\frac{D_s}{y_c} = 8\left(\frac{V_c}{w_o}\right)^{3/4} - \frac{\frac{6+V_c}{w_o}}{\sqrt{1+\frac{2\Delta WS}{y_c}}} \qquad (3)$$

where

D_s = the scour depth measured from the downstream tailwater, ft
y_c = the critical depth of flow, ft
V_c = the critical velocity, ft/s
w_o = the fall velocity of a quartz sphere of median diameter d of the material being scoured, ft/s
ΔWS = the drop in water surface across the structure, ft

The Laursen-Flick equation provided a scour depth of 30 ft, almost twice that of the Veronese Equation. A conservatively weighted average of the two results was taken resulting in the adopted scour depth for a single drop structure, measured from the original stream bed downstream of the drop to the bottom of the scour hole, of 25 feet.

Channel and Bank Protection - Riprap placement would be necessary both upstream and downstream of the structure. The limits of placement are indicated in Figure 2. Computations using COE criteria (U.S. Army, 1994) showed that a nominal 24-inch gradation placed in a layer two feet thick would provide sufficient bank and channel protection. The standard U.S. Army Corps of Engineers gradation meeting this criteria, for rock of unit weight not less than 150 lb/ft^3, was:

Percent Lighter by Weight	*Stone Weight, lbs.*	
	Minimum	Maximum
W_{100}	250	630
W_{50}	125	185
W_{15}	40	90

Rock downstream of the structure would be placed as shown in Figure 2 and would be launched as the scour hole developed. Because the riprap would be launched, and not placed, enough rock was included to provide a thickness 1.5 times the calculated thickness (U.S. Army, 1994). A launch slope of 2 horizontal to 1 vertical (2H to 1V), based on Corps of Engineers model and prototype data (U.S. Army, 1994), was used to calculate the volume of rock needed per unit foot of length.

Sloping Grade Control Structure

The sloping grade control structure is comprised of stone and is similar to rock stabilizers used by the Los Angeles District of the U.S. Army Corps of Engineers throughout the Southwest United States. The sloping sill is constructed with a 2H to 1V approach slope and a 4H to 1V exit slope, as shown in Figure 3. A filter fabric is overlain by a layer of bedding material 1 foot thick, which is in turn overlain by two feet of rock (24-inch maximum stone size, gradation same as shown above for vertical structure). At the upstream and downstream ends of the structure, additional stone is provided to prevent undermining. The stone extends down to the expected scour depth.

Scour Depth - The depth of the scour hole formed by the water downstream of the structure was calculated by the equation developed for sloping sills with a 4H to 1V slope by Laursen and Flick (1983). This equation is

$$\frac{D_s}{y_c} = 4\left(\frac{y_c}{d}\right)^{0.2} - 3\left(\frac{d_{rr}}{y_c}\right)^{0.1} \tag{4}$$

where D_s = the scour depth measured from the downstream tailwater, ft
       y_c = the critical depth of flow, ft

d = the size of the material being scoured (or the riprap blanket in the bottom of the scour hole), ft

d_{rr} = the size of the riprap layer protecting the surface of the sloping sill, ft

In the calculations, the rock sizes were both taken as two feet diameter. The critical depth at the sill is $y_c = (q^2/g)^{1/3} \approx 4$ feet. Equation (4) then yields a scour depth of approximately eight feet.

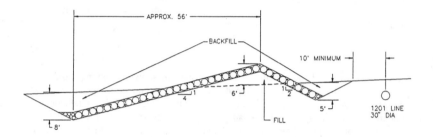

Figure 3. Sloping Grade Control Structure

Channel and Bank Protection - Riprap would be placed for protection along either the earthen berm or vertical wall. Either of these options would be acceptable and the preferred option would be based on cost comparisons.

Protection on Sloping Sill - The report prepared by Laursen and Flick (1983) suggests that any rock on the downstream 1V to 4H sill be surrounded by wire mesh and anchored in place as it was in the experiments they performed in preparing their report. Therefore, the rock placed on the sill of the structure should either be grouted in place or used in gabions or reno mattresses that are anchored in place.

An additional option for protection of the sloping sill, the use of concrete-filled fabric mats, was considered. The particular type of mat considered was the articulating block mat (ABM) which has reinforcing cables inserted between two layers of fabric prior to mortar injection. This type of mat is available commercially from several companies. The cables allow the mat to flex slightly (articulate) to adjust to changing foundation conditions. A six inch ABM was designed and would be placed over a gravel bedding.

Effects on Downstream Pipelines

Construction of a grade control structure per the designs was expected to have no effect on the amount of scour that other downstream pipelines would otherwise experience. Scour resulting from water plunging over a structure is expected to be localized in the area of the plunge pool. During the design event, the plunge pool was not expected to exceed 60 feet in length. The closest pipeline crosses the wash approximately 300 feet downstream of the proposed grade control structure. Since the velocities exiting the plunge pool would not be greater than preproject conditions (as determined by use of a hydraulic model, HEC-2), local downstream scour would not increase. Thus, the downstream pipelines were not expected to be affected at all by construction of a grade control structure at the selected location.

Local scour was calculated for the downstream pipelines where they cross the wash using three methods described in USBR's technical guideline publication (Pemberton and Lara, 1984): the Lacey, Blench and "envelope curve" methods. The results indicated six feet of scour at the pipelines furthest downstream.

Summary

Different methods of preventing both lateral erosion (channel migration) and vertical erosion (scour or degradation) are presented and briefly summarized. Given expected scour parameters obtained through assessment procedures such as those outlined in Part 1 of this paper, proper countermeasures may be designed to protect pipelines from expected erosion. A design summary for a grade control structure, as an alternative to lowering of pipelines, was also presented showing some of the basic parameters to be considered in the design process.

References

Brown. S.A. (1985). "Design of Spur-Type Streambank Stabilization Structures." *Report No. FHWA/RD-84/101*, prepared for the Federal Highway Administration, Washington, D.C.

Brown, S.A., and Clyde, E.S. (1989). "Design of Riprap Revetment." *Hydraulic Engineering Circular No. 11, Report No. FHWA-IP-89-016*, prepared for the Federal Highway Administration, McLean, Virginia.

Copeland, R.R. (1983). "Bank Protection Techniques Using Spur Dikes." *Miscellaneous Paper HL-83-1*, U.S. Army Engineer Waterways Experiment Station, Vicksburg, Mississippi.

Henderson, J.E. and Shields, F.D., Jr. (1984). "Environmental Features for Streambank Protection Projects." *Technical Report E-84-11*, U.S. Army Corps of Engineers, Washington, D.C.

Kehe, S.M. (1984). "Streambank Protection by Use of Spur Dikes." Project Report, Department of Civil Engineering, Oregon State University, Corvallis, Oregon.

Laursen, E.M., and Flick, M.W. (1983). "Scour at Sill Structures." *Report Number FHWA/AZ83/184*, prepared for Arizona Department of Transportation, Tucson, Arizona.

Odgaard, J.A., and Mosconi, C.E. (1987). "Streambank Protection by Submerged Vanes." *J. of Hydraulic Engineering*, Vol. 113, No. 4. ASCE, New York, New York.

Pemberton, E.L. and Lara, J.M. (1984). "Computing Degradation and Local Scour." *Technical Guideline for Bureau of Reclamation*, Denver, Colorado.

Portland Cement Association. (1984a). "Soil-Cement for Facing Slopes and Lining Channels, Reservoirs, and Lagoons." *PCA Publication IS126.05W*, Skokie, Illinois.

Portland Cement Association. (1984b). "Soil-Cement Slope Protection for Embankments: Planning and Design." *PCA Publication IS173.02W*, Skokie, Illinois.

Richardson, E.V., Harrison, L.J., Richardson, J.R., and Davis, S.R. (1993). "Evaluating Scour at Bridges." *Hydraulic Engineering Circular No. 18, Report No. FHWA-IP-90-017, Rev. April 1993,* prepared for the Federal Highway Administration, McLean, Virginia.

Soil Conservation Service. (1984). *Engineering Field Manual*. 4th printing. Washington, D.C.

U.S. Army Corps of Engineers, Hydrologic Engineering Center. (1990). *HEC-2, Water Surface Profiles*. Davis, California.

U.S. Army Corps of Engineers. (1992). "Design and Construction of Grouted Riprap." *Engineer Technical Letter 1110-2-334,* Washington, D.C.

U.S. Army Corps of Engineers. (1994). "Hydraulic Design of Flood Control Channels." *Engineer Manual 1110-2-1601,* Washington, D.C.

U.S. Bureau of Reclamation. (1977). *Design of Small Dams*. Washington, D.C.

Criterion for Predicting Failure of Corroded Linepipe

Bin Fu[1] and Mike G Kirkwood[2]

ABSTRACT

This paper describes an analytical study of the failure behaviour of corroded linepipe. The study is based on an elastic-plastic, large-deformation finite element analysis of simulated pipeline corrosion shapes. Corrosion pits and narrow corrosion grooves in pressurised linepipe were analysed. A failure criterion, based on the local stress state at the corrosion and a plastic collapse failure mechanism, is proposed. Failure stress predictions obtained for simulated corrosion defects are compared with predictions using the existing ANSI/ASME B31G code and a modified B31G method. It is concluded that corrosion geometry significantly affects the failure behaviour of corroded pipe and categorisation of pipeline corrosion should be considered in the development of new guidance for integrity assessment.

INTRODUCTION

Predicting the failure of damaged oil and gas pipelines is essential for the determination of design tolerances, post inspection integrity assessment and effective maintenance action. A pipeline may experience significant internal and external corrosion defects which reduce its strength and resistance to fatigue cracking, local buckling, leakage and bursting. The existing criterion, detailed in ANSI/ASME B31G (ASME 1985), was developed on an empirical basis over 20 years ago. The code was based on an extensive testing of pipe vessels with narrow machined slots in the external surface, and the majority of the vessel materials were low grade steels (Maxey et al 1972, Kiefner et al 1973). The corrosion assessment methods in the B31G code have been successfully used in the oil and

---
[1] Senior Consulting Engineer, Billington Osborne-Moss Engineering Ltd, Ledger House, Fifield, Maidenhead, SL6 2NR, UK.
[2] Principal Engineer, British Gas plc, Engineering Research Station, Newcastle upon Tyne, NE99 1LH, UK.

gas industries but it has been recognised that they can be over-conservative (Coulson and Worthingham 1990). This is mainly due to the simplifications embodied in the methods and through its application to a complex variety of defect shapes which have an inherently different failure mechanism. Additionally, the accuracy of the method in corroded pipeline with higher grade steels has not been fully justified.

High-resolution on-line inspection techniques developed during the last decade have enabled the accurate location and sizing of pipe wall corrosion. In parallel, modern numerical analysis methods have enabled the modelling of realistic defect shapes and nonlinear material behaviour. Backed by experimental validation, these are proving a powerful and accurate tool in predicting critical condition against failure. In order to reduce unnecessary repair/replacement actions and optimise pipeline design, research aimed at developing a new failure criterion and guidelines, is being undertaken worldwide. Limited numerical studies using thin shell models (Wang 1991, Bubenik et al 1992), plane strain models (Jones et al 1992) and general 3D models (Chouchaoui et al 1992) have recently been published, in which failure pressures are predicted using either an elastic limit state criterion, plastic limit state criterion or allowable plastic strain criterion. These studies have improved failure predictions for the defect groups considered.

This paper discusses the limitations in the use of the existing code and presents a numerical study of the failure behaviour of corroded linepipe. The study is based on an elastic-plastic, large-deformation finite element (FE) analysis of simulated pipeline corrosion shapes, external corrosion pits and external corrosion grooves. A failure criterion, based on the local stress state in the corroded region and failure due to plastic collapse, is proposed.

EXISTING CRITERION

The original B31G criterion for predicting the failure of corrosion was derived from an extensive database of flawed-pipe burst tests (axial through wall and part-through wall crack-like slots). A fracture mechanics analysis, based on a Dugdale crack-tip plasticity model and a Folias bulging factor representation of the stress intensity factor for an axial crack in pressurised cylinder, resulted in the following failure equation:

$$\sigma_f = \bar{\sigma} \left[\frac{1 - (A/A_o)}{1 - (A/A_o) M^{-1}} \right] \tag{1}$$

This equation was validated by a series of corroded pipe burst tests, the materials ranging from API 5L Grade A-25 to 5LX Grade X-52. The failure equation was then modified by (i.) limiting the maximum hoop stress by the material's yield strength and (ii.) characterising a corrosion geometry by a projected parabolic shape for relatively short corrosion and a rectangular shape for long corrosion. Simple failure equations are embodied in the B31G code as below:

$$\sigma_f = 1.1\, SMYS \left[\frac{1-(2/3)(d/t)}{1-(2/3)(d/t)M^{-1}} \right] \quad \text{(for } \sqrt{0.8\left(\frac{L}{D}\right)^2\left(\frac{D}{t}\right)} \leq 4 \text{)} \tag{2a}$$

$$\sigma_f = 1.1\, SMYS\, [1-(d/t)] \quad \text{(for } \sqrt{0.8\left(\frac{L}{D}\right)^2\left(\frac{D}{t}\right)} > 4 \text{)} \tag{2b}$$

where $SMYS$ is the specified minimum yield strength, and the bulging factor was defined by:

$$M = \sqrt{1 + 0.8\left(\frac{L}{D}\right)^2\left(\frac{D}{t}\right)} \quad \text{(for } \sqrt{0.8\left(\frac{L}{D}\right)^2\left(\frac{D}{t}\right)} \leq 4 \text{)} \tag{3}$$

M is infinity otherwise.

Kiefner and Vieth (1989, 1990) argued that the excess conservatism in the original B31G criterion is due to the expression for flow stress ($\bar{\sigma} = 1.1 \times SMYS$), which under-estimates the true ultimate strength of the pipe materials, the parabolic representation of the metal loss, and the approximation of the bulging factor, M, They proposed a new flow stress value $\bar{\sigma} = SMYS + 69$ (MPa), an "effective area" for projected corrosion shape, i.e. A/A_o, and a new bulging factor:

$$M = \sqrt{1 + 0.6275\left(\frac{L}{D}\right)^2\left(\frac{D}{t}\right) - 0.003375\left(\frac{L}{D}\right)^4\left(\frac{D}{t}\right)^2} \quad \text{(for } \left(\frac{L}{D}\right)^2\left(\frac{D}{t}\right) \leq 50 \text{)} \tag{4a}$$

$$M = 3.3 + 0.032\left(\frac{L}{D}\right)^2\left(\frac{D}{t}\right) \quad \text{(for } \left(\frac{L}{D}\right)^2\left(\frac{D}{t}\right) > 50 \text{)} \tag{4b}$$

Bubenik et al (1992) compared the failure predictions obtained using the original and this modified B31G criteria and showed that the change in flow stress definition and bulging factor formulae do not give significant improvement, but the use of effective area considerably reduces the conservatism.

Linepipe corrosion often has anomalous shapes which are normally classified by corrosion pitting, corrosion bands (or grooving) and general corrosion. A corrosion band is usually formed by an array of pitting, and a general corrosion may be a group of pits which spread over a wide circumferential extent on the pipe wall. However, the existing codes use a single simple corrosion geometry, i.e. either a parabolic or a rectangular shape and the corrosion width is not considered.

Figure 1 shows a comparison of failure predictions using the B31G code for a specific linepipe geometry with a range of corrosion lengths. It shows large differences between the predictions assuming the parabolic and the rectangular corrosion shapes and inconsistent expansions of the failure pressure functions for long corrosions. Kiefner and Vieth (1990) stated that the rectangular shape simplification gives too conservative predictions by comparison with the burst results of corroded pipe vessels. This implies that the failure pressure function for longer corrosion (L/D>0.65), shown in Figure 1, may have the same level of conservatism, according to the comparison. Test results from a British Gas study of long pipe wall defects (Hopkins and Jones 1992) indicate that the above failure function is adequate if the flow stress is defined by the material's UTS. However, the corrosion models considered in the tests were crack-like slots.

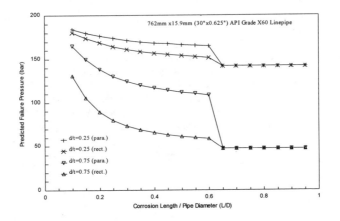

Fig.1 Comparison of Failure Predictions Assuming Different Corrosion Shapes.

An unpublished British Gas numerical study of corrosion models shows that the local stress state and the failure behaviour of the corrosion models are affected by their circumferential widths in addition to their lengths and depths. As the width increases from a crack-like slot to a wide extent of metal loss (more than 40 degree in angle), the failure behaviour changes from one extreme (controlled by fracture) to another extreme (controlled by plastic collapse after three plastic hinges develop). Therefore a single shape representation for all type of corrosions has an inherent limitation in application, and a simple fracture mechanics approach could result in more conservative predictions. This implies that failure criteria should be developed for various corrosion shapes, such as isolated pitting, interactive pitting, narrow/wide band grooves and interacting grooves. The following sections describe a development of failure criterion for isolated corrosion pits and narrow band corrosion grooves.

ANALYTICAL STUDY OF FAILURE BEHAVIOUR

Corrosion Model

Two types of external corrosions, isolated pits and narrow band grooves, have been studied using the FE method. The pit models have a semi-spherical shape with a diameter of 60 mm and the groove models have a semi-cylindrical shape with a width of 60 mm and a length of 190.5 mm (L/D=0.25). Three corrosion depths, d/t=0.25, 0.50 and 0.75, have been considered, these represent shallow, intermediate and deep defects respectively. The geometry of the pipe considered in this study is of 762 mm (30 inch) outside diameter and 15.88 mm (0.625 inch) wall thickness. Material properties of the FE models were calibrated from a database of standard tensile tests of API grade X60 linepipe material. A true stress-strain relationship of the X60 linepipe material was used in the FE analysis. Figure 2 shows 3D FE meshes for local regions around the pit and the groove models.

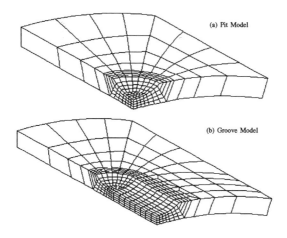

Fig.2 Local Details of the 3D FE models.

Stress Analysis

The FE analysis was carried out using the ABAQUS v5.3 FE software. Both material non-linearity and large displacements were considered in the elastic-plastic stress analysis by means of a geometrically nonlinear static analysis procedure (ABAQUS 1993). Internal pressure was modelled as a static loading condition in which the pressure load increases to a level at which equilibrium requires a load increment less than a specified minimum value, 10^{-5} of the total load specified.

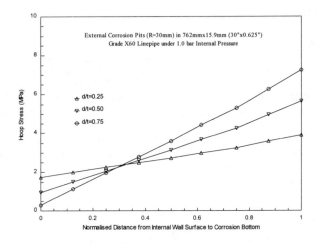

Fig. 3 Distributions of Hoop Stress through Remaining Ligament of the Corrosion Pit Models.

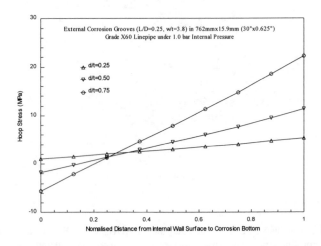

Fig.4 Distributions of Hoop Stress through Remaining Ligament of the Corrosion Groove Models.

Figures 3 and 4 show hoop stress distributions through the remaining ligament of the corrosion models. Although the corrosion shape changes from shallow to deep in depth and from extremely short to relatively long in length, linear through thickness stress distributions were obtained for all these models.

These types of stress distributions indicate that the local stress states are controlled by a membrane stress and a bending moment, instead of a stress concentration. Both the membrane stress and the bending stress in the hoop direction increases as the corrosion depth and the corrosion length increases. Such local stress states result in a localised bulging deformation and failure occurs in the manner of plastic collapse as the pressure load exceeds a critical level.

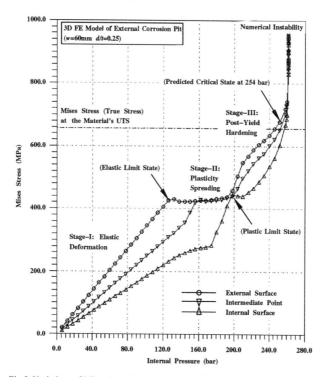

Fig.5 Variations of Mises Stress State at an External Corrosion Pit (w/t=3.8, d/t=0.25).

Figures 5 and 6 show the variation of local Mises stresses, at the corrosion surface, internal wall surface and an intermediate point, with an increase in the pressure load for the corrosion pit model and the corrosion groove model respectively. Variations of the stress states indicate that, before structural failure occurs, the remaining ligaments under the corrosions experience three distinct loading stages, i.e. the elastic deformation stage; the plasticity-spreading stage and the post-yield hardening stage. In the first stage, the local ligament deforms elastically. After the stress state at the corrosion bottom exceeds the material's yield strength, the plasticity spreads through the remaining ligament while the Mises stress level remains approximately constant until the plasticity reaches the opposite wall surface. The third stage shows the ultimate capacity of the corroded pipe at which point the whole ligament deforms plastically. This feature is due to the structural constraint provided by the surrounding pipe wall. At the stress level (true stress) corresponding to the material's ultimate tensile strength (engineering stress), the through ligament stress distribution becomes nearly uniform and a steep increase in the local stress level occurs, this leads to the stress level corresponding to that at the material's final elongation in a standard tensile test with a very small increment in pressure load.

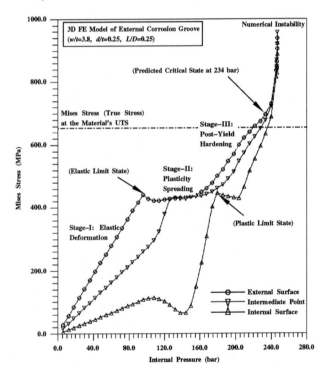

Fig.6 Variations of the Mises Stress State at an External Corrosion Groove (d/t=0.25, L/D=0.25).

Figures 7 and 8 show the FE results for all the corrosion models considered. They indicate that, for corrosions with higher local bending, e.g. deeper and longer corrosions, the first and the second stages reduce. These results demonstrate that changes in corrosion depth and in corrosion length affect the failure pressure level but not the failure behaviour characterised by the three stages. The failure behaviour is primarily affected by the corrosion width.

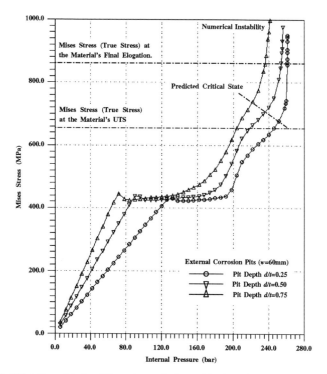

Fig.7 Variations of the Local Stress State at the External Surface of Corrosion Pit Models.

Failure Prediction

The FE results in Figures 5 and 6 show that a failure criterion based on the elastic limit state, at which the maximum stress at the corrosion bottom reaches the material's yield strength, is over-conservative and a failure criterion based on a plastic limit state, at which the minimum stress at the opposite wall surface reaches the material's yield strength (excluding the post-yield hardening stage), may also considerably under-estimates the ultimate strength of corroded pipe. However, failure predicted by the state of numerical instability may over-estimate the remaining strength. The results obtained suggest a failure criterion, which incorporates both the material's strength (i.e. the UTS) and the structural constraint (i.e. the plastic spreading stage and the post-yield hardening stage) and relates to the failure mechanism of plastic collapse.

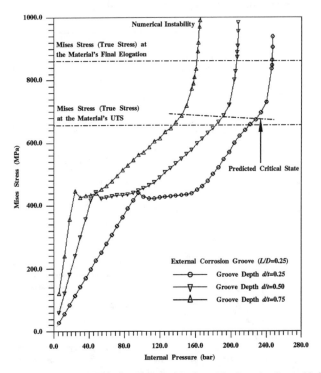

Fig.8 Variations of Local Stress State at External Surface of the Corrosion Groove Models.

The FE analysis indicates a critical condition that can be approximately determined by a point, which indicates the end of the post-yield hardening stage, i.e. the beginning of the acceleration in stress variation leading to an instability, in the plot of Mises stress vs pressure load, as shown in Figures 7 and 8. Although this includes an uncertainty in determining the end of the post-yield hardening, the margins of such uncertainty in the pressure values, as shown in Figs 7 and 8, are small (few percentage of the predicted failure pressure values). Unpublished British Gas hydro-test results for linepipes with a different material grade and a different pipe geometry (D and t values) show that actual burst pressures measured are few percentage higher than numerical predictions using the above criterion.

The failure pressures predicted, by the FE analysis, the original B31G and the modified B31G (MB31G), for the three pit models and the three groove models are compared in Table 1.

The comparison of predicted failure pressures, listed in Table 1, shows that (i.) the modified B31G method gives only a few percent increase in failure pressures for the pit models and the shallow corrosion groove models, but lower failure pressure values for the intermediate and the deep corrosion groove models than the predictions obtained using the methods in the original B31G code. The present failure criterion gives higher failure pressure values from both the original and the modified B31G criteria.

d/t	Predicted Failure Pressure (bar)					
	The Corrosion Pit Models			The Corrosion Groove Models		
	F.E.	MB31G	B31G	F.E.	MB31G	B31G
0.25	254.0	198.0	186.0	231.0	180.1	173.8
0.50	237.0	193.3	180.6	195.0	151.3	154.4
0.75	213.0	186.0	172.3	143.0	109.8	130.2

Table 1 Comparison of Predicted Failure Pressures for the Pit Models and the Groove Models.

Although there is no test data to support the analytical results for the specific corrosion models presented here at this stage, the validity of the proposed failure criterion has been partially justified by unpublished British Gas experimental studies (vessel burst tests and pipe ring tests) of similar corrosion models with different pipe sizes and material grades, which show reduced conservatism and increased accuracy. For instance, an externally machined pit model with a depth of 70% of wall thickness survived in the burst tests when the nominal hoop stress level reaches 95% of material's UTS. The predicted failure pressure for the pit model, $d/t=0.75$, in the present study is related to the nominal hoop stress level which is 87% of the UTS of the X60 linepipe material. In general, the proposed criterion gives conservative failure predictions when compared to the British Gas test data.

The proposed failure criterion, which is based on a rigorous analysis of corrosion models, has an advantage that enables the complexity of defect shapes, the variety of pipe geometries and material grades to be considered. A group sponsored research project for linepipe corrosion is currently undertaken by British Gas. The project includes a large number of full-scale burst tests and pipe ring tests and a comprehensive numerical study. The proposed criterion will be validated by the test results.

CONCLUSIONS

This paper has described a rigorous analytical study of the failure behaviour of corroded linepipe. A new failure criterion, based on plastic collapse mechanism, is proposed. This criterion can be applied to advanced numerical analysis of realistic corrosion models for the development of new guidelines on corrosion in transmission pipelines.

ACKNOWLEDGEMENTS

The authors wish to thank British Gas plc for permission to publish this paper. The authors also wish to thank their colleagues at British Gas plc, Engineering Research Station, who have contributed to the paper.

APPENDIX I, REFERENCES

ABAQUS v5.3, (1993). Hibbit, Karlsson and Sorensen Inc. Providence.

ANSI/ASME B31G-1984. (1985). *Manual for Determining the Remaining Strength of Corroded Pipelines - A Supplement to ASNI/ASME B31 Code for Pressure Piping*. The American Society of Mechanical Engineers, New York.

Bubenik, T. A., Olson, R. J., Stephens, D. R. and Francini, R. B. (1992). Analyzing the pressure strength of corroded line pipe. *Proc., 11th Int. Conf. Offshore Mech. & Arctic Eng.*, Calgary, Canada, V. 225-232.

Chouchaoui, B. A., Pick, R. J. and Yost, D. B. (1991). Burst pressure predictions of line pipe containing single corrosion pits using the finite element method. *Proc., 11th Int. Conf. Offshore Mech. & Arctic Eng.*, Calgary, Canada, V. 203-210.

Hopkins, P. and Jones, D. G. (1992). A study of behaviour of long and complex-shaped corrosion in transmission pipelines. *Proc., 11th Int. Conf. Offshore Mech. & Arctic Eng.*, Calgary, Canada, V. 211-218

Jones, D. G., Turner, T. and Ritchie, D. (1992). Failure behaviour of internally corroded linepipe. *Proc., 11th Int. Conf. Offshore Mech. & Arctic Eng.*, Calgary, Canada, V. 219-224.

Kiefner, J. F., Maxey, W. A., Eiber, R. J., and Duffy, A. R.. (1973). Failure Stress Levels of Flaws in Pressurized Cylinders. *Progress in Flaw Growth and Fracture Toughness Testing, ASTM STP 536*. American Society for Testing and Materials. 461-481.

Kiefner, J. F. and Vieth, P. H. (1989). *A Modified Criterion for Evaluating the Remaining Strength of Corroded Pipe*. Project PR 3-805 Pipeline Search Committee, American Gas Association.

Kiefner, J. F. and Vieth, P. H. (1990). New method corrects criterion for evaluating corroded pipe. *Oil and Gas Journal*, 88(32), 56-59.

Maxey, W. A., Kiefner, J. F., Eiber, R. J., and Duffy, A. R. (1972). Ductile Fracture Initiation, Propagation and Arrest in Cylindrical Vessels. *Fracture Toughness, ASTM STP 514*, American Society for Testing and Materials, 70-81.

Wang, Y. S. (1991). An elastic limit criterion for the remaining strength of corroded pipe. *Proc., 10th Int. Conf. on Offshore Mech. & Arctic Eng.*, Stavanger, Norway, V, 179-18 .

APPENDIX II, NOTATIONS

The following symbols are used in this paper:

A = projected cross-sectional area of corrosion (mm^2);
A_o = $L \times t$ (mm^2);
D = outside diameter of pipe (mm);
d = maximum corrosion depth (mm);
L = longitudinal length of corrosion (mm);
M = bulging factor (dimensionless);
p_f = failure pressure (bar);
t = nominal wall thickness (mm);
w = circumferential width of corrosion (mm);
$\bar{\sigma}$ = flow stress of the pipe material (Ma);
σ_f = failure stress (MPa);

EARTH PRESSURE ON PIPELINES IN CENTRIFUGED MODELS

Li L.[1] and Tohda J.[2]

Abstract

Total values and circumferential distributions of earth pressures acting on pipelines transiting through differential settlement grounds were investigated through 2-D centrifuge model tests, where model pipes buried in different grounds were pulled up or pulled down during centrifuge flight. The pull-up tests showed that: 1) effects of both pipe size and stress level were not observed, 2) the similarity law was satisfied, and 3) the earth pressures depend on the shearing properties of the soils. The similarity law was also satisfied in the pull-down tests when the size of the model grounds was widened.

Introduction

The safety of buried pipelines transiting through differential settlement grounds is usually examined by a beam theory on an elastic foundation, in which the ground is replaced with linear or non-linear elastic springs fixed to the pipelines, and the differential equation is solved by applying either loads acting on the upper-halves of the pipelines or relative displacements between the pipelines and grounds. Since analyzed results depend on the intensities of the spring coefficients, the main concern in the study has been how to estimate them to adjust the calculated results to actual data. Many researchers have pointed out that the spring coefficient is not an inherent modulus of the soil, because it depends on either structure size or stress level (e.g., Hyodo 1991).

The authors believe through both their studies on the longitudinal behavior of pipelines (Tohda 1991) and liter-

....
[1]Graduate student, [2]Associate prof., Osaka City University, 3-3-138, Sugimoto, Sumiyoshi-ku, Osaka, 558, Japan

ature survey (Trautman 1985, Dickin 1988, Ng 1994) that: 1) the loads and reaction forces acting on the upper-halves and lower-halves of the pipelines are earth pressures that are equally produced as a result of 3-D soil-pipe interactions involving various factors such as boundary conditions at the surfaces of the pipelines, and therefore, 2) their behavior can not be expressed through the simple spring coefficients. Accordingly, earth pressures acting on the pipelines should be directly investigated to obtain the realistic safety factors of the pipelines. As a first step to do this, 2-D centrifuge model tests were carried out under various conditions, in which 3-D behaviors of the actual pipelines along their longitudinal axes were replaced with two ideal 2-D conditions that the settlement of the pipelines, δ_P, was either smaller or greater than that of the ground, δ_G, as illustrated as zones A and B in Fig. 1(a). In the tests, model pipes buried in the grounds were pulled up or pulled down during centrifuge flight to simulate these two conditions, as shown in Fig. 1(b), and total earth pressures and circumferential distributions of normal and tangential earth pressures acting on the model pipes were measured.

The following three effects, expressed as three axes in Fig. 2, should be considered in model tests including centrifuge model tests: size effect, stress effect and modeling effect. They are denoted here as D-effect (D: external diameter of pipes), H-effect (H: cover height), and N-effect (N: scaling ratio of models; model scales are 1/N), respectively. Although both D-effect and N-effect have been called as the scaling effect without distinction (Ovesen 1981), they are clearly distinguished in this paper to avoid the confusion. D-effect and H-effect mean that the phenomena, earth pressures in this case, are dependent on both the size of structures and the stress level. Fig. 3 shows a simple slip model to define both D-effect and H-effect in the pull-up tests using non-cohesive soils. If both effects exist, either the gradients of the slip planes, α, or internal friction angles of

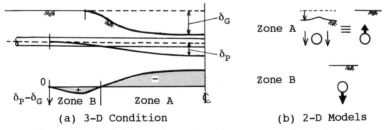

(a) 3-D Condition (b) 2-D Models

Figure 1. Modeling of A Pipeline Transiting through Differential Settlement Grounds

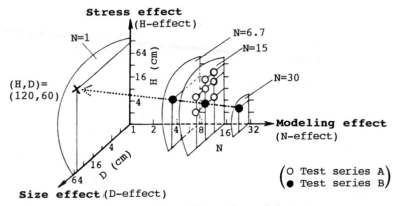

Figure 2. Three Effects in Model Tests

soils, ϕ, will vary for different D and H, resulting in non-constant coefficient $f(\alpha, \phi)$ of H/D in the equation of $P_v/\gamma HD$ shown in the figure, in which P_v: total earth pressures acting on the model pipes, and γ: unit weight of the soil under centrifugal accelerations. Thus, existence of both effects can be examined by investigating $f(\alpha, \phi)$. Furthermore, N-effect means if the phenomena in scaled models are similar to those in prototypes or not; the similarity law governs their similarity.

Thus, the existence of both D-effect and H-effect was examined first in the pull-up tests using dry sand grounds by measuring the total earth pressures for different D and H, whose combinations are illustrated as the marks ○ on the N=15 plane in Fig. 2. Second, N-effect in the pull-up tests was examined for six ground conditions by using a technique of modeling of models, in which different model scales of 1/N and centrifugal accelerations of N G (G: gravitational acceleration) were combined, as illustrated as the marks ● on the planes of

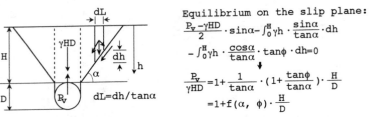

Figure 3. A Slip Model for the Pull-up Tests

N=30, 15 and 6.7 in Fig. 2, to produce an identical prototype having D=60 cm and H=120 cm (cf. the mark × on the plane of N=1 in Fig. 2). After examining these three effects, the distributions of the normal and tangential earth pressures were measured in the pull-up tests for the six ground conditions. On the other hand, it was found in the pull-down tests that the similarity law was not satisfied owing to both the friction acting on the wall of the test container and small size of the container. As to the pull-down tests, therefore, only the data, obtained in the tests using a revised method to satisfy the similarity law, are reported.

Experiment Representation

Fig. 4 shows models tested and their setup. Model pipes having 1-9 cm in external diameters and 15-37 cm in lengths were made of aluminum alloy or stainless steel, whose surfaces were finished to be smooth. They were connected by a pair of steel bars with diameters of 4-8 mm to a hydraulic cylinder mounted on the container. The steel bars were inserted into thin tubes to cut friction forces transmitted from the model ground. The edges of the steel bars were not fixed to the model pipes, except in Test series A described later, so that the model pipes could settle as the ground settled during centrifugal loadings. A pair of load cells was mounted at the other edges of the steel bars to measure the total earth pressures acting on the pipes. The test container was made of hard aluminum alloy, whose inside dimensions are 45 cm in width x 45 cm in height, its thickness, W, being varied to be 15-37 cm by inserting dummy plates between its main body and the front or back plate. The container was reinforced by steel angles in the pull-down tests.

Three soils with dense and loose densities, whose properties are shown in Tables 1 and 2, were used to construct the model grounds. Dense and loose grounds using

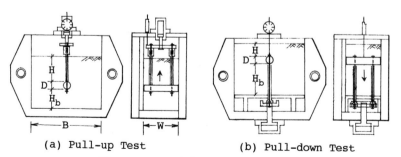

(a) Pull-up Test (b) Pull-down Test

Figure 4. Models and Test Setup

Table 1. Primary Properties of Soils

Soil*	G_s	Grain Size Distribution Max. mm	Sand %	Silt %	Clay %	U_c	ρ_{dmax} g/cm³	ρ_{dmin} g/cm³	w_{opt} %
S	2.65	1.4	100	0	0	1.75	1.58	1.32	–
G	2.71	2.0	84	9	7	70	1.92	1.37	11.4
F	2.67	2.0	70	17	13	115	1.86	1.18	13.5

* S: Dry Sand, G: Decomposed Granite, and F: Silty Sand.

Table 2. Secondary Properties of the Grounds

Soil	Density	w %	ρ_d g/cm³	D_r 1) %	D_c 2) %	S_r %	c_d 3) tf/m²	ϕ_d 3) degree	c_p 4) tf/m²	ϕ_p 4) degree
S	Loose	0	1.43	47	91	0	0	37	0	16
S	Dense	0	1.55	83	97	0	0	43	0	17
G	Loose	10	1.50	30	78	34	0.9	38	0.1	24
G	Dense	10	1.70	68	89	46	2.3	38	1.1	25
F	Loose	12	1.50	58	81	41	3.0	32	0	25
F	Dense	12	1.70	84	91	56	4.6	32	1.1	26

1) D_r: relative density, 2) $D_c = \rho_d / \rho_{dmax}$, 3) shear strength parameters under drain condition, and 4) friction parameters against aluminum alloy pipes.

a dry silica sand, S, were prepared by pouring from 50 cm and 1 cm heights by means of a small hopper having a circular hole of 1 cm diameter. The other two soils, G and F, are a decomposed granite (w=10 %) and a silty sand (w=12 %) containing fine fractions of 16 % and 30 % in weight, respectively; dense and loose grounds using these two soils were constructed by heavy and light compaction in each layer 2 cm thick. The direction of both pouring and compaction was vertical as in actual construction.

The models were subjected to centrifugal acceleration fields, and the model pipes were pulled up or pulled down in constant displacement rates by means of the hydraulic cylinder. The earth pressures and displacements of the pipes were recorded by a digital strain meter, and photographs were taken to observe the ground deformation.

Table 3 shows conditions of the four test series, A-D. Contents of each test series are as follows:

Test series A (the pull-up test): Tests using dense and loose dry sand grounds were carried out in a 15 G

Table 3. Conditions of Four Test Series

Test series[1]	Scale N	Pipe D cm	Pipe Material	Soil[2]	Model H cm	Model H_b cm	Model W[3] cm	Model B[4] cm
A	15	2-8	Al[5]	S	8-32	4-16	22	45
B	6.7-30	2-9	Al	S,G,F	4-18	4-10	15-22	33-45
C	6.7	9	Al	S,G,F	18	10	15-22	33-45
D	15-60	1-4	Al,St[6]	S	2-8	4.5-18	20-37	45

[1]Test series A-C: pull-up test and Test series D: pull-down test, [2]S: dry sand, G: decomposed granite, and F: silty sand; they were used under both dense and loose conditions, [3]thickness of the ground, [4]width of the ground, [5] aluminum alloy, and [6] stenless steel.

field with a combination of different H and D to investigate the existence of both D-effect and H-effect. Three model aluminum pipes having D=2 cm (bar), D=4 cm (wall thickness t: 6.5 mm), and D=8 cm (t=10 mm) were used to measure the total earth pressures by means of the load cells, their length L being 21.9 cm. The rate of pulling up the pipes was 1 mm/min..

Test series B (the pull-up test): The satisfaction of the similarity law was investigated by using three model pipes of D=2 cm, 4 cm, and 9 cm under six ground conditions when H/D=2; D and H in the corresponding prototype were 60 cm and 120 cm. The model pipes of D=2 cm and D=4 cm were the same as those used in Test series A. Another model pipe of D=9 cm is shown in Fig. 5. Its length is 15.1 cm, except for one case when Soil F was used under the dense condition; in this case, its length was extended to 21.9 cm by using a pair of dummy pipes, as shown in the figure, to minimize the effect of the friction force acting on the container wall. The rate of pulling up the pipes was 0.42 mm/min..

Test series C (the pull-up test): Tests in this series are common to those using the model pipe of D=9 cm in Test

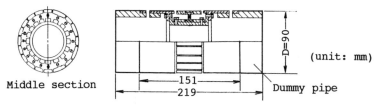

Figure 5. A Model Pipe Used in Test Series B and C

series B. Distributions of normal and tangential earth pressures, σ and τ, together with the total earth pressures, were measured in a 6.7 G field for the six ground conditions.

Test series D (the pull-down test): The satisfaction of the similarity law was investigated in 15 G, 30 G, and 60 G fields for only dry sand grounds. Model pipes were aluminum and stainless steel bars having D=1-4 cm and L=20-37 cm; different pipe materials were selected to minimize the pipe deflections along their longitudinal axis within 0.5 mm. H and D of the corresponding prototype were coincident with those in Test series B; H_b/D (H_b:thickness of the sand below the pipes) was 4.5. The rate of pulling down the pipes was 0.23 mm/min..

Preliminary tests using dry sand grounds showed that measured results when D=4 cm were independent of the following factors: 1) roughness of the pipe surface in the pull-up tests, 2) displacement rate of the pipes in the pull-up tests within 0.17-1.55 mm/min., 3) H_b in the pull-up tests (within H_b=4-24 cm) and in the pull-down tests (within H_b=18-28 cm), and 4) lubrication on the wall of the test container in the pull-up and pull-down tests.

<u>Measured Results in the Pull-up Tests (Test Series A-C)</u>

Measured results in Test series A are shown in Figs. 6 and 7. Fig. 6 shows changes in $P_v/\gamma HD$ against the pipe displacement δ, where P_v: the total earth pressures acting on the model pipes, and γ: unit weight of the soil under a 15 G condition. P_v were not zero at δ=0, because the

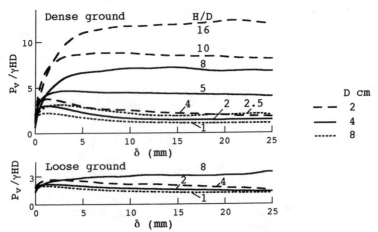

Figure 6. Change in $P_v/\gamma HD$ against δ (Test Series A)

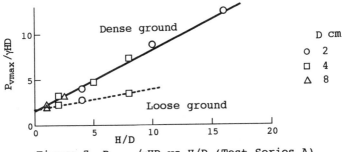

Figure 7. $P_{vmax}/\gamma HD$ vs H/D (Test Series A)

model pipes and the edges of the steel bars were fixed in this test series. Nevertheless, the other tests without fixing them showed that the intensities of P_v were almost coincident with those in Fig. 6. Fig. 7 shows $P_{vmax}/\gamma HD$ (=the maximum values of $P_v/\gamma HD$) against H/D, indicating that the dense sand ground generated $P_{vmax}/\gamma HD$ greater than the loose ground. The figure also indicates that $P_{vmax}/\gamma HD$ increase linearly with an increase of H/D in both dense and loose grounds. According to the definition of both D-effect and H-effect in the pull-up tests (cf. Fig. 3), the constant gradient of $P_{vmax}/\gamma HD$-H/D in Fig. 7 confirmed that both effects do not exist in Test series A, and as a result, they do not affect the test results, at least, in the dry sand grounds.

Fig. 8 shows $P_{vmax}/\gamma HD$ against the scaling ratio of the models, N, for six ground conditions, obtained in Test series B, where γ: unit weight of the soils under N G conditions (γ mean γ_t for Soils F and G). $P_{vmax}/\gamma HD$ for each ground condition is almost identical at any model scale, indicating that the pull-up tests satisfied the similarity law for different ground conditions including partially saturated grounds.

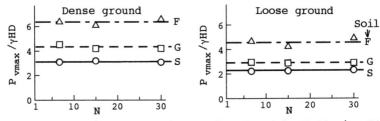

Figure 8. $P_{vmax}/\gamma HD$ vs Scaling Ratio of Models N (Series B)

Measured results in Test series C are shown in Figs. 9-12 and Table 4. Fig. 9 shows changes in $P_v/\gamma HD$ against δ for six ground conditions; the thick and thin lines denote the data for the dense and loose grounds for each soil, respectively. The dense condition generated $P_{vmax}/\gamma HD$ greater than the loose condition for each soil ground. Fig. 10 shows changes in earth pressure distributions at five stages during the tests, measured for three grounds under the dense condition. The earth pressures in the figure are illustrated by using the polar coordinates; compressive σ and downward τ are taken as positive. σ and τ on the lower-half of the pipe became to be zero at the beginning of pulling up the pipe; σ on the upper-half of the pipe changed complicatedly their distribution shapes with an increase of δ. The loose ground for each soil generated the earth pressure distributions in the manner similar to those in Fig. 10. Fig. 11 shows the earth pressure distributions measured when $P_{vmax}/\gamma HD$ was recorded (at the stage ③); the P_v values calculated by using the earth pressure distributions, shown as curves in Fig. 11, were almost coincident with the measured P_{vmax} values, indicating high accuracy in the earth pressure measurement. The figure showed that: 1) the measured τ are small in any ground, and 2) σ in the dry sand grounds concentrate onto the pipe top, whereas in the other grounds σ at the pipe top are smaller than those in the neighboring areas. Table 4 shows the measured $P_{vmax}/\gamma HD$ and soils' strength, τ_f, calculated by $\tau_f = c_d + \gamma H \cdot \tan\phi_d$. These values except one data indicate that the $P_{vmax}/\gamma HD$ increases with an increase of τ_f. Furthermore, the difference between the curves of $P_v/\gamma HD$-δ shown in Fig. 9 could be explained through the difference in dilatancy properties of the soils. Thus, it was confirmed that the mechanism in the pull-up tests is governed by the soil's shearing properties. This conclusion was supported by the ground deformation at $\delta=18$ mm (at $\delta/D=0.2$) shown in Fig. 12, in which clear slip planes were observed except in the loose dry sand ground.

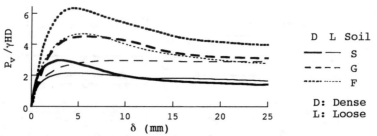

Figure 9. Change in $P_v/\gamma HD$ against δ (Test series C)

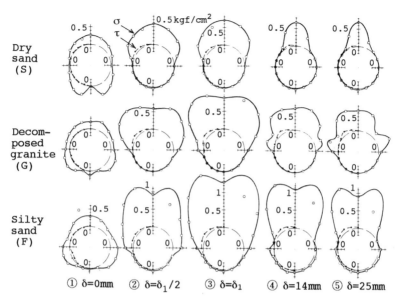

Figure 10. Change in Distributions of σ and τ in Test Series C (P_{vmax} was recorded at the Stage ③)

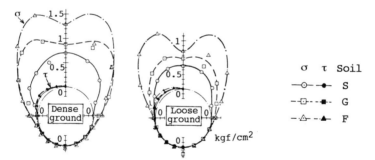

Figure 11. Distributions of σ and τ at the Stage ③

Table 4. Measured $P_{vmax}/\gamma HD$ and Soil's Strength τ_f at the Depth of the Pipe Top

	Dry Sand		Decomposed Granite		Silty Sand	
	Dense	Loose	Dense	Loose	Dense	Loose
$P_{vmax}/\gamma HD$	3.1	2.1	4.5	3.0	6.5	4.7
τ_f (tf/m^2)	1.8	1.3	4.0	2.5	6.1	4.4

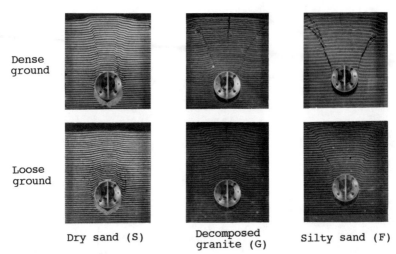

Figure 12. Deformation of Grounds at δ=18mm (Test Series C)

Measured Results in the Pull-down Tests (Test Series D)

Preliminary tests showed that the effects of both the friction acting on the container wall and the small size of the container could not be neglected for the dense sand ground in the pull-down tests. The lubrication applied on the container wall was not sufficient. Thus, the widening of both thickness, W, and width, B, of the ground was employed as a countermeasure for the dense sand ground. Fig. 13(a) shows the measured $P_r/\gamma HD$ (P_r: total reaction earth pressure) against δ/D for different sizes of the model ground when D=4 cm. This figure indicates the ef-

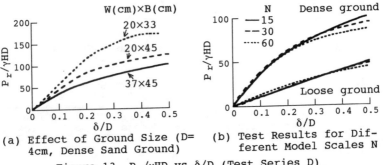

(a) Effect of Ground Size (D= 4cm, Dense Sand Ground)
(b) Test Results for Different Model Scales N

Figure 13. $P_r/\gamma HD$ vs δ/D (Test Series D)

fectiveness of this countermeasure. Fig. 13(b) shows changes in $P_r/\gamma HD$ against δ/D, measured for different scaling ratios of the models, N, in Test series D, where the size of the dense sand ground was widened as W x B=37 cm x 45 cm (W x B=20 cm x 45 cm in the loose sand ground). The figure shows that: 1) the similarity law was satisfied within the range of $\delta/D<0.3$, as expected, and 2) the relationship between $P_r/\gamma HD$ and δ/D was almost linear in the loose ground, while it was non-linear in the dense ground.

Conclusion

Main conclusions drawn from this study are as follows:

1. Both D-effect and H-effect can be neglected in the pull-up tests using the dry sand grounds. Both effects either for the other soil grounds or for the pull-down tests should be investigated further.
2. The similarity law was satisfied in the pull-up tests using six ground conditions including partially saturated soils. It was also satisfied in the pull-down tests using the dry sand, when the size of the dense ground was widened. Thus, the test results can be extrapolated to the prototype cases with the similar ground conditions.
3. The intensities of the total earth pressures and their changes against the pipe displacement in the pull-up tests were governed by the shearing strengths and dilatancy properties of the grounds.
4. The distributions of both normal and tangential earth pressures measured in the pull-up tests were dependent on the types and densities of soils used as the ground materials, and they changed their shapes complicatedly with an increase of the pipe displacement.

References

Dickin E. A. 1988. Stress-Displacement of Buried Plates and Pipes. Centrifuge 88. Balkema. pp.205-212.
Hyodo M., Shimamura K., and Takagi N. 1991. Field Test and Analysis for Evaluating a Coefficient of Subgrade Reaction for Buried Pipes. Tuchi-to-Kiso (Journal of JSSMFE). Vol. 39-4 (399). pp.3-8. (in Japanese)
Ng C. W. W and Springman S. M. 1994. Uplift Resistance in Granular Materials. Centrifuge 94. Balkema. pp.753-758.
Trautman C. H. and O'Rourke T. D. 1985. Lateral Force-Displacement Response of Buried Pipe. ASCE. Journal of Geotechnical Engineering. Vol.111. No.9. pp.1077-1092.
Tohda J. and Yoshimura H. 1991. Failure of Buried Gas Pipeline Crossing a Trench. Proc. of ASCE Specialty Conf. on Pipeline Crossings. pp. 190-201.
Tohda J. etc. 1991. Centrifuge Model Tests of Uplift Capacity of Buried Pipes. Proc. of the 26th JSSMFE Annual Meeting. pp. 1769-1772. (in Japanese)

Large-scale model test of leachate pipes in
landfills under heavy load

Helmut Zanzinger[1]
Erwin Gartung[1]

Abstract

A large-scale model test is used to examine the behaviour of HDPE leachate pipes to be used in full-scale landfill construction. Waste coverage heights of up to 60 m are simulated as static load above the pipe.

A comprehensive test program determines tension and deformation in the soil, as well as elongation and deformation of the pipe. The results of the measurements make it possible for current conventional analytical models to be modified.

Introduction

The question of the structural stability of leakage pipes at the base of landfills, particularly under very high coverage heights, has not been adequately treated. Mere simulations of static load with numerical calculations do not resolve the uncertainties: in situ measurements should be carried out. However, because of the need to leave the mineral basal liner undamaged, and because of the operating conditions at a landfill, it is not possible to carry out measurements which record with certainty all the basic parameters at the landfill site.

Accordingly, tests are carried out on a leachate pipe in the LGA Geotechnical Institute's model test bay, under the conditions actually obtaining in landfills. The construction, the materials and the size of the model correspond to the conditions in a landfill. The model was subjected to a uniformly-distributed vertical load of up to 800 kN/m^2. The following parameters were recorded in a comprehensive measuring program:

---

[1] LGA-Grundbauinstitut, Nuremberg, Germany

- radial deformations of the leachate pipe
- surface strains of the pipe
- stresses in the soil around the pipe
- soil deformation in the pipe area

The objective is to use the measurement results to check, and possibly to calibrate existing analytical models, and if necessary to develop new processes of calculation.

The tests were financially supported by the Bavarian Office for Environmental Protection.

Scale of the model test

The size of the pipe cross section is prescribed by directives, e.g. DIN 19667 (1991). In accordance with this, pipes with a minimum diameter of DN 250 must be used. Most commonly, DN 300 HDPE pipes, of pressure stage PN 10 or higher, are chosen for use in landfill construction. These have an external diameter of da = 315 mm or 355 mm. For hydraulic reasons, the perforations of the pipes should consist of slots. From a static point of view, however, perforated pipes with circular openings can cope with higher loads. The coverage height of the drainage gravel should be more than two times the external diameter. Outside the pipe area, the drainage gravel layer is approximately 30 cm thick.

In accordance with the German regulations for solid waste landfills, it has become customary for many landfill constructions to install a composite lining, with a geomembrane laid on with a compacted clay liner. To make it possible for pipes to be bedded onto composite linings, it is necessary for the pipes to be laid on a sand/bentonite bearing surface.

Numerical calculations (Zanzinger et al. 1992) have shown that the pipe's static loading capacity is greater if a sand/bentonite bearing surface is provided, than when installed directly on the clay liner. For the pipe, therefore, direct bedding on a mineral basal liner is the less favourable option. As little research has been carried out on conditions in the pipe area of HDPE leachate pipes in landfills, directly bedding on clay was investigated in the first phase of this research project. We report on the results obtained. Further work to study conditions in the pipe area on composite linings is under way.

The model size was established on the basis of finite element analyses, while varying the geometric boundaries. This resulted in the following minimum dimensions where no further marginal disturbances were to be expected:

Model height ≥ 1.6 m
Model width   ≥ 4.4 m

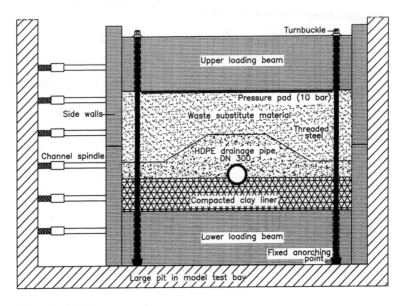

Figure 1. Model cross-section

The model is cased in a construction of pre-fabricated, reinforced concrete units, which can be assembled on a modular basis (Figure 1). Load was applied by pressure pads.

Clay

Some 15 m³ of compacted clay material were necessary to install the mineral basal liner, representing approximately 29 tonnes of clay. Accordingly, several batches of milled and watered material were delivered from the nearby landfill construction site currently being built at Nuremberg, and immediately installed in the LGA's model test bay. The parameters relevant to deformation behaviour are given in Table 1.

Table 1. Soil parameters

		Clay	8/16 gravel	Waste substitute material
Density ρ	t/m³	2.02	1.65	0.87
Dry density ρ_d	t/m³	1.68	1.65	0.82
Water content w	%	20.5	0	6.1
Loosest state max. n	1	-	0.42	-
Densest state min. n	1	-	0.33	-
Relative density I_D	1	-	0.35	-
Proctor density ρ_{pr}	t/m³	1.69	-	-
Optimal water content w_{pr}	%	20	-	-
Degree of compaction D_{pr}	%	100	-	-
Confined modulus (for 200 kPa)	MN/m²	15	50	3
Confined modulus (for 400 kPa)	MN/m²	20	65	5
Confined modulus (for 800 kPa)	MN/m²	22	80	14

Drainage gravel

Based on experience gained in tests with gravel from different places, we chose drainage gravel with qualities and mineral composition well suited for a landfill basal drainage layer. A fraction of 8/16 mm was chosen to match the size of the measuring devices. The characteristic soil mechanical parameters were determined in the LGA Geotechnical Institute's laboratory.

Waste substitute material

When the mechanical properties of solid waste are discussed, it rapidly becomes apparent that they vary in wide ranges. They differ from region to region and the composition in future will be different from what it is today or what it used to be. In any case, waste is so heterogeneous that it is almost impossible to set up a representative sample model in the context of a laboratory test.

In addition, its handling causes hygienic problems, e.g. in the case of municipal waste. For these and other reasons, it is necessary to use a material with the mechanical properties of waste, which can easily and safely be handled in model tests any time.

After experiments with different materials, we used mixtures of sand and wood shavings. These can be made into very stable, homogeneous mixtures, which are easy to control in their load-settlement behaviour. Table 1 shows the soil-mechanical coefficients of the selected mixture.

Leachate pipe

For the tests described below, pipes with the following specifications were used:

Material:	HDPE Hostalen GM 5010 T2
External diameter:	315 mm
Wall thickness:	28.7 mm
Slot width:	7 mm
Slot interval:	8 cm

Test program

For the load test to check the structural stability of leachate pipes under conditions specific to landfills under static load, our aim was to place the pipe under stress in as realistic a way as possible. The load of $p = 800$ kN/m² corresponds to a waste coverage of approximately 60 m. This represents a fully realistic and adequate situation.

Plastics such as the HDPE pipes mainly used in landfill construction tend to react to constant load with stresses and deformations that vary with time. They may creep or relax. So that the pipe in our model test would not be placed under stress for too short a period of time, but tested over as long a period as possible, a test duration was planned of over 1,000 hours. Published details confirm that the modulus of elasticity of HDPE drops to approximately 30% of its short-term value after 1,000 hours load.

Even with a disproportionate effort, e.g. a test duration of one year, no really different information can be expected, as the modulus of elasticity would only have a slight further drop to approximately 25% of the temporary value. On the basis of these specifications for the size and duration of load, a test program was developed to determine the following:

Radial deformation of the leachate pipe

For this purpose, a powered measuring device was developed, which is pushed into the pipe on runners and causes a laser head to rotate automatically. This laser is able to measure precisely the distance between light source and the wall of the pipe (accuracy 0.1 mm).

Surface strain measurement of the leachate pipe

To establish the normal forces and bending moments in the pipe, the leachate pipe was fitted with strain gauges. These were placed at a number of cross-sections on the external and the internal surfaces of the pipe.

Soil stress measurement around the pipe

Earth pressure cells were placed in the waste substitute material, in the drainage gravel and in the clay liner, to measure vertical soil stresses. In addition, earth pressure meters were placed in the gravel and clay to measure horizontal stresses.

Soil deformation measurement around the pipe

Vertical soil movements were measured at different levels by fibreglass rod extensometers. Multiple extensometers were also available to measure horizontal movement beside the pipes.

All measurements were sent by measurement amplifiers and multi-position measurement equipment directly to a computer for storage and evaluation. In addition, the following measurements were recorded:

- pressure in the loading pads
- force in the tendons steel
- temperature in the leachate pipe and in the clay.

Load

The model was placed under pressure by applying constant water pressure in all pressure pads. The internal pressure in the pressure pads was constantly controlled by a pneumatically powered hydraulic pump. In addition, the load caused by the pressure pads was measured by pressure meters at selected tendons steel.

The system was planned in such a way that if the pressure in the pads dropped because of soil settlement, this was automatically adjusted, thus keeping the internal pressure of the pad constant. Water consumption was constantly registered. Even when, towards the end of the test program, one pad showed slight signs of leakage, it was possible to pump in sufficient water to compensate for the water loss and retain the pressure in the pad. This made it possible for life-like conditions to be achieved. As a result of the leak in a pressure pad towards the end of the test, a brown leachate fluid emerged from the pipe, and the bearing surface softened to a certain degree. It was possible to establish that clay had been pressed through the slots near the basal liner.

Up to this point it had been possible to measure pipe deformations with the laser measuring device. Afterwards, as there was a constant flow of water along the base, it was no longer possible to carry out complete optical measurements.

The load lasted for a total period of four months, with load being continually increased in single steps, up to a pad inner pressure of p = 12 bar. The duration of the individual load levels is given in Table 2. Readings were constantly taken and evaluated by a computer-controlled data registration program, either manually or automatically.

Table 2. Load Levels

Internal pressure (bar) in pressure pads	0.5	1.0	2.0	3.5	5.0	7.0	8.0	10.0	12.0
duration (days)		11	17	6	13	61	7	3	1	1

Results of measurements

Soil stress

Vertical soil stress was registered at the following heights: measuring level 2 (lower edge of pipe base), measuring level 4 (upper edge of pipe top) and measuring level 6 (upper edge of drainage layer). The vertical stresses show the effect of the so-called "shear beam". When load was first applied (Figure 2), the stress was evenly spread. From a load of approximately 100 kN/m^2, however, stress moves to the vertex, thus relieving the load on the pipe.

An arch is produced in the body of gravel, which remains when load is increased. Only the sizes of the redistributed stresses change when the load is increased (Figure 2). The maximum vertical stresses are measured approximately 0.6 m from the pipe axis. The vertical stress is more evenly distributed below the base of the pipe. The archforming effect is not as noticeable in the clay (Figure 3). Load above the drainage layer is practically constant. However, the "shear modulus beam" attracts load from the waste layer, as a result of its greater stiffness.

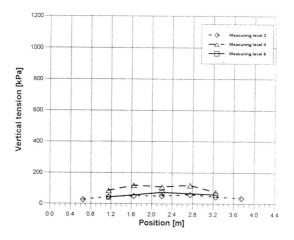

Figure 2. Vertical stresses at the measuring levels with a pad pressure of 50 kN/m²

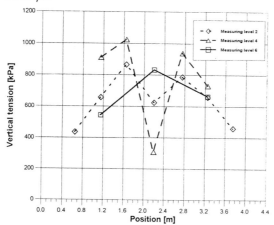

Figure 3. Vertical stresses at the measuring levels with a pad pressure of 1,200 kN/m²

The horizontal stresses measured at the abutment height of the pipe were on average approx. $\sigma_H = 200$ kN/m², under maximum load. The lowest horizontal stresses were measured at a distance of approx. 1 m from the pipe axis.

Pipe wall deformations

Measurement of elongation or compression at the inside and outside of the pipes did not confirm the assumption that the most critical point for the pipe would be in the transition from clay bearing surface to gravel layer. Instruments were concentrated very densely in the lower half of the pipe, between the polar angle of $\varphi = 0°$ (base) and 90° (abutment), and between $\varphi = 270°$ (abutment) and 0°, in order to measure any maximum load there accurately.

However, the greatest compression was measured in the abutments, with up to 3.5% at the inside. The maximum elongation of 1.5% was measured at the top surface, on the outside.

The boundary bending strain for the material was not exceeded, in the pressure or tensile area, even using a creep modulus of $E_c = 300$ N/mm².

Figure 4 shows typical elongation curves in 3 measurement cross-sections for the extent of the pipe. The surface strains were measured at different points, at 3 cross-sections of the outer surface (solid line in Figure 4). Strains were also measured on the inside at cross-section H.

Pipe deformations

The measurements showed very large deformations of the pipes. Figure 5 shows a comparison between the maximum deformation and the undeformed pipe. The relative diameter changes, related to the axis, were $\delta_v = 14\%$. After the load has been relieved, this value drops to 8,5%.

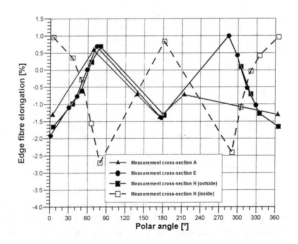

Figure 4. Surface strains in pipe at pad pressure of 700 kN/m² in three different measurement sections.

Calculatory results producing figures of approx. 6% deformation under these conditions are obviously on the unsafe side. Accordingly, it is necessary to verify the calculatory models by using the measurement results from the loading tests.

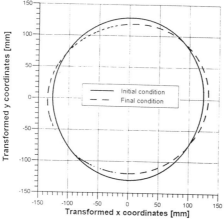

Figure 5. Pipe deformation in the final loading stage of the test

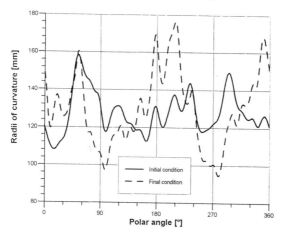

Figure 6: Radii of curvature of the deformed pipe in the final loading stage, example of measurement

For example, the pipes have undergone pre-deformation before installation. Each further increase in load resulting from the waste covering places strain on a pre-deformed system. Accordingly, it would seem sensible to carry out analysis in a pre-deformed status.

The deformation measurements confirm the test results obtained by the strain gauges. After entering the radii of curvature of the inside pipe wall, it is seen that the largest radii (up to 200 mm) were measured in the area of the pipe invert and crown ($\varphi = 0°$ and $180°$). In contrast, the contortion of the pipe in the abutments ($\varphi = 90°$ and $270°$) was at its lowest (90 mm). Figure 6 shows a typical comparison at maximum load and at zero measurement in unloaded state.

Summary

The trial construction of a pipe area at a landfill base, as a large-scale model, equipped with extensive recording meters, showed the following:

- The gravel covering above the piping, corresponding to twice the pipe diameter, is sufficient for an arch to form above the piping.

- The horizontal stress distributions show, in conjunction with vertical stress distributions, the changes of direction of the principal stresses in the area of the arch.

- By considering the horizontal displacements of soil at pipe height, it was established that the gravel was pressed away laterally from the pipes, in an area of up to 30 cm from the outer wall of the pipe. Outside this area, it moved in the direction of the pipes.

- Compressive strain is greatest at the pipe abutments.

- Pipe deformations (changes in diameter) of up to 14% were measured. This shows that the dimensioning of the pipes seems to be a problem of deformation rather than stress.

In future, pipe durability will play a decisive role. As it is in the nature of flexible pipes to deform, it is proposed that deformations of up to 12% be permitted for conditions in use. This also lets the load-bearing behaviour of these pipes be utilized in a way suitable to their material. Calculatory models to forecast deformation, while taking into consideration the necessary safety coefficients, have still to be developed for design purposes.

References

(1) DIN 19667
Dränung von Deponien, Technische Regeln für Planung, Bauausführung und Betrieb, 1991.

(2) Zanzinger, H., Gartung, E. and Hoch, A.
Grundsatzuntersuchung über die statische Berechnung von Rohrleitungen in Sickerwasserentsorgungssystemen bei Abfalldeponien, Veröffentlichungen des LGA-Grundbauinstituts, Heft 61, 1992.

INSPECTING BURIED PLASTIC PIPE
USING A ROTATING SONIC CALIPER

Ted Price

ABSTRACT

Direct-buried plastic pipe is frequently used in sewer and wastewater applications. The SONEX ROTATOR has the ability to accurately inspect buried plastic pipes and to measure and display any deflection of the pipe walls. This paper describes the technique of the ROTATOR Sonic Caliper and presents examples of field and laboratory demonstrations of the tool.

INTRODUCTION

Plastic pipe, made by many manufacturers, has become an accepted resource for new and replacement pipes. Plastic pipes are used as inserts inside existing damaged pipes, or can be directly buried in the ground.

With any pipe materials, installation inspection is important to verify long and useful life. This is particularly true with the direct-buried pipes where improper installation methods can negate structural properties of the pipe. Further, measurement of the deflection of plastic pipe within the first year after installation can be used to predict the rate of deflection over the desired life of the pipe.

Until recently, there have been no satisfactory methods of measuring pipe-wall deflection. Single-point and single-axis tool are limited in use; using a fixed-size size mandrel to provide a go-no-go result gives no information about degree of deflection or the reasons for deflection. The best tool would be one to describe the inside circumference of the pipe for 360 degrees. The SONEX ROTATOR was designed to do just that.

----
1. Owner - SONEX, 608 Williams Boulevard, Richland, WA 99352, USA - 509-375-1096

INSPECTING BURIED PLASTIC PIPE

BACKGROUND

In the early 1980's cities realized that the collapse of concrete sewer pipes had become a serious and recurring problem. Hydrogen-sulfide corrosion of the pipe walls was the most common cause. If the pipe breaks, the soil around the pipe can wash in, creating a cavity above the pipe which can eventually break through to the surface.

SONEX learned of this problem in 1984 when a client city contacted us to see if we could adapt our geo-technical sonic probes to measuring concrete thickness at the crown of their buried sewer pipes. We took the approach of measuring inside vertical diameter and comparing it with as-built diameter to determine the corrosion loss.

The simple but elegant concept was to use sonic-distance measurement methods to measure the distance from a tool in the pipe to the wall of the pipe. We were primarily concerned with the crown of the pipe since that is where the television inspection of the pipes showed the worst corrosion.

The tool that we developed for this purpose we named the Sonic Caliper. This tool had several sonic transducers mounted in fixed positions to measure the distance from the tool to the crown, to the sidewalls, and to the bottom so that we could also measure debris in the pipe.

From 1986 to 1993 versions of the Sonic Caliper were used to inspect pipes ranging from 68 cm in diameter to three-by-four meters. The results from a typical Sonic Caliper program are shown below in Figure 1.

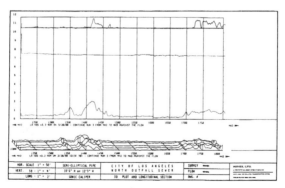

Figure 1. SR1000 Sonic Caliper inspection of a section of the North Outfall Sewer for the City of Los Angeles - 1987.

In 1992 we built an experimental model of a rotating Sonic Caliper with a single transducer on a motorized shaft that nutated from side to side, covering about 120 degrees of arc. This was tested on a program that we were running using the SR2000 Sonic Caliper. Information from the tool was interesting, but not as complete as the information from the SR2000, so we filed the concept and went forward with the SR2000 tools.

At several conferences in 1993 we were asked if the Sonic Caliper could be used to inspect plastic pipe as well as concrete pipe, and of course it could - with the caveat that we were only looking at the top of the pipe. The rotating-transducer concept came back, and we then preceeded with the development of the ROTATOR to inspect all of the pipe wall above the waterline. The first field application of the ROTATOR was in October 1993, inspecting sections of 30- and 36-inch plastic pipe.

THEORY

The concept of measuring distances with sound involves measuring the time needed for a burst of sound to travel from a source to a target, or in the case of distance ranging, from a source to a target and back. The travel time multiplied by the velocity of sound in the given transmission medium equals the total travel distance.

The velocity of sound in any medium can be determined by comparing the travel time of a sonic pulse over several known distances. The velocity changes with the density and elasticity of the medium.

Typical velocities in various media are shown below. The listed velocities can vary greatly with temperature or changes in the chemical makeup of the medium.

TABLE I

Material	Approximate Velocity of Sound	
	(mm/second)	(ft/second)
Iron/Steel	5900	19350
Aluminum	6400	21000
Gold	3240	10600
Granite	5400-6000	17700-19700
Basalt (Hanford)	6550	21500
PVC Plastic	2400	7870
Wood (along grain)	3000-4000	9800-13000
Wood (across grain)	1000-1400	3300- 4600
Air	345	1130
Water	1430	4700
Single Malt Scotch	1350	4430

The sonic velocities with which we are concerned are the velocity in air (about 350 mm/sec), and in water (about 1430 mm/sec). These are the media through which we will have to transmit the sonic pulses to reflect from the walls of the pipe.

Sound is generated by piezoelectric transducers. These have the ability to convert electrical pulses into mechical pulses (to transmit pressure waves), and also to convert pressure waves into electrical pulses (to act as receivers). The ROTATOR transducers can have a resonant frequency of 120 KHz, 200 KHZ, 350 KHz or 500 Khz. The choice depends upon the application.

The ROTATOR, as with all Sonic Calipers, records the first return signal only. We are not concerned with signal amplitude or phase, but only the time of the first detected signal. This will be the time for the sonic pulse to travel by the most direct path to the pipe wall and be reflected back to the transducer.

Crystal clocks in the Sonic Caliper electronics circuits produce pulses used to time the signal returns. We start to count pulses when the transducer is fired, and stop counting when the return echo is detected. This count is converted into time and used to calculate the distance using the formula $D = 1/2\ VT$.

We simplify that by using the clock frequency and the sonic velocity to calculate a count factor which is the number of clock counts per inch of range. This will vary slightly with the ambient temperature but is generally about 14.76 counts per centimeter for ranging in air, and 13.98 counts per centimeter for ranging in wastewater. The count factor is calculated at the start of each inspection run, after the calibration process. Calibration consists of recording the time count for a series of known distances, and running a linear progression analysis of the data.

EQUIPMENT

The ROTATOR Sonic Caliper can be mounted on skids or can be mounted on a floating platform. The motorized section of the ROTATOR can accept any one of a series of sonic tranducers - designed for specific inspection needs. A four-conductor wireline cable supports the ROTATOR, sending control signals to the ROTATOR, and bringing the sonic returns to the control computer in the van. The ROTATOR can operate on cables up to 3000 meters long. Standard SONEX trucks have up to 1000 meter of cable.

The first design of the rotating Sonic Caliper had the transducer rotate at 12 RPM. By the time we were ready for the first commercial application, the rotation speed had been increased to 36 RPM, and we recorded nineteen readings for a 260-degree arc. Todays ROTATOR can rotate at 120 RPM, and record more than fifty distance readings over the full 360 degrees.

The ROTATOR tool is pictured in Figure 2. Electronics for the ROTATOR are in the lower tube and the rotating motor, sliprings and transducer are in the upper tube. The legs can be adjusted to place the tool center close to the center of the pipe.

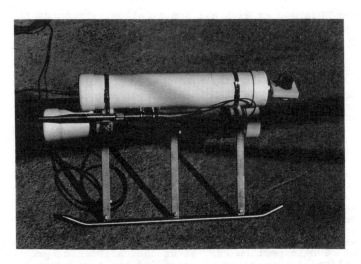

Figure 2. SONEX ROTATOR Sonic Caliper

The ROTATOR normally fires its single transducer fifty times in one rotation. Fifty discrete distance readings are taken. The distances are calculated, and used to plot the circumference of the pipe on a computer monitor in real time as the inspection is taking place. All of the distance readings are stored in the computer and transferred to disk at the end of the run. The data are used for later analysis and profile plotting.

We can control the speed at which the ROTATOR travels through the pipe. We normally set the speed so that we record at least one cross section for each 0.3 meters of pipe inspected. For some applications, the speed can be reduced to provide more detail.

The data from a ROTATOR inspection are used to recreate the run on a computer monitor so that the responsible engineer can see cross sections of the pipe. A proprietary program supplied with the data allows the viewer to step through the sections one by one, to automatically scan through the file, or to jump to a desired station and continue scanning at that point.

The data are also used to plot a vertical axis profile, or a combined invert and crown profile. This plot gives an overview of more than two hundred meters of pipe on a single page so that the engineer can readily see where there are problem areas. A printed summary of the data can also be prepared, showing the average and maximum deflection or corrosion for each three-meter segment of the inspection run.

INSPECTION RESULTS

The ROTATOR had been used to inspect plastic pipe and concrete pipe. It is equally useful for brick or clay pipe. We can operate in pipes as small as one-half meter or large as four meters.

In this section are examples of a buried 760mm plastic showing deflected and undeflected sections; a corroded 600mm concrete pipe; an uncorroded 1.14m concrete pipe; and a controlled field test where a 760mm plastic pipe was deformed in a hydraulic press while the ROTATOR recorded the change in realtime. Figures illustrating these examples were printed from the actual screen displays using data from the various jobs or tests. These show the ability of the ROTATOR to capture cross section data.

In November, 1994 we were able to conduct a field test of the ROTATOR. A short section of 760mm plastic pipe was obtained and placed, unconstrained, in a hydraulic press. The press could apply more than 1400 kilograms per square meter on the pipe. The setup used for the test is shown in Figures 3a and 3b. Figure 3a shows the pipe in the press before pressure was applied. Figure 3b shows the pipe when the vertical axis had been compressed from 760mm to about 560mm.

As the pipe was being compressed, we ran continuous data with the ROTATOR. Figures 4a, on the following page, shows the ROTATOR screen display soon after the pipe was placed in the press. Figure 4b shows the screen display near the end of the test when the pipe had been fully compressed to 560mm (twenty-two inches). All dimensions on the displays are given in inches or feet to conform to U.S. industry standards.

A 760mm diameter pipe was placed in the hydraulic press so that controlled pressure could be applied. We took the vertical axis down in one-inch increments. Data were recorded continuously. At the end of the field test the vertical axis was reduced from 760mm to 560mm using about 1353 kilograms per square meter pressure.

Figure 3a. Test Equipment for ROTATOR Field Test

Figure 3b. 760mm Pipe at Conclusion of the Field Test

INSPECTING BURIED PLASTIC PIPE 133

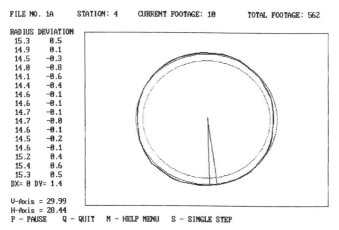

Figure 4a. ROTATOR Display at the Start of Field Test

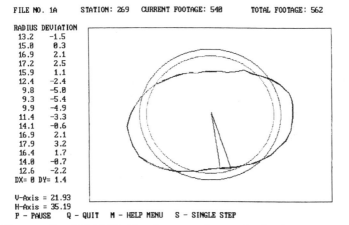

Figure 4b. ROTATOR Display at the End of Field Test

The ROTATOR screen shows that the pipe originally was slightly higher than wide, about 760mm high by 720mm wide. A cross section was drawn for every two seconds as the pipe was deformed. In a standard pipe inspection three sections per meter would be recorded. Applying 1353 kg per sq m, the pipe was deformed to 560mm on the vertical axis. The screen print shows this cross-section of the deflected pipe. The readings at the left show the measured radii in inches and the deflection in inches for every third reading around the pipe.

The top line shows the file number; the station number, which is the sequential number of the dataset; the footage at that station; and the total footage for the file. Since the ROTATOR did not move we assigned each station a value of two feet for plotting purposes. The file had 281 datasets, with fifty distance readings per set; a total of 14,050 separate readings to define the pipe.

Each display starts with two "ideal" circles. These are the designed inside circumference of the pipe; and, in the case of plastic pipe, the five-percent deflection limit. For concrete pipe, the second circle drawn is the outer circumference showing pipe-wall thickness. The cross section drawn by the ROTATOR is superimposed on these two ideal circles so that the viewer has a scale to estimate the deviation better.

The numbers at the left show the sonic distance from the tool to the pipe wall for sixteen readings around the pipe (every third reading) and the deflection in inches at that orientation. The values for "DX" and "DY" show the tool's offset from the center of the pipe in inches. V-axis and H-axis are the calculated values for vertical and horizontal axis.

The first commercial use of the ROTATOR was in Houston in October, 1993. We inspected several kilometers of 760mm and 910mm buried plastic pipe. Figure 5 shows the smaller pipe with less than one-half percent deflection.

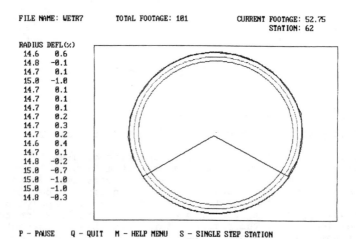

Figure 5 ROTATOR Display of a 760mm Plastic Pipe.

INSPECTING BURIED PLASTIC PIPE   135

For this early use of ROTATOR, the transducer rotated
at 36 RPM, and we recorded 19 distance readings around
the roughly 260 degrees of the pipe wall that was not
under water. The low was about 20cm deep. The bottom of
the pipe can not be seen. The ROTATOR can be used in air
or in water, but not both simultaneously. Different
electronics and transducers are needed.

The pipe was buried seven meters beneath a residential
street, with heavy traffic and poor soil conditions.

In Figure 6, the same pipe is seen at a location about
seven meters away. There is nearly four-percent deflection at the left shoulder. This inward deflection is
accompanied by bulging of the right and left walls.
This was the worst-case deflection in the pipe that we
inspected. It is our understanding that based on this
inspection, that manufacturer's pipe was accepted for
use in the city's wastewater programs.

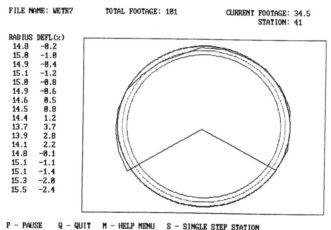

Figure 6. Four-percent Deflection of a 760mm Pipe.

Figures 7 and 8 on the following pages are examples of
the ROTATOR used to inspected concrete pipe for damage
caused by hydrogen-sulfide corrosion. Video inspection
of concrete line confirms the presence of corrosion, but
cannot tell how much material has been lost. The Sonic
Caliper can measure corrosion loss and also measure the
volume of debris in the invert.

The ROTATOR was rotating at 36 RPM, and we were firing
the transducer fifty times per rotation. The printed
data at the left side of the screen display shows the
radial distance and the corrosion loss in inches. The
values for DX and DY show the position of the tool in
the pipe measured in inches.

Figure 7 shows the ROTATOR display from a 600mm concrete pipe with 1 to 1.5 inches of corrosion. The flow
was about 15cm deep. There was no debris.

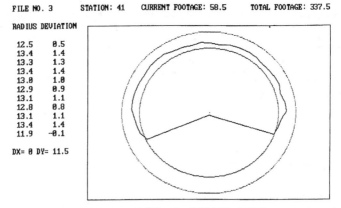

Figure 7. 600mm Concrete Pipe with Severe Corrosion.

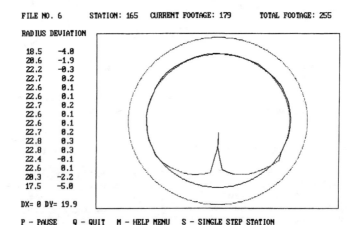

Figure 8. 1140mm Concrete Pipe without Corrosion Damage.

For comparison with the corroded pipe, Figure 8 is a cross section of a 1140mm concrete pipe inspected the following day. The lower part of the trace shows the water level at about 12cm.

It is clear that there is no corrosion. The abovewater part of the pipe is within 0.5cm (0.2 inches) of the as-built radius, and the arc of the wall is smooth.

All of the above ROTATOR projects were conducted with a skid-mounted ROTATOR tool. The ROTATOR may be mounted on a floating platform, in pipes where the flow level is too deep to operate with the skid.

CONCLUSION

Sonic Caliper tools have proved useful for corrosion and debris mapping. The ROTATOR with its ability to portray the entire abovewater circumference of the pipe, can now inspect plastic pipes. It is now possible to provide accurate inspection data showing the deflection in pipes as small as one-half meter and as large as two meters or more in diameter.

In addition to deflection monitoring, the ROTATOR can be used to measure corrosion in concrete pipes, physical damage to brick pipes, displacement of cracked clay pipes, and debris volumes in large-diameter tunnels.

Field Performance of PVC Water Mains
Buried in Different Backfills

Balvant Rajani,[1] Member, ASCE, and Senro Kuraoka[2]

Abstract

During the autumn of 1993, several sections of PVC pipes were instrumented in an ongoing water mains renewal project in Edmonton, Alberta. These PVC water mains were buried in several different types of backfill materials in order to evaluate the thermal and mechanical performance of these materials. Concurrently, PVC water mains at a shallow depth, which function as a bypass, were installed in parallel to the water mains at normal depth (renewal). The intent of this arrangement was to attempt to purposefully freeze the water in the bypass water mains to gain a better understanding of possible failure modes of the PVC water mains. PVC water mains at both depths were instrumented with strain gauges to monitor longitudinal and circumferential strains.

This paper describes the strain histories at selected locations on the PVC pipes during a 15-month period beginning November 1993. The strains developed during the period when no water flow was permitted in the bypass water mains are compared with those measured for the renewal water mains. The backfill-PVC water mains interaction and its implications to the PVC water mains structural behaviour are discussed.

Introduction

PVC water mains are frequently used in North America whether it is for renewal or new construction projects. Its introduction in the water distribution systems began about 20 years ago in Canada but many water utility authorities feel reluctant to use PVC pipes instead of ductile iron pipes because of limited experience on its long term performance.

The City of Edmonton has been using PVC water mains to replace existing deteriorated metallic water mains for over 15 years in its annual replacement program. Concurrently the city has had various initiatives to use different backfill

---

[1] Research Officer, National Research Council Canada, Institute for Research in Construction, Ottawa, Ontario. Canada K1A 0R6

[2] Research Associate, National Research Council Canada, Institute for Research in Construction, Ottawa, Ontario. Canada K1A 0R6

materials instead of the native backfill soils. One of the principal purpose of using alternative backfill materials is to reduce the risk of freezing water mains, as well as to reduce their burial depth, whereby obtaining a substantial reduction in construction costs.

Unshrinkable fill or controlled low-strength material (CLSM) is one of the backfill materials currently used by the City of Edmonton. Unshrinkable fill consists of materials with controlled density and is a mixture of Portland cement, water, fine and coarse aggregates and may contain an air-entraining admixture. Unshrinkable fill has extremely low-strength and its maximum 28-day compressive strength does not exceed 0.40 MPa. Thermal finite element analyses by Goodrich and Sepehr (1993) indicated that it was possible to freeze PVC water mains buried in unshrinkable fill under severe winter conditions. The City of Edmonton also has a good local source of bottom ash and consequently expressed an interest to explore its possible utilization as a suitable backfill material.

Field tests of seven backfill materials were tested by the National Research Council of Canada and the City of Edmonton in order to assess their thermal performance and to examine the structural performance of the PVC water mains when subjected to freezing conditions. These backfill materials were: light-weight aggregate (LWA), native clay (Edmonton clay), clean sand, unshrinkable fill (CLSM), sand with and without insulation, fly ash and bottom ash. In this paper, only the performance of PVC water mains buried in native clay (Edmonton clay), clean sand and unshrinkable fill are discussed.

Scope of present study

The major purpose of this field study was to evaluate the thermal performance of the different backfills that are either being used or may be under consideration in the near future. The study was extended to monitor the longitudinal and circumferential strains at selected points on the PVC water mains, as well as to observe the eventual failure modes of the buried PVC pipes.

This paper only addresses the aspects related to structural performance of the PVC water mains even though this field study was extensive and addressed a number of concerns. The evaluation of the structural performance is accomplished by monitoring strains in PVC pipes buried in selected backfills exposed to freezing ground conditions. Another important objective was to observe modes of failure of the PVC water mains by intentionally inducing failure in the PVC pipes by permitting the water in the PVC water mains to freeze.

Details of field installation and the layout of PVC water mains at 77 Avenue in Edmonton

The City of Edmonton identified a stretch of water mains on 77 Avenue that required replacement and, yet, where installation of temporary facilities for data logging equipment would not present a problem. It was clear from the early planning stages of the project that additional water mains would have to be constructed so that normal water service would not be interrupted in the event that the PVC water mains failed. This arrangement provided flexibility in the operation of the water mains to the extent that water flow could be shut off and purposefully frozen in an attempt to induce failure of the PVC bypass water mains. Concurrently, PVC bypass water mains at a shallow depth of 1 m were installed in parallel to the water mains at a normal depth of 2.4 m (Fig. 1). PVC water mains at both depths were instrumented with strain gauges to monitor longitudinal and circumferential strains.

The details of a typical layout for the bypass and the renewal PVC water mains are shown in Fig. 2. The site was divided into seven test sections with each test section backfilled with different backfill material. It is to be noted that even though reference is made to different backfills, the renewal PVC pipe is always embedded in clean sand in order to conform to current installation standards. The different trench backfills are only placed above the 50 mm extruded polystyrene insulating board above the renewal PVC water mains. The 200 mm (8") PVC pipe installed at 77 Avenue conforms to AWWA standard C900. Pipe lengths of 6.13 m and 3.05 m were used. The principal characteristics of the 200 mm PVC water mains are summarized in Tablé 1.

Table 1. Description of PVC water mains at 77 Ave. in Edmonton, Alberta.

pipe geometry		mechanical properties	
nominal diameter, D_n	200 mm	elastic modulus, E_p (-30°C)	2 550 MPa
outside diameter, D	229.95 mm	elastic modulus, E_p (0°C)	2 300 MPa
wall thickness, t	12.78 mm	elastic modulus, E_p (+20°C)	2150 MPa
typical pipe length, L	20 ft or 6.1 m	Poisson's ratio	0.45
		coefficient of thermal expansion	79×10^{-6}/°C
		tensile strength	50 MPa
		strain at tensile failure	18 250 µε
		strain at burst pressure failure	2 600 µε

Data acquisition system

Data acquisition at the site was accomplished under a contract to Campbell Scientific (Canada) Corp. by linking three CR-10 measurement and control modules. The data acquisition equipment monitored all sensors with the exception of the soil moisture probes and the thermal conductivity probes.

Instrumentation of PVC water mains

Strain gauges were installed on three of the seven test sections to monitor the deformation of renewal and bypass water mains. Micro-Measurements strain gauges type EA-30-500BL-350 with a range of 0 to 50 000 micro-strains (µε) were used. The strain gauges were mounted at mid-span and quarter-span as shown in Fig. 2 to monitor circumferential (hoop) strains, axial and flexural strains.

An additional six dummy strain gauges, forming independent Wheatstone bridges, were installed to permit temperature corrections. These dummy gauges were mounted on small rectangular coupons prepared from PVC pipe of identical specifications to the pipe installed. These gauges were sealed in canisters to isolate them from the surrounding soil, while being responsive to temperatures comparable to those experienced by the active gauges.

Earth pressure cells (Roctest EPC-6) were also placed 200 mm above the crown of renewal water mains in these three sections. Reference to the time histories of earth pressures are not specifically discussed in this paper, except when it may influence other measurements.

PVC water mains

The longitudinal strains measured at crown and bottom of the pipe can be resolved to identify the components of pure flexure (bending) and axial deformations. If it is assumed that the neutral axis remains at the centre of the pipe, flexural (ε_b) and axial strains (ε_a) are given by:

[1] $$\varepsilon_a = (\varepsilon_T + \varepsilon_B)/2, \qquad \varepsilon_b = (\varepsilon_T - \varepsilon_B)/2$$

where ε_T and ε_B are the strains measured at the crown and bottom of the pipe, respectively. Although all the strain gauges used have self temperature-compensation feature, additional corrections were required because of the difference in the temperature strain property of PVC and the gauge. The additional corrections were made using the temperature-strain response of the dummy gauges. It is also noted that the strains indicate a change in magnitude relative to the initial values measured after backfilling.

Axial strains

If the 6.13 m (20 ft) standard lengths of PVC pipes were free to expand or contract longitudinally, then a strain of 79 µε would be noted for each degree change of temperature. Since the pipes were installed during the first week of October 1993, it is reasonable to assume that the temperature at the time of installation was in the range of 8°C to 10°C. As mentioned previously, the water temperature during mid-November was about 7.5°C. However, the water temperature did drop to as low as 2°C. Consequently, based on these considerations alone, the cumulative longitudinal strain would be expected to be in the range of 474 µε.

Figure 3 shows the measured axial strain histories for both bypass and renewal PVC pipes embedded in native clay, clean sand and unshrinkable fill. The following comments can be made on these axial strain histories:

- During the winter months, absolute levels of axial strains in the renewal pipes (RN) are consistently lower than in the bypass pipes (BP), independent of the backfill material.

- The axial strains indicate that the bypass pipes are almost always in compression and strains in the renewal pipes fluctuate from tensile to compressive within a range of ± 400 µε.

- The axial strains in the renewal pipes which are all embedded in clean sand do not exceed 300 µε during the first 120 days of observations.

- The axial strains in the bypass pipes embedded in native clay and clean sand did not exceed 300 µε until after the water valve was closed to initiate freezing of the water in the pipes. After the strain gauges had registered the maximum strains, the axial strains did decrease but not to the values corresponding to the state of strain prior to the closing of water valves. As anticipated, the renewal pipes did not produce any significant response to the action of freezing water in the bypass pipes.

- In the order of increasing strains, the maximum axial strains registered during the first 180 days of observations in the PVC bypass pipes were 750 µɛ (native clay), 600 µɛ (clean sand) and 650 µɛ (unshrinkable fill).
- The axial strains in the bypass water mains after valve closure were notably higher than those registered before closure. Coincidentally, the temperatures were also warmer and thawing of uppermost backfill had already initiated. After valve closure, axial strains measured at mid-length of span (each section of bell and spigot pipe is 6.13 m long) were consistently higher than those at quarter-length, except for pipes embedded in native clay (section B).
- The axial strains in the bypass pipes embedded in unshrinkable fill during the warmer months are considerably higher than those measured during colder periods with strains reaching as high as 1 250 µɛ.

The above observations indicate that the unshrinkable fill provides less restraint to axial movement than either native clay or clean sand. This is probably due to the fact that unshrinkable fill shrinks after placement and, thus, forms a bridge over the pipe. This behaviour is congruent with the earth pressure measurements where the registered earth pressures were significantly reduced.

Though the axial compressive strains did not exceed the elastic limit when the water valve was closed, these strains did not decrease to values prior to valve closure indicating that frictional restraint provided by the backfill has an influence on the axial behaviour of water pipes. Nonetheless, axial strain recovery for water pipes embedded in unshrinkable fill was substantially more than that for pipes buried in native clay and clean sand. This difference in behaviour reconfirms the assumption that there is reduced contact between the unshrinkable fill and PVC pipes. The development of axial strains in the pipe is a consequence of complex interactions between water temperature, frictional restraint and shrinkage and swelling characteristics of the backfill. Except for PVC pipes embedded in unshrinkable fill, field observations indicate that the axial strains at the half-span length (centre of 6.13 m standard pipe-length) is higher than at the quarter-span length of the pipe.

Flexural strains

Ideally, flexural bending should be minimal if each of the 6.13 m (20 ft) standard lengths of PVC pipes were installed in well-compacted trench bedding and backfill. Since we are dealing with flexural strains, there is no particular significance as to whether the strains are positive (tension) or negative (compression). The following points can be summarized on flexural strain histories (Fig. 4) keeping in mind that renewal pipes are always embedded with sand while the soil surrounding the bypass pipes depends on the particular sections of native clay, clean sand or unshrinkable fill:

- Overall the flexural strains in bypass and renewal pipes embedded in clean sand (section C) and unshrinkable fill (section D) are minor in comparison to those that developed in pipes embedded in native clay (section B).
- The flexural strains in the renewal pipes which were all buried in clean sand did not vary significantly and these strains did not exceed 150 µɛ.

- Flexural strains at both locations in the bypass pipe embedded in native clay had a cyclic variation between ± 350 µε.

The above observations indicate that significant flexural strains should not develop if the pipe bedding material is easy to place and compact. Since native clay is neither easy to place nor easy to compact, as well as the fact that fine-grain soils are more frost susceptible, flexural strains are considerably higher than for pipes buried in clean sand or unshrinkable fill.

It is also interesting to note that flexural strains for bypass pipe buried in native clay decreased after 150 days with the increase in ground surface temperatures (thawing). In the same time period, the variation in the flexural strains for pipes buried in clean sand or unshrinkable fill was insignificant.

Circumferential (hoop) strains

Hoop strain variation in PVC bypass pipe would be in the range of 930 µε to 2 360 µε based on the maximum (7.7 bars or 111 psi) and minimum (3 bars or 44 psi) water pressures registered in the bypass pipe, and assuming that the PVC bypass pipes were not surrounded by either unfrozen or frozen backfill. Any external restraint offered by the unfrozen or frozen surrounding soil will undoubtedly tend to counteract the effects of internal water pressure.

Figure 5 shows the measured hoop strain histories for PVC bypass pipes embedded in native clay, clean sand and unshrinkable fill, respectively. Similar hoop strain histories for PVC renewal pipes are not shown since neither significant strains nor variations in strain levels with surface temperatures were observed. The following comments can be made on the hoop strain histories:

- The trends in the time histories (Fig. 5) for hoop strains in the bypass pipes (BP) are similar for all three backfill materials but distinct from the time histories for renewal pipes (RN). Similar signatures of strain histories at the time of valve closure are observed with the development of strains as high as 2 250 µε.

- A sharp increase in the tensile strains on the day of valve closure noted in all bypass pipes buried in the three backfills are due to the sudden increase in water pressure associated with the valve closure. The subsequent increase in the tensile strains is attributed to the freezing of water. As expected, the tensile strains rapidly decreased during the thawing period but abruptly changed to compressive strains after bypass pipes were drained.

- The time histories (not shown) for the hoop strain for renewal pipes (RN) buried in native clay, clean sand and unshrinkable fill have similar trends. Hoop strains as high as ± 510 µε developed during the first year in the renewal pipes with all three backfills.

- Typically, hoop strains at the crown of the renewal pipes (RN) were either near zero or tend to be tensile during the first 120 days. Meanwhile, the hoop strains at the spring-line of the renewal pipes (RN) were tensile except for the pipe in the trench with native clay backfill.

The fact that hoop strains did develop nearly as high as those predicted (2 360 µε) for unrestrained pipe confirms that the surrounding soil does not effectively provide a radial restraint.

The intent of the closure of valves on March 7, 1994 was to promote freezing of the water in the pipe so that the pipe would fail. Prior to valve closure, the flow of water in the pipe did not permit the freezing to take place. Since the valves were closed rather late in the cold season, it is most likely that all the water in the pipe did not freeze. Though high hoop strains in the range of 2 000 µε were registered in the pipe, a CCTV (closed circuit television) inspection on May 27, 1994 did not indicate any serious damage to the pipe.

Water expands by 9% when it freezes to ice. It is possible to determine an upper limit on the expansion of water to ice that is necessary to induce hoop strains that would lead to the failure of the PVC pipes. If the free expansion of water to ice, α_{ice}, were to be suppressed by the buried pipe, the pipe would be subjected to an internal pressure given by:

[2] $$P_{internal} = -\frac{E_{ice}\alpha_{ice}}{2(1-v_{ice})}$$

where E_{ice} is elastic modulus for ice under plane strain condition, v_{ice} is Poisson's ratio for ice under plane strain condition. The hoop stress, σ_θ, that is generated assuming there is no interaction between the pipe and surrounding soil is given by:

[3] $$\sigma_\theta \quad \sigma_\theta = \frac{Dp_{internal}}{2t}$$

where D and t are pipe diameter and thickness, respectively. A hoop stress of 47 MPa would result if it is assumed that water to ice expansion is limited to 0.10 %, and if reasonable estimates for elastic modulus of ice (7 GPa) and Poisson's ratio for ice (0.33) are used. This hoop stress corresponds to an upper bound strain of 17 000 µε. It is important to emphasize that this calculation ignores the effects of soil-structure interaction between the pipe and the surrounding frozen soil, but it is possible that high stresses in the pipe are induced if the water to ice expansion is greater than 0.10 %.

Conclusions and recommendations

This field experiment was intended to investigate the rupture behaviour of a live PVC water line during freezing. The relatively moderate weather conditions combined with the high flow rates and warm water temperatures kept the active bypass water mains from freezing and although flow was stopped in late winter, complete freezing was not achieved in the bypass water mains.

The axial strains in the bypass PVC water mains buried in the native clay backfill and sand backfill increased substantially but the corresponding response for the pipe in the unshrinkable fill was subdued. Unshrinkable fill provides less restraint to axial movement than either native clay or clean sand.

Flexural strains were found to be considerably higher in pipes buried in native clay than in pipes buried in clean sand or unshrinkable fill. This is primarily due to the fact that native clay is neither easy to place nor easy to compact as well as due to the fact that fine-grain soils are more frost susceptible. Nonetheless, these observations are based only on the short-term data available to date.

Hoop strains in the bypass water mains did not exceed 2 360 με even when the valves were closed to induce water to freeze. The most likely reason why higher strains were not registered is because of the possibility that not all the water had time to freeze. Since this study is an on-going, alternative operating procedures have been suggested for the second year of the field experiment. For these purposes, it is proposed that the water mains be operated with stagnant air until the third week of January 1995. At this time, in order to accomplish the objectives of the experiment concerned with the rupture failure behaviour of the PVC pipe, the line should be charged but the flow closed off to allow complete freezing.

Acknowledgments

The authors extend their appreciation to Mr. John Ward, Director of Engineering Services at The City of Edmonton and all the technical staff at Infrastructure Laboratory for their valuable assistance in making this project a success.

References

American Water Works Association. (1990). "PVC Pipe - Design and installation. AWWA Manual M23." vi + 89p.

Goodrich, L.E. and Sepehr, K. (1993). "Frost protection of buried water mains: Results from a finite element model study." Proceedings of 1993 CSCE Annual Conference.

Moser, A.P. (1990). Buried Pipe Design. McGraw-Hill, NY.

Zhan, C., Goodrich, L. and Rajani, B. (1995). "Thermal Performance of Trench Backfills for Buried Water Mains." Proceedings of 2nd International Conference on Advances in Underground Pipeline Engineering, ASCE, Seattle, June 25-27.

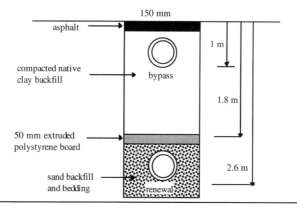

Fig. 1. Typical cross-section of PVC bypass and renewal water mains buried in native clay backfill (section B, Edmonton clay).

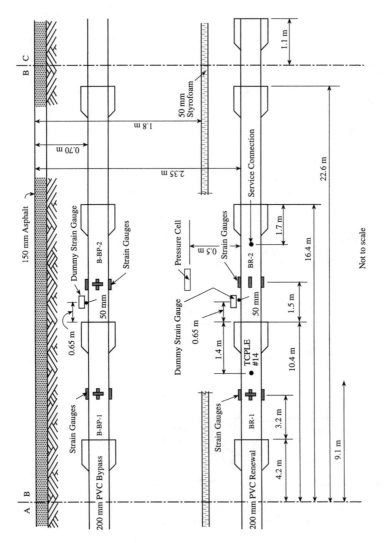

Fig. 2. Longitudinal section of bypass and renewal pipes with native clay backfill (section B, Edmonton clay).

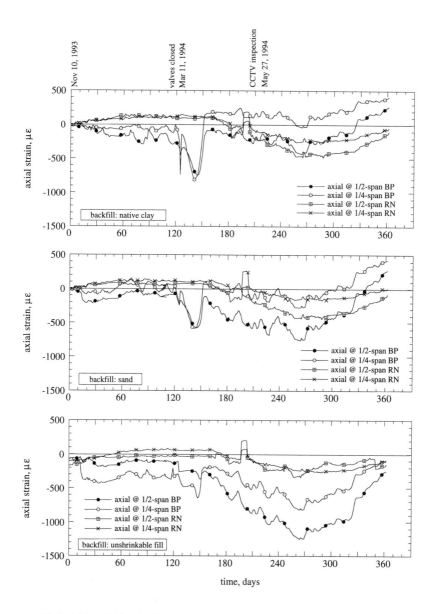

Fig. 3. Axial strain histories of bypass (BP) pipes buried in native clay (section B), clean sand (section C) and unshrinkable fill (section D) and renewal (RN) pipes buried in clean sand.

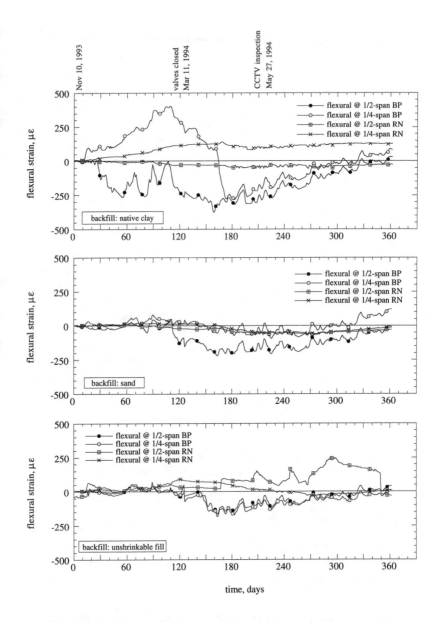

Fig. 4. Fexural strain histories of bypass (BP) PVC pipes buried in native clay (section B), clean sand (section C) and unshrinkable fill (section D) and renewal (RN) PVC pipes buried in clean sand.

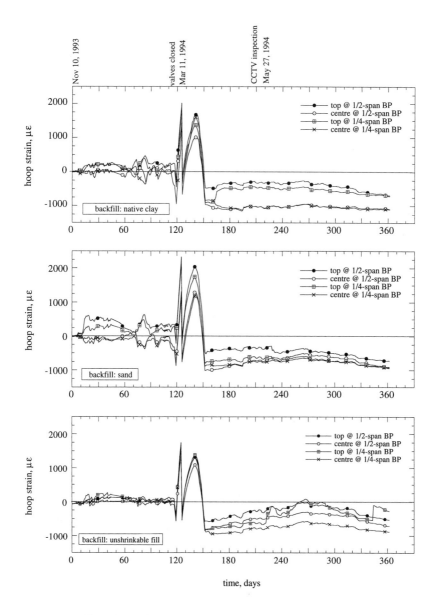

Fig. 5. Hoop strain histories of bypass (BP) PVC pipes buried in native clay (section B), clean sand (section C) and unshrinkable fill (section D).

Improved Joint System for Concrete Pipe

Kenneth W. Anderson[1]

This article describes the Anderson Seal System. It is an EPDM rubber joint liner mechanically interlocked to the tongue and bell of concrete pipe joints to improve the seal provided by the O'ring pipe. These seals are designed to allow the producer to produce a premium pipe product using his existing O'ring equipment.

The Anderson Seal Company was founded in 1982 as a result of twenty years of effort to improve the quality of concrete pipe. It was felt that a consistently better concrete pipe product could be produced. It was felt that the need for joint painting and touch up for concrete sanitary sewer pipe could be eliminated.

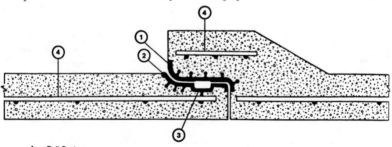

1 — Bell Seal
2 — Tongue Seal
3 — Rubber O-Ring Gasket
4 — Reinforcing Steel

Figure 1.

----
[1] President, Anderson Seal Company, Inc., P.O.Box 351, Rochester, Illinois 62563

JOINT SYSTEM FOR CONCRETE PIPE

It appeared that it was possible to develop a liner for the joint surfaces that would provide a slick smooth resilient sealing surface upon which to seat the O'ring greatly improving the seal of the joint as well as the ease of installation.

The prototypes of joint seals were made using urethane to test the concept of resilient joint seals. These were actually molded on the pallets and headers. Tests made using these seals did perform well, but it was found the urethane was not weather resistant and that molding the seals would not be a practical process for mass production.

Upon further study, tests were made using EPDM hydrocarbon rubber that had been extruded through a die having the profile of the pipe joints to be used. It was found that this approach was very practical and that EPDM material was an ideal product to use for manufacturing the seals.

The Anseal (Anderson Seal System) was designed to give concrete pipe producers the ability to produce a premium jointed pipe using their existing equipment.

With the Anseal, precast reinforced concrete pipe can be manufactured with a resilient hydrocarbon rubber seal mechanically interlocked with this concrete in the tongue and bell ends of the pipe. This allows use of existing o'ring equipment to provide a much improved sealing connection.

The Anseal can be used in the manufacture of concrete pipe with wet cast, dry cast/vibration, or packerhead production configurations.

The Anseal centers the reinforcing steel, keeping it from being exposed at the sealing surfaces. With the seal, the reinforcing steel can be extended up into the seal area in both the tongue and bell ends, and the seal acts as a protective bumper on the tongue end, reducing damage during shipping and storage of the finished concrete pipe.

The Anseal helps to insure smoother ends during production, thus reducing the need for added labor to repair damaged sealing surfaces.

The Anseal eliminates the differential expansion problem associated with concrete pipe utilizing steel end rings. The slick, smooth surfaces help prevent gasket roll-out.

Sunlight and weather have little adverse effect on the hydrocarbon (EPDM) rubber used in the fabrication of the seal. Rubber samples exposed for over ten years in Florida have been found free from blemishes and display excellent retention of their physical properties.

The rubber compound also is flexible at temperatures as low as $-58°$ to $300°$. It will resist attack by many acids and alkalies, detergents, phosphate esters, ketones, alcohols and glycols.

These seals are custom designed for each plant's existing equipment and fitted to the pallets and headers. The seals arrive in boxes cut to length and vulcanized in a ring ready to be placed on the base pallet or in the header ring. The production crew can be quickly taught to produce premium pipe using the Anderson Seal System.

During pipe production, the only change required is a reduction of pallet rotation time as the rubber provides the slick, smooth surface in the bell; thus, rotation is only used to trowel off the end surface of the bell. The pallets and headrings should come off easier than when producing pipe without the seal system. The cleanup time for headers and base pallets is reduced, for with the use of the Anseal, the rubber seal separates the concrete from the metal surface. Patching of the sealing surface is eliminated.

It is highly recommended that all pipe joints should be field tested as installed. This can be done while excavating for the next section and will confirm a properly positioned gasket assuring outstanding final line tests.

Even if the bell end of the pipe should be damaged during installation the seal provides a continuous embedded rubber surface protecting the integrity of the bell.

Several hundred thousand feet of Anseal pipe have been installed. Projects using the Anseal include those in Illinois, Indiana, areas near Nashville, Tennessee, Memphis, Tennessee, Little Rock, Arkansas, and in areas around Toronto, Ontario, Canada. There are upcoming projects throughout the United States.

Patents for this system have been obtained in Canada, Australia, U.K., and U.S. with patents pending in Japan.

Tests performed by Illinois Environmental Protection Agency in Springfield, Illinois, resulted in approval of the use of Anseal pipe in areas previously requiring pressure pipe.*

Pressure tests have been performed exceeding 45 PSI. Differential load tests have been performed pressurizing the pipe sections to 15 PSI while supporting the bell of the pipe and applying a 4400 lb. load to the tongue of the pipe. No leakage was observed.

In a project for the City of Indianapolis known as the Camby Sanitary Sewer Interceptor including approximately 6000 feet of 30" pipe ranging in depth from 20 to 40 feet, tests were performed by engineers for the City with exfiltration results in the range of 10-15 inch gallons throughout the line. All concerned were pleased with the results. The pipe was produced by Independent Concrete Pipe, Inc. of Indianapolis, Indiana, and installed by Thompson Construction of Indianapolis, Indiana. When exfiltration results fall in this range, it is just water being absorbed by the concrete pipe.

One of the most recent projects is in Memphis. Tennessee, and consists of approximately one mile of 42" pipe buried to depths of 50 feet. This pipe was produced by the Choctaw plant in Little Rock, Arkansas, where it was air-tested prior to shipment to the job site in Memphis and joint-tested as installed. We have been informed that test results have been excellent. Approximately 20,000 feet of Anseal pipe of 42" and 54" diameter were used in a project in Jackson, Mississippi, with depths up to 50 feet and 30 feet of water above it at times. Tests on this project well exceeded all requirements.

Pipe using the Anseal has been produced in sizes ranging from 12" to 120" using wet cast, dry cast, and packerhead systems. We have found many specifiers of pipe products who have stopped using concrete pipe will go back to concrete pipe when they are made aware of the availability of the improved joint system.

----
*Ill. EPA Letter, Nov.10, 1981, W.Akers, C. Fellman, R. Selburg

Since the development of the basic Anseal, an additional patent has been obtained for a method of connecting the Anseal to a P.V.C. corrosion protection liner for concrete pipe.

The use of this approach would eliminate the requirement to heat weld the overlapping liner in the pipe as done in the past and allow for the natural settling of the pipe line without loss of the continuity of the corrosion protection that could occur with the present method.

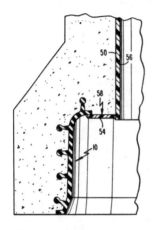 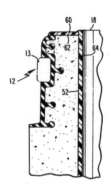

Figure 2.

Repeatedly, surveys of concrete pipe specifiers have indicated they wanted better joints for concrete pipe. We feel that the Anderson Seal System has met this need and added easier installation thus providing concrete pipe users with the high quality products they should receive.

The Anseal system is available on a license basis to producers of high quality concrete pipe.

Figure 3. 21" Anseal Tongue

Figure 4. 36" Anseal Bell

WHEN SHOULD PCCP WITH INTERPACE CLASS IV WIRE BE REPLACED?

Bryan M. Bradish[1], Roger J. Cronin[2], Richard O. Lewis[3]

ABSTRACT

The City of Newport News's raw and treated water systems include 12 miles of pipelines constructed of prestressed concrete cylinder pipe (PCCP) manufactured with Interpace Class IV prestressing wire. Because of City concerns about the service life of PCCP with Class IV wire, the City developed a plan for scheduled replacement of this problem pipe. The replacement schedule was based upon a testing program that provides a basis for determining the wire deterioration rate.

CITY OF NEWPORT NEWS'S WATER SYSTEM

The Newport News Department of Public Utilities (Waterworks) is among the 100 largest water utilities in the United States. It provides potable water over a 250 square mile area to a population of more than 340,000 people living and working on the lower Virginia Peninsula.

[1] Chief Engineer, City of Newport News, Newport News Waterworks, 2600 Washington Avenue, Newport News, Virginia 23607

[2] Partner, Greeley and Hansen, 2116 W. Laburnum Avenue, Suite 100, Richmond, Virginia 23227

[3] Gould, Lewis & Proctor, 6712-1 NW 18th Drive, Gainesville, Florida 32606

PROBLEMS EXPERIENCED WITH PCCP MANUFACTURED BY INTERPACE CORPORATION

During the mid to late 1970s, failures across the nation began occurring in PCCP and continued thereafter. The vast majority of these failures involved pipe manufactured by Interpace Corporation. Investigations ensued and the failures were initially attributed to a variety of suspected causes, generally contractor damage or hydraulic transient effects. After the failures became chronic in several pipelines, the investigations came to focus on a common denominator, Class IV prestressing wire. In 1982, the scope of the problem was made widely known in an article appearing in the May 13th edition of "Engineering News Record." Since then numerous investigations, often connected with litigation, have led to an understanding of the history and failure mechanism of the Class IV wire problem.

Until the mid-1960's, prestressing wire utilized in the manufacture of PCCP under AWWA Standard C301 was Class I or II tensile strength. By 1968, a higher strength wire, Class III, emerged and became accepted in the industry. Interpace, however, pushed the tensile strength class yet higher and produced wire at tensile strengths in excess of any then existing, or current, AWWA or ASTM prestressing wire specification. This wire, in 8, 6 and 1/4-inch gauges, was designated Class IV by Interpace, and was manufactured by Interpace from 1972 to 1979.

Prestressing wire is made by taking hot-rolled, mill-produced rod, and pulling it through a succession of dies. This is a type of cold working that increases strength and some other mechanical properties, while reducing the wire diameter.

The important manufacturing variables are control of the amount of diameter reduction at each die and control and dissipation of the considerable heat generated by the drawing process. Interpace purchased the Solon, Ohio, wire mill in the mid-1960's and began drawing its own prestressing wire without having a metallurgist on staff. The evidence obtained during the Richmond, Virginia litigation in 1989 and later investigations and testing of Class IV prestressing wire manufactured by Interpace indicates that the wire was drawn at excessively high temperatures. Wire drawing temperatures exceeding about 400°F, while increasing tensile strength, compromises ductility, and causes the wire to become dynamically strain aged (DSA) which in turn makes the wire susceptible to a form of stress corrosion cracking called hydrogen induced delayed brittle fracture, or hydrogen embrittlement. Since this condition may lead to failure of the wire by higher forms of corrosion, the protection of the wire from potential sources of corrosion becomes more critical concerning the expected service life of the wire which is in excess of 100 years.

Drawing the wire too hot also results in the development of longitudinal splits. Split wire has been found in all investigations of Interpace Class IV wire to date, varying in amount from as little as 15 percent to as much as 85 percent of wire samples examined. The hotter the wire drawing temperature, the more the wire is both dynamically strain aged and the higher the amount of split wire. Split wire is a major problem because of the inability of the mortar coating to protect the wire within the split from corrosion. Water can migrate by capillary action into the split and initiate corrosion regardless of the quality of the mortar coating. This corrosion then charges the wire with hydrogen which causes it to become embrittled and then fail. Split wire is a major factor in the Class IV wire problem.

The Richmond, Virginia litigation revealed that Interpace was using 3/8-inch rod and a five die wire drawing machine at Solon to make the Class IV wire and that it was not possible for them to make Class IV wire which is not dynamically strain aged as shown on Figure 1. The tensile strength gained by the combination of chemical composition and thermal treatment and drafting (cold work) without dynamic strain aging the wire is only 278,000 psi which is less than the minimum strength of 293,000 psi that was used by Interpace in the design of the pipeline.

The problems associated with defective Class IV wire have been exacerbated in many cases by poor mortar coatings. The mortar coating is the primary protection of the wire from corrosion. Permeable voids in the coating can facilitate the movement of ground water and gases through the coating to the steel elements. This can permit the entrance of deleterious compounds such as chlorides, which in sufficient concentration can initiate corrosion, or carbon dioxide in one form or another, which can result in carbonation of the cement paste, thereby reducing the pH and compromising the alkaline protection of the coating.

WATERWORKS' APPROACH

The City's system contains 12 miles of PCCP with Interpace Class IV wire which will need to be replaced prior to the originally expected service life of 100 years. Most of the pipelines are very critical elements of the City's transmission system. The replacement of all the PCCP with Class IV wire would have cost the City over $23,900,000 based on 1994 cost levels.

The Waterworks' approach to respond to the question **"When Should PCCP with Interpace Class IV Wire Be Replaced?"** was developed based on information obtained from the various litigations against Interpace. The approach development team included experts with diverse backgrounds and Waterworks' staff members with detailed knowledge on the City's transmission system.

Defining the approach to address the pipelines was a challenge because several issues had to be dealt with at one time. These include technical issues, customer service issues, and legal issues. Each of these issues had to be analyzed in both the long term and the short term. The first action was briefing the Council so that a decision on legal action could be made and that Council would be aware of the potential for major pipeline failure.

The following short term actions were taken:

- Development and implementation of an emergency repair procedure
- Installation of alternative supply system to critical customers such as major manufacturing plants and hospitals
- Testing of all pipelines with Class IV wire as described in the following section
- Installation of additional isolation valves on critical pipelines

The long term action to address this problem pipe was considered an opportunity to develop a new and improved transmission master plan for the water system. A Class IV wire pipeline replacement plan was developed and made part of the system's long term master plan. The Class IV replacement costs were minimized by reducing the size of the 54-inch replacement pipe. The size of the 54-inch pipeline was able to be reduced to a 48-inch because of the location of a new water treatment plant that was not part of the initial pipeline sizing plan. This system change is an improvement because it fulfills all the system's needs and reduces the replacement costs. The replacement plan assessed the hydraulic value of each pipeline and the expected remaining service life.

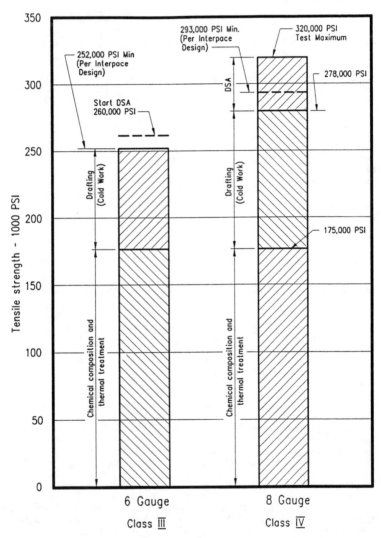

Figure 1. Degree of "Dynamic Strain Aging" (DSA) in PCCP wire

PCCP TESTING PROGRAM

Inspection and Sampling Procedures

After reviewing the original Interpace design sheets and pipe laying schedules, individual pipe lengths representing various design classes were located for sampling along the pipelines at intervals, which varied between 1,000 and 5,000 feet based on the overall length of each pipeline. The sample pipe lengths were located, where feasible, in unrestrained sections and outside of street pavements.

A total of 17 excavations were made to expose, examine and sample PCCP in the four pipeline facilities investigated. Each excavation was centered on a pipe joint, thereby allowing prestressing wire samples to be taken from two adjoining lengths of pipe. Prior to removing mortar coating and wire samples each pipe was cleaned, sounded and inspected for cracks, delaminations and damage. Photographic and verbal recording logs were made at each stage of the work.

Lightweight pneumatic chipping hammers were used to carefully remove the mortar coating. Mortar coating samples were taken from one pipe at each site. Wire samples measuring approximately one-half the pipe circumference in length were carefully removed from both pipe lengths, tagged and visually examined. When lined cylinder pipe was sampled, the steel cylinder was exposed when the mortar coating was removed. This afforded the opportunity to examine the condition of the cylinder. Replacement Class III wire was installed and restressed to the proper tension by the use of unique equipment designed solely for this purpose. A cement slurry followed by a correct mix of cement-rich, dry-pack mortar was then applied over the sampling sites to protect the replacement wire. After completion of sampling, the excavation was carefully backfilled and compacted.

Laboratory Testing and Analyses

Testing of prestressing wire included microscopic examination, tests for ultimate tensile strength and reduction in area ductility properties, torsion tests, "as received" hydrogen content analysis, and hydrogen charging tests. The mechanical tests (tensile and torsion) are indicators of brittleness and dynamic strain aging in the wire. All wire samples were subjected to dissolved hydrogen analysis, which indicates the content of atomic hydrogen in the wire at the time of the test. Direct tension hydrogen charging

(DTHC) tests were performed on selected wire samples to assess the relative sensitivity of the Class IV wire to embrittlement by atomic hydrogen. This is a time-related test to failure.

Petrographic analysis and testing were performed on all mortar coating samples to determine quality of the coating (i.e. compositional and textural features, porosity, absorption, specific gravity and permeable pore space), chloride concentrations in the mortar and the extent of carbonation.

Summary of Results

The field examinations and laboratory testing resulted in a substantial amount of data on the condition of these pipelines. Examinations of the exposed exterior of the PCCP indicated no evident cracks or other distress in the mortar coating at any of the sites. Nearly all of the wire samples had some degree of surface corrosion; however, it was generally light to occasionally moderate. On the 54-inch treated water pipeline, it was found that 6-gauge wire had been substituted for the specified 8-gauge wire which is of concern because 6-gauge wire is more dynamically strain aged than 8-gauge wire.

The mechanical properties and dissolved hydrogen sensitivity of the wire were characteristically similar to other Interpace Class IV wire tested in other investigations in Florida, Minnesota, and Virginia. A brief summary of the Class IV wire test results is presented in Table 1. In the tensile strength tests, all but a few wire samples met Interpace's minimum ultimate tensile strength for Class IV wire of 283,000 psi and 293,000 psi for wire gauge 6 and 8 respectively.

The torsion test is especially sensitive to wire drawn at temperatures greater than 400°F, above which dynamic strain aging becomes prevalent. The results of the torsion tests confirmed that all of the wire sampled was dynamically strain aged. The wire was brittle and "splitty" and had very poor torsional properties. Pre-existing splits existed in some of the wire samples, and some broken wire was found under sound coating.

		TENSILE TESTS		TORSION TESTS	HYDROGEN TESTING		
						CHARGING TEST	
Pipeline Facility	Test Result Range	Ultimate Tensile Strength (ksi)	Red'tn in area (%)	Revolutions to Failure	Hydrogen Content as Rec'vd (ppm)	Time to Failure (hrs)	Hydrogen Content at Failure (ppm)
54-Inch Raw Water (6-gauge, Class IV)	Low Avg. High	289.2 299.3 315.4	1.5 32.5 47.3	0.85 2.57 7.08	0.43 0.84 1.91	1.80 4.66 8.51	3.62 6.51 7.43
42-Inch Raw Water (8-gauge, Class IV)	Low Avg. High	300.5 313.5 320.5	11.1 36.0 45.3	0.61 2.90 3.08	0.49 0.67 1.20	1.44 2.20 2.73	2.97 4.06 4.92
54-Inch Finished Water (8-gauge, Class IV)	Low Avg. High	282.8 286.6 293.7	14.9 34.2 47.0	1.14 5.88 11.38	1.31 2.54 5.87	2.00 4.94 9.39	3.94 4.93 7.51

Table 1. Summary of Class IV Wire Test Results

The as-received hydrogen analyses indicated that atomic hydrogen has been charged into the wires to varying degrees. Wire that has been maintained completely free from corrosion on PCCP will normally contain, on average, 0.15 to 0.40 ppm dissolved hydrogen. Any concentration of dissolved hydrogen greater than that is generally due to anoxic corrosion occurring on the surface of the wire. The hydrogen charging process is irreversible.

The DTHC tests confirmed the relative sensitivity of the Class IV wire to atomic hydrogen. All of the wire samples exhibited a strong sensitivity to hydrogen embrittlement.

In general, the testing and analysis of the mortar coating samples indicated that the coatings were of moderate quality and appear to have provided acceptable protection for the prestressing wire during the 14 to 18 years these pipelines have been in service. Nearly all the coatings exhibited stratified bands alternating between low and moderately high porosity and water-cement ratio. Carbonation did not penetrate deeper than 1/16 to 1/8 inch below the surface in any of the samples.

It is important to note that corrosion can cause prestressing wire failure by more than one mechanism. While corrosion induced hydrogen charging and hydrogen embrittlement have been the focus of most Class IV prestressing wire investigations and failures, oxygen induced pitting corrosion can also occur. In fact, dissolved oxygen reduction competes cathodically with hydrogen ion reduction in the corrosion process, and is more energetically favorable. If dissolved oxygen is available at the wire level, pitting corrosion will dominate the corrosion process, effectively inhibiting hydrogen ion reduction and hydrogen embrittlement.

Pipeline Deterioration Rate Assessment

Methods of quantifying hydrogen effects are limited, but a graphical method was devised for this project to give some indication of the potential for time-related increase in the concentration of atomic hydrogen in the prestressing wire. The extrapolation of a straight line from the estimated hydrogen concentration at the time of manufacture through a point representing the as-received hydrogen concentration to the intersection of the concentration at failure in the DTHC test may provide an indication as to whether or not the apparent historic increase in hydrogen concentration, if it were to continue at the same apparent rate, could result in possible near-term failure of the wire. A typical graphical illustration of time-related potential for increased hydrogen concentration in the Class IV wire taken from the 54-inch raw water pipeline and the 54-inch finished water pipeline are shown on Figures 2 and 3 respectively. This graphical approach assumes a constant rate of increase in the hydrogen content with time. The actual rate of hydrogen charging of the wire could be faster or slower than predicted by the graphs. Such graphical illustrations are intended as a first approximation only of the hydrogen charging process that could lead to wire failure, but not as a prediction of the time to failure of the pipeline itself. Nonetheless, a wire sample which exhibits a high as-received hydrogen concentration and appears to be rapidly approaching the hydrogen concentration resulting in failure in the DTHC test gives rise to legitimate concern as the possibility of pipe failure in the near future such as the 54-inch finished water pipeline. The wire in this pipeline indicates a high sensitivity to hydrogen failure and the mortar coating has not been able to prevent a buildup of hydrogen in the wire. This pipeline has been scheduled for replacement in the next few years.

In contrast, the 54-inch raw water pipeline data indicates that the wire is not as sensitive to hydrogen failure and the mortar coating has been able to prevent hydrogen buildup in the wire. This pipeline has been scheduled for long-term replacement (15 to 20 years) with retesting in 5 years. The problem with the 54-inch

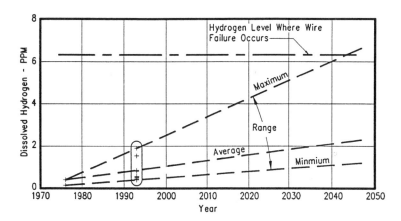

Figure 2. Time Related Corrosion Failure Potential
54" Raw Water Pipeline Wire

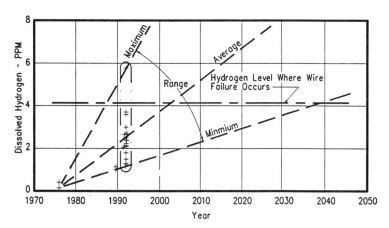

Figure 3. Time Related Corrosion Failure Potential
54" Treated Water Pipeline Wire

raw water pipeline is that the mortar quality is poor and is carbonated to a moderate extent. Fortunately, to this time, the coating has been able to protect the wires from hydrogen buildup.

PIPELINE LOCATION RISK ASSESSMENT

A pipeline location risk assessment was conducted for each pipeline segment. The following criteria were used in the risk assessment:

- Effect of a potential failure on overall system
- Transmission main segment assessment and need for additional isolation valve evaluation
- Duration of failure event and ability of system to be operated without the pipeline segment during failure repair
- Private property risk assessment
- Personal injury risk assessment
- Risk to critical customers

INTERNAL INSPECTION OF MAJOR PIPELINES

The City conducted internal inspections of available portions of the 54-inch finished water pipeline and the 42-inch raw water pipeline. The 54-inch finished water pipeline was inspected in the segment which showed the highest levels of hydrogen in the Class IV wire. This inspection was performed during the installation of an additional isolation valve on the pipeline. This inspection revealed one pipe which had a 40-inch long horizontal crack which is an indicator of a large number of broken wires. This is a typical wire failure, characterized by longitudinal cracking and a hollow sound in the internal concrete wall. Repair straps were installed to replace the structural support which had been provided by the broken wires. The internal inspection confirmed the accuracy of the wire testing.

The City is scheduled to conduct additional internal inspections of the 54-inch raw water pipeline, because this pipeline is critical to the operation of the system during high demand periods. The inspection will also provide additional information on the condition of the pipeline. The pipeline replacement schedule will be adjusted based on the results of the internal inspection and other inspections concerning the condition of the pipelines.

PCCP REPLACEMENT ROUTE STUDY

A replacement route study was conducted for all of the PCCP mains which included Interpace Class IV wire. The replacement alternatives included consideration of alternative systems which would make this pipeline less critical to the operation of the entire system.

The result of the PCCP replacement study was the development of a phased program which provides a higher level of redundancy in the finished water system and reduced the total net cost of the combined replacement and expansion program.

PHASED REHABILITATION/REPLACEMENT PROGRAM

At this point, Newport News has not had a PCCP emergency failure due to Class IV wire. They are currently scheduled to replace the treated 54-inch pipeline with a 48-inch and 42-inch pipeline in 1998. All the other Class IV wire pipelines are scheduled to be replaced by the year 2030.

DETAILED CITY COUNCIL BRIEFINGS

A total of three detailed City Council briefings were held to explain the potential problem, provide information concerning potential risks and to outline the proposed approach to the problem. The Waterworks' approach to this problem allows them to more fully define and quantify the problem. The proposed approach to the problem was accepted by City Council as a reasonable balance between the potential risks and the cost for defective pipeline replacement.

SUMMARY

Like many cities, Newport News faced the challenge of ascertaining when to replace transmission pipelines that had prematurely reached the end of their service life, how to explain the problem to their elected officials, and how to pay for the replacement costs. The City's approach allowed them to fully define and quantify the problem. This resulted in development and implementation of a staged replacement program for the defective PCCP pipelines. The Waterworks is currently on schedule with its short and long term programs for replacement of PCCP pipelines with Interpace Class IV wire. They now have a firm and proactive reply to the question **"When Should PCCP with Interpace Class IV Wire Be Replaced?"**

Cathodic Protection Requirements of Prestressed
Concrete Cylinder Pipe

Sylvia C. Hall[1]
Ivan Mathew[2]

Abstract

Concrete pressure pipe (CPP) is used in water and waste water systems that serve virtually every city in North America. Due to the passivating (corrosion inhibiting) properties of the highly alkaline portland cement, the cement mortar coating provides the only protection that CPP normally requires. Under certain conditions, such as high chloride environments, the steel can depassivate, leading to corrosion. Under these conditions, cathodic protection (CP) can be used to protect the encased steel elements. The purposes of this project were to determine current density requirements of CPP with and without supplemental barrier protection and to determine the effect of cathodic over-protection on the properties of the encased prestressing wire.

This project consisted of installing a CP system on a 73 m (240') long by 1.22 m (48") diameter prestressed concrete cylinder pipe (PCCP) line and measuring current, polarization, and depolarization of the pipeline during two to three months of system activation. It also consisted of subjecting prestressing wire to cathodic over-protection and determining the effect of CP on time-to-failure, hydrogen content, tensile strength, and reduction of area of the prestressing wire.

Introduction

PCCP is used in water and waste water systems that serve virtually every city in North America. It is primarily used for distribution of water for industrial, agricultural, and residential use. The pipe is typically designed and manufactured in accordance with AWWA Standards C304 and C301, respectively. It is manufactured in sizes from 410 mm (16") to 6.4 m (21') in diameter.

----
[1]Director, [2]R&D Engineer, Engineering Development Center, Ameron, Inc., 8627 S. Atlantic Ave., S. Gate, CA 90280-3501

Approximately 29,000 km (18,000 miles) of PCCP have been installed in North America during the past 50 years (Clift 1991).

Due to the passivating (corrosion inhibiting) properties of the highly alkaline portland cement, the cement slurry and mortar coating provides the only protection that CPP normally requires. One survey showed that concrete pipe had the lowest problem occurrence rate and that the average level of satisfaction was highest for more than 185,000 km (115,000 miles) of pipe surveyed (AWWA Research Foundation). Another survey stated that the overall performance of PCCP has been excellent. Only about 1 project out of 700 (25 out of 17,400 projects) has had any type of problem with external corrosion and that, in most cases, only 1 or 2 pipe sections were affected (Clift 1991).

In unusual circumstances, such as in high chloride environments, the passivating properties of the highly alkaline cement may be compromised. In such environments, it may be necessary to provide supplemental protection for PCCP. Supplemental protection is usually in the form of barrier coatings, barrier membrane encasement, or, in rare cases, CP.

Since extreme conditions are required to cause corrosion of PCCP, CP has rarely been used. One investigator reported that it appears that less than 0.5% of all PCCP in the United States is under CP (Benedict 1989). Another report indicated that over 20 projects of a total of 28,900 PCCP projects (less than 0.1%) are under CP (Clift 1991).

Since CP is rarely required, potential criterion and current density requirements for buried PCCP are generally not available. The authors have seen current density design requirements of 10.8 to 21.6 mA/m^2 (1 to 2 mA/ft^2) in project specifications for buried PCCP. This appears to be derived from the design current densities used for CP of reinforced concrete bridge components. It is expected that buried pipelines will require considerably less current density than an atmospherically-exposed bridge.

The potential criterion required to achieve protection of underground organically-coated oil and gas pipe lines has been under considerable debate for the past 10 years. The most accepted criterion of -850 mV versus a copper-copper sulfate electrode (CSE) is often used on bare or organically coated steel (RP0169 1992). A potential of -500 mV (CSE) was reported as a criterion to protect uncorroded steel in an alkaline environment in the presence of high levels of chloride ions (Hausmann 1969). A potential of -710 mV (CSE) was found to prevent further corrosion once corrosion was initiated (Hausmann 1969). A 100 mV polarization or depolarization shift is another criterion that is used to protect steel from corrosion of either organically coated steel or steel in concrete (Benedict 1989; RP0169 1992; RP0290 1990). Polarization shifts of only 20 mV have been found to effectively protect corroding steel in mortar (Hall et al. 1994).

One of the authors (Hall) obtained current and potential data for a 4.5 mile long, 60" and 66" coal-tar-epoxy-coated PCCP under CP from a water district. The supplemental barrier coating of coal tar epoxy and CP was specified due to the possibility that stray current from a nearby electric rail system may cause stray current interference on the pipeline. Based on 12 years of available data, current density requirements were approximately 120 $\mu A/m^2$ (11 $\mu A/ft^2$) to achieve polarization potentials ranging from -600 mV to -800 mV and approximately 43 $\mu A/m^2$ (4 $\mu A/ft^2$) to achieve polarization potentials ranging from -400 mV to -450 mV. This is at least a 100 to 150 mV shift from the baseline potential.

Due to the use of high strength prestressing wire in the pipe, the effect of high levels of CP at the potential required to cause hydrogen embrittlement must also be addressed. The most probable reaction occurring on the pipe under excessive CP at pH greater than 7 is the electrolysis of water, $2H_2O + 2e^- \rightarrow H_2 + 2OH^-$. During the formation of H_2, hydrogen atoms (H°) are produced on the metal surface. Prior to combining, atomic hydrogen may penetrate the steel. This entry causes a loss of ductility, or embrittlement, of the prestressing wire.

Based on this reaction, the potential at which hydrogen evolution occurs can be calculated using the Nernst Equation (eq. 1) below:

$$E = E^\circ - (RT/nF)\ln([H_2O]^2/[H_2][OH^-]^2) \quad (1)$$

where E = Hydrogen evolution potential, V
E° = Oxidation potential = +0.828 V (SHE)
R = 8.314 J/°K·mole
T = 298.2°K
F = 96,500 coulombs/equivalent
n = 2 equivalents/mole (electrons in reaction)
$[H_2O]$ = Activity of water = 1
$[H_2]$ = Activity of hydrogen = 1 atmosphere
$[OH^-]$ = Hydroxide ion activity = antilog(pH minus 14)

Therefore, at a pH of 12.5, E = +0.739 volt (SHE) for the oxidation reaction. For the reduction reaction (hydrogen production), the sign is changed and -0.316 volt is added to convert from the standard hydrogen electrode (SHE) to the copper-copper sulfate electrode (CSE). Thus at typical pHs of portland cement mortar of 12.5 to 13.5, the hydrogen evolution potential is -1055 mV to -1114 mV (CSE), respectively. The calculated potential for hydrogen evolution is more positive by approximately 59.2 mV for each decrease in one pH unit. In carbonated concrete, under CP, the pH at the wire surface will increase rapidly to a value greater than 12.4 due to the production of hydroxide ions or consumption of hydrogen ions in accordance with the following reactions:

$2H_2O + 2e- \rightarrow H_2 + 2OH^-$ at $E < -0.0592pH-0.316$ and $pH>7$
$O_2 + 2H_2O + 4e- \rightarrow 4OH^-$ at $E > -0.0592pH-0.315$ and $pH>7$
$2H^+ + 2e- \rightarrow H_2$ at $pH < 7$

Since hydroxide ions, but no hydrogen, are produced in the second reaction, low levels of current at the indicated potential can be used to increase the pH without producing hydrogen.

The purpose of this project was to determine the current density requirements of PCCP with and without supplemental barrier protection, such as coal tar epoxy coatings and polyethylene encasement, at polarization and depolarization shifts of 100 mV. The project also consisted of determining the effect of CP on time-to-failure, hydrogen content, tensile strength, and reduction of area of prestressing wire.

Test Setup

Cathodic Protection of PCCP Line

The project consisted of installing an impressed current CP system and applying current to attain a 100 mV shift to a 73 m (240') long by 1.22 m (48") diameter embedded cylinder PCCP line. Ten 7.3 m (24') long PCCP sections were manufactured and installed. The pipe was manufactured with two 2.54 cm (1") wide shorting straps, 180° apart to reduce the electrical attenuation along the prestressing wire in each pipe section. CP requires that all steel elements within the pipe and between each pipe section be electrically continuous (or bonded). In most cases, the pipe joints have a bell and spigot configuration with a rubber gasket. This configuration requires that the joints be bonded for CP to be effective. Most PCCP lines installed in the Western United States in the last 15 years have been bonded. In the Eastern United States, most pipelines are not bonded. The prestressing wire was made electrically continuous to the steel cylinder at each end of each pipe. The steel joints were specially manufactured with oversized bells, epoxy coated, and installed with oversized gaskets to ensure electrical discontinuity between adjacent pipe sections so that only the joint bonds provided electrical continuity between joints. Additional details of the pipeline are reported elsewhere (Hall 1994).

Two of the pipe sections were coated with a 660-micron (0.026") thick supplemental coal tar epoxy (CTE) coating and two of the sections were encapsulated in a 200-micron (0.008") thick polyethylene (PE) encasement. Pinholes and holidays were present in both coating systems. Six sections had no additional supplemental protection beyond the highly alkaline cement slurry and mortar coating.

The pipe sections were installed with 1.8 m (6') of cover in an arid environment in Palmdale, California. Mortared night caps were provided at each end with two access manholes. The site was selected to be representative of arid environments.

The native soil at the site is a sandy gravel. Soil-box resistivity ranged from 100,000 to 200,000 ohm·cm dry and 16,000 to 30,500 ohm·cm saturated. The pH of the soil samples ranged from 7.8 to 8.2. Water-soluble chloride and sulfate contents were less than 10 mg/kg which indicates that the soil is non-corrosive. The pipeline was backfilled with sand from an adjacent aggregate pit which had similar chemical properties. The Wenner four-pin soil resistivity values of the backfilled area at 1, 1.5, and 3.0 m (3', 5', and 10') spacings ranged from 13,400 to 63,200 ohm·cm when wet or dry.

Provisions were made to allow electrical connection or disconnection between adjacent pipe sections to simulate bonded and unbonded pipelines. This was done by connecting insulated 4/0 copper cables to the joints and bringing them to a test station at the surface above each joint.

Permanent copper-copper sulfate reference electrodes (CSE) and soil moisture cells were installed. The reference cells were buried 0.6 m (2') from the pipe at the pipe top, both springlines, and bottom, and 1.2 m (4') above the pipe top. They were placed at and midway between each joint. The lead wires were brought to a common junction box.

A 10 cm (4") diameter by 114 cm (45") long steel pipe buried 6.1 m (20') perpendicular to the last pipe was used as the anode. This is expected to simulate CP systems where anodes are installed close to the pipeline because of space limitations. Gypsum was placed around the anode as backfill. A variable power supply was used to supply the current.

A baseline potential survey was taken prior to activating the CP system. Potentials were measured approximately every 1.5 m (5 feet) along the centerline of the pipeline. Upon activation, an over-the-line potential survey was taken at intervals from 90 minutes to up to 3 months of CP. Current-on and polarization (current-interrupted) potentials and current were recorded. The polarization potentials were measured manually approximately

1 second after current interruption in the first series of surveys with the line at 4640 to 5180 $\mu A/m^2$ (430 to 480 $\mu A/ft^2$) and in the second series at 540 $\mu A/m^2$ (50 $\mu A/ft^2$). The polarization potentials were measured using a datalogger 300 ms after the current was interrupted in the remaining surveys. At selected intervals, the CP system was deactivated and an over-the-line survey was performed at intervals from 4 hours to 1 month during pipeline depolarization. The current density was calculated based on the mortar coating surface area. The actual surface area of the prestressing wire and steel cylinder is 57% and 88%, respectively, of the mortar coating surface area.

Cathodic Over-protection of Prestressing Wire

Prestressing wire was subjected to no CP and to CP polarization potentials of -850 mV and -1000 mV as well as cathodic over-protection values of -1200 mV to determine the effect of cathodic protection on the performance of prestressing wire. The wire was stressed to 60% of its specified minimum tensile strength in a saturated calcium hydroxide solution. This is the approximate stressed value of the wire on PCCP.

Cantilever-type wire tensioning apparatus with plastic enclosures for immersing approximately 32" of 60" long wire specimens were used. Potentiostats were used to control the potential. Twenty-four specimens were exposed at any one time. Six and eight gage, ASTM A648, class III prestressing wire specimens were used during the investigation.

The specimens were immersed in a saturated calcium hydroxide solution. Calcium hydroxide is the principle soluble compound of portland cement which provides for the high alkaline, corrosion inhibiting environment. The solution was prepared using reagent grade calcium hydroxide and ASTM D1193-77 type IV reagent water. Calcium hydroxide was periodically added to ensure that a pH of 12.45 was maintained. The pH was determined periodically.

At 6 and 12 months, two wire specimens from manufacturer "A" at each CP level were removed from test. At 3 months, two wire specimens from manufacturer "B" at each CP level were removed from test. Reduction of area, normal-load-rate tensile strength, and presence of longitudinal splits of the immersed, air-exposed, and control specimens were determined in accordance with ASTM A648-90a.

A 2" long piece from each removed specimen and a control, non-exposed specimen were immediately cleaned, immersed in liquid nitrogen to prevent hydrogen diffusion, placed in a plastic bag in an ice chest containing dry ice, delivered by overnight delivery to a laboratory, and analyzed for hydrogen content using a LECO DH103 Total Hydrogen Determinator.

Two specimens each from coils with different performance properties in continuous torsion remained under exposure at no CP, -850 mV, -1000 mV, and -1200 mV and the time-to-failures of the specimens were recorded. Continuous torsion of the wire specimens prior to exposure was determined in accordance with ASTM A648-90a.

Test Results and Discussion

Cathodic Protection of Uncoated PCCP

The current-on and polarization (current-interrupted) potentials of the entire pipeline at 4 hours and 5 days at 1.74 A (5180 $\mu A/m^2$; 480 $\mu A/ft^2$) and 1.56 A (4640 $\mu A/m^2$; 430 $\mu A/ft^2$), respectively, are shown in Figure 1. The baseline potentials prior to CP are also shown and indicate that the pipeline was passive prior to CP activation. The actual current flowing to each type of pipe was measured and the average current density based on the mortar coating surface area is shown. This current density is excessive since the pipe polarized 400 mV on the uncoated pipe sections, more than 600 mV on the CTE-coated sections, and exceeded even the commonly-used -850 mV criterion on the PE-encased sections during the initial 4 hours of CP. The potential of the end PE-encased section exceeded -1000 mV. This indicates possible hydrogen embrittlement of the prestressing wire on this section. Potentials taken at 5 days indicate that the pipeline was continuing to polarize. The potentials of the CTE-coated sections exceeded -850 mV at 5 days of CP. In addition, the difference in potential between the current-on and polarization (I-interrupted) potentials indicate that the IR drop in the CTE-coated and PE-encased sections is excessive at high current density.

The current was then reduced by approximately a factor of 8 to 200 mA (540 $\mu A/m^2$; 50 $\mu A/ft^2$). The current-on and polarization potentials and the resulting average current density on each coating type five weeks after reducing the current are shown in Figure 2. The polarization (I-interrupted) potentials were greatly reduced but still exceeded the 100 mV polarization shift criterion. The IR drop on the CTE-coated and PE-encased system was still excessive. At 5 weeks of CP, the system was deactivated. The 4-hr and 1-week depolarization potentials of the pipeline are shown in Figure 2. The average depolarization shifts at 4 hours were 220 mV on the PE-encased sections, 180 mV on the CTE-coated sections, and 120 mV on the uncoated sections which is greater than the 100 mV depolarization required. At 1 week, the potential had depolarized to approximately the baseline potential. The average depolarization shifts at 1 week were 420 mV on the

PE-encased sections, 400 mV on the CTE-coated sections, and 280 mV on the uncoated sections.

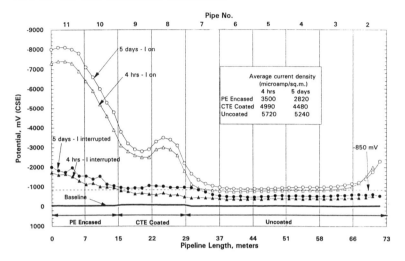

Figure 1. Current-on and Polarization Potentials of Entire PCCP Line At Approximately 4640 to 5180 $\mu A/m^2$ (430 to 480 $\mu A/ft^2$)

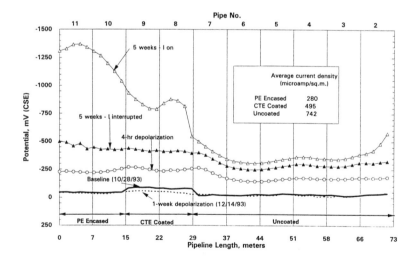

Figure 2. Current-on, Polarization, and Depolarization Potentials of Entire PCCP Line at 540 $\mu A/m^2$ (50 $\mu A/ft^2$)

Since the various coating systems polarized to varying levels, it was decided to protect each type of coating system separately. The CTE-coated and PE-encased pipe sections were electrically disconnected from the uncoated pipe sections. Since the uncoated sections had polarized 280 mV at 540 $\mu A/m^2$ (50 $\mu A/ft^2$), the current to the uncoated sections was reduced by a factor of two to 270 $\mu A/m^2$ (25 $\mu A/ft^2$). Polarization potentials at 9 and 13 weeks of CP are shown in Figure 3 and are approximately the same. This indicates that the pipeline was not continuing to polarize during the 13 weeks of CP. At 13 weeks, the CP system was deactivated. The depolarization at 4 hours and 1 week are also shown in Figure 3. The depolarization shift at 4 hours was approximately 100 mV indicating that the 100 mV depolarization shift criterion was met. The pipeline continued to depolarize during the week until it depolarized another 100 mV and returned to a slightly more passive potential than its baseline potential. Since 540 $\mu A/m^2$ produced a depolarization shift of 280 mV and 270 $\mu A/m^2$ produced a shift of 200 mV, a lower current density is expected to produce a 100 mV shift.

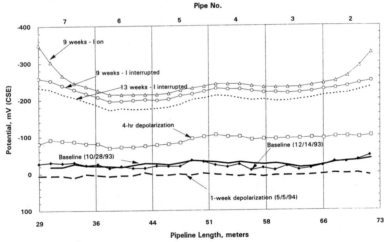

Figure 3. Current-on, Polarization, and Depolarization Potentials of Uncoated PCCP Line at 270$\mu A/m^2$ (25 $\mu A/ft^2$)

The CTE-coated sections were cathodically protected at 108 $\mu A/m^2$ (10 $\mu A/ft^2$). The polarized potentials 90 minutes after CP activation are shown in Figure 4. The polarization shift at 90 minutes was greater than 100 mV so the current density was reduced to 54 $\mu A/m^2$ (5 $\mu A/ft^2$). The polarized potentials 2 months at 54 $\mu A/m^2$ are given in Figure 4. The polarization shift is still greater than 100 mV so the current density was further reduced to 32 $\mu A/m^2$ (3 $\mu A/ft^2$).

The polarized potential at 2 months is shown in Figure 4. The polarization shift from the baseline potential was approximately 150 mV at 32 µA/m² at 2 months of CP.

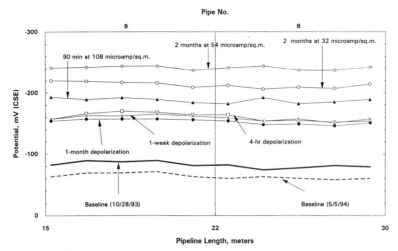

Figure 4. Polarization and Depolarization of CTE-Coated PCCP Line

Depolarization of the CTE-coated sections after 2 months at 32 µA/m² is shown in Figure 4. The pipeline depolarized 65 mV during the first 4 hours with less than 15 mV occurring during the following month. Although depolarization to the baseline potential did not occur, it is believed that complete depolarization of the CTE-coated sections would eventually occur.

The PE-encased sections were cathodically protected at 32 µA/m² (3 µA/ft²). Polarization at 3 hours, 1 week, 5 weeks, and 3 months of CP are shown in Figure 5. Polarization of approximately 100 mV was achieved at 3 hours and remained essentially unchanged for the following 3 months.

Depolarization of the PE-coated sections after 3 months at 32 µA/m² (3 µA/ft²) is shown in Figure 5. The pipeline depolarized 140 mV during the first 4 hours and 200 mV at 1 week. It depolarized to levels more positive than the initial baseline.

Polarized potentials of the uncoated section using the permanent reference cells buried around the pipe every 3.5 m (12') are shown in Figure 6. The potentials were within 25 mV of the over-the-line potentials. Differences in potential around the pipeline or at different depths are not apparent, indicating uniform distribution of current around the pipe circumference.

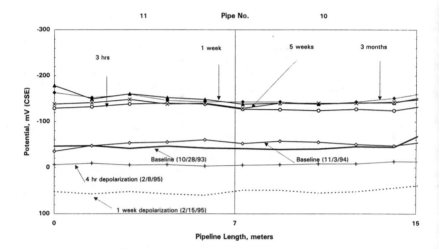

Figure 5. Polarization Potential of PE-Encased PCCP Line at 32 $\mu A/m^2$ (3 $\mu A/ft^2$)

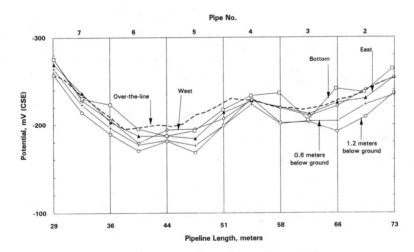

Figure 6. Effect of Reference Electrode Position on Polarization Potential of Uncoated PCCP Line at 270 $\mu A/m^2$ (25 $\mu A/ft^2$)

Current density requirements of uncoated, CTE-coated, and PE-encased PCCP to achieve or exceed a polarization or depolarization shift of 100 mV were determined to be 270, 32, and 32 $\mu A/m^2$ (25, 3, and 3 $\mu A/ft^2$), respectively.

Cathodic Over-protection of Prestressing Wire

The effect of CP on the tensile strength and reduction of area of 6 gage prestressing wire from manufacturers "A" and "B" maintained at 60% of its specified minimum tensile strength in a saturated calcium hydroxide solution for 3, 6, and 12 months is shown in Figure 7. For plotting purposes, the potentials of the non-cathodically protected specimens were arbitrarily selected to be -200 mV which is within the passive range of 0 to -300 mV (CSE) measured during the exposure.

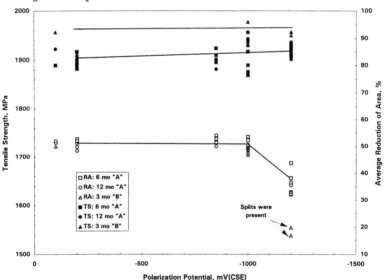

Figure 7. Reduction of Area and Tensile Strength of Prestressing Wire Loaded at 60% of Its Tensile Strength & Subjected to CP in a Saturated Calcium Hydroxide Solution

For comparison purposes, the reduction of area of the control specimens are plotted at -100 mV even though they were never loaded, immersed, or subjected to CP. Reduction of area of 16% and 19% of wire "B" at -1200 mV was due to splits found in the wire. The splits opened up at the ends resulting in a lower but erroneous reduction of area value. A significant decrease in reduction of area is evident at polarization potentials of -1200 mV with no significant decrease at -1000 mV. These results were consistent with the authors' unpublished results using unloaded prestressing wire which showed that excessive CP at -1100 mV (CSE) or more negative values decreases the ductility of prestressing wire. Normal-load-rate tensile strength did not change at any level of CP.

The effect of CP on the hydrogen content of the immersed specimens is plotted in Figure 8. For comparison purposes, the hydrogen contents of the control specimens are plotted at -100 mV even though they were never loaded, immersed, or subjected to CP. A significant increase in hydrogen content is evident at polarization potentials of -1200 mV. This is consistent with the observation that copious amounts of fine bubbles were produced on the specimens at -1200 mV. No bubbles were seen at -1000 mV. This is consistent with the calculated hydrogen evolution potential of -1055 mV (CSE) at a pH of 12.45. The increase in hydrogen content is consistent with the decrease in reduction of area shown in Figure 7.

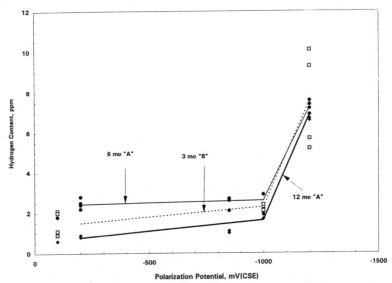

Figure 8. Hydrogen Content of Prestressing Wire Subjected to CP

Time-to-failure of prestressing wire specimens held at -1200 mV at 60% of its specified minimum tensile strength is shown in Figure 9. It is plotted as a function of its initial ductility, as expressed by turns to break in continuous torsion. The time-to-failure varied from 9 to 41 months and is dependent on the number of turns to break in continuous torsion determined on non-exposed (control) specimens. The fractures were all brittle in nature and the reduction of area at the failure point was less than 2%. This indicated that embrittlement of the wire occurred. Longitudinal splits terminating near the center of the wire were found on some of the specimens held at -1200 mV. Hydrogen contents of the fractured end were similar to those shown in Figure 8. It appears that wire with lower turns to break in continuous torsion has a greater susceptibility to hydrogen embrittlement. This supports the continuous torsion requirement in AWWA C301.

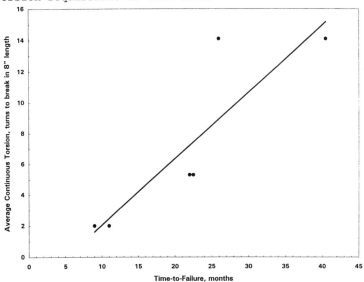

Figure 9. Time-to-Failure of Prestressing Wire Held at -1200 mV as a Function of Continuous Torsion

Prestressing wire held at -1000 mV or more positive potentials, where hydrogen is not produced, have not failed after up to 53 months of exposure. Exposure of the specimens at these more positive potentials is continuing.

It is evident that prestressing wire is susceptible to hydrogen embrittlement at potentials negative enough to generate hydrogen. Polarization potential should be maintained at potentials less negative than -1000 mV to avoid hydrogen embrittlement.

Conclusions

- Current density required to achieve or exceed a 100 mV polarization or depolarization shift of uncoated, CTE-coated, and PE-encased PCCP in a non-corrosive environment was approximately 270, 32, and 32 $\mu A/m^2$ (25, 3, and 3 $\mu A/ft^2$), respectively.

- If CP of PCCP is necessary, the polarization potential should be maintained at potentials less negative than -1000 mV to avoid subjecting PCCP to hydrogen embrittlement which could lead to failure of the pipe.

Acknowledgment

The authors wish to acknowledge the American Concrete Pressure Pipe Association for supporting portions of this project.

Appendix I. References

"AWWA Standard for Prestressed Concrete Pressure Pipe, Steel-Cylinder Type, for Water and Other Liquids." (1992). C301-92, AWWA, Denver, CO.

"AWWA Standard for Design of Prestressed Concrete Cylinder Pipe." (1992). C304-92, AWWA, Denver, CO.

Benedict, R. L. (1989). "Corrosion Protection of Concrete Cylinder Pipe," *CORROSION/89*, paper no. 368, NACE, Houston, TX.

"Cathodic Protection of Reinforcing Steel in Atmospherically Exposed Concrete Structures." (1990).RP0290,NACE,Houston.

Clift, J. S. (1991). "PCCP - A Perspective on Performance", *1991 ACE Proceedings*, AWWA, Denver, CO.

"Control of External Corrosion on Underground or Submerged Metallic Piping Systems." (1992). RP0169, NACE, Houston,TX.

Hall, S. C. (1994). "Analysis of Monitoring Techniques for Prestressed Concrete Cylinder Pipe," *CORROSION/94*, paper no. 510, NACE International, Houston, TX.

Hall, S. C., Carlson, E. J., and Stringfellow, R. G. (1994). "Cathodic Protection Studies on Coal Tar Epoxy-Coated Concrete Pressure Pipe," *Materials Performance*, 33(10), 29.

Hausmann, D. A. (1969). "Criteria for Cathodic Protection of Steel in Concrete," *Materials Protection*, 8(10), 23.

"Review of Water Industry Plastic Pipe Practices." AWWA Research Foundation - Research Report, Distribution Systems.

Cathodic Protection Retrofit of a 35-Year-Old Concrete Pressure Pipeline

Stephen P. Turnipseed[1]

Richard I. Mueller, P.E., Member, ASCE[2]

Abstract

This paper presents a case history of a CP (cathodic protection) retrofit on a 35-year-old concrete pressure pipeline. Corrosion control was selected over pipeline replacement to curb the increasing leak frequency. The CP project required excavation of approximately 3,800 joints and electrically bonding across each connection to ensure the 37 km (23 miles) of 840 mm (33 inch) and 910 mm (36 inch) diameter AWWA C303-style pipe was electrically continuous. Nine 107 m (350 ft) deep impressed current anode beds with rectifiers were installed to provide the current sources. Surveys taken before and after energizing the CP system indicated that the pipe was meeting the NACE -100 mV potential shift criteria. No additional leaks have occurred in over two years since the CP system was energized.

Introduction

The CCWIS (Crane County Water Injection System) pipeline includes approximately 37 km (23 miles) of 840 mm (33 inch) and 910 mm (36 inch) diameter AWWA C303-style pipe. The pipeline carries brine water from source wells to an oilfield 64 km (40 miles) south of Odessa, Texas. The pipe was placed in service in 1958, but was considered for replacement due to an increasing frequency of leaks. The leaks in the pipeline resulted from a combination of external corrosion caused by long-term exposure to moist, well-aerated, chloride-laden soils along the pipeline and inadequate surge control. Surges are thought to cause cracks in the external mortar coating, which would tend to accelerate the corrosion process. Cathodic interference with foreign pipelines was another contributing factor to the external corrosion. Internal corrosion from the brine being transported is not suspected to be a cause of the failures.

---

[1]Lead Materials Engineer - Corrosion, Chevron Research and Technology Company, 100 Chevron Way, Richmond, CA 94802-0627

[2]Vice President, Engineering, Gifford-Hill-American, Inc., 1003 Meyers Road, Grand Prairie, Texas 75050

Samples taken from the pipeline wall during hot taps and observations during repairs indicated that, in spite of the failures, pipeline integrity was generally good. It appeared failures were occurring in scattered areas of active corrosion along the length of the pipeline.

A formal decision to make the pipeline electrically continuous and install cathodic protection, rather than replace large sections of the pipeline, was implemented. This decision was an economic choice, as the cost of CP was 1/10th the cost of replacement. Vacuum breaking valves were also added to minimize surges. The improved operating systems have eliminated the problem with on-going pipeline leaks. The cathodic protection system has since been reviewed to determine current attenuation and minimum current densities necessary for this pipeline at various locations along its length.

Failure Analysis

The retrofit project began with a failure analysis of the pipe at one of the leak sites. Figures 1 and 2 are cross-section photomicrographs that clearly show external corrosion was the cause of this failure. Although there was a very thin layer of iron sulfide present on the ID, internal corrosion at this location was not significant.

Figure 1. Cross Section of Steel Can at a Failure (50 X)

Figure 2. Cross Section of Steel Can at a Failure (100 X)

Pilot Project

The major obstacle to installation of CP was that the pipeline must be made electrically continuous. Due to the joining method used with C303 pipe, it was necessary to excavate and electrically bond each joint to its neighbor. The magnitude of the bonding work indicated that a pilot project would be prudent. In November of 1991, two methods of excavation and electrical bonding were evaluated on 550 m (1,800 ft) of the pipeline as follows:

    Excavation Techniques:

        1. Vacuum Excavation
        2. Trackhoe

    Electrical Bonding Techniques:

        1. Thermite Welding Copper Jumper Wires
        2. Electric Arc Welding Steel Bonding Clips

Based on the pilot study results, it was decided to electrically connect the joints by excavating with a trackhoe and welding a U-shaped steel bonding clip (electrical resistance of about 56 micro-ohm) between the bell and spigot. Figure 3 shows a detailed drawing of the bonding clip installed between two joints.

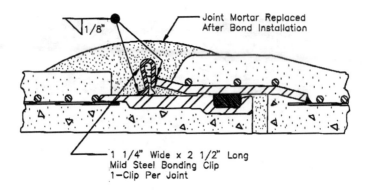

Figure 3. Retrofit Electrical Joint Bond Detail

To provide current to the 550 m (1,800 ft) long section of bonded pipe, a single 122 m (400 ft) deep anode bed and rectifier was installed. An estimate of the number of anode beds required for the rest of the pipeline was made from the average current density on this section, 645 to 970 u-A/m^2 (60 to 90 u-A/ft^2). Similar values have been reported. (Benedict 1989).

Electrical Bonding Project

The pilot proved that excavation, electrical bonding, and regrouting could be performed at a reasonable cost. In the summer of 1992 approximately 3,800 joints (37 km or 23 miles) of pipe were electrically bonded. During the bonding work, test stations with two lead wires (thermite welded to the bonding clip) were installed every 305 m (1,000 ft) and at every major pipeline crossing. The construction crew used the following equipment during the excavation and bonding operation:

- Trackhoe - to excavate a bell hole every 9.8 m (32 ft)
- Compressor & pneumatic chipping hammer - to clean off mortar at each joint
- Electric arc welder - to attach the steel bonding clip
- Bulldozer - to cover the bell holes.

Because the top of the pipe averaged only 0.9 to 1.2 m (3 to 4 ft) from the surface and digging was relatively easy, a high rate of bonding was possible. Unobstructed by line crossings, it was possible for the crew to excavate, bond, grout, and backfill as many as 70 joints per day.

Close-Interval Pipe-to-Soil Survey

After the pipeline had been electrically bonded, but prior to the application of CP, a true native-state P/S (pipe-to-soil) potential survey was performed. The close-interval over-the-line survey technique was selected to obtain the maximum amount of potential data on the pipeline. A native-state potential survey provides a baseline reference for meeting the NACE -100 mV potential shift criteria. It also allows areas of high corrosion activity to be identified (potentials more negative than 300 mV vs. Cu/CuSO$_4$). Figures 4 and 5 are examples of survey data showing areas of various levels of corrosion activity. Should future pipe replacement ever be considered, the areas of highest corrosion activity will be considered a first priority.

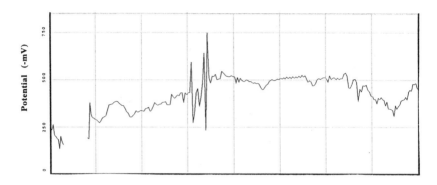

Figure 4. Native-State Potential Survey Showing Area of Active Corrosion

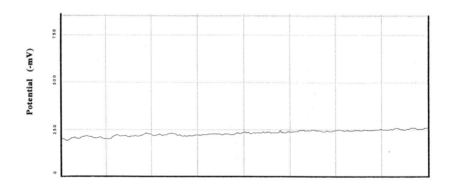

Figure 5. Native-State Potential Survey Showing Area with Minimal Corrosion

Resistance Survey and Location of Discontinuities

After the bonding work was completed, the resistance of 305 m (1,000 ft) spans of pipe was measured between the two-wire test stations. This was performed by impressing current through the pipe and measuring the voltage drop across the span. Figure 6 is a circuit diagram which illustrates how the resistance test was conducted.

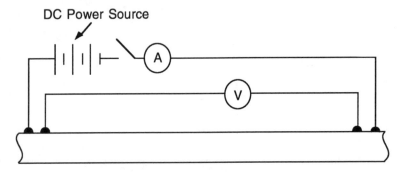

Figure 6. Pipe Resistance Test

A statistical analysis of the pipe resistance indicates an average resistance of 29.5 to 30.8 u-ohm/m (9.0 to 9.4 u-ohms/ft). Figure 7 is a distribution of the pipe span resistance data collected on pipe without unresolved discontinuities.

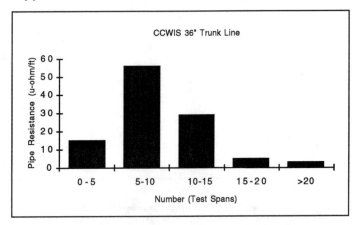

Figure 7. Distribution of Measured Pipe Resistances

Spans of pipe with resistance values greater than twice the average usually indicate an electrical discontinuity. A more detailed survey was necessary to pinpoint the actual high resistance joint(s) for repair. To locate the problem joints, a close-interval over-the-line survey was performed while 10 amps of direct current was being circulated through the pipe span. Most of the electrical discontinuities were a result of short repair joints that were missed during the initial bonding operation. As illustrated in figure 8, areas of high pipe resistance show up as sharp changes in potential.

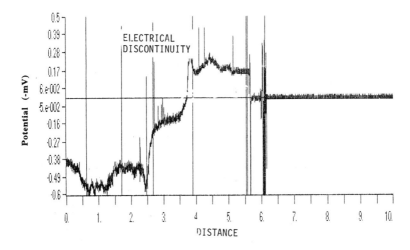

Figure 8. Survey Showing Electrical Discontinuity

Deep Anode Beds

While the electrical bonding work was proceeding, one 107 m (350 ft) deep anode bed was installed on the 840 mm (33 in) gathering system and two additional 107 m (350 ft) deep anode beds were installed along the 910 mm (36 in) trunk line. After energizing the rectifiers, it was found that poor current attenuation along the pipeline caused sections of the pipeline between anode beds to remain unprotected. Five additional deep anode beds were installed in 1993 to protect the lengths of pipe with low potential shifts.

Test Station P/S Potentials and Rectifier Current Output

In July of 1994 both on and instant-off potential measurements were taken at each test station. Comparing the instant-off potentials with the native-state potentials taken before CP was applied showed the majority of the pipe met the NACE -100 mV potential shift criteria. Figure 9 is a graphical plot of the on and instant-off potential data on one section of the pipeline at the test stations.

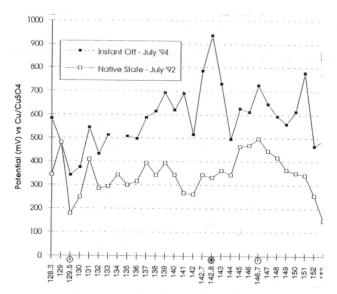

Figure 9. Test Station Potentials On and Instant-Off

The rectifier current outputs ranged from 16 to 29.5 amps each, with an average of 22 amps on the 910 mm (36 in) diameter pipe. This equates to an average current density of 1,400 u-A/m^2 (130 u-A/ft^2). It should be noted that electrical bonds exist with several foreign pipelines. The exact amount of current being returned through these bonds is not known.

<u>ACPPA Funded Testing</u>

The ACPPA (American Concrete Pressure Pipe Association) became interested in the project and funded additional tests on three 1,220 m (4,000 ft) sections of pipe to further define the CP requirements on concrete pressure pipe. The additional tests included:

- Soil Analysis

    + Soil resistivity at 30 m (100 ft) intervals using the 4-pin method
    + Classification of soil from 12 locations
    + Laboratory analysis of 14 soil samples and 1 surface-water sample to determine:

        - Resistivity
        - pH
        - Chloride
        - Sulfate
        - Carbonate
        - Moisture

- Close-Interval Over-the-Line Potential Surveys
    + On-potential survey - All rectifiers energized
    + Slow-cycle instant-off potential survey - all rectifiers simultaneously interrupted
    + After 12 to 16 hours of depolarization

- Resistance measurement of each 305 m (1,000 ft) pipe span surveyed.

- Current flow and direction in each 305 m (1,000 ft) pipe span surveyed.

- Plot of potential depolarization with time at one location.

Soil and Surface Water Data Analysis

The soil was classified as poorly-graded sand, tan to brown in color. Of the twelve samples, only two passed more than 5 percent by weight through a 200 mesh sieve. The soil samples had moisture contents which ranged from 1.6 to 4.9 percent by weight.

The field soil resistivities taken at 1.5 m (5 ft) pin spacing averaged 37,500 ohm-cm. Laboratory soil resistivities averaged about 100,000 ohm-cm in the as-received condition, but dropped to 15,000 ohm-cm in the saturated minimum condition.

Soil pH values ranged from 6.9 to 8.5. Water soluble chlorides were typically less than 10 mg/kg. Chlorides content was higher in areas where leaks had occurred. Water soluble sulfate contents were typically less than 30 mg/kg. Carbonate content varied dramatically, ranging from 0.1 to 43.1 percent as $CaCO_3$. Except for isolated locations where leaks may have occurred, the literature suggests that the soil be considered non-corrosive to concrete pressure pipe.

Flowing water from a small creek was also sampled. The surface water contained 2,140 mg/l chloride and 366 mg/l sulfate. This chloride content, combined with the oxygen in the well-aerated soil, is considered corrosive to concrete pressure pipe.

Potential Survey Data Analysis

Because the -0.85 volt minimum potential vs. $Cu/CuSO_4$ criteria from NACE RP0169-92 would result in excessive over-protection of the concrete pressure pipe, the -100 mV polarized potential shift criteria was used as a basis. This criteria requires two measurements, an instant-off potential and a depolarized or native-state potential.

An initial close-interval over-the-line on-potential survey was made in November of 1994 to determine if the interrupted surveys would cause any substantial depolarization during the work. Afterwards, a slow-cycle interrupted survey was made to obtain the instant-off potentials. Finally, a third survey was made with all rectifiers turned off overnight to obtain a partially depolarized measurement with which to compare the instant-off potentials.

The surveys revealed the majority of the pipeline had achieved a negative polarization shift of about 150 mV. In addition, there was typically an additional 25 mV of difference between the original native state potentials taken in 1992 and the off-potentials taken after leaving the rectifiers down overnight. The depolarization plot confirmed that the pipe potentials had not reached equilibrium, even after 40 hours with the CP current turned off.

Pipe Resistance and Current Flow

The average pipe span resistances were very close to those measured during the original survey two years earlier (25 to 34 u-ohm/m or 7.5 to 10.4 u-ohm/ft). Using the cylinder, bond, and fringing resistance formulas in the AWWA M9 Manual, a theoretical pipe resistance of 45 u-ohm/m (13.7 u-ohm/ft) was estimated. The lower resistance measured in the field may be attributed to some of the impressed current flowing through the soil or brine, rather than through the steel reinforcement. Most importantly, no electrical discontinuities developed during the two years since the original measurements.

During the measurement of pipe resistance, current flowing along each 305 m (1,000 ft) span of pipe was also measured. By taking the difference in the magnitude of current flow from adjoining spans, the average current pick up can be calculated. It was found that a typical 305 m 1,000 foot span picked up about 0.9 to 1.2 amps of current. This equates to a current density requirement of 970 to 1,400 u-A/m^2 (90 to 130 u-A/ft^2) of total pipe surface area.

Reduction in Leak Frequency

Ultimate success of a corrosion mitigation project can be measured in terms of leak frequency. There have been no leaks in over two years of operation. A cumulative leak versus time is shown in figure 10.

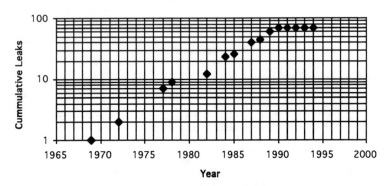

Figure 10. Cumulative Leaks vs. Time

Conclusions

External corrosion was determined to be the major cause of leaks on the 35-year-old concrete pressure pipeline even though the majority of the soil conditions were non-corrosive. Cathodic interference, well-aerated soil with locally-high chlorides, and inadequate surge controls were contributing factors to the corrosion.

A CP system was successfully installed on the 37 km (23 mi) long pipeline which allowed the potential of the pipe to be polarized more than -100 mV. The cost of bonding and adding CP was less than 1/10th of cost of pipe replacement.

In the open field, excavation and bonding between joints to achieve electrical continuity was most economically accomplished using a trackhoe and attaching steel bond clips using electric arc welding.

Test stations placed every 1,000 feet enhanced the ability to verify pipe continuity and locate bond problems.

Field measurements taken between test stations indicated that the average bonded pipe resistance is about 30 u-ohm/m (9 u-ohms/ft).

Typical CP current densities required to achieve a minimum -100 mV potential shift have ranged from 970 to 1,400 u-A/m^2 (100 to 130 u-A/ft^2) in these cohesionless soils having moisture contents from 2 to 5 percent.

Acknowledgments

The authors thank Chevron U.S.A. Production - South Permian Profit Center for permission for and assistance with the follow-up study on the CCWIS pipeline. Without their support, this paper would not have been possible. Special thanks goes to the ACPPA for funding the detailed CP surveys and soil analyses. And finally thanks to the management of Chevron Research and Technology Company, Gifford-Hill-American, Inc., and Ameron, Inc. for supporting this effort.

References

Benedict, Risque L., "Corrosion Protection of Concrete Cylinder Pipe", NACE Corrosion '89, paper #368, 1989.

NACE Standard RP-0169-92, "Control of External Corrosion on Underground or Submerged Metallic Piping Systems", 1992.

AWWA Manual M9, Concrete Pressure Pipe. American Water Works Association, Denver, CO (1995).

Impact Factors for Estimating Vehicle Live Loads

Mark C. Gemperline,[1] Member, ASCE,
and Thomas J. Siller,[2] Member, ASCE

Abstract

The maximum vertical surface force exerted by a pneumatic tired vehicle on a road surface is related to the surface roughness, vehicle speed, and vehicle physical characteristics. A relationship is developed between dimensionless forms of these variables and presented as design charts. The design charts may be used to develop restrictions on pipe crossing conditions and/or to estimate vehicle live loads for buried pipe design.

The relationship between variables is developed by numerical modeling. Road surfaces are characterized as self-similar fractals and vehicles are characterized as single-degree-of-freedom linear mass-spring-dashpot systems. Ground surface roughness is quantified as the fractal dimension of the pathway profile. Numerical model results are shown to compare favorably with vehicle impacts measured in a field study.

[1] Civil/Geotechnical Engineer, U. S. Bureau of Reclamation, Earth Sciences and Research Laboratory, Denver, CO 80225.

[2] Professor, Department of Civil Engineering, Colorado State University, Fort Collins, CO 80523.

Introduction

Estimating live vehicle loads on rough road surfaces is paramount to determining live loads on buried pipe. Surface loads are dependent on vehicle stiffness, damping characteristics, and speed; as well as ground surface roughness. Efforts to relate these variables to vehicle impact is handicapped by the lack of a simple quantitative measure of road roughness from which a similar surface profile can be recreated.

In this paper unpaved road profiles are characterized as deterministic fractals. The fractal dimension, representing road roughness, is related to maximum expected vehicle impact factor.

A one-dimensional, single-degree-of-freedom, quarter-car-simulation model is used to calculate forces a vehicle imparts on computer generated self-similar fractal surfaces. These forces are divided by vehicle static weight to obtain impact factors. Vehicle surface impact data for the condition of a fully loaded construction scraper traveling on an unpaved pathway has been collected in a previous study by Gemperline (1984, 1985). Maximum Impact factor values for the fractal surfaces are compared with maximum impact factors calculated from the field measurements.

A relationship between problem variables is determined using quarter-car-simulation and is presented in dimensionless design charts. The potential use of these charts is developed and discussed.

Vehicle Model

A single-degree-of-freedom quarter-car-simulation model characterizes a vehicle as a mass, spring and dashpot. Three variables, quarter vehicle mass, M; spring constant, K; and damping coefficient, D_c; are required to define the problem. The model, illustrated in Figure 1, is one-dimensional and has one degree-of-freedom; i.e., only the displacement of the mass from the neutral position, $y_m(t)$, is unknown.

The neutral position is the static position the mass takes if only the force of gravity is acting on it. The mass represents one quarter of the total vehicle mass. The surface displacement is given by y(t). Vehicle surface velocity, V_t, is used with the surface profile data to define y(t). Viscous damping is assumed; i.e., the damping force is opposite in direction but proportional to vertical velocity.

A free body diagram describing dynamic equilibrium of the system is shown on Figure 2. The equation for dynamic equilibrium is:

$$k\ (\ y(t) - y_m(t)) + c\ (\frac{dy(t)}{dt} - \frac{dy_m(t)}{dt}) - M\frac{d^2 y_m(t)}{dt^2} = 0 \quad (1)$$

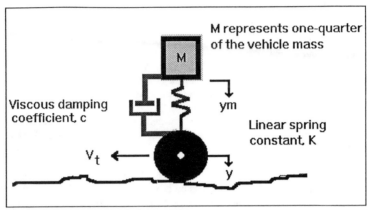

Figure 1. Quarter-car-simulation model.

A finite difference solution is applied in program QCAR.BAS (Gemperline 1994). Prior to implementing the solution the surface profile is smoothed to reflect tire envelopment of small surface asperities. This is accomplished by using the average elevation of the surface profile in contact with the tire to define y(t).

Fractal Dimension as a Measure of Roughness

Mandelbrot has demonstrated the ability of fractals to portray real land surfaces (Mandelbrot 1983; Barnsley et al. 1988; Feder 1988). Fractals are statistically self-similar or self-affine at all scales. Statistically self-similar means that each small portion, when magnified, can reproduce exactly the statistical characteristics of the larger portion. Statistically self-affine means that each small portion, when magnified using axis transformations, can reproduce exactly the statistical characteristics of a larger portion. Self-similar fractals are used in this study to represent road surface profiles. Figure 3 shows a series of computer generated fractal profiles of various roughness and the associated fractal dimension.

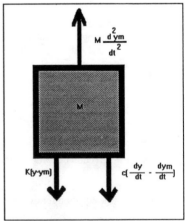

Figure 2: Quarter-car-simulation model dynamic equilibrium.

Fractal geometry provides simple techniques for measuring the fractal dimension of rough surfaces as well as simple methods of generating fractal profiles representing specified fractal dimensions. A computer program COMP.BAS was developed by Gemperline (1994) to determine the fractal dimension of profiles defined by few points.

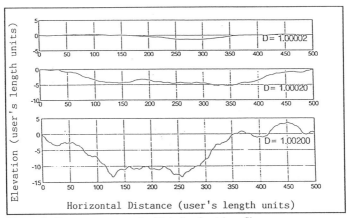

Figure 2: Computer generated self-similar fractal profiles.

Recently the International Roughness Index (IRI) has been developed as a measure of ground surface roughness (Sayers et al. 1986). The IRI is the average of the absolute value of the pathway profile slope measured over a wide range of wavelengths. It is correlated to dynamic roughness measurement instruments using quarter-car simulation. A self-similar fractal profile, by definition, is assured to have a constant absolute value of the profile slope at all wavelengths. This slope can be easily calculated for each fractal dimension and related directly to the IRI. Consequently, it is expected that methods used to estimate the IRI may be used to estimate the fractal dimension of a road profile.

Comparing Field Study Results with Quarter-car-simulation

Subsurface vehicle induced pressures were measured for the condition of a fully loaded construction scraper traveling on an unpaved haul road leading to an embankment dam (Gemperline 1984, 1985). The collected data permits estimation of spring constant, damping ratio, and quarter-vehicle mass required to represent the scraper for use in the quarter-car-simulation model. This study was performed to determine the magnitude of forces imparted to a buried concrete slab by moving fully loaded construction scrapers. Eight soil stress meters were imbedded at 1.4 m (4.5 ft) centers along the length of a 0.6 m-thick (2 ft) by 1.2 m-wide (4 ft) by 12.6-m-long (41.5 ft) concrete slab. The slab was located adjacent to a construction haul road leading to Red Fleet Dam near Vernal, Utah. Scrapers traversed the buried slab, traveling along its length with the tires on one side passing directly over all eight load cells.

The scrapers used in the study are Baker, model No. 657. The tires are 2.4 m (8 ft) in diameter and have a tread width of approximately 0.9 m (3 ft). The tire envelopment length, the length of tire in contact with the ground in the direction of travel, is approximately 0.46 m (1.5 ft). The scraper design fully loaded weight is 10.9 kN (240,000 lbs). The scraper width is 2.59 m (8.5 ft) measured from tire center to tire center. The front and rear tire centers are separated by 9.75 m (32 ft) (Gemperline 1984, 1985).

The scrapers traveled at speeds as great as 54.1 km/h (33 mi/h). The ground surface roughness and concrete slab burial depth were varied. One hundred and seventy five passes were recorded, 64 of these were at speeds of less than 3.2 km/h (2 mi/h). The low speed passes provided pressure records in a quasi-static condition. Low speed passes are subsequently referred to as "static" runs. Maximum recorded cell pressures for high speed, "dynamic" runs were divided by the maximum pressure on the same cell during a corresponding static run yielding a calculated impact factor. Impact factor profiles were generated; an example is shown on Figure 4. The natural frequency and damping ratio of the scraper is estimated from impact factor records.

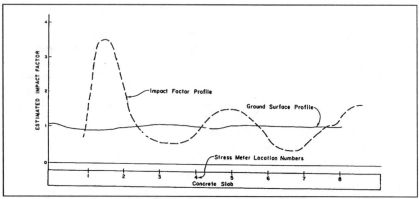

Figure 4.. Impact Factor Profile (Gemperline 1984, 1985).

The fractal dimension is determined for profiles representing the roughness of the field study surfaces using the computer program COMP.BAS (Gemperline 1994).
A comparison is presented of the maximum impact factor calculated by quarter-car-simulation with maximum impact factor calculated from field study measurements. This comparison gives an indication of the uncertainty expected in impact factor calculation.

Impact factor profiles were calculated for a model scraper traversing half-mile fractal pathways using quarter-car-simulation discussed earlier. Maximum impact factors were calculated for conditions representing field study pathway roughness (expressed as fractal dimension) and vehicle speeds. A comparison of the maximum impact factor calculated by quarter-car-simulation with the field observation is presented on Figure 5. The line in this figure represents the ideal condition of equality between the calculated and observed values.

The maximum impact factor observed in the field study is expected to be less than, or equal to, the maximum impact factor calculated by quarter-car simulation. Computer program QCAR.BAS calculates impact factors at approximately 0.015 m (0.05 ft) intervals and models a scraper traversing a 0.8 km section (0.5 mi) (Gemperline 1994). The field study measurement permits the calculation of impact factors at 1.4 m (4.5 ft) intervals over a length of 12.6 m (41.5 ft). Large vehicle impacts may occur between field measurement points or beyond the end of the test section. Consequently, larger impact factors are expected to be predicted by the numerical method than are observed in the field study. This is observed in 88 of the 92 comparisons presented on Figure 5. The four inconsistencies are explained by the uncertainty inherent in fractal dimension estimates representing field study surface roughness and also the uncertainty in estimates of quarter-car characteristics.

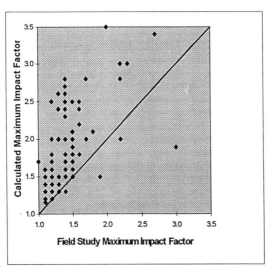

Figure 5. A comparison of the maximum impact factor calculated by quarter-car-simulation with field observations.

The agreement between calculated and observed impact factors supports the validity of using the quarter-car model with fractal representation of pathway roughness to calculate model vehicle impact on rough pathways.

Method Application in Engineering Practice

Vehicle impact factors used in design of buried structures may be calculated using the presented method for a reasonable worse case vehicle type, speed, and pathway roughness. Conversely, the model may be used to develop controls which limit vehicle type, speed, and ground surface roughness so that vehicle impact during construction does not govern design.

The following parameters are required to define the problem.

1. Fractal dimension representing pathway roughness, D.
2. Vehicle Speed, V.
3. Quarter-vehicle mass, M.
4 Quarter-vehicle spring constant, K.
5. Quarter-vehicle damping ratio, D_c.
6. Tire envelopment length, T.
7. Impact Factor, I.

Dimensional analysis yields the following complete set of dimensionless terms to define the problem.

1. Fractal Dimension, D.
2. Dimensionless Velocity, $V\dfrac{\omega}{g}$
3. Dimensionless tire envelopment length, $T\dfrac{\omega^2}{g}$
4. Damping ratio, D_c.
5. Impact Factor, I.

ω is the natural frequency of the scraper associated with vertical bouncing and g is the acceleration of gravity.

Figures 6 and 7 show the relationship between these variables for conditions which result in impact factors of 3.0 and 1.5 respectively. These relationships were developed using program QCAR.BAS. The lines represent limits to the condition $0.00 \le D_c \le 0.05$. If it is assumed that the vehicle of concern has at least 5 percent critical damping then the upper bound line may be used conservatively to estimate conditions which result in the impact factor represented by the figure. Note that tire envelopment length appears to have an immeasurable effect and for practical purposes can be ignored.

If the vehicle is loaded, as was the case with the scrapers used in the field study, the natural frequency must be estimated for this condition. Vehicle

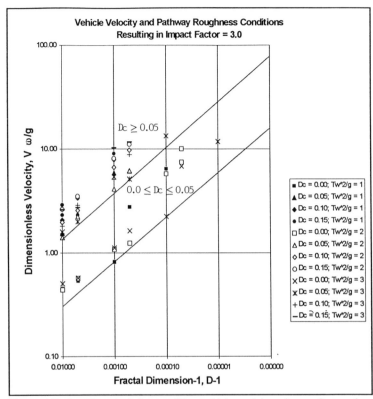

Figure 6 Design Chart depicting conditions which result in an impact factor of 3.0; data points represent the results of quarter-car-simulation.

manufacturers do not routinely measure and publish vehicle natural frequency and damping ratio. In this study, considerable data was available from field studies to assess these values. Data from other studies might provide useful information about other vehicles. A possible alternative method of evaluation is to use video tapes of the vehicle of concern encountering surface anomalies. Visually observe the bouncing frequency and changes in subsequent cycle bouncing amplitude and then estimate the natural frequency and damping coefficient. Validating this approach is beyond the scope of this work.

After the design criteria and vehicle natural frequency are defined, Figures 6 and

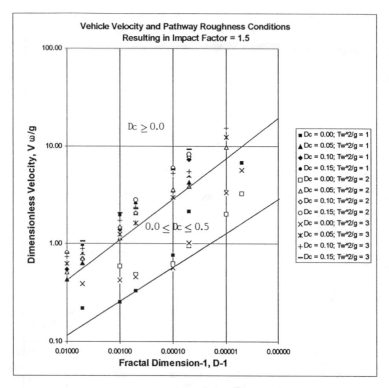

Figure 7. Design Chart depicting conditions which result in an impact factor of 1.5; individual data points represent the results of quarter-car-simulation..

7 may be used to determine the conditions of velocity and surface roughness which will result in unacceptable impact factors. The following example describes the process.

Example Use of Design Charts

The following example simulates design chart use for a case familiar to the authors. The objective is to set safe limits on vehicle speed and ground surface roughness at a pipe crossing.

It is desired to construct a haul road from a barrow area to a dam currently under construction. The route traverses a quarter mile section of 3.0 m (10 ft) diameter concrete pipe which, in some locations, has only .9 m (3 ft) of cover. Fully loaded scrapers, observed to have a natural frequency of approximately 6.3 rad/s (1 c/s), are expected to use the haul road. It is determined that the pipe could be safely traversed if quarter-vehicle impacts do not exceed 1.5 times the static weight of the fully loaded scraper. Limits must be placed on vehicle velocity and ground surface roughness to assure this condition is not exceeded.

The dimensionless variables describing the problem are:

1. Fractal Dimension (representing road roughness), D = unknown

2. Dimensionless Velocity, $V\frac{\omega}{g}$ = unknown

3. Dimensionless tire envelopment length, $T\frac{\omega^2}{g}$ = effect is insignificant and will be ignored.

4. Damping ratio, D_c assumed greater than 0.05
5. Acceptable Impact Factor, $I = 1.5$

The area above the upper line on Figure 7 represents unfavorable conditions, i.e., conditions expected to correspond to impact factors greater than 1.5.

A smooth road surface is difficult to maintain and monitor. It is decided to determine the acceptable maximum vehicle velocity for a relatively rough surface and grade the haul road as needed to assure this roughness condition never occurs. A road surface characterized by a fractal dimension of 1.0005 is rough as indicated by the fractal profiles shown on Figure 3. Normal grading efforts are expected to yield a smoother surface. The maximum value for the dimensionless velocity term, from Figure 7, is approximately 1.5. Therefore the maximum scraper velocity permissible at the pipe crossing is approximately 5 mi/h (8 km/h). Road roughness can be measured by developing a profile and calculating its fractal dimension using program COMP.BAS (Gemperline 1994).

Conclusion

This paper presents a new method of predicting vehicle surface impact which is necessary for estimating live loads on buried pipe. Understanding the mechanism of impact suggests methods of control. For example, if a buried pipe is designed using an impact factor equal to 1.5 to account for construction vehicle loading, then the maximum pathway roughness of the pipe crossing, expressed as a fractal dimension, and maximum vehicle speed may be identified as construction controls.

It is shown that fractal representations of real road profiles may be used with numerical quarter-car-simulation to yield reasonable estimates of maximum vehicle impact. It is demonstrated that the impact factors measured in a field study can be reasonably explained by numerical quarter-car-simulation while characterizing the pathway profile as a fractal shape.

It is suggested that the International Roughness Index is directly related to the fractal dimension. This is recommended as a topic for future study.

Appendix. References

Barnsley, M. F., Devaney, R. L., Mandelbrot, B. B., Peitgen, H-O, Saupe, D., and Voxx, R.F. (1988). "The Science of Fractal Images." Springer-Verlag, Inc., New York, N. Y., 312.

Feder, J. (1988). "Fractals." Plenum Press, New York, N. Y., 283.

Gemperline, M. C. (1984). "Results of the Currant Creek Pipeline Impact Loading Study,Central Utah Project." *REC-ERC-83-16*, U.S. Bureau of Reclamation, Engineering and Research Center, Denver, Co., 93.

Gemperline, M.C. (1985). "Construction-induced Dynamic Pressure and Corresponding Impact Factors for Pipelines." *Proc., Advances in Underground Pipeline Engineering*, ASCE, University of Wisconsin, Madison, Wisc., 1-10.

Gemperline, M. C. (1994). "Vehicle Induced Loads on Fractal Road Surfaces." *Dissertation*, Colorado State University, Fort Collins, Co.

Mandelbrot, B. B. (1983). "The Fractal Geometry of Nature." W. H. Freeman and Company, New York, N. Y., 468.

Sayers, M. W., Gillespie, and Queiroz, C.A.V. (1986). "Transportation Research Record 1084." Transportation Research Board, National Research Council, Washington, D.C., 76-85.

Earthquake Loss Estimation Techniques for Pipelines

Donald Ballantyne, P.E.[1]

Abstract

Estimates of pipeline losses for earthquake scenarios are useful evaluation tools in assessing post-earthquake pipeline system function. People responsible for both system planning and emergency response use loss estimates.

Different types of permanent ground deformation, PGD, hazard information is required for continued development of pipeline loss estimation methods. This paper summarizes approaches for pipeline earthquake loss models used over the past 20 years considering wave passage and PGD (liquefaction, settlement, and landslide). Current trends in loss modeling are explored such as:

- PGD net displacement for segmented pipelines.

- Soil block dimension influence on vulnerability.

- Methodology to define areal extent of PGD.

The need for PGD information associated with non-lateral spread and non-tectonic related ground movement is discussed. Approximately 900 water transmission and distribution pipeline failures occurred in the Northridge earthquake with almost no liquefaction and no fault expression.

Introduction

Continued development of pipeline earthquake loss estimation methods is important; loss estimation is a useful tool in assessing the risk of pipeline system failure in earthquakes, developing earthquake mitigation programs, and developing emergency response programs.

---

[1]Associate, Dames & Moore, 2025 First Avenue, Suite 500, Seattle, Washington 98121

This paper briefly reviews pipeline earthquake loss estimating methods used over the past 20 years. It then presents a proposal for a methodology for earthquake pipeline loss estimation associated with liquefaction induced lateral spreading. Finally, the proposed methodology is discussed, and the future direction of pipeline earthquake loss modeling is posed.

One of the objectives of having such a methodology available is for application in regional loss studies. With that in mind, it is important to minimize the number of parameters required to achieve a meaningful result. The methodology proposed herein tries to limit the number of those variables.

This proposed methodology has been developed working with the Fragility Task Committee of the ASCE Technical Council on Lifeline Earthquake Engineering. The objective of the committee is to establish a methodology for earthquake loss modeling of lifelines. Once an approach is established, it will provide a format for acquisition of damage data from future earthquakes. The methodology is designed to allow upgrading of components as new information is developed.

Summary of Pipeline Loss Estimation Approaches

This section summarizes the development of pipeline earthquake loss estimation as well as methods used to estimate PGD in support of pipeline loss estimation. Empirically based water pipeline damage algorithms reviewed in this paper were initiated in Japan and refined in both the United States and Japan as described below.

Initial Japanese Efforts - A method for estimating pipeline earthquake losses was introduced by Professor Katayama in the mid-1970s. He developed pipeline damage algorithms relating pipe failures and earthquake peak ground acceleration, PGA. His damage algorithm enveloped loss estimates with specific estimates dependent on the soil characteristics, such as liquefaction susceptibility (Katayama, 1975).

Segregation of Wave Propagation and PGD Effects - Pipeline damage from wave passage, fault rupture, and liquefaction was segregated by Eguchi in the early 1980s. He gathered empirical damage data from over 20 earthquakes worldwide, but was able to develop the most significant relationships based on damage data from the 1971 San Fernando Earthquake (Eguchi, 1982). He assumed that Modified Mercalli Intensity, MMI, was an indicator of wave propagation effects on pipelines. For cast iron pipe, the relationship between MMI and the failure rate was established. He then related damage rates for other pipe materials to cast iron for one intensity, establishing a family of damage algorithms. For that same earthquake, he also developed pipeline damage rates for liquefaction conditions for a family of pipe materials, but did not relate them to PGD from liquefaction. Finally, he developed damage rates based on the proximity to and displacement of fault offset.

Differentiation of Pipeline Failure Consequences - In the late 1980s, Ballantyne segregated pipeline damage into pipeline breaks and pipeline leaks (Ballantyne, 1990). This information became valuable for use in deterministic post-earthquake water system hydraulic modeling. As part of the same study, the question of areal extent of liquefaction along a pipeline corridor or in a microzone had a very significant effect on loss estimation results. It became clear to this author in that study that PGD-related pipeline damage would often control the overall system performance, and that pipeline unit damage rates for liquefaction and PGD were an order of magnitude greater than for wave passage.

Quantification of Liquefaction Related PGD - In 1987, Youd and Perkins published the Liquefaction Severity Index, LSI, approach to estimate the maximum PGD at a given site for a particular earthquake scenario. Initially, this information was not applied as a pipeline damage estimation tool (Youd, 1987). More recently, Bartlett and Youd have refined the LSI method with the Multiple Linear Regression analysis method, MLR, for estimating maximum PGD from liquefaction related lateral spreading (Bartlett, 1992).

PGD Soil Block Geometry - In 1992, M. O'Rourke identified the significance of lateral spread block geometry on the extent of continuous (welded steel) pipeline vulnerability (O'Rourke, 1992). It still remains that segmented pipe PGD-related damage is controlled by net PGD. A major problem related to this approach was being able to estimate the block size and ground breakup pattern.

San Francisco Liquefaction Study - Following the 1989 Loma Prieta Earthquake, the City and County of San Francisco selected a project team to estimate utility losses that might occur in liquefiable soil areas around the periphery of the city for a magnitude 8.3 San Andreas Earthquake (Harding Lawson, 1992). The project team developed damage algorithms relating pipeline damage to PGD using empirical damage data from the 1971 San Fernando Earthquake, the 1989 Loma Prieta Earthquake (including the San Francisco Marina District and the City of Santa Cruz data), and the 1983 Nihonkai Chubu, Japan Earthquake. It was found to be very difficult to find damage data that included a record of PGD.

The San Francisco project team geotechnical engineers used Tohata's (1990) method to estimate the extent of lateral PGD. The assumption was made that the entire soft soil area would liquefy based upon the soil properties, large peak ground accelerations, and long duration. While this assumption is for the most part valid for large, near-field earthquakes, it becomes inappropriate for smaller or more distant events. This issue is further discussed later in this paper. This PGD displacement and areal extent information was then passed along to the earthquake lifeline pipeline project team members for use in estimating pipeline damage.

Strain-Related Pipeline Vulnerability Assessment - Studies of the Greater Vancouver Regional District, GVRD, water system in British Columbia (Kennedy/Jenks Consultants, 1993), and BC Gas transmission system (Honegger, 1994) have used pipe strain induced by PGD as a indicator of vulnerability. In the GVRD study of their welded steel pipe system, PGD was estimated using the LSI approach. Pipeline strain was calculated considering the length along the pipeline where PGD was expected and the pipe wall thickness to radius ratio, t/R. This approach estimates the relative pipeline vulnerability but does not estimate the expected total number of failures.

The BC Gas study applied a similar approach. Segments of pipe were identified where there was a high liquefaction susceptibility and lateral spread potential (slope greater than 0.5 percent or a location near a free face). In these locations, an estimate was made of the expected PGD. This information was then input into a finite element analysis of the pipeline segment. The findings indicated that pipelines with 90 degree bends in areas where significant PGD was expected were the most vulnerable. Again, an estimate of the number of failures could not be made using this method.

Uncertainty of Areal Extent of Liquefaction - The San Francisco project had an extensive geologic data set available. Pipeline earthquake loss estimation projects for the City of Everett, Washington (Ballantyne, 1991) the Greater Vancouver Regional District in Vancouver, British Columbia (Kennedy/Jenks Consultants, 1993), and the Portland Bureau of Environmental Services, Portland Oregon, (Dames & Moore, 1994) made pipeline loss estimates considering liquefaction as the primary damage mechanism. In these three cases the project team geotechnical engineers were asked to make estimates on the areal extent of liquefaction. In this author's opinion there was a significant level of uncertainty associated with those estimates because of the lack of available methods. These three studies all considered the liquefaction susceptibility of alluvial deposits in similar geotechnical settings along rivers ultimately discharging into the Pacific Ocean. The estimates for areal extent of liquefaction varied by a multiple of seven. The areal extent of liquefaction estimate is directly related to the damage estimate, so the degree of certainty is very important.

Pipe Type Categorization Issues

Damage algorithms are developed for each pipe type category. Each pipe "type" category may include a range of materials and designs that will perform differently when subjected to similar earthquake hazards. Each pipe type category would likely include a range of pipe classes because 1) empirical pipe damage data is usually not gathered in enough detail to identify pipe class, and 2) there is too little data to develop a algorithm for each pipe class. Similarly, joint types may differ. For example, for welded steel pipe, in both the 1971 Sylmar and 1994 Northridge earthquakes oxyacetylene gas welded pipe performed poorly. Following the 1971 earthquake, a separate damage algorithm had been developed for oxyacetylene welded pipe. In the Northridge earthquake, modern welded steel water transmission pipe performed poorly. This issue is not addressed in existing pipeline damage algorithms. Pipe damage algorithms are only as good as the data used for development. The resulting loss estimates are only as good as the algorithm, and the appropriate application of that algorithm.

Liquefaction Areal Extent Methodology

This section presents a methodology for estimating pipeline losses from PGD. It includes consideration for estimating liquefaction susceptibility, probability, and areal extent, and applies that information to pipeline loss estimation.

Liquefaction Susceptibility - This reduced set of variables has been selected so that the methodology is applicable to conduct regional loss studies. First, establish three levels of liquefaction susceptibility: none or low, medium, and high. From a loss estimation perspective, none or low can be ignored, and medium susceptibility is usually ignored as it typically only represents less than 1 percent of estimated losses. In general, liquefaction susceptibility would take into account (criteria in parenthesis for high susceptibility) groundwater table depth (< 12 feet below grade), blow count N_1 (< 12), and depth to liquefiable deposit (< 25 feet below grade).

Liquefaction Probability and Magnitude Scaling Factor - The curves shown in Figure 1 define the probability of liquefaction, P_L, as a function of peak ground acceleration, PGA. These curves are defined as the probability of liquefaction occurring at a point in a soil mass in controlled field test conditions.

A family of curves is proposed for a range of earthquake magnitudes. Ultimately, separate curves would be required for a range of soils such as clean sands, silty sands, etc.. These curves could likely be developed from existing information such as Liao (1986). The PGA scale has purposely not been provided. These curves show the form expected in a finalized methodology. Additional curves can be added for earthquakes of other magnitudes. These curves are included to address the magnitude scaling factor compensating for the number of earthquake cycles for earthquakes of varying magnitude.

Liquefaction Areal Extent, P_{AE} - Apply a factor to estimate areal extent of liquefaction, P_{AE}, using the probability of liquefaction, P_L. P_{AE} is the conditional probability that an arbitrary surface location will exhibit the results of liquefaction below grade.

$$P_{AE} = A_L / A_T \tag{1}$$

where:

A_L = Area of liquefaction where it is evident that liquefaction has occurred by field observation such as where sand boils appear, or ground has subsided or moved laterally as evidenced by cracking.

A_T = Total area with same susceptibility to liquefaction (low, moderate, or high as defined above), subjected to the same approximate PGA.

It is the intent that this factor use the probability of liquefaction developed for a volume of soil in a controlled condition and correct it to estimate the areal extent of liquefaction occurring in an earthquake. The occurrence of liquefaction is identified by field observation, as that is the basis on which most pipeline damage data has been generated.

The family of curves relating P_L and P_{AE} is expected to take the general form shown in Figure 2. Surface expression of liquefaction is expected to be influenced by the thickness of the layer of liquefiable (and liquefied) material. Other conditions may also have an effect such as variability of the liquefiable deposit. The final family of curves should reflect as many variables as applicable.

<u>Maximum Permanent Ground Deformation</u> - Apply the Liquefaction Severity Index methodology, LSI, to estimate the maximum PGDs within the study area. Correct the LSI for slope or proximity to a free face. If there is adequate information, apply the Multiple Linear Regression, MLR, analysis technique to estimate the maximum PGD. (Note that it is suggested only to correct LSI for slope/free face proximity, and not subsurface data because slope/free face proximity information is more readily available in Geographic Information System, GIS, format.)

Map the mean PGD, PGD_m (one-half times the maximum PGD as a starting point) based on the LSI or the MLR.

<u>Pipe Parameters and Permanent Ground Deformation Pipe Exposure</u> - Measure the pipe length, L_P, of each pipe type category (defined by material and joint type) within each area with a defined range of PGD_m (such as 0-2 cm; 2-10 cm, etc.). Note that pipe type category may include more than one type of pipe.

<u>Pipeline Damage Algorithms</u> - Read the failure rate for the average of the range of PGD_m for each pipe type using the appropriate pipe damage algorithm, presented in terms of percent of length requiring replacement P_R(as a function of PGD), or failures per km F_K(as a function of PGD). Pipe damage algorithms are shown in Figures 3 and 4.

<u>Pipeline Repair / Replacement</u> - Calculate the pipe length to be replaced:

$$P_{AE} \times L_P \times P_R = \text{Pipe Length to be Replaced} \qquad (2)$$

for each PGD_m range/pipe type category for each range of PGD. Alternatively, calculate the total number of expected pipe failures:

$$P_{AE} \times L_P \times F_K = \text{Expected Pipe Failures} \qquad (3)$$

for each PGD_m/pipe type category.

Further Corrections - Provide further corrections for: 1) pattern of liquefaction deformation, 2) pipe orientation to PGD, 3) corrosion condition/maintenance history, 4) number of connections per unit length, and 5) for welded steel pipe, wall thickness/pipe radius ratio. Additional research is required to develop and expand upon these correction factors.

Discussion and Unresolved Issues

This proposed methodology identifies two concepts that will have to be developed with time, areal extent of liquefaction, and mean PGD, PGD_m. As it is defined, areal extent of liquefaction would have to be developed using empirical field data gathered following earthquakes, in conjunction with liquefaction hazard maps for the same areas that have defined liquefaction susceptibility relationships.

PGD_m is selected as being representative of the PGD which is seen by pipelines in the immediate area. Trying to develop a density function for the distribution of PGD, using the LSI as a maximum, was not thought to be useful due to the uncertainties associated with the density function as well as the pipeline damage algorithm itself. Using one-half the PGD is considered only a starting point.

It is recognized that pipeline damage associated with liquefaction is not only related to PGD, but to the breakage pattern and size of soil blocks that develop when lateral spreading occurs. Methods are needed to enable lifeline earthquake engineers to determine likely soil block patterns, and then to relate pipeline damage to those patterns. Other parameters will also affect pipe strain as it is related to soil block patterns and movement including the coefficient of friction between the soil and pipe.

Ultimately, it may be appropriate to revert back to the generalized damage pipeline damage estimation approach originally proposed by Katayama. It is very difficult to clearly define soil parameters and sources of PGD along every length of pipeline. Techniques to quantify liquefaction, lateral spread, landslide, and fault displacement-associated PGDs are currently available. However, there are two examples where mapping of these hazards have failed. First, in the study of the Seattle water system (Ballantyne, 1990), there were a number of clusters of pipeline failures following the 1949 and 1965 Seattle earthquakes in areas that today are mapped as competent non-liquefiable soils. Based on evaluation of leak repair records, it is likely that there was localized liquefaction in those areas resulting in some PGD and pipeline failure. There was no reported indication of liquefaction on the surface.

Second, in the Northridge Earthquake, there was significant surface cracking throughout the San Fernando Valley. There was on the order of 700 transmission and distribution pipeline failures in the valley (as well as an equal number of service failures). However there was only limited liquefaction reported, and not widely distributed. The cracking may be related to either lurching associated with wave passage, slope instability, or tectonic movement. Until these ground cracking mechanisms can be understood and quantified, it will be difficult to generate meaningful pipeline estimates.

One of the objectives of estimating earthquake damage is to appraise post-earthquake system serviceability. A program to develop a standard for earthquake loss estimation, funded by the Federal Emergency Management Agency, proposes a relationship between average break rates for a system for a given earthquake scenario, and a serviceability index (RMS, 1994). Hazard and pipeline vulnerability information is still required to calculate the average break rate system wide. However, uncertainty in estimating pipeline failures to a small section of the system would likely not have a significant effect on the average system break rate, and the resulting serviceability index. This approach may resolve some of the issues discussed herein. The approach is currently being beta tested in the Portland, Oregon metropolitan area.

Post-Earthquake Soil Failure/Damage Data Needs

If we focus on areas of likely liquefaction-related PGD, to make the proposed methodology a reality, specific soil failure and pipeline damage data will be required such as the following:

- Inventory of areas where liquefaction has occurred
- Liquefaction susceptibility maps
- Definitive mapping of PGD
- Pipeline locations relative to PGD
- Pipeline damage mechanisms, and damage descriptions (leak, break, leakage rate)

Post-earthquake investigators are urged to gather this type of data. Some of this information currently exists such as that developed from the Loma Prieta and Nihonkai Chubu earthquakes. The Japanese lifeline community has been more aggressive than their U.S. counterparts in gathering this type of data.

Conclusions

Pipeline earthquake loss estimation has developed from simple damage algorithms incorporating damage from "all" earthquake hazards, to more sophisticated algorithms that segregate damage mechanisms.

A methodology has been proposed to enable the lifeline community to better estimate pipeline damage from PGD in future earthquakes as a damage data base is developed using the proposed parameters.

Acknowledgements

We thank the National Center for Earthquake Engineering Research for support of expenses associated with meetings of the ASCE TCLEE Fragility Task Committee. Members of the ASCE TCLEE Fragility Task Committee are acknowledged, listed alphabetically after the chair include: Professor Anshel Schiff, Chair, Mr. Donald Ballantyne, Mr. James Clark, Dr. C.B. Crouse, Mr. John Eidinger, Professor Anne Kiremidjian, Professor Michael O'Rourke, Dr. Douglas Nyman, Mr. Alex Tang, and Dr. Craig Taylor.

References

Ballantyne, D.B.; April 4, 1994; *Development of Liquefaction Areal Extent Evaluation Methodology*; Memorandum to the ASCE TCLEE Fragility Task Committee.

Ballantyne, D.B.; Heubach, W.F.; Crouse, C.B.; Eguchi, Wong, F.; R.T.; Werner, S.; Ostrom, D.; 1991, *Earthquake Loss Estimation for the City of Everett, Washington's Lifelines*, funded by USGS award number 14-08-0001-G1804, Kennedy/Jenks/Chilton Report No. 906014.00, Federal Way, Washington.

Ballantyne, D.B.; Taylor, C.; 1990; *Earthquake Loss Estimation Modeling of the Seattle Water System*, USGS Grant Award 14-08-0001-G1526, Kennedy/Jenks/Chilton.

Bartlett, S.F.; and T.L. Youd; 1992; "Empirical Prediction of Lateral Spread Displacement"; *Proceedings of the Fourth Japan-U.S. Workshop on Earthquake Resistant Design of Lifeline Facilities and Countermeasures for Soil Liquefaction*, Hawaii, Report No. NCEER-92-0019.

Dames & Moore; 1994; *Seismic Vulnerability Assessment of the Portland Bureau of Environmental Services Sewer System*; being prepared for Brown & Caldwell and the Portland Bureau of Environmental Services; Project currently underway.

Eguchi, R. T.; *1982; Earthquake Performance of Water Supply Components During the 1971 San Fernando Earthquake*; Prepared for the NSF, J. H. Wiggins Company, CA.

Harding Lawson Associates, Dames & Moore, Kennedy/Jenks Consultants, EQE Engineering, 1991, *Liquefaction Study, San Francisco, California*, prepared for the City and County of San Francisco, Department of Public Works, San Francisco, California.

Honegger, Douglas, 1994, *Assessing Vulnerability of BC Gas Pipelines to Lateral Spread Hazards*, presented at the National Center for Earthquake Engineering Research Fifth US-Japan Workshop on Earthquake Resistant Design of Lifeline Facilities and Counter-Measures Against Soil Liquefaction, Snowbird, Utah, September.

Katayama, T.; Kuho, K.; Sato, N.; 1975; "Earthquake Damage to Water and Gas Distribution Systems," *Proceedings of the 1st U.S. National Conference on Earthquake Engineering*. Berkeley, CA: Earthquake Engineering Research Institute.

Kennedy/Jenks Consultants in association with EQE Engineering and Design; 1993; *A Lifeline Study of the Regional Water Distribution System*; Prepared for the Greater Vancouver Regional District, Vancouver, British Columbia.

Liao, S.S.C.; 1986; *Statistical Modeling of Earthquake-Induced Liquefaction*; Ph.D. Thesis, Massachusetts Institute of Technology, Department of Civil Engineering.

Risk Management Solutions, 1994, *Development of a Standardized Earthquake Loss Estimation Methodology*, Draft Technical Manual, 95% Submittal, prepared for NIBS..

O'Rourke, M. J.; C. Nordberg; 1992, *Longitudinal Permanent Ground Deformation Effects on Buried Continuous Pipelines*, NCEER-92-0014, Buffalo, N.Y..

Towhata, I., K. Tokida, Y. Tamari, H. Matsumoto, and K. Yamada.; 1990; "Prediction of Permanent Lateral Displacement of Liquefied Ground by Means of Variational Principle", *Proceedings of the Third Japan-U.S. Workshop on Earthquake Resistant Design of Lifeline Facilities and Countermeasures for Soil Liquefaction;* San Francisco.

Youd, T. L., D. M. Perkins, 1987, "Mapping of Liquefaction Severity Index" *Journal of Geotechnical Engineering, Vol. 113, No. 11*, pp. 1374-1392.

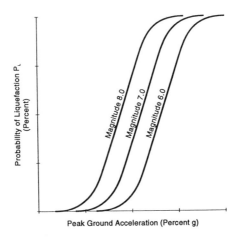

Figure 1
Probability of Liquefaction, P_L, Versus PGA for a Range of Earthquake Magnitudes

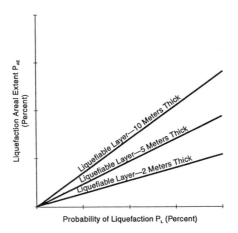

Figure 2
Areal Extent of Liquefaction, P_{AE}, Versus Probability of Liquefaction, P_L, for a Range of Liquefiable Layer Thicknesses

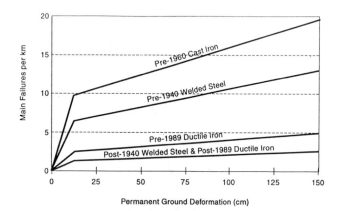

Figure 3
Failures Versus Permanent Ground Deformation
for Water Pipelines
(Harding-Lawson, 1990)

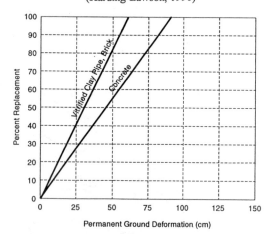

Figure 4
Percent Replacement versus Permanent Ground Deformation
for Gravity Sewer Pipelines
(Harding-Lawson, 1990)

ADVANCES IN PIPELINES FOR DESALINATED WATER IN SAUDI ARABIA

Abdullah Al-Shaikh, Mohammed Al-Amry and Saleh Al-Najdi.[1]

Abstract

The industrialization in Saudi Arabia, which began in 1975, in order to catch up with the industrial world, has resulted in major construction of large Seawater Desalination and Power Plants, and water transmission pipelines. Saline Water Conversion Corporation (SWCC) is the Government Agency responsible for constructing, operating and maintaining Desalination Plants and Water Transmission Pipelines.

The Kingdom of Saudi Arabia is a world leader in the production of desalinated seawater. There are 23 desalination plants at 14 different locations on the eastern and western coasts. The total production of these plants exceeds 500 MIGD of water and 3,600 MW of electrical power and the total length of water transmission pipelines is about 3,722 km, with diameters varying from 32 to 84 inches (SWCC 1993).

The goal of this paper is to describe the selection of material of construction for a new pipeline based on the experience gained by the SWCC during the last two decades in the operation of transmission pipelines, and with construction for unique hydraulic conditions and geotechnical concerns which have been identified and addressed during this period.

---

[1]Saline Water Conversion Corporation, P.O.Box. 5968, Riyadh - 11432.

Introduction

The pipeline transmission system plays a very important role in the desalination industry especially in a country such as Saudi Arabia where it is necessary to transport water for hundreds of kilometers across the country. Selection of the suitable type of pipeline material is influenced by many factors, including:

a) The availability of the required pipe diameters.
b) The pressure rating.
c) The type and number of joints.
d) Transportation and handling problems.
e) The effect of the external environment on the pipe.
f) The cost of the pipe.

SWCC water transmission systems relevant for this paper can be divided basically into two groups as follows:

A) Pipelines under Operation
B) Pipelines under Construction

Coating and Lining

A - Coating

Coating is a very important item for the protection of pipes especially in a country like the Kingdom of Saudi Arabia, where the environment is very aggressive, and the climate is hot and humid.

Since 1978 SWCC has been installing pipes in Saudi Arabia using various types of coatings following the international developments in coating techniques such as coal tar epoxy for concrete pipe, extruded Polyethylene, sintered Polyethylene, liquid epoxy, fusion-bonded epoxy and finally, three-layer Polyethylene for steel pipes.

B - Cement Mortar Lining

Almost all of SWCC pressurized water transmission systems are made of high-tensile steel AP1 (Grade B to X60). These pipelines are internally lined with cement mortar lining using either spinning process or centrifugal process for application.

Cement mortar lining for straight pipes is shop applied using the spinning process where the cement mortar is filled in while the pipe is rotated and the mortar is compacted due to centrifugal force according to AWWA C-205 (1985). The pipe specials like bends, field joints and pipe spools are cement lined using the centerline process which is accomplished by spraying mortar in small particles from a rapidly revolving dispensing head against stationary pipe wall according to AWWA C-602 (1976).

Case Study

Shoaibah - Jeddah Water Transmission System

The water transmission system between Shoaibah and Jeddah forms part of the development of the Shoaibah desalination complex. The water output capacity of the desalination complex is being increased by the addition of a new plant currently under construction. The additional output will be used to supply up to 100 MIGD of water to Jeddah, and to increase by 40 MIGD to a maximum of 94 MIGD the water supplied to Makkah and Taif from the complex.

New supply to Jeddah

For the new supply to Jeddah, the works to be implemented include the following:

- Provision of one new pumping station at Shoaibah which contains 8 booster and 8 main pumps, and associated pipework, valves, surge protection, scraper launching station, control and instrumentation systems and ancillaries.
- Supply, lining, coating and installation of approximately 122 km of 1524 mm outside diameter steel twin buried pipelines.
- Three line/cross connection installations, pipeline washouts, air-valves and cathodic protection system.
- Terminal works at Jeddah including a scraper receiving station.

In the future the output from the Jeddah pumping station will be increased to 120 MIGD, and the twin pipelines have been designed for this flow.

Route options

Eleven route options have been considered and investigated. Finally, one route was preferred based on these advantages:

a) Shortest route:
b) Lowest total pumping head:
c) Lowest capital and operating costs.

Ground conditions

The depth of soil cover to the top of the pipe will generally be 1.30 m, made up of 0.8 m below ground and a 0.5 m berm above ground. Sabkha deposits are encountered north of the Shoaibah site. Along the coast, Sabkha deposits are present across a zone some 2 km to 5 km wide extending inland. These deposits comprise calcareous, gypsiferous and saline silts, clays and sands. Saline ground water is present either close to or above ground surface level. The pipelines in Sabkha areas are constructed on embankment fill materials over the Sabkha.

At the coast, groundwater levels are at or above the ground surface. Inland, as the desert plain rises, groundwater is encountered at greater depths below the surface. In Jeddah, ground water tables are variable.

Selection of pipeline material

Five (5) primary pipe materials were considered and evaluated for technical and price suitability, viz. reinforced concrete (cylinder), prestressed concrete, ductile iron, steel and glass reinforced thermosetting resin. Steel pipe was selected for technical, operation, and maintenance considerations and because the material is the cheapest for the duty considered.

Selection of pipeline size and pumping arrangements

Options for pipeline size and pumping arrangements were identified and compared technically and economically. The results of the evaluation indicated that a single pumping station at Shoaibah, with no intermediate pumping station between Shoaibah and Jeddah, and twin 1524 mm outside diameter steel pipelines, would satisfy the required duty of transferring 120 MIGD to Jeddah.

Hydraulic gradient

The pipeline was designed for the worst hydraulic profile. The hydraulic gradient was established using the Colebrook-White equation for friction gradient based on a Ks(roughness) of 0.15 mm as being approximate to a 50-year design life when compared to a pipe roughness at commissioning of 0.03 mm.

Design hydraulic profile

The design flow was taken as 120 MIGD, 60 MIGD per pipeline. For this flow, 3 duty pumps would feed each pipeline with 1 standby pump per pipeline. An increase in flow rate through the system is possible if 4 pumps feed each pipeline, i.e., 3 duty pumps plus the standby pump. Under this condition, the flow in each pipeline will increase to about 65 MIGD, i.e. 130 MIGD total.

This possible (but unlikely) operating condition was used as a reasonable basis on which to base the design hydraulic profile.

Surge protection

Surge analyses were carried and the air vessel volumes were determined to prevent execessive negative and positive pressures occurring in the pipeline during surge events.

Pipe pressure ratings

The pressure ratings of the pipeline were matched to the design hydraulic profile taking into account the ground levels. By increasing the pipe pressure rating, a more economic and efficient pipe design was achieved. This is because thinner steel wall thickness in higher grades of steel are cheaper overall in comparison to thicker plates in lower grades of steel.

Pipe design

A maximum D/t ratio not exceeding 150 was selected to limit flexibility for transportation and laying. Pipe thicknesses as used in the pipe design were standard thicknesses to API 5L. The pressure ratings selected that fitted to the pipeline and decreased in a logical order were:

Grade X 46	t = 15.88 mm	Grade X 42	t = 15.88 mm
Grade X 42	t = 14.28 mm	Grade X 42	t = 12.70 mm
Grade X 42	t = 11.91 mm	Grade X 42	t = 11.13 mm
Grade B	t = 11.13 mm		

The structural adequacy of the pipes was checked for various loading conditions.

Manual calculations

Manual Calculations were carried out for each pipe section. The yield stress of each steel grade was selected from API 5L. The allowable percentage yield stress at working pressure was taken as 50% of yield stress (Fy). The minimum wall thickness was checked based on tolerance for grade B of 10% for welded pipes and for X42 and higher of 8% for welded pipes. The hoop stress was checked based as an allowable design wall stress of 50% Fy. Deflection and buckling of the pipe were calculated for loads and pressures on the pipelines arising from soil, uniform and live loads (from wheel loads) and ground water table based on AWWA M11 design manual and CIRIA 78 report. The allowable deflection of 2% for mortar lined and coated pipe was not exceeded.

Line/cross connection installations

Three line/cross connection installations are incorporated, one at Shoaibah and at 40 km and 80 km from Shoaibah. Their inclusion permits increased flows in comparison to that possible with two single pipelines if one has to be taken out of service for maintenance. In order to allow more throughput during an abnormal operating condition when cross connections are being used and a length of pipe is out of service, up to 60% of yield stress was allowed.

Conclusion

1. SWCC has adopted the three-layer Polyethylene system as preferable protection for steel pipes against mechanical damages and corrosion attacks.
2. Surge vessels and pressure relieve systems are still used by SWCC as means for pipeline protection against transient pressure conditions during power failure and sudden valve closures.

References
1. Annual Report 1993, Saline Water Conversion Corporation.
2. "Cement Mortar Lining and Coating for Steel Water Pipe - 4 inches and larger, Shop Applied". American Water Works Association, ANSI/AWWA C-205-85, USA, 1985.
3. "Standard for Cement-Mortar Lining of Water Pipelines - 4 inches and Larger - In place" AWWA C-602-76, American Water Works Association, 1976.

A) PIPELINES UNDER OPERATION (SWCC, 1993)

Project Name	Pipe Length km	Pipe Diameter mm	Pipe Material, coating and lining	Pumping Stations	Reservoirs	Remarks
Yanbu-Madina Water Transmission System	A. 175 to Madina B. 51 to Yanbu	A. 800 B. 600	A. Steel pipes with polyethylene coating and epoxy lining. B. Asbestos cement pipes.	2 Pumping stations at Desalination Plant and Al-Missayjed (90 km from Desalination Plant)	2 Reservoirs at Desalination Plant, Capacity 20,000 M³ each.	Supplying Water to Medina and Yanbu
Riyadh Water Transmission System	Double line 2x466	1524	Steel pipes with polyethylene coating and cement mortar lining.	6 Pumping stations at Al-Jubail, Al-Dharan, Shedgum, Al-Hofuf, Khurais and wasia	6 Concrete reservoirs at high point Terminal (Near Riyadh), with Capacity of 50,000 M³ each.	Supplying Water to Riyadh City.
Riyadh city feeder lines.	Three lines A. 43 B. 36.5 C. 53	A. 1600 B. 2000 C. 2000	Pre-stressed concrete cylinder pipes	Gravity System	3 Concrete Reservoirs: • Al-Rouda 150,000 M³. • Northern 7500 M³. • Southern 100,000 M³.	Supplying Water to Riyadh City.
Makkah-Taif Water Transmission System.	A. 96.4 Double Line to Makkah B. 41 to Taif.	A. 1400 B. 1050	Steel pipes with polyethylene coating and cement mortar lining.	4 Pumping stations.	4 Reservoirs at Arafat, 50,000 M³ each. 4 Reservoirs at Taif, 25,000 M³ each.	Supplying Water to Makkah and Taif

A) PIPELINES UNDER OPERATION (SWCC, 1993) (continued)

Project Name	Pipe Length km	Pipe Diameter mm	Pipe Material, coating and lining	Pumping Stations	Reservoirs	Remarks
Al-Khobar Water Transmission System.	A. 56 B. 82 C. 98	A. 1100 B. 1000 C. 900	Concrete cylinder pipe.	8 Pumping Stations at Al-Khobar-I, Al-Khobar-II, Dahran, Dammam, Sayhat, Safwa, Qateef and Rahemah.	3 Reservoirs each at: Al-Khobar 3x12,700M^3 Dahran 3x25,000M^3 Dammam 3x25,000 M^3 Sayhat 3x25,000 M^3 Qateef 3x11,000 M^3 Safwa 3x25,000 M^3 Rahema 3x2,500 M^3	Supplying Water to: Al-Khobar, Dammam, Dhahran, Sayhat, Safwa, Qateef and Rahemah
Assir Water Transmission System.	A. 121 B. 36 C. 10 D. 19 E. 29	A. 1100 B. 900 C. 800 D. 600 E. 500	A. Steel pipes with polyethylene coating and cement mortar lining. B,C,D,E. Concrete cylinder pipe with coal tar epoxy coating.	4 Pumping Stations at Shuquyq Wadi Maraba At km 81 Near Abha. (km 96)	2 Reservoirs at Abha, Capacity 50,000 M^3 each. 1 Reservoir at A-Bin Numan Capacity 50,000M^3 1 Reservoir at Ukad Capacity 20,000 M^3. 1 Reservoir at Ahad Rafidah Capacity 8,000 M^3	Supplying Water to: Abha, Khamis-Mushayt, Ahad - Rafidah and Military City.

B) PIPELINES UNDER CONSTRUCTION (SWCC, 1993)

Project Name	Pipe Length km	Pipe Diameter mm	Pipe Material, coating and lining	Pumping Stations	Reservoirs	Remarks
Al Jubail-Riyadh Line 'C' Project.	384	1524	Steel pipes with polyethylene coating and cement mortar lining.	4 Pumping Stations at Al Jubail, Al-Lidam, Ashuaib and Al-Wasia	6 Reservoirs capacity 50,000 M^3 each.	To supply Water to Sudair, Washam, and Qassim cities
Riyadh - Qassim Water Transmission System	Double line 389 Single line 135	A. 2000 B. 1600 C. 200 D. 500	A. *PCCP 17 km. B. PCCP 418 km. C. Steel 316 km. D. **CCP 41 km.	2 Pumping Stations, At Al-Hissi and Al-Majmaáh.	17 reservoirs as follows Ghunman 3x20,000 M^3 Al-Majmaáh 2x20,000 M^3 Shaqra 3x20,000 M^3 Al-Ghat 3x20,000 M^3 Buraidah 6x50,000 M^3	
Yanbu - Madina Water Transmission System. (Phase II)	A. 150 B. 40 C. 51 D. 66 E. 47	A. 1524 B. 1320 C. 800 D. 600 E. 300	Steel pipe with three-layer polyethylene coating and cement mortar lining.	2 Pumping Stations: One at Desalination Plant and Another at Al-Missayjed (90 km from Desalination Plant)	14 reservoirs with the capacity of 50,000 M^3 each.	To Supply Water to Madina and Yanbu cities.

* PCCP. Pre stressed concrete cylinder pipe.
** CCP. Concrete cylinder pipe.

LANDSLIDES, LIQUEFACTION, AND ARTESIANS PIPELINE CROSSING OF THE SNOQUALMIE RIVER VALLEY

Roger Beieler[1], Karen Dawson[2], and Stephanie Murphy[3]

Abstract

The Seattle Water Department is currently designing a second water transmission pipeline between the Tolt River Watershed in the Cascade Mountains and its customer service area. The pipeline extends approximately 19 miles and varies in diameter from 81-inches to 54-inches. The pipeline crosses several waterways, including the Snoqualmie River. The paper briefly discusses the Seattle Water Department's water transmission system and the need for a second supply line from the Tolt Reservoir. The majority of the paper discusses the results and recommendations of the geotechnical investigation and preliminary design associated with crossing the Snoqualmie River Valley. Several unique geological features were confirmed as part of the investigation: a steep slope with landslide and severe erosion hazards, liquefaction potential near the river, and artesian conditions at the base of the slope. The recommended construction technique for the river crossing and artesian area is microtunnelling.

Introduction

Approximately 1.2 million Seattle-area residents receive water from the Seattle Water Department (SWD) system. Surface water sources provide 94 percent of the average annual demand and groundwater accounts for the remaining six percent. The surface water originates in two protected watersheds located in the foothills of the Cascade Mountains. The Cedar River Watershed is 143 square miles in size and the South Fork Tolt River Watershed is 21 square miles in size (see Figure 1). Annual rainfall is more than 160 inches in the upper portions of the watersheds.

---
[1] Design Engineer, CH2M HILL - Bellevue, WA
[2] Geotechnical Engineer, CH2M HILL - Bellevue, WA
[3] Project Manager, Seattle Water Department - Seattle, WA

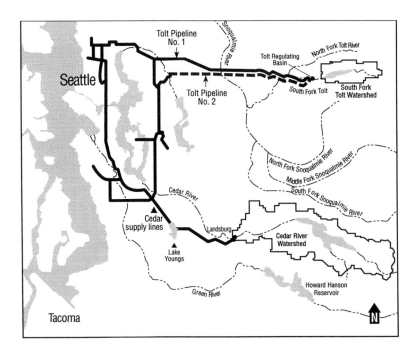

Figure 1. Seattle's Regional Water Supply System

The Cedar system began operation in 1901, twelve years after the "Great Seattle Fire" of 1889 destroyed the entire 64-acre business district. The lack of water available from the patchwork of private water companies was a major contributor to the widespread destruction. The Tolt system, which began operation in 1964, provides about 28 percent of the average annual system demand. The Tolt system currently relies on a single transmission pipeline to convey water from the watershed to the Tolt service area. The capacity of the existing Tolt Pipeline No. 1 is approximately 85 mgd.

The proposed second transmission main, Tolt Pipeline No. 2, in conjunction with the North Fork Tolt Project will improve the reliability of the supply system and increase the system's capacity to 240 mgd. The pipeline will be 19 miles in length and vary in diameter from 81-inches to 54-inches. The proposed pipeline alignment is shown in Figure 2. The pipeline will cross several waterways, including the Snoqualmie River. At the Snoqualmie River, the pipeline will be 75 inches in diameter with a static water pressure in the range of 300 psi.

Description of Snoqualmie Valley

The Snoqualmie Valley is approximately two miles wide at the location of the proposed pipe crossing. The valley floor is relatively flat and approximately one-mile wide with steep slopes on either side. The river occasionally floods in the spring, essentially covering the entire valley floor. A cross section of the valley showing the slopes and location of the river is included as Figure 3. The western and central portions of the valley include several unique features. For purposes of analyzing pipe installation techniques and comparing alternatives, the valley has been separated into three areas: the west slope area, the western valley floor, and the river channel. Standard construction techniques will be used in the eastern portion of the valley.

West Slope

Slopes along the western edge of the Snoqualmie Valley are mapped as erosion hazards. Increased runoff, most likely due to upslope development, has resulted in recent formation of significant gullies (up to 40 feet deep) less than a mile south of the proposed pipeline alignment. The alignment on the slope is heavily forested. Removal of vegetation in this area, as required for pipeline construction, will increase the opportunity for erosion. The cleared area will need to be carefully protected, reseeded, and monitored until vegetative growth has reestablished. It may be necessary to install a temporary piped drain system with storm water inlets to minimize overland flow during construction.

Several springs were observed on the lower half of the slope. These springs will need to be controlled and the area dewatered prior to trench excavation in order to allow placement of pipe and safeguard the stability of the hillside. A permanent pipe system, installed parallel to the transmission main, will likely be required to collect and discharge subsurface flows and prevent internal erosion, or piping (the loss of fine grained soils from the subsurface). The permanent pipe system will probably include perforated collection pipes in the bottom of the trench and cutoff collars placed around the transmission main. The pipe collars would extend into the native soils on the bottom and sides of the trench to prevent water from flowing in the pipe trench backfill material.

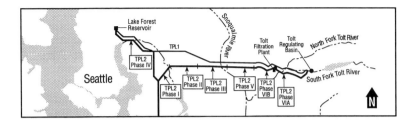

Figure 2. Tolt Pipeline No. 2 Alignment

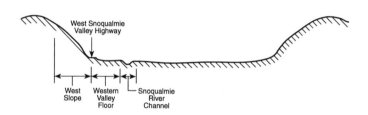

Figure 3. Valley Cross Section

The subsurface in this area is composed of till overlying lacustrine and outwash sediments. The till consists of a dense, silty sand with gravel and cobbles. The lacustrine sediments consist of hard clay and silt. The clay and silt were likely deposited in waters ponded some distance from the glacial ice front during the most recent Ice Age. The advance outwash consists of fine sand deposited in slow moving meltwater streams at the front of the glacier. In this area, it appears that there have been multiple advances and retreats of the glacier. Sediments that were deposited were eroded away locally. As a result, the stratigraphic profile is complex and varies in both lateral and vertical extent. Test borings, which were spaced as closely as 100 feet in some locations, could not fully define the complex interfingering between the various units present.

Confined groundwater was observed in some of the sand layers or fingers at depths ranging from about 35 to 45 feet. Though these observed depths are below the bottom of the standard trench section, water-bearing layers could extend into the trench zone at locations between the borings. If these layers are present near the excavation, dewatering will need to be undertaken to prevent uncontrolled artesian flow into an open trench and possible progressive landsliding of the hillside. The well points would have to be closely spaced because of the variability in stratigraphy along this portion of the alignment.

Western Valley Floor

High artesian groundwater conditions exist primarily in the most westerly 500 feet of the valley. Piezometric heads up to 17 feet above the ground level have been measured. Piezometer readings indicate that total water head increases with depth and decreases with distance eastward from the base of the slope. It appears that the fines content, which generally increases closer to the ground surface, provides a graded cap to confine groundwater.

The West Snoqualmie Valley Highway runs parallel to the base of the slope. The area covered by the West Snoqualmie Valley Highway is anticipated to have the greatest artesian pressures and would require substantial dewatering. It is desirable, however, to keep the roadway operational during construction. Several miles of detour using other roads or a large construction easement for a temporary bypass will be required if this road is closed.

Snoqualmie River Channel

The channel width of the Snoqualmie River at the proposed pipeline crossing is about 300 feet and the approximate elevation difference between the top of bank and the high water mark is about 23 feet. Soils within the upper 50 feet from the top of bank are typically sand with some silt. Below a depth of 50 feet, the sand is more dense and slightly more coarse.

The loose sands below and adjacent to the river are subject to liquefaction during a design seismic event. Based on estimated undrained residual soil strength during liquefaction, "flow"-type failures extending up to approximately 170 feet beyond the top of the river banks and down to an elevation of -15 feet (approximately 20 feet below the existing river bed) could occur. The approximate limits of a flow failure during a design earthquake are shown in Figure 4.

A "flow"-type failure was designated as one involving several feet of movement, with the potential for complete loss of pipeline support. The limits of the flow failure were initially estimated by using undrained residual strength in a static stability analyses. Analyses using a seismic load and residual strength resulted in failures extending well over 1000 feet away from the river, which were judged to be unrealistic. The initial flow failure boundaries were verified by estimates of lateral spreading displacement, evaluated by several methods (Dobrey and Bazier, 1992; Mabey and Youd, 1992; Newmark, 1965). The displacement calculations showed that movements within the bounds determined from static stability analyses could be measured in feet, while those well outside averaged between about 4 and 8 inches.

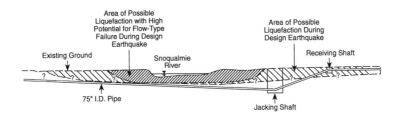

Figure 4. Approximate Limits of a Flow Failure

Analysis of Alternatives for Constructing the River Crossing

Environmental concerns dictate that any excavations near the river must be shored, and no soil or excavation debris can be discharged into the river. The high river flows and resulting scour potential preclude a buried pipeline with minimal cover. High flows and environmental constraints also preclude simple support of the pipeline on piles over the water. A bridge supported pipeline with clearance for high flows could be designed, but the cost would be high and environmental impact would still be an issue.

The loose soil conditions, susceptibility to liquefaction, high groundwater, and the 300-foot width of the Snoqualmie River make an open trench construction method unfeasible, even if depths of cover were shallow and the risks of liquefaction during an earthquake could be accepted. If the profile is lowered below the potential flow failure area caused by liquefaction, the costs and technical difficulties associated with an open excavation multiply. Because earlier studies ruled out a bridge-supported pipeline, microtunneling is the obvious alternative to trenching. Preliminary estimates indicate that microtunneling will be less costly than an open trench with minimal cover. Though microtunneling will involve environmental impacts associated with construction of the jacking and receiving shafts and disposal of the cuttings, the total amount of soil disturbed would be less than with the open trench option and there would be no direct impacts to the river. For the above reasons, microtunneling has been selected as the preferred method for crossing the Snoqualmie River.

The jacking shaft needed for microtunneling beneath both the river and the potential flow failure zone will need to be at least 60 feet deep. High groundwater and loose soils make it unfeasible to construct an open trench for the pipeline between the jacking shaft and the open-cut sections of the alignment on either side of the river. Therefore, at least one deep jacking shaft and two shallow receiving shafts will be needed to microtunnel beneath the river and raise the pipeline to a depth shallow enough to allow open trench construction for the remainder of the alignment. Because a tunneling machine would be on site for the river crossing and the mobilization costs paid, microtunneling becomes a viable option for dealing with the technical and environmental difficulties in the western valley floor and the west slope.

Development of Alternatives

To determine the extent that microtunneling should be considered as an option, three alternative microtunneling scenarios, as shown in Figure 5, were developed for the area between the top of the steep lower hillside of the western slope and the east side of the Snoqualmie River. All three alternatives have the same 2,300-foot total length, but involve different distance combinations of microtunneling and open trench.

The lengths selected for possible microtunneling were based on topography, minimum cover requirements for microtunneling, and a maximum grade of approximately 36 percent where possible to keep the interior of the pipeline more accessible to inspectors and maintenance personnel. The limits of the area of possible liquefaction shown in Figure 4 also controlled the profiles for crossing beneath the Snoqualmie River.

Trenching Across the Western Valley Floor (Alternative A)

If the western valley is open cut, extensive dewatering will be needed to provide stable excavations. West of the highway, it may be possible to drive sheet piling through the artesian layers, into hard, low permeability, glacially overconsolidated silts. These silts could provide a cutoff to groundwater and allow excavation within the sheet piling with moderate dewatering efforts. However, the glacially overconsolidated silt unit is either nonexistent or deeper than the practical depth of sheet piles on the eastern side of the highway. Therefore, east of the highway, sheet piles will do little to slow or limit flows into an excavation.

RIVER VALLEY PIPELINE CROSSING

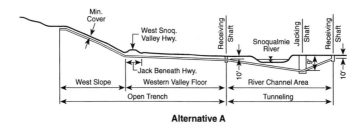

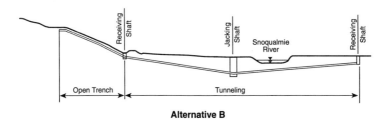

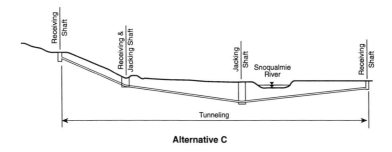

Figure 5. Construction Alternatives

Replacement of Confining Layer. Trench excavation across the artesian portions of the western valley, as shown in Alternative A, will remove the existing layer of finer grained soils that presently tend to cap the area. Three approaches were analyzed for replacement of the confining layer.

1. Backfilling the excavation with fine-grained or impervious material so that the existing state is approximated and upward water flow is minimized or halted, allowing excess upward groundwater pressures to be applied to the fine-grained fill cap.

2. Backfilling the entire trench section with granular material that would act as a graded filter and allow water to flow upward to the ground surface, relieving confined groundwater pressure at the depth of the pipe, but exerting an upward seepage force.

3. Providing free-draining backfill in the pipe zone and perforated drain pipes below the pipe to relieve pressures on the pipe and overlying soil, and backfilling the trench zone with fine-grained material to provide a cap and eliminate seepage forces.

The first option will be the least expensive, but the risk of piping and eventual destabilization of the trench is much higher than in the other options. Even with careful compaction under dry conditions, it will be very difficult to assure a bond between the backfill and the sides of the excavation. If water is allowed to seep along this contact, piping of fines could occur. This piping could eventually lead to instability.

The second option involves the costs of disposing of all excavated material and importing granular backfill. In addition, seepage forces require 16 feet of soil cover to prevent floating of the pipe. The depth of cover could be reduced by anchoring the pipe in the ground with concrete collars, but this option is also expensive. In addition, the water escaping to the ground surface will disrupt the current farming operation.

In the third option, the cap restricts permanent flow to the surface and the drains reduce the upward pressures on the cap, controlling the risk of piping. Eight feet of cover is recommended to prevent floating of the empty pipe with this option. Only 4 feet of cover are needed if the drains function perfectly. The additional recommended depth is intended to provide a factor of safety against temporary blockage or clogging of the drains. The drains empty into a slough which connects to the Snoqualmie River. At times of seasonal high water or floods, the elevation of water in the slough is higher than the drain pipe invert; however, the drains still provide pressure relief in the pipe zone. This third option is recommended for open cuts in this area and was used as a basis for the cost comparison.

Disadvantages of Open Trench Construction. There are several disadvantages of open trench construction across the western valley floor. First, the permanent pipe zone drainage system would alter the groundwater flow regime, permanently diverting water toward the alignment and lowering the piezometric head. This may or may not cause changes in vegetation in the surrounding field and hillside. The system would probably raise the water level in the slough and increase the flow of water through the slough into the Snoqualmie River. The impact to local groundwater wells is unknown.

The second disadvantage is the environmental impact and cost of the dewatering, which would be required even if sheet piling is used. Based on preliminary trench depths, groundwater levels in the area would have to be lowered at least 17 feet below the ground surface, with a total possible decrease in water head of 25 to 35 feet. If a deep well system with wells extending into the deeper, more permeable sands and gravel is used, the radius of influence of dewatering could extend a substantial distance. Discharge volumes are estimated to be very high.

Trench excavation would cause additional environmental impacts. If 1:1 slopes are used for the excavation, the trench width at the top would be roughly 40 feet. If well points are used, a two-stage pumping system would be required, making the total trench width larger. Additional area would be disturbed for equipment access and stockpiling of material. At least 5.5 cubic yards of material per linear foot of pipe would have to be hauled offsite for disposal. The remaining 10.5 cubic yards excavated per linear foot could be placed as trench backfill, but the moisture content is estimated to be too high to achieve adequate compaction. Spreading and drying to reduce the moisture content is probably impractical.

Finally, the open trench option would require either jacking under the West Snoqualmie Valley Highway, or disruption of traffic while a trench is cut. The highway is believed to be on the transition zone where the glacial till cap overlying the artesian layers ends and loose valley and landslide deposits begin. It is expected to be the most difficult and unpredictable section to dewater on the project. Problems during construction could cause loss of the roadway or disruption of traffic for an extended period of time.

Microtunneling Across the Western Valley Floor (Alternative B or C)

Microtunneling allows the pipe to be placed at a depth that will prevent floating. Disturbance to soil above and around the pipe is minimal, so concentration of groundwater and the potential for piping are of little concern. Long-term groundwater levels should not be altered. There will be minimal disruption to traffic on the highway and environmental impact to the Valley will be minimized. Construction dewatering will be required at the shaft locations. The proposed pit location just west of the highway will probably intersect confined groundwater. The primarily granular soils at shaft locations outside of the artesian zone are likely to require partial or complete dewatering in conjunction with a cutoff wall/shoring system. Though this is expected to be a large dewatering effort, the total water volumes discharged are anticipated to be orders of magnitude less than those involved with an open cut through the western valley because of the comparatively small area involved.

Microtunneling also requires space to settle cuttings from the slurry and subsequent disposal or use of the cuttings elsewhere. Ideally, a large basin could be excavated in the vicinity of the pit and cuttings allowed to settle and remain permanently in the basin. Cuttings could also be allowed to settle in temporary lined ponds, with the cuttings periodically removed and hauled offsite, or the slurry could be pumped into several large portable tanks. A typical size of basin or pond for slurry recycling at each jacking pit is anticipated to be about 200 cubic yards.

Trenching (Alt. A or B) vs. Microtunneling (Alt. C) the West Slope

If the pipe is installed by open trenching in areas underlain by artesian layers on the west slope, the weight and soil strength which counteract the upward pressures exerted by any underlying confined groundwater will be reduced. If the confined groundwater is close enough to the base of the excavation and under enough pressure, the base of the excavation could fail, either by sudden heaving or by gradual flow of water through the confining soil into the excavation, causing uncontrolled piping on the steep slope.

Information from the subsurface exploration indicates that artesian layers will be intersected or come close enough to the bottom of the trench to cause flow into the excavation near the base of the lower hillside. Because a catastrophic failure is possible if these artesian zones are intersected without dewatering, the observational method of dewatering after groundwater is encountered should not be applied. Therefore, a dewatering system, to maintain hillside stability, should be installed.

Microtunneling the hillside would eliminate most erosion concerns since vegetation would be left in place except at the jacking and receiving pits. Hillside soils include glacial till that is very dense and contains gravel and scattered cobbles and boulders. Cutting heads in the 88-inch-diameter range can typically pass boulders up to 2 or 2.5 feet in diameter. It is possible that boulders larger than 2.5 feet could be encountered along the alignment. If these larger boulders are encountered, they could be broken up in place or excavated from the surface so that tunneling could continue. Access for an air track rig to drill and blast a boulder into smaller pieces could probably be done with minimal damage to surrounding landscape by clearing brush from a trail and working around large trees. Microtunneling in this section does involve construction of a receiving pit, but artesian conditions are not expected at the top of the slope and any necessary dewatering would be concentrated at the pit location.

Conclusions

Based on the preliminary opinions of cost, the least expensive method for constructing the pipeline between the west slope and the east side of the Snoqualmie River is Alternative B, which entails open trenching the west slope and microtunneling between the base of the slope and the east side of the river. Installing the pipe by microtunneling across the western valley floor would also cause less environmental damage than placing the pipe in an open trench. Although there are risks to slope stability and road stability associated with constructing a receiving shaft at the base of the lower hillside, these risks are believed to be similar to or less than the risks involved with an open trench or conventional jacking across the highway. Therefore, microtunneling is recommended between the eastern side of the Snoqualmie River and the base of the west slope.

Microtunneling appears to be more expensive than open trenching for the west slope. In order to realize the anticipated savings associated with open trench construction, SWD has chosen to proceed with design and permitting activities based on open trench construction. Microtunneling is likely to cause less disturbance along the alignment, even if the presence of large boulders requires clearing of localized areas on the hillside for surface access, eliminating a portion of the environmental benefit. Microtunneling will be included in the bid documents as an acceptable alternative for installing the pipeline on the west slope.

References

Dobry, R. and M.H. Bazier, 1992. Modelling of Lateral Spreads in Silty Sands by Sliding Soil Blocks. In: Stability and Performance of Slopes and Embankments, ASCE Geotechnical Special Publication No. 31, p. 625-652.

Mabey, M.A. and T.L. Youd, 1991. Liquefaction Hazard Mapping for the Seattle, Washington Urban Region Utilizing LSI. Final Technical Report, U.S.G.S. Grant No. 14-08-0001-G1695.

Newmark, N.M., 1965. Effects of Earthquakes on Dams and Embankments. Geotechnique, Vol. 15, No. 2, January, 1965.

Stress Relaxation in Constantly
Deflected PE Pipes

Lars-Eric Janson[1]

Abstract

In 1982 a long-term study was commenced in Sweden on constantly deflected polyethylene (PE) pipes. A first report was presented 10 years ago at the ASCE International Conference in Madison, USA, 1985. Since then the main study has continued up to 1990 (8 years of constantly deflected pipes). At that time the pipe samples were released from the forces which had kept the samples deflected, and the time depended recovery of the deflection was observed and registered. The latest recovery record was made in November 1994, 4 years after the release. In the following paper, the new findings are discussed as a complement to the first report in 1985.

Introduction

A long-term study of constantly deflected polyethelene (PE) pipes started in 1982 as a VBB commission for Neste Chemicals (now Borealis Polythene) and the Swedish KP-council. The laboratory tests were performed at the Swedish National Testing and Research Institute (SP) in Gothenburg, Sweden. A first report was presented in 1983 after one year of investigation. This report was used as a basis for a paper presented at the ASCE international conference in Madison, USA [1 Janson 1985]. Since then, the study has continued during totally 72,000 h (8.2 years) after which time the main test was terminated. The following paper provides complementary information on the findings presented in [1] and [2 Janson 1991].

---

[1] Dr. Sc., Professor, M. ASCE.
Senior Advisor of the VBB/SWECO Consulting Group.
P.O. Box 34044, S-100 26 Stockholm, Sweden

Test Samples

Three different PE grades have been studied, all from Neste Chemicals (Sweden). One is the HDPE grade DGDS 2467 (Type 1), another the modified HDPE grade DGDS 2467 BL (Type 2) and the third is the MDPE grade DGDS 2418. (The letters DGDS are today recognized as NCPE). As test samples extruded pipes have been used with an external diameter of 315 mm. The length of the sample is the same as the diameter. The wall thickness of the PE samples corresponds in one series to the pressure class PN4 (nominal wall thickness 12.1 mm) and in another series to the pressure class PN6 (nominal wall thickness 17.8 mm). Both classes are based on a long-term tensile design stress of 5 MPa. The pipe samples were deflected as to give constant bending strains in the pipe wall between 1.1 % and 3.5 %. The test samples were deflected to min. 4.3 % and max. 13.6 % by use of two linear loads parallel to the pipe axis and acting against each other through the vertical diameter plane. The temperature was constantly +23°C ±0.5°C. The force which was needed to keep the pipe sample deflected was successively measured in the coarse of time in a load machine, type Instron 1195.

The following theoretical expressions have been used in the study. Eq. (1) gives the relation between the pipe deflection δ (the decrease of the vertical pipe diameter) and the undeflected mean diameter D_m (D_m = D-s where D is the external pipe diameter and s is the wall thickness). In the equation, P stands for the linear load per length of the pipe sample, E for the relaxation modulus and I for the moment of inertia.

$$\delta/D_m = 0.0186 \frac{P \times D_m^2}{E\ I} \tag{1}$$

The relative strain ϵ in the pipe wall caused by the deflection can be expressed according to eq. (2) provided that the deflection is moderate and the pipe is thin-walled:

$$\epsilon = \frac{\sigma}{E} = 4.28\ (\delta/D_m)\ (s/D_m) \tag{2}$$

The nominal ring stiffness S_R of the pipe expressed in kPa is given by the formula:

$$S_R = EI/D_m^3 \tag{3}$$

In [1] the deduced E-modulii were presented in graphs giving the logarithmic E values versus the logarithmic loading time up to 0.9×10^4 h (1 year). In [2] test results have been achieved for almost one more time decade, or up to 0.7×10^5 h (8.2 years).

Evaluation of E-modulus and Bending Stress

The study during this long testing period has gradually taught us a lot about how to evaluate the long-term behaviour of various polymer materials subjected to mechanical stress. Thus, it has been found that the tests performed so far have much more to tell us than was originally understood. This concerns particularly the approach to the study of the compliance C (compare [3 Struik, 1978 and 1989]) defined as the inverted value of the E-modulus:

$$C = 1/E \qquad (4)$$

As the E-modulus according to the classical Hooke's law is defined as the relation between stress σ and strain ϵ, the following equation is also valid:

$$C = \epsilon/\sigma \qquad (5)$$

For viscoelastic materials the E-modulus and consequently the compliance C described the change of the relation between stress and strain in the course of the loading time, independent of whether the study concerns stress relaxation or strain creeping. Thus, as the strain ϵ is constant in the relaxation studies of constantly deflected pipe samples, an increase of C describes the decrease (or relaxation) of the stress σ in the course of time according to eq. (5). In the same way, an increase of C in the case of constant stress σ, giving rise to free creeping, describes the increase procedure of the strain ϵ.

In [4 Janson, 1988] various types of relations are discussed between the strain in a polymer material and the loading time, if the stress is constant or alternatively is subjected to relaxation. The analysis is connected to the discovery of a physical ageing process going on in stressed amorphous polymer materials such as PVC, but to a significant degree also in semi-crystalline polymers such as polyethylene. The physical ageing implies a strengthening of the material in the course of time due to the fact that a successive consolidation of the molecular structure is taking place. As a consequence of this ageing process, it has been found that the strain in a constantly stressed material will increase linearly as a function of the logarithmic loading time. In a similar way the stress in a constantly strained material will decrease linearly as a function of the logarithmic loading time. This means that the compliance C according to eq. (5) can give good advice about the long-term behaviour of the loaded material. Thus, as C has a rectilinear course in a lin C/log t graph, it opens up the possibility to perform a linear extrapolation of C despite the fact that the time is given in a logarithmic mode. Then it is easy to find the relaxation modulus E according to eq. (4) by simply calculating the inverted value of C.

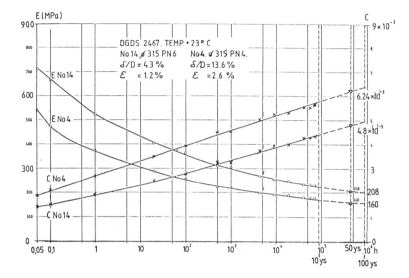

Figure 1. Relaxation E-modulus for the HDPE (Type 1) pipe.

This type of behaviour of the polymer material has been applied when evaluating the long-term studies of constantly deflected PE pipes now terminated after more than 8 years. Thus *Figure 1* shows an example of the lin C/log t curves and the corresponding E-modulus curves for the HDPE (Type 1) grade. As the compliance C stands for the inverted value of the E-modulus the rectilinear part of the lin C/log t-curve illustrates the relaxation behaviour of the physically aged PE material.

It has been found that the rectilinear course of the curve occurs very early, which means more or less immediately for highly strained material and after 10 to 100 h for less strained material. (This finding has recently been applied as a basis for a new field method for tightness testing of thermoplastics pressure pipelines [5 Janson 1994]). Another practical positive effect of this observation is that the long-term (50-100 years) E-value can be found for most PE materials after a testing time of less than 1000 h. This can be recognized in Figure 1 from which it can be seen that the measurements after 10,000 h concur well with the rectilinear lin C/log t curves drawn up at a testing time of only 1000 h. Thus, the more than eight years of constant pipe deflection have given no reason from a practical point of view to change the long-term E-values that could be determined after only approximately six weeks of testing.

One important finding from the long-term studies is that the E-modulus after 50 years is significantly higher than was previously assumed. Thus, for moderately deflected pipes (4-5 %) the 50-year E-modulus is approximately 200 MPa, while previously 100 MPa was often used. - An extrapolation of the C-curves to 100 years and after inverting the C-values, corresponding E-values are given which are only slightly less than the 50-year value. This supports the statements made in [6 Janson 1987], implying that the service period for standardized gravity thermoplastics sewer pipes can certainly be 100 years or more, provided properly installed so that buckling does not occur. The present long-term investigation also supports the statement already made in [1], that there is no practical upper bending strain limit for the design of buried gravity polyethylene pipes to be used up to 100 years. The condition is, however, that the internal pipe wall has not been thermally oxidized during manufacture, and moreover that the stabilization system of the grade is chosen to resist the chemical ageing process.

The lin E/log t diagrams of the type illustrated in Figure 1 have been transformed to σ/t diagrams with the loading time as a parameter. *Figure 2* illustrates the relations for the HDPE (Type 2) grade and *Figure 3* for the MDPE grade. The difference between the two grades is insignificant. Thus, it can be stated that the short-term bending stress (3 min value) will relax to approximately 30 % of the original value after 50-100 years of constant bending strain caused by pipe deflection.

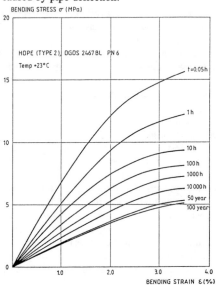

Figure 2. Stress/strain diagram for the HDPE (Type 2) pipe

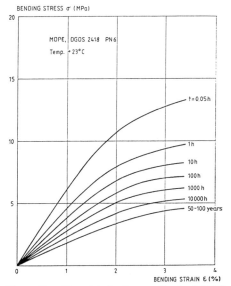

Figure 3.   Stress/strain diagram for the MDPE pipe

Deflection Recovery of Released Samples

All samples were doubled in the investigation. The study of the first series was terminated in 1983, and at that time these samples were released and the deflection recovery measured for 1000 h. *Figure 4-6* present the results of the study for this first series of samples (full lines). As can be seen, a remarkable rectilinear course of the curves is recognizable for all samples kept deflected for one year. If a rectilinear extrapolation had been performed, the deflection would have recovered fully after some thousand years. However, a long-term curve configuration of this kind should not be true, as the viscous part of the strain would give a residual deflection even after an infinitively long time. But it is certainly true that the shorter the time of forced deflection, the faster the recovery process will take place after releasing the sample.

This is well illustrated in Figures 4-6 by the recovery curves (dotted lines) valid for the 'daughter' samples released after 8.2 years of deflection. The recovery measurements have here been made for approximately four years. (This means 12 years after start of the testing.) Hence, it can be found that the immediate deflection recovery is somewhat less after many years of constant deflection than after only one year of deflection. This is to be expected, as the bending stress in the pipe wall, which has to force the pipe deflection to recover, has further decreased in the course of time due to relaxation.

244 UNDERGROUND PIPELINE ENGINEERING

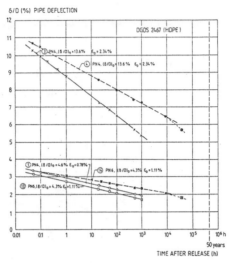

Figure 4. Pipe deflection recovery when releasing the HDPE (Type 1) pipes after 10,000 h of constant deflection: Full lines. - The same after 8.2 years of deflection: Dotted lines

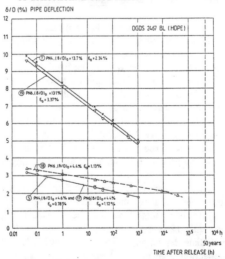

Figure 5. Pipe deflection recovery when releasing the HDPE (Type 2) pipes after 10,000 h of constant deflection: Full lines. - The same after 8.2 years of deflection: Dotted lines

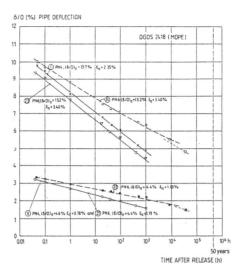

Figure 6. Pipe deflection recovery when releasing the MDPE pipes after 10,000 h of constant deflection: Full lines. - The same after 8.2 years of deflection: Dotted lines

A more unexpected finding is that the rectilinear course of the lin $(\delta/D)/\log t$ curves also takes place during the prolonged time of observation up to one year. Thus, no retardation of the recovery rate can be found. On the contrary, the recently performed control four years after the release indicates a smaller increase of the recovery rate. This should mean that the PE pipes may have an ability to more or less fully revert to their original circular shape within a definable space of time. This in turn would mean that the elastic strain and the so called retarded elastic strain dominate and that the viscous part of the strain is insignificant.

Ring Stiffness after Long-Term Pipe Deflection

As discussed, the initial bending stress in the pipe wall caused by the constant deflection of the pipe will decrease in the course of time, due to stress relaxation. As the deflection is constant, the bending strain in the pipe wall caused by the deflection is also constant. The E-modulus as defined according to eqs (4) and (5) will consequently decrease in the course of the loading time in the same way as the bending stress does. However, it must be understood that the decrease of the E-modulus is only fictive and has nothing to do with the capability of the polymer material to resist new initial forces after a long loading time, a capacity represented by the short-term relation between stress and strain. As the ring stiffness according to eq. (3) is a linear function of the E-modulus, it consequently means that even after a long loading time, the ring stiffness continues to retain its short-term value for each new impulse of loading [7 Janson 1990].

In order to demonstrate this theoretical fact practically, the HDPE (Type 2) pipe sample constantly deflected to 4.4 % during 8.2 years, was after release subjected to another short-term loading procedure. The short-term ring stiffness calculated according to eq. (3) was then 18.1 kPa, while the original ring stiffness eight years earlier was 15 kPa. This clearly confirm that the short-term E-modulus has not declined after long-term loading of the pipe. On the contrary, it has in fact increased significantly, most probably due to the physical ageing effect mentioned above. This fact is of utmost importance for an adequate understanding of the deflection process for buried thermoplastics gravity pipes as discussed in detail in [7].

Appendix

References

[1] *Janson, L-E.* 1985. Investigation of the long-term creep modulus for buried polyethylene pipes subjected to constant deflection. - Proc. Int. Conf. Advances in the Underground Pipeline Engineering. ASCE. Madison, USA, 1985, p 253-262.

[2] *Janson, L-E.* 1991. Long-term studies of PVC and PE pipes subjected to forced constant deflection. - Report No. 3 from the Swedish KP-Council. Stockholm Dec. 1991. (Distribution VAV, Regeringsgatan 86, S-111 39 Stockholm, Sweden).

[3] *Struik, L.C.E.* Physical ageing in amorphous polymers and other materials. - Ph.D. Thesis, Technical University, Delft, 1977. Elevier. Amsterdam (1978) and *Struik* 1989. Mechanical behaviour and physical ageing of semi-crystalline polymers: 3. Prediction of long-term creep from short-time test. - Polymer (1989) Vol. 30, p 799-814.

[4] *Janson, L-E.* 1988. Physical ageing of buried PVC sewer pipes as affecting their long-term behaviour. - Int. Conf. Plastics Pipes VII, Bath, England, 1988, p 28/1-28/10.

[5] *Janson, L-E.* 1993. Method for tightness testing of plastics pressure pipelines. - Construction and Building Materials, (7), No. 4, p 241-244.

[6] *Janson, L-E.* 1987. Hur gammalt kan ett plaströr bli? (How old can a plastic pipe become?) - Report No. 1 from the Swedish KP-Council, Oct. 1987. In Swedish.

[7] *Janson, L-E.* 1990. Short-term Versus Long-term Pipe Ring Stiffness in the Design of Buried Plastic Sewer Pipes. - Proc. Int. Conf. Pipeline Design and Installation. ASCE. Las Vegas, USA. 1990, p 160-167.

BURIED PLASTIC PIPE - PERFORMANCE VERSUS PREDICTION

CDF Rogers[1], PR Fleming[2], MWJ Loeppky[3] & E Faragher[4]

Abstract

The design criteria for buried flexible pipe in the UK have traditionally been based on a conservative application of the Iowa formula. The UK Transport Research Laboratory (TRL) has produced an alternative, theoretically sound, method. The former has been proved through extensive worldwide experience, whereas the latter suffers from lack of experience in its use (i.e. in parameter definition).

A critical review of the two design methods is presented in the context of prediction for small diameter corrugated twin-wall HDPE pipe. Full-scale laboratory box loading tests of such pipe under static and cyclic loads is described. The results, deflection and pipe wall strain data, are analyzed with reference to the two design methods using classical prediction and parameter back-calculation. The Iowa formula proved superior due to better (particularly soil stiffness) parameter definition.

Introduction

Performance prediction and assessment of performance when constructed remain two issues of crucial importance to engineers who are responsible for flexible pipeline construction. Novel pipe materials and structural forms make performance criteria difficult to define, which, combined with different prediction methods, results in confusion. This paper aims to clarify the situation for corrugated, twin-wall high density polyethylene (HDPE) pipe by presenting experimentally derived performance data and analyzing (via prediction and back-calculation) these data using currently favoured methods.

For the relatively shallow buried flexible pipes, often subject to surface traffic loading, that are used in the UK, it is failure by excessive deformation (as opposed to pipe wall buckling or excessive hoop strain) that must be guarded against. Long experience of flexible pipes has demonstrated that a more substantial (i.e. stiffer) pipe will

---

[1] Senior Lecturer, [2] Lecturer, [4] Research Assistant, Department of Civil and Building Engineering, University of Technology, Loughborough, Leicestershire, LE11 3TU, UK.
[3] Product Manager Pipes, B&H (Leics) Ltd, Union Works, Bishop Meadow Road, Loughborough, Leicestershire, LE11 0RE, UK.

deform less under the same conditions as a lighter pipe. It rapidly became apparent, however, that the surrounding soil stiffness was of greater importance, both in the vertical direction (controlling load shedding by arching) and horizontal direction (controlling pipe wall support hence deformation). Thus any method of deflection prediction must take account of both the pipe and soil stiffness. This is true whether the pipe has to come into equilibrium under cyclic or static loading.

Theories to predict the deformation of flexible pipes were developed by Marston, Spangler and others in Iowa, USA in the early 20th century. The development of the Iowa formula (Spangler, 1941) was based on the behaviour of large (typically greater than 600mm) diameter corrugated steel culverts having Standard Dimension Ratios (SDR, the ratio of internal diameter to wall thickness) ranging from 300 to 600. The method developed has been in worldwide use since then, and has been accepted as a reliable predictor of pipe performance. The main reason for this is that much work has been done to define the soil stiffness parameter (E') for use in the design equation, notably by Howard (1977).

In the UK, the use of flexible pipes has become more common in the last twenty years. This is because of their light weight and high toughness (compared to traditional materials such as clay and concrete), factors which contribute to quicker, and hence cheaper, installation. The most common material of construction for these is plastic (PVC-U, polyethylene or polypropylene). Maximum diameters do not generally exceed 600mm, and SDRs tend to be less than 30. Multi-walled pipes of various profiles are now in widespread use (because of their higher strength to weight ratio). Until the early 1980s, the Iowa formula was applied to these types of pipes in deflection calculations. However, it became apparent that the pipes under consideration differed fundamentally from those for which the method was intended, and the UK Department of Transport commissioned research to find a more suitable method for these changed circumstances. The method was developed by the TRL (Gumbel et al, 1982 and Gumbel, 1984).

The Design Methods

The Iowa formula is used to calculate the horizontal diametral increase δh using the equation:

$$\delta h = \frac{k_b W_c D_L}{E_p I / r^3 + 0.061 E'}$$

where: δh = horizontal deflection (m)
k_b = bedding factor (dimensionless)
W_c = vertical load per unit length of pipe (kN/m)
D_L = deflection lag factor (dimensionless)
$E_p I$ = bending stiffness of the pipe ring per unit length (kNm)
r = radius of pipe = D/2 (m)
E' = Modulus of Soil Reaction (kN/m^2)

The vertical diametral reduction (δv) is determined by assuming that the pipe deforms to an ellipse, and hence δv = 0.91δh. However, for practical purposes the two are often assumed equal due to the lack of precision in parameter definition (especially E') and the known deviation from elliptical deformation (Rogers, 1988). The first term in the denominator is commonly substituted by $8S_{fp}$, where S_{fp} is the flexural stiffness of the pipe (E_pI/D^3) calculated according to UK standard tests. For the short-term case (including back-calculation from laboratory tests) $D_L=1.0$ and S_{fp} is that immediately derived from the parallel plate loading tests, whereas for long-term deformation D_L varies upwards depending on the pipe surround and an extrapolated 50 year stiffness value is used for S_{fp}. As Spangler admitted in his later work, the equation is semi-empirical and thus relies upon its extensive use for adequate parameter definition.

Gumbel's (TRL method) is a more rigorous method in which the pipe-soil system is treated as the basic structural unit. The method defines excessive pipe deflection and buckling of the pipe wall as the two primary failure criteria. The idealised analysis considers a linear elastic system consisting of a long, thin-walled cylinder deeply buried in a uniform weightless soil medium and subjected to two-dimensional loading in the plane of cross-section. Young's Modulus (E_p, E_s) and Poisson's Ratio (ν_p, ν_s) are defined for the pipe and soil respectively, from which a plane strain soil stiffness ($E_s^*=E_s/(1-\nu_s^2)$) is derived. A flexural stiffness ratio ($Y=E_s^*/S_{fp}$) is used to describe system behaviour as rigid ($Y<10$), flexible ($Y>1000$) or intermediate, as is the case for the buried plastic pipes described herein ($10<Y<1000$). Buckling is only likely for very thin-walled flexible pipes and/or deeply buried pipes.

A dead load lateral pressure ratio (k_d) is used to relate the free-field horizontal (p_{hd}) to vertical pressure (p_{vd}), and this in turn is used to determine uniform (p_z) and distortional (p_y) components of stress acting on the pipe. Uniform deformation (i.e. hoop compression) is small and generally ignored for design. Distortional, or out-of-round deformation (δy) is calculated as the sum of initial deformation (i.e. that after installation, δy_o), first-order (δy_1) and second-order (δy_2) deformation and long-term deformation (δy_3) to allow for creep (equivalent to deflection lag). The first-order component, derived from elastic theory, is given by:

$$\delta y_1 = \frac{4p_y}{108S_{fp}+E_s^*}$$

and has the important difference from the Iowa formula of the distortional pressure as numerator. Second-order deformation relates to p_z acting on the deformed pipe ring. An "arching factor", α, is introduced to account for the proportion of p_z carried by the pipe. For design purposes, α is safely equated to a uniform thrust coefficient α_z and generally assumed to be unity for intermediate behaviour. A set of design charts has been produced to determine δy directly.

While the method has been proved by limited laboratory studies, difficulties arise in accurately defining E_s^*. E_s^* is acknowledged to depend upon the soil surround, its placement and compaction, trench wall geometry, and the methods of insertion and withdrawal of trench supports. Gumbel acknowledges that E_s^* can only be defined by back analysis and, since little experience of applying the technique is available, this makes the method difficult to use.

Experimental Work Carried out at Loughborough

An extensive programme of experimental research at Loughborough University has included testing of twin-walled, annular corrugated HDPE pipe ranging in internal diameter from 100mm to 375mm. LVDTs were used to measure vertical and horizontal pipe deformations (Figure 1). Circumferential pipe wall strains were measured with uniaxial, foil/epoxy strain gauges bonded to the walls in the locations shown in Figures 2 to 4.

Testing was carried out in a 1.8m long x 1.5m wide x 1.8m high steel-sided test box. The cover to the pipe was 1m in all cases, virtually the minimum allowed in the UK (DoT, 1989). Three types of surround were used: an uncompacted and a well compacted river sand ($c_u=4.37$, $c_c=0.65$, $D_{10}=0.19$mm, relative densities 80% and 90% respectively), and a rounded river gravel ($c_u=1.55$, $c_c=0.96$, $D_{10}=5.5$mm, relative density 94%), this latter being the most commonly used in the UK.

Each test comprised four phases:
(i) installation, in which the pipe surround was placed and the test tank filled,
(ii) a static stress of 70kPa, to simulate a stationary heavily laden vehicle above the trench, or an overburden pressure equivalent to an approximate burial depth of 4m,
(iii) 1000 cycles of sinusoidal loading (amplitude = 70kPa, period = 120s), to represent the repeated passing of a heavy vehicle over the trench, and
(iv) a static stress of 140kPa, to simulate an equivalent burial depth of 8m.

The static loads were applied for a period of 12 hours. A recovery period of four hours was allowed following removal of the load after all phases. A fuller description of the test facilities is given elsewhere by Rogers et al (1994).

Discussion of Results

The results are presented in Table 1 and Figures 1 to 4. Deformations (measured by vertical diametral strain, VDS) are generally low in relation to the maximum permissible value in the UK, which is 5% (BSI, 1980). Clear trends can be discerned. The first of these is the influence of the surround material on deflection, a poor surround (uncompacted sand) producing the largest deflection in all cases. This is because of its poor arching ability (smaller vertical stiffness) and its lower resistance to lateral movement of the pipe wall (smaller horizontal stiffness) when compared with the well compacted sand. The latter phenomenon is evinced by the data for horizontal diametral strain (HDS). The gravel surround produced deflections between these extremes, showing that some settlement of the (uncompacted) gravel takes place on loading, but that the material (because of its uniformity and roundness) quickly develops a more load-resistant structure.

The lowest recorded deflections were for the 100mm diameter pipe in a heavily compacted sand surround. This is partly attributed to the pipe being somewhat stiffer than the larger diameter pipes because of its design, but also to the small "span" (i.e the external horizontal pipe diameter) being an advantage in terms of the geometry of the soil arch.

Considering the various stress phases, it can be observed that negligible positive deformations occurred during installation (the maximum recorded VDS being 0.31%). Negative values were common after installation, the increase in vertical diameter being caused by sidefill placement, and particularly compaction, in layers. Once filling progressed above the springline, the effect was to increase load on the upper half of the pipe, thus compensating (either partly or wholly) for the reduction in horizontal diameter.

The 70kPa static stress phase produced similar trends in terms of the effect of surround conditions. The maximum recorded VDS was 2.7% (300mm pipe in uncompacted sand). Since this phase simulates a heavy construction vehicle parked on the trench with minimum cover to the pipe and no load-alleviating effects from a pavement structure (which would normally be the case) the performance, being within the 5% VDS limit, must be considered good.

The 70kPa cyclic phase produced some interesting trends. The deflection (both transient and permanent) builds up rapidly during the first passes of loading, then tends towards a residual value (see Figure 1). This phenomenon is the result of soil movements and densification within the surround as it forms a more stable, load-resistant structure. These laboratory results were confirmed in field work carried out on larger (600mm diameter) pipes, in which similar trends were noted.

The 140kPa static stress, representing an abnormally deep burial (for the UK), produced predictably higher deflections than the previous phases. For only one case was the 5% limit marginally breached (300mm pipe in uncompacted sand). This must be viewed in the context of the high stress, the poor state of the surround (not allowed under UK specifications) and the fact that the deflection was a cumulative effect of the three previous stress phases applied to the pipe by that stage of the test. A single application of such a stress at the start of the test would not have produced such a large deformation. The effect on the same pipe in a gravel surround (a more typical situation) produced an acceptable VDS of 2.8% under a stress of 140kPa.

Consideration of strain gauge data gives an insight into the deformed shape of the pipe. An elliptical deformation, assumed in the Iowa formula, produces the parallel plate strain profile shown in Figure 2. The parallel-plate loading test is untypical of normal installation conditions, as the surround (air) has zero stiffness and the pipe is supported only at two points (crown and invert). Profiles similar to this have been produced, mainly for uncompacted surrounds. Figure 2 also shows the strains produced by the 375mm diameter pipe in a gravel surround. The top of the pipe shows an approximately heart-shaped deformation (Rogers, 1988), but the lower half exhibits large tensile strains at 135° and 225°, and large compressive strains at 180°, the pipe invert. This is explained by the installation procedures, which involve feeding surround material to the haunches in order to provide support at these points. In this case, too much material was placed, resulting in a slight lifting of the pipe off the bedding layer, thus transferring the main contact area from the invert to the haunches. The reduction of support caused at the invert allowed it to deflect downwards and become more rounded (and hence producing compression at the inside wall).

An interesting effect was noticed for a stiff 600mm pipe in very highly compacted sand (see Figure 3). The high state of compaction produced a very stiff, load-resistant structure, and this led to an all-round distribution of stress, with very small deformations. The strains were slightly compressive, but virtually identical at all points where they were measured. This indicates an approximately hydrostatic stress distribution in dense surrounds under high loads.

The effects of pipe geometry were investigated, and can be seen on Figure 4. The strains at the "single wall" of the pipe were less than those at the twin wall section, confirming that the structural form of the pipe affects the stress distribution within it (i.e. that the corrugated profile resists the applied stresses). Shortening of the circumference may be estimated from strain gauge data. Under the 140kPa stress, the circumference of the 375mm pipe decreased by approximately 0.25mm.

Application of Deflection Theories to Test Results.

The two methods (Iowa and TRL) have been used both to predict deflections from known or estimated pipe and soil stiffness parameters, and to back-calculate soil stiffness parameters using measured deflections. The results are presented in Tables 2-5.

Rigorous predictions using the Iowa method (Tables 2 and 4) have been obtained using E' values of 7 (arguably high) and 20MPa for the lightly (LC) and heavily (HC) compacted sand respectively, together with a bedding factor of 0.103 (representing the mid-point between full underlying support and line loading). The predictions for LC sand are good for the two smaller (150mm and 225mm) pipes yet underestimate the deformations for those with a larger span. The mean of the back-calculated E' values for the 70kPa static stress was 7.0MPa (3.4-9.3MPa) and for the 140kPa stress was 5.3MPa (3.3-7.5MPa). The poorer performance of the pipe-soil system at higher imposed loading perhaps reflects the degree of "locked-in" stress caused by installation, thus reducing the effect of the smaller surface stress. The predictions for the HC sand are consistently overestimates, with back-calculated E' values of 94MPa (38-127MPa) and 74MPa (32-117MPa) for the 70kPa and 140kPa surface stresses respectively.

For the uncompacted river gravel, using E'=10MPa, the predictions were generally acceptably good, with some underestimation (1.7 vs. 2.6% being the greatest) and overestimates for the 375mm pipe. Back-calculated E' values were 13.3MPa (6.5-30.4MPa) and 10.7 (6.0-19.9MPa) for the 70 and 140kPa surface stresses respectively. The common perception (often quoted in design) that rounded gravel is essentially self-compacting and thus a higher design value (14 or 20MPa) can be used has been shown to be invalid in these tests. However, in UK practice the gravel is used as sidefill only and heavy compaction is applied to an overlying 300mm thick layer of (typically selected as-dug) backfill, thus causing some densification of the gravel and better performance of the installation. The use of E'=10MPa in practice would thus seem justified.

Both prediction and back-calculation using the TRL method were more problematic. The analysis is limited to the two larger pipes due to SDR considerations (α_z is quoted only for SDR>20). For prediction, a range of k_d and E_s^* has been used according to Gumbel's recommendations (and lack of data to refine them). The measured VDS in almost all cases lies in the upper portion of the range or exceeds the predicted VDS (by up to a factor of 1.9, see Tables 3 and 5). This implies the need for caution in parameter definition, perhaps by choosing the lowest values of k_d and E_s^* in the suggested ranges and even adjusting downwards the E_s^* values for uncompacted surrounds.

Back-calculation required interpretation from charts with logarithmic scales, thus was imprecise. The major difficulties, however, arose due to the interrelationship of the parameters. The value of α (required to define which chart to use) is dependent on Y, hence on E_s^* (the object of the back-calculation). In addition a value of k_d had to be fixed if a single value of E_s^* was to result. For this exercise values of k_d were fixed according to preconceptions of soil behaviour, yielding values of E_s^* that appear to be low and lack distinction between pipe surrounds. It should be noted that higher E_s^* values would result from the use of lower k_d values and that an iterative process (using E_s^* to give Y, hence α, hence E_s^*, etc) can be adopted in order to produce values that are closer to preconceived ideas of the parameter.

The back-calculated E_s^* values, which are not numerically dissimilar to E_s according to the above equation, would indicate a material of low stiffness (e.g. a soft clay). Gumbel et al (1982) state that "E_s^* is a measure of the bulk response of the backfill around the pipe (which) cannot be measured directly from laboratory tests on small soil samples." Thus E_s^* is not equivalent to a traditionally perceived Young's Modulus, due to the abnormal loading applied and the proximity of a flexible boundary (the pipe wall) since it is the soil immediately adjacent to the pipe that controls behaviour. The value is perhaps closer to being a bulk modulus (K), and one for a material that is being placed or compacted next to a flexible boundary, thus achieving only a low stiffness.

Conclusions

Deflection data for twin-wall corrugated HDPE pipe when buried in uncompacted river gravel and heavily and lightly compacted river sand demonstrated consistently good performance. Pipe wall strain data indicated the mechanism of structural resistance to both cyclic and static stresses and demonstrated that the corrugated section provides the main structural element (as might be expected). Installation procedures have a marked influence on the deformed pipe profile, especially in terms of feeding material under the pipe haunches to provide the main underlying support. A good, uniformly dense installation proved to cause an almost uniform compressive strain all around the pipe. Prediction and back-calculation techniques demonstrated that the Iowa formula (after Spangler, 1941) remains superior to the TRL method (after Gumbel, 1984) by virtue of its simplicity and, most importantly, its long period of use allowing judgements on parameter definition. Nevertheless, the TRL method provides a good insight into pipe performance and, with time and use, it is expected that this analytical technique will prove valuable.

References

British Standards Institution (1980). "Plastics pipework (thermoplastics materials) Part 6. Code of Practice for the installation of unplasticized PVC pipework for gravity drains and sewers", BS5955:1980.

Department of Transport (Highways and Traffic)(1989). "Highway Advice Note HA40/89 Determination of Pipe and Bedding Combinations for Drainage Works", HMSO, London.

Gumbel, JE, O'Reilly MP, Lake LM and Carder DR (1982). "The Development of a New Design Method for Buried Flexible Pipes", Proceedings of the Europipe 82 Conference, Basel, Switzerland.

Gumbel, JE (1984). "Analysis and Design of Buried Flexible Pipes", PhD thesis, University of Surrey, Guildford, UK.

Howard, AK (1977). "Modulus of Soil Reaction (E') Values for Buried Flexible Pipe", Journal of the Geotechnical Division of the ASCE, Volume 103.

Rogers, CDF (1988). "Deformed Shape of Flexible Pipe Related to Soil Surround Stiffness", Transportation Research Record 1191, TRB, Washington DC, USA.

Rogers, CDF, Loeppky, MWJ and Faragher, E (1994). "The Performance of Profile Wall Drainage Pipe in Relation to UK Specifications", Proceedings of the 4th International Conference on Pipeline Construction, Hamburg, Germany.

Spangler, MG (1941). "The Structural Design of Flexible Pipe Culverts", Bulletin 153, Engineering Experiment Station, Iowa State College, Iowa, USA.

Table 1. Experimental data at critical points of tests

Pipe Size	Soil	Sidefill Compaction		I	70S	70C	140S	END
100	RS	Not compacted	δv	0.07	0.8	2.7	2.8	2.6
			δh	-0.12	-0.8	-2.4	-2.4	-2.4
100	RS	Heavily compacted	δv	-0.17	-0.11	0.08	0.10	0.04
			δh	-0.01	-0.03	-0.19	-0.16	-0.17
100	RG	Not compacted	δv	0.03	0.4	1.2	1.2	1.2
			δh	-0.03	-0.3	-0.9	-0.9	-0.9
150	RS	Not compacted	δv	0.10	1.6	2.6	3.3	3.0
			δh	-0.10	-1.0	-1.9	-2.1	-2.1
150	RG	Not compacted	δv	0.15	1.3	1.9	2.3	2.1
			δh	-0.05	-0.7	-1.1	-1.3	-1.3
225	RS	Heavily compacted	δv	-0.12	0.1	0.1	0.2	0.1
			δh	0.06	0.0	-0.1	-0.1	-0.1
225	RS	Not compacted	δv	0.04	1.2	2.0	2.7	2.1
			δh	-0.07	-1.0	-1.9	-2.1	-1.8
225	RG	Not compacted	δv	0.14	1.0	1.5	1.9	1.6
			δh	-0.11	-0.7	-1.1	-1.3	-1.2
300	RS	Heavily compacted	δv	-0.31	0.0	0.1	0.4	0.2
			δh	0.40	0.3	0.2	0.1	0.2
300	RS	Not compacted	δv	-0.01	2.7	4.2	5.1	4.1
			δh	-0.05	-2.6	-4.2	-4.7	-4.2
300	RG	Not compacted	δv	0.18	1.5	2.2	2.8	2.3
			δh	-0.08	-1.2	-1.9	-2.5	-2.4
375	RS	Not compacted	δv	0.14	1.3	3.9	4.4	4.0
			δh	-0.03	-0.6	-2.6	-2.7	-2.6
375	RS	Heavily compacted	δv	-0.70	-0.6	-0.6	-0.5	-0.7
			δh	0.80	0.7	0.7	0.7	0.8
375	RG	Not compacted	δv	-0.30	0.0	0.3	0.6	0.2
			δh	0.30	0.1	-0.1	-0.2	-0.1

Legend and Sign Convention

Soil Types   RS = Well graded river sand
RG = Relatively uniform, rounded 10 mm gravel

Deflection δv =   Vertical diametral strain (% of mean external diameter)
δh =   Horizontal diametral strain (% of mean external diameter)

Table 2. Calculations for 70kPa static stress using the Iowa method

Pipe	k	E' (MPa)	VDS (T) (%)	VDS (M) (%)	E'(bc) (MPa)
150 RS LC	0.1	7	1.5	1.5	6.6
150 RG	0.08	10	0.9	1.1	6.5
225 RS LC	0.1	7	1.5	1.2	9.3
225 RS HC	0.1	20	0.6	0.2	126.5
225 RG	0.08	10	0.9	0.9	9
300 RS LC	0.1	7	1.5	2.7	3.4
300 RS HC	0.1	20	0.6	0.3	38.1
300 RG	0.08	10	0.9	1.3	7.3
375 RS LC	0.1	7	1.4	1.2	8.6
375 RS HC	0.1	20	0.6	0.1	116.9
375 RG	0.08	10	1	0.3	30.4

Key: RS = river sand   RG = river gravel   LC = lightly compacted
HC = heavily compacted   VDS(T) = theoretical VDS
VDS(M) = measured VDS   E'(bc) = back-calculated E'

Table 3. Calculations for 70kPa static stress using the TRL method

Pipe	Constants	VDS (T) (%)	VDS (M) (%)	E_s^* (bc) (MPa)
300 RS LC	k_d 0.3-0.5 E_s^* 5-20MPa	1.5 1.2 0.4 0.3	2.7	2.1 (k_d=0.4)
300 RS HC	k_d 0.3-1.0 E_s^* 15-35MPa	0.5 0 0.3 0	0.3	4.2 (k_d=0.9)
300 RG	k_d 0.4-1.0 E_s^* 20-30MPa	0.7 0 0.3 0	1.3	2.1 (k_d=0.7)
375 RS LC	k_d 0.3-0.5 E_s^* 5-20MPa	1.5 1.2 0.4 0.3	1.2	4.3 (k_d=0.4)
375 RS HC	k_d 0.3-1.0 E_s^* 15-35MPa	0.6 0 0.3 0	0.1	11.2 (k_d=0.9)
375 RG	k_d 0.4-1.0 E_s^* 20-30MPa	0.4 0 0.3 0	0.3	10.2 (k_d=0.7)

Key: k_d = dead load lateral pressure ratio E_s^*=soil plane strain modulus
E_s^*(bc)=back calculated soil plane strain modulus.
VDS(T) values are in order of increasing k_d (left to right) and increasing E_s^*

Table 4. Calculations for 140kPa static stress using the Iowa method

Pipe	k	E' (MPa)	VDS (T) (%)	VDS (M) (%)	E'(bc) (MPa)
150 RS LC	0.1	7	2.8	3.2	6
150 RG	0.08	10	1.7	2.2	7.5
225 RS LC	0.1	7	2.8	2.6	7.5
225 RS HC	0.1	20	1.1	0.3	72
225 RG	0.08	10	1.7	1.8	9.4
300 RS LC	0.1	7	2.8	5.1	3.3
300 RS HC	0.1	20	1.1	0.7	32
300 RG	0.08	10	1.7	2.6	6
375 RS LC	0.1	7	2.8	4.3	4.2
375 RS HC	0.1	20	1.1	0.2	117
375 RG	0.08	10	1.8	0.9	19.9

Key: RS = river sand   RG = river gravel   LC = lightly compacted
HC = heavily compacted   VDS(T) = theoretical VDS
VDS(M) = measured VDS   E'(bc) = back-calculated E'

Table 5. Calculations for 140kPa static stress using the TRL method

Pipe	Constants	VDS(T) (%)	VDS(M) (%)	E_s^*(bc) (MPa)
300 RS LC	k_d 0.3-0.5 E_s^* 5-20MPa	4.7 3.0 0.8 2.4	5.1	2.6 (k_d=0.4)
300 RS HC	k_d 0.3-1.0 E_s^* 15-35MPa	0.7 0 0.3 0	0.7	3.2 (k_d=0.9)
300 RG	k_d 0.4-1.0 E_s^* 20-30MPa	1.5 0 1.3 0	2.6	3.7 (k_d=0.7)
375 RS LC	k_d 0.3-0.5 E_s^* 5-20MPa	3.0 3.5 0.8 2.4	2.8	3.4 (k_d=0.4)
375 RS HC	k_d 0.3-1.0 E_s^* 15-35MPa	0.7 0 0.3 0	0.9	14.0 (k_d=0.9)
375 RG	k_d 0.4-1.0 E_s^* 20-30MPa	1.2 0 0.5 0	0.7	9.4 (k_d=0.7)

Key: k_d = dead load lateral pressure ratio E_s^* = soil plane strain modulus,
E_s^*(bc) = back calculated soil plane strain modulus.
VDS(T) values are in order of increasing k_d (left to right) and increasing E_s^*

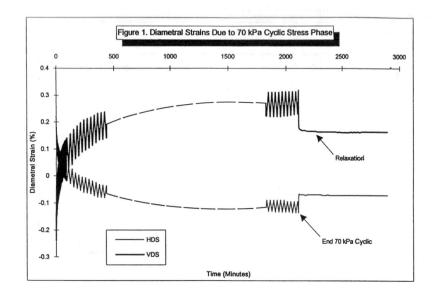

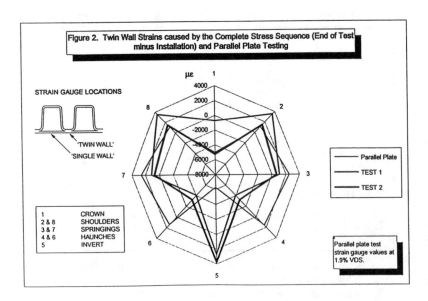

BURIED PLASTIC PIPE

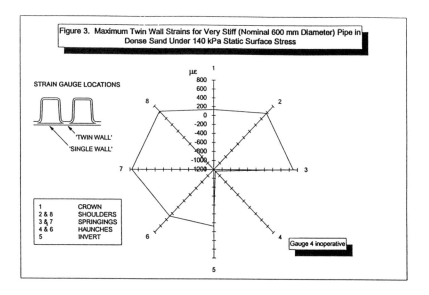

Figure 3. Maximum Twin Wall Strains for Very Stiff (Nominal 600 mm Diameter) Pipe in Dense Sand Under 140 kPa Static Surface Stress

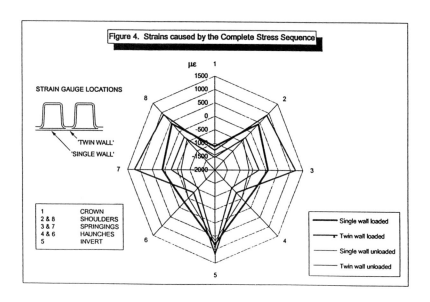

Figure 4. Strains caused by the Complete Stress Sequence

INSTALLATION TECHNIQUE AND FIELD PERFORMANCE
OF HDPE, PROFILE PIPE
Larry J. Petroff [1]

ABSTRACT

Designers are often surprised to discover that the field performance of their buried, flexible pipes differs markedly from the performance they anticipated when designing according to the "textbook". Quite often these differences are attributed to variances in the embedment and the insitu soil properties as most design equations and design parameters were developed through analytical mechanics, laboratory studies, or computer models based on ideal conditions. But, the root of the unanticipated performance often lies in the installation technique. Attempts have been made to modify design equations to account for construction practice by using "add-on" deflection factors. However, after reviewing nearly 15 years of buried applications of large-diameter, HDPE profile pipes, it seems that in addition to these modifications, the designer must understand the effect local construction methods have on pipe performance. With this knowledge, the designer can produce a specification that will minimize unexpected variances from the design.

This paper reports on deflection data from 28 installations of HDPE, profile pipe and shows how construction techniques effect soil parameters such as the modulus of soil reaction, E', and deflection variability. HDPE, profile pipe is an ideal pipe for this study as its large diameter allows for easy access and it reacts immediately to external load changes. It will be argued that construction considerations often may have more influence on deflection performance than the embedment type, insitu soil stiffness, or pipe stiffness.

---

[1] Engineering Supervisor, Technical Services, Chevron / PLEXCO, 1050 Busse Highway, Suite 200, Bensenville, IL 60106

INTRODUCTION

If the pipe could be inserted into the ground without any disturbance to the insitu soil, then elastic solutions, i.e. Hoeg's [1], would produce predictable designs. Returning excavated soil to its undisturbed conditions is a goal even the most dedicated and conscientious contractor cannot achieve. Empirical design methods, such as Watkins-Spangler's Iowa Formula, account for some construction effects in the deflection calculation by using field determined parameters, ie. E'. However, this is not always sufficient. Howard [2] and Molin [3] have suggested "add-on" factors to apply to deflection. This paper attempts to improve on these approaches by suggesting a statistical approach to design coupled with specifying construction techniques that maximize embedment performance.

EVALUATION OF FIELD PERFORMANCE

Construction procedures were examined on 28 field installations [4]. Data were gathered from specifications, profile drawings, geotechnical reports, compaction records, and field logs. Information obtained on each project included pipe properties, soil type, groundwater levels, trench geometry, compaction methods, shoring techniques, inspection levels, construction quality, and vertical diameter measurements.

Deflection Data Collection and Evaluation

For the pipes in the study, vertical diameter measurements were collected using an extension ruler or a deflection caliper followed by a closed circuit television. Some end-user supplied measurements were included. The simplest and most accurate method was to enter the pipe and measure with an extension ruler. Allowing error for reading and for locating the vertical diameter, accuracy ranged from one-eighth to one-quarter inch, which is less than 1 percent of the diameter for the smallest pipes measured. The accuracy of the calipers was not quite as good due to limitations in reading the caliper's scale and its tendency to rotate.

Given that measurements can be made with a reasonable degree of accuracy, there are limitations to their interpretation. The purpose of measuring is to establish the change in diameter in response to the earth loading. On a flexible pipe, this can be obtained only if the diameter prior to backfill above the crown is known. Except in a research study, it is difficult to convince the installing contractor to wait while initial measurements are taken. For most of the projects reported herein, no readings were taken prior to backfilling. Except where initial diameters were known such as strutted pipes, deflection was calculated from an assumed reference diameter equal to the nominal diameter less manufacturing tolerances. Any out-of-roundness in the pipe is "averaged" out by random orientation when laying. (The compaction levels observed in the projects were rarely sufficient to cause significant "rise".) While this method gives a fairly accurate value for the average deflection caused by earth loading, it gives misleading results for maximum deflections, as diameter changes may be the result of out-of-roundness rather than a response to earth load.

Time-dependent Deflection

Interpreting measurements is further complicated by the fact that deformations are time-dependent. The pipe begins deflecting from the time the first lift of backfill is placed over it. The type of backfill and its compaction level are significant factors in determining the deflection of the pipe. This process is not effectively represented by Spangler's lag factor. The rate of change of deflection with time is better characterized by viscoelasticity. Fig. 1 shows deflection versus time data for a 54" diameter pipe. The data plots as a straight line on a semi-log scale, like almost all deflection data this writer has plotted. Linearity on the semi-log scale is a result of the soil's viscoelastic nature [5]. This linearity has two useful results. First, long-term deflections can be predicted by simply extrapolating short-term data. Second, deflection data from two different projects can be adjusted to the same time period for comparison.

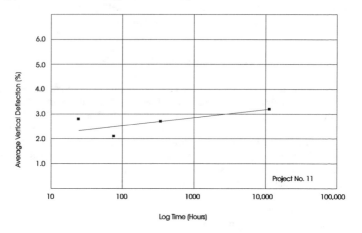

Figure 1. Deflection versus Time

Reduction of Data

The study produced a very large number of individual deflection measurements. Statistics were used to evaluate the data. For each line measured, the vertical diameters were grouped into 1/8 inch to 1/4 inch increments. The frequency of points within each increment was determined. From this grouping, the frequency distribution of the various increments were plotted against the cumulative relative frequency on probability paper. A straight line relationship indicates a normal distribution, as shown by the upper curve in Fig. 2. Visual observation of these graphs for a large number of projects indicate that the distribution of deflection points in the majority of cases approached the normal distribution, with the exception of a few curves which had "outliers" toward the high deflection end. This may be a result of inconsistent construction techniques.

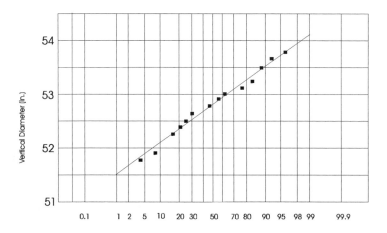

Figure 2. Deflection Data Distribution

Deflection Data

Deflection data gathered in this study are summarized in the tables given in the Appendix. Tables A1 through A4 are pipes in compacted embedments in the following insitu soils; (1) dry sand, (2) rock or dry, cemented soils, (3) medium to stiff clay (wet), and (4) wet, loose sands or soft clays. Tables A5 and A6 are for uncompacted embedments in medium to stiff clay and in wet, loose sands or soft clays. Table A7 gives miscellaneous conditions.

Statistical Analysis

The mean and standard deviation characterize the normal distribution. These values were calculated for each project. The standard deviation is reported as a coefficient of variation (COV), which is given as a percentage of the diameter.

$$COV = \frac{STD.DEV.}{AVG.DIAM.} (100) \quad (1)$$

For example, on Project 09, the average of the vertical diameter measurements at 10 months was 46.68" with a standard deviation of 0.42". The average vertical deflection, calculated from the nominal I.D. less manufacturing tolerances (47.75"), was 2.2% with a COV of 0.9%.

Statistical Maximum Deflection

Most designers limit maximum deflection. Therefore, information regarding maximum deflections is useful. As discussed above, diameter changes found without a reference reading prior to backfilling are misleading, therefore the maximum deflections obtained in this study are of little value for understanding the pipe's response to earth load. However, a good approximation of the actual maximum deflection due to earth loading can be obtained by statistically analyzing the data. As it turns out, on the 28 projects studied the actual observed maximum deflection values were typically within a percent or less than the maximum values arrived at statistically. This is consistent with the fact that most projects had normally-distributed deflection data.

The statistical maximum deflection can be determined from Eq. 2, by selecting the desired confidence level. To determine the maximum deflection with a 96 percent confidence level, Z equals 1.76. For other confidence levels consult a statistic text.

$$Max\ Defl = Avg\ Defl + (Z)(COV)(Avg\ Defl) \quad (2)$$

For instance, Project 01a had an average deflection of 2.8 percent and a COV of 1.6 percent. The statistical maximum deflection equals 5.6 percent while the measured maximum deflection was 5.4 percent.

CONSTRUCTION VARIABILITY AND INSTALLATION TECHNIQUES

Most plastic pipes are flexible and are designed to use external soil support to resist deflection and buckling. The chief objective of installation is to construct an embedment that provides and maintains firm, continuous support to the pipe. The following factors strongly influence the support; embedment material quality, placement, and densification, trench shoring techniques, groundwater, and workmanship. Variability in these factors results in changes in deflection.

Densification of Embedment

Probably, the most significant factor of the above factors is the embedment material placement and densification. The projects studied were all installed with granular embedment with the exception of Project 16 (Table A7). Specifying a good quality granular embedment alone was found not to be sufficient to control deflection. A wide range of performance was seen with granular embedments depending primarily on their method of placement. Back-calculated E' values ranged from just under 300 psi for the dumped sand on Project 23a to over 5000 psi for the mechanically compacted crushed stone on Project 04.

The most effective means of maximizing the supporting resistance of an embedment was found to be mechanical compaction with an impact tamper or a plate vibrator. It is often thought that crushed stone will achieve a near maximum density when dumped into the trench. On site density testing found that for dumped crushed stone the density ranged from about 84 to 88 percent Std. Proctor. Whereas lightly vibrating or tamping the same material produced densities excess of 90 percent. This small difference in density was found to be very significant in controlling deflection. Compare the average deflection values in Table A4 (compaction) with those in Table A6 (no compaction).

Other means of effecting densification such as by using concrete stingers were found not as effective. On a project not reported herein, the use of stingers to densify pea gravel for a 54" pipe with 25 ft of cover produced an E' of approximately 1000 psi. Stingers seem to perform slightly better than shovel slicing, which is probably the least effective method of compaction and herein is not considered mechanical compaction. Both stingers and shovel slicing are very sensitive to effort. Where shovel slicing is carried out in small lifts on angular stone, and thoroughly done, it may approach an E' of 1000 psi. On Projects 12 and 20 which had wet ground and soft soils the E' for shovel slicing stone was found to be about 500 psi.

Placing granular material, including crushed stone, without mechanical compaction or shovel slicing results in large deflections, except in very shallow cover. Compare Projects 23a with 23b. Even with permanent sheeted trench walls, large deflections were experienced with dumped crushed rock. See Project 19. Projects 8a and 8b show the improvement gained from shovel slicing dumped stone. The projects studied suggest that the E' for dumped or lightly shovel sliced crushed rock did not exceed 500 psi.

Trenching Methods

Trenching is important because it often determines the stability of the embedment. Improper installation techniques can disturb the insitu soil and create voids or loose zones into which the embedment can move, thus reducing side support. A key objective then, is to excavate and place embedment material while minimizing disturbance to the insitu soil.

Unshored Trenches

Generally, unshored trenches are untilized in stable ground, which tends to make installation easier. Except for excavations in soft rock or stiff clays, some sloughing or raveling of sidewall materials may occur that reduces support. The embedment soil is placed directly against the insitu soil. In this case, there are extra reasons for mechanical compaction; (1) it stiffens the trench sidewall and (2) densifies sloughed in materials. Dewatering of wet, soft ground, often creates stable trenching conditions.

Tables A1 through A3 show that for unshored trenches with mechanically compacted embedment in either dry ground or firm to stiff wet ground, the average deflection is quite low, less than 3 percent. (The absolute minimum diameter found within these 866 measurements from these twelve projects ranging in cover depths up to 35 ft was only 6.6 percent smaller than the nominal diameter.)

In Table A4, Projects 06, 17, 22, 23, and 25 were installed in soft ground either in unshored trenches or with trench shields kept essentially above the top three-quarter point of the pipe diameter. The average deflections and COV's of these pipes do not vary significantly from the respective values reported in Tables A1 through A3. This indicates that the soft ground probably offers as much support to the embedment material as stiffer ground, when the embedment is mechanically compacted. (This has not been verified for very soft grounds such as loose organic silts or bay muds.) For uncompacted embedments, it appears that deflections may increase in soft ground. Compare Tables 5A and 6A.

Permanent Shoring

Permanent shoring normally consists of piles with timber lagging or wooden boxes. Recently concrete "sleds" have been used. The shoring eliminates many of the construction problems involved with soft ground, however, embedment densification is still critical. For instance, on Project 19 crushed stone was dumped around the pipe and the resulting back-calculated E' was 400 psi.

Temporary Shoring

Both sheet piling and trench shields are used for temporary shoring. If not properly used, large deflections may result from disturbance to the insitu soil. The biggest problem with sheet piling is its removal after placement of the embedment. When corrugated piling is used, extraction often removes insitu soil. This void will fill with embedment material, thus reducing side-support. Use of a vibratory extractor may cause not only loss of ground but liquefaction followed by the pipe deflecting excessively. Restrict piling to thin, non-corrugated sections, unless a test section has been successfully installed.

The use of the trench shield in many parts of the country is almost mandatory. Proper use of this device is essential to protect life. When installing rigid pipes, that have been designed without any benefit from soil/structure interaction, the installer will normally set the shield on pipe grade and drag the shield along the trench line as pipe is installed. This practice is discouraged with flexible pipes for two reasons: (1) The soil outside of the shield is excavated down to trench grade, which remains disturbed after the shield is moved. (2) A large void between the embedment and the insitu soil is created when the shield is moved. Many flexible sewer pipes have experienced larger than design deflections because the installer followed techniques used for rigid pipes.

On the projects reported herein, excavation below the pipe crown elevation was done from within the shield, except for Project 15. That is, the mechanical excavator removed soil from within the shield and forced the shield down as it dug. This minimized the disturbance between the shield and the insitu soil.

When shield is moved forward on pipe grade elevation, any densification of materials within the shield is totally lost. To minimize deflection, embedment must be compacted directly against the undisturbed insitu soil. One technique is to lift the shield in stages as the embedment is placed. Another is to place thin plates extending beneath the shield. This technique was used on Project 11 and 14 with excellent results. (An engineer should review the anticipated technique to assure its safety.)

STATISTICAL DESIGN METHOD FOR MAXIMUM DEFLECTION

So far, this paper has discussed field reported deflections, variability and how various installation techniques affect predicted deflection. This section will discuss how this information may be used for design. A designer can predict maximum deflection statistically, if the COV is known. The designer can calculate the predicted deflection using average values of design parameters, such as the E'. Then the maximum deflection can be found using Eq. 2 with a Z value for the desired confidence level. COV's observed on the reported projects are listed in the Tables. However, these values may be unique to HDPE, profile pipes or pipes of similar stiffness and ductility. Table 1 can be used to help the designer select a COV. Table 1 gives estimated COV values as a fraction of the design average deflection. The designer should be aware of the limited scope of the data base used to develop Table 1. Therefore, it is suggested that these values be used only for qualitative analysis until more experience is gained using this method. The range of the variability in Table 1 is consistent with the range reported by Molin [3]. Since his study was for stiffer pipes than reported herein, pipe stiffness is probably not a significant factor in determining variability in the pipe stiffness range of plastic pipes.

Table 1. Ratio of COV to Average Deflection[1]

		Std. Proctor Density	COV / Avg.Defl.
Dry Soil, Rock or Firm Saturated Soil	Mechanical Compaction	≥ 90	0.4 - 0.6
	Dumped or Shovel Slice	< 90	0.4 - 0.6
Wet, Loose Soil or Soft Soil	Mechanical Compaction	≥ 90	0.5 - 0.8
	Dumped or Shovel Slice	< 90	0.5 - 0.8

[1] Factor does not account for out of roundness of pipe. Therefore use Base I.D. for reference diameter for acceptance inspection.

CONCLUSIONS

Statistical design will account for some of the construction and soil variability, however it assumes reasonable construction techniques are used to install the pipe. To achieve field deflection close to the design (predicted) deflection, the designer must determine, then specify, construction techniques that maximize performance of the embedment materials by achieving proper densification and minimizing disruption to insitu materials.

REFERENCES

[1] Hoeg, K. (1968) "Stress Against Underground Structural Cylinders". J. Soil Mech. and Found. Div., ASCE, No. SM1, July, pp.833-858.
[2] Howard, A.K. (1981) "The USBR Equation for Predicting Flexible Pipe Deflection", Proc., Conf. on Underground Plastic Pipe, ASCE, New Orleans.
[3] Molin, J. (1985) "Long-Term Deflection of Buried Plastic Sewer Pipes, Proc., Intl. Conf. on Advances in Underground Pipe Line Engineering, ASCE, Madison, WI.
[4] Chua, K.M. and Petroff, L.J. (1988) "Predicting Performance of Large-Diameter Buried Flexible Pipes", Proc., 2nd Intl. Conf. on Case Histories in Geotechnical Engrg., St. Louis.
[5] Chua, K.M. and Lytton, R.L. (1989) "Viscoelastic Approach to Modeling Performance of Buried Pipes." J. Transp. Engrg. Div., ASCE, 115(3), 253-269.

APPENDIX A1. Deflection Measurements

Table A1. Summary of Measurements for Dry Sandy Soil with Mechanically Compacted Crushed Stone Embedment. Unshored Trench.

Project No.	DIA (in)	PS (psi)	Depth (ft)	Time	Avg. Defl. (%)	COV (%)	N
01a	36	7.0	15	4y	2.8	1.6	66
02	18	39.2	17	1.5y	1.3	1.6	120
07	36	12.2	15	3d	0.9	2.5	17

Table A2. Summary of Measurements for Rock or Dry, Cemented or Lightly Cemented Insitu Soil with Mechanically Compacted Crushed Stone Embedment. Unshored Trench.

Project No.	DIA (in)	PS (psi)	Depth (ft)	Time	Avg. Defl. (%)	COV (%)	N
01b	36	12.2	22	4y	1.8	1.4	378
03	42	7.8	10	1m	1.7	1.1	42
04	42	41.6	35	2w	1.0	0.4	16
05	36	19.4	24	10d	1.0	0.5	29
13	54	7.0	20	10m	2.8	1.3	29
21	72	6.4	22	3m	2.5	1.1	120

Table 3A. Summary of Field Measurements in Medium to Stiff Clay with Mechanically compacted Crushed Stone Embedment.

Project No.	DIA. (in)	PS (psi)	Depth (ft)	Time	Avg. Defl. (%)	COV (%)	N	Trench Shield
09	48	20.8	24	10m	2.2	0.9	83	Yes
10	36	12.2	11.5	2y	2.3	1.2	47	No
26	24	12.7	17	1.5y	2.7	1.1	12	No

d = day, m = month, y = year, N = no. of measurements

TABLE A4. Summary of Field Measurement in Wet, Loose Sands or Soft Clays with Mechanically Compacted Embedment.

Project No.	DIA. (in)	PS (psi)	Depth (ft.)	Time	Avg. Defl. (%)	COV (%)	N	Trench Shield (Box)
06	42	7.8	15.5	2.1y	2.5	1.5	39	yes
11	54	14.7	12.5	15m	3.7	1.0	34	yes
14	60	4.3	12	3y	2.1	1.0	50	yes
15	24	26	13	1y	1.5	1.9	548	yes
17	30	20.7	20	2w	1.3	1.1	13	yes
22	24	17	15	1d	2.1	1.3	5	no
23a	30	13.7	16	7d	1.5	0.8	14	yes
25	36	24.2	10	1d	0.3	0.7	10	no

Table A5. Summary of Measurements in Rock or Medium to Stiff Clay with Crushed Stone Embedment without Mechanical Compaction.

Project No.	DIA. (in)	PS (psi)	Depth (ft)	Time	Avg. Defl. (%)	COV (%)	N	Box	Compact
09	48	20.8	20	11m	2.0	.8	17	yes	SS
13	54	5.8	25	3w	5.4	2.3	22	no	Dump
21	48	8.3	10	5m	4.4	2.2	68	no	?
27	36	7	15	3m	2.6	1.3	10	no	SS
28	24	17	7	1m	2.5	1.0	77	?	Dump SS

d = day, m = month, y = year, N = Number of measurements, Depth = Depth to Springline, SS = Shovel Slice

Table A6. Summary of Measurements in Wet, Loose Sands or Soft Clays without Mechanically Compacted Embedment.

Project No.	DIA (in)	PS (psi)	Depth (ft)	Time	Avg. Defl (%)	COV (%)	N	Box	Compact
08a	48	5.2	11	18d	2.3	1.2	30	no	SS
08b	48	5.2	11	1m	4.1	3.5	112	no	Dump
12	36	7	13.5	5m	4.2	2.1	59	no	SS
19	48	8.3	15	14m	6.0	0.6	8	P	Dump
20	36	7	11.5	6m	4.6	2.7	135	yes	SS
23b	30	13.7	16	5d	6.4	2.4	20	yes	Dump

Table A7. Summary of Measurements for Miscellaneous Applications in Stiff Clay with Sand Seams.

Project No.	DIA (in)	PS (psi)	Depth (ft)	Time	Avg. Defl (%)	COV (%)	N	Box	Compact (Bed)
16	18	39.2	7.5	27m	2.5	2.8	41	no	Dump Clay
18	48	5.2	12	6m	3.3	-	-	no	Dump (Cem.Sa)
24	36	12.2	15	1d	2.2	2.1	12	yes	? (Cem.Sa)

D = Diameter, d = day, m = month, P = permanent sheeting, N = number of measurements, Cem.Sa = Cement stabilized sand, Bed = Bedding Type

Time-Deflection Field Test of 120-cm Steel, Fiberglass, and Pretensioned Concrete Pipe

Amster Howard [1]

Abstract

In 1971 the Bureau of Reclamation constructed a special test section of 120-cm diameter reinforced plastic mortar (RPM), plain steel, and pretensioned concrete (PT) pipe at the Denver Federal Center. During construction, measurements were made of pipe deflections, soil pressures on the pipe, soil strain around the pipe, soil properties, and in-place densities. The elongation of the pipe (increase in vertical diameter) was carefully monitored during compaction of soil at the sides of the pipe. The pipe deflections have been continually measured over the past 22 years.

Introduction

The load-deflection behavior of three types of 120-cm diameter flexible pipe was evaluated using a special field test installation which was constructed in 1971. Sections of reinforced plastic mortar (RPM) [a type of fiberglass pipe], plain steel, and pretensioned concrete (PT) pipe were installed in an existing drainage channel on the Denver Federal Center (DFC) in Colorado. Two different soil types were used as embedment material. The installation was done in accordance with the United States Bureau of Reclamation (USBR) specifications that were in effect in 1971.

This paper is one of a series by the author to document the results of field studies so the resulting data may be compared with other studies to gain a better understanding of the behavior of buried flexible pipe (Howard 1974) (Howard 1990) (Howard, Spridzans, and Schrock 1994).

Purpose of Study

Flexible pipe is designed so the soil at the sides of the pipe helps support the backfill load on the pipe. As the load on the pipe increases, the vertical diameter of the pipe decreases and the horizontal diameter increases. The increase in the horizontal diameter

---
[1]Consulting Civil Engineer, 1562 S. Yank St., Lakewood CO 80228

is resisted by the soil strength at the sides of the pipe. The side soil support must be strong enough to carry this load without the pipe deflecting significantly. At the time of construction of this test section, a single value for soil stiffness (E prime) was used in the Iowa Formula for predicting the deflection of buried flexible pipe (Spangler 1941). This study helped show that the soil support depends on the type of soil and its degree of compaction (Howard 1977) (Howard 1981a).

After the initial deflection occurs, a flexible pipe continues to deflect with time. The full weight of the backfill load does not reach the pipe immediately, and the soil at the sides of the pipe will compress with time. At the time of this study, minimal data was available on this deflection increase with time. Accordingly, the study had two objectives:

> 1. The first objective was to evaluate the initial load-deflection behavior of RPM, steel, and PT pipe installed in two different types of soil placed at a high degree of compaction. Deflections of the pipe were measured during the construction sequence up until the time the final backfill was placed over the pipe.

> 2. The second objective was to evaluate the long-term "deflection-lag" behavior of the three types of pipe. Following the final backfilling over the pipe, the pipe deflections were measured periodically for 22 years.

The test installation was 50 m in length on the bottom of an existing drainage channel and was divided into two sections, with each containing 6.1 m lengths of PT, steel, and RPM pipe laid end to end. Two additional sections of RPM pipe were used as inlet and outlet transitions. The joints were sealed by wrapping them with a 300-mm-wide sheet of 6-mm thick plastic, vinyl-backed asphalt membrane. The installation drawings are shown on figure 1.

The bottom of the existing channel was excavated down to firm *Denver Formation,* a siltstone. The trench bottom was then covered with a 0.6-m thick layer of 19-mm to 37.5-mm clean gravel followed by a 0.3-m thick layer of 9.5-mm to 19-mm clean gravel to provide a uniform foundation for the pipe bedding material and to carry the anticipated flow of intercepted groundwater beneath the pipe. Each lift was compacted by a crawler-tractor. A 0.3-m thick layer of concrete sand was placed above the gravel to provide the bedding for the pipe. The sand was compacted by the crawler-tractor to 82 % *relative density.* A cradle for the pipe was then trimmed in the sand. After the pipe was placed on the sand, the pipe in one-half of the test section was installed using a clayey sand (SC) compacted beside the pipe as embedment while the pipe in the other half was installed using a poorly graded sand (SP). The soil was compacted by mechanical tampers using 15 cm compacted lifts up to a height of 0.7 of the pipe outside diameter. The clayey sand was compacted to an average density of 95 % Proctor and the sand was compacted to an average of 75 % *relative density.*

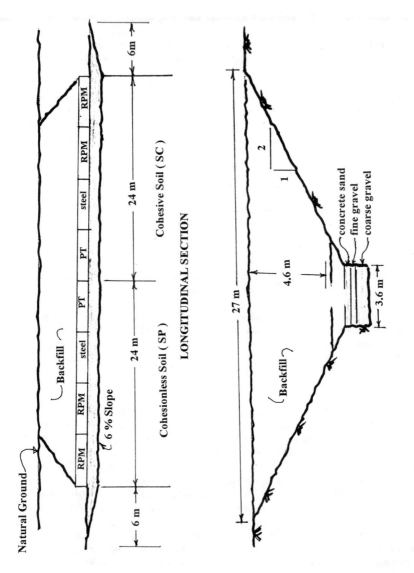

Figure 1. Installation Details

Uncompacted backfill was then placed above the pipe in approximate 0.6-m to 0.8-m thick lifts until a total cover depth of 4.6 m was reached. During this time, periodic measurements were made of the pipe deflection and of the soil pressures on the pipes for each increment of fill load applied. Other data obtained included deflection measurements (horizontal and vertical) of the RPM pipe joints and of the inlet-outlet transition pipe.

Pipe diameter measurements were made at the 1/3 points of each pipe (two measurements stations for each pipe) at 15 degree intervals around the inner circumference at each location; a total of 12 diameter measurements were obtained at each location.

The properties of the pipe used in this study are given in table 1.

Table 1. - Pipe Properties

Description of Pipe	Inside Diameter cm (inch)	Wall thickness mm (inch)	EI m-kg (in-lb)	EI/r^3 kN/m^2 (psi)
RPM pipe class I type II gasket (USBR design RPM 325 for H = 15 feet)	120 (48)	13 (1/2)	300 (27,000)	14 (2.0)
Steel pipe	120 (48)	5 (3/16)	190 (16,500)	8.4 (1.2)
Pretensioned concrete pressure pipe	120 (48)	50 (2)	8800 (760,600)	380 (55)

Cylinder gage = 10   bar size = 3/8 at 1.16 in spacing
Pretension stress = 12 kips/in²   (AWWA C303 - 70)

Concrete sand, a cohesionless free-draining material, was used as the embedment for the pipe installed in the upstream half of the test section. The sand was classified as a *POORLY GRADED SAND* (SP) and contained only 2 % fines. Laboratory tests (Designation E-12, Earth Manual, 1963) indicated a minimum dry density of 1.54 Mg/m³ (96.0 lb/ft³) and a maximum dry density of 1.88 Mg/m³ (117.4 lb/ft³). The USBR compaction test is considered equivalent to the standard Proctor test (ASTM D 698).

Nineteen in-place density tests were performed on the embedment material using the *water balloon* method. The in-place dry densities ranged from 1.68 to 1.88 Mg/m^3 (104.7 to 117.3 lb/ft^3) with an average value of 1.78 Mg/m^3 (111.2 lb/ft^3). The average in-place density corresponds to a *relative density* of 75 %.

A cohesive soil was used as the embedment material for pipe installed in the downstream half of the test installation. This material was classified as a *SANDY CLAY* (SC). The material (sample No. 53K-5) contained 56 % sand and 44 % fines having a liquid limit of 39 and plasticity index of 23. The laboratory maximum dry density was 1.81 Mg/m^3 (112 lb/ft^3) at an optimum moisture content of 14.7 % (Designation E-11, Earth Manual, 1963).

During placement of the embedment, 14 in-place field density tests were performed using the *water balloon* method. Values of dry density ranged from 1.58 to 1.82 Mg/m^3 (98.8 to 113.5 lb/ft^3) with an average value of 1.71 Mg/m^3 (106.9 lb/ft^3), which corresponds to 94.7 *percent compaction* (sufficiently close to specifications requirements of 95 *percent compaction)*. The placement moisture content ranged from 13.8 % to 19.0 % with an average of 16.0 %, which is about 1.3 % above optimum moisture content (within spec. limits of plus or minus 2 % of optimum moisture).

Pipe Elongation During Embedment

Flexible pipe can elongate (increase in vertical diameter and decrease in horizontal diameter) due to compaction of the soil alongside the pipe. The diameters (horizontal and vertical) of the pipe were measured with the pipe resting in place on the trench bottom before any embedment soil was placed. Diameter measurements were again made when the compacted embedment had been completed; that is, compacted up to 0.7 of the outside diameter of the pipe. The next diameter measurements were made when 0.5 m of backfill had been placed over the pipe. Usually, any elongation ceases when the compacted bedding reaches 0.7 o.d. and then any backfill placed over this level results in deflection of the pipe (vertical diameter starts decreasing). However, in this study, placement of the first 0.5 m of backfill resulted in additional vertical elongation of the pipe. The construction equipment was not permitted to work over the pipe until at least two feet of backfill covered the pipe. Additional pressure was placed on the sides of the pipe since the construction equipment could only work along the side, which resulted in additional elongation of the pipe. Subsequent layers of backfill over the pipe did result in deflection. Thus, the point where elongation changed to deflection occurred at 0.5 m of backfill. All deflections were calculated assuming the diameter measurement at 0.5 m of cover as the initial reading. Table 2 summarizes the elongation of the pipe at 0.5 m of cover.

As the pipe stiffness (EI/r^3 value) increased, the elongation decreased. The percent vertical elongation values appear to be typical based on other reported values (Howard 1981b) (Howard, Spridzans, and Schrock 1994).

Table 2. - Pipe Diametral Elongation

Type of Pipe	% average elongation with soil at 0.5 m over pipe	
	Vertical ΔY	Horizontal ΔX
COHESIVE EMBEDMENT (SC)		
RPM	- 1.2 %	- 1.0 %
Steel	- 4.0	- 3.7
PT	- 0.2	- 0.3
COHESIONLESS EMBEDMENT (SP)		
RPM	- 0.8	- 0.9
Steel	- 2.2	- 2.1
PT	- 0.2	- 0.2

Pipe Deflection During Backfilling

Flexible pipe deflects (decreases in vertical diameter and increases in horizontal diameter) due to backfill load on the pipe. The initial diameter (or zero) reading for calculating deflection was the pipe diameter measured when the backfill had been placed to 0.5 m over the top of the pipe (see discussion on pipe elongation). From this zero point, any changes in pipe diameters are due to backfill placed over the pipe. Pipe deflection versus height of cover is shown in figure 2. Table 3 summarizes the average vertical and horizontal deflection at 4.6 m of cover:

Table 3. - Percent Deflection at 4.6 m of Cover

Type of Pipe	Vertical ΔY	Horizontal ΔX	Ratio $\Delta X/\Delta Y$
COHESIVE EMBEDMENT (SC)			
RPM	0.7 %	0.7 %	1.0
Steel	1.2	0.7	0.6
PT	0.5	0.6	1.2
COHESIONLESS EMBDMNT (SP)			
RPM	0.4	0.3	0.8
Steel	0.4	0.2	0.5
PT	0.1	0.1	1.0

278 UNDERGROUND PIPELINE ENGINEERING

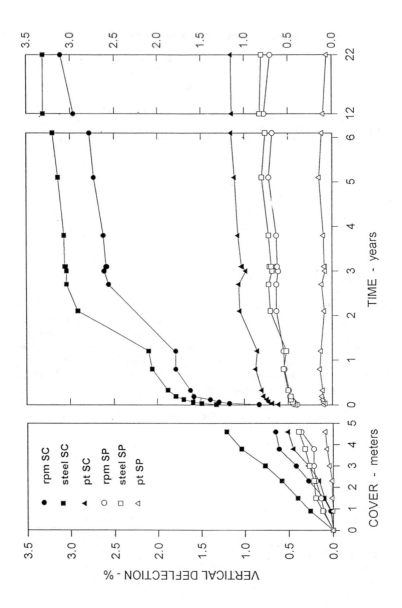

Figure 2. Pipe Deflection due to Backfill and Time

The percent deflection was directly related to the type of embedment soil and to the pipe stiffness. The deflection of the pipe with the cohesionless embedment was about one-half or less of the deflection of the pipe with the cohesive embedment. As the pipe stiffness (EI/r^3 value) increased, the deflection decreased.

However, the main conclusion is that regardless of the type of pipe, or the soil type, the initial deflections were all less than 2 %. The degrees of compaction used in this study were the minimum accepted values in Bureau of Reclamation specifications. Thus, the study verified that the specification requirements would result in acceptable pipe deflections of flexible pipe, even when the results are extrapolated to 6 m of cover (maximum depth of cover for USBR standard pipe design in 1971).

The ratio of $\Delta X / \Delta Y$ shows that the RPM and PT pipe deflected in an elliptical pattern while the steel pipe was deforming rectangularly (Howard, 1981b). Flexible pipes have been assumed to deform elliptically under load with the horizontal deflection roughly equal to the vertical deflection. In a *three-edge bearing test* or *parallel plate test*, flexible pipe deform elliptically with the horizontal deflection equal to 91 % of the vertical deflection. However, when buried, if the soil stiffness at the sides of the pipe is large compared to the pipe stiffness, pipe can deform rectangularly and the horizontal deflection is much less than the vertical deflection. In this study, the vertical and horizontal deflections were about the same for the RPM and the PT pipe; however, the horizontal deflections for the steel pipe were about 50 - 60 % of the vertical deflections.

Net change in pipe diameter

The net change in pipe diameter from measurements made when the pipe was in place on the trench bottom before embedment to measurements made after backfilling was completed is shown on table 4.

Table 4. - Net Pipe Deflection During Installation

Type of Pipe	Vertical Elongation %	Vertical Deflection %	Net Change %
COHESIVE EMBDMNT (SC)			
RPM	- 1.2	0.7	- 0.5
Steel	- 4.0	1.2	- 2.8
PT	- 0.2	0.5	0.3
COHESIONLESS EMBDMNT (SP)			
RPM	- 0.8	0.4	- 0.4
Steel	- 2.2	0.4	- 1.8
PT	- 0.2	0.1	- 0.1

On the day that the 4.6 m of cover was completed, only one pipe out of six had returned to its original diameter. The other five pipe were still vertically elongated.

Saturation of Pipe Backfill

Three years following the completion of the test section, an attempt was made to increase the load on the pipe to see what corresponding increase in deflection might occur.

To increase the load on the pipe, the backfill over the pipe was wetted since more conventional methods of adding surcharge were not readily available. A series of gravel-packed infiltration wells and interconnecting trenches were constructed in the center of the test installation. Water obtained from the pond immediately upstream from the test section was pumped into the trenches, keeping the infiltration wells continuously full. Pipe deflection measurements were obtained after a 30-day period of soaking.

From calculations performed to determine available void space in the backfill, it was determined that the backfill above the pipe was about 75 % saturated. Water was entering the pipe at several locations through the pipe sleeves used for installation of the soil pressure cells.

Pipe deflection measurements made during a 24 day period after wetting the backfill indicated no effect of the added load of the water. The weight of the water added to the backfill was probably not enough to cause an increase in pipe deflection. The added water could possibly affect pipe support by reducing the stiffness of the soil placed beside the pipe, but since the pipe embedment was well compacted, no effect was seen.

Time Lag of Pipe Deflections

A flexible pipe continues to deflect over time for two reasons (Spangler 1941):
 1. Increase in the soil load on the pipe
 2. Compression and consolidation of the soil at the sides of the pipe

Diameter measurements were made at various times following completion of the backfill (day 0) up through 22 years. Plots of vertical pipe deflections versus time are shown on figure 2.

Table 5 summarizes the vertical deflection measured at selected time intervals:

Table 5. - Deflection Over Time

Type of Pipe	Average vertical deflection - %						
	0 day	2 mo	1 yr	3 yrs	6 yrs	12 yrs	22 yrs
COHESIVE EMBDMNT (SC)							
RPM	0.7	1.6	1.8	2.6	2.8	3.0	3.1
Steel	1.5	1.8	2.1	3.0	3.2	3.3	3.3
PT	0.5	0.8	0.9	1.1	1.2	1.2	1.2
COHESIONLESS EMBDMNT (SP)							
RPM	0.4	0.5	0.6	0.6	0.7	0.8	0.7
Steel	0.4	0.5	0.5	0.7	0.8	0.8	0.8
PT	0.1	0.1	0.1	0.1	0.1	0.1	0.1

The following observations can be made from Table 5:

>1. For the period from 3 years to 22 years, all the pipe with cohesionless embedment showed very slight increases in deflection and the deflection can be considered as being relatively constant since year 3. The PT pipe with cohesive embedment had similar behavior. The vertical deflections in the RPM pipe and the steel pipe with the cohesive embedment increased over the period of 3 to 22 years from 2.6 % to 3.1 % and 3.0 % to 3.3 %, respectively. If the 22 year measurements are considered to be the final deflection of the pipe, then the 3 year deflections of the pipe were about 80 percent to 90 percent of the final.
>
>2. Except for the pretensioned concrete pipe with cohesionless embedment, the vertical deflections of the pipe have at least doubled for most of the pipe since the day that the backfilling over the pipe was completed. This emphasizes the concept that designing pipe for deflection must consider the long-term deflection of the pipe.
>
>3. While the initial deflections were about 1 % or less for any of the pipes, the RPM and steel pipe had a final deflection of 3 % in the cohesive bedding.
>
>4. The long-term deflections were still less than the maximum allowable deflections (5% for RPM and steel and 2% for PT - Reclamation requirements in 1971), emphasizing the need for proper embedment of flexible pipe.

Time Lag is defined as the ratio of the deflection measured at some time period following completion of backfill to the deflection measured at completion of backfill. [Note: Time-Lag is not the same as Deflection Lag as used in Iowa Formula.

Deflection lag is the increase in deflection *after* the maximum load occurs on the pipe, which doesn't happen until about 3-6 months after completion of backfill (Spangler 1941)]. Table 6 gives *time lag* factors for average vertical deflections measured over time. The *time lag* factors for the cohesionless embedment material are less than those for the cohesive material embedment material.

Table 6. - *Time Lag* Factors

Type of pipe	3 days	2 mo	1 yr	3 yrs	6 yrs	12 yrs	22 yrs
COHESIVE EMBDMNT (SC)							
RPM	1.3	2.4	2.8	4.0	4.3	4.6	4.8
Steel	1.1	1.5	1.7	2.5	2.6	2.7	2.7
PT	1.2	1.5	1.7	2.2	2.3	2.2	2.3
COHESIONLESS EMBDMNT (SP)							
RPM	1.1	1.3	1.5	1.7	1.9	2.1	1.9
Steel	1.1	1.2	1.3	1.7	1.9	2.1	2.1
PT	1.0	1.0	1.0	1.0	1.0	1.0	1.0

Net Change in Pipe Diameter After 22 Years

The net change in pipe diameter from the measurements made when the pipe was in place on the trench bottom before embedment and measurements made after 22 years is shown on table 7.

Table 7. - Net Vertical Diameter Change After 22 Years

Type of Pipe	Elongation %	Deflection %	Net Change %
COHESIVE EMBEDMENT (SC)			
RPM	- 1.2	3.1	1.9
Steel	- 4.0	3.3	- 0.7
PT	- 0.2	1.2	1.0
COHESIONLESS EMBMNT (SP)			
RPM	- 0.8	0.7	- 0.1
Steel	- 2.2	0.8	- 1.4
PT	- 0.2	0.1	- 0.1

The two steel pipe have not yet returned to their original diameter and are still vertically elongated. The RPM pipe and the PT pipe in the cohesionless bedding have returned to their same diameter with about zero net change. In the cohesive embedment, the RPM pipe and the PT pipe have diameters less than the initial diameter by about 2 % and 1 %, respectively. The allowable long-term average deflection for the Bureau of Reclamation in 1971 was 5 % for RPM and steel (flexible coating and lining) pipe and was 2 % for PT pipe (note: current AWWA standard limits deflection to $D^2/4000$ or 1.2 % for 48 inch pipe). The 22-year deflections are less than the allowable limits verifying that Reclamation specifications result in acceptable flexible pipe deflections.

Summary

Diameter measurements were made over a 22-year period on 120-cm diameter RPM, steel, and PT pipe buried under 4.6 m of cover. Two types of soil were used as the pipe embedment. The results of the diameter changes are summarized as follows:

1. Flexible pipe installed using Reclamation specifications (circa 1971) will deflect less than the allowable deflection limits.
2. Significant elongation (increase in vertical diameter and decrease in horizontal diameter) due to compacting soil alongside flexible pipe can occur during installation. As the pipe stiffness increased, the elongation decreased.
3. The pipe deflected less with a cohesionless soil (SP) embedment compacted to 75 % *relative density* than with a cohesive soil (SC) embedment compacted to 95 *percent compaction.*
4. Since the 3-year reading, diameter measurements of all pipe with cohesionless embedment and of the pretensioned concrete pipe with cohesive embedment have been relatively constant.
5. The steel and RPM pipe with cohesive embedment have continually shown increases in deflection over time. About 80 percent to 90 percent of the deflection occurred in the first three years.
6. Except for the pretensioned concrete pipe with the cohesionless embedment, the vertical deflections of the pipe have at least doubled in the 22 years since installation. The deflections of the pipe in the cohesionless soil (SP) embedment showed less increase over time than the pipe in the cohesive (SC) embedment.

REFERENCES

Earth Manual. (1963) Bureau of Reclamation, Denver Colorado, Second Edition

Howard, Amster (1974), "Field Test Deflections of Reinforced Plastic Mortar Pipe," Transportation Research Board Record No. 518, Washington DC

Howard, Amster (1977), "Modulus of Soil Reaction Values for Buried Flexible Pipe," *J. Geotechnical Engineering* , ASCE, vol 103, No GT1, Proc Paper 12700

Howard, Amster (1981a). "The USBR Equation for Predicting Flexible Pipe Deflection." *Proc. International Conference on Underground Plastic Pipe*, ASCE, New Orleans Louisiana

Howard, Amster (1981b), "Diametral Elongation of Buried Flexible Pipe," *Proc. International Conference on Underground Plastic Pipe*, ASCE, New Orleans Louisiana

Howard, Amster (1990), "Load-Deflection Field Test of 27-inch (675-mm) PVC (Polyvinyl Chloride) Pipe," *Buried Plastics Pipe Technology, ASTM STP 1093*, American Society for Testing and Materials, Conference held in Dallas Texas

Howard, Amster; Spridzans, J.B.; and Schrock, B.J.(1994), "Latvia Field Test of 915 mm Fiberglass Pipe," *Buried Plastic Pipe Technology: Second Volume, ASTM STP 1222* ASTM, Conference held in New Orleans Louisiana

Spangler, M.G. (1941), "The Structural Design of Flexible Pipe Culverts," *Iowa Engineering Experiment Station Bulletin No. 153*, Iowa State Univ., Ames Iowa

"DEVELOPMENTS IN DUCTILE IRON PIPE"
RANDALL C. CONNER[1]

Abstract

The history of welding gray cast iron and ductile cast iron products, which spans more than half a century, is briefly discussed. Welding effects on gray and ductile cast irons and slides and explanations illustrating pipe welding applications and welded products are presented. A new National Standard for welding of ductile cast iron materials, ANSI/AWS D11.2 is also discussed. In addition to providing valuable engineering information concerning weldability parameters, welding processes, and welding materials, this standard also provides criteria for the establishment of recognizable welding procedure, welder qualification, and welding Q.A. practice. Current "field welding" situations and concerns are also presented.

Design and field conditions in many cases dictate alternatives to some welded products, including the need for "field adaptable" restrained joints. While various types of "retainer glands" used to restrain "stuffing box" or mechanical joints, have been utilized in the iron pipe industry for several decades, beginning roughly 20 years ago a new generation of self-actuating, push-on joint restraint devices were introduced and became very popular for use with small diameter ductile iron pipes (DIP) in Europe. This type of product later spread to the USA. Design and performance aspects of various types of current gripping, self-actuating restraint devices using high-strength stainless steel or ductile iron elements which "grip" on to the DIP spigot are presented and discussed. In general, these devices are much less labor intensive and offer strength, dependability, and availability advantages (in addition to their obvious convenience and field-adaptable simplicity) over prior joint restraint alternatives.

---

[1]Research Engineer, Technical Division, American Cast Iron Pipe Co. (Acipco), P.O. Box 2727, Birmingham, AL 35202

Welded Products

We have records indicating Acipco began welding gray cast iron bossed outlets on gray cast iron pipes prior to 1939, or over half a century ago. Early boss welding was by the "gas bronze", "Tobin bronze", or "oxyacetylene torch bronze" welding process. On or about 1969 (or 25 years ago), Acipco began electric arc welding bosses and other appurtenances on DIP pipe using a revolutionary new welding electrode including approximately 55% nickel and 45% iron and originally developed by the International Nickel Company (Inco, incidentally, was the original co-developer of "ductile iron"). While former bronze-welded boss pipes had generally provided good service for decades, they were rapidly supplanted by the new, more "strength dependable" process and material, which controlled weld shrinkage and microstructure by innovative metallurgical means.

A general discussion of the challenges presented in welding cast irons is appropriate. Cast irons are not readily welded with the same steel welding electrodes and the same techniques used in widespread fashion to satisfactorily weld-join structural steel members. The worst problems in this regard are seen when welders attempt to arc weld on gray cast iron structures with conventional covered steel electrodes. For example, cast irons contain much more carbon than mild steel, which can contribute to the formation of massive iron carbides (cementite) with accompanying hardness and brittleness effects in the weld joint. The metallurgy of cast irons predicts the formation of other hard constituents or phases, such as martensite, upon the rapid cooling and solidification inevitable in arc welding processes. Gases emitted from castings, contaminants, filler metals, and/or atmospheres can also cause porosity in the weld deposit. Furthermore, the tensile strengths and ductility of gray cast irons are low compared to ductile cast iron and mild steel, and inevitable shrinkage stresses due to the volume changes in solidification also can cause cracking in the gray cast iron heat-affected zone and/or weld and/or parent metals. While ductile iron is much more conducive to welding than gray iron, mild-steel arc welding electrodes are also not recommended for welding DIP.

It is for these reasons that cast iron welding processes have traditionally been geared toward "low heat input" processes (such as brazing or welding with lower melting point materials such as aluminum-bronze, etc.). A major advance in the ductile iron pipe industry was the development and introduction of the aforementioned special nickel-alloyed iron welding electrodes which became widely employed by Acipco in the 1970's. An example of such class manual electrodes (now AWS A5.15 Class ENi-C1) is Inco Alloys International (Huntington Alloys) "Ni-Rod 55", or for later developed semi-automatic (wire welding) applications "Ni-Rod FC 55 Cored Wire" from the same manufacturer.

Deposited metal from such electrodes possesses a carbon content well above the "solubility limit", and excess carbon is rejected as spheroidal graphite during solidification of the weld metal. In truly innovative fashion also, this process minimizes weld porosity in welding cast irons by absorbing hydrogen safely in an austenitic weld deposit, and it also causes an increase in volume which tends to minimize the effects (residual stresses, tendencies to crack, etc.) of weld shrinkage during solidification. With good joint design, proper quality control of parent metal, control of heat input (amperage ranges, travel speeds, etc.), and the use of such welding electrodes with good welding practice, dependable and dramatic improvements in microstructure and joint strength are realized. Figures 1 and 2 are metallurgical photomicrographs of comparative weldment cross-sections (welded on DIP) showing the dramatic differences between conventional, low hydrogen mild steel electrode microstructure and that of Ni-Rod FC 55, respectively.

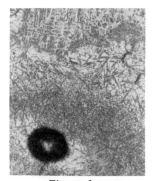
Figure 1

Figure 2

It is evident in examining these photomicrographs that the heat affected zone (HAZ) of the steel weld is significantly larger, with a substantial, continuous band of iron carbides present in the joint. On the other hand, the Ni-Rod FC 55 weld contains a lesser HAZ with a very narrow, substantially discontinuous band of carbides. Figures 3 and 4 depict cross sections of circumferential "weld beads" deposited on the outside of DIP for the purposes of joint restraint. Figure 3 is of a carbon steel, semi-automatic arc weld (containing some manganese) applied with a "low heat" metal-arc inert gas (MIG) welding process, and Figure 4 is of a Ni-Rod FC 55 weld applied with a submerged arc-wire welding process.

Figure 3

Figure 4

Even though an effort has been made to control weld heat input in the steel weld, it is obvious that greater heat effects and a wider band of carbides are present (compared to the nickel-iron weld). Also, the steel weld exhibits noticeably greater weld zone hardness. Vickers hardness 255-429 (Bhn 243-404) and 184-299 (Bhn 175-283) were measured across the welds and HAZ of the steel and nickel-iron welds, respectively. In another interesting comparison of these weldments, two each comparative weld-pipe samples were milled to identical pipe wall thicknesses and lengths, and the welds were then shear-load tested in a specially designed fixture. While the shear strengths of all weldments were substantial and the averages similar, the two steel weldments had widely divergent results and also failure modes strikingly different than that of the nickel-iron beads. Graphic illustrations of the failure modes in these tests are shown in Figure 5 and 6. It should be noted that the steel samples broke through the HAZ and the pipe wall, whereas the nickel-iron samples sheared thru the weld metal.

Figure 5

Figure 6

DUCTILE IRON PIPE DEVELOPMENTS

Another example of the capabilities of the nickel-iron weld material and ductile iron pipe can be seen in special and demanding, pipe butt-weld, "transverse" tensile tests. In such testing procedures, two actual DIP coupons are butt-welded together with a "back-gouged" 45° V-groove weld, then a conventional DIP tensile sample is machined from the welded panel transverse to the weld joint, with the weld joint at the center of the "gauge length". In such testing, Ni-Rod 55 pipe-weld samples will consistently exceed the minimum tensile and yield strength requirements for ductile iron pipe metal as per ANSI/AWWA C151/A21.51, and the apparent elongation, ductility, and toughness exhibited by the Ni-Rod 55 pipe-weld samples is also better than that offered by alternative welding materials. What follows are example results of such transverse tensile testing:

Electrode	**Type**	**T.S.(KSI)**	**Y.S.(KSI)**	**Elong(%)**	**R.A.(%)**
Ni-Rod 55	Pipe-Weld	70.2	57.1	6.7	8.2

In 1989, the first edition of voluntary consensus standard ANSI/AWS D11.2, <u>Guide for Welding Iron Castings</u>, was published. This new standard for the first time codified suggested qualification of welding procedures and welders who perform these procedures in welding iron castings. Guidance in quality control practice and weldability parameters was also included. Recognizing the value of such a standard in fulfilling a corporate mission objective of providing premium quality products to our customers, Acipco embarked on an extensive program involving cooperative effort of our production and technical divisions to formally document our welding production and Q.A. efforts in producing all our various weld-fabricated ductile iron pipe products in accordance with this new standard.

This effort was completed, and the procedures for the production of all Acipco DIP welded product lines and all welders who perform these procedures have been properly qualified in accordance with ANSI/AWS D11.2. The result is a well documented Q.A. system, involving welding procedure specification (WPS), procedure qualification (PQR), and welder performance qualification test (WPQR) records, all with terminology familiar to welding engineers.

The WPS, of course, defines specific material requirements, basic processes and techniques, and operating parameters for the product weld class involved. The PQR documents the precise make-up and parameters of an actual test pipe and weld applied in accordance with a specific WPS by a specific welder, then it documents the inspection (and inspector) of this weld as per Class I (critical service) visual, dye penetrant, and (cross-section) macroetch requirements of Appendix D Table D2 of ANSI/AWS D11.2 (thus "qualifying" the WPS). The WPQR forms(for every welder and welding operator at Acipco) document that individual welders have produced referenced test welds and are qualified to produce fabricated products per the welding procedure specification, and they also document the inspection and inspector of these test welds. This system of Q.A. forms and their development represent considerable investment in effort and expense on the part of Acipco, and these documents are considered proprietary; however, along with other quality control documents, records, etc., these resident documents are maintained in the Production units and Quality Assurance Department of Acipco for the perusal of our Customers, their Engineers, and/or their third party inspecting agents as they visit our factory.

This documentation is part of an overall Acipco quality control system which has received a Certificate of Registration from BSI Quality Assurance to quality system standards ISO 9002-1987, EN 29002-1987, and BS 5750/Part 2-1987. The scope of this registration applies to the manufacture of ductile iron pipe and fittings including machining and fabrication to national and international standards and specifications. Produced pipe and fittings up to 64" diameter (1600mm) are included.

Design, Testing, and Routine Q.C. of Welded-Outlet Pipes

In addition to qualification of welding production to all suggested requirements of ANSI/AWS D11.2, Acipco also further assures the quality and performance of these products by other means. For example, the wall thickness and weld reinforcement design of outlet pipes is governed by a proprietary computer program design wherein a safety factor of at least 2.5 is available in the weakest (largest) outlet configurations at maximum pressure rating versus the minimum yield strength of DIP per ANSI/AWWA C151/A21.51. Weld reinforcement in this design process is based on a method similar to that which is depicted in Section 13 of AWWA Manual M11 for similar welded outlets on steel water pipe (which in turn refers to Sec. VIII of the ASME Unfired Pressure Vessel Code for detailed design method details).

In the development of the welded-on outlet product line, multiple (weakest) outlet pipe configurations were proof-of-design, hydrostatically tested to failure to verify the computer design procedure. In addition to this developmental design testing, Acipco initially also directed that each and every outlet pipe of all sizes and all lengths and all joint configurations be first air tested at 15 psi(one bar) pressure and then hydrostatically tested to a pressure of 400 psi (27.5 bars) prior to shipment. While this early rigorous testing program involved several logistical problems and represented significant developmental expense in itself, impressively there were absolutely no failures in the 400 psi (27.5 bar) hydrostatic testing program for its entire duration over a period of approximately two years. At this point the routine high-pressure hydrostatic testing requirement was discontinued, but the requirement to air leakage test all outlet pipe fabrication welds remains. The one bar air test is also a current requirement for metric ductile iron fitting and special pipe fabrications per ISO 2531, although no routine leakage testing is required for AWWA standard fittings.

Field Welding

A discussion of welding DIP would not be complete without addressing the issue of "field welding" (welding accomplished at site). There are many field situations that suggest the application of field-cut and/or field-weld fabricated ductile iron pipe. These situations include normal installed pipe length variations in project easements,(which is affected by variations in manufactured lengths and by purposeful and inadvertent pipe joint deflections and other factors), errors in design, lay-out, survey, and/or installation, unforeseen underground obstacles or required pipeline off-sets, and many other situations. These situations generally "cannot wait" for "custom closure pipes" from the factory.

In the mid-1980's, amid a rapidly expanding demand and use of restrained joint DIP, Acipco received a great many requests for suggested procedures for field-welding of ductile iron pipe. After weighing all factors involved, Acipco published a field-welding brochure in 1986, suggesting how to cut and weld-fabricate our restrained joint pipes in the field. Since that time, we have shipped literally hundreds of restrained joint weld rings per year to our customers, "loose" for field welding.

It is natural to assume that some past problems with field welding DIP have been due to a lack of welding "skill" on the part of the field welders. In my experience I have found this to be only a small "part of the story". What are the most frequent problems I have seen in 20 years of pipeline work are use of totally inappropriate and non-recommended welding materials, often combined with a total lack of understanding of the magnitude of the forces involved in restrained pipeline work. Small welds, even skillfully applied by a welder using his or her "favorite" steel electrode (for welding the steel members with which he is most familiar), will not "hold the world" when it comes to DIP restraint matters, and there have been some failures in such cases. In other words, what is apparently most lacking is knowledge of appropriate materials and the size of weldments that must be applied to handle the service loads. The fact that we have been advised of no problems to date with the proper application of the Acipco field welding brochure over several years appears to reinforce this theory.

Practical DIP Welding Considerations

With regard to weldment design, it is generally preferable that welds and weldments on DIP be placed so as to be subjected to predominantly shear or bearing loads as opposed to predominantly direct tensile loads in service. This is the case with all contemporary welded restrained joints for DIP, but is not true of most designs familiar to some pipeline engineers and contractors for weld-restraining some steel component pipes.

Where some tensile loadings on DIP welds or HAZ of the welds cannot be avoided, as in the case of welded-on branch bosses or welded-on pipe branch outlets, it is generally good practice to "oversize" such welds so as to minimize the level of tensile stress applied to the HAZ. It should be noted that while oversizing DIP welds increases fabrication costs over lesser welded steel structures from the standpoint of labor and material (for example, just the recommended DIP welding materials typically cost per pound 10-15 times those used to weld steels), we feel these costs are justified due to the reinforcement and lesser stress conditions afforded by the larger welds.

While the welded products offered by Acipco possess considerable strength in bending or when subjected to large bending moments, reasonable care in design and installation should be exercised to avoid great relative movements or deformations of members which could cause damage to the welded structure. For example, when welded-on bosses or welded-on outlets are used to make smaller branch line connections underground to a larger pipeline, it is advisable to locate one or more flexible (outlet) joints near the welded branch connection to allow for relative motion of the two lines without imposing potentially damaging bending loads on the weld joint. It is also not advisable to specify or make connections to large underground lines utilizing relatively rigid flanged or restrained mechanical piping connections that could impart deleterious beam loads to the joints or outlets. It should be noted that the appendices to ANSI/AWWA standards C115/A21.15 and C110/A21.15 for ductile iron flanged pipe and fittings, respectively, dissuade the use of underground flanged pipe or fitting joints because of the rigidity of the joints. Small branch lines off much longer transmission lines are sometimes not installed at the same time as the mainline pipe, inviting unequal settlement of the pipes after the smaller piping is installed.

Acipco recommends underground use of flexible Fastite* (push-on) or mechanical outlet joints, and where restraint is necessary the use of flexible, Fastite (push-on) outlets provided with Fast-Grip gasket* restraints (to be discussed later) or flexible, welded push-on restrained joints such as American Flex-Ring* or Field Flex-Ring*. Other DIP manufacturers offer joints with similar claimed flexibility and restraint features. Supporting our recommendations is a record of no reported problems with outlet pipes* employing flexible, Fastite, Flex-Ring, or Fastite with Fast-Grip gasket joints at or near the outlets. Engineers are becoming increasingly aware of the possibilities (or in some cases probabilities) of relative movements between structures and buried pipelines, or at junctions of buried pipelines, and they are subsequently specifying flexible connections at these locations.
* *These products are detailed in the American Pipe Manual.*

"Gripping" Field-Adaptable DIP Joint Structures

With the undeniable need for rapid field adjustment for many aforementioned reasons, and in spite of proper available materials and instructions for field-welding, many Engineers are uncomfortable with, or at least want to minimize, field welding of DIP. I suspect that some have had poor experience with past improper iron or steel field welds. Recognizing these realities, there has been considerable interest for many years in means of field adaptability that do not require field welding of the pipe. Prior to the 1970's, these means generally consisted of masses of plain and reinforced concrete, myriad types of (often threaded) steel joint "rodding", many different types of mechanical joint "retainer glands" various types of "friction clamps", and many innovative combinations of these restraints. In some cases, engineers have had poor experiences with such restraints, and there has been a "yearning" in this field for more dependable alternatives.

In the 1970's, a revolutionary new type of joint restraint device was developed in Europe for use with common proprietary push-on joint DIP and fittings. This invention incorporates multiple strengthened, "series 400" or like chemical analysis stainless steel inserts or teeth embedded at short, regular intervals in the front or "hook" end of the rubber gaskets. These devices utilize standard push-on pipes and assemble with essentially no more joint preparation and little more effort than do ordinary, non-restrained DIP joints, yet they generally provide exceptional restraint when even large joint separating forces are applied. This is accomplished with gripping, "camming", and/or wedging action of these elements between the socket and spigot pipe elements as the joints try to separate.

In early 1980, the writer assisted in an extensive laboratory testing program involving actual samples of these proprietary push-on joint restraint gaskets and pipes. While some slight technical problems were noted in these tests, the exceptional and non-obvious restraint capabilities of these products were well exhibited. Later, Acipco actually received the aid of the European developers of this original product in research efforts to adapt this product to the standard Fastite (push-on) DIP joint structures furnished by Acipco. This effort was unsuccessful; therefore, Acipco independently developed a new gasket device referred to as "Fast-Grip" in the late 1980's and presently markets this in the 4"-16" (102mm-406mm) size range. This device, which utilizes pairs of cooperating wedging "400 series" stainless steel elements in the front portion of the gasket, has provided quite excellent service to our customers, and to date we have provided approximately 80,000 of these gaskets to pipeline projects throughout the United States.

In addition to hundreds of developmental tests in our research laboratories, Fast-Grip gaskets are currently Underwriter's Laboratory listed in 4"-16" sizes(based on extensive testing witnessed by a U.L. inspector) for pressure ratings of from 200-350 psi (14-24 bars), depending on size, when utilized with minimum pressure class DIP per ANSI/AWWA C151/A21.51. Fast-Grip gaskets are also Factory Mutual Systems approved(also based on extensive testing witnessed by an F.M. inspector).

Soon after the development of the Fast-Grip gasket, Acipco also independently developed a new restraint device for larger pipes(though separated from the Fastite gasket in a Flex-Ring socket) called "Field Flex-Ring". This device, which is capable of even greater load restraint capabilities than Fast-Grip gaskets, is offered in 14"-36" (356mm-914mm) size ranges. The Field Flex-Ring also utilizes elastomer-actuated wedging restraint, although in this case the biting elements are large, strengthened ductile iron segments which wedge directly in a heavy cross-section, modified push-on pipe bell.

Durability of Gripping and Welded Products

Acipco has received a few inquiries over the years concerning the corrosion resistance of gripping and welded products, particularly those that employ metals other than "ductile iron". It has been previously noted that our products have been and are variously fastened with alloy bronze welds, (high) nickel-iron welds, and strengthened, "400 series" stainless steel gripping wedges. All of these materials are dependably cathodic to DIP (or are more "noble" in galvanic series than DIP).

Although buried metal corrosion processes cannot be reduced to simple models in some cases, the basic mechanism of galvanic and other corrosion are well documented and well understood. The essential details of many of these processes were actually worked out by Sir Humphrey Davy as early as 1824. Corrosion will generally occur or be accelerated at the anode of a corrosion cell. In general (and though not well explained in some corrosion textbooks), it is advantageous in aggressive environments to utilize where needed relatively highly stressed fasteners with small relative surface area which are cathodic to the larger surface area members they fasten.

In some instances and in aggressive environments, designers have not paid heed to these Engineering design principles and problems have quickly resulted. Examples are the use of anodic carbon steel fasteners bolting cathodic cast iron or stainless steel members on ships in saltwater service, or anodic unalloyed and stressed cast iron or carbon steel bolts or rods in some buried cast iron piping systems. These failures are predicted by conventional corrosion principles.

On the other hand, the standard use of dependably anodic metal elements with DIP creates a desirable condition known conducive to successful performance. This design results in a very small exposed cathode area in contact with a large anode area, in some cases also in a relatively protected or shielded environment up inside a pipe socket. The very limited exposure of our restrained joint welds and wedges in service is admittedly very similar to the elements used in others' similar products, and to the best of our knowledge the stainless steel used in these gripping products have similar chemistry, "400 series" stainless steel wedging elements(we have subjected a few samples of same to chemical analyses dating back more than 10 years). These similar European and domestic products have been around for more than 20 years, and while we listen closely for all problems with products available in the marketplace (for obvious reasons), we have heard of no problems attributed to corrosion. Of course, it is claimed that there are literally millions of these similar products in service in varied soils around the world, covering this time span of roughly two decades.

Our known experience with 400 series stainless steel elements in contact with cast iron and ductile iron pipe is quite excellent. As noted in the historical section of the American Pipe Manual, the standard bolting material for the venerable Molox critical service ball joint pipe (Molox has recently been totally replaced by the boltless Flex-Lok) was "American Stainless" or Type 416 stainless steel. Figures 7(as-received) and 8(after wire brushing) document customer retrieval of a bolt which had been in service for 17 years in a brackish river crossing application coupled to cast iron pipe in Charleston, S.C. Both the pipe and the bolt were in good shape. As is also suggested by the excellent "after" condition of this bolt in difficult service, we have no known fastener or pipe corrosion complaint associated with the use of American stainless or "400 series" stainless steel elements used with cast iron and ductile iron pipe.

Figure 7

Figure 8

This record spans several decades experience with American Molox, Lok-Fast, and Mechanical Joint T-head bolts and nuts, as well as approximately five years experience with Fast-Grip gasket joints.

Similar appropriate cathodic-anodic relationships are also represented in ANSI/AWWA C111/A21.11 for underground, T-head watermain bolts. Alloy steel T-head bolts are required to contain small amounts of nickel, copper, and chrome so as to result in dependable proper galvanic relationships.

Summary

While welded products and field adaptable joint structures have been around for a long time in the ductile iron pipeline field, extensive developments in technology and manufacturing quality systems have greatly extended the capabilities and utility of these products for the benefit of design engineers, contractors, and ultimately the system owners and users who derive long-term economic and practical benefit from these developments. Engineers are urged to examine these products in detail and utilize them where their engineering attributes are appropriate for the intended service.

This paper is not intended to be an all inclusive primer concerning welding of ductile iron pipe, gripping products available to be used with DIP, or other discussed matters. Readers are encouraged to contact Acipco or other ductile iron pipe manufacturers for further details and are also encouraged to obtain referenced standards and other documents for further information.

References

American Cast Iron Pipe Company(Acipco), American Pipe Manual, 17th Edition, 1994

American Welding Society(AWS), ANSI/AWS D11.2-89, "Guide for Welding Iron Castings", 1989

Bishel, R.A., "Flux-Cored Electrode for Cast Iron Welding", Welding Journal, 52(6), June 1973

Ductile Iron Pipe Research Association(DIPRA), "Field Welding and Cutting Ductile Iron Pipe", 1992

Inco Alloys International(Inco), "NI-ROD and NI-ROD 55 Welding Electrodes", Data Sheet and Technical Info, 1983

Inco Alloys International(Inco), "NI-ROD FC 55 Cored Wire", Data Sheet and Technical Info, 1987

Pressure Class Ductile Iron Pipe and Its Design

L. Gregg Horn, P.E.[1], member, ASCE

Abstract

The latest revisions to ANSI/AWWA C150/A21.50 and ANSI/AWWA C151/A21.51 changed the way ductile iron pipe is designated. While the standard reference to this product has changed from Thickness Class to Pressure Class, the design and manufacturing procedures have not changed.

This paper discusses the reasons for the change and its significance for current and future designs and specifications. The thickness design procedure is also discussed in some detail, focusing on the various design parameters, models and assumptions that go into the routine design of ductile iron pipe. The external load models, the significance of the defined laying conditions, internal pressure design and the tolerances that are applied are covered.

The reader of this paper will receive a basic understanding of the new Pressure Class designation and the standard approach to the design of ductile iron pipe. It will give the designer and owner a fundamental understanding of how a routine application is addressed in the standard and it will provide the information necessary to know when special or non-routine designs are required.

Introduction

The performance record for iron has been an impressive one. Long respected for their strength and durability, both ductile iron and its predecessor gray cast iron pipe have provided outstanding service to those who have selected them for their

---

[1]Director of Regional Engineers, Ductile Iron Pipe Research Association, 245 Riverchase Parkway, East, Suite O, Birmingham, Alabama 35244

PRESSURE CLASS DUCTILE IRON PIPE   299

piping systems. Cast iron pipe has served more than 400 utilities for over 100 continuous years, and at least ten communities for 150 years or more (Reference 5). Since ductile iron pipe is stronger and more flexible than cast iron, it would be reasonable to expect similar long-term utility.

The performance history of ductile iron pipe results from a design approach that takes advantage of the pipe's inherent strength. This allows design models to be extremely conservative in assumptions that are used to quantify the loadings on and the level of support provided to the pipe. Understanding these assumptions, the engineer can quickly and easily accomplish a pipe wall thickness design for most applications without the need for rigorous definition of the factors that are used to calculate the loads the pipe will be required to sustain or the level of soil support that can be counted on from the backfill. When addressing typical installations, the loads that are modeled are generally assumed to be as high as they can be for that model and the sidefill soil support relied upon is reasonably established at practical levels. It is not necessary to "fine tune" the design based on limitations of the pipe.

Because of this approach the ANSI/AWWA design standard (Reference 3) can incorporate many tables that compile the results of these "worst case" assumptions. Such tables afford the designer the opportunity to quickly look up results rather than go through step-by-step calculations. Further, to allow flexibility, the tables have been developed to accommodate certain possible special design criteria. The engineer may recognize that all of the standard assumptions are valid and go to a table that accomplishes the design directly, or, if some assumptions require modification, the design may be accomplished by individual steps. The standard also allows for intermediate situations and provides reference tables to help the designer along.

For example, it may be that while the earth load assumption is valid, the traffic load may be greater than the standard AASHTO H-20 load. In this case, the earth load may be found in Table 50.1, the traffic load calculated and added to the earth load, and the appropriate external load tables used to find the required diameter-to-thickness ratio (D/t or D/t_1) for that load.

Although one would now specify a pressure class instead of a thickness class, the wall thickness design procedure is unchanged from that traditionally used for ductile iron pipe. Consistent for all diameters, the design procedure is the most demanding of all piping materials commonly used in water and sewer applications.

Revisions to Standards

The standards that apply to Ductile Iron pipe for water or force main applications are ANSI/AWWA C150/A21.50, "Thickness Design for Ductile Iron Pipe," and ANSI/AWWA C151/A21.51, "Ductile Iron Pipe for Water or Other

Liquids" (References 3 and 4). As noted, these two ANSI/AWWA standards have been revised so as to change the preferred designation of the product from "Thickness Class" to "Pressure Class." Since most Ductile Iron pipe projects involve internal pressure applications, it was appropriate to have the pipe designation reflect its pressure design. The thickness class designation simply referred to an available thickness, but gave no indication of its ability to support loads or hold pressure. Further, the thickness of a given class of pipe is dependent on the pipe's diameter. Six-inch Class 50 is not as thick as 12-inch Class 50, for example. Thickness class pipes had "Rated Water Working Pressures," however, these ratings were a function of the thicknesses of pipe, rather than the reverse. In other words, to determine the pressure capacity, one would start with a thickness and solve for pressure, up to a maximum of 350 psi. For these and other reasons, the meaning of thickness class was often confusing. In the current revision, the thickness of pressure class pipe is a direct function of the pressure capacity. One begins with a working pressure and calculates the resulting thickness. This is more directly related to function and is, therefore, more easily appreciated.

Additional revisions included adding 60- and 64-inch diameter pipe, changing the outside diameter of 54-inch pipe to correspond to International Standards Organization (ISO) standards, and introducing Pressure Class 150 pipe in certain larger diameters. However, while the primary designation has been changed, thickness class pipes remain in the standard as "Special Thickness Classes." This allows for a transition, should the market dictate such, into pressure classes and recognizes that for some installations, such as those involving extreme depths of cover, a thickness exceeding that corresponding to a pressure class 350 psi design might be required. Thus, if the design indicates that a nominal thickness greater than that available with Pressure Class 350 is required, an appropriate Special Thickness Class would be specified. This product would have the additional thickness required to accommodate the larger load, but would still have a Rated Water Working Pressure of 350 psi.

Design Criteria

Ductile iron differs from gray cast iron in general terms by the difference in the character of the graphite. In gray cast iron, the graphite is in a flake form. In this form, the graphite plays a more dominant role influencing the mechanical behavior of the pipe. As a result, gray cast iron is more brittle and has a lower tensile strength. In ductile iron, the graphite is spheroidal rather than in a flake, which allows the characteristics of the surrounding iron to dominate. This translates into an increase in the ultimate tensile strength and in the introduction of a yield strength and corresponding flexibility (Reference 6).

This is why ductile iron pipe uses the flexible conduit design approach. In modeling external loads, separate stress analysis is utilized. No support of external loads that might result from internal pressure is assumed. Also, since ductile iron pipe is a flexible conduit, design addresses both ring bending stress and deflection.

Pipe Design Rationale

In the ANSI/AWWA design standard, assumptions are made for what are considered "routine" installations that address pressure design and modeling of external loads. Because they are safe assumptions, they also make routine design of ductile iron pipe a simple procedure. For example, the standard considers working pressures of 350 psi and below "routine," although such high working pressures are rare. A further pressure design assumption recognizes that surge pressures are possible. In addressing internal pressure design, a surge allowance of 100 psi is nominally added to the working pressure before a factor of safety is applied and the thickness calculation is accomplished.

Since ductile iron pipe is a flexible conduit, the standard external load design models the earth pressure bearing on the pipe by assuming a prism load. This is the highest calculable earth load for a flexible conduit in a trench condition. For the traffic load, the standard uses the American Association of State Highway and Transportation Officials (AASHTO) model H-20 (Reference 1). This is a 16,000-pound point load that is distributed over a three-foot length of pipe at pipe depth. An impact factor of 1.5 is used regardless of the pipe depth and the surface reduction factor is based on an unpaved road. These last two assumptions provide maximum contribution of dynamic impact and minimal support of the load by the road surface.

All of these criteria apply to a routine installation and they are conservative. However, if the predicted surge pressure is above 100 psi, or the earth load should be calculated for an embankment condition, or the pipe is to be laid under a railway, the loads may be modeled to reflect such differences. However, in most cases, a routine installation is easily designed under the given assumptions.

As noted above, a flexible conduit design theory is used in modeling external loads. This approach recognizes that the backfill surrounding a flexible conduit will be mobilized to provide some measure of support for the external load (Reference 9). When the external load is applied, the pipe deflects. The sides of the pipe move outward against the passive resistance of the soil resulting in its assisting in supporting the load. This reduces the stresses in the pipe wall that occur from ring bending, and lessens the ring deflection the pipe experiences.

Obviously, the nature of the backfill determines the degree of sidefill soil support that will be available. The type of soil, the nature of the bedding, and the amount of compaction imparted are all factors. The more effort applied, the more support that results. Both the bedding angle, which is the angle subtended by that arc of the bottom of the pipe that is bedded in the trench, and the modulus of soil reaction (E'), are used. E' is not measurable, but is used to describe the level of support the soil would provide.

Research and literature are available that allow a theoretical prediction of E' based on soil classifications and compaction. In ductile iron pipe design a prudent approach makes this typically unnecessary. Unless the soils to be encountered are unusually weak, the E' suggested for a given laying condition is conservatively assumed (Reference 8) and is easily accomplished in the field.

For example, work done by the Bureau of Reclamation (Reference 7) resulted in a table that predicts the value of E' based on the soil's ASTM classification and the level of compaction to be applied. Therefore, if crushed stone were compacted to 95 percent Proctor density, an E' of 3,000 psi would be predicted. This is most closely approximated by a Type 5 laying condition as defined in ANSI/AWWA C150/A21.50. The ductile iron pipe design would, in this approach, assume an E' of just 700 psi. This assumption takes advantage of the strength of ductile iron while recognizing that the actual installation may not occur precisely as specified. While an E' of 3,000 psi may be realized at 95 percent compaction, it is sometimes a challenge to achieve such a high level of compaction during installation. Ductile iron pipe design takes a practical approach by not relying so much on the support that the sidefill soils might actually provide or on a contractor's ability to achieve an explicit and high level of compaction.

The same approach is found in all five laying conditions. This allows design values for E' that are reasonably achieved in the field without having to classify the soil. Of course, some soils require special approaches. For example, if the installation is through a marsh it may not be practical to rely on sidefill soil support at all and a special design is required. However, this would be unusual.

Pressure Class versus Pressure Rated: Internal Pressure Design

In pressure class design, the wall thickness is calculated so as to accommodate a surge pressure over and above the working pressure. Thus, a pressure class product design would add a surge allowance to the working pressure before applying a factor of safety. In ductile iron pipe design, the surge allowance is nominally 100 psi, and the factor of safety is two. Therefore, for a working pressure of 150 psi, the design pressure would be:

$$2(150 \text{ psi} + 100 \text{ psi}) = 500 \text{ psi}$$

In a pressure rated pipe, however, no surge allowance is incorporated into the design pressure. Therefore, for a the same working pressure and factor of safety, the design pressure would be:

$$2(150 \text{ psi}) = 300 \text{ psi}$$

Both pipes would be referred to as 150 psi, but the pressure class 150 pipe would have a design pressure 200 psi higher than the pressure rated 150 product. Therefore, for a given pipe material and size, pressure class pipe would have a thicker wall than pressure rated pipe, for the same working pressure.

Design Steps

The procedure for calculating a wall thickness for ductile iron pipe involves the separate stress analysis discussed above. Designs are performed for internal pressure and external load, exclusive of one another. Two thicknesses are calculated that are based on the material strength of the pipe; one for internal pressure and one for the ring bending stress.. A third thickness, the deflection thickness, is also calculated. This is a second external load design, but is based on the flexibility of the lining rather than the material strength of the iron.

In designing for internal pressure or ring bending stress, a nominal service allowance is added to the calculated thickness. This service allowance reflects an added factor of safety over and above the nominal factor of safety used in all ductile iron pipe design calculations. This total thickness is called the "minimum manufacturing thickness" and is denoted as t_1. On the other hand, the deflection design criterion is a function of the pipe lining's flexibility, not the pipe wall's mechanical strength. Therefore, t_1 is calculated directly.

The three resulting t_1's are compared to one another and the largest of the three is selected. Then, a foundry casting tolerance is added based on the diameter of the pipe. The result is called the "total calculated thickness." This thickness is then compared to the nominal thicknesses corresponding to the various available pressure classes for that given diameter. The next higher nominal thickness is selected and the corresponding pressure class for that pipe is specified.

Internal Pressure Design

The hoop stress equation is used in ductile iron pipe internal pressure design. The design pressure is as described above, incorporating a working pressure and surge allowance. The tensile stress against which we design is limited to the yield strength of ductile iron, 42,000 psi at a minimum.

The formula is:

$$t = \frac{F_s(P_w + P_s)}{2S} \quad (1)$$

where:
- t = net thickness, inches
- F_s = nominal factor of safety (2.0)
- P_w = working pressure
- P_s = surge allowance (100 psi, nominally)
- S = minimum yield strength of Ductile Iron (42,000 psi)

The design pressure should incorporate the actual working and surge pressures (if surges in excess of 100 psi are anticipated) to calculate the required Pressure Class.

Design for External Load

The prism load is used to calculate the magnitude of the earth load that will bear on the proposed pipe. As noted above, for a flexible conduit in a trench condition, this is the most conservative earth load model. In this model, the pressure of the soil prism directly above the pipe is considered to be bearing fully upon the pipe, without shear support from soils immediately outside the prism. In the ANSI/AWWA C150/A21.50 standard, the soil is nominally assigned a density of 120 pounds per cubic foot (pcf).

It would be rare for the trench condition model to be inappropriate, but if, for example, an embankment condition were a better approximation, such a model could be used. Even so, it would be rare for the embankment condition to result in a significantly higher external load. Similarly, it would be unusual to encounter soils with greater densities than 120 pcf, but if they are, their actual densities should be used as well.

The formula for calculating the prism earth load is:

$$P_e = \frac{wH}{a} \quad (2)$$

where:
- P_e = earth load, psi
- w = soil density, pcf (120 pcf, nominally)
- H = depth of cover, feet
- a = conversion factor (144, for pounds/square-foot to psi)

The AASHTO H-20 traffic load is used in the design standard. This model distributes a single wheel load of 16,000-pounds over an effective length of three feet of the pipe at pipe depth. The nature of the roadway, whether it is paved or unpaved, is accounted for by incorporating a surface load factor. Also, the impact of the dynamics of a traffic load (impact factor), and the fact that the load is supported in part by immediately adjacent portions of the pipeline (reduction factor) are incorporated into the calculation. Except for the impact factor, all other considerations are a function of the diameter of the pipe and the depth of cover.

The formula used is:

$$P_t = \frac{CRPF}{bD} \qquad (3)$$

where:
- P_t = traffic load, psi
- C = surface load factor, based on an assumed unpaved road
- R = load reduction factor
- P = wheel load (16,000 pounds)
- F = impact factor (1.5 for all depths of pipe)
- b = conversion factor (12 for pounds-per-linear-foot to psi)

The total external load, P_v, would simply be the sum of the earth and traffic loads:

$$P_v = P_e + P_t \qquad (4)$$

Ring Bending Stress Design

Once the magnitude of the external load is determined, two thicknesses are calculated. The first, for ring bending, calculates the thickness required to minimize the stresses that are allowed to build in the pipe wall when the external load is applied. A second thickness calculation ensures that the flexibility of the lining in the pipe will not be compromised as the pipe deflects under that load. In each design, the formulas used incorporate constants that are functions of the laying condition to be used or of ductile iron as a pipe material.

The ring bending design formula is:

$$P_v = \frac{f}{3\left(\frac{D}{t}\right)\left(\frac{D}{t} - 1\right)\left[K_b - \frac{K_x}{\frac{8E}{E\left(\frac{D}{t} - 1\right)^3} + 0.732}\right]} \qquad (5)$$

where:
- P_v = external load, psi
- f = allowable ring bending stress (48,000 psi)
- D = outside pipe diameter, inches
- t = net thickness, inches
- K_x = deflection coefficient
- K_b = bending moment coefficient
- E = modulus of elasticity of Ductile Iron (24 x 10^6 psi)
- E' = modulus of soil reaction, psi

All values of the constants and variables are known except t, the net thickness of the pipe wall. Bending stress is limited to 48,000 psi, one-half (again, the factor of safety is two) of the minimum ultimate bending stress of ductile iron pipe. K_x and K_b are constants that are defined by the laying condition. Their values may be found in Table 50.2 in the design standard. E' is the modulus of soil reaction described above and is also defined by the laying condition.

Deflection Design

The deflection of the pipe under the application of the external load is not a concern with regard to stresses in the pipe wall, as are the internal pressure and ring bending designs. Rather, the magnitude of the deflection is controlled so as to prevent damage to the lining. Linings control internal corrosion and have the added benefit of improving the flow characteristics by lessening the "roughness" of ductile iron pipe. In the ANSI/AWWA design standard, the lining is assumed to be the standard cement mortar lining described in ANSI/AWWA C104/A21.4 (Reference 2).

The deflection design formula is:

$$P_v = \frac{(\Delta x/D)}{12K_x} \left[\frac{8E}{\left(\frac{D}{t_1} - 1\right)^3} + 0.732E' \right] \quad (6)$$

where:
- $\Delta x/D$ = allowable deflection (3% for cement mortar linings)
- t_1 = minimum manufacturing thickness, inches

As with all ductile iron pipe design criteria, a nominal factor of safety of two is used, in this case based on the flexibility of the lining.

General Discussion of Design Criteria

When designing pipe wall thickness for ductile iron pipe, it is important to be aware of the assumptions made in developing the standards. Their purpose is to make pipe wall thickness design for typical pipeline installations easily and safely accomplished. The prism earth load, the 120 pcf soil density, the H-20 traffic load and associated factors, the low E' values, pressure class rather than pressure rating design, the 100 psi surge allowance, the nominal factor of safety of two for all design models are all conservative for most installations. One simply goes to the appropriate table in the design standard and finds the pressure class that is required. There are tables to look up the magnitude of the external load; either earth or traffic as well as their total. There is a table that compiles the required wall thicknesses for either external load or internal pressure. If a standard assumption is not appropriate there are still tables to use. For example, if the soil density is higher than 120 pcf, or if the traffic load is greater than the H-20 load, each load may be calculated using the appropriate method and the total external load derived from the resulting sum. The values of D/t for bending stress design and for D/t_1 for cement-mortar lining deflection design for a given magnitude of external load may be found in the appropriate table, regardless of how the external load was calculated. And if all assumptions hold true, there is a table (see Table 1) that allows direct selection of the appropriate pressure class based on diameter, laying condition, depth of cover, and internal pressure.

Table 1. Excerpt from Table 50.14 in ANSI/AWWA C150/A21.50-91, Rated Working Pressure and Maximum Depth of Cover.

Size in.	Pressure Class* psi	Nominal Thickness in.	Laying Condition				
			Type 1 Trench	Type 2 Trench	Type 3 Trench	Type 4 Trench	Type 5 Trench
			Maximum Depth of Cover - ft†				
24	200	0.33	**	8§	12	17	25
	250	0.37	**	11	15	20	29
	300	0.40	**	13	17	24	32
	350	0.43	**	15	19	28	37

*Ductile Iron pipe is adequate for the rated working pressure indicated for each nominal size plus a surge allowance of 100 psi. Calculations are based on a 2.0 safety factor times the sum of working pressure and 100 psi surge allowance. Ductile Iron pipe for working pressures higher than 350 psi is available.
†An allowance for a single H-20 truck with 1.5 impact factor is included for all depths of cover.
§Minimum allowable depth of cover is 3 ft. (for this pressure class and laying condition).
**For pipe 14 in. and larger, consideration should be given to the use of laying conditions other than Type 1.

For example, if a 24-inch pipe is required for a gravity sewer, the depth of cover is eight feet, the laying condition specified is Type 3, and the external load assumptions discussed above are appropriate, Pressure Class 200 Ductile Iron would be selected from Table 50.14 (see Table 1) of ANSI/AWWA C150/A21.50. In fact, with a Type 3 laying condition, 24-inch Pressure Class 200 pipe will support 12 feet of cover, assuming a prism load, a soil density of 120 pcf, and the AASHTO H-20 traffic load. And although not required in this installation, it will also be sufficient to sustain a 200 psi internal pressure plus a 100 psi surge pressure. Further, these levels of performance all incorporate a nominal factor of safety of two. If the depth of cover were increased to 14 feet, either the specified laying condition could be improved to Type 4, or Pressure Class 250 pipe could be selected.

Conclusion

The most recent revisions to the manufacturing and design standards for ductile iron pipe change the standard way this product is to be specified. By incorporating a pressure class designation, the standards have removed the confusion that was associated with thickness classes. By referencing a pressure capacity, the designer references a performance capability. Further, the addition of 60 and 64-inch diameters and the introduction of a pressure class 150 product in larger diameters give owners and engineers more variety and better conformance to the smaller pressures commonly found in transmission mains.

However, the approach to design of ductile iron pipe has not changed. In ductile iron pipe design standards "worst case" assumptions are used to model the magnitude of the loads that may occur and the support that is given by pipe backfill. The result is a safe design that is easy to accomplish for routine installations. Special installations will require special design considerations, but these are rare. For most applications the design engineer can confidently select an appropriate Pressure Class of ductile iron pipe without rigorously defining the specific environment into which the pipe will be installed.

Appendix

References:

(1) American State Highway and Transportation Officials, Standard Specifications for Highways and Bridges, 14th ed., 1989, AASHTO, Washington, D.C.

(2) ANSI/AWWA C104/A21.4-90, "American National Standard for Cement-Mortar Lining for Ductile-Iron Pipe and Fittings for Water," AWWA, Denver, Colorado.

(3) ANSI/AWWA C150/A21.50-91, "American National Standard for the Thickness Design of Ductile-Iron Pipe," AWWA, Denver, Colorado.

(4) ANSI/AWWA C151/A21.51-91, "American National Standard for Ductile-Iron Pipe, Centrifugally Cast, for Water or Other Liquids," AWWA, Denver, Colorado.

(5) Ductile Iron Pipe Research Association, Ductile Iron Pipe News, Fall/Winter 1993.

(6) Ductile Iron Pipe Research Association, Handbook of Ductile Iron Pipe, 6th ed., DIPRA, Birmingham, Alabama.

(7) Howard, A.K., Modulus of Soil reaction (E') Values for Buried Flexible Pipe, Denver, Colorado, United States Department of the Interior, Bureau of Reclamation, Technical Report REC-ERC-77-1, (January 1977).

(8) Kennedy, H., Jr., "The New Ductile Iron Pipe Standards," Journal American Water Works Association, Vol. 68, No. 11, pp. 622-626, (November 1976).

(9) Spangler, Merlin G. and Handy, Richard L., Soil Engineering, Harper & Row, 1985.

The Development and Installation Of
Ductile Iron Microtunneling Pipe

Ralph R. Carpenter, M. ASCE[1] and Richard Croxton[2]

Abstract

Domestically, ductile iron pipe has not had a presence in the microtunneling pipe market. With the increased demand forecasted for the trenchless installation of pressure pipelines (water mains and force mains) and the limited choices of pipe materials available for this application, the inherent characteristics of ductile iron pipe make it a natural candidate for the development of a microtunneling pipe. This paper discusses American's development, testing and pilot installation of the first domestically designed and manufactured ductile iron microtunneling pipe.

Introduction

American Cast Iron Pipe Company (ACIPCO) has always been perceived as an innovator in the ductile iron pipe industry. Several industry "firsts" include: pioneered the first application of cement lining for cast iron water pipe, the mechanical joint, the parallel development of ductile iron, introduction of large diameter ductile iron pipe, and the first in the industry to receive ISO 9002 certification (August 1994), just to mention a few.

Transitioning into our tenth decade brings forth the responsibility to service American's customers with superior products and product innovation.

---

[1]Marketing Specialist, American Ductile Iron Pipe, a Division of American Cast Iron Pipe Company, P.O. Box 2727, Birmingham, AL 35207-2727
[2]Vice President, Kinsel Industries, P.O. Box 750160, Houston, TX 77275-0160

Development

ACIPCO, like most progressive companies, has a product development committee whose responsibility is to perpetuate new product development. American has a formalized critical path for the development of a new product or refinement of an existing product. This critical path has several detailed tasks that must be satisfied before a product is released for production. In the body of this paper several of these tasks will be discussed in detail as they relate to "*The Development of Ductile Iron Microtunneling Pipe*."

ACIPCO was first made aware of the potential growth for the microtunneling industry in the late 1980's. At that time it appeared that most, if not all, of the microtunneling installation work was for gravity sewer applications. American's standard cement or polyethylene lined ductile iron pipe has been used extensively in gravity sewer applications installed by traditional cut and cover, as well as, some carrier pipe and slip lining applications. The evaluation of the market demand for microtunneling pipe (at that time) suggested that growth would be slow.

In early 1992, ACIPCO began to reexamine the market potential for ductile iron microtunneling pipe. The evaluation this time showed that microtunneling construction was being promoted for both gravity and pressure applications. Additionally, the value added features of microtunneling were beginning to be exploited by some of the industry's promoters with some success.

In August 1992, Dr. Tom Iseley, one of our Country's most recognized authorities on microtunneling, stated that in his opinion, "in five to ten years, this (microtunneling) will be the standard way of installing pipe in congested areas or below the groundwater table."[1] Microtunneling was now being viewed as a means to limit the socio-economic impact of standard cut and cover installation in urban areas. This was an obvious and powerful value added feature of microtunneling for any piping infrastructure (gravity, pressure, utility, et al). There was a tremendous need to enhance the understanding that owners, engineers, and the general public have about the true costs of a project, "that is, direct bid cost plus indirect or the socio-economic cost to society of a proposed design and construction method."[2] American began the product development process in late 1992.

American Cast Iron Pipe Company, with a staff of research engineers assigned to the Technical Division, was committed to the development of the microtunneling pipe. An early assessment of the requirements for a microtunneling pipe dictated that the pipe have the ability to transmit the columnar load across or through the joint, and that the pipe, through its entire length, must have an uniform outside diameter.

When developing a new product it is foolhardy to neglect the experiences learned in the development of products that may have reached maturity and/or have completed their product's life-cycle.[3] Several of American's products contributed to the preparation of the preliminary joint concepts.

Integral-Bell Joint

The initial joint concept is the *integral bell* joint, see Figure 1 (right). This joint, utilizes American's standard Fastite joint with the application of cement mortar to the exterior of the pipe to form a uniform outside diameter for its entire length. Columnar loads from jacking forces would be distributed to this coating and transferred across the joint by a compressible filler.

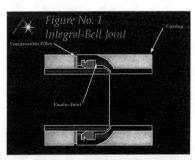

Integral-Bell Joint with Push Bar

The second joint concept is the *integral bell joint with push bar*, see Figure 2 (right). This joint is similar to the first, with the addition of a push bar for transmitting the columnar load from the face of the bell to the pipe wall. American has supplied this joint (absent the cement mortar exterior) for rehabilitation of deteriorated pipes by slip lining.

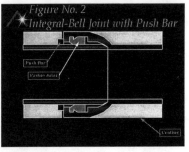

An efficiency ratio was used to analyze the actual inside diameter of the pipe to the theoretical bore inside diameter (maximum pipe outside diameter). For example: a 406mm (16in) integral bell ductile iron microtunneling pipe has an actual inside diameter of 427mm (16.80in). This same pipe has a maximum outside diameter (at the bell face) of 504mm (19.86in). The ratio or the efficiency of the flow area to the bore hole area is, therefore, 0.85 or 85%.

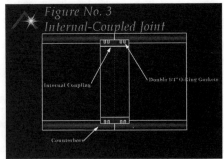

Internal-Coupled Joint

Designers were optimistic that a more efficient joint could be developed using an inside diameter coupling. Our optimism and ingenuity brought about the third and final design, the *internal-coupled joint*, see Figure 3 (left). This design has an efficiency ratio of 92%.

The joint is formed by counterboring both ends of a ductile iron pipe. An independent, internal steel coupling with double 6.35mm (1/4in) O-ring grooves, is utilized in this joint configuration, see Figure 4 (below).

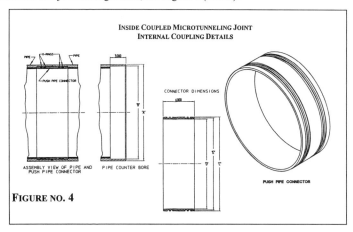

FIGURE NO. 4

In proceeding through the processes established for new product development, the next step was to produce a prototype for each of the three (3) joint concepts. The design team then made a presentation to the Product Planning Committee (PPC) suggesting that we continue with the development of a ductile iron microtunneling pipe. The PPC concurred with the team's recommendation and allowed phase I testing of the inside coupled joint to begin.

Testing -- Phase I

Chronologically parallel with Phase I testing, and in concert with the National Utility Contractors Association (NUCA), American was assisting the NUCA Microtunneling Specification Task Group in the preparation of microtunneling pipe specifications. Specifically, our contribution was the preparation of a specification for *Ductile Iron Microtunneling Pipe*. The results of this Task Group have been forwarded to the North American Society for Trenchless Technology (NASTT), who in turn will forward them to ASCE for publication.

Phase I testing began in early 1993 and involved a battery of four (4) tests. They included:

   1. Finite Element Analysis (FEA),
   2. Columnar Load Test,

3. Hydrostatic Test of Joint Assembly, and
4. Hydrodynamic Flow Test for Determining Head Loss.

All tests, except the hydrodynamic test (completed using 102mm pipe), were performed using 406mm (16in) *internal coupled joint*.

A preface to any discussion about the finite element analysis (FEA) requires the review of the physical properties of ductile iron. The following are representative minimum values for the physical properties of ductile iron for use as microtunneling pipe for pressure or gravity service.

1. Tensile Strength: Min. 413.7 MPa (60,000 psi)
2. Tensile yield strength: Min. 289.59 MPa (42,000 psi)
3. Compressive Strength Properties: The compressive yield strength of ductile iron is 10%-20% higher than the tensile yield strength. The ultimate strength in compression is not normally determined for ductile metals, though apparent strength in tests may be several times the tensile strength value.
4. Elongation: Min. 10%
5. Modulus of Elasticity: 16.548×10^4 MPa (24×10^6 psi) -- tension or compression
6. Poisson's Ratio: 0.28

Finite Element Analysis

Finite element analysis was accomplished by analyzing the pipe model and internal coupling model independently. This was viewed as being a more conservative approach, as the deformation of the counterbore of the pipe and deformation of the internal coupling would be constrained by the geometry of the assembled joint.

The pipe model analysis was based on the following design criteria:

Two dimensional(2D), axisymmetric elements, coordinate system x (radial), y (axial).
- Outside diameter = 441.96mm (17.40in)
- Thickness = 12.45mm (0.49in)
- Counterbore diameter = 427mm (16.80in)
- Counterbore depth = 76.2mm (3.00in)
- Ductile Iron properties

The internal coupling model analysis was analyzed in the same axis as the pipe model and with the dimensions and properties listed herein below:
- Outside diameter = 425.96mm (16.77in)
- Inside diameter = 406mm (16.00in)
- Steel properties (Young's Modulus = 20.685 x 10^4 MPa, Poisson's Ratio = 0.29)

In both models, the total thrust load was converted to pounds per radians and applied to the nodes. Three (3) different scenarios were constructed, each accepting a simulated total end thrust load of 4,448,000 N (1,000,000 lbs.). For each scenario, three plots were created; Von Mises Stress, resultant displacement, and an exaggerated overlay of the deformation.

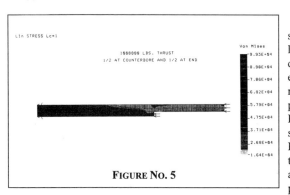

FIGURE NO. 5

The first case simulated the thrust load (4,448,000 N) distributed about the end of the pipe, resulting from pipe-to-pipe contact only, see Figure 5 (left). The second case 227,272 Kg (500,000 lbs.) thrust is distributed about the end of the pipe and 2,224,000 N (500,000 lbs.) is distributed about the base of the counterbore, resulting in the equal contact between the pipe and connector, see Figure 6 (below). In the final case, a 2,224,000 N (500,000 lbs.) thrust is distributed about the end of the internal coupling, resulting from equal contact between the pipe and the coupling (Figure 7).

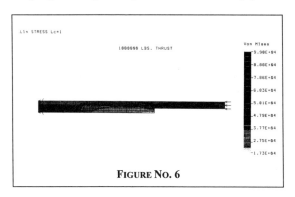

FIGURE NO. 6

Columnar Load Testing

The end thrust loading that was previously discussed was selected to correspond more closely with the results of an actual columnar load test conducted at American's Research facility. American's hydraulic test press

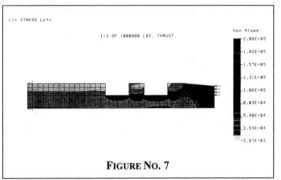

FIGURE NO. 7

has an ultimate design capacity of over 44,480,000 N (10 million lbs.) and was configured with an assembly of two (2) 3.5m (10ft) joints of 406mm (16in), *internal-coupled* joint pipe in a non-confining state (allowed to deform without restriction). The load at which failure (buckle failure at counterbore) was observed and recorded was 4,465,792 N (1,004,000 lbs.).

Hydrostatic Testing

American constructed a 18.29m (60ft) hydrostatic test platform for the sole purpose of hydrostatically testing an assembly of three (3) *internal-coupled* joints (nominal length 5.94m). The 406mm (16in) assembly was tested to 51 BAR (750 psi) and held for 5 minutes without leakage. Tests at reduced pressures and changes of pressure did not result in leakage.

Hydrodynamic Flow Rate Testing

Hydrodynamic flow testing and measurements have been conducted on a standard Fastite push on joint and for the proposed *internal-coupled* joint. These tests were conducted with a 102mm (4in) diameter, cement lined pipe with a nominal inside diameter of 99.57mm (3.92in). This diameter was chosen because it represents the worst case flow condition. The internal coupling for the 102mm design has an actual inside diameter of 91.95mm (3.62in) which is a larger percentage reduction in cross sectional area than would occur in an *internal-coupled* joint of larger diameter. Results show that the head loss for the *internal-coupled* joint was slightly higher at low flow rates, and that for higher flow rates the head loss is less (see Table No. 1). Larger diameter *internal-coupled* pipe joints should have lower head losses as a result of little to no intrusion of the coupling sleeve into the waterway.

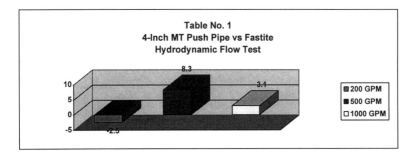

Testing -- Phase II, Field Testing

With Phase I testing completed, the Product Planning Committee recommended that an actual field test of American's *Internal-coupled* Ductile Iron Microtunneling Pipe (now renamed and shortened to *American's MT Push Pipe*) be secured. *MT* in the product name denotes the pipes primary functionality, microtunneling. The *Push Pipe* portion suggests that the pipe has the versatility to be installed by other construction methodologies.

The primary stimulus for arranging the inaugural field test of American's MT Push Pipe was provided by Kinsel Industries (Kinsel). Kinsel was established in 1976 as primarily a paving and utility construction company. Kinsel, recognizing the need for trenchless pipe installation, entered the trenchless construction market early in 1990. They now are recognized as one of the premier trenchless contractors in the Gulf Coast area. Kinsel has been averaging approximately 80 km of trenchless pipe installation per year.

Kinsel Industries was the successful bidder on a project for a water main installation paralleling Memorial Drive in Houston, TX. The bulk of this project was specified to be a bore and jack installation, as the alignment affronts some very affluent areas in the Western portion of Houston. Surface disruption on the heavily traveled Memorial Drive corridor and the economic impact to business along the right-of-way also impacted the decision to use this trenchless installation method.

Kinsel arranged several meetings with City of Houston personnel to present the option of providing a 305m (1,000 linear feet) test section of American's MT Push Pipe on the Memorial Drive project. After resolution of some design questions presented to American by City of Houston engineers, authorization was given to Kinsel Industries to proceed with the test installation.

Kinsel immediately began planning and organizing for the installation which would be scheduled for June 1994. Kinsel planned on using the Hydrahaul 225, push/pull unit, manufactured by Trenchless Replacement Systems (TRS) of Calgary, Alberta, Canada, for the phase II field test installation. As a part of the preplan, Kinsel (TRS licensee for State of Texas) and TRS would be involved in designing new equipment that would be needed to install the MT Push Pipe. It must be noted that a considerable investment of time, effort, and money was made by both firms.

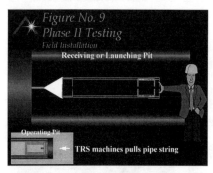

Figure No. 9
Phase II Testing
Field Installation

The Hydrahaul 225 equipment, installed in a secured 2.4m by 2.4m (8ft x 8ft) operating pit (see Figure No. xx, insert represents operating pit), was used to push a 114mm (4.5in) OD, heavy steel, drive-string (1.2m sections of "oil field type" drill rod) through a 250mm (10in) auger bored hole to a receiving pit, approximately 122m (400ft) East. Existing soils were silty-clay at the installation depth of two and one-half meters (8ft). After completion of the "push" a cone shaped installation tool (560mm O.D. and approximately 2.5m in length) was attached to the drive-string. This tool facilitated the expansion of the hole from 250mm (10in) to 560mm (22in). Kinsel followed instructions provided by American on the proper assembly of the MT Push Pipe joint. American's MT push pipe was placed onto a custom designed and constructed assembly and guide frame. Utilizing this frame to support and align the pipe sections, a 38mm (1.5in), heavy steel, drill pipe, called the *Tow Pipe*, was connected to the installation tool and was extended through the entire 5.9m (19.5ft) length of the MT Push Pipe. At the opened end of the pipe a specially designed pushing head was slipped over the exposed Push Pipe connector. With the tow pipe extending through the center of the MT Push Pipe and pushing head, a quick release gripper tool (see Figure 10) was pressed into place by the assembly and guide frame, securing the completed pipe string. The Hydrahaul 225 equipment then began to pull the pipe string (114mm drive-string, installation tool, 38mm tow pipe, MT Push Pipe, Pushing head, and gripper tool) back toward itself. As the pipe string moved, the pipe and pusher head assembly was supported by a small traveling gantry crane (built onto the assembly and guide frame). After movement of the pipe string, equivalent to one standard length, was completed, the quick release gripper tool and pushing head were removed. The process was then repeated. Installation of the 442mm (17.4in) pipe was completed at a rate of one joint every 8 - 10 minutes.

Prior to the field test (described above), Kinsel Industries pretested the procedure by installing 25m (80ft) of MT Push pipe at a remote test sight. The installation was identical to the final test, except that the 114mm (4.5in) drive string was simply pushed through the insitu soil with no pilot (auger bored) hole. The 442mm (17.4in) pipe was then pushed/pulled back through a hole expanded from 114mm (4.5in) to 559mm (22in). Maximum surface displacements were as great as 300mm (12in) following the installation of the 400mm MT Push Pipe.

Conclusions and Observations

Both American and Kinsel shared the opinion that the installation was superb, reflecting the excellent preparation of both companies. This opinion was substantiated by the successful hydrostatic test conducted on the completed pipeline in August 1994.

The installation of the MT Push Pipe challenged neither the capability of the TRS equipment nor the columnar load capabilities of the pipe. The TRS installation minimized the loading on the pipe. Basically, the only load the pipe endures is the total of the weight of the pipe and the frictional drag of the pipe being pulled through the soil. This field test produced several positive outcomes including:

- The ability of the MT Push Pipe to perform under the rigors of trenchless and microtunneling installations.
- The ease of assembling the close tolerance internal-coupled joint.
- A successful partnership between City of Houston, TRS, Kinsel, and American.
- The first domestic installation of Ductile Iron, bell-less, microtunneling pipe.

Acknowledgments

The authors are grateful for the time and efforts given by: The City of Houston: Mr. Tim Lincoln, Mr. Showri Nadagiri, P.E., and Mr. Jack Wells, P.E.. Kinsel Industries: Mr. Bubba Bland, Mr. Glen Crawford, Mr. Mutt DeBorde, and Mr. Richard Kinsel. Trenchless Replacement Systems: Mr. Reginald Handford. Purdue University: Dr. Tom Iseley, P.E. American Cast Iron Pipe Company: Mr. Tom Adams, Mr. Randy Conner, Mr. Brett Mitchell, and Dr. Gene Oliver.

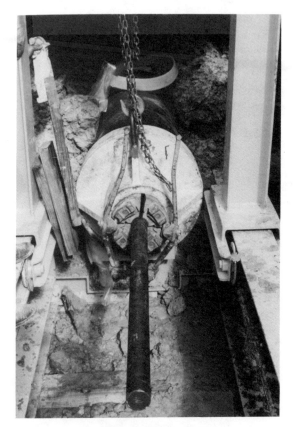

**FIGURE NO. 10
QUICK RELEASE GRIPPER TOOL**

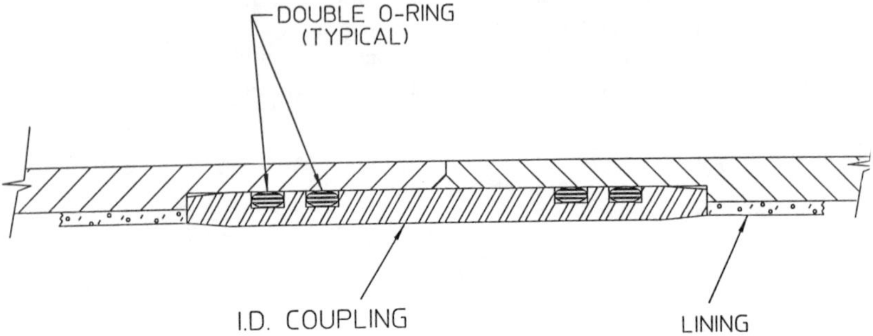

**FIGURE NO. 11
MT PUSH PIPE -- JOINT CROSS SECTION**

Appendix

[1] Kosowat and Powers, "Microtunneling forges ahead", Engineering News-Record, August 17, 1992, pp33.
[2] Budhu and Isley, "The Economic of Trenchless Technology vs. the Traditional Cut and Fill in High-Density-Activity Urban Corridors -- A Research Concept in a Real World Environment", presented at No-Dig 94, Dallas, Texas, 1994
[3] Kotler, "Principles of Marketing," 3rd Edition, Prentice-Hall, 1986, pp334.

POLYETHYLENE ENCASEMENT
AND THE 1993 REVISION OF ANSI/AWWA C105/A21.5 STANDARD

Richard W. Bonds, P.E.[1]

Abstract

The external corrosion protection afforded iron pipe by polyethylene encasement has continued to be extensively researched by the Ductile Iron Pipe Research Association (DIPRA) in all types of corrosive environments since 1952, and has been successfully used for in service iron piping systems since 1958. This paper discusses the evolution of this corrosion protection system, how it works, and the research related to the 1993 revision of ANSI/AWWA C105/A21.5 Standard "Polyethylene Encasement For Ductile Iron Pipe."

Introduction

The earliest recorded installation of cast iron pipe was in 1455 at the Dillenburg Castle in Germany. In 1664, French King Louis XIV ordered the construction of a cast iron pipeline extending 15 miles from a pumping station at Marly-On-Seine to Versailles to supply water for the fountains and town. This pipeline is still functioning after more than 300 years of service. (DIPRA, 1984)

Cast iron pipe was installed in the United States as early as 1804 in Philadelphia. Today, more than 400 utilities in the United States have had cast iron pipe that has provided service for 100 years or longer. At this writing, at least ten utilities in the United States have cast iron mains that have served continuously for 150 years or more. (DIPRA, 1994)

For centuries, utilities installed cast iron pipelines with little thought to the possible need for corrosion protection. The causes of corrosion on underground pipelines were not well understood, and affordable, reliable methods of methods of protection were virtually nonexistent. Fortunately, cast iron pipe

---
[1]Research & Technical Director, Ductile Iron Pipe Research Association, Birmingham, Alabama

proved to be a very durable conduit in most soil environments. In those instances where soil conditions or stray currents promoted extensive corrosion on cast iron pipelines, repair and replacement were accepted as common remedies.

Ductile Iron Pipe

The first ductile iron pipe was cast experimentally in 1948 and introduced into the marketplace in 1955. Since 1965 ductile iron pipe has been manufactured in accordance with ANSI/AWWA C151/A21.51 by centrifugal casting methods that have been in the process of commercial development and refinement since 1925. (DIPRA, 1984)

Ductile iron is a material that has castibility and chemical properties similar to gray iron and mechanical properties more like steel. Both gray iron and ductile iron contain approximately the same amount of carbon. In gray iron, the carbon is present in the form of flakes that are interconnected throughout the metal matrix. These flakes create planes of weakness within the metal that limit the physical properties that can be achieved and thus, are responsible for the characteristic low ductility of gray iron. In ductile iron, the carbon is present in the form of discrete nodules or spheroids that do not create planes of weakness, which results in better continuity of the ferritic structure (see Figure 1). The spheroidal graphite is accomplished by an additional treatment of the molten iron, usually with magnesium, and exercising greater control over certain embrittling agents during the manufacturing process. The result is a form of cast iron with much greater strength as well as ductility. The C151/A21.51 Standard requires the metal to meet minimum physical characteristics of 60,000 psi ultimate tensile strength, 42,000 psi yield strength, and 10 percent elongation.

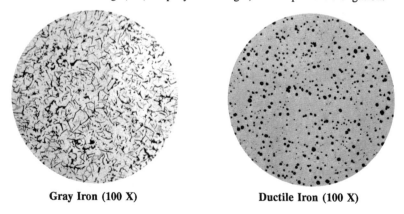

Gray Iron (100 X)   **Ductile Iron (100 X)**

Figure 1

Ductile iron pipe is manufactured in standard sizes ranging from 3 inches to 64 inches in diameter. For water applications, it is typically furnished with cement-mortar lining to prevent internal corrosion. The exterior of the pipe is normally provided with an asphaltic coating that offers some protective value but is not intended to provide long-term corrosion protection.

The pipe is manufactured in nominal 18- and 20-foot lengths and employs a rubber-gasketed jointing system. Although several types of joints are available for ductile iron pipe, the push-on joints and, to a lesser extent, the mechanical joint are the most prevalent.

Corrosion

Corrosion is an electrochemical process involving both a chemical reaction and the flow of electrical current. In order for corrosion to occur, there must be an anode and a cathode, electrically connected by a metallic path, with the anode and cathode immersed in an electrically conductive electrolyte that is ionized. There must also be an electrical potential between the anode and the cathode. If any of these components are eliminated, corrosion cannot occur.

Basically there are two types of corrosion that occur on underground pipelines: galvanic corrosion and electrolytic corrosion. In a galvanic cell, the electrical energy required to drive the corrosion reaction is supplied within the cell due to differences in electrical potential resulting from two dissimilar metals exposed to a single, uniform electrolyte or from a similar metal(s) exposed to a common electrolyte of uneven composition. Electrolytic corrosion is similar to a galvanic cell except that the electrical energy is supplied by a direct current source originating outside the cell. Impressed current cathodic protection systems are an example of direct current sources that can cause electrolytic corrosion. Methods of corrosion prevention can vary depending upon the type of corrosion encountered.

Polyethylene Encasement

Polyethylene encasement was first used experimentally in the United States in 1951 to protect cast iron pipe in a highly corrosive cinder fill in Birmingham Alabama. In conjunction with the Cast Iron Pipe Research Association study initiated to investigate methods of protecting mechanical joint bolts from corrosion, several 6-inch mechanical joint pipe samples were buried with 4-mil thick polyethylene wrapped around the joint assembly. In 1952, similar samples were buried in a highly organic muck test site in the Florida Everglades and later, additional samples were buried in a tidal salt marsh in Atlantic City, New Jersey.

The first inspection of samples in 1953 revealed that the polyethylene had provided excellent protection not only for the bolts but also for the entire pipe surface under the wrap. Successive inspections at later dates revealed the same degree of protection being provided underneath the polyethylene wrap while severe corrosion was evident elsewhere on the pipe (see Figure 2).

Polyethylene Encasement - Bolt Protection Study Specimen
Figure 2

As a result, in the early 1950's, CIPRA began an ongoing testing program, burying bare and polyethylene encased cast iron pipe specimens in highly corrosive muck in the Florida Everglades and later in a tidal salt marsh in Atlantic City, New Jersey.

The success of these early installations led to the development of an extensive ongoing research program that determined polyethylene encasement's efficacy in providing a high degree of corrosion protection for cast and ductile iron pipe in all types of soil environments.

By the late 1950's, successful results of CIPRA's research program led to the first use of polyethylene encasement in operating water systems in Lafourche Parish, Louisiana, and Philadelphia, Pennsylvania. And, in 1963, CIPRA continued its research with the burial of the first polyethylene encased ductile iron pipe specimens in test sites in the Everglades, and Wisconsin Rapids, Wisconsin. Millions of feet of polyethylene encased cast and ductile iron pipe have since been installed in thousands of operating water systems across the United States and throughout the world.

How Polyethylene Encasement Works

Ductile iron pipe is encased with a tube or sheet of polyethylene encasement at the trench immediately before installation. The polyethylene acts as an unbonded film, which prevents direct contact of the pipe with the corrosive soil. It also effectively reduces the electrolyte to any moisture that might be present in the thin annular space between the pipe and the polyethylene film.

Polyethylene encasement is not designed to be a watertight system. However, when surrounded by compacted soil, the polyethylene is tightly pressed against the pipe, thereby preventing any significant exchange of groundwater between the wrap and pipe.

Under the film, this minimal amount of water loses its dissolved oxygen content through a reaction with the iron surface, thereby becoming deficient in the chemical constituents necessary to sustain further corrosion. The water enters a state of stagnant equilibrium and a uniform environment exist around the pipe, which effectively halts the oxidation process long before any damage occurs. The polyethylene film also retards the diffusion of additional dissolved oxygen to the pipe surface and the migration of corrosion products away from the pipe surface. (STROUD, 1989)

Polyethylene Encasement Standards Development

In 1926, the American Standards Association (ASA) (now known as American National Standards Institute [ANSI]) Committee A21, Cast-Iron Pipe and Fittings, was organized under the sponsorship of the American Gas Association (AGA), the American Society for Testing Materials (ASTM), the American Water Works Association (AWWA), and the New England Water Works Association (NEWWA). In 1958, Committee A21 was reorganized where subcommittees were established to study each group of standards in accordance with the review and revision policy of ASA (now ANSI). In 1984, the committee became AWWA Standards Committee A21 on Ductile-Iron Pipe and Fittings.

Due to polyethylene encasement's excellent success under actual field conditions, the first national standard, ANSI/AWWA C105/A21.5, was adopted in 1972. The American Society for Testing and Materials issued a standard for polyethylene encasement (ASTM A674) in 1974. Today, the USA (ANSI/AWWA C105/A21.5 & ASTM A674), Japan (JDPA Z 2005), Great Britain (BS6076), Republic of Germany (DIN 30 674, Part 5), and Australia (A.S. 3680 & A.S. 3681) have adopted national standards for polyethylene encasement of ductile iron pipe. An international standard for polyethylene sleeving (ISO 8180) was adopted in 1985.

ANSI/AWWA C105/A21.5 covered material specifications and installation procedures in the body of the standard, and soil investigation procedures were detailed in an appendix to the standard. The material specified was 8-mil low-density polyethylene for which three different installation methods were included: Methods A and B use polyethylene tubes and Method C uses polyethylene sheets.

The soil evaluation procedure in Appendix A is based upon information drawn from five tests and observations: soil resistivity, pH, oxidation-reduction (redox) potential, sulfides, and moisture (see Table 1).

Table 1
Soil-Test Evaluation*

Soil Characteristics	Points
Resistivity--ohm-cm (based on single-probe at pipe depth or water-saturated soil box):	
< 700	10
700-1000	8
1000-1200	5
1200-1500	2
1500-2000	1
> 2000	0
pH:	
0-2	5
2-4	3
4-6.5	0
6.5-7.5	0**
7.5-8.5	0
> 8.5	3
Redox potential:	
> +100 mV	0
+50 to +100 mV	3.5
0 to +50 mV	4
Negative	5
Sulfides:	
Positive	3.5
Trace	2
Negative	0
Moisture:	
Poor drainage, continuously wet	2
Fair drainage, generally moist	1
Good drainage, generally dry	0

*Ten points--corrosive to gray or ductile cast-iron pipe; protection is indicated.
**If sulfides are present and low or negative redox-potential results are obtained, three points shall be given for this range.

For a given soil sample, each parameter is evaluated and assigned points according to its contribution to corrosivity. The points for all five areas are totaled, and if the sum is 10 or more, the soil is considered corrosive to ductile iron pipe and protective measures should be taken. (ANSI/AWWA, 1993)

In addition, potential for stray direct current corrosion should also be considered as part of the evaluation. Notes on previous experience with underground structures in the area are also very important in predicting soil corrosivity. (AWWA, 1987)

It is important to note that the 10-point system, like any evaluation procedure, is intended as a guide in determining a soil's potential to corrode ductile iron pipe. It should be used only by qualified engineers or technicians experienced in soil analysis and evaluation.

Polyethylene Encasement Field Investigations

To complement CIPRA's own test sites in evaluating the effectiveness of polyethylene encasement to protect cast and ductile iron pipe in corrosive soils, a program was initiated in the early 1960's to excavate and inspect cast and ductile iron pipelines in operating systems protected with polyethylene encasement. It is DIPRA's practice to invite interested consultants and/or utilities to participate in the inspection.

The usual procedure for performing the inspection starts with the utility providing a backhoe and crew to uncover the pipe. The location is chosen by the utility. Typically, approximately 10 feet of pipe is uncovered and the polyethylene removed. The pipe is then cleaned and washed with a stiff wire brush and inspected around its total circumference with a sharp geologist's hammer. A sample of the polyethylene is sent to the DIPRA laboratory for analysis and a sample of soil adjacent to the pipe is collected for evaluation. Pictures of the dig-up are taken and a detailed report is prepared and distributed to the utility and participants.

Polyethylene Encasement Investigations
Figure 3

Figure 3 shows the locations of numerous investigations of polyethylene encased pipelines that have been conducted since the early 1960's. Photographs of representatives dig-ups, which are included in Figure 4 and 5 are typical of the protection provided by polyethylene encasement at each location.

LaFoush Parish, LA - 35 Years
Resistivity: 520 ohm-cm
Figure 4

Charleston, SC - 21 Years
Resistivity: 560 ohm-cm
Figure 5

4-Mil, High-density, Cross-laminated (HDCL) Polyethylene

During the 1982 revision of C105/A21.5, the A21 Committee considered a manufacturer's request that high-density, cross-laminated (HDCL) polyethylene be incorporated into the standard. The A21 Committee concluded that while logic would indicate that the material should perform as well as the standard 8-mil, low-density polyethylene, there were insufficient field tests to verify that fact. Consequently, the material was not incorporated into the 1982 revision of the standard.

To investigate the corrosion protection afforded ductile iron pipe by 4-mil HDCL polyethylene encasement, DIPRA initiated a comparative study at its highly corrosive Logandale, Nevada test site in 1982. Soils analysis of the Logandale test site are shown in Table 2. Since that time, 3-, 6-, and 10-year removals of specimens have been conducted in conjunction with that study. Results of the study show that the corrosion protection afforded ductile iron pipe by the 4-mil, HDCL polyethylene encasement is comparable to that afforded by the standard 8-mil polyethylene encasement. Figures 6 and 7 compare the average of the ten deepest surface variations and weight loss of specimens encased in both polyethylenes after 3, 6, and 10 years exposure. The 10 year exposure specimens are shown in Figure 8.

HDCL polyethylene was first installed in an operating pipeline in Aurora, Colorado in 1981. Dig-up investigations of this pipeline were conducted in 1986 and in 1992. Both investigations revealed that the 4-mil HDCL polyethylene had effectively protected the ductile iron pipe. Since that first installation, 4-mil HDCL polyethylene has been used on several pipelines in operating systems.

Table 2
Soils Analysis
Logandale, Nevada Test Site

Resistivity	40 to 100 ohm-cm
pH	7.1
Redox potential	+100 mV
Sulfides	Negative Reaction
Chlorides	9,000 ppm
Sulfates	40,000 ppm
Description	Red & brown, silt & clay, very moist, plastic

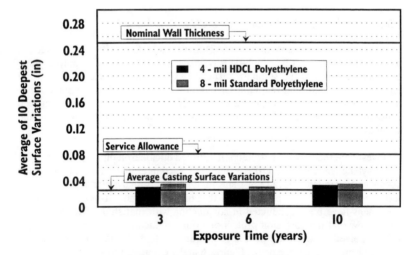

Average of 10 Deepest Surface Variations
Logandale, Nevada Test Site
Figure 6

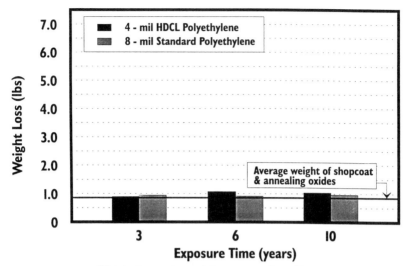

Weight Loss - Logandale, Nevada Test Site
Figure 7

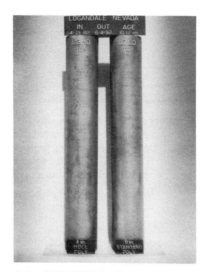

8 mil LD and 4 mil HDCL Polyethylene Encased Specimens
10 Years Exposure
Figure 8

During the 1993 revision of ANSI/AWWA C105/A21.5 Standard, the A21 Committee reviewed that test data on 4-mil HDCL polyethylene and concluded that from all indications it provided comparable protection of ductile iron pipe to that afforded by the standard 8-mil, low-density polyethylene. Based on that conclusion, the A21 Committee elected to incorporate the 4-mil HDCL polyethylene into the standard.

Laboratory tests indicate that the 4-mil HDCL polyethylene may be more resistant to construction damage than the standard 8-mil low-density polyethylene. Tensile strength, impact strength, and puncture resistance of the 4-mil HDCL polyethylene are typically greater because of inherent differences in the materials (Table 3).

Table 3
Typical Physical Properties
8-mil LD and 4-mil HDCL Polyethylene

	8-mil Low Density	4-mil HDCL
Tensile Strength (psi)	2,260	7,990
Elongation (%)	750	415
Dielectric Strength (V/mil)	2,280	3,420
Falling Ball Impact (kg/cm)	60	200
Beach Puncture Resistance (kg/cm)	40	78

Based on DIPRA's laboratory and field research, either the 8-mil, low-density or the 4-mil HDCL polyethylene material is recommended in accordance with ANSI/AWWA C105/A21.5 Standard for corrosion protection of ductile iron pipe in aggressive environments. There are also other types of polymeric materials available that may provide equally suitable protection, some of which are presently being researched by DIPRA.

Conclusion

Since 1958, polyethylene encasement has been used extensively in the waterworks industry to protect cast and ductile iron pipe in corrosive environments. The advantages of this system of corrosion protection that utility managers and engineers find most attractive are its low initial cost, ease of installation, ease of repair, and lack of requirement for monitoring and maintenance once installed. As with all corrosion protection systems, however, polyethylene encasement must be installed properly in order to provide the desired level of protection. Extensive usage history has proved this system of protection to be both economical and effective in a broad range of soil conditions.

References

ANSI/AWWA 1993, "Polyethylene Encasement For Ductile Iron Pipe", American Water Works Association, Denver, CO (1993)

AWWA 1987, "External Corrosion: Introduction to Chemistry and Control", Manual M27, 1st ed., American Water Works Association, Denver, CO (1987)

DIPRA 1984, "Handbook of Ductile Iron Pipe", 6th ed., Ductile Iron Pipe Research Association, Birmingham, AL (1984)

DIPRA 1994, "Cast Iron Pipe Century and Sesquicentury Club Records and Correspondence", Ductile Iron Pipe Research Association, Birmingham, AL (1994)

STROUD/NACE 1989, "Corrosion Control Measures For Ductile Iron Pipe", Paper No. 585, CORROSION 89, National Association Of Corrosion Engineers, Houston, TX (April 17-21, 1989)

Reexamining Seismic Risk Assessment for Buried Pipelines

Douglas G. Honegger[1] M. ASCE

Abstract

Seismic risk assessment often plays a key role in emergency planning of pipeline systems and justification capital projects. The methodology for performing seismic risk assessments has changed considerably over the past 10 years to try to meet the demands for more specific information on the location and extent of potential earthquake damage. One consequence of this is the promotion of detailed probabilistic approaches as a means to assess risks and system performance. However, there has been little improvement in the quality of data available on past earthquake performance of pipeline systems. Pipeline performance remains based on a limited set of earthquake data expressed in terms of breaks, repairs, or leaks per unit length of pipeline. Use of historical pipeline damage to establish vulnerability is judged appropriate only when gross estimates of damage are needed. Where more detailed information is required, analytically developed vulnerability relationships are preferable because of their capability to distinguish performance based on specific pipe parameters.

Introduction

The earliest efforts to assess pipeline performance focused on estimation of the overall level of damage to buried pipelines. Planners and pipeline operators used this information primarily as a basis for estimating the level of effort associated with post-earthquake recovery. With each earthquake, it seems that there is more pressure on pipeline system operators to have more specific information on pipeline damage and to have detailed plans in place to rapidly respond to this damage.

---

[1] Technical Manager, EQE International, Inc., 18101 Von Karman Ave, Suite 400, Irvine, California 92715-1032 ph. 714-833-3303

This pressure is most evident for natural gas pipeline systems because of the public and political perception of the potential hazard of these pipelines. However, the increasing cost and public opposition to large water projects can produce similar pressures on the risk assessment procedures used to justify the project.

Engineers have responded by incorporating more sophisticated analysis techniques into the seismic risk assessment. These techniques include probabilistic definition of regional earthquake activity and incorporation of qualitative estimates of seismic hazards (e.g., ground shaking, liquefaction, landslide).

Seismic risk assessment of pipeline systems are generally regional in nature and are characterized by four basic components:

1. Identification of earthquake hazard
2. Determination of pipeline vulnerability to the seismic hazards
3. Quantification of damage sustained by the pipeline system
4. Assessment of pipeline system performance in the damaged state

There is considerable uncertainty in each of these components. However, this paper focuses on assignment of pipeline vulnerability using existing data on earthquake damage data.

Current Approaches for Assessing Pipeline Vulnerability

The most commonly used damage models for regional studies are based upon past observations of earthquake damage that have been correlated with a general earthquake parameter such as Modified Mercalli Intensity, peak ground acceleration, occurrence of liquefaction, or distance from fault rupture. Typically, damage models relate these general measures of earthquake hazard to a rate of repair or break per unit of pipeline length (e.g., breaks per 1000 feet, repairs per kilometer).

Damage information from past earthquakes is very sparse in terms of detail and variety of pipe manufacturing technologies represented. A review of reported damage from past earthquakes reveals the following characteristics:

1. Most of the data concerns cast-iron and jointed pipe
2. Past experience is largely limited to pipe less than 24 inches in diameter
3. An extremely limited amount of information on pipe strength characteristics has been collected (i.e., wall thickness, joint type, failure mode)
4. Collected data are poorly differentiated between various contributors to pipe damage such as corrosion, ground

shaking, subsidence, liquefaction, lateral spread movement, and surface faulting.

Regardless of these shortcomings, past damage data statistics continue to be regularly used to identify potential damage to pipeline systems. Use of this poorly defined set of data is typically argued as acceptable for several reasons. The most common justification is that a great deal of accuracy in not warranted because of the considerable level of uncertainty in estimating seismic hazards, especially those related to permanent ground deformations. Other bases for justification include references to previous assessments utilizing similar approaches and lack of any better statistical data. In cases where an engineer is faced with trying to meet the needs of his client, the latter justification may often carry great weight.

The current approach for performing risk assessment of pipeline systems is to utilize engineering mechanics principles to scale trends in past earthquake damage. Indeed, this is the general approach recommended in Taylor (1991). Application of this approach in practice include Harding and Lawson et al. (1991, 1992) and Honegger and Eguchi (1992). The scaling process requires several assumptions regarding critical failure mode and the relative impact of such pipe characteristics as material strength and ductility, joint strength, diameter, and burial depth.

Selection of a Baseline Damage Relationship

Unfortunately, the approach is highly dependent upon the baseline damage relationship selected. To provide an indication of the degree of variability in pipeline damage statistics, data reported from several sources has been compiled in Figure 1. Note that Figure 1 does not employ a logarithmic scale for the damage axis as has been adopted by most investigators of earthquake-related pipeline damage. Development of Figure 1 required the use of an expression relating Modified Mercalli Intensity (MMI) to peak ground acceleration (PGA). The relationship used for Figure 1 is the same as that used by Taylor (1991):

$$\text{Log PGA} = -2.94 + 0.286 \text{ MMI}$$

There are clearly two trends indicated in the data set. These two trends indicate the impact of segregating damage data between areas of competent soils versus areas subjected liquefaction and sources of permanent ground deformation. Eguchi (1983) first demonstrated the significant impact of such data segregation. On the other hand, much of the data presented in Katayama et al. (1975) is known to have occurred in liquefaction areas and areas where permanent ground deformation was present.

Also plotted in Figure 1 are examples of damage relationships that have been developed to represent the variation of pipeline damage with earthquake severity. It is clearly evident that developing pipeline damage algorithms using a logarithmic formulation has severe practical limitations in highly seismic areas. This problem is addressed by Taylor (1991) and Wang et al. (1991) by artificially limiting the upper damage level to that corresponding to replacement of 10% of the pipeline.

In reviewing the data and the published vulnerability relationships, several key observations can be made:

1. There is considerable scatter in the trend of historical data (as much as an order of magnitude)
2. There is very limited data for ground shaking above about MMI 9 to verify existing vulnerability relationships
3. The data used to derive vulnerability relationships is highly variable depending upon whether damage from liquefaction or other sources of permanent ground deformation are excluded.
4. There seems to be little practical advantage to using logarithmic expressions for vulnerability relationships

It is important to understand the impact of observation 3. More careful review of the data suggests that the data segregation carried out by Eguchi (1983) had a significant impact on lowering the resulting vulnerability relationship. Much of the damage data presented in Katayama et al. (1975), Miyajima and Kitaura (1991), O'Rourke and Pease (1992) and O'Rourke and Ballantyne (1992) is from areas in which liquefaction occurred. A limited investigation of the available earthquake damage data identified many sources of data that contain a large percentage of data from soft sites, liquefied soils or soils that experienced some permanent ground deformation.

Given the above discussion, an analyst choosing to use past damage data for a regional assessment of pipeline damage has two options. The analyst can develop two baseline damage relationships, one for regions susceptible to liquefaction or permanent ground deformation and one for regions with competent soils. Examples of how such relationships could be developed from the data presented in Figure 1 are presented in Figure 2. Alternatively, the analyst can use a damage relationship based on damage data from competent soils and develop a damage relationship for regions susceptible to permanent ground deformation based on analytical approaches.

In any case, the use of a damage relationship in which the logarithm of the rate of damage increases with ground shaking severity is judged inappropriate.

Damage relationships developed by Katayama et al. (1975) are all believed to be only applicable to sites with a high liquefaction potential or where varying amounts of permanent ground deformation are expected. Rather than an indication of soil condition, the "good" and "poor" soil bounds presented in Katayama et al. (1975) simply reflect the variability in the damage data used in the correlation study. In competent soil conditions, these relationships may overpredict damage by an order of magnitude or more.

Alternative Approaches

At the present time, the best alternative to the reliance on questionable earthquake damage data is to develop vulnerability relationships analytically. Such an approach is described in Honegger (1994) for application to welded steel natural gas pipelines . By performing a set of non-linear analyses on a variety of pipeline configurations, it is possible to differentiate levels of pipeline performance accounting for variation in material, diameter, wall thickness, configuration (e.g. straight section, ell, tee), and soil type. Pipeline vulnerability can then be related to the magnitude and size of ground movements associated with lateral spread deformations. It should be noted that the approach of Honegger (1994) was limited to steel pipelines with welded connections that have strengths similar to that of the pipe body.

Pipelines constructed of less ductile materials such as cast iron, concrete, asbestos cement or very weak joints are not as amenable to analysis. This class of construction is typical for water and sewer pipelines. Without a positive joining mechanism, the deformation capacity of bell-and-spigot type joints is limited to the rotational and extensional capacity of a single joint. Overall pipeline vulnerability is therefore a function of the variation in individual joint strength and deformation capacity.

Mechanical joining can improve the situation by assuring that multiple joints are activated. This permits a section of pipeline to conform to the imposed ground deformation. With mechanical couplings, it is possible to develop analytically based estimates of pipeline vulnerability of segmented pipelines. An example of an dapproach for estimating the deformation capacity for jointed pipelines is that developed by O'Rourke and Trautman (1981).

Reliance on analytically derived pipeline vulnerability relationships requires considerable refinement in the definition of seismic hazards related to permanent ground deformation. Typical practice is to simply provide qualitative rankings on the susceptibility to liquefaction. One means to perform such a qualitative ranking is provided by Youd and Perkins (1987). However, analytically derived vulnerabilities require some quantitative description of the amount and extent of ground deformation. The geotechnical community has made significant advancement in the area of ground deformation hazard estimation.

One approach for obtaining estimates for the amount of lateral movement that may be associated with development of a lateral spread is that developed by Bartlett and Youd (1992). Implementation of this approach requires some additional site information such as the slope of the ground surface, grain size distribution, thickness of liquefiable layer and normalized blow count. This information can generally be obtained in a developed area from boring logs performed as part of pre-construction site investigations.

An example of how to obtain estimates on the dimensional characteristics associated with lateral spread displacement is described in Honegger (1994). The approach relies on interpretation of detailed ground displacement data collected for two Japanese earthquakes with site characteristics judged to be generally similar to those of the study region in question. The resulting distributions of lateral spread dimensions parallel to and perpendicular to the predominant direction of lateral spread movement are shown in Figure 4.

An alternate approach to interpreting the same set of data is described in Suzuki and Masuda (1991). Their approach focused on relating lateral spread dimensions to to the amount of observed peak ground displacement.

Improving the State of Practice

Analytical approaches similar to those developed by Honegger (1994) have been demonstrated to be far superior to approaches utilizing poorly defined vulnerability relationships based on past earthquake damage. This approach is felt to be most readily implemented for relatively modern (post-1950) steel pipelines with good-quality butt-welded joints. However, in cases where significant joint strength exists, significant portions of a segmented pipeline can be counted on to respond to imposed ground deformations. Analytically developed vulnerability relationships should be strongly considered under such circumstances.

Analytical approaches to determine pipeline vulnerability are generally not considered applicable to most water or sewer pipeline systems because of the inherent weaknesses in their construction. These weaknesses include the use of brittle materials such as cast iron or asbestos-cement, low ductility mechanical pipe joints or non-structural welded joint construction. If past performance is an indication of future developments, the construction practices for water and sewer pipelines in seismically active areas will not improve to eliminate well-known weaknesses. As a result, scaling of vulnerability relationships as discussed in Taylor (1991) and Honegger (1994) will continue to be one of the few practical approaches to assessing seismic risk in the near future.

Improvement of the state-of-the-art in assessing pipeline risk from earthquake hazard requires research to better quantify the variability in establishing pipeline vulnerability. Suggested areas of investigation include the following:

1. Verification of the adequacy of current analytical approaches for both welded steel pipelines and segmented pipelines with restrained joints through near full-scale tests
2. Experimental verification of failure modes (e.g., joint pull-out, buckling) and failure criteria (e.g., compressive strain, curvature) for various types of pipelines
3. Variability in constructed versus design conditions and quantification of the impact on seismic ruggedness
4. Quantification of the influence of a built-up urban environment on idealized representation of pipeline conditions and earthquake hazard definition

Implications for the Risk Assessment Practice

The current state-of-practice in assessing seismic risks to pipelines has evolved to include mathematical formulations that incorporate probability concepts into estimates of expected pipeline damage. Despite the fact that the risk assessment community has been aware of the shortcomings presented in this paper for some time, there has been little effort to clearly identify the high degree of variability in the results from such numerical exercises.

Typical practice is to utilize previously developed pipeline vulnerability relationships from past earthquake damage data. These relationships are often scaled to account for differences in the mechanical characteristics of the pipeline being evaluated and those included in the historical damage database. From the author's experience in a recent project, using this approach can easily have a 5-fold increase between a mean and 16% exceedance estimate of damage. This high variability generally undermines the value of carrying out a detailed probabilistic seismic risk assessment. If areas of high ground shaking potential and poor soil conditions are fairly limited and well-defined, estimates of damage based on expert opinion and judgement can be as reliable and much less costly.

Use of past pipeline damage as the basis for assessing earthquake damage is most useful when providing an order of magnitude estimate of earthquake damage potential. This very approximate information is appropriate for supporting regional emergency planning activities. Using the information in other applications such as justifying large capital projects, acquisition and prepositioning of materials to be used in emergency response, prioritization of post-earthquake inspection activities, and assessing post earthquake system performance should be questioned whenever the impact of variability in the resultant risk assessment is not explicitly accounted for.

Given the preponderance of historical damage data related to areas of poor soil or subjected to liquefaction, it is likely that the risk of earthquake damage may have been greatly overestimated in past risk assessments for regions with less than a high potential for liquefaction. Even in areas with large regions of high liquefaction potential, past studies have likely greatly overestimated the damage potential by using logarithmic damage relationships and overestimating the occurrence of liquefaction induced displacements. Furthermore, relatively modern steel pipelines with butt-welded joints, pipeline vulnerability should be established using analytic methods. Advancements in analytic approaches to pipeline risk assessment are available as a result of information and research that has become available in the past several years. Past studies of such pipelines using scaled vulnerability relationships developed from historical data should be viewed as potentially flawed.

References

1. Ayala, A.G. and M.J. O'Rourke, 1989, "Effect of the 1985 Michoacan Earthquake on Water Systems and Other Buried Lifelines in Mexico," NCEER-89-0009.

2. Bartlett, S.F. and T.L. Youd, 1992, "Empirical Analysis of Horizontal Ground Displacement Generated by Liquefaction-Induced Lateral Spreads", NCEER-92-0021

3. Eguchi, R.T., C. Taylor, and T.K. Hasselman, 1983, "Seismic Component Vulnerability Models for Lifeline Risk Analysis," prepared for the National Science Foundation under grant No. PFR-8005083, J.H. Wiggins Company Technical Report No. 82-1395-2c.

4. Eguchi, R.T., 1983, "Seismic Vulnerability Models for Underground Pipes," Proceedings of the 1983 International Symposium on Lifeline Earthquake Engineering, PVP-Volume 77, American Society of Mechanical Engineers.

5. Harding and Lawson, Dames and Moore, Kennedy/Jenks/Chilton, and EQE Engineering, 1991, "Liquefaction Study for the Marina and Sullivan Marsh Areas," final report prepared for the City of San Francisco.

6. Harding and Lawson, Dames and Moore, Kennedy/Jenks/Chilton, and EQE Engineering, 1992, "Liquefaction Study for North Beach, Embarcadero Waterfront, South Beach, and Upper Mission Creek Areas," final report prepared for the City of San Francisco.

7. Honegger, D.G., 1994, "Assessing Vulnerability of BC Gas Pipelines to Lateral Spread Hazards," Proceedings of the 5th U.S.-Japan Workshop

on Earthquake Resistant Design of Lifeline Facilities and Countermeasures Against Liquefaction, Snowbird, UT, September 29-October 1, (in press)

8. Honegger, D.G. and R.T. Eguchi, 1992, Determination of Relative Vulnerabilities to Seismic Damage for San Diego County Water Authority Water Transmission Pipelines, report prepared for San Diego County Water Authority.

9. Katayama, T., K. Kubo and N. Sato, 1975, "Earthquake Damage to Water and Gas Distributions," Proceedings of the U.S. National Conference on Earthquake Engineering, EERI.

10. Miyajima, M. and M. Kitaura, 1991, "Performance of Pipelines During Soil Liquefaction, Proceedings of 3rd U.S. Conference on Lifeline Earthquake Engineering, Los Angeles, ASCE-TCLEE Monograph No. 4.

11. O'Rourke, M.J. and D. Ballantyne, 1992, "Observations on Water System and Pipeline Performance in the Limon Area of Costa Rica Due to the April 22, 1991 Earthquake", National Center for Earthquake Engineering Research, Report NCEER-92-0017.

12. O'Rourke, T.D. and J.W. Pease, 1992, "Large Ground Deformations and Their Effects on Lifeline Facilities: 1989 Loma Prieta Earthquake", Case Studies of Liquefaction and Lifeline Performance During Past Earthquakes, O'Rourke, T.D. and M. Hamada editors, Vol. 2, United States Case Studies, Technical Report NCEER-92-0002

13. O'Rourke, T.D. and C.H. Trautmann, 1981, "Earthquake Ground Rupture Effects on Jointed Pipe", in Lifeline Earthquake Engineering: The Current State of Knowledge, D.J. Smith ed. ASCE.

14. Suzuki, N. and N. Masuda, 1991, "Idealization of Permanent Ground Movement and Strain Estimation for Buried Pipes", Proceedings from the Third Japan-U.S. Workshop on Earthquake Resistant Design of Lifeline Facilities and Countermeasures for Soil Liquefaction, NCEER 91-001.

15. Taylor, C.E. (editor), 1991, Seismic Loss Estimates for a Hypothetical Water System, ASCE-TCLEE Monograph No. 2.

16. Wang, L.R.L., I. Ishibashi, and J.C.C. Wang, 1991, "Inventory and Seismic Loss Estimation of Portland Water/Sewer Systems - GIS Application to Buried Pipelines," Proceedings of 3rd U.S. Conference on Lifeline Earthquake Engineering, Los Angeles, ASCE-TCLEE Monograph No. 4.

17. Youd, T.L. and D.M. Perkins, 1987, "Mapping of Liquefaction Severity Index," Journal of Geotechnical Engineering, vol. 103, no. 11, ASCE, November.

SEISMIC RISK ASSESSMENT

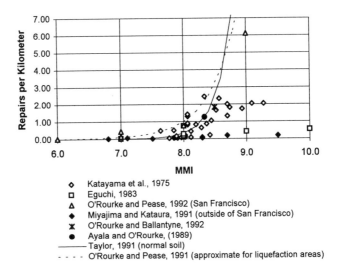

- ◇ Katayama et al., 1975
- ☐ Eguchi, 1983
- △ O'Rourke and Pease, 1992 (San Francisco)
- ◆ Miyajima and Kataura, 1991 (outside of San Francisco)
- ✕ O'Rourke and Ballantyne, 1992
- ● Ayala and O'Rourke, (1989)
- ——— Taylor, 1991 (normal soil)
- - - - - O'Rourke and Pease, 1991 (approximate for liquefaction areas)

Figure 1. Plot of Historical Pipeline Damage Data From Various Sources

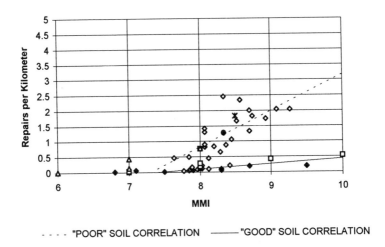

- - - - "POOR" SOIL CORRELATION   ——— "GOOD" SOIL CORRELATION

Figure 2. Example of "Good" and "Poor" Soil Vulnerability Relationships

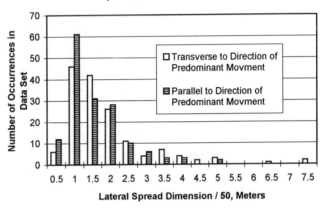

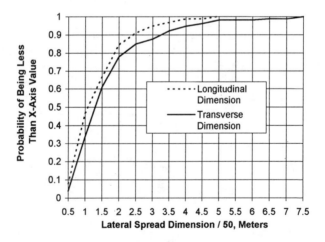

Figure 3. Distribution of Estimated Lateral Spread Dimensions

Influence of Tertiary Creep on the Uplift Behaviour of a Pipe Embedded in a Frozen Soil

J. Hu[1] and A.P.S. Selvadurai[2]

Abstract

The mechanics of the interaction between a pipeline and soils which are susceptible to freezing and frost action can be influenced by the creep characteristics of the frozen soil. Such effects can be particularly important in regions where the frozen soil exhibits discontinuous heave processes. The creep behaviour of a frozen soil has three characteristic stages involving primary, secondary and tertiary phenomena. This paper investigates the uplift behaviour of a section of pipe which is embedded in a frozen soil which exhibits all three creep processes. The finite element technique is used to examine the time-dependent displacements of the pipe which is subjected to constant uplift loads and the time dependent evolution of creep failure zones within the frozen soil. The modelling is viewed as a methodology for identifying time-dependent peak loads that can be sustained by frozen soil, particularly at zones involving discontinuous heave.

Key words

Creep, creep damage, frozen soil, pipeline, uplift load, finite element method.

Introduction

Pipelines in northern environments which convey chilled gas may be located in discontinuous frost susceptible regions. The presence of chilled gas further enhances frost action around the pipeline. When the frost susceptibility of the soil varies along the pipeline this causes discontinuous heave which can create

---

[1] Ph.D student, Department of Civil Engineering and Applied Mechanics, McGill University, Montreal, QC, Canada H3A 2K6
[2] Professor and Chair, Department of Civil Engineering and Applied Mechanics, McGill University, Montreal, QC, Canada H3A 2K6

uplift zones where the heave processes are markedly different (see Figure 1).

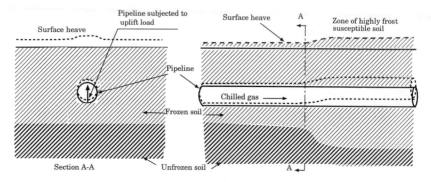

Figure 1: Uplift force on pipeline induced by discontinuous frost heave

The soil-pipeline interaction in such discontinuous regions is influenced by the mechanical behaviour of the frozen soil. Creep characteristics feature dominantly in the mechanical behaviour of a frozen soil owing to the presence of ice and unfrozen water within it. Creep behaviour of a frozen soil has three characteristic stages involving primary, secondary and tertiary creep. The creep rate strongly depends upon the temperature and applied stress. Structures embedded in frozen soils can undergo movements as a result of the creep process. The allowance for tertiary creep will allow micro-mechanical failure of the frozen soil and such processes influence the soil-pipeline interaction process. In this paper, a unified creep model is developed to characterize the complete creep behaviour of a frozen soil which exhibits all three phases of creep behaviour. The creep model is implemented in a finite element code which is used to examine the time-dependent displacements of the pipe which is subjected to a constant uplift load. This paper also examines the time-dependent evolution of creep failure zones within the frozen soil. Such creep failure effects can influence the behaviour of the interaction between a frozen soil and a pipeline which intersects a discontinuous frost heave zone and influences the curvature of the pipeline located at the transition zone.

Constitutive Modelling of Frozen Soil

During the past four decades, the modelling of creep in frozen soils has received considerable attention. Interests in such creep processes stem from the involvement of engineering in the northern environments where geomaterial such as permafrost can be predominant. Many models and empirical relationships have been put forward to describe the creep behaviour of frozen soil. In the early stages of research, these models focused on the description of mainly the creep behaviour in the primary stage. Examples of such models include the power function relationships proposed by Vyalov (1963), Assur (1963), Sayles (1973) and the exponential function representations given by Andersland and AlNouri (1970). In later studies, creep behaviour in the secondary stage was taken into consideration in modelling. A purely secondary creep model which uses a power function was proposed by Hult (1966) and Ladanyi (1972) to represent creep responses in which the stage of the secondary creep dominates the complete creep curve. The creep strain rate was approximated by a linear function. Ting and Martin (1979) employed Andrade's equation to represent creep in the secondary stage. Complete creep models were proposed by Fish (1980, 1983), Ting (1983) and Gardner et al. (1984). These models all utilized a single equation and they belong to the family of creep curves characterized by a rate process theory. It is noted that these complete creep models can only represent primary and tertiary creep stages and the secondary stage degenerates to an inflexion point of the complete creep curve. However, the secondary stage does exist and it is a recognizable component in the constitutive behaviour of many frozen soils.

Comparisons of creep models mentioned above were made by Hampton et al. (1985) and Sayles (1988). These investigations found that a power function is able to predict the creep behaviour of frozen soil at primary stage, but it is not applicable to the latter two stages. The engineering model proposed by Ladanyi (1972) is only applicable in situations where the secondary creep stage dominates the complete creep curve. The models by Fish (1980) and Gardner et al. (1984) greatly overestimate the creep strain in the early portion of the curve but gives better agreement in other stages of the creep response. The model by Ting (1983) fits well the early part of the creep curve, but underpredicts the strain for other ranges of creep. Based on the above observations, it would appear that the available models are only capable of describing, at most, only one or two stages. There is no single complete model that can accurately predict all three stages of creep behaviour. It may even be questioned whether it is reasonable to use a single equation to represent the three distinct mechanisms inherent in a typical creep curve.

In this study, a unified creep model is proposed as a method for characterizing the complete creep behaviour of a frozen soil which exhibits all three stages of creep behaviour. It is well known that the initiation of tertiary creep implies the generation of creep failure within the frozen soil. It is assumed that the occurrence of tertiary creep in a frozen soil does not imply the complete collapse of the region, however, it can induce relatively large creep deformations. In the proposed model, instead of using a single equation to describe a complete creep curve, three separate creep equations are used to characterize individually primary, secondary and tertiary stages of creep behaviour. The model is made complete by prescribing criteria for transition from one stage of creep behaviour to another.

Primary Creep Stage

The power laws representation has been one of the most popular procedures for characterizing constitutive phenomena and as such has been widely accepted for the description of primary creep. i.e. the primary creep behaviour is modelled by the power law:

$$\dot{\epsilon}_{ij}^{cp} = \frac{3}{2} A C \sigma_e^{B-1} t^{C-1} s_{ij} \tag{1}$$

where $\dot{\epsilon}_{ij}^{cp}$ is the creep strain rate in the primary stage; $\sigma_e = \sqrt{\frac{3}{2} s_{ij} s_{ij}}$ is equivalent stress ; $s_{ij} = \sigma_{ij} - \frac{1}{3}\sigma_{kk}\delta_{ij}$ is the stress deviator tensor and A, B and C can be temperature dependent material parameters.

Secondary Creep Stage

The secondary creep stage is also described by a power law of the form

$$\dot{\epsilon}_{ij}^{cs} = \frac{3}{2} A_2 \sigma_e^{B_2-1} s_{ij} \tag{2}$$

where $\dot{\epsilon}_{ij}^{cs}$ is the creep strain rate in the secondary stage; A_2 and B_2 can also be temperature dependent material parameters.

Primary creep parameters A, B, C and secondary creep parameters A_2 and B_2, can be evaluated from the creep curves of a uniaxial constant stress creep test by a graphical method. The detail of these procedures are given by Andersland and Anderson (1978).

Tertiary Creep Stage

In this study, tertiary creep is modelled using a phenomenological theory of creep damage mechanics. The theory of damage mechanics has been used to examine the stress induced progressive deterioration of engineering materials such as concrete, ice, composites, etc. (Kachanov, 1986, Lemaitre and Chaboche, 1974, Boehler and Khan, 1991, Selvadurai and Au, 1991). In the phenomenological theory of creep damage, the damage is identified as the time-dependent stress induced accumulation and growth of micro-voids within a material. The strain rate acceleration in the tertiary stage and the process of creep rupture are explained by appeal to the degradation of the material. Consider a material element in a body, with undamaged original area A_0. As damage evolves, the effective load-bearing area diminishes with the creation of voids. The area of voids induced by damage is defined by A_D (Figure 2). A damage variable (ω) applicable to a uniaxial stress state can be:

$$\omega = \frac{A_D}{A_0} \tag{3}$$

The damage process reduces the load-bearing area of material. Consequently the net stress $\bar{\sigma}$ in a damaged material is defined as:

$$\overline{\sigma} = \frac{\sigma_0}{1-\omega} \qquad (4)$$

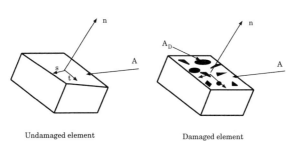

Figure 2: Evolving damage in element

The parameter ω is understood to be an evolving internal variable of the system. For a given stress state at a given time, the evolution of damage will be a property of the material.

Such damage evolution criteria can either be postulated or determined by recourse to creep experiments which are carried out to include the tertiary stage. In the current study we assume that the rate of evolution of damage can be determined by a power law of type:

$$\dot{\omega} = D\left(\frac{\sigma_0}{1-\omega}\right)^k \qquad (5)$$

where D and k are temperature dependent constants.

Damage evolution is generally anisotropic (Lemaitre and Chaboche, 1974, Chow and Wang, 1987). Here we adopt a simplified *isotropic* damage model. The creep strain rate in the tertiary stage $\dot{\epsilon}_{ij}^{ct}$ derived from Equation (2) is therefore defined by:

$$\dot{\epsilon}_{ij}^{ct} = \frac{3}{2}A_2\left(\frac{\sigma_e}{1-\omega}\right)^{(B_2-1)}\left[\frac{s_{ij}}{(1-\omega)}\right] \qquad (6)$$

where A_2 and B_2 are temperature dependent constants referred to the *secondary* creep stage.

A new creep equation for the tertiary stage is derived from Equation (5) and Equation (6)

$$\dot{\epsilon}_{ij}^{ct} = \frac{3}{2}A_2(\sigma_e)^{(B_2-1)}[1-(k+1)D(\sigma_e)^k t]^{\frac{-B_2}{k+1}} s_{ij} \qquad (7)$$

All four parameters A_2, B_2, D and k in Equation (7) can be determined from creep curves obtained from uniaxial constant stress creep tests. The procedures of determining D and k are as follows: (i) two sets of uniaxial creep tests are conducted at different but constant stress levels (σ_1, σ_2), (ii) the creep curves

$\dot{\epsilon}_1$ vs t and $\dot{\epsilon}_2$ vs t are obtained, and (iii) the result (7) is applied to each creep test giving rise to the following equations. The unknown parameters D and k can be obtained by solving the non-linear equations.

$$\left. \begin{array}{l} \dot{\epsilon}_1 = A_2 \left(\dfrac{\sigma_1}{[1-(k+1)D\sigma_1^k t]^{\frac{1}{k+1}}} \right)^{B_2} \\ \dot{\epsilon}_2 = A_2 \left(\dfrac{\sigma_2}{[1-(k+1)D\sigma_2^k t]^{\frac{1}{k+1}}} \right)^{B_2} \end{array} \right\} \quad (8)$$

Transition of Creep Stages

In order to complete the creep model it is now necessary to identify the *inflexion times* at which transition takes place from one creep process to another. In this study, the function of *inflexion times* T_m and their corresponding equivalent stresses σ_e are established by a power law approximation $T_m = E\sigma_e^{-F}$ with a least square method (E and F are material parameters). By the simulation of creep stage transition, the unified creep model is able to examine, depending on the stress levels, both individual and combinations of the creep processes (e.g. the primary stage, the secondary stage, the primary and the tertiary stages, secondary and tertiary stages and the combination of all three stages).

Finite Element Implementation

Details of the finite element modelling of media susceptible to creep are given by Zienkiewicz (1977) and in this section a brief account of the important elements are outlined.

A general incremental constitutive relationship can be developed by assuming that the incremental strain rate is the sum of the incremental strain rate components associated with the respective stages, ie:

$$d\dot{\epsilon}_{ij} = d\dot{\epsilon}_{ij}^e + d\dot{\epsilon}_{ij}^{cp} + d\dot{\epsilon}_{ij}^{cs} + d\dot{\epsilon}_{ij}^{ct} \quad (9)$$

where $d\dot{\epsilon}_{ij}$ - total incremental strain rate; $d\dot{\epsilon}_{ij}^e$ - incremental elastic strain rate; $d\dot{\epsilon}_{ij}^{cp}$ - incremental creep strain rate in the primary stage; $d\dot{\epsilon}_{ij}^{cs}$ -incremental creep strain rate in the secondary stage; and $d\dot{\epsilon}_{ij}^{ct}$ -incremental creep strain rate in the tertiary stage;

A fully implicit algorithm is used for creep analysis (Zienkiewicz, 1977). The scheme is outlined as follows:

$$\epsilon_{c,n+1} - \epsilon_{c,n} = \Delta t_n \dot{\epsilon}_{c,n+\alpha} = \Delta t_n \frac{\partial \dot{\epsilon}_c}{\partial \sigma}(\sigma_{n+\alpha} - \sigma_n) = \Delta t_n \frac{\partial \dot{\epsilon}_c}{\partial \sigma}\alpha\Delta\sigma_n \quad (10)$$

where $\epsilon_{c,n+1}$ and $\epsilon_{c,n}$ are creep strain vectors at time t_{n+1} and t_n respectively; $\Delta t_n = t_{n+1} - t_n$; σ_n is stress vector at time t_n; σ is current stress; α is a conditioning parameter.

Governing equation for this algorithm is given by:

$$\left(\int [\mathbf{B}]^T[\mathbf{D}]^0[\mathbf{B}]dV\right)\Delta\delta_{n+1} + \Delta F_n - \int [\mathbf{B}]^T[\mathbf{D}]^0 \Delta t_n \frac{\partial \dot{\epsilon}_c}{\partial \sigma}\alpha\Delta\sigma_n = 0 \quad (11)$$

where $\Delta\delta_{n+1}$ is the vector of incremental nodal displacement; ΔF_n is incremental loads applied during the time interval Δt_n.

The matrix $[\mathbf{D}]^0$ in Equation (11) can be derived as follows; we have

$$\Delta\sigma_{e,n} = [\mathbf{D}](\Delta\epsilon_n - \Delta\epsilon_{c,n}) \quad (12)$$

where $\Delta\sigma_{e,n}$ is the increment of elastic stress vector during the interval Δt_n and $[\mathbf{D}]$ is the elastic matrix; Consequently

$$\Delta\epsilon_n = [\mathbf{D}]^{-1}\Delta\sigma_{e,n} + \Delta\epsilon_{c,n} = ([\mathbf{D}]^{-1} + \Delta t_n \frac{\partial \dot{\epsilon}_c}{\partial \sigma}\alpha)\Delta\sigma_{e,n} \quad (13)$$

$$\Delta\sigma_{e,n} = \frac{1}{([\mathbf{D}]^{-1} + \Delta t_n \frac{\partial \dot{\epsilon}_c}{\partial \sigma}\alpha)}\Delta\epsilon_{e,n} \quad (14)$$

$$[\mathbf{D}]^0 = \frac{1}{([\mathbf{D}]^{-1} + \Delta t_n \frac{\partial \dot{\epsilon}_c}{\partial \sigma}\alpha)} \quad (15)$$

If $\alpha \geq \frac{1}{2}$ the scheme is unconditionally stable.

Post Failure Analysis

Consider the problem of a pipeline which is embedded in a frozen soil. Due to the process of discontinuous heave the pipe section can be subjected to an uplift force. Such forces can initiate tertiary creep damage in limited zones of the frozen soil and the failure process can attain their peak values. With further loading there can occur distinct zones of failure or localized damage which ultimately result in the development of cracking or separation. Consequently, the stress dependent tertiary creep failure can influence failure that can result from inelastic or plasticity phenomena. In this section we shall address the problem of failure development as a consequence of creep phenomena which includes all three stages. In order to achieve this, new constitutive equations which describe failure has to be introduced.

The initiation of failure of frozen soil is prescribed by appeal to the Mohr-Coulomb failure criterion with a tension cut off condition. The Mohr-Coulomb failure criterion, F in the compression range takes the form:

$$F = \frac{I_1}{3}\sin\phi + \sqrt{J_2}(\cos\theta - \frac{1}{\sqrt{3}}\sin\theta\sin\phi) - C_0\cos\phi = 0$$

where
$\theta = \frac{1}{3}\sin^{-1}\{-\frac{3\sqrt{3}}{2}\frac{J_3}{J_2^{\frac{3}{2}}}\}, \quad (-\frac{\pi}{6} \leq \theta \leq \frac{\pi}{6})$;
$I_1 = \sigma_{kk}, \quad J_2 = \frac{1}{2}\sigma_{ij}^D\sigma_{ij}^D, \quad J_3 = \frac{1}{3}\sigma_{ij}^D\sigma_{jk}^D\sigma_{ki}^D; \quad \sigma_{ij}^D = \sigma_{ij} - \frac{1}{3}\sigma_{kk}\delta_{ij}$;

C_0 is the cohesion and ϕ is the angle of internal friction.

For the tensile failure criterion, F is given by

$$F = \sigma_i - R_T = 0$$

where R_T is tensile strength.

Incremental theory of plasticity is used to develop the constitutive relationships which describe post failure processes. We assume that the total incremental strain in the medium undergoing elasto-plastic failure takes the form

$$\{d\dot{\epsilon}_{ij}\} = \{d\dot{\epsilon}^e_{ij}\} + \{d\dot{\epsilon}^p_{ij}\} \tag{16}$$

where elastic strain $\{d\epsilon^e_{ij}\} = [\mathbf{D}^e]^{-1}\{d\sigma^e_{ij}\}$ and $[\mathbf{D}^e]$ is an elastic matrix. It must be noted that the elasticity matrix $[\mathbf{D}^e]$ should correspond to that which relates to elastic behaviour at the particular level of damage $[\mathbf{D}^e_\omega]$. The limits of elastic behaviour are such that $[\mathbf{D}^e_\omega] \to [\mathbf{D}^e]$ when $\omega \to 0$ and $[\mathbf{D}^e_\omega] \to 0$ as $\omega \to 1$. In this study we assume that, since $\omega \in (0,0.2)$, the elasticity matrix in the damaged state $[\mathbf{D}^e_\omega]$ is approximated by the initial elasticity matrix $[\mathbf{D}^e]$. The plastic strain increment $\{d\epsilon^p_{ij}\}$ can be obtained for a flow rule of the type

$$\{d\epsilon^p_{ij}\} = d\lambda \frac{\partial G}{\partial \sigma_{ij}} \tag{17}$$

where G is the plastic potential function and $d\lambda$ is the positive scalar factor of proportionality.

The general elastic-plastic constitutive matrix $[\mathbf{D}^{ep}]$ can be written as

$$[\mathbf{D}^{ep}] = [\mathbf{D}^e] - [\mathbf{D}^p] = [\mathbf{D}^e] - \frac{[\mathbf{D}^e]\{\frac{\partial G}{\partial \sigma_{ij}}\}\{\frac{\partial F}{\partial \sigma_{ij}}\}^T[\mathbf{D}^e]}{A + \{\frac{\partial F}{\partial \sigma_{ij}}\}^T[\mathbf{D}^e]\{\frac{\partial G}{\partial \sigma_{ij}}\}} \tag{18}$$

where A is the hardening parameter.

The initial stress algorithm for non-linear analysis, used as the iteration procedure, will be summarized in the following: (i) Determine if an element has failed by applying the failure criteria at each time interval. (ii) For an element which has experienced failure, adjust the state of stress based on the corresponding post-failure constitutive relations and compute the initial stress $\{\sigma_0\}$. The initial stress of each element can be evaluated by the difference of stress before and after adjustment. (iii) Compute the equivalent force induced by the initial stress σ_0 and repeat the above scheme till convergence is established. Calculate next time interval for creep until the process is complete.

Verification of Computational procedure

The deflection analysis of a frozen beam is considered for the purpose of verification. The frozen beam of length=1 m, height=0.15 m and width=0.1 m used for calibration is shown in Figure 3 (also see Klein, 1979). Due to the symmetry of the problem, only half of the beam was discretized.

We examine a complete creep curve which includes primary, secondary and tertiary stages. Creep parameters used for the simulation are derived from the experimental data (Gardner et al., 1984). The creep parameters are listed in Table 1.

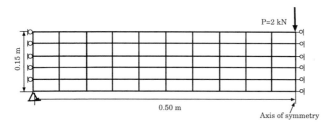

Figure 3: Finite element model of frozen beam

Table 1

Creep model	Parameters
Primary creep model	$A = 0.00002115,\ B = 1.05,\ C = 0.29$
Secondary creep model	$A_2 = 0.00014611,\ B_2 = 1.111$
Tertiary creep model	$D = 0.00013712,\ k = 0.7649$

Figure 4 illustrates the complete creep deflection curve at midspan at the neutral axis of the frozen beam. Creep damage can reduce the load-bearing area of material and eventually yield the structure to failure. The frozen beam is assumed to fail when ω approaches 1. From Equation (7), the time to failure (i.e. $\dot{\epsilon}_{ij}^{ct} \to \infty$) from the initiation of tertiary stage is:

$$t_f = [(k+1)D\sigma_e^k]^{-1} \tag{19}$$

The steady-state stress distributions in a beam susceptible to creep are derived by Odqvist (1966). The expression is given in the form:

$$\sigma_x = (4 + \frac{2}{B})(\frac{M}{bh^2})(\frac{2z}{h})^{\frac{1}{B}} \tag{20}$$

Combining Equation (19) with Equation (20), the analytic solution of failure time of the beam from the initiation of tertiary stage can be written as

$$t_f = \left\{(k+1)D[(4+\frac{2}{B})(\frac{M}{bh^2})(\frac{2z}{h})^{\frac{1}{B}}]^k\right\}^{-1} \tag{21}$$

Since stress distributions are non-uniform, the times for failure of the beam subjected to bending will vary at different locations. Creep failure initiates at the point $z = \frac{h}{2}$, of cross-section at the midspan of the beam. The time for failure derived via Equation (21) is 91.9 hours. The corresponding time obtained via the FEM is 97.25 hours (Figure 4).

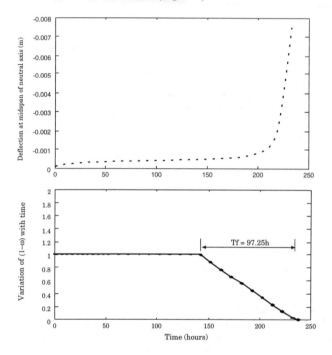

Figure 4: Creep displacement for frozen beam

A Pipeline Subjected to Uplift Load

Pipelines which convey chilled gas which are located in discontinuous frost susceptible regions can be subjected to non-uniform displacements. The non-uniform displacements can create uplift in transition regions in which the frozen soils are either frost susceptible or non-frost susceptible. The mechanics of the soil-pipeline interaction in such discontinuous regions is influenced by the creep characteristics of the frozen soil. In this paper, the finite element technique is used to examine a two dimensional plane strain problem of the time-dependent displacement of the pipe which is subjected to constant uplift loads. The modelling also focussed on the time-dependent evolution of creep failure zones within the frozen soil.

A pipeline of diameter 50 cm and wall thickness 0.5 cm is located at a depth of 175 cm below the surface of a frozen layer. The mechanical parameters of

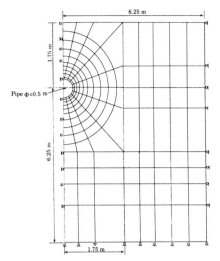

Figure 5: Discreted finite element mesh of pipeline and frozen soils

pipe and frozen soil are assumed as follows: Pipe: $E = 200 \ GPa$, $\nu = 0.3$; Frozen soil: $E = 520 \ MPa$, $\nu = 0.3$, $C_0 = 1.5 \ MPa$, $\phi = 30^0$. The tensile strength of frozen soil is assumed to be $0.5 \ MPa$. The weight of the frozen soil is taken as $19 \ kN/m^3$. The mesh discretization with 152 elements and 182 nodes is shown in Figure 5. The creep parameters used for simulation are listed in Table 1.

Case 1: An uplift load $P = 80 \ kN$ is applied to the pipe and this load is maintained constant with time. The magnitude of load was chosen in order to initiate detachment of elements in the vicinity of the base of the pipe, upon application of the load. Figure 6 illustrates the time-dependent variation in the displacement of the pipe which is subjected to the 80 kN of loading. It indicates that the simulation which couples tertiary creep with post failure development gives a much larger value of total displacement after 6300 hours than for simulation where post failure is not included. Figure 6 also indicates the time-dependent evolution of failure in the frozen soil which couples tertiary creep with failure development.

Case 2: In this problem, a 60 kN load is applied to the pipe and this load is maintained constant with time. At this load level there is no initiation of detachment of the pipe from the frozen soil. At the time of application of the 60 kN load, however failure was allowed to develop by the initiation and evolution of tertiary creep. The elements in the vicinity of the base of the pipe enter the tertiary stage at $t = 1010$ hours. The maximum tensile stresses in these elements are attained and failure occurs at $t = 3100$ h, due to material degradation and reduction of loading-bearing area within them. Figure 7 shows the total displacement and creep displacement with time. Figure 7 also illustrates the contour of the progressive failure zone. As can be observed, the

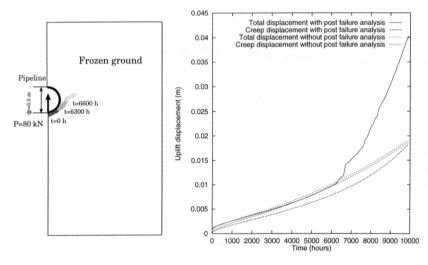

Figure 6: Pipe under uplift loading P=80 kN

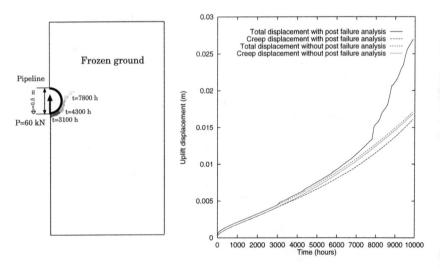

Figure 7: Pipe under uplift loading P= 60 kN

failure zone initiates at 3100 hours after the application of the uplift load. The failure zone gradually expanded thereafter and the displacement curve accelerated markedly at 7800 hours. Further computations also indicate that the frozen medium remains in the primary stage, without evolution of tertiary creep and failure, when the uplift load is 40 kN. The 40 kN load can be identified as the peak loading in the range for a duration of 10000 hours (approximately 1.14 years).

Conclusions

A finite element procedure which incorporates a unified creep model and an algorithm for post failure analysis is applied to examine the uplift behaviour of a section of pipe which is embedded in a frozen soil. The creep behaviour of the frozen soil is modelled by a continuum theory which exhibits three characteristic stages involving primary, secondary and tertiary creep phenomena. In particular the tertiary effects are modelled by consideration of damage mechanics. The finite element algorithm was verified by comparing the numerical results with the analytical solution for a frozen beam. The technique is applied to examine the time-dependent displacement of a section of pipe which is subjected to different magnitudes of uplift loads which remain constant with time. The time-dependent evolution of creep failure zones within the frozen soil are also examined. The numerical procedure can be applied to estimate the maximum uplift loads that can be sustained by a frozen soil in a region where an embedded pipe is subjected to uplift loads.

References

Andersland, O.B. and AlNouri, I. (1970). " Time - dependent strength behaviour of frozen soils." *J. Soil Mech. Found. Div.*, ASCE, 96(4), 1249-1265.
Andersland, O.B. and Anderson, D.M. (1978). *Geotechnical Engineering for Cold Regions*. McGraw-Hill Book Company.
Assur, A. (1963). Discussions. *Proc. 1st Int. Conf. Permafrost*, Purdue University, Lafayette, 339-340.
Boehler, J.P. and Khan, A.S. (1991) *Proc. of PLASTICITY 91: 3rd Int. Symp. Plasticity and Its Current Applications*, Elsevier Applied Science, London.
Chow, C.L. and Wang, J. (1987). "An anisotropic theory of elasticity for continuum damage mechanics," *Int. J. of Fracture*, Vol 33, 3-16.
Fish, A.M. (1980). "Kinetic nature of the long-term strength of frozen soils," *Proc. 2nd Int. Symp. on Ground Freezing*, Trondheim, Norway, 95-108.
Fish, A.M. (1983). "Thermodynamic model of creep at constant stresses and constant strain rates," *U.S. Army Cold Regions Research and Engineering Laboratory, CRREL Report*, 83-33.
Gardner, A.R., Jones, R.H., and Harris, J.S. (1984). "A new creep equation for frozen soils and ice," *Cold Regions Science and Technology*, Vol 9, 271-275.
Hampton, C.N. Jones, R.H. and Gardner, A.R. (1985). "Modelling the creep behaviour of frozen sands," *Proc. 4th Int. Symp. on Ground Freezing*, Sapporo, Japan, 27-33.
Hult, J.A.H. (1966). *Creep in Engineering Structures*. Blaisdell Publ. Co., Waltham, Mass.

Kachanov, L.M. (1986). *Introduction to Continuum Damage Mechanics*, Martinus Nijhoff Publishers.

Klein, J.A.H. and Jessberger, H.L. (1979). "Creep stress analysis of frozen soils under multiaxial states of stress," *Eng Geology*, Vol 13, 353-365.

Klein, J (1979). "The application of finite elements to creep problems in ground freezing," *Proc. 3rd Int. Conf. on Num. Methods in Geomechanics*, Vol 1, 493-502.

Ladanyi, B. (1972). "An engineering theory of creep of frozen soils," *Canadian Geotech. J.*, Vol 9, No 1, 63-68.

Lemaitre, J. and Chaboche, J.L. (1974). "A nonlinear model of creep-fatigue damage accumulation and interaction," *Proc. IUTAM Symp. Mech. of Viscoelasticity Media and Bodies*, Gothenbourg, Springer-Verlag.

Odqvist, F.K.G. (1966). *Mathematical Theory of Creep and Creep Rupture*, Clarendon Press, Oxford, England.

Puswewala, U.G.A. and Rajapakse, R.K.N.D. (1990). "Numerical modelling of structure-frozen soil/ice interaction," *J. of Cold Regions Eng.*, Vol 4, No 3, 133-151.

Sayles, F.H. (1988). "State of the art: Mechanical properties of frozen soil," *Proc. 5th Int. Symp. on Ground Freezing*. University of Nottingham, England, 143-165.

Selvadurai, A.P.S. and Au, M.C. (1991). "Damage and visco-plasticity effects in the indentation of a polycrystalline solid, " *Proc. of PLASTICITY 91: 3rd Int. Symp. on Plasticity and Its Current Applications*, (J.P. Boehler and A.S. Khan Eds.), Elsevier Applied Science, London, 405-408.

Selvadurai, A.P.S. and Shinde, S.B. (1993). "Frost heave induced mechanics of buried pipelines," *J Geotech. Eng. Proc. ASCE*, Vol 119, 1929-1951.

Ting, J.M. (1983). "Tertiary creep model for frozen sand," *J. Geotech. Eng.* , ASCE, Vol 109, No 7, 932-945.

Ting, J.M. and Martin, R.T. (1979). "Application of the Andrade equation to creep data for ice and frozen soil." *Cold Regions Science and Technology*, 1. 29-36.

Vyalov, S.S. (1963). " Rheology of frozen soils," *Proc. 1st Int. Conf. on Permafrost*, 332-339.

Zienkiewicz, O.C. (1977). *The Finite Element Method*. 3rd Ed. McGraw-Hill Co., London, United Kingdom.

DUCTILE IRON PIPE IN EARTHQUAKE/SEISMIC ACTIVITY

Michael S. Tucker[1], P.E., Member ASCE

Abstract

Earthquakes have shown the vulnerability of pipelines at locations of ground rupture due to soil liquefaction, landslides or fault slippage. Pipelines constructed of brittle materials and/or rigid joints have experienced major damage during earthquakes.

Technological advances in the iron pipe industry introduced ductile iron pipe into the marketplace in 1956. Since then, there have been significant improvements in joint designs that are sufficiently flexible to prevent damaging stress build up resulting from seismic excitation and movement. Good jointing techniques and practices allow pipelines to increase or shorten in length and to rotate or bend without leakage or failure.

This paper reports on recent earthquakes and describes the technological advances in ductile iron pipe and jointing systems, as well as special design considerations for pipeline installations in the vicinity of active earthquake faults.

Introduction

Worldwide, of the 50,000 earthquakes of sufficient magnitude to be felt or noticed without aid of instruments, about 100 are large enough to cause substantial damage if their epicenters are near populated areas. Intense earthquakes (generally 6.5 to 7.5 on the Richter scale) occur on the average of one per year. This paper is addressed at these quakes and their effect on underground water and sewer pipelines. See Appendix A for a listing of the most intense earthquakes (greater than 8.0 on the Richter scale) that occurred during the 1900's.

Recent earthquakes in North America have shown the vulnerability of pipelines at locations of ground rupture due to soil liquefaction, landslides or fault slippage. Although earthquake-induced ground shaking is known to cause major damage to

---

[1]Michael S. Tucker, P.E., Senior Regional Engineer, Ductile Iron Pipe Research Association, 1016 E. Rosewood Ave., Orange, California 92666.

pipelines constructed of brittle materials and/or rigid joints, most present day pipeline construction materials employ rubber gasketed joints that are sufficiently flexible to prevent damaging stress build up resulting from seismic excitation and movement. For distribution and transmission piping, through 64-inch diameter, ductile iron pipe has become the leader in water and sewer piping by providing a strong, resilient pipe with flexible joints, as well as recent technological advances in restrained, flexible expansion/contraction couplings for seismic areas. It has become evident that pipelines subject to seismic forces must have high strength and flexibility and that good jointing techniques must be practiced, thereby enabling the pipeline to increase or shorten in length and to rotate or bend without leakage, failure or interruption to service.

But first, let's take a look at the earthquake phenomenon itself and how it affects underground pipelines due to shaking, fault displacement and ground failure. To address this, the paper is divided into the following sections:

1. The Earthquake Phenomenon
2. Earthquake Measurement
3. History of Recent Earthquakes
4. Designing Pipelines for Seismic Resistance
5. Pipe Materials and Jointing Systems
6. Conclusions

The Earthquake Phenomenon

Of all of nature's phenomena, the earthquake is one of the most awesome and devastating. The Western/Pacific Ocean edge of the United States is located on what is known as the "Ring of Fire," the volcanic belt that roughly coincides with the circum-Pacific seismic belt (see Figure 1). Eighty percent of the world's earthquakes occur along this belt.

Figure 1

Damage from earthquakes may result from three separate affects, listed here and described below:

1. Shaking (strong ground motion)
2. Fault Displacement (surface rupture)
3. Ground Failure (landslides, settlement, liquefaction)

1. Shaking - The extent of damage is directly related to the maximum horizontal and vertical ground accelerations and the duration of the seismic tremors. During an earthquake, a structure may be bent or distorted. If its maximum flexure is exceeded, the structure will fail. Above ground structures are generally more affected by shaking than underground structures.

2. Fault Displacement - Surface ruptures occur in three primary forms: a) normal faults (downward dip-slips) occur when overlying fault blocks slip vertically downward in relation to underlying blocks; b) reverse faults (upward dip-slip) are the same as normal faults but the direction of movement is upward; c) lateral faults (strike-slips) occur when fault blocks slip in the horizontal plane.

   Slips can vary from a few inches to nearly 50 feet. The largest known slip occurred in Yakutat, Alaska in 1899. This fault had a vertical slip of 47 feet. When a lateral fault occurs in combination with either a normal or reverse fault, it is called an oblique normal or oblique reverse fault. An example is the San Fernando earthquake of 1971. There, fault movement was in a left reverse oblique direction with major ground shaking lasting about seven seconds.

   When displacement occurs over extended periods, it is known as fault creep or tectonic creep. A well known example is the San Andreas fault in Southern California, which moves an average of 0.8-inches annually.

   In addition to normal, reverse and lateral fault displacements, surface faulting is often accompanied by extension or compression of the earth, occurring approximately perpendicular to the fault. The width of this zone may vary from only a few feet to several hundred feet.

3. Ground Failure - Landslides and subsidence may result from causes other than earthquakes, but their effect upon structures above and below ground are the same. However, in earthquakes, ground failure occurs instantaneously and without warning. Ground failure that results from earthquakes may vary depending upon the type of soil involved. Silts, sands and gravels are usually susceptible to rapid settlement and liquefaction under shaking motions, while rigid clay soils may undergo ground cracking. Damaging effects occur least in bedrock and most in fine-grained sediments. Group ruptures can occur from lateral spreading due to landslides or soil liquefaction.

Earthquake Measurement

The first scale to reflect earthquake intensities was developed by deRossi, of Italy, and Forel, of Switzerland, in the 1880's. The need for a more refined scale increased with the advancement of the science of seismology. In 1902, the Italian seismologist, Mercalli, devised a new scale that was then modified in 1931 by American seismologists, Wood and Neumann, to take into account modern structural features.

Currently, earthquakes are most commonly measured using a seismograph and the Richter magnitude scale. This scale gives an indication of the energy at the source (epicenter) of the quake. The Richter magnitudes can be confusing and misleading unless the mathematical basis for the scale is understood. The magnitude varies logarithmically with the wave amplitude of the quake recorded by the seismograph. Each whole number step of magnitude on the scale represents an increase of ten times in the measured wave amplitude. This corresponds to an increase of thirty-one times in the amount of energy released by the quake.

Strong motion instruments, called accelerographs, record only when they are triggered by strong earthquake motions; in contrast, a seismograph records continuously. Earthquake motions that trigger an accelerograph will generally drive a seismograph off the scale.

History of Recent Earthquakes

The most violent earthquake to occur in North America this century took place in Alaska in 1964. The magnitude of this quake measured approximately 8.5 on the Richter scale. The unusually long duration of strong ground motion (3 to 4 minutes by eye witness reports) intensified the effects of the shock. Spectacular damage was caused by landslides, both subsurface and submarine.

Another recent earthquake occurred in San Fernando, California in 1971. The magnitude of this quake measured 6.6 on the Richter scale. Several surface ruptures occurred with slip displacements of 6-feet reverse and 6-feet left lateral. Separated joints were very evident in areas experiencing extension and elongation. Most failures were at pipeline joints as a result of horizontal ground motion that pulled welded and caulked joints apart. Some failures resulted from compressive action of the spigot within the bell. Over the total area, there were 829 main breaks, 19 shattered mains and 647 service leaks. The greatest number of failures were in cement caulked joints and the least were in flexible joints. Also, there were many failures in nearby steel pipelines connected to, or parallel to, an existing 20-inch ductile iron pipeline (4,250 feet) on Bradley Avenue. There were no leaks or breaks in the ductile iron pipe.

The Managua, Nicaragua earthquake of December 23, 1972 had a devastating effect on the capital city and its inhabitants. This earthquake measured 6.2 on the Richter scale and caused extensive loss of life. This earthquake erupted from five faults. Of these, the left lateral fault displacement was about 4 times greater than the downward displacement that occurred along the four major planes that moved during this event. Asbestos cement (AC) pipe was most affected with 391 breaks, while gray cast iron pipelines suffered 90 breaks. Ductile iron pipe and older cast iron pipe in greater than or equal to 14-inch diameter sizes experienced only eleven joint separations/gasket displacements (fishmouthing - see Figure 2) with no structural failures.

The Miyagi-Ken-Oki earthquake (Japan, 1978) was unique in that it provided the most significant data to date on the performance of ductile iron pipe in an earthquake. The damage reports to water supply pipes in the 29 utilities in the Miyagi area compared ductile iron pipe, cast iron pipe, polyvinyl chloride (pvc) pipe, AC pipe and steel pipe. Of the 215 breaks, steel pipe with turnbuckle joints had the highest rate, AC pipe "which is considered to be the most vulnerable against earthquake ground motions"[2] was next. Ductile iron pipe showed the best performance in this 7.4 magnitude quake.

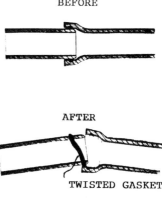

Figure 2

In 1980, three 6.0 magnitude earthquakes occurred in Mammoth Lakes, California. The predominant water pipes in use were steel pipe with threaded couplings and cast iron pipe. Steel pipe experienced the most failures. There were no failures reported on any of the 25,000 feet of ductile iron pipe in the system.

In Hawaii in 1983 a tsunami (tidal wave) generated as the result of an earthquake caused one death and injury to 28. There were no reported breaks or failures of ductile iron pipe as a result of this earthquake.

In 1991 an earthquake occurred near a treatment plant in Limon, Costa Rica. Two pipelines connect the treatment plant to Limon; an 18 km (59,000 feet) long, 500 mm (20-inch) concrete cylinder pipeline installed in 1930 (with a variety of joints including mechanical and leaded) and an 18 km (59,000 feet) long, 300 mm (12-inch) cast iron pipeline installed in 1990 (with rubber-gasketed bell-and-spigot joints). Additionally, there was a secondary supply line consisting of 12 km (39,300 feet) of 300 mm (12-inch) ductile iron pipe. There were 36 failures in the concrete pipe (30 joint separations), 20 to 30 failures in the cast iron pipe (mostly joint separations) and only 4 joint separations with no structural failures in the ductile iron pipe.

California has become known as the earthquake state. With its most recent earthquakes (Coalingua 1983, Whittier 1987, Loma Prieta 1989, Landers 1992 and

---

[2]"Effect of the Miyagi-Oki, Japan Earthquake on June 12, 1978 on the Lifeline Systems", K. Kubo, Professor, University of Tokyo, Tokyo, Japan.

Northridge 1994), the need for a high strength, flexible pipe with flexible joints has been demonstrated over and over again. Ductile iron pipe has performed best with only minimal structural damage (mostly joint separation), while AC pipe exhibited the highest number of failures, both in shear and in compression (spigot end driven into bell) modes. Plastic pipe has performed adequately with its flexible joints but is limited in strength, resulting in some failures. Steel pipe, especially with welded joints, suffers from crimping or buckling in the steel cylinder because of its lower rigidity and section modulus due to its thin walls.

Designing Pipelines for Seismic Resistance

Pipelines used to transport water and sewer may be considered "lifelines" because they transport essential services. Avoiding damage to theses underground "lifelines" during an earthquake is especially important.

Many important parameters control the soil-pipe interaction during an earthquake, including the nature of the soil, the pipeline material and the pipeline jointing system. The nature of the soil affects the load transfer from the soil to the pipe. Additionally, the deformation and stress in the pipeline are significantly influenced by the lateral support provided by the fill material. The pipeline material, whether it is flexible or rigid, as well as the jointing systems determine the ability of the pipeline to resist the motion and energy associated with earthquakes.

One of the first considerations in properly designing a water or sewer line to cross a fault zone is to determine the shortest possible link. When crossing active faults, special couplings with enough flexibility to withstand displacements and to facilitate prompt repair should be used. Additionally, restrained connections that allow flexibility and expansion/contraction should be employed. If rigid body valves or fittings are used, they should be designed to withstand expected high surge forces. Pipe materials must be of sufficient strength and durability to resist earthquake tremors without rupture or cracking. A standard practice is to increase the pipe wall thickness in the fault zone area. In most cases, for water lines, isolation valves should be installed on each side of the crossing.

Most water and sewer systems employ segmented (jointed) piping systems. Segmented pipelines are considered less sensitive to the affects of seismic shaking than continuous pipelines, because they tend to allow some stress relief at each joint. Also, stress amplifications at fault crossings that usually result from confining soil friction around the pipe are not as likely on discontinuous pipelines, because joint separations tend to relieve these stresses. Fault displacements will generally cause pipe to shorten or lengthen at pipe/fault intersections. Pipes can then fail by compressional shear or buckling, or by joint pull-out or separation.

Pipe Materials and Jointing Systems

"Man will continue to live and work in areas of seismic hazards and that first consideration, therefore, should be given to safety factors in structural design and in patterns of use and occupance."[3] In general, the more flexible the pipeline is, the better it can handle the large ground strains associated with earthquakes. Ductile iron pipe has considerable advantage in ground movement situations because of its well known ability to bend without fracturing. Typical is the recommendation by Richard Hazen in his report on the Managua earthquake given at the AWWA Conference in 1973, stating "Non-brittle pipe should be used for water mains and services, with special attention to flexible connections. We should like to see all future mains of ductile iron, even down to 4-inch diameter."[4]

From a review of the limited statistics regarding pipe failure records from the modern-day earthquakes, it is apparent that ductile iron pipe has fared the best. The greatest vulnerability to ductile iron systems is pipe joint separation. However, present day rubber gasketed joints, which have been in use since 1955 in both cast and ductile iron pipe systems, show adequate engagement to resist normal ground shaking displacement. Additionally, in areas where severe displacements are expected, restrained push-on and restrained mechanical joints are available. Asbestos cement pipe has had the worst record of failures with gray cast iron pipe and steel pipe with welded or coupled (threaded) joints next.

Recent developments in jointing and expansion hardware offer additional promise of economic solutions to seismic design problems. The use of restrained expansion couplings in lifelines should solve most problems associated with extension and compression movements along faults. However, shear forces can develop on underground piping at the fault surface (see Figure 3). By properly locating the pipeline at an oblique angle across the fault, the shearing action on the pipeline can be reduced and the pipe joint will tend to elongate or compress. If a right angle crossing is necessary, or if added flexibility is deemed important, standard ductile iron ball joint pipe connections can be connected to the restrained expansion couplings.

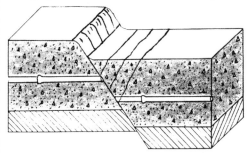

Figure 3

Further considerations in design may include encapsulation or backpacking to protect the pipe. Backpacking material has the capability to crush and deform in the event of

---

[3]National Academy of Sciences, The Committee of the Alaska Earthquake, "Toward Reduction of Losses from Earthquakes", Washington, D.C., 1969.

[4]"Managua Earthquake - Some Lessons in Design and Management", by Richard Hazen, New York, presented at AWWA Conference, Las Vegas, NV, May 14, 1973.

earth movements, yet has sufficient strength to support normal loads. The main purpose of softening the trench is to increase the effective width of the fault zone and to reduce or eliminate soil friction. Multiple layers of loose polyethylene encasement (AWWA C105) may be applied to reduce pipe to soil friction and maintain clean connections. Rigid foam-fiberglass coverings are also available to encapsulate the pipe from the soil.

Rigid connections aboveground to water storage tanks present a highly vulnerable design based on seismic activity. An economic and effective solution can be achieved using a flexible expansion/contraction coupling. This type of device provides strength and flexibility at the joints, as well as the ability to lengthen or shorten. Similar devices are available from ductile iron pipe manufacturers.

Within the past few years, significant progress and technological improvements have been made in the ductile iron industry with regard to jointing systems that withstand seismic activity; telescoping sleeves and flexible couplings are used, as are standard restrained joints and high deflection ball and socket joint pipe. These unique couplings and joints are specially designed for use in fault zone areas. Prudent design using ductile iron pipe along with these specific jointing/coupling systems will significantly promote continuity and performance during seismic activity.

Conclusions

Up to now, geologic hazards have largely been ignored or left to the probabilities of nature. However, a new pipeline usually represents considerable replacement costs at a future time. The time to make the most economical impact on the system reliability is during the initial planning and design phase through application of proper lifeline earthquake engineering principles.

Utilities faced with earthquake hazards should take a long, hard look at the lifelines that are critical to system operations and their customers. These lifelines should receive top priority for retrofitting or replacing to a greater seismic resistant capability. Redundancy, or the availability of parallel systems, will increase chances for survival or allow repairs to continue on damaged sections while others operate.

Ductile iron pipe has thus far achieved an enviable record of service in areas that have experienced moderate and large earthquakes. The 1978 Miyagi-Ken-Oki, Japan earthquake provided an excellent comparison of the earthquake-resistance of ductile iron with other pipe materials. The strength of ductile iron pipe and the flexibility of its joints with various capabilities to rotate, extend, contract or deform without fracture are essential in fault movement resistant pipe design.

The characteristics of pipeline strength, joint flexibility, joint restraint and expansion/contraction capability are top priority. Modern-day ductile iron pipe and connections meet these criteria.

APPENDIX A

Most Intense Earthquakes (>8.0 Richter): 1900 to Present

Date	Location	Richter
1906	San Francisco, California	8.2
1922	Central Chile	8.5
1923	Russia	8.5
1932	Mexico	8.1
1933	Pacific Ocean floor, near Japan	8.4
1938	Banda Sea floor, near Indonesia	8.5
1946	Pacific Ocean floor, near Japan	8.1
1949	British Columbia	8.1
1950	India	8.6
1952	Japan	8.1
1952	Russia	9.0
1957	Aleutian Islands	9.1
1957	Southwestern Mongolia	8.1
1958	Kuril Islands	8.3
1960	Southern Chile	9.5
1963	Kuril Islands	8.5
1964	Southern Alaska	8.6
1965	Aleutian Islands	8.7
1968	Pacific Ocean floor, near Japan	8.2
1977	Indonesia	8.3
1985	Mexico City	8.1
1989	Southern Pacific Ocean floor, near Australia	8.2

APPENDIX B

References

American Society of Civil Engineers, "Lifeline Reconnaissance Report on Costa Rica Earthquake", ASCE News, July 1991.

Blair, William, "Earthquake Engineering for Water Utilities", CA-NV Section AWWA, Reno, Nevada, October 1984.

California Division of Mines and Geology, "Earthquakes - Be Prepared!", Special Publication 39, California Geology, November 1971.

California Division of Mines and Geology, "How Earthquakes are Measured".

Earthquake Engineering Research Institute, "Managua, Nicaragua Earthquake of December 23, 1972", Conference Proceedings, Volume 1, San Francisco, California, November 29-30, 1973.

Earthquake Engineering Research Institute, The EERI Reconnaissance Team, "Engineering Aspects of the Lima, Peru Earthquake on October 3, 1974", Oakland, California, January 1975.

Engineering News Record, "Managua Dozes Wreckage, Sorts Out What's Salvageable", January 11, 1973.

Ford, Duane B., Effects of Ground Failure on Ductile Iron Pipe, American Society of Civil Engineers (undated).

Ford, Duane B., "Seismic Resistant Connections for Water and Sewer Lifelines", Cast Iron Pipe Research Association, March 1974.

Ford, Duane B., "Design Considerations for Underground Pipelines in Geologically Hazardous Areas", Cast Iron Pipe Research Association, Cast Iron Pipe News, Spring/Summer 1975.

Ford, Duane B., "Joint Design for Pipelines Subject to Large Ground Deformations", American Society of Mechanical Engineers, Earthquake Behavior and Safety of Oil and Gas Storage Facilities, Buried Pipelines and Equipment, 1977.

Ford, Duane B., "Earthquake Experiences with Ductile Iron Pipe", ASCE Convention, St. Louis, Missouri, 1981.

Hazen, Richard, "Managua Earthquake - Some Lessons in Design and Management", AWWA Conference, Las Vegas, Nevada, May 1973.

Kachadoorian, Reuben, "Earthquake: Correlation Between Pipeline Damage and Geologic Environment", American Water Works Association Journal, March 1976.

Kubo, K., "Effect of the Miyagi-Oki, Japan Earthquake on June 12, 1978 on Lifeline Systems", University of Tokyo, Tokyo, Japan.

National Academy of Sciences, The Committee on the Alaska Earthquake, "Toward Reduction of Losses from Earthquakes", Washington, D.C., 1969.

Singhal, Dr. Avi, "How to Design Pipelines for Earthquake Resistance", Pipeline and Gas Journal, July 1983.

Wang, Leon R. and Holly A. Cornell, "Evaluating the Effects of Earthquakes on Buried Pipelines", American Water Works Association Journal, April 1980.

Wang, Leon R. L. and Robert V. Whitman, "Seismic Evaluation of Lifeline Systems - Case Studies", American Society of Civil Engineers, October 1986.

World Book Encyclopedia, "Earthquakes", Volume 6, pages 33-39, 1993.

An Economic Approach to Sewer Pipe Selection

by E.C. LAMB[1], M. ASCE

When calling for bids for a new gravity flow sanitary sewer, agencies often specify alternate pipe materials in order to obtain the lowest competitive bid price. However, the lowest bid is not always the best bid or in the best interests of the owner. The difference in durability of materials must be weighed against differences in initial bid price. The additional cost of replacing limited-life sewer pipe must be taken into account at the time of the initial investment.

Shouldn't your sanitary sewer system be expected to last at least 100 years? When one considers that sanitary sewers are usually the most expensive utilities to replace because they are laid beneath water and gas lines, telephone cables, electric services, sidewalks, curbs and gutters and paved roads, selection of pipe material must be based upon a carefully calculated least-cost analysis and not solely on the initial bid price.

"Revenue Shortfall   The Public Works Challenge of The 1980s", a report of a task force (of The American Public Works Association) on revenue shortfall indicated the problems that existed in the 1980s in obtaining funds for infrastructure. Funding infrastructure has not become easier in the interim. The desire to balance the federal budget; the vast sums necessary to pay the interest on the current debt; the cost of entitlement programs; and health and welfare programs take so much of the public money, there is even less money available today for infrastructure construction and/or rehabilitation. Planning, engineering and material selection must be based upon long-term factors and the selection choice must be cost-effective. The goal is to select the pipe material which will provide the greatest value when compared in accordance with accepted principles of engineering economics. Least-cost analysis has been used by Engineers for many years to determine present worth, or Total

----
[1]District Manager-Technical Services, National Clay Pipe Institute

Effective Cost in current dollars.

Often it is suggested that an owner can invest the difference in bid price, between a 100-year pipe and a limited-life pipe, at some assumed or "guessed-at" interest rate. It is anticipated that the funds, at interest, will be adequate to cover the replacement costs, including inflation, when replacement is required. It should be noted that most agencies don't have a vehicle for "investing" the difference. There is also a difficulty in predicting the future rates of inflation and interest over an extended period. There is, however, a relationship between the inflation rate and the interest rate which, due to market factors, is reasonably fixed over long periods of time. That is to say, as interest rates rise or fall, inflation rates also rise or fall, such that the differential over an extended period of time remains somewhat constant. This relationship was researched and reported in Taking the Guesswork Out of Least-Cost Analysis by W.O. Kerr Ph.D. and B.A. Ryan, Arthur Young & Company and updated by W.O.Kerr for The National Clay Pipe Institute in April 1993.

Although the inflation rate may be similar for different levels of government or the private sector, the interest rate can vary according to the borrower's source of funds. The differential between interest and inflation rates for Federal projects is derived from long-term U.S. Treasury instruments. The differential for state and local projects is generally related to municipal bond rates, and the private sector differential is based upon the prime rate. Historical interest/inflation differentials are shown in Table 1 for each of the borrowing sectors.

TABLE 1 - HISTORICAL INTEREST-INFLATION DIFFERENTIALS

Time Period	Sector		
	Federal (Treasury Bonds)	State/Local (Municipal Bonds)	Private (Prime Rate)
1954-1963	2.74	2.08	3.29
1964-1973	1.69	0.81	2.48
1974-1983	0.55	-1.32	2.81
1984-1992	6.31	4.90	6.40
1954-1992	2.73	1.53	3.68

Note: Differentials represent average differences between stated interest and the Producer Price Index for the indicated period.

The Total Effective Cost (Current Dollars) of the limited life pipe is determined from Equation 1.

Equation 1

$$EC = P\{1+[(1+I)/(1+i)]^n +[(1+I)/(1+i)]^{2n} .....[(1+I)/(1+i)]^{mn}\}$$

Where:
 EC = Total Effective Cost (Current dollars)
 P = Bid Price (Current dollars)
 I = Inflation Rate over the period of the Project Life(percent)
 i = Interest Rate over the period of the Project life (percent)
 n = Service Life of the material (years)
 m = Number of times the material with the limited life must be replaced to equal the longer service material

The Average Inflation-Interest Rate Factors shown in Table 2 are based on the term $(1+I)/(1+i)$ in Equation 1 for the stated differentials from Table 1.

TABLE 2 - AVERAGE INFLATION-INTEREST RATE FACTORS

Sector	Factor (F)
Federal Projects	0.9760
State/Local Projects	0.9864
Private Projects	0.9679

Note: The Average Inflation-Interest Factor is computed for the differentials shown in Table 1 for the period 1954-1992.

Using these historic average inflation-interest rate factors in Equation 1 provides a very practical basis for least cost comparisons and greatly simplifies the analysis. The material with the lowest Total Effective Cost is the most economical.

Example 1. A project specified 7,963 feet (2,429 meters) of 15 inch (375 millimeter) sanitary sewer. Project service life is 100 years. Depth of cover was approximately 18 feet (5.5 meters). Both VCP and a limited life pipe were bid. The Bid Price plus engineering and administration totaled $704,661 for VCP. The total for limited life pipe was $634,587.

SEWER PIPE SELECTION

Determine the most economical bid using Least Cost Analysis. The service life of VCP is 100 years based upon demonstrated service. The service life of the limited life pipe has been extrapolated to 50 years. Assuming that the project was funded by state/local government, an average inflation/interest rate factor of 0.9864 is chosen from Table 2.

Since VCP will not require replacement for 100 years, the Effective Cost in today's dollars is $704,661. Using Equation 1, the Effective Cost of limited life pipe is $954,582.

Solution for Example 1

$$EC = P\{1+[(1+I)/(1+i)]^n\}$$
$$= \$634,587\{1+(0.9864)^{50}\}$$
$$= \$954,582$$

VCP has a cost advantage of $249,921. Stated differently; in addition to the $634,587 required to construct the project with the limited life pipe, an additional $319,905 must be set aside today to rebuild the sewer in 50 years.

Example 2.

The Project specified the pipe sizes and quantities shown below. While the project was local with funding by municipal bonds, calculations were also made based on private and federal funds to illustrate how project costs vary with different funding sources.

Pipe Size	Quantity
42" (1050 mm)	4,951' (1,510 M)
36" (900 mm)	1,613' (492 M)
33" (825 mm)	7,411' (2,260 M)
30" (750 mm)	1,039' (317 M)
27" (675 mm)	11,865' (3,619 M)

	State/Local	Private	Federal
Project Design Life, years	100	100	100
Inflation/Interest Factor	0.9864	0.9679	0.9760
VCP Pipe			
Bid Price (Incl Eng/Adm)	$4,078,511	$4,078,511	$4,078,511
Service Life, years	100	100	100
Limited Life Pipe			
Bid Price (Incl Eng/Adm)	$3,692,511	$3,692,511	$3,692,511
Service Life, years	50	50	50
Effective Cost: VCP	$4,078,511	$4,078,511	$4,078,511
Effective Cost: Limited Life Pipe	$5,554,644	$4,415,135	$4,788,448
Cost Advantage			
VCP Over Limited Life Pipe	$1,476,133	$ 336,624	$ 709,937

Sewers are expensive to build, <u>but they are even more expensive to replace!</u> In Example 1, if limited life pipe is selected; and if funds are not set aside to rebuild the project after 50 years; and if the inflation rate over the period is 6% (approximate average inflation rate 1952-1992) the cost of replacing the sewer will be $11,689,190.

It makes good economic sense to use pipe materials which are the most cost-effective. This method of making a least-cost analysis facilitates the pipe selection because the use of historic data simplifies the analysis, and avoids the pitfalls that will likely result from trying to predict future, long-term inflation and interest rates.

PROTECTIVE COATINGS - - CURRENT PRACTICES

WILLIAM C. ROBINSON, MEMBER[1]

ABSTRACT

This paper is presented as an overview of current protective coating practices for buried pipelines.

Protective coating and lining technologies improve as materials and methods are improved or changed to satisfy economic and societal needs. Industry's most common pipeline coatings: the dielectrics (tape, enamel, paint), cement mortar, and metallic are discussed.

Cement mortar coatings and linings are most common to the water and waste industries. Zinc, as a galvanizing material, is the most widely used metallic coating. More recently, aluminum alloy coatings have replaced zinc for some applications.

INTRODUCTION

One of our least expensive construction materials, steel, is the one most susceptible to degradation from the natural environment. Much effort goes into isolating the steel from the environment with protective coatings of various types to prevent corrosion.

Environmental regulations issued in the past few years have had an impact on the use and application techniques of some common protective coatings. This paper will address current practices and materials used in the protective coating industry today. The coatings and linings addressed are common to the pipeline industry and are selected for use in various environments. Coatings are selected by owners or designers based on cost, durability, and environment. Due to the decreasing demand for field applied coatings, plant applications are increasing and coatings are being

---

[1] President, Intermountain Corrosion Service, Inc., 4624 16th St. East, #A6, Tacoma, WA 98424-2664.

developed to answer the demand for plant application. Table I compares some of the important features of coatings addressed in this paper. The data contained in the table is based on coating manufacturer and coating applicator data.

DIELECTRIC

New coatings and composite coating combinations, as well as application methods, are continuing to be developed, while trusted coatings are being improved. Some popular dielectric coatings in use today are:

Epoxies	Urethanes
Coal Tar Enamel	Tape
Coal Tar Epoxy	Extruded Polyolefins
Coal Tar Urethanes	

Dielectric coatings in the form of mineral filled asphalt were developed to protect steel pipe. Originally asphalt was applied by dipping, brushing or mopping. Asphalt coatings, however, are losing favor with the pipeline industry due to their porosity and low adhesion qualities, which result in low resistance to soil stress and are not discussed.

Epoxy

The most popular epoxy for underground steel pipelines, both interior and exterior is fusion bonded epoxy (FBE). A fusion bonded epoxy coating is a one part, heat curable, thermosetting powdered resin designed as a protective coating.

After the pipe is blast cleaned to the specified finish, it is then uniformly heated to approximately $+232°C$ ($+450°F$). At this pipe temperature, the powdered epoxy resin is electrostatically sprayed on the surface of the pipe. When the resin comes in contact with the heated steel surface, it melts, flows out to the required thickness, gels and then cures by means of the residual heat in the pipe. The normal average thickness of the cured coating is 12 to 14 mils. Application thickness up to 30 mils are used for special conditions, such as directional boring.

The FBE coatings can withstand large temperature variations. Due to the hardness of the cured material, field bending tends to overstress the coating and results in cracking and disbonding. The FBE coatings have a high resistance to soil stress and disbondment. Soil stresses are greatest in the clay soils where volume is affected by moisture content, and least in the gravel soils.

The two component liquid epoxies are quite common, but tend to be less durable than FBE. Liquid epoxies are used where epoxy is the desired coating, but the structures are too large, or awkward, for the FBE process. Pipelines that are pre-lined with cement mortar or other materials that can not withstand the heat required for FBE are commonly coated with liquid epoxy. Surface preparation for FBE and liquid epoxy are similar. Like materials can be used for field repairs for the FBE and sprayed coatings.

Coal Tar Enamel

Coal tar enamel has been in use as a pipeline coating for over 60 years and has one of the longest performance records of any pipeline coating. Coal tar has been distilled from coal for nearly 100 years. During distillation of coal tar, as many as 190 chemicals have been identified, some of which are: benzene, toluene, xylene, phenol, naphthalene and creosote, which makes it's use difficult in some of today's industries. The extensive use of coal tar in the past can be attributed to it's good corrosion protection properties, availability and cost.

Early application of coal tar for pipelines was by hot mopping or dipping. Over time, improved techniques led to application by spray or drip flooding and reinforcement was added to exterior coating. The reinforcement allows application of more layers of enamel resulting in a thicker, more durable coating. Coal tar is also popular for over-the-ditch coating operations when used with an outerwrap for mechanical strength.

The use of coal tar has declined in the past few years due to the development of new products and the decrease of over-the-ditch coating application. Environmental regulations have also affected the use and handling of coal tar. At the time of this writing, coal tar has not received approval of the National Sanitation Foundation for contact with potable water. Coal tar is not used for potable water service in most of Europe because of leaching of polyaromatic hydrocarbons, some of which are known to be carcinogenic. Some owners have discontinued the use of coal tar for exterior pipe surfaces due to potential contaminant disposal problems in the future, should the coating be removed from the pipe surface.

Minimum surface preparation for coal tar is a commercial blast (NACE No. 3/SSPC-SP6)[2], which is less restrictive than the thinner dielectric coatings. Over the ditch surface preparation can be accomplished by wire brushing.

---

[2] NACE No. 3/SSPC-6 "Commercial Blast Cleaning"

Coal Tar Epoxy / Coal Tar Urethane

Coal tar epoxy and coal tar urethanes have been used successfully for many pipeline coating projects since their development in the mid 1950's. Coal tar epoxy and coal tar urethanes are two component, chemically cured coal tar coatings. They owe their popularity to their ease of application to irregular surfaces, good corrosion protection and ease of field repairs. They are also known for their versatility due to wide variations in formulations producing different cured properties.

These coatings can be applied in a single coat of 5 to 15 mils dry film thickness. Multiple coats with high build qualities up to 60 mils have been used successfully for buried pipelines.

Both have excellent resistance to temperature variations and petroleum products with the exception of coal tar derivatives, some of which are listed above.

Both coatings are sensitive to surface preparation and require a white metal finish (NACE No. 1/SSPC-SP5)[3] for good performance.

Urethane

Urethane coatings are increasing in popularity as a protective coating. The most common urethane types are solvent release, 100 percent solids, and moisture cured. Gaining popularity and receiving much attention to improvement are the moisture cured urethanes. Moisture cured urethanes require atmospheric moisture to initiate the chemical reaction necessary to cure the applied coating. Spray application methods are used, but care must be taken because moisture cured urethanes have been known to set up in feed lines where condensed moisture from the air mixes with the urethane. Most urethanes have good resistance to soil stress. For buried service, exterior coating thickness must be increased from that required for above ground to resist damage from abrasion. Urethanes can be obtained in different colors, where most buried pipeline coatings are the basic black.

The solvent release and 100 percent solids are two component coatings while the moisture cured are single component. Urethane primers can be applied over commercial blast surfaces. Other types of coatings are commonly used as a primer for the urethane top coats.

Tapes

Pipeline tape has been in wide use for nearly 40 years. Most tapes used today have moderate resistance to soil stress and are fabricated with a self-release adhesive and

---

[3] NACE No. 1/SSPC-SP5 "White Metal Blast Cleaning"

a plastic backing material to provide mechanical strength. An advantage of tapes is that the primers can be supplied with an additive that will resist stress corrosion cracking. Tapes have a high resistance to disbondment and stand up well during field bending, but do not perform well in high temperature service.

One of the features of tape is that it is applied under tension and controlled temperature. When applied under tension, the backing compresses the adhesive, producing a "gasket" effect that increases it's resistance to disbonding and soil stress. Popularity for plant application of tapes is increasing, due to it's improved quality compared to over-the-ditch application.

A disadvantage of tape is the extra effort required to coat irregular surfaces on straight pipe sections. Irregular surfaces, such as weld seams, will require grinding or a special filler tape. Rough surfaces, such as that on ductile cast iron, can be coated with tape; but a thicker, more pliable adhesive is necessary for the inner wrap than would be required for steel pipe. Primers for tape can be applied to a commercial blast surface.

Extruded Plastics

Extruded polyolefins are the common choice for many applications up to 20-inches in diameter. More plants that will handle larger diameters are becoming a demand where an extruded coating is desired. The extruded coatings are typically applied over a primary coating or an adhesive. The prepared pipe is routed through an extrusion die where a high density coating is applied in a uniform thickness, then water quenched to shrink and cure the coating, resulting in a hard moisture resistant seal.

Many users of extruded coatings prefer their use over other coatings because they are good performers in many different conditions. The extruded coatings can withstand extreme temperature variations and can be formulated to resist deterioration by ultra violet light. They have high chemical resistance, high impact resistance, adapt well to field bending, and have a high resistance to disbonding. Some users are experimenting with various under coatings, including epoxies that are top coated with extruded coatings, with great success. However, some multi-layer systems are cost prohibitive when comparing performance characteristics to other less costly coatings. Field joints and repairs are commonly completed with compatible heat shrink sleeves or pipeline tape.

CEMENT MORTAR

Cement mortar has been used as a protective coating and lining for steel pipe since the early 1940's. The corrosion protection quality of cement mortar is due to the passivation of the steel surface. When cement mortar is held in intimate contact with the steel pipe surface, the highly alkaline environment with pH of 12.5 to 13, forms an iron oxide film that inhibits corrosion. Although there are various configurations of pipe manufactured with cement mortar such as mortar coated steel, concrete cylinder pipe, prestressed concrete pipe, and reinforced concrete pipe, the cement mortar is the protective medium. Because the continued contact between the mortar and the steel is crucial, pipe design must limit deflection to prevent excess cracking of the mortar. Depending on soil chemistry, the designer has the option to select Type II (moderate sulfate resisting) or Type V (sulfate resisting) cement where water soluble sulfates could cause rapid deterioration of Type I cement. Field repairs and field joints are commonly repaired by a dry-pack procedure with rich cement grout or pouring liquid grout into a special design form.

METALLIC

Metallic coatings are commonly zinc and aluminum. Other types of metallic coatings are also employed where sour environments (H_2S, CO_2) are encountered. Zinc has been used as a protective coating for more than 100 years in the form of galvanizing. More recently aluminum has been adapted as a galvanizing material. One important matter to keep in mind is that galvanizing material is sacrificial and when its gone, so is the corrosion protection.

The most common application method for application of zinc is by hot dipping, where the steel is prepared in an acid bath and dipped in molten zinc. A less effective method is electrodepositing zinc on the steel. Recently, equipment has been developed that allows zinc and aluminum to be sprayed on large steel structures. This procedure shows great promise since it allows larger and awkward materials of construction to be coated.

Although zinc and aluminum can be used alone for protective coating, the zinc formula for sheet steel contains 15 percent aluminum. The aluminum prevents the zinc/iron alloy from layering on the steel surface that would interfere with bending and shaping of the sheets.

Hot dip zinc contains a small amount of aluminum to provide brightness, but not enough to interfere with the alloy layering necessary to provide abrasion resistance and long service life.

SPECIFICATIONS

Many projects have been successfully completed with specifications written in reference to a standard or guide specification only. This procedure will work many times over, but one failure can, in some cases, be catastrophic. Standards or guide specifications such as those prepared by AWWA, API and APWA, are guides only and should be treated that way. It is not uncommon for the Engineer or owner to reference a guide specification and prepare 10 to 15 sheets of definitive specifications addressing guide specification options and specific site conditions.

AWWA C214[4] contains a minimum of 15 options that must be addressed by the purchaser for a product that exceeds minimum requirements. Although AWWA C214[4] was developed as a guide specification for steel pipe, it can also be used for ductile cast iron.

As a safety measure, the Engineer or owner should request comments from the coating material supplier on the application specifications.

CONCLUSIONS

Protective coating selection has advanced to a highly technical procedure. Pipeline owners and coating manufacturers are constantly driving to develop the "perfect" coating. Many of the older proven coatings are being replaced by newer products designed with new technology and recently developed materials.

Reduction of failures, by isolation of pipelines from the soil, can be achieved through proper coating selection. Specifications can be written to accept alternative coatings with equal performance and equal service life.

---

[4] AWWA C214, "Tape Coating Systems For the Exterior of Steel Water Pipelines"

TABLE 1
PROTECTIVE COATING* COMPARISON TO APPLICATION & USE

	Surface Preparation	Temperature Variations	Soil Stress	Ease of Field Repairs	Bending	Disbonding	Adhesion
Epoxies, (FBE)	4	1	1	5	3	1	5
Epoxies, Liquid	4	2	2	4	3	2	5
Coal Tar Enamel	3	5	3	1	1	4	4
Coal Tar Epoxy	5	3	2	2	2	3	4
Coal Tar Urethanes	5	3	2	2	2	3	4
Urethanes	5	2	2	4	5	1	4
Tape	3	5	4	1	2	4	3
Extruded Plastics	4	2	2	2	2	2	4
Metallic	2	1	1	5	1	1	5
Cement Mortar	1	1	1	2	N/A	5	5

* Plant applied coatings.

WEIGHT RATING

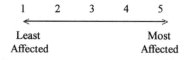

1   2   3   4   5

Least Affected   Most Affected

Specify the Right Steel for your Steel Water Pipe
by George J. Tupac[1]

Abstract

The properties of steel are governed by their chemical composition, by the processes used to transform the base metal into a shape and by their heat treatment. The effects of these on the properties of steel are discussed, such as chemical composition, casting, killed steels, heat treatment, mechanical properties, ductility, toughness and charpy v-notch fracture toughness. Examples of charpy v-notch values for commonly used steels in steel water pipe are given.

Introduction

All steel materials used in the fabrication of pipe, including pressure carrying components and non-pressure carrying attachments such as ring girders, stiffeners rings, thrust rings, lugs, support systems, and other appurtenances are manufactured and tested in strict accordance with appropriate specifications.

The properties of steels are governed by their chemical composition, by the processes used to transform the base metal into the shape, and by their heat treatment. The effects of these parameters on the properties of steels are discussed in the following sections.

---
[1] Engineering Consultant, G.J. Tupac and Assoc., Inc.
904A, 1150 Bower Hill Road
Pittsburgh, PA 15243
Telephone (412) 276-4136

Chemical Composition

Constructional steels are a mixture of iron and carbon with varying amounts of other elements - primarily manganese, phosphorus, sulfur, and silicon. These and other elements are either unavoidably present or intentionally added in various combinations to achieve specific characteristics and properties of the finished steel products. The effects of the commonly used chemical elements on the properties of hot-rolled and heat-treated carbon and alloy steels are presented in Table 1. The effects of carbon, manganese, sulfur, silicon, and aluminum are of primary interest to the present discussion.

Carbon is the principal hardening element in steel where each additional increment increases the hardness and tensile strength of the steel. Carbon has a moderate tendency to segregate and increased amounts of carbon cause a decrease in ductility, toughness, and weldability.

Manganese increases the hardness and strength of steels but to a lesser degree than does carbon. Manganese combines with sulfur to form manganese sulfides thus decreasing the harmful effects of sulfur.

Table 1 -- Effects of Alloying Elements

Carbon (C)
* Principal hardening element in steel.
* Increases strength and hardness.
* Decreases ductility, toughness, and weldability.
* Moderate tendency to segregate.

Manganese (Mn)
* Increases strength.
* Controls harmful effects of sulfur.

Phosphorus (P)
* Increases strength and hardness.
* Decreases ductility and toughness.
* Considered an impurity, but sometimes added for atmospheric corrosion resistance.
* Strong tendency to segregate.

Sulfur (S)
* Considered undesirable except for machinability.
* Decreases ductility, toughness and weldability.
* Adversely affects surface quality.
* Strong tendency to segregate.

Table 1 (contd.)

Silicon (SI)	*Used to deoxidize or "kill" molten steel.
Aluminum (Al)	*Used to deoxidize or "kill" molten steel. *Refines grain size, thus increasing strength and toughness.
Vanadium (V)	*Small additions increase strength.
Columblum (Nb)	*Small additions increase strength.
Nickel (NI)	*Increases strength and toughness.
Chromium (Cr)	*Increases strength. *Increases atmospheric corrosion resistance.
Copper (Cu)	*Primary contributor to atmospheric corrosion resistance.
Nitrogen (N)	*Increases strength and hardness. *Decreases ductility and toughness.
Boron (B)	*Small amounts (0.0005%) increase hardenability in quenched and tempered steels. *Used only in aluminum-killed steels. *Most effective at low carbon levels.

Sulfur is generally considered an undesirable element except where machinability is an important consideration. Sulfur adversely affects surface quality, has a strong tendency to segregate, and decrease ductility, toughness, and weldability.

Silicon and aluminum are the principal deoxidizers used in the manufacture of carbon and alloy steels. Aluminum is also used to control and refine grain size.

Casting

The traditional steelmaking process is the one where molten steel is poured (teemed) into a series of molds to form castings known as ingots. The ingots are removed from the molds, reheated, then rolled into products with square or rectangular cross sections. This hot-rolling operation elongates the ingot and produces semi-finished products known as blooms, slabs, or billets. All ingots exhibit some degree of nonuniformity of chemical composition known as segregation, which is an inherent characteristic of the cooling and solidification of the molten steel in the mold.

The first liquid steel to contact the relatively cold walls and bottom of the mold solidifies very rapidly having the same chemical composition as the liquid steel entering the mold. However, as the rate of solidification decreases away from the mold sides, crystals of relatively pure iron solidify first. Thus, the first crystals to form contain less carbon, manganese, phosphorus, sulfur and other elements than the liquid steel from which they were formed. The remaining liquid is enriched by these elements that are continually being rejected by the advancing crystals. Consequently, the last liquid to solidify, which is located around the axis in the top half of the ingot, contains high levels of the rejected elements and has a lower melting point than the poured liquid steel. This segregation of the chemical elements is frequently expressed as a local departure from the average chemical composition. In general, the content of an element that has a tendency to segregate is greater than average at the center of the top half of an ingot and less than average at the bottom half of an ingot.

Certain elements tend to segregate more readily than others. Sulfur segregates to the greatest extent. The following elements also segregate, but to a lesser degree, and in descending order: phosphorus, carbon, silicon, and manganese. The degree of segregation is influenced by the composition of the liquid steel, the liquid temperature and ingot size. The most severely segregated areas of the ingot are removed by cropping, which is cutting, discarding sufficient material during rolling.

Continuous Castings

The direct casting of steel from the ladle into slabs. This steelmaking development bypasses the operations between molten steel and the semi-finished product which are inherent in making steel products from ingots. In continuous casting, molten steel is poured at a regulated rate into the top of an oscillating water-cooled mold with a cross-sectional size corresponding to the desired slab. As the molten metal begins to solidify along the mold walls, it forms a shell that permits the gradual withdrawal of the strand product from the bottom of the mold into a water-spray chamber where solidification is completed.

The solidified strand is cut to length and then reheated and rolled into finished products as in the conventional ingot process. The smaller size and higher cooling rates for the strand result in less segregation and greater uniformity in composition and properties for steel products made by the continuous casting process than for ingot products.

Killed and Semi-Killed Steels

The primary reaction involved in most steelmaking processes is the combination of carbon and oxygen to form carbon monoxide gas. The solubility of this and other gases dissolved in the steel decreases as the molten metal cools to the solidification temperature range. Thus, excess gases are expelled from the metal and, unless controlled, continue to evolve during solidification. The oxygen available for the reaction can be eliminated and the gaseous evolution inhibited by deoxidizing the molten steel using additions of silicon or aluminum or both. Steels that are strongly deoxidized do not evolve any gases and are called killed steel because they lie quietly in the mold. Increasing amounts of gas evolution results in semi-killed, capped or rimmed steels.

In general, killed steel ingots are less segregated and contain negligible porosity when compared to semi-killed steel ingots. Consequently, killed steel products usually exhibit a higher degree of uniformity in composition and properties than semi-killed steel products.

Heat Treatments For Steels

Steels respond to a variety of heat treatments that can be used to obtain certain desirable characteristics. These heat treatments can be divided into slow cooling treatments and rapid cooling treatments. The slow cooling treatments, such as annealing, normalizing and stress relieving, decrease hardness and promote uniformity of structure. Rapid cooling treatments, such as quenching and tempering, increase strength, hardness and toughness. Heat Treatments of base metal are generally mill options or ASTM requirements.

Annealing -- Annealing consists of heating the steel to a given temperature followed by slow cooling. The temperature, the rate of heating and cooling, and the time the metal is held at temperature depends on the composition, shape and size of the steel product being treated and the desired properties. Usually steels are annealed to remove stresses; to induce softness; to increase ductility and toughness; to produce a given microstructure; to increase uniformity of microstructure; to improve machinability; or to facilitate cold forming.

Normalizing -- Normalizing consists of heating the steel to between $1650°F$ and $1700°F$ ($899 - 927°C$) followed by slow cooling in air. This heat treatment is commonly used to refine the grain size, improve uniformity of microstructure, and improve ductility and fracture toughness.

Stress Relieving -- Stress relieving of carbon steels consists of heating the steel in the range 1000 to 1200 F (538 - 649°C) and holding for a proper time to equalize the temperature throughout the piece followed by slow cooling. The stress relieving temperature for quenched and tempered steels must be maintained below the tempering temperature for the product. Stress relieving is used to relieve internal stresses induced by welding, normalizing, cold working, cutting, quenching and machining. It is not intended to alter the microstructure of the mechanical properties significantly.

Quenching and Tempering -- Quenching and tempering consists of heating and holding the steel at the proper austenitizing temperature (about 1650 F) (899°C) for a significant time to produce a desired change in microstructure, then quenching by immersion in a suitable medium (water for bridge steels). After quenching, the steel is tempered by reheating to an appropriate temperature usually between 800 and 1200°F, (427 - 649°C), holding for a specified time at temperature, then cooling under suitable conditions to obtain the desired properties. Quenching and tempering increases the strength and improves the toughness of the steel.

Controlled Rolling -- Controlled rolling is a thermo-mechanical treatment at the rolling mill that tailors the time-temperature-deformation process by controlling the rolling parameters. The parameters of primary importance are (1) the temperature at start of controlled rolling in the finishing stand, (2) the percentage reduction from start of controlled rolling to the final plate thickness, and (3) the plate finishing temperature.

Hot-rolled plates are deformed as quickly as possible at temperatures above about 1800°F (982°C) to take advantage of the hot workability of the steel at high temperatures. In contrast, controlled rolling incorporates a hold or delay time to allow the partially rolled slab to reach a desired temperature before start of final rolling. Controlled rolling involves deformations at temperatures in the range from 1500 and 1800°F (815 - 982°C). Because rolling deformation at these low temperatures increases the mill loads significantly, controlled rolling is usually restricted to less than two inch thick plates, (50 mm). Controlled rolling increases the strength, refines the grain size, improves the toughness, and may eliminate the need for normalizing.

Controlled Finishing-Temperature Rolling -- Controlled finishing - temperature rolling is a less severe practice than controlled rolling and is aimed primarily at improving notch toughness of plates up to 2 1/2 inch thickness, (63.5 mm). The finishing temperatures in this practice (about 1600°F), (871°C), are higher than required for controlled rolling. However, because heavier plates are involved than in controlled rolling, mill delays are still required to reach the desired finishing temperatures. By controlling the finishing temperature, fine grain size and improved notch toughness can be obtained.

Ductility and Toughness -- Stress - strain curves are explained in the references.

The total percent ELONGATION and the total percent REDUCTION OF AREA at fracture are two measures of ductility that are obtained from the tension test. The percent elongation is calculated from the difference between the initial gage length and the gage length after fracture. Similarly, the percent reduction of area is calculated from the difference between the initial and the final cross-sectional area after fracture.

Ductility is an important material property because it allows the redistribution of high local stresses. Such stresses occur in welded connections and at regions of stress concentration such as holes and changes in geometry.

Toughness -- The ability of a material to absorb energy prior to fracture and is related to the area under the stress-strain curve. The larger the area under the curve the tougher is the material.

Fracture Toughness -- Most constructional steels can fracture either in a ductile or in a brittle manner. The mode of fracture is governed by the temperature at fracture, the rate at which the loads are applied and the magnitude of the constraints that would prevent plastic deformation. The effects of these parameters on the mode of fracture are reflected in the fracture toughness behavior of the material. In general, the fracture toughness increases with increasing temperature, decreasing load rate and decreasing constraint. Furthermore, there is no single unique fracture toughness value for a given steel even at a fixed temperature and loading rate.

Traditionally, the fracture toughness for low and intermediate strength steels has been characterized, primarily by testing Charpy V-notch (CVN) specimens at different temperatures. However, the fracture toughness for materials can be established best by using fracture mechanics test methods. The following presents a few aspects of fracture toughness of steels by using CVN and fracture mechanics test results.

Charpy V-Notch Fracture Toughness - The Charpy V-notch specimen has been the most widely used specimen for characterizing the fracture toughness behavior of steels. These specimens may be tested at different temperatures and the impact fracture toughness at each test temperature may be determined from the energy absorbed during fracture, the percent shear (fibrous) fracture on the fracture surface or the change in the width of the specimen (lateral expansion). At low temperatures, constructional steels exhibit a low value of absorbed energy (about 5 ft.-lb.), (6.77 joules), and zero fibrous fracture and lateral expansion. The values of these fracture toughness parameters increase as the test temperature increases until the specimens exhibit 100 percent fibrous fracture and reach a constant value of absorbed energy and of lateral expansion. This transition from brittle to ductile fracture behavior usually occurs at different temperatures for different steels and even for a given steel composition. Consequently, like other fracture toughness tests, there is no single unique CVN value for a given steel, even at a fixed temperature and loading rate. Therefore, when fracture toughness is an important parameter, the design engineer must establish and specify the necessary level of fracture toughness for the material to be used in the particular structure or in a critical component within the structure.

Materials

The most common grades of steel used are listed in Table 2:

Table 2 Grades of Steel

Specifications for Fabricated Pipe	Minimum Yield Point psi	(Mpa)	Minimum Ultimate Tensile Strength psi	(Mpa)
ASTM[1] A36	36,000	(248.2)	58,000	(399.9)
ASTM A283 GR C	30,000	(206.8)	55,000	(379.2)
GR D	33,000	(227.5)	60,000	(413.7)
ASTM A516 GR 60	32,000	(220.6)	60,000	(413.7)
ASTM A516 GR 70	38,000	(261.9)	70,000	(482.6)
ASTM A570 GR 30	30,000	(206.8)	49,000	(337.8)
ASTM A570 GR 33	33,000	(227.5)	52,000	(358.5)
ASTM A570 GR 36	36,000	(248.2)	53,000	(365.4)
ASTM A570 GR 40	40,000	(275.8)	55,000	(379.2)
ASTM A570 GR 45	45,000	(310.3)	60,000	(413.7)
ASTM A570 GR 50	50,000	(344.7)	65,000	(448.2)
ASTM A572 GR 42	42,000	(289.6)	60,000	(413.7)
ASTM A572 GR 50	50,000	(344.7)	54,000	(448.2)
ASTM A572 GR 60	60,000	(413.7)	75,000	(517.1)
ASTM A907 GR 30	30,000	(206.8)	49,000	(337.8)
ASTM A907 GR 33	33,000	(227.5)	52,000	(358.5)
ASTM A907 GR 36	36,000	(248.2)	53,000	(365.4)
ASTM A907 GR 40	40,000	(275.8)	55,000	(379.2)

Specifications for Manufactured Pipe	Minimum Yield Point psi	(Mpa)	Minimum Ultimate Tensile Strength psi	(Mpa)
ASTM A53, A135, and A139 GR A	30,000	(206.8)	48,000	(330.9)
GR B	35,000	(241.3)	60,000	(413.7)
ASTM A139 GR C	42,000	(289.6)	60,000	(413.7)
GR D	46,000	(317.2)	60,000	(413.7)
GR E	52,000	(358.5)	66,000	(455.1)

[1] American Society of Testing Materials

Impact Tests, Toughness Requirements

Many specifications call for toughness requirements. Typical requirement is 25 ft.-lbs. @ 30°F (33.9 joules @ -11.1°C). Reference No. 3 does not require impact testing for thickness of 5/8 inch (15.8 mm) or less thickness for a yield strength of 55,000 psi (379.2 MPA) or less. Higher strengths, then require impact testing. These requirements assume killed-fine grain material. Special carbon equivalent (CE) criteria applies. The are: for a yield strength stress near 55,000 psi (379.2 MPA), the CE is .45. For a yield strength over 75,000 psi (517.2 MPA), the CE is .53. This may be obtained from the following formula:

$$CE = C + \frac{Mn}{6} + \frac{Cr + Mo + V}{5} + \frac{Ni + Cu}{15}$$

Steels have been performing safely and reliably under normal operating conditions. This observation is supported by years of satisfactory field experience. Present continuous cast steels with improved chemical and mechanical properties over ingot cast further enhance this performance. Recent impact test results from some current projects are as follows:

Project No.1

Requirement   15 ft. lbs. @ 0°F (20.4 joules @ -9.44°C)

Thickness .625 inches (15.8mm) A139 GRC.
Tests in transverse direction

Results - 75 tests:

Outside coil	-	53.2 ft. lbs. @ 0°F (72.1 joules @ -9.44°C)
Center Coil	-	40.8 ft. lbs. @ 0°F (55.3 joules @ -9.44°C)
Inside Coil	-	47.0 ft. lbs. @ 0°F (63.7 joules @ -9.44°C)

Project No. 2

Requirement   25 ft. lbs. @ 30°F (33.9 joules @ 1.11°C)
                      15 ft. lbs. @ 0°F (20.4 joules @ -9.44°C)

Thickness .875 inches (22.3mm) ASTM A516GR70

Results - 10 tests:

42 ft. lbs. @ 30°F (56.9 joules @ -1.11°C)
32 ft. lbs. @ 0°F (43.4 joules @ -9.44°C)

Thickness .8125 inches (20.6 mm) ASTM A GR70

45 ft. lbs. @ 30°F (61 joules @ -1.11°C)
35 ft. lbs. @ 0°F (47.5 joules @ -9.44°C)

Thickness .723 inches (18.4 mm) ASTM A516 GR70

49 ft. lbs. @ 30°F (66.5 joules @ -1.11°C)
31 ft. lbs. @ 0°F (42 joules @ -9.44°C)

Project No. 3

Requirement: 25 ft. lbs. @ 30°F (33.9 joules @ -1.11°C)

Thickness varies from .365 inches (9.28 mm) to .615 inches (15.6 mm) ASTM A139 GR42

Results, 1 test per coil:

.365 inches (9.28 mm), 35 tests, average 75 ft. lbs. @ 30°F (101.7 joules @ -1.11°C)

.427 inches (10.8 mm), 5 tests, average 95 ft. lbs. @ 30°F (128.8 joules @ 1.11°C)

.490 inches (12.5 mm), 6 tests, average 100 ft. lbs. @ 30°F (135.6 joules @ 1.11°C)

.552 inches (14.02 mm) 4 tests, average 94 ft. lbs. @ 30°F (127.5 joules @ -1.11°C)

.615 inches (15.62 mm), 5 tests, average 75 ft. lbs. @ 30°F (101.7 joules @ -1.11°C)

Conclusion

Properties of steel are mainly governed by their chemical composition. Continuous casting, because of the smaller size and higher cooling rates for the strand, results in less segregation and greater uniformity in composition and properties. With both of the above, the ductility and toughness is assured. One only needs to specify the allowable stress, minimum thickness, allowable deflection and toughness requirements because of the many choices of steel available. Then, allow the pipe fabricator to choose the right steel.

References

1. Steel Pipe - "A Guide for Design and Installation, Manual M11", American Water Works Association, Third Edition, 1989.

2. "Buried Steel Penstocks, Steel Plate Engineering Data - Volume 4", American Iron and Steel Institute, 1992.

3. "Steel Penstocks, ASCE Manual and Reports on Engineering Practice No. 79", American Society of Civil Engineers, 1993.

4. Barsom, J.M., "Steel for Welded Water Pipe - Fracture Toughness and Structural Performance", Steel Plate Fabricators Association and American Iron and Steel Institute, 1993.

THE STRAIN EQUATION. --
Incidence on Buried Steel Pipe Design.

R. C. Prevost, M. ASCE .[1]

Abstract.

The strain equation (1) below for buried pipes is based on deformation geometry, to the exclusion of any other consideration; it is therefore valid regardless of the deformations' origin, whether from loads or installation, regardless of the stress/strain status of the pipe material (whether its behaviour is "elastic" or not; whether yield or creep proceeds); regardless of the soil. It has therefore particular importance.

Actual strains are determined by actual deformations, not calculated ones, and the permitted maximum strains by the permitted maximum deformations.

The D_f factor of equation (1) is determined by the shape of the deformed pipe; in elastic pipe and environment, D_f is a simple function of the loadings' parameters.

Application of the strain equation to steel pipe shows that, inevitably, large areas of all steel pipelines are plastically strained to an extent which, normally, only marginally impacts steel's considerable ductility. Therefore, application of the classical elastic stress analysis to steel pipelines is inappropriate. Design methods which deal separately with internal pressure and external load, and with each exclusively, should be used in all circumstances, whether for ordinary pipelines, or crossings, etc.; and "risk" assessments should focus on steel quality (ductility), not on stresses.

Key words: Buried pipe, strain equation; strain; stress; D_f; curvature; steel pipe; steel quality.

The Strain Equation.   Introduction.

The conventional strain equation [2] $\varepsilon = D_f \dfrac{\Delta}{D} \dfrac{t}{D}$ (1)

is usually derived by means of the classical methods of Strength of Materials

---

[1]/ Dr. Ir. 2402 Route des Vallettes, 06140 Tourrettes sur Loup, France.

[2]/ D_f is the strain factor; Δ the diametral deformation; t the thickness, to be replaced by 2v in profile wall pipes, v the distance of the extreme fiber to the central axis; D is the diameter of the ring, the symbol within which inside, outside and mean diameters are confounded, and R the corresponding radius.

techniques applied to a thin elastic circular ring, embedded in an elastic medium. Its validity limits are never spelled out.

This ring model universally represents the buried pipe.

The strain equation (1) is generalized below solely in terms of geometry which only requires the assumption that plane ring sections remain plane during deformation; this assumption is fundamental and applies throughout Strength of Materials' methodologies. No other assumption is needed, particularly about material behaviour (eg, elastic properties) untill the section "Application to the Thin Elastic Ring Model" below, which expresses D_f in function of the ring's parameters. The strain formula is therefore valid regardless of: (a) the origins of the deformations, whether they are induced by the pipe loading or installation (which causes random deformations)

(b) the loading or the stress/strain status, whether "elastic" conditions prevail or creep or yield proceeds, and thus even past the proportionality limit, where the elastic relationship between stresses σ and strains ε, $\sigma = E \varepsilon$, becomes invalid[3];

(c) regardless of the surrounding soil characteristics.

The Generalized Strain Equation.

Deformation moves a point A (Fig. 1), of radial coordinates $\overline{R}$, and angle x with the axis of reference (usually the vertical or crown axis), to A', of coordinates $\overline{r}$ and virtually the same angle x. In A, the radius of curvature was R which is constant; it becomes ρ at A', which varies with x. The deformation $\overline{AA'} = \overline{\delta}$ is admittedly small -- i. e., of the order of a few %. The angle between $\overline{R}$, $\overline{r}$, and $\overline{\delta}$ is small and is neglected[4]. We can therefore address only their projections on $\overline{R}$, or their arithmetic values r and δ, both functions of x, and the constant R.

We can write the following equation for the functions of x :

$$\rho - R = a\,\delta\,. \qquad (2)$$

where "a" is an appropriate function of x.

---

[3]/ E is the pipe material's modulus of elasticity.

[4]/ Indeed, the angle between $\overline{\delta}$ and $\overline{R}$ is small everywhere on the ring save on the "diagonal" (x≅45°) where the deformations change sign and are about nil; this approximation is thus acceptable.

The ring's bending strains are considered exclusively; strains of other origin (i.e. thrust as, for example,. in pressure pipe) can be dealt with separately and added when needed. In such conditions, the length of elementary arcs along the central axis of the ring does not change, neither does that of the deformed ring. The bending deformation is thus "inextensional".

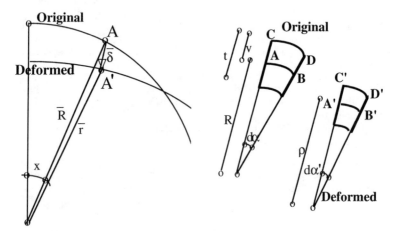

Fig. 1 The Ring's Deformation   **Fig.2 Ring's Segment Deformation**

The length of the elementary segment $AB \cong R\, d\alpha$ on the central axis anywhere on the ring (Fig 2) is thereforegiven by:
$A'B' \cong \rho\, d\alpha'$, and, since $AB = A'B'$:

$$R\, d\alpha = \rho\, d\alpha'$$

Within the ring's thickness, deformations and strains vary linearly with the distance from the neutral axis; at the neutral axis, they are zero (bending only is considered). We address the absolute value of the maximal strain which in plain (solid) wall pipe is either on the inside or outside. In profile-wall pipe, it is where "v", the distance between neutral axis and inside or outside wall is a maximum. Let us assume this happens along CD [5].

---

[5]/ Positive strain for decrease in radius of curvature.

The maximum strain in the ring segment that spans $d\alpha$, is thus:

$$\varepsilon = \frac{C'D' - CD}{CD} = \frac{(\rho+v)d\alpha' - (R+v)d\alpha}{(R+v)d\alpha} = \frac{\frac{\rho}{R} + \frac{v}{R}}{1 + \frac{v}{R}} \frac{R}{\rho} - 1$$

or, $\quad \varepsilon \cong \dfrac{R}{\rho}\left[\left(\dfrac{\rho}{R} + \dfrac{v}{R}\right)\left(1 - \dfrac{v}{R}\right)\right] - 1 = \dfrac{R}{\rho}\left[\dfrac{\rho}{R} + \dfrac{v}{R} - \dfrac{\rho}{R}\dfrac{v}{R} - \left(\dfrac{v}{R}\right)^2\right] - 1$

With R and ρ being of the same order of magnitude, and neglecting $\left(\dfrac{v}{R}\right)^2$ against $(\dfrac{\rho}{R}\dfrac{v}{R})$ in thin pipes:

$$\varepsilon = \left[1 + \frac{R}{\rho}\frac{v}{R} - \frac{v}{R}\right] - 1 = v\left[\frac{1}{\rho} - \frac{1}{R}\right]$$

which is a classic equation. Taking into account equation (2):

$$\varepsilon = v\frac{R-\rho}{R\rho} \cong v\left[\frac{R-\rho}{R^2}\right] = (-) a \frac{v}{R}\frac{\delta}{R} \tag{3}$$

For plain-wall pipe, v is to be replaced by t/2, hence, disregarding signs:

$$\varepsilon = a \frac{t}{2R}\frac{\delta}{R} = a \frac{t}{D}\frac{\delta}{R}$$

Comparing the latter equation to (1), we identify "a" to D_f and write:

$$\varepsilon = \mathbf{D_f} \frac{\mathbf{t}}{\mathbf{D}}\frac{\delta}{\mathbf{R}} \tag{4}$$

which generalizes, and is the <u>radial version</u> of equation (1). equation (4) can also be written:

$$\varepsilon = D_f \frac{v}{R}\frac{\delta}{R}.$$

and equation (2) also becomes:

$$\rho - R = D_f \delta \tag{2a}.$$

providing a relationship between curvature radii and deformations, which is valid at any point on the ring, along which D_f also varies.

Application to the Thin *Elastic* Ring Model.

The formulas for strains ε, moments M and radial deformations δ can be applied to loading cases such as that of Fig. 3 (an elastic ring under vertical loading, Spangler's or otherwise distributed lateral earth pressures), and generalized as follows:

$$\varepsilon = \frac{M}{EI/v} = Mv/EI$$
$$M = mWR$$
$$\delta = \delta_r W/S = \delta_r WD^3/EI$$

Standards symbols and stiffness $S=EI/D^3$ are used; m and δ_r are the coefficients for moments and <u>radial</u> deformations. Both are functions of x, and include the effects of earth lateral pressures; they vary along the ring.

The strains are thus:

$$\varepsilon = \delta \frac{mRv}{\delta_r D^3} = \frac{m}{2\delta_r} \frac{v}{D} \frac{\delta}{D}$$

Comparison of this equation to equation (3) yields, since $a \equiv D_f$:

$$\mathbf{D_f} = \frac{m}{8\delta_r} \qquad (5)$$

which applies **only** to a thin <u>elastic</u> ring,, subjected to above types of loading.

These equations can be extended to complex loading cases through summation of the strains due to each loading, each with its own D_f and deformation parameters.

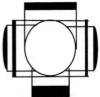

Fig. 3. Typical Earth Pressures Distribution

For plain, solid, wall pipe, the average "<u>diametral</u>" strains (where the c, i, s, v, h subscripts apply, respectively, at crown, invert, spring line, and on the vertical, and horizontal axes) are:

$$\varepsilon_{mv} = 1/2\ (\varepsilon_c+\varepsilon_i)$$
$$= (\ [\ D_{fc}\ \delta_c + D_{fi}\ \delta_i\ ]\ /\ 2R)\ (t/D)$$
$$= ([D_{fc}\ \delta_c + D_{fi}\ \delta_i]\ /\ D)\ (t/D)$$
$$= D_f\ \Delta/D\ t/D \quad \text{and}$$

$$\varepsilon_{mh} = D_{fs}\ (2\ \delta_s\ /\ D\ )\ (t/D) = D_{fs}\ (\ \Delta_h\ /\ D\ )\ (t/D)$$

<u>D_f in equation (1) is thus a "weighted" average when the ring shape is asymmetric.</u>

When the deformation is symmetric with respect to the major axes, $\delta/R \equiv \Delta/D$ and the two versions are identical; when the deformation is not symmetric, equation (1) may be applied on the main diameters, but this may hide higher maximal strains.

The Pipe Case.

Deformation equations applicable to pipe should include the Poisson's correction $1/(1-\nu^2)$, where ν is Poisson's ratio. This however has not been done above, because it is presumed that values of S measured according to standards such as ASTM D 2412 or AWWA C 950. are used; such standards de facto include the Poisson's correction for the measured value of S.

Pressure Pipes. Pressure significantly alters the deformation parameters m and δ_r; it stiffens pipe, and increases D_f when positive, though decreasing δ much more; the reverse happens under negative (external) pressure -- See Prevost/Kienow, 1994.

D_f Values.

The maximum value of D_f equals 3 for an ellipse (small deformations); it remains about 3 for an elliptically-deformed pipe, as long as the deformation remains small.

In the parallel-plate test, the values reach 4.27 and 2.7 on the vertical and horizontal axes respectively.

For flexible pipe, the calculated value of D_f can reach 8, depending on their flexibility

For the loading case in Fig 3 -- vertical loading combined with Spangler's earth horizontal side reactions to deformations -- the calculated value of D_f is about 7 for the ratio of modulus of earth side reactions to pipe stiffness E'/S =2000 which should be about the normal maximum limit.

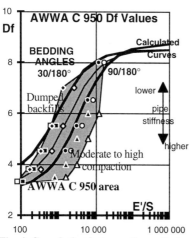

Fig. 4. Correlation between theoretical and AWWA C95O D_f values.

Correlation of the above theoretical values with AWWA C950 data is quite good (Fig. 4). Values up to 40 have been measured in the field.

COMMENTS, CONSEQUENCES AND CONCLUSIONS.

a/ It must be emphasised (i) that the deformation input into the generalized equation (4) is not the diametral deformation Δ, but the **radial deformation** δ or the relative deformation δ/R; and (ii) that equation (4) delivers the maximal strains anywhere on the ring, while equation (1) computes a diametral average (most often on the vertical axis). Because it is usually derived after the calculation of the (vertical) diametral deformation, as in Gerbault (1985)[6], equation (1) may wrongly appear valid only in the context of the conventional "elastic" and other assumptions of Strength of Materials.

b/ It must also be stressed that the radial version (4) of the strain formula shows the <u>actual</u> bending strains all along the ring, when the <u>actual</u> deformations δ are put in, regardless of pipe's strain/stress status, loading, installation, or embedding soil's conditions. When <u>calculated deflections</u> are used, equation (4) or the averaging diametral (1) <u>yields only calculated strains, which may be significantly different from those achieved in practice.</u>

c/ The correct evaluation of strains requires actual deformations to be input into the equation, regardless of their origin, whether from loading or installation, or whether creep or yield occurs.

Therefore, whenever a maximal deformation is specified, as in a standard for example, <u>the calculation of the maximal strain must be based on this maximal deformation</u> -- which is what AWWA C 950 rightly prescribes --, and <u>not on calculated values</u>.

d/ Of course, D_f and δ are determined by the shape of the deformed pipe determined by installation, loads, and pressure.

The strain factor D_f is sometimes called the "shape factor", which may well be a consequence of equation (2a). This is not appropriate because, if well-defined D_f values are attached to a given shape, the reciprocal is not true: D_f does not define the pipe's shape.

e/ Measuring strains is relatively easy using strain gauges, or through measurements of the variations of the radius of curvature -- though with some risk of error -- and the use of the equation:

$$\varepsilon = v \left[\frac{1}{\rho} - \frac{1}{R} \right].$$

On the contrary, measuring the vertical or horizontal diametral deflection can be readily made -- this is indeed what has been done in practice[7]. The diametral version (1) of the strain equation is thus more convenient to use and will remain so. The available data are those required for equation (1), but it must kept in mind that the D_fs so found are only approximate averages.

All this means that the <u>D_f data are uncertain</u>, and evaluating strains using equations (1) or (4) is volatile.

D_f, it must be remembered, is an approximate, though most worthwhile, indicator.

---

[6] In Gerbault, 1985, the vertical pressures on crown and invert are also symmetrical, hence same strains in both locations, and a symmetrically deformed ring.

[7] It is unclear where the scarce strain measurements have in the past been made whether at crown or invert.

APPLICATION TO STEEL PIPE.
Stresses/strains in steel.

Buried, all welded, steel pipelines are highly stressed structures. In short, depending on usage, API, ASME, AWWA and other standards or codes of practice allow "Barlow" or hoop $\sigma = pD/2t$ stresses as high as 50 to 72, even 75 % of the specified minimum yield strength (SMYS) [8], depending on the sites of the lines and the pipes manufacturing process, *regardless of other stresses of any kind.*

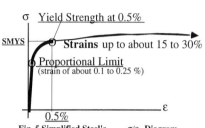

Fig. 5 Simplified Steel's σ/ε Diagram

There is a trend to increasing the SMYS of pipeline steels: grade 30, 60, and even, now, 80, are used, which have minimum yield strength of, respectively, 210, 400 and 560 MPa (30 000 to 80 000 psi).

These values of the SMYS are the stresses that cause 0.5% strain.

The proportional limit is somewhat lower than the SMYS, and its corresponding strain is close to 0.1% for the mild steel generally used in water-supply practice, and up to about 0.25% for the higher-strength steels often used for higher pressure gas and oil transmission lines (strain ε = stress σ/E, where E is the modulus of elasticity).

Beyond the proportional limit, steel enters the plastic domain; the equation $\varepsilon = \sigma/E$ no longer applies; stresses that exceed the yield strength level off, redistribute and are in the whole or in part "erased" through yield; and steel somewhat hardens. But strains add as yield proceeds. This is sketched in the simplified Steel's σ/ε diagram shown Fig. 5, which is sufficient for our purpose; see also, for example, AWWA M11 Manual,1989.

Beyond the proportional limit, the analytical or stress design method no longer applies.

Level of stresses/strains.

Let us try to assess the overall order of magnitude of stresses or strains in buried steel pipelines, while reviewing the loadings they are subjected to, and estimating the stresses/strains they cause.

---

[8] There is a weld joint factor to be taken into account, but it most often is 1.

Pressure. The Barlow stress consumes up to 50 to 75% of the permitted stress; it is unaffected by any other loading or deformation, whatever they may be and cannot be "erased" by yield; it is uniform stretch. This is the primary pressure effect which, alone, brings the steel "barrel" near the proportional limit.

Pressure also induces "secondary" strains/stresses associated with the rerounding of pipe's deformations or of its "shape defects" (i.e., any out of roundness). *These stresses/strains may equal and even exceed Barlow's* (Prevost, 1990),as shown in Fig 6, one example among several. Barlow and secondary stresses/strains compound; pressure alone can thus bring the pipe steel well above the proportional limit, causing yield to proceed, in a random manner, since pipe never has a uniform circular shape. [9]

Let us remember that the theory of pressure action on near-circular structures is about a century old; it was well established by Lazard in 1933, and is satisfactorily verified by experience (Prevost, 1990); shape defects rerounding stresses/strains and deformations can be predicted with good accuracy once the "original" shape is known.

Transverse deflections up to 3, or 5%, or more, are allowed; the strains they cause can be estimated by using equation (4). As the D/t ratios of steel pipe range, normally, between 50 and 100 and their D_f is at least 3, equation (1) shows that the transverse bending strains to be expected in a buried pipe are *at the very least* about $\varepsilon \cong 3 \times 0.03 / 100 \cong 0.1\%$, which about equal the proportional limit of many steels. Under the permittted maximal deflection alone, plain wall steel pipe can thus be subjected to yield. The parameter "v" of profile wall pipe that replaces, and is bigger than, t/2 in equation (1) aggravates the condition of corrugated pipe; Spangler did note the fact.

Longitudinal load variations, any departure from a straight alignment cause deflections, strains and stresses, which may have the same order of magnitude as the transverse ones. Their origins are diverse: uneven bedding; joints or welding pits more or less well backfilled; live loads or load variations along the lines. All cause shears, which in turn deflect the transverse sections, flattening the pipe when bending is vertical, elongating it when horizontal; and lesser important thrusts. Pressure tends also to straighten-out pipes, hence giving rise to further strains and stresses.

---

[9] GRI/Cornell (1991) experiments, and the Batelle tests they cite, measured straying strains as high as 30% of the Barlow values on high pressure pipes, but do not explain them. They deny pipe pressure strengthening.

High pressure pipes are mechanically expanded and pressure tested close to, or at, the yield strength at the end of the manufacturing process; they are thereby well rounded, which is not the case of ordinary fabricated pipe; they therefore show lower shape defects strains.

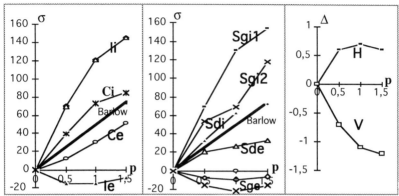

Fig. 6. Calculated Barlow (Hoop), and Actually Measured, Stresses σ (MPa), and Rerounding Deformation Δ (mm) vs Internal Pressure p (MPa) in a Tested Pressure Pipe, 500 mm Diam., 5mm Thick
C=Crown; I=Invert; S=Springline; i=Inside; e=Outside; g=Left; d=Right
No External Load; 2 measuring sections

External accidental damages, in the form of groves (possibly from welding) or other kind of sharp departure from circular shape, may add their own particular strains and stresses. Such shape defects are less rerounded, if at all, by pressure than "ovals". Such stresses are not "ironed out", and they may generate complex stress/strain fields. Tri-dimensional stresses might also arise, possibly embrittling the best of steels; their magnitude may be quite high, and is, of course, unpredictable.

There may also be residual stresses/strain from the manufacturing process.

There is thus a series of stresses and strains -- out of which only the Barlow and transverse-bending are usually considered--. They compound, and, each, can bring the pipe material close to, or above, the proportional limit, and often into yield. Each strain have the magnitude of a tenth, or of a couple of tenths, of 1 %; their aggregated strains could thus reach, roughly, 1% which is harmless when compared to steel's large ductility value, normally significantly higher than 14%. Most such strains are randomly distributed in a pipeline.

Consequences.

a/ It is inevitable that, during construction or operation or both, large areas of all steel pipelines are brought into, and/or remain in, yield domain.

This is in principle incompatible with *the stress or analytical design method* which *is thus inappropriate for buried steel pipe;*.

b/ ***The above implies thorough, though implicit, reliance on steel's quality, on its ductility*** -- yield should not damage pipes -- which is not often, but should be, emphasized. The use of the analytical method, by limiting stresses, tends to create a false sense of security, oblivious of the real problem.

c/ However, the analytical design method is still the preferred method in Continental Europe for use on all sorts of pipes (with maximum steel stresses often a factor of the ultimate limit), and the prevailing one in the USA for steel pipe crossings of highways and railroads.

d/ Obviously such crossings (as well as urbanized areas) are sensitive structures, and the agencies who own the right-of-way at crossings are obviously right to be concerned about safety.

A distinction was established, long ago, in the USA, between steel pipeline crossings at highways and railroads and those installed elsewhere. The common method is to design "pipeline crossings" according to the analytical method, and API 1102 is now strengthened in its 1993 revision by its enshrining of the GRI/Cornell, 93, proposals. As said above, this could appear safe because limits are set to the stresses and yet is deceptive It is further faulty because analytical methods have to consider all loading, but, in this case, only the Barlow, and few of the bending stresses actually are.

The fact however is that the external and internal loading and the responses of all pipes, whether at crossings or elsewhere, are identical; therefore all should be designed according to identical principles.

e/ The method of AWWA Manual M11 (1989) or similar documents should thus supersede the analytical method for buried steel pipe design, and do so indeed, successfully and economically, in many parts of the world.

Accordingly, steel pipe design should separately consider:
i/ transverse bending forces, deflection being a major criterion; and
ii/ internal pressure, exclusively on the basis of the Barlow stresses.

Other loadings need not be taken into account (see Prevost/Kienow, 1994).

f/ Deformations and strains would normally spread, but might concentrate, e. g., in areas where uneven settlements proceed, which could create hazards. This has very low probability but is not to be overlooked.

g/ Generally, analytical method increases pipe thickness. Contrary to base common sense, this in turn would increase strains whenever thickness does not affect transverse or longitudinal deformations.

h/ Documents assessing risks associated with pipelines (e. g. Murray, 1990) also rely on stress analysis -- and, similarly, overlook the series of generally-disregarded strains listed above. They thus do not focus enough on a real issue, that is steel quality.

However, Dauby, et. al., 1993, in their "Gas Pipeline Fitness for Service Investigation" go a long way towards the positions developed above (yield status was found at several locations in a high pressure gas line).

It appears also to be acknowledged that offshore lines in the North Sea are subjected to yield in spanning bedding unevenness.
Further developments should be watched.

Appendix. References.

API 1993 Steel Pipelines Crossing Railroads and Highways. Recommended Practice 1102. April

AWWA (1989) Steel Pipe Design and Installation, AWWA Manual M 11 AWWA

Dauby F. A. 1993 Lee Chih-Hung, Vollbrecht G. M. Gas Pipeline Fitness-for-Service Investigation Proceedings ASCE Pipeline Infrastructure II International Conference, San Antonio, Aug.

Gerbault M, 1985. Calcul des Canalisations Circulaires Semi-rigides. Annales de l'Institut du Batiment et des Travaux Publics N°439 Série Théories et Méthodes de Calcul 278.

GRI / Cornell Univ. Ingraffea A. R., O'Rourke T. D., Stewart H. E., Behn H. T., Barry A., Crossley C. W., El-Gharbawi S. L., 1991 Final Report Technical Summary and Data Base for Guidelines for Pipelines Crossing Railroads and Highways Gas Research Institute, Transport and Storage Department, Dec.

Lazard R. (1933) "Le calcul de ouvrages circulaires" Le Génie Civil, Tome CIII, N° 13, 14, and 15, respectively 23 and 30 september. and 7 october.

Murray A. G. 1990 Controlling danger from a hidden source. Pipes & Pipelines Int. July-August

Prevost R. C., 1990 Flexible Pipe Design Revisited, ASCE. Pipeline Design and Installation Conference, Las Vegas, March

Prevost R. C., Kienow K. K. 1994. Basics of Flexible Pipe Structural Design. ASCE Journal of Transportation Engineering. July

THE EFFECTS OF HIGH-STRENGTH STEEL
IN THE DESIGN OF STEEL WATER PIPE

Robert J. Card, P.E.[1]

&

Dennis A. Dechant, P.E.[2]

Abstract

In steel water pipe design, two variables must be considered, namely the steel's thickness and yield strength. Traditionally, varying the wall thickness of the steel pipe cylinder was the principle means of designing pipe for internal pressure. Specifiers gave relatively little consideration to using steels with a higher yield strength mainly because such grades were not readily available. However, the steel industry has modernized significantly over the last decade, and higher grade steels are now commonplace. Today with the availability of higher grade material, steel pipe can be specified with lesser wall thicknesses than what were historically used in the past. While this is particularly advantageous in cases where a pipeline must operate at high internal pressures, a reduction in wall thickness

---

[1] Co-Chairman, Pipe Committee
Steel Plate Fabricators Association

6400 South Fiddler's Green Circle, Suite 580
Englewood, Colorado 80111-4955
(303) 290-9490 Fax: (303) 290-0662

[2] Co-Chairman, Pipe Committee
Steel Plate Fabricators Association

12005 North Burgard
P.O. Box 83149
Portland, Oregon 97283-0149
(503) 285-1400 Fax: (503) 285-2913

also has an important bearing on other elements of steel water pipe design.

This paper focuses on the inter-relationship between pipe wall thickness and higher grade steels and its effects on other common steel water pipe design parameters. These parameters include internal pressure, external loading, pipe handling stiffness and resistance to vacuum.

Introduction

In this age of improving technology and rapid advancements in material design, engineers are faced with many new and improved options. Many are great advancements and are worthy of all the claims of greatness, but anyone who has been burned by new products that have failed prematurely may become somewhat leery of claims of improvement. Compounding this concern is the fact that in the waterworks field "new" (be it design, manufacture, or material) may take 20 years to test out.

The typical water pipeline in America has working pressures less than 400 psi and fill heights less than 15 feet over the top of the pipe. Pipelines with greater internal pressures are typically penstocks and the design procedure for such pipelines is outlined in American Society of Civil Engineers (ASCE) or Steel Plate Fabricators Association (SPFA) manuals. Pipelines with greater fill heights would require a careful review of the installation practices, native soils, and bedding and backfill soils. This paper addresses the above mentioned typical water pipeline project.

Steel Making Historical Overview

With the development of the Bessemer process in 1855 and the open hearth process 1861, steel, the strongest and most versatile refinement of iron, became available for water pipe. Pipe mills sprang up in many cities and steel pipe started its' long history of use in North America. Looking at a historical listing of projects prior to 1900[1], it is easy to recognize steel wall thicknesses in similar diameters of pipe that are still being utilized today.

The quality of steel produced by the Bessemer process was dependent on the incoming iron ore quality. As high quality iron ore was depleted, the impurities (such as high amounts of sulfur or phosphorous) started to cause problems for both producers and users. The negative pipe properties of being brittle and having poor impact resistance started to appear.

The open hearth process afforded the opportunity to partially remove sulfur and phosphorous during the refining process. This process also yielded a more economical means of reducing iron ore. This process proved to be so very popular that there was still an open hearth shop built in the United States as late as 1958. Additionally, there are still several furnaces in operation today.

The basic oxygen process was introduced to the steel making process in the late 1950's. This process differed from the open hearth process by the use of commercially pure oxygen. The main advantage of the basic oxygen process over the open hearth process is the speed of the melting. Steel can be refined in less than 45 minutes in the basic oxygen process while a comparable heat may take up to 10 hours by the open hearth process.

In the late 1960's, the continuous cast process was also introduced. Up until this time, steel was cast into ingots with an intermediate step to produce slabs. These slabs can now be produced directly from the continuous cast process. This process also reduces reoxidation and greatly improves steel cleanliness.

In the last 15 years, an intermediate step has been introduced between the basic oxygen process and the continuous casting process. This process is known as ladle metallurgy. This process can and often does include degasing, desulfurization, temperature control and alloy additives and trimming. It can produce very clean steels tailored to exacting end user requirements, be it chemical or physical. This means that the end product is steel made for the requirements needed by the project conditions.

Pipe Manufacturing Improvements

The greatest manufacturing advancement in the production of steel pipe since it was first utilized in the 1850's is the method of jointing the plate to form the pipe. The first joints were single rows of rivets. Then multiple rows of rivets were used to increase the joint efficiency (to hold the steel thickness down). In 1905 the Lock-bar process was developed. Pipe was formed by rolling half circles and the circles were then joined by an H shaped member which was pressed closed to obtain a joint which was considered at that time to be 100% efficient.

Finally, in the 1920's automatic welding was developed and refined. Since the 1920's, the welding of plates, sheets, or coils of steel has shown drastic improvement, decade by decade as more and more automatic machinery is developed. Multiple methods of testing the strength of these welds were concurrently developed. Destructive testing such as hydrotesting the cylinder, tension testing of a reduced section sample, or face and root bend tests were developed to test the strength and ductility of the welds.

Non-destructive testing followed later. These tests include magnetic particle, dye penetrant, and ultrasonic examination to find any kind of defect in the weld that a destructive test would not uncover. Lastly, for the highest quality pipe in the most severe applications, x-ray testing of the welds is now sometimes utilized.

Design Philosophy Development

It is interesting to note that the history of design development very closely follows with the improvements in the steel making practices and manufacturing processes. In the early years of riveted steel pipe, the design of the pipe was based on a safety factor of approximately 4:1 against ultimate tensile which resulted in a design stress of 10,000 psi. As the steel quality improved and the manufacturing processes were refined, confidence grew and the design was based on 12,500 psi then 13,750 psi, and finally 15,000 psi. The safety factor was still 4:1 which means the ultimate tensile value of the steel was 60,000 psi. Pipe sizes of 4" through 144" with steel thicknesses of 16 gage (0.0598") to 1-1/4" were being produced. The joint efficiency of the single rivet applied to design was 45%; quadruple rivets raised the efficiency to 90%.

The pipe design in this time period was totally based on the internal pressure that the pipe needed to resist. Marston's theory of earth loads on rigid pipe was not developed until approximately 1913. Deflection, vacuum, or other external type loadings were not part of the design procedure for steel pipe.

Even with lock-bar pipe becoming more prevalent in the early 1900's, the design stresses were still limited to 13,750 psi to 15,000 psi as the steel making process still only produced 55,000 psi to 60,000 psi ultimate tensile values.

When welding of the seams finally took over the steel pipe manufacturing process, the design philosophy gradually changed to a 2:1 safety factor against the yield strength of the steel instead of the tensile value. This was primarily due to the more refined testing of the joint that welding allows to show a truly 100% efficient seam.

During World War II, the United States suffered from extreme steel shortages and alternate concrete pipe products were developed which severely cut down the amount of steel in pipe. Due to the rigid coatings on these products, design stresses were typically limited to 15,000 psi in order to prevent over straining and cracking the concrete or mortar coatings.

Marston's earth load theory was now being utilized to predict loads on pipe, both rigid and flexible pipe, therefore, the design process evolved to include the effect of external load on pipe. In 1941, Dr. Spangler published his first work on deflection of flexible pipe which utilized the Marston earth loads on rigid pipe. The first revision of the Iowa deflection equation, as it was called, was shown to not adequately predict flexible pipe deflection.

Shortly after World War II, the gas and petroleum industry began using steels with yields up to 50,000 psi designed at 50% of yield and steels of 42,000 psi yield often designed at 60 to 65% of yield. Even today, gas and oil pipelines may be designed for up to 72% of yield.

During this same time frame (the late 1940's), many very high pressure penstocks were designed and built using fine grain, killed and/or heat treated steel with yields as high as 100,000 psi. The design was based on stresses equal to 1/3 the tensile value or 2/3 the yield, whichever was less. The United States Bureau of Reclamation published their first penstock design manual at this time.

The decade of the 1950's brought about a revision to the original Iowa formula by Dr. Spangler and Dr. Watkins. Watkins had rederived the Iowa formula and found the original assumptions regarding the soil property term, er, were dimensionally incorrect. The term was rewritten with a correct interpretation of the property of material and was introduced as E'. The formula was now called the Modified Iowa Formula and was first published in 1958. Actual field measurements corresponded reasonably well with the predicted values and the Formula quickly found widespread use.

This widespread use of the Modified Iowa Formula soon began a change in design philosophy from internal pressure design control to external load barrel deflection design control. Because of more awareness of external load design, much deflection testing of pipe and soil testing to determine E' values was undertaken in the 1960's and 1970's. The deflection testing was mainly performed on mortar coated steel pipe to control the cracking of the mortar coating. Utah State University published a report in 1966[2] which stated allowable levels of deflection for cement mortar lined and cement mortar coated steel pipe. In 1977, Amster Howard of the USBR published his report on the modulus of soil reaction, E', which received immediate widespread attention.

With all the attention being given to external load design, internal pressure design, the original proper design procedure, was neglected for years. The allowable stress level in the pipe more or less stayed in the 15,000 psi to 16,500 psi range, matching the design stress level of the concrete pipe.

All of this history brings us to where the state of design of steel pipe is at today. Helpful information that the modern steel pipe designer should consider prior to design is as follows:

1. Steel utilized today in the manufacture of steel pipe is of a much higher quality and are most probably a fine grain continuous cast material. The typical steels of today match the quality of pressure vessel steels of the 1960's.

2. Thicknesses of steel coils can be ordered to the nearest thousandth of an inch. There is no reason to use incremental thicknesses (3/16", 1/4", etc.) that were originally developed for plates. The use of 52,000 psi yield strength steel should be analyzed as this yield strength can now be produced with excellent ductile properties.

3. If the design is controlled by internal pressure then steels with yield strengths up to 52,000 psi should be considered. Typically, within reason, the design control should be the handling requirements. The most recent edition of AWWA M-11 gives time tested and proven handling thickness requirements.

4. Dr. Watkins has written numerous papers on steel pipe design regarding deflection and vacuum design. These papers are based on actual laboratory testing, field investigations, and performance history. The design procedures that are developed in these papers leads the designer from utilizing the Modified Iowa Formula (which is based on the unmeasurable soil characteristic of E') to a more quantifiable soils driven procedure. As the design procedure for deflection control matches actual field installations the closest, they should not be ignored.

5. Above ground or pipe in a fluid environment vacuum conditions must be analyzed. However, buried buckling conditions will not control if the proper deflection levels are maintained (less than 3%), and the minimum handling thickness criteria is not violated (D/t=288).

Summary

Designers often overlook the pipe design and specify an arbitrary minimum steel thickness. This relatively new practice of specifying the steel wall thickness is counter to the long time history of designing the wall thickness for steel pipe. By designing the wall thickness of the steel pipe, the utility does not "overpay" for their pipe and obtains the most cost effective steel pipe for their money spent. A complete review of design procedures and information performed prior to issuing the project plans and specifications most often pays great dividends in the end.

References:

1. Steel Plate Engineering Data Volume 3, *"Welded Steel Pipe"*

2. *"Ring Deflection of Buried Pipes"*, Engineering Experiment Station, Utah State University

Newton Meets Darcy and Colebrook in a Programmable Calculator

Jorge A. Garcia, Ph.D., P.E.[1], Assoc. Member, ASCE

Abstract

Pipeline design requires the computation of friction losses in the conduit and minor losses in fittings, valves and other appurtenances. Depending on the equations used, these computations can be performed manually, using tables or nomographs, using a computer or simply a programmable calculator.

Modern programmable calculators have incorporated tremendous computing power in relatively small and inexpensive machines. Hydraulic computations such as friction and fitting losses which in some instances are of implicit nature, can be implemented in these new machines.

The Darcy-Weisbach equation is frequently used to compute friction losses in pipelines. The equation requires the pipe flowrate, length, diameter and friction factor. The friction factor can be obtained from the Colebrook equation, which is implicit, and therefore requires an iterative procedure to find a solution. The Newton-Secant method, which is essentially a numerical approximation to a function derivative, has proved to be very effective in solving for the friction factor in the Colebrook equation, which is then used in the Darcy-Weisbach equation.

Depending on the design or analysis application being addressed, the unknown in the friction equation can be the friction loss only, the total friction plus minor loss, or maybe the diameter, the flow, or possibly the pipe length. This adds another level of complexity in design and analysis which is usually overcome by manual trial-and-error approach or more sophisticated computer programs or models.

---

[1]Design Engineer, Utilities Engineering Dept., City of Las Cruces, P.O. Drawer CLC, Las Cruces NM 88004.

This paper presents a technique to solve for any unknown in the Darcy-Weisbach equation utilizing the implicit Colebrook equation for friction factor and an option of two other explicit friction factor equations. The solution is implemented using the Newton-Secant method and is incorporated in an HP 48GX programmable calculator.

Introduction

Fluid flow between points A and B along a closed conduit can be described by the general energy equation. The total energy at point A is given by the sum of the pressure head (P_a/W), the elevation head above a given datum (Z_a) and the velocity head ($V_a^2/2G$). Energy additions (H_{ad}), energy removals (H_r) and energy losses due to friction (H_l) are added or subtracted from the energy at point A to find the energy at point B given by the sum of pressure head (P_b/W), elevation head (Z_b) and velocity head ($V_b^2/2G$). In equation form we have:

$$\frac{P_a}{W} + Z_a + \frac{V_a^2}{2G} + H_{ad} - H_r - H_l = \frac{P_b}{W} + Z_b + \frac{V_b^2}{2G} \quad (1)$$

where: P_a, P_b are pressures in KN/m² (lb/ft²)
W is the specific weight of fluid in KN/m³ (lb/ft³)
Z_a, Z_b are elevation heads in meters (ft)
V_a, V_b are the flow velocities in m/s (ft/s)
G is the acceleration of gravity in m/s² (ft/s²)
H_{ad}, H_r are energy additions and removals in meters (ft)
H_l is the friction loss in meters (ft)

The flow of a fluid in a closed conduit generates internal friction losses (H_l) which are proportional to the velocity head and to the ratio of pipe length to diameter. Several equations have been proposed in the literature to quantify these friction losses, the most commonly used being the Darcy Weisbach equation for most fluid applications. Other equations such as the Hazen Williams equation is commonly used in water distribution systems.

Computational difficulties when utilizing the Darcy Weisbach equation forces some designers to use other more simple equations which may have important limitations. These difficulties are associated with the implicit nature of the friction factor which accounts for the type of flow regime and to the computation of flow velocity.

The technique proposed in this paper solves for friction loss, flow, length of conduit or diameter in the Darcy Weisbach equation. This methodology uses a numerical approximation to a function derivative to iteratively solve for the implicit equation of the friction factor nested within a built-in calculator function.

Friction Loss Equation

The Darcy Weisbach equation relates friction loss in a conduit (H_l) to the velocity head ($V^2/2G$) and to the ratio of pipe length (L) to pipe diameter (D), accounting for fluid properties, conduit properties and type of flow regime with the friction factor (F). In equation form the expression is:

$$H_1 = (F \frac{L}{D} + M) \frac{V^2}{2G} \quad (2)$$

where:
H_l is the friction loss in meters (feet)
F is the dimensionless friction factor
L is the pipe length in meters (feet)
D is the conduit diameter in meters (ft)
M is a dimensionless minor loss coefficient
V is the velocity head in m/s (ft/s)
G is the acceleration of gravity in m/s^2 (ft/s^2)

Depending on the flow regime, the friction factor is a function of the Reynolds number for laminar flow, and both the Reynolds number and the relative roughness of the pipe for turbulent flow. The Reynolds number is defined as the ratio of the inertia force of an element of fluid to the viscous force. This dimensionless number depends on the fluid density, fluid viscosity, pipe diameter and average flow velocity. The expression for the Reynolds number, denoted by R_y in this paper is as follows:

$$R_y = \frac{\rho VD}{\mu} \quad (3)$$

or

$$R_y = \frac{4Q}{\pi DV} \quad (4)$$

where:
R_y is the Reynolds number (dimensionless)
ρ is the fluid density in kg/m^3 (slugs/ft^3)
μ is the fluid dynamic viscosity in N.s/m^2 (lb-s/ft^2)
v is the kinematic viscosity (μ/ρ) in m^2/s (ft^2/s)
V is the flow velocity in m/s (ft/s)
Q is the flowrate in m^3/s (ft^3/s)
D is the pipe diameter in meters (feet)

The relative roughness is given by ratio of average wall roughness (e) to pipe diameter (D). Average wall roughness values for new commercial pipe have been widely published in the literature and will not be included in this paper.

Probably the most widely used method of evaluating the friction factor employs the Moody diagram which relates the friction factor (F) to the Reynolds number (R_y) and the relative roughness (e/D). This diagram was experimentally developed by L.F. Moody in 1944. Computerized calculation of the friction factor, however, requires equations for different flow regimes. For laminar flow conditions, where Reynolds numbers are below 2000, the friction factor (F) is given by the equation:

$$F = \frac{64}{R_y} \qquad (5)$$

where: F is the dimensionless friction factor
         R_y is the dimensionless Reynolds number

Reynolds numbers between 2000 and 4000 indicate flow is in a critical range and values of the friction factor cannot be accurately predicted. Reynolds numbers above 4000 indicate a region of flow called turbulent. The turbulent zone is further divided into a "smooth pipe" boundary, a transition region and a region of complete turbulence. Along the "smooth pipe" boundary the friction factor is a function of the Reynolds number only and in the fully turbulent region, the friction factor is a function of the relative roughness. In the transition zone, however, the friction factor is a function of both the Reynolds number and the relative roughness.

C.F. Colebrook developed a relation for the friction factor in the transition zone. This equation, which is of implicit nature, includes two terms inside the log function to account for both the "smooth pipe" boundary and the fully turbulent flow region. The implicit form of Colebrook's equation can be written as follows:

$$\frac{1}{\sqrt{F}} = 1.14 - 2LOG(\frac{e}{D} + \frac{9.35}{R_y\sqrt{F}}) \qquad (6)$$

where: F is the dimensionless friction factor
         R_y is the dimensionless Reynolds number
         D is the pipe diameter in meters (feet)
         e is the wall roughness in meters (feet)

Equation (6) above accounts for fully turbulent flow as the second term inside the parenthesis becomes very small. Also, very small values of e/D reduces the equation to that of the "smooth pipe" boundary.

Explicit equations have been developed to compute the friction factor as an alternative to Colebrook's equation. P.K. Swamee and A.K. Jain developed the following equation:

$$F = \frac{0.25}{[LOG(\frac{1}{3.7(D/e)} + \frac{5.74}{R_y^{0.9}})]^2} \quad (7)$$

where: F is the dimensionless friction factor
R_y is the dimensionless Reynolds number
D is the pipe diameter in meters (feet)
e is the wall roughness in meters (feet)

A second frequently used explicit equation to compute the friction factor was developed by Churchill. The expression is given as follows:

$$F = 8[(\frac{8}{R_y})^{12} + \frac{1}{(A+B)^{1.5}}]^{\frac{1}{12}} \quad (8)$$

$$A = [2.457 LN((\frac{7}{R_y})^{0.9} + 0.27(\frac{e}{D}))^{-1}]^{16} \quad (9)$$

$$B = [\frac{37530}{R_y}]^{16} \quad (10)$$

where: F is the dimensionless friction factor
R_y is the dimensionless Reynolds number
D is the pipe diameter in meters (feet)
e is the wall roughness in meters (feet)

The explicit equations described above solve the problem of having to implement an iterative process in finding the friction factor when the unknown in Darcy's equation is the friction loss H_l. If the unknown in Darcy's equation is the flowrate, however, an iterative process is required because the Reynolds number cannot be explicitly computed since it is a function of velocity and therefore flowrate. This problem is further complicated when the diameter is the unknown in Darcy's equation since both the Reynolds number and the relative roughness cannot be calculated. Regardless of the unknown being considered, a technique is available to the design engineer to compute any of the three scenarios just described by treating Darcy's equation as an implicit function as follows:

$$f(x) = \left(F\frac{L}{D} + M\right)\frac{V^2}{2G} - H_l = 0 \quad (11)$$

$$V = \frac{4Q}{\pi D^2} \quad (12)$$

where: $f(x)$ is an implicit function
x is a dummy variable representing H_l, Q or D
H_l is the friction loss in meters (feet)
L is the pipe length in meters (feet)
D is the conduit diameter in meters (feet)
M is a dimensionless minor loss coefficient
V is the flow velocity in m/s (ft/s)
Q is the flowrate in m^3/s (ft^3/s)
G is the acceleration of gravity in m/s^2 (ft/s^2)

The technique commonly used to solve implicit equations is called the Newton-Raphson method. This method in modified form can be easily programmed in BASIC or any programmable calculator. Some modern programmable calculators have a built-in function which implements the Newton routine. The section that follows will address this method in detail. A nested Newton procedure will be introduced to solve the implicit form of Darcy's equation in conjunction with the implicit Colebrook equation for the friction factor.

Newton-Secant Method

The traditional Newton-Raphson method takes a seed value of the unknown being solved (x), finds the derivative of the implicit function evaluated at "x", then finds a new value of the unknown by correcting the initial value by the ratio of the function to the function derivative. In equation form the iterative process is as follows:

$$x_{i+1} = x_i - \frac{f(x_i)}{f'(x_i)} \quad (13)$$

where: x_i is the current value of the unknown
x_{i+1} is the new value of the unknown
$f(x_i)$ is the implicit function evaluated at x_i
$f'(x_i)$ is the derivative of the function evaluated at x_i

The iterative process is continued until the absolute value of the function is within a specified error limit. One problem with this method is that it requires the computation of the function derivative with respect to the unknown being solved. Each different unknown will require the formulation of a new derivative.

The Newton-secant method is a more simplified iterative procedure where the function derivative is numerically approximated. In this manner, only two functions are evaluated regardless of the unknown. Care must be exercised, however, in supplying the correct seed value for the iterations. The equation for this technique is as follows:

$$x_{i+1} = x_i - f(x_i) \left[\frac{x_i - x_{i-1}}{f(x_i) - f(x_{i-1})} \right] \qquad (14)$$

where: x_i is the current value of the unknown
x_{i+1} is the new value of the unknown
x_{i-1} is the previous value of the unknown
$f(x_i)$ is the implicit function evaluated at x_i
$f(x_{i-1})$ is the previous value of the implicit function

As mentioned earlier, the Newton-secant method can be programmed very easily in BASIC or a programmable calculator. Modern calculators such as the HP 48GX have a built-in function called Solver which basically implements this iterative technique. The implicit function is programmed in the calculator and all pertinent variables are defined. Values are entered for all known variables and a seed or guess is supplied for the unknown variable. The solver function then utilizes this seed value and iterates to find the solution or root.

This paper utilizes a combination of both the built-in Solver function and a programmed Newton-secant procedure to solve two implicit equations. The built-in function is used to solve for Darcy's equation in implicit form (11) accounting for friction as well as minor losses. The programmed Newton-secant procedure is utilized within the built-in iterative procedure to compute the friction factor from Colebrook's equation (6). This is a situation where there are two implicit functions, one nested within the other. Nested Newton iterations programmed in BASIC to solve these same equations have been previously developed and tested by the author. The use of the built-in function mainly facilitates the process and does not offer any mathematical advantage over complete programming.

Figure 1 is a schematic representation of the Newton-secant method. The curve represents an arbitrary function with a single root at a function value of zero. The two initial seed values provide two function points through which a secant line can be projected. This secant line is extended to intersect the x-axis, providing a new point at which the function can be evaluated. Next a second secant line is drawn through the new point and the previous point, and again extended to intersect the x-axis. The process is continued until a predetermined error criteria is satisfied.

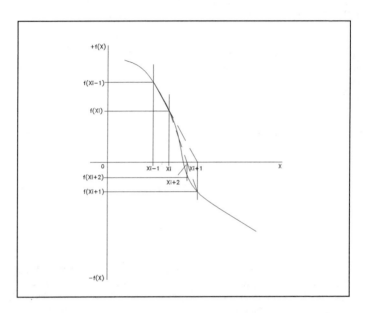

Figure 1. Graphical representation of the Newton-secant method.

The Program

A structured approach was used in the development of the program, utilizing separate procedures for each equation. The program consists of five procedures and the main body. The main program called DWE is executed after all data has been entered. This portion of the program calls for the computation of the Reynolds number and depending on the option of friction factor equation, it invokes the appropriate procedure. After the friction factor is computed, the implicit form of Darcy's equation is solved.

Three options are offered for the computation of the friction factor in addition to the laminar friction factor equation (OPT=0). The first option (OPT=1) utilizes the Newton-secant method which solves Colebrook's equation (6). The second option (OPT=2) uses the Churchill's explicit computation of the friction factor (8,9,10). The third and last option (OPT=3) of the friction factor computation is Swamee and Jain's equation (7) which is also explicit.

Figure 2 shows a process flowchart for program DWE. The flowchart shows an input block followed by a call to compute the Reynolds number. Based on the magnitude of the Reynolds number, a decision is made on whether flow is laminar. Reynolds numbers less than or equal to 2000 set the friction factor option variable to zero (OPT=0). Next, a CASE branch structure selects the friction factor computation option based on the variable OPT. This branch structure calls the laminar friction equation, the Newton-secant method for Colebrook's equation, Churchill's equation or Swamee and Jain's equation. The CASE branch structure is followed by the solution of Darcy's equation in implicit form.

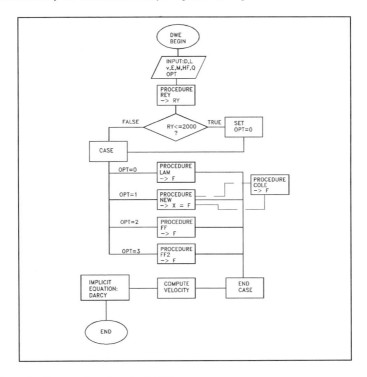

Figure 2. Process flowchart for DWE program

Figure 3 shows a process flowchart for the Newton-secant method used to solve Colebrook's equation. The flowchart shows an initialization block for seed values and allowable error. Colebrook's equation is first solved using one seed value x_{i-1} to obtain an initial function value $f(x_{i-1})$. Next a WHILE loop starts the iteration process which continues until the absolute value of the change in function $f(x_i)$ becomes smaller than the allowable error. Inside the WHILE loop, Colebrook's equation is solved for x_i and the Newton-secant equation is calculated for a new value of the unknown x_{i+1}. After the WHILE loop is terminated, the last value of the variable x_i is assigned to the friction factor (F).

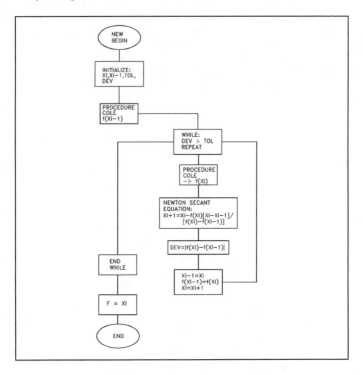

Figure 3. Process flowchart for Newton-secant method.

NEWTON MEETS DARCY AND COLEBROOK   423

The code for the complete calculator program is shown in the Appendix. The following procedures are listed:

DWE: Main procedure which calls other procedures for friction factor and solves Darcy's equation.
REY: Computation of the Reynolds number.
FF: Churchill's equation for friction factor.
COLE: Colebrook's implicit equation for friction factor.
NEW: Newton-secant method.
FF2: Swamee and Jain's equation for friction factor.
LAM: Laminar flow equation for friction factor.

Example Problem

An 18-inch ductile iron transmission pipeline delivers water from a tank to a subdivision. The tank has an overflow elevation of 4472 feet MSL and is located 4.24 miles north east of the subdivision. The high end of the subdivision has an elevation of 4330 feet MSL. Combined minor losses of bends and butterfly valves amount to a factor of 5.74 and approximate water temperature is 70 deg. F. Compute the maximum amount of water that can be delivered to the subdivision if a minimum pressure of 50 PSI is to be maintained at the high end. Use the Darcy Weisbach equation for friction losses. Try Colebrook's, Churchill's and Swamee and Jain's equations for the friction factor (F).

Data: L = 22,387 ft.
         D = 18.76 in. (from tables)
         M = 5.74
         $v = 1.05 \times 10^{-5}$ ft^2/s (from tables)
         $e = 4.8 \times 10^{-3}$ in. (from tables)

         H_l = 4472 - (4330+50(2.31)) = 26.5 ft.

Solution:

Friction Equation	Flowrate (GPM)	Reynolds Number- R_y	Friction Factor-F
Colebrook	2307.3	3.987×10^5	0.016215
Churchill	2300.8	3.9761×10^5	0.016310
Swamee & Jain	2300.8	3.9762×10^5	0.016309

Summary

Computation of friction losses is an essential component of pipeline design. The most commonly used equation in a variety of fluid applications is the Darcy Weisbach equation which in most situations is implicit in nature. The Darcy Weisbach equation requires the computation of the friction factor utilizing the implicit Colebrook equation or other explicit equations such as Churchill's or Swamee and Jain's equations.

A methodology has been introduced in this paper to solve for friction loss, flow or pipe diameter in Darcy's equation with the choice of either implicit or explicit friction factor equations. The methodology uses a nested Newton-secant method and is implemented in a programmable calculator.

Appendix

References

1. Mott, Robert L., (1994). "Applied Fluid Mechanics", Fourth Edition, Merrill an imprint of Macmillan Publishing Company.

2. Streeter, Victor L. and E. Benjamin Wylie, (1985). "Fluid Mechanics", Eighth Edition, McGraw Hill Book Company.

Calculator Program

```
DWE
« REY
  IF 'RY≤2.0000E3'
  THEN 0.0000E0 'OPT'
STO
  END
  CASE 'OPT==0'
    THEN LAM
    END 'OPT==1'
    THEN NEW
    END 'OPT==2'
    THEN FF
    END 'OPT==3'
    THEN FF2
  END
  END 'Q/QCON/(π*(D/DCON
)^2/4)' →NUM 'V' STO '(F
*(L/(D/DCON))+M)*(V^2/(2
*G))-HF' →NUM
»

REY
« '4*(Q/QCON)/(π*(D/DCON
)*VIS)' →NUM 'RY' STO
»

FF
« '(3.7530E4/RY)^16'
→NUM 'B' STO '(2.4570E0*
LN(((7/RY)^9.0000E-1+
2.7000E-1*(E/D))^-1))^16
' →NUM 'A' STO '8*((8/RY
)^12+(A+B)^-1.5000E0)^(1
/12)' →NUM 'F' STO
»

COLE
« → F '1.1400E0-2*LOG(E/
D+9.3500E0/(RY*√F))-1/√F
' →NUM
»

NEW
« '64/RY' →NUM 'XN' STO
2.5000E-2 'X' STO
1.0000E-4 'TOL' STO
1.0000E0 'DEV' STO 'XN'
RCL COLE 'FXN' STO
  WHILE 'DEV>TOL'
  REPEAT 'X' RCL COLE
'FX' STO 'X-FX*((X-XN)/(
FX-FXN))' →NUM 'XP' STO
'FX-FXN' →NUM ABS 'DEV'
STO 'X' RCL 'XN' STO
'FX' RCL 'FXN' STO 'XP'
RCL 'X' STO
  END 'X' RCL 'F' STO
»

FF2
« '2.5000E-1/LOG(1/(
3.7000E0*D/E)+5.7400E0/
RY^9.0000E-1)^2' →NUM
'F' STO
»

LAM
« '64/RY' →NUM 'F' STO
»
```

Hydraulic Transient Analysis in a Large Water Transmission System

by

Awni Qaqish, P.E.
David E. Guastella, P.E.
James H. Dillingham, Ph.D.
Donald V. Chase, Ph.D., P.E.

Abstract

Hydraulic transients have caused sensational failures in pipelines due to the transient-induced pressures exceeding the capacity of the pipe. Hydraulic transients occur when there are sudden changes in flow and usually take place in a matter of seconds. They can be caused by the starting and stopping of pumps or the opening and closing of valves that result from operational errors or equipment failure. Transients that result from the stoppage of one or more pumps (pump trip) because of power failure often cause the most severe hydraulic transient problems. This causes pump discharge pressure to drop rapidly and may lead to pressures below atmospheric and even to vapor cavity formation immediately downstream of the pumps and at intermediate high points in the pipeline. But more likely, the returning pressure wave causes the separated water columns to "slam" together causing a sharp rise in the pressure that may exceed pipe design capacity. This is a serious problem and must be carefully examined. Due to the importance of its largest transmission mains, the Detroit Water and Sewerage Department (DWSD) is currently involved in a project to identify facility improvements that would control transient pressures within prescribed limits.

Mr. Awni Qaqish, Detroit Water and Sewerage Department, Detroit, MI; Mr. David E. Guastella, Tucker, Young, Jackson, Tull, Inc., Detroit, MI; Dr. James Dillingham, Metcalf & Eddy, Wakefield, MA; Dr. Donald Chase, University of Dayton, Dayton, OH.

Introduction

DWSD owns and operates one of the largest water supply systems in the nation. The system consists of five water treatment plants (1.6 BGD combined capacity) and twenty booster pumping stations with a total storage capacity of 363 MG. The Lake Huron Water Treatment Plant (presently rated at 240 MGD) is connected to the transmission system by a single, 120-inch main. This main connects to two other mains at the Imlay Station. One of these mains is a 72-inch line that feeds one of DWSD's major water users (40 mgd average day demand). The other main (96-inch) supplies water to the Detroit metropolitan area.

These mains have experienced hydraulic transients in the past due to power failures to the treatment plant high-lift pumps. To date these transients do not appear to have damaged the transmission mains. However, DWSD is planning to increase the flowrate from the treatment plant. Since an increase in flow may cause larger transients to occur, a project was initiated by DWSD to model the transient effects on the system due to the projected flowrates. This paper will describe the results of the computer simulation used to analyze the hydraulic transient effects of water hammer/column separation in the 120-inch transmission main during the operation of the system at the maximum anticipated flow rate. Recommended facilities and operational procedures to control the hydraulic transients in the 120-inch main will also be identified.

DWSD Water Transmission System

The City of Detroit is located in the southeastern part of Michigan with Canada as its border to the south and east. The area served by DWSD is shown in Figure 1. As shown in this figure, the DWSD system provides water to a very large area which includes a total of 122 communities. With a population of approximately four million people served, DWSD provides water to almost half Michigan's population. The current average day demand is 640 MGD and has peaked at over 1.5 BGD in the summertime. DWSD operates the entire system from one central location in downtown Detroit called the System Control Center (SCC).

Due to the size of the system and the topography, the transmission system is divided into three pressure districts - low, intermediate and high. The different pressure districts are interconnected to facilitate the continuity of service and to meet the flow and pressure requirements occurring during peak load periods or emergencies.

The area being studied for hydraulic transients is the transmission system north of the North Service Center (NSC). Existing facilities and the pipe centerline profiles are shown in Figure 2. The system is comprised of 26 miles of 120-inch main from Lake Huron Water Treatment Plant (WTP) to Imlay Station, 24 miles of 72-inch main from Imlay Station to Flint, and 32 miles of 96-inch main from Imlay Station to NSC. The City of Flint and its surrounding environs is one of DWSD's largest customers.

Operation of the Lake Huron WTP

The Lake Huron WTP began supplying water to DWSD customers in 1974. At that time, it was envisioned that base flow rates for the plant would be 210 MGD or 345 MGD, dependent on how the transmission system was to be operated within the Detroit metropolitan area. At a flowrate of 210 MGD, Imlay Station would repump Lake Huron water to Flint and the Lake Huron WTP would pump directly to NSC. At the base rate of 345 MGD, all the water in the 120-inch main would be diverted into the Imlay Station and repumped from Imlay to Flint and NSC. Unfortunately, these flowrates have not materialized as originally anticipated.

Currently, the flowrate from the Lake Huron WTP averages 125 MGD. The recent largest maximum hour flowrate for the plant was 230 MGD. These flowrates are much lower than originally envisioned. Because of the low flows, the head loss in the 120-inch line is small enough that no pumping at Imlay is necessary; therefore, the Imlay Station is completely bypassed.

However, water demands in the northern and western portion of the DWSD service area have been increasing rapidly due to a population shift to the area. The construction of a large transmission main south of the NSC will be completed by 1996. Additionally, the Lake Huron WTP is being expanded to a capacity of 400 MGD. Completion of this main and the plant expansion will allow more water to be supplied by the Lake Huron plant to the rapidly growing northern and western areas. These recent events will require the Lake Huron plant to supply water at rates close to those originally anticipated in the 1970's.

Hydraulic Transient Concerns

Although the project identified earlier will examine all three transmission mains shown in Figure 2 for hydraulic transients, the focus of this paper is on the 120-inch main. The 120-inch transmission main is a prestressed concrete embedded steel cylinder pipe with a rubber and steel joint configuration. The main has been designed for maximum operating pressures of 225 psi near the Lake Huron WTP and 175 psi upstream of the Imlay Station. Between these two sections the pipe is designed for a maximum pressure of 200 psi.

The magnitude of the pressure due to hydraulic transients is directly proportional to the velocity in the pipe. Therefore, to prevent hydraulic transients in the 120-inch main, the water plant operators provide very good coordination with SCC when high-lift pumps are brought in or taken out of service. This procedure typically takes from five to thirty minutes as the operators closely monitor the pressure in the 120-inch main. Though this procedure has been successful in preventing hydraulic transients during the starting and stopping of the pumps, transients have occurred due to the loss of power to the pumps. In fact, three such events have occurred in the past year alone.

The hydraulic transients due to the power failures have not appeared to have caused any damage in the 120-inch main. This is probably attributed to relatively low velocities in the main and the design pressure of the pipe sections. Based on the average day and maximum hour flowrates, the velocity in the transmission main varies from 2.5 ft/s to 4.5 ft/s, respectively. Additionally, the pipe sections are designed to withstand transient pressure surges of 40% greater than the design operating pressures described earlier.

Based on hydraulic analyses performed on the 120-inch transmission main, a maximum flowrate of 400 MGD from the Lake Huron WTP is possible. At this flowrate, the transmission main would experience a velocity of 8.0 ft/s. This velocity is almost twice that of the current maximum velocity. Additionally, there are many high and low points in the transmission main due to the surrounding topography. This fact coupled with the higher velocity may yield column separation at the high points of the transmission main. To evaluate the effects of hydraulic transients at the maximum flowrate it was decided to analyze the 120-inch main using a computer model.

Computer Model Description

In order to analyze the transient behavior of flow through the 120-inch transmission main, a mathematical model of the main was developed and used with an existing computer simulation model. The water hammer computer program, WHAM, was developed by Metcalf & Eddy using well-established methods and procedures (Wylie and Streeter, 1993). The program uses the method of characteristics to solve the characteristic equations which are developed using momentum and continuity concepts. The characteristic equations are solved within a finite difference framework for pressure head and flow throughout a hydraulic network as a function of space and time. The program includes complex pump rotational relationships. This allows time-varying pump characteristics resulting from transient-induced changes in pump speed to be predicted. WHAM can handle column separation and subsequent column collapse or "slam." This program takes account of the actual profile of each pipe in the network and friction losses are considered.

Each pipe in the network is divided into pipe sections of equal length and the characteristic equations are applied to adjacent pipe sections to compute the flow and head at each internal point. The length of the pipe increments is dependent on the computational time step and there is usually a remaining fractional section. The size of the time step, and therefore the size of the pipe section, will have an effect on the precision of the calculations.

WHAM determines the appropriate length of pipe section and the corresponding time increment to be used in the calculations. Smaller values of pipe length and time can also be specified by the user. Computational results are verified by using successively smaller time increments to reduce the effect of the short section.

Four quadrant pump characteristic curves are stored in WHAM for each of five values of specific speed. Interpolation between curves is used to estimate pump characteristics. Actual pump curves can be used, if available, when more accurate values are needed.

Water column separation (cavity formation) is taken into account fully by computing the cavity volume in accordance with the continuity principal during each time step. On the rejoining of the water columns, the resulting pressure surges are computed throughout the network. In this way, WHAM helps to evaluate maximum and minimum pressures in water systems with potential for column separation.

The assumption of cavities concentrated at intermediate high points along the pipeline where pockets can coalesce is reasonable. However, the number of pipe sections have an affect on the results. Because of this, results obtained by varying the number of pipe sections modeled were compared to each other.

The conditions analyzed are for maximum anticipated flow of 400 mgd with a pumping head of nearly 200 psi from Lake Huron WTP to the Imlay Station. A detailed piping profile for the 120-inch main containing nearly 150 pipe segments is shown in Fig. 3a. This actual profile has several intermediate high and low points along its 26 mile length. Seven 60 mgd pumps pumping 400 mgd at 450 ft of head were specified for the simulation to represent maximum anticipated pumping conditions for future flow. The transient condition specified was simultaneous power failure to all pumps.

For comparison purposes, pump check valves (PCV) that close at the time of flow reversal were assumed for some of the simulations. Additionally, the existing 42-inch timed closure valves (TCV) at the Lake Huron WTP that close

430   UNDERGROUND PIPELINE ENGINEERING

in 15 to 30 seconds were also simulated. Four quadrant pump curves were selected automatically by the WHAM program based on rated pump flow, head and speed. Values for pump inertia were estimated by the formula based on horsepower and pump speed. Subsequent sensitivity analyses indicated that these estimates were sufficiently accurate. Existing eight inch air vacuum valves (AV) were also included in some of the simulations.

Results and Recommendations

Extensive sensitivity analyzes were performed to develop confidence in the parameters and the modeling. Many WHAM simulations were run varying parameter values for the time increment used in the calculations, wave speed, pump inertia and other characteristics, valve characteristics, and alternative check valve locations. Also, several simulations were run using least, intermediate, and most detailed pipe profiles with favorable comparisons between the intermediate and most detailed profiles. The results of the WHAM simulations using the most detailed pipe profile are shown on Figures 3 and 4.

Fig. 3a shows the pipe profile and the HGL for approximately 400 mgd flow from the Lake Huron WTP to Imlay. Prior to the low point at Sta. 317 (Black River), there are intermediate high points at stations 47, 151, 198, and 241. The low point at Sta. 317 is the Black River. Sta. 718 is Mill Creek. There are three more major intermediate high points (Sta. 329, 434, 611) before reaching the highest point (Sta. 932).

Fig. 3b shows the pump discharge pressure and the pump flow (in % of total flow) for power failure to all pumps at t=0. The pump flow drops to zero in about 40 seconds, which is the time when flow would normally reverse if there were no valves to prevent backflow. The transient pressure at the pumps show sharp rises after power failure with the largest more than 300 psi at about 350 seconds. The results of the computer simulation indicate that the pressure spikes are caused by the returning pressure transients from the column separation and subsequent slam at intermediate high points. The columns do not rejoin at the highest point (Sta. 932).

The largest pressure is caused by the slam at Sta. 241. The cavity volume grows to about 17,500 CF (about 220 ft of pipe) and then collapses to zero at about 300 seconds causing the pressure to rise sharply. Cavity growth and slam also occur at Sta. 198 and Sta. 151, but are less severe.

Fig. 3c shows the pump discharge pressure for the same conditions except that check valves to prevent backflow are included in the main (MCV) at Sta. 329. The MCV prevent the driving force downstream (Sta. 329 to 923) from causing slam at Sta. 241. Several other simulations (not included herein) were run for alternative MCV locations along the pipeline to determine the best location. Sta. 329 is the apparent optimum location from a hydraulic basis, but this is subject to further analysis and verification. The search for an optimum location did not include consideration of accessibility of the site and other factors. The cavity at Sta. 241 grows but does not collapse. The rise in pressures at the pumps are caused by slam at the other intermediate high points such as Sta. 198. The maximum transient pressure for this simulation is below the 200 psi operating pressure.

Fig. 4a shows the simulation for actual pump timed closure cone valves (PTV). In this case, the 42-inch PTV closes in 15 seconds, which is considerably less than the time of normal flow reversal that would occur if there were no pump check valves. The results indicate that the PTV closing before the normal time of flow reversal (such as the results from 3b) aggravates the transient problem; the maximum transient pressure is about 230 psi. However, the PTVs may need to close quickly for cases when only some of the pumps trip in order to prevent excessive backflow through the pumps.

Fig. 4b shows the same simulation with actual air valves (AV) included. Each pump has a quick open / quick close eight inch diameter AV installed in the discharge line. The AV substitutes an air cavity for the vapor cavity (shown in Fig. 4a from 0 to 45 seconds). This delays the slam because the air cavity grows larger with the introduction of air and the air cushion delays cavity closure. However, the results indicate that the magnitude of the slam resulting from air cavity (250 psi maximum) may be as large or larger than slam from vapor cavity.

Fig. 4c conditions are the same as Fig. 4b except that the PTV closes in 30 seconds and the maximum pressure is reduced to about 220 psi. However, proper operations prefer maximum transient pressures below 200 psi.

Further analysis and verification are currently underway to determine optimal timing and MCV closing characteristics. It is desirable to install an MCV structure that closes to prevent backflow at the instant of flow reversal. However, a reverse velocity will likely be required to close the MCV, and this might create transient problems downstream as the valve closes. Because there are several apparent advantages for using the MCV structure, oppose to other surge relief devices, further investigation is warranted. These advantages include: no water quality problems introduced, less likely to stick or freeze than other devices, closing at the optimal time (flow reversal), relatively minimal maintenance, exercised each time the system is shut down, and electrical service is not required. These are important considerations, especially since the 120-inch main runs through remote areas where electrical service is not readily available and access is difficult.

Additionally, in the event of an upstream main failure for any reason, the MCV would prevent downstream water entering the failed area. The volume of water is significant. The 120-inch main contains about 3.1 MG/mile, about 14 MG from the Lake Huron WTP to Sta. 241, and about 20 MG from Sta. 500 to Sta. 932 (the downstream portion of the pipe above the elevation at Sta. 241).

Further analysis is also warranted to investigate potential remedies at the Lake Huron WTP to reduce the maximum transient pressure below 200 psi. Results shown on Fig. 3c indicate that relatively inexpensive remedies are likely. One alternative is the quick opening, slow closing air valve that expels water after cavity closure to control column slam. Another potential remedy is the anticipating surge relief valve (SRV) that opens quickly on pump trip to allow water downstream of the pumps to prevent column separation. This valve remains open to receive the return surge, and then closes slowly. These devices are more practical at the Lake Huron WTP than at remote locations along the pipeline because they can be more easily exercised and maintained.

Conclusions

1. Pump trip of all seven pumps causes column separation at several intermediate high points along the main with subsequent slam exceeding 300 psi.

2. An in-line main check valve (MCV) structure can be effective in controlling transient pressures upstream of the MCV (Lake Huron WTP side) by eliminating the driving force of the water column downstream of the MCV (Imlay side).

3. Pump timed closure cone valves (PTV) closing before normal time of flow reversal in the main aggravates the column separation problem near the pumps resulting in greater transient pressure.

4. Use of quick closing air valves (AV) at the pump may cause higher transient pressures than no AVs. Substitution of an air cavity for a vapor cavity does not necessarily decrease slam.

5. Pump valves (PTV) closing in 30 seconds rather than 15 seconds alleviates column separation near the pumps because the closing time is closer to the normal flow reversal time (40 seconds).

HYDRAULIC TRANSIENT ANALYSIS

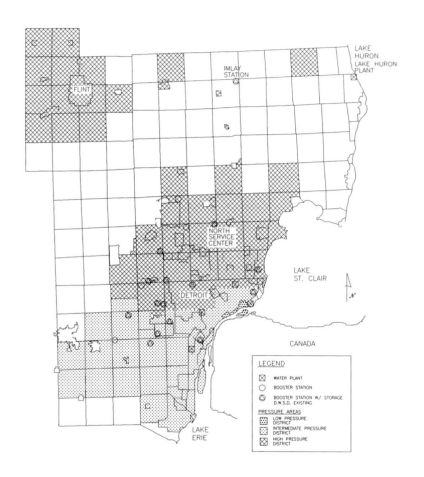

FIG 1. AREA SERVED BY DWSD WATER TRANSMISSION SYSTEM

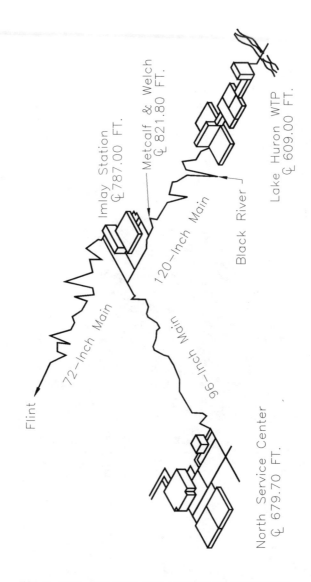

FIG 2. AREA ANALYZED FOR HYDRAULIC TRANSIENTS

3a. PIPE PROFILE AND HGL AT 400 MGD

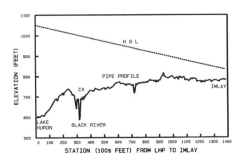

3b. NO CHECK VALVES (MCV) IN WATER MAIN

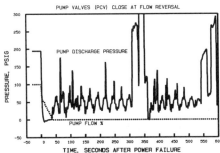

3c. CHECK VALVES (MCV) AT STATION 329

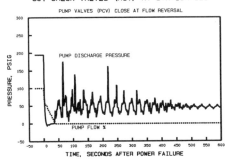

FIG 3. 120" MAIN, 7 PUMPS WITH SWING CHECK VALVE (PCV)

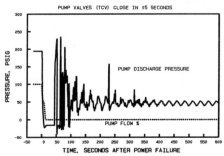

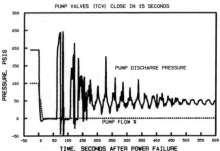

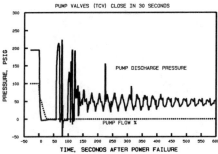

FIG 4. 120" MAIN, 7 PUMPS WITH TIMED CLOSURE VALVE (TCV)

SLEEVE VALVES FOR A WASTEWATER APPLICATION

By

David Timmermann, Black & Veatch
Arne Sandvik, Black & Veatch
Hans Torabi, City of San Diego

Abstract

The Point Loma Wastewater Treatment Plant, in San Diego, California resides on a bluff approximately 30 meters (100 feet) above the Pacific Ocean. The view from the perch is spectacular. However, it is a challenge to dispatch the treated wastewater from the plant to the ocean mixing zone approximately 7 kilometers (4 miles) offshore. The plant has operated with a single effluent facility consisting of a vortex structure for more than thirty years. Recognizing the need to provide a second effluent facility to allow rehabilitation of the existing facilities, in 1989 the City of San Diego, California embarked on a program to provide an alternative effluent outfall system. After evaluation of several alternatives including drop structures and helix ramps, it was decided to pursue the construction of the South Effluent Outfall Connection (SEOC) project. The new effluent outfall system is designed to carry 18.92 cubic meters per second (432 million gallons per day) of primary treated wastewater and dissipate approximately 30.5 meters (100 feet) of head prior to discharge into a deep ocean outfall. The SEOC utilizes three 137 centimeter (54-inch) angle pattern sleeve valves, a 214 centimeter (84-inch) in-line pattern sleeve valve, and one 214 centimeter (54-inch) ball valve. All of these valves are among the largest ever constructed, especially for an effluent application.

This application of sleeve valves in a wastewater environment is the first of its kind. The sleeve valves will be protected from clogging through the use of 3 millimeter mesh screens upstream of the valves. The new system significantly reduces the amount of air currently entering the outfall pipeline through an existing vortex structure. Surge protection will be provided by the existing vortex structure and through control of the rate of operation of the throttling valves.

While all of the facilities were constructed on-shore, adjacent to the shoreline, many of the facilities were constructed below sea level and shoring and dewatering facilities were required. Piping materials consisted of 137 centimeter (54-inch) and 214 centimeter (108-inch) cathodically protected steel pipe and 214 centimeter (108-inch) American Water Works Association (AWWA) C302 Low Head Concrete Pressure Pipe.

Background and Existing Facilities

The Point Loma Wastewater Treatment Plant, constructed in 1961, is located on a bluff overlooking the Pacific Ocean. Figure 1 shows the general layout of the Plant. The treatment facilities are approximately 30.5 meters (100-feet) above sea level with the plant effluent discharged through a vortex structure into a 214 centimeter (108-inch) diameter ocean outfall. A hydro-electric plant consisting of a turbine and needle valve was constructed in 1983. The needle valve is designed to serve in parallel with the existing hydro-electric plant. With the need to reduce air entrapment from the vortex, the needle valve has served as the primary effluent flow control element for the past two years. The needle valve is experiencing excessive cavitation and vibration due to low downstream pressures.

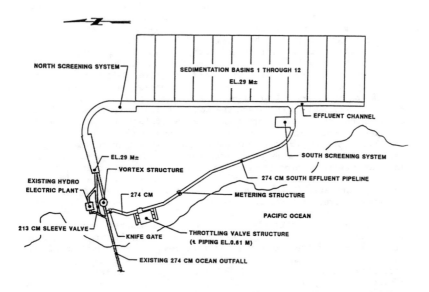

PROJECT OVERVIEW
FIGURE 1

While the existing vortex structure has proven to be effective at dissipating head before the effluent enters the outfall, it has also caused air to become entrapped and enter the outfall.

New Facilities

While the use of mechanical systems is typically not desirable if there are non-mechanical solutions available, there are several factors which made construction of the SEOC attractive from both an economic and engineering standpoint. First, the City of San Diego was concerned with the amount of air being carried into the existing outfall. This accumulation of air was allowing the formation of hydrogen sulfide corrosion on the crown of the outfall pipe. Since the plant has only one outfall pipe, it was imperative that the amount of air entering the pipe be limited to reduce corrosion. The second factor was the City's use of 3 millimeter mesh screens on the plant effluent. These screens allowed a degree of protection from plugging for the small orifices of the sleeve valves. The third factor, and probably the most important factor, was the presence of the vortex structure. While the City desired not to use the vortex structure as the primary means of energy dissipation, it provided an ideal back-up system for the SEOC facilities as well as an excellent system for surge protection. If the control system for the SEOC facilities fails for any reason and water begins to rise in the sedimentation basin effluent channel, an isolation gate with an emergency battery backup power supply, will automatically open, allowing water to be discharged through the vortex structure.

The control philosophy for the SEOC facilities is to maintain an operating water level in the sedimentation basin effluent channel within an 18-inch operating band. The valves are opened or closed, according to their priority and the water level in the effluent channel. The operating water level is monitored through the use of a bubbler system which is tied into the valve control system.

Throttling Valves

The layout of the throttling valve structure for the three 137 centimeter (54-inch) sleeve valves and 137 centimeter (54-inch) ball valve is shown on Figure 2.

Sleeve valves were chosen as the primary energy dissipation valves due to their superior head reduction characteristics, superior control characteristics, and they work well with low backpressure. In order to reduce the possibility of air entering the outfall it was necessary to locate the valves below sea level. This eliminated air entrapment and increased the back pressure on the valves. The valves will be subjected to a cavitation index ranging from 0.30 or greater. Butterfly valves, a commonly used throttling valve in less severe applications, can safely operate at conditions with a cavitation index

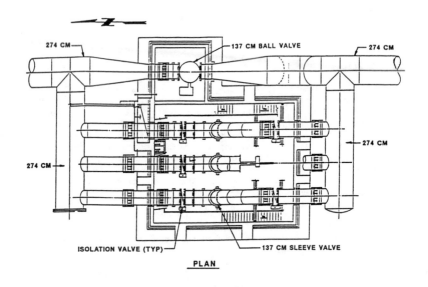

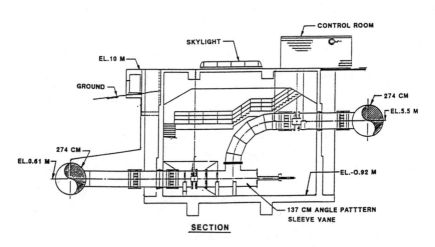

THROTTLING VALVE STRUCTURE
FIGURE 2

of around 2.5 or greater. In-line pattern sleeve valves were initially considered for use since this style of valve is not as subject to mechanical wear, but excavation for construction of the Throttling Valve Structure would have been significantly increased. Space limitations and increased headless, due to more piping bends, also restricted their use.

While 24.4 meters (80 feet) of head is required to be dissipated at 3.33 cubic meters per second (76 mgd), the maximum headloss which can be tolerated at 18.9 cubic meters per second (432 mgd) is 1.2 meters (4 feet). Since headloss needed to be limited at maximum flowrate, a ball valve was selected for operation in conjunction with the sleeve valves at flows exceeding 14.45 cubic meters per second (330 mad). The ball valve is brought on-line only after the capacity of the sleeve valves has been exceeded. Once the ball valve is opened, there is a large change in flow from the sleeve valve circuit to the ball valve circuit. The ball valve is opened in 10 degree increments. A sleeve valve is then modulated to fine tune the hydraulics to maintain the set point in the operating range. When the three sleeve valves and the ball valve are all fully open, the ball valve circuit carries approximately 76 percent of the total flow. Figure 3 shows the available head versus flow capabilities of the 137 centimeter (54-inch) sleeve valves and ball valve.

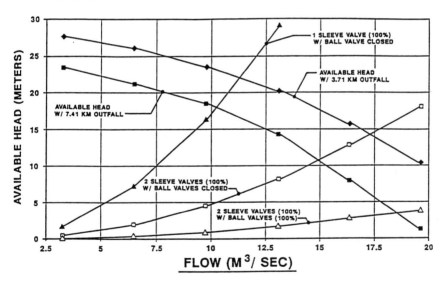

HYDRAULIC CHARACTERISTICS
FIGURE 3

Pipe Material

The piping materials used for this project consisted of AWWA C302 Low Head Concrete Pressure Pipe and concrete lined and coated steel pipe. Concrete pipe was used from the connection to the South Screening Structure to immediately upstream of the Throttling Valve Structure influent header. Pipe lengths are primarily 4.88 meter (16-feet). A former sea cove is located through the middle section of the concrete piping. This area was constructed with uncompacted, undocumented fill material in 1961 when the original plant was constructed and therefore has a high potential for differential settlement. To compensate for this, the concrete pipe through this section is constructed with 1.22 meter (4-foot) pipe lengths and the existing fill material was overexcavated 0.61 meters (2-feet) below the bottom of the pipe and replaced with compacted granular embedment material. The typical cover depth for the concrete pipe is one meter (3-feet).

Piping material was changed to concrete lined and coated steel pipe from the connection to the Throttling Valve Structure influent header to the connection with the existing ocean outfall due to the significant number of fittings and to simplify connection for a future outfall. A significant portion of the steel pipeline is constructed below sea level, therefore, an impressed current cathodic protection system is utilized to protect the pipeline. Flange isolation kits are used at the connection between the steel and concrete pipe, at the connection to the existing outfall, and where the piping enters and exits the Throttling Valve Structure. The steel pipe was further protected from hydrogen sulfide corrosion by increasing the cement lining to five centimeters (2-inches) and coating the lining with an epoxy coating. Cover depth varies from one meter to nine meters (3-feet to 30-feet).

Flow Metering

A magnetic probe insert flow meter was installed on the effluent piping. Because the meter may be used only in the future to pace dechlorination chemical feed, it was determined that only about 5 percent accuracy would be needed. Venturi or magnetic flow meters would be very accurate but also very expensive because of the large pipe size. The meter vault, however, was designed to accommodate alternative meter types if a change was desired in the future. The design also included a section of steel pipe without concrete lining to allow for alternative meter types.

Connection to the Existing Ocean Outfall

One of the biggest challenges of this project was the connection of the SEOC facilities to the existing ocean outfall. The Plant currently treats an average 8.32 cubic meters per second (190 million gallons per day). The plant operates 24 hours per day but it is possible to shut the plant down for approximately 3 hours between the hours 2:00 a.m. and 5:00 a.m. on certain days of the week. The hydroelectric circuit can be used for short-term flow diversion around the vortex structure, but it does not have adequate capacity to handle maximum day flow conditions, therefore, the vortex structure and connection point of the SEOC could not be isolated by a "permanent" bulkhead. Operating pressures in the outfall prohibited insertion of a balloon-type plug in the existing outfall.

With these limitations in mind, the connection was made by fabricating a two-piece 244 centimeter (96-inch) diameter tee and welding it to the outside of the existing steel pipe outfall pipe. A vault was then constructed around the pipe, up to ground level at elevation 6.1 meters (El. 20 feet). A 244 centimeter (96-inch) flanged knife gate valve was then attached to the two piece saddles. With this construction complete, it was then possible to make the wet tap during several early morning shut-down periods. The wet tap was accomplished by filling the vault with fresh water to improve visibility, opening the knife gate valve, and then having a diver cut out the 244 centimeter (96-inch) coupon from the existing outfall. Upon completion of each shutdown period, the knife gate valve was closed and flow restored through the outfall. Two plant shutdowns were required to remove the coupon. Damaged areas were repaired by epoxy coating.

Construction Considerations

The construction of the SEOC facilities involved some unusual construction considerations. Due to the narrow access of the Lower Hydro Access Roadway, the construction of the SEOC facilities had to follow a very sequential order. Work had to begin at the Ocean Outfall Connection Vault and proceed up the Lower Hydro Access Roadway. Work had to be nearly complete on one facility before work could begin on the next upstream facility.

Construction of the Ocean Outfall Connection Vault and the Throttling Valve Structure were both helped and hindered by the in-situ soil conditions. Both of these structures are constructed partially in a sandstone material and undocumented fill comprised primarily of large riprap. The sandstone formation in the area of these structures, was primarily located at below El. 0.00, but along the eastern side of these structures, it was extended to much higher elevations. This proved to be invaluable since it allowed almost 12 meter (40 foot) deep unshored, near vertical excavations. It also significantly reduced the costs of dewatering. Flow of seawater through the undocumented fill was retarded by shotcreting the face of the excavation.

Conclusion

Construction of this project allows the City of San Diego the opportunity to perform needed maintenance on the existing effluent system. The SEOC will reduce the amount of air entering the existing outfall and will increase the effluent disposal system reliability. The use of sleeve valves in this application proved to be an innovative means of meeting the unique constraints at this plant.

Trench Widths for Buried Pipes

by Reynold King Watkins[1]

Abstract

What is the best trench width for a buried pipe? The trench only needs to be wide enough for alignment of the pipe and for placement of a soil envelope that fills the space between pipe and trench walls. Ideally, the pipe would be bored into place. A flexible pipe is stable if soil embedment is uniform all around the pipe, and if the pipe remains nearly circular. Even if the trench is narrow and is excavated in poor soil, soil support is usually adequate. Determination of trench width by back-calculating Spangler's ring deflection formula can lead to ambiguous conclusions. If ring deflection is excessive, or if the pipe has inadequate soil cover when surface loads pass over, the soil at the sides of the pipe can slip and allow ring inversion. Soil slip analyses make it possible to quantify these limits of performance. Soil slip analyses show that a wide trench is seldom justified.

Introduction

In general, the trench should be narrow -- just wide enough to align the pipe and to place soil around the pipe in full contact with the pipe. If there is any possibility of soil liquefaction, the embedment should be denser than critical void ratio density. Including a safety factor, 85 percent standard density is often considered to be minimum. Soil compaction may or may not be required depending upon the quality of the embedment soil. Good embedment, such as pit-run gravel or pea

---

[1]Professor Emeritus, Department of Civil and Environmental Engineering, Utah State University, Logan, UT 84322-4110 Telephone: (801) 797-2864 Fax: (801) 797-1185

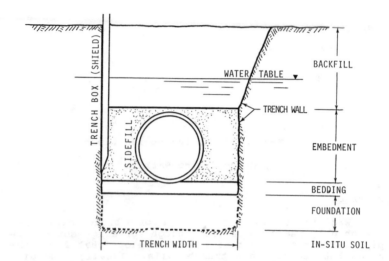

Terminology for trench cross section.

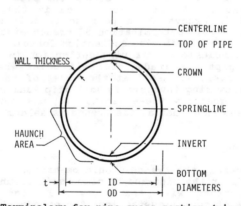

Terminology for pipe cross section (ring).

Figure 1. Typical cross sections of a buried pipe. The foundation is specified to be select soil wherever it is found necessary to overexcavate and replace unacceptable native (in situ) soil.

gravel, falls into place at densities greater than 85 percent. Liquefaction can be caused by earth tremors if the soil is saturated and is looser than critical. See Figure 1 for terminology and Appendix II for notation. Loss of embedment should be prevented. Loss of embedment (piping) is usually the wash-out of soil particles by groundwater flow.

In the 1920's, Marston proposed a design theory for a rigid pipe buried in a trench. (Marston 1930) He assumed that the load on the pipe was the entire weight of backfill in the trench, reduced by frictional resistance of the trench walls. The narrower the trench, the lighter the load on the pipe. The pipe had to be strong enough to support the backfill. Marston neglected the strength contribution of the sidefill -- both horizontal support of the pipe, and vertical support of backfill. Trenches are kept narrow for rigid pipes.

In the 1930's, when flexible steel pipes began to impact the market, M.G.Spangler observed that a flexible ring depends upon support from sidefill soil. (Spangler 1941) (Watkins 1958) That observation, although correct, led to the inference that, if the trench is excavated in poor soil, there is a possibility that trench walls can not provide adequate horizontal support for the pipe. The remedy seemed to be wider embedment -- a wider trench -- especially in poor native soil. However, the need for a wide trench was seldom verified by experience or by basic principles of stability.

Figure 2a shows a flexible pipe in equilibrium when subjected to uniform external pressure P. Shearing stresses between the soil and pipe reduce pressure on the pipe. But shear is undependable and, therefore, is ignored for worst-case analysis. Figure 2b is equivalent to 2a. Horizontal pressure, P_x, is equal to vertical pressure, P_y. If the ring is nearly circular, it is stable. The only requirement of the soil is that sidefill soil strength must be greater than P_x.

Figure 2c shows a pipe that is perfectly flexible, but is deflected out-of-round. It is stable if,

$P_x r_x = P_y r_y$ UNIFORM CIRCUMFERENTIAL WALL THRUST   (1)

If the deflected ring is an ellipse,
$r_y/r_x = (1+d)^3/(1-d)^3$.
When ring stiffness is taken into account, $P_x r_x$ is less than $P_y r_y$. The contribution to the support of load P by

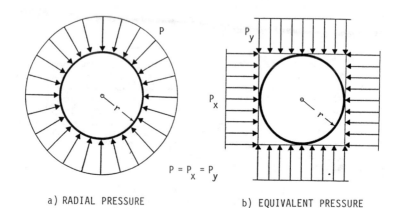

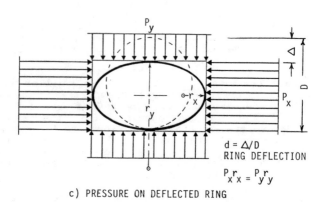

Figure 2. External radial pressure on a perfectly flexible pipe ring at equilibrium of the ring, showing:
a) the uniform pressure for equilibrium of a circle,
b) the equivalent pressure diagrams for a circular ring,
c) the equivalent pressure diagrams for an ellipse.

stiffness of the ring can be included in the analysis, but is often ignored in conservative pipe design.

Theoretically, as long as the ring is nearly circular, the embedment does not need high strength. A flexible pipe would be stable in poor soil. Practically, good sidefill adds a margin of safety. See Figure 3 where the infinitesimal soil cube B is in equilibrium as long as pipe pressure P_x does not exceed sidefill soil strength, σ_x. For stability,

$$P_x < \sigma_x = K\sigma_y \qquad (2)$$

where $K = (1+\sin\phi)/(1-\sin\phi)$, and ϕ is soil friction angle at soil slip. If sidefill soil is granular and denser than critical, its soil friction angle is no less than 30°, for which $K = 3$. From Equation 2, because $P_x = P$, the safety factor against soil slip is no less than three.

Suppose the trench walls are poor soil; with blow count less than four. What should be the trench width? See Figure 4. If the friction angle of sidefill is $\phi = 35°$, then the soil shear plane is at angle $(45°-\phi/2) = 27.5°$ and P_x is transferred to the trench wall by a soil wedge with 1:2 slopes as shown. If ring deflection is less than five percent and the width of sidefill is half the pipe diameter, the pressure on the trench wall is about P/2 as shown, and can be supported by trench walls with a blow count less than four. The trench width in poor soil does not need to be more than twice the diameter of the flexible pipe. The margin of safety is increased by: stiffness of the ring, shearing resistance of soil on pipe, and arching action of the soil. Both ring stiffness and ring deflection can be included in the analysis of P_x, if greater accuracy is required.

<u>Narrow Trench</u>

In general, it is prudent to keep the trench narrow -- just wide enough to align the pipe and to assure good placement of the embedment. As long as the pipe is fairly circular, trench widths need not be more than twice the pipe diameter -- for all pipes (rigid and flexible) -- even in poor soils. If ring deflection of flexible pipes is no more than five percent, the effect of ring deflection can be neglected. The pipe is designed by wall crushing according to the equation:

$$\sigma = PD/2t = Pr/t \qquad \text{RING COMPRESSION STRESS} \qquad (3)$$

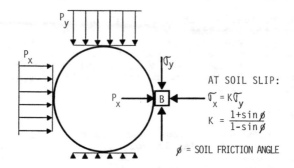

Figure 3. Infinitesimal soil cube at spring line, B, showing conditions for soil slip when $P_x = K\sigma_y$.

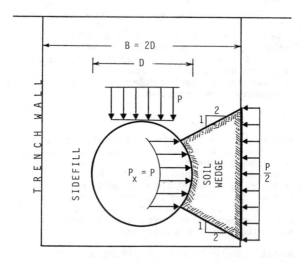

Figure 4. Soil wedge at incipient slip of sidefill soil, showing how pressure P is transferred to the trench wall where it is distributed and reduced by half if the trench width is 2D.

where σ is the compressive stress in the pipe wall. P is vertical pressure on the pipe; i.e., $P = P_l + P_d$ where P_l is pressure due to a surface live load and P_d is the dead load pressure due to backfill. On a rigid pipe P_d is the Marston load. (Marston 1930) On a flexible pipe, P_d is more nearly the prism load, γH. (Spangler 1941 and 1973, page 679) Consequently, the dead load on the flexible pipe is about two-thirds as great as the dead load on the rigid pipe because the flexible pipe deflects enough to relieve itself of soil pressure concentration. A flexible pipe is less of a hard spot in the soil.

The height of soil cover, H, is not a pertiment variable in the analysis of trench width. As soil load is increased, the pressure on the pipe increases; but the strength of the sidefill soil increases in direct proportion. See Equation 2.

A good rule of thumb for width of sidefill for buried flexible pipes is, "In poor soil, specify a minimum width of sidefill of D/2 from the pipe to the walls of the trench, or from the pipe to the windrow slopes of the embedment in an embankment." In good soil, sidefill widths can be less. However, it must be possible to place a good embedment.

Following are four concerns for all buried pipes:
1. Wheel loads over a pipe with minimum soil cover,
2. Water table above the pipe, and/or vacuum in the pipe,
3. Removal of soil particles from the embedment,
4. Voids left by a trench shield, sheet piling, etc.

1. Wheel Loads

Figure 5 shows a wheel load on a pipe with minimum soil cover. The angle of punch-through for cohesionless soil is approximately 1h:2v. The punch-through is approximately a truncated cone for a single tire and a truncated pyramid for a dual wheel. For the dual wheel load shown, vertical pressure on the pipe is $P = P_d+P_l$ where $P_d = \gamma H$, is dead load, and P_l is live load. $P_l = Q/(B+H)(L+H)$ for load, Q, on a rectangular tire print area, LxB [about 0.18x0.56 m (22x7 inches) for HS-20 dual wheel load]. Sidefill soil strength must support the pipe under this additional live load. However, a minimum cover of compacted granular soil is about H = 0.3m (1 ft) for HS-20 dual wheel, and H = 3 ft 0.9m (3 ft) for the single wheel of a scraper. Manufacturers of large steel pipes with mortar linings like a margin of safety of 0.46m (1.5 ft) added to the minimum cover such that minimum cover is 0.75m (2.5 ft) for HS-20 loads and 1.4m (4.5 ft) for

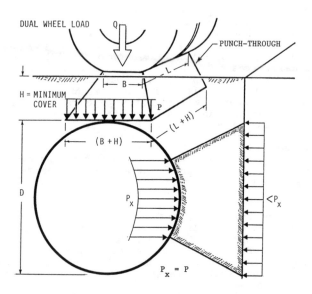

Figure 5. Dual wheel load passing over a pipe buried under minimum soil cover, showing a sidefill wedge at incipient soil slip.

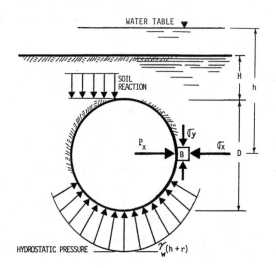

Figure 6. Free-body-diagram of an empty flexible pipe buried below the water table.

scraper wheel loads. With soil cover greater than minimum, arching action of the soil supports wheel loads. No punch-through is felt by the pipe. Trench width could be critical only if the sidefill embedment -- not trench wall -- were poor soil.

2. Water Table

See Figure 6. When the water table is above the top of the pipe, the sidefill soil strength is effective strength, $\sigma_x = K \sigma_y$. The effective vertical soil stress is $\sigma_y = \sigma_y - u$, where u is the pore water pressure of the water table; i.e. $u = \gamma_w h$, where γ_w is the unit weight of water and h is the height of the water table (head) above the spring line of the pipe. If the pipe tends to float, critical P is hydrostatic buoyant pressure on the bottom of the pipe, $P = \gamma_w(h+r)$, rather than the soil pressure on top. This is noteworthy because water pressure is maximum at the bottom of the pipe.

3. Soil Particle Migration

Soil particle migration is generally a function of either: a) groundwater flow that washes trench wall fines into the voids in a coarser embedment; or, b) wheel loads and earth tremors that shove or shake coarser particles from the embedment into the finer soil of the trench wall. If fines migrate from trench wall into embedment, the trench wall may settle, but the pipe is unaffected. If embedment particles migrate into the trench wall, the shift in sidefill support may allow a slight ring deflection. This could occur only if the trench wall soil is loose enough, or plastic enough, that the embedment particles can be shoved into it. The conditions for soil particle migration are unusual. Nevertheless, they must be considered. Remedies for soil particle migration include: a) embedment with enough fines in it to filter out migrating particles in groundwater flow; and, b) trench liners. Geotextile trench liners may be specified -- but only in very severe cases.

4. Trench Box (Voids in the Embedment)

Soil should be in contact with the pipe in order to avoid piping (channels of groundwater flow) under the haunches. Voids can be avoided in the bedding and under the haunches if the embedment is a soil cement slurry with slump of 0.25m (10 inches) minimum compressive strength of 1400 kPa (200 psi). The slurry may be a good idea under difficult installation conditions -- for example, when trench widths are too narrow for placement of soil under the haunches. Flexible pipes tend to conform with loose soil under the haunches by squatting down slightly into the soft bedding. However, the increased radius of the bottom, r_y, must neither be great enough to cause spalling of the mortar lining, nor great enough that the relationship $P_y r_y = P_x r_x$ causes P_x to exceed sidefill support capability. Width of trench is not an issue. Constructors have some clever ways to "chuck" soil down under the haunches: J-bar it, or vibrate it, or flush it, or pond it, etc. It is sometimes proposed that the bottom of the trench be shaped to fit the pipe. But the effort usually does not justify the cost and does not assure uniform support.

Voids left by the withdrawal of sheet piling or trench shield do not affect the pipe if the tips of the piles or shield are above the spring line of the pipe, or if the shields are pulled ahead of the backfilling.

Appendix I. References

Marston, Anson (1930). The Theory of External Loads on Closed Conduits in the Light of the Latest Experiments. Bul. 96, Engr. Exp. Sta., Iowa State College.

Spangler, M.G. (1941). The Structural Design of Flexible Pipe Culverts. Bul. 153, Engr. Exp. Sta., Iowa State College.

Spangler, M.G., and R.L. Handy (1973). Soil Engineering, 3ed. Intex Educational Publishers.

Watkins, R.K., and M.G. Spangler (1958). Some Characteristics of the Modulus of Passive Resistance of Soil: A Study in Similitude. Proceedings, Highway Research Board 37, 576.

Appendix II. Notation

Lengths:

B	=	width of trench at spring line of the pipe
B	=	width of dual wheel tire print
L	=	length of dual wheel tire print
d	=	ring deflection = Δ/D
Δ	=	decrease in vertical diameter
D	=	diameter of the pipe = 2r
r	=	radius of curvature of the ring
H	=	height of soil cover over the pipe
h	=	height of groundwater table above spring line of the pipe

Properties of Materials:

- E = modulus of elasticity
- K = $(1+\sin \phi)/(1-\sin \phi)$ = ratio of normal principle stresses at soil slip at a point in the soil
- α = bedding angle for Spangler's derivation of the Iowa Formula
- ϕ = soil friction angle

Loads, Pressures, and Stresses:

- γ = unit weight of soil
- γ_w = unit weight of water
- σ = normal stress
- P = pressure
- Q = surface wheel load
- W = Marston load on a rigid pipe

Subscripts indicate directions. A bar over a soil stress indicates effective soil stress (intergranular stress).

Mill Woods Sanitary Storage Tunnel

Ken Chua[1], P.Eng. and John Kelly[2], P.Eng.

Abstract

Pre-cast segmented and shotcrete liner construction of a 22 m deep (72.2 ft.), 2,125 m (6,972 ft.) long tunnel of approximately 3.0 m (9.8 ft.) diameter is presently underway to resolve Mill Woods basement flooding at a cost of $8.8 million CDN.

In the design of the sewer relief schemes, four storage concepts were considered: subsurface storage reservoir, subsurface storage pipe, on-line storage, and distributed storage. Various options and alignments for storage alternatives were evaluated and pros and cons were identified. The subsurface storage pipe alternative was selected to detain the required 15,630 m^3 (4,129,000 U.S. gallons) wet weather flow.

A TBM capable of constructing a 3.48 m (11.42 ft.) inside diameter tunnel was chosen to construct 480 m (1,575 ft.) of tunnel. The primary and secondary liner (two pass system) utilized shotcrete. A second TBM was chosen to construct 1,645 m (5,397 ft.) of a 2.92 m (9.58 ft.) inside diameter tunnel utilizing a "single pass" method consisting of pre-cast concrete segments.

Introduction

The Mill Woods district of Edmonton is located in the southeast portion of the City of Edmonton (Figure 1).

---

[1]General Supervisor, Drainage Branch, Transportation Department, City of Edmonton, 9803 - 102A Avenue, Alberta, Canada T5J 3A3

[2]Vice-President, Bel·MK Engineering Ltd., 10532 - 110 Street, Edmonton, Alberta, Canada T5H 3C5

It encompasses an area of 2,270 hectares (5,600 acres) and has a population of approximately 79,000. Over 21,000 dwellings are serviced by 206 km (128 miles) of sanitary sewers. Since 1978, there have been numerous incidents of reported basement flooding.

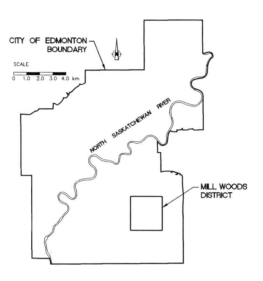

Figure 1   Location Plan

Various studies have been undertaken focusing on the cause of basement flooding and other related issues. Findings of the studies confirmed that the major cause of the sanitary sewer surcharging and backing up into resident's basements was inflow/infiltration (I/I) of rainwater into the sanitary sewer system. In particular, the settlement of backfill around the basements of dwellings is directing surface run-off to the foundation walls and then into the weeping tile pipes. In Mill Woods, as in much of Edmonton, the weeping tile pipes are connected to the sanitary sewer, a practice that is no longer permitted.

Previous studies evaluated various amounts of sewer upgrading and storage options[1]. It was determined that a combination of upgrading particular sewers and constructing storage facilities at key locations (to detain flows until there was a capacity available in the downstream sewers) was the most cost effective.

Evaluation of Storage Alternatives

Computer hydraulic modelling was used to confirm key points in the sewer network where relief of the sanitary trunk sewers had to be provided.

Storage concepts evaluated consisted of "off-line" and "on-line" underground storage tunnel and tanks, and a "distributed storage" concept.

The subsurface tank concept consisted of excavating soil to appropriate depths below the elevation of the sanitary sewers, and constructing large tanks "off-line" to detain the excess flows from the sanitary sewers until the rain storm subsided and there was capacity available in the sewer to pump the stored flows. To provide "failsafe" (i.e. gravity) filling of the storage facility, the required volume must be provided below the elevation of the trunk sewers.

The subsurface storage tunnel option consisted of constructing pipes ranging from 2.0 m (6.6 ft.) to 5.0 m (16.4 ft.) diameter of appropriate lengths to provide the required volume "off-line". Due to the pipe storage requirement of being below the elevation of the trunk sewers, excavation by tunnelling methods was more cost effective. Stored flows are later removed by pumping stations when sufficient sewer capacity becomes available.

On-line storage, sometimes referred to as "super pipe" storage, can be described as installing pipes of diameters greater than that required to convey the flows and using a restriction device, typically an orifice, to allow storage to occur and release the flows at a rate that will not exceed the capacity of the downstream sewers.

Distributed storage is somewhat of an unproven concept within the City of Edmonton. It consists of installing an additional pipe in areas prone to flooding. When the sanitary sewer surcharges and begins to flow up the service pipe to the house, a check valve arrangement directs the sanitary flows to a new pipe designated to store the flows. Once the surcharging has subsided, the storage pipe drains into the existing sewer system.

The evaluation process consisted of not only determining costs, but undertaking a public consultation program and using a Kepner Tregoe (K-T) Analysis to quantify some of the other criteria that would lead to a successful storage system most suited to the Mill Woods community. Typical criteria consisted of various performance and technical aspects, community relations and disruptions, and operation/maintenance aspects. One particular performance aspect that was very important was "How failsafe is it?". The subsurface storage reservoir, storage pipe and on-line storage concepts did not rely on any mechanical valving, pumping or other equipment to direct flows to storage. However, the distributed storage option relied on valves to operate correctly to provide storage.

The subsurface storage pipe "off-line" option was selected as the best option to store the required 15,630 m^3 (4,129,000 U.S. gallons). In determining the alignment of the new subsurface storage pipe, a previous study regarding a regional trunk sewer to service a nearby City and other communities to the south was reviewed. It was discovered that a subsurface storage pipe approximately 3.0 m (9.8 ft.) in diameter and 2,100 m (6,890 ft.) long would connect to the three key trunk sewer diversion points and could be aligned and sloped to potentially serve as a portion of the regional trunk sewer.

Geotechnical Investigation

A geotechnical investigation was undertaken to evaluate the sub-surface soil and groundwater conditions. 25 test holes were drilled along the proposed tunnel alignment to a maximum depth of 25.9 m (85 ft.) below ground surface. Standard penetration tests were conducted at selected locations and disturbed samples were collected for laboratory testing and analysis. Ten groundwater monitoring wells were installed to establish static water levels. The depth to water ranged from 2.2 m (7.2 ft.) to 8.9 m (29.2 ft.) below ground surface. The progression of soil types with increasing depth in the test holes generally consisted of fill and organic soils, glaciolacustrine clay with silty/sandy zones, glacial clay till with sand layers, and shale/siltstone/sandstone bedrock. Tunnel construction will occur entirely within the shale/siltstone/sandstone bedrock zone which was considered suitable for tunnelling in general, and in particular, very suitable for using Tunnel Boring Machine (TBM) equipment.

Tunnel Design Criteria

Design loadings were based on the following:

1) Initial Loading Conditions
   a) Vertical Soil Pressure: $3.0\ \gamma_s D$
   b) Lateral Soil Pressure: $K_o = 0.8$ to 1.2

2) Final Loading Conditions
   a) Vertical Soil Pressure: $\gamma_s z$ (full overburden)
   b) Lateral Soil Pressure: $K_o = 1.0$
   c) Full Hydrostatic Pressure: $\gamma_w z$

Where: D = excavated diameter of tunnel, m
γ_s = 21 kN/m³, dry unit weight of soil
γ_w = 9.8 kN/m, unit weight of water
z = depth of overburden, m

Subsurface Tunnel Storage

At this point in time, it was necessary to evaluate various diameters and construction methods. Early discussions with the City of Edmonton Drainage Branch and City of Edmonton Public Works Construction Branch confirmed that the City's own forces had the required resources available and could undertake the work economically. To meet the schedule, it was determined that the tunnel should be constructed in two sections, using two tunnel boring machines (TBM's). Review of available equipment and various techniques determined that construction of a 3.48 m (11.42 ft.) inside diameter, 480 m (1,575 ft.) long shotcrete lined tunnel and a 2.92 m (9.58 ft.) inside diameter, 1,645 m (5,397 ft.) pre-cast concrete segment lined tunnel was the preferred option. Other construction methods considered included steel rib and timber lagging primary liner and cast-in-place concrete final liner. Figure 2 shows the tunnel alignment.

Shotcrete Lined Tunnel

As shown on Figure 2, the portion of tunnel to be constructed as a shotcrete lined tunnel is 480 m (1,575 ft) long. The equipment selected to excavate the tunnel is an "open face" type TBM. This type of TBM is typically suitable for stable ground conditions.

As with most tunnel construction, both the initial soil loading conditions and the final (long term) loading conditions must be addressed. With this particular method, shotcrete will be applied 100 mm (4 inches) thick to the excavation walls immediately following the cutting head of the TBM. The purpose of this layer is to take the initial loadings of soil and to deflect, and crack as necessary to create equilateral circumferential loads. When the primary liner is in place, excavation can continue. Once the tunnel is entirely excavated and the primary liner is in place, the final 100 mm (4 inch) layer of shotcrete can be applied. A welded wire steel mesh will be placed prior to applying the final shotcrete layer to provide protection against shrinkage cracking. A cross-section of the shotcrete lined tunnel is shown as Figure 3.

SANITARY STORAGE TUNNEL

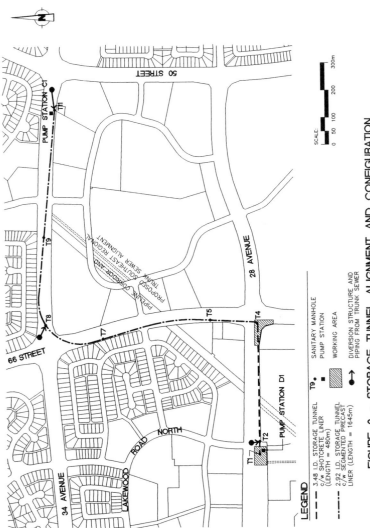

FIGURE 2 - STORAGE TUNNEL ALIGNMENT AND CONFIGURATION

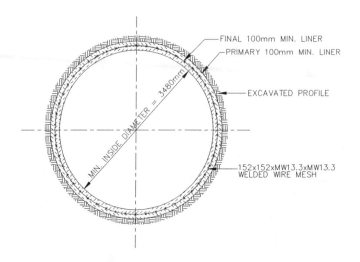

Figure 3   Shotcrete Lined Tunnel Cross-Section

The shotcrete mix design was based on information and product available from Lafarge Fondu International of France. This admixture had been successfully used in Europe but not in North America. The admixture is an accelerator called Shotax, a cementicious material consisting of highly reactive calcium aluminate. Shotcrete mix accelerated with this admixture usually has the same 28 day strength as unaccelerated mixes but the set time is quickened. It gels on impact and then stiffens within minutes. During these critical minutes, the shotcrete provides a mattress effect for the subsequent shotcrete, cutting rebound to a minimum.

Construction of the shotcrete tunnel portion is proceeding eastward from T1 to T4 (see Figure 2). T1 was constructed as the main working shaft for this portion of the tunnel and consists of a 4.57 m (15 ft.) diameter shaft. Since the TBM was assembled to begin excavation at T1 and the spoil material is being removed at this point, an "undercut" and "tail tunnel" were excavated at this location in advance of the tunnelling. The undercut east of the shaft was predominantly excavated by hand and supported by steel rib and timber lagging. This area allows for TBM assembly and alignment set-up. A tail tunnel (west of T1) was excavated (by hand with rib and lagging) to allow movement/switching and storage of the rail cars used to haul spoil material. A hoist at this location lifts the rail cars of spoil material and dumps them into awaiting trucks.

The production rate for excavation and installation of the primary liner is approximately 3 m (10 ft.) per 8 hour shift. Application of the shotcrete secondary liner is expected to take 1 month. On this particular project, 2 shifts/day have been utilized.

The construction cost of the shotcrete lined tunnel is estimated to be $2,598,000 CDN. Unit costs are $5,410 CDN per linear metre of tunnel ($1,650 CDN/ft.) or $569 CDN/m^3 of storage ($2.15 CDN/ U.S. gallon).

Pre-Cast Concrete Segment Lined Tunnel

The portion of tunnel being constructed with a pre-cast concrete segment liner method is 1,645 m (5,397 ft.) long and is shown in Figure 2.

This particular method is known as a "one-pass" system as the pre-cast segments are capable of withstanding both the initial loadings and the final loadings.

To undertake this portion of the project, the contractor, Public Works Construction Branch, decided to retrofit one of their closed face moles with a hydraulic arm to place the segments as tunnelling proceeds (see Plate 1). To undertake the retrofit, the TBM and some pre-cast "test" panels were sent to the manufacturing plant in eastern Canada where the erector arm was installed and tested. Manufacture of the pre-cast segments was carried out locally. A cross section of the pre-cast segment lined tunnel is shown as Figure 4.

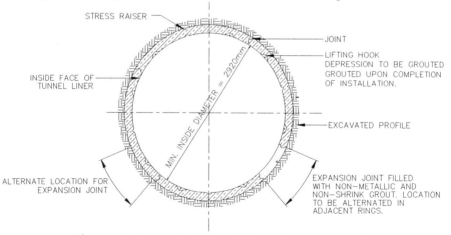

Figure 4   Segmental Lined Tunnel Cross-Section

Construction of the pre-cast segment tunnel is proceeding northward and then eastward from T4 to T11 (see Figure 2).

T4 was constructed in advance of tunnelling as the main working shaft and consists of a 4.57 m (15 ft.) diameter shaft. Similar to T1, a tail tunnel was constructed south of T4 and an undercut (see Plate 2) north of T4 was built to facilitate installation of the TBM and soil removal. Due to the length of this portion of the tunnel, the TBM will be removed for refurbishing at T8 and reinstalled. Also, due to the length of this portion, the tunnel will be over-excavated in the vicinity of T8 to construct a crossover (parallel track) that will allow the full and empty rail cars to pass one another.

The production rate for excavation and installation of pre-cast liner is approximately 9.0 m (30 ft.) per 8 hour shift.

The construction cost of the pre-cast segment lined portion of the tunnel is expected to be $6,217,000 CDN. On a unit cost bases, this is $3,780 CDN per linear metre of tunnel ($1,152/ft.) or $564 CDN/$m^3$ of storage ($2.14 CDN/U.S. gallon).

Related Storage Information

To divert sanitary flows into the storage tunnel, diversion structures with adjustable weirs will be constructed. The three optimal diversion locations and weir heights were confirmed by computer modelling of the sanitary trunk network. When the level of flow in the trunk sewer reaches the critical height, it overflows the weirs in the chambers and is piped to the storage tunnel. Trunk sewer diameters were increased to ensure the flows reached the diversion structures before they caused flooding.

Two pump stations will be constructed to pump out the storage tunnel when capacity is available in the sanitary trunk sewers. A level control system consisting of level sensors in the storage tunnel, diversion chamber and downstream sewer will detect the need to pump and confirm acceptable flow conditions in the trunk sewer.

Tunnel maintenance (cleaning in particular) was an important design parameter. Manholes will be installed at a maximum of 473 m (1,552 ft.) spacings to allow access and, in particular, cleaning with high pressure sewer cleaning trucks. Cleaning after each use (9 - 13 times/yr.) is expected.

Conclusions

Due to the depth of the sanitary sewers requiring relief, the construction of a storage tunnel was found to be economical. The selection of this method was further justified by input received during the public consultation program.

It was important to obtain geotechnical information to evaluate storage options and determine construction methods and equipment selection. Based on the geotechnical findings and local expertise and equipment available, it was determined that construction of a 3.48 m (11.42 ft.) inside diameter shotcrete lined tunnel and a 2.92 m (9.58 ft.) inside diameter tunnel was cost effective.

Even if methods somewhat less costly than tunnelling are proposed, the impact on the community, public relations costs and the ultimate restoration costs must be considered.

References

1. UMA Engineering Ltd., Mill Woods Drainage Improvements Phase III, Preliminary Analysis, November 1992

Closed faced TBM ready for lowering down working shaft T-4 (foreground)

Pre-cast concrete segments

Plate 1

Undercut at main working shaft T-4

Closed face TBM cutting head

Plate 2

POINT LOMA TUNNEL OUTFALL
F. Stuart Seymour[1]; Thomas Willoughby[2] and Richard Trembath[3]

Abstract

This paper provides an overview of the Pt. Loma Tunnel Outfall Project. This City of San Diego project will involve a 16.7 kilometer (10.5 mile) tunnel system for conveying up to 26 cubic meters per second (614 million gallons) per day of treated wastewater to offshore disposal.

The system is described in general terms followed by a discussion of system hydraulics. The geological setting is reviewed and two tunnel design issues are discussed; selection of the specified tunnel boring machine (TBM) and tunnel liner system. In addition, a review of project benefits and environmental issues is provided.

Introduction

The Point Loma Tunnel Outfall is a 16.7 kilometer (10.5-mile) tunnel system which will connect existing and new city of San Diego wastewater treatment facilities to the Point Loma Wastewater Treatment Plant (PLWTP) outfall extension pipeline, some 3,500 meters (11,500-feet) offshore. Upstream, the 3.7 meter (12-foot) diameter tunnel will be connected to a new headworks facility serving the planned North City Water Reclamation Plant and Central Region Pipeline. Downstream, the tunnel will connect at the shoreline via a shaft to a new Tie-In Structure at the PLWTP and extend offshore to intercept the existing seafloor pipeline through a new riser shaft.

Treated wastewater effluent will flow in the tunnel and pipeline, and discharge into the ocean, approximately 7,300 meters (24,000-feet) offshore, via a diffuser system. The

---

[1] F. Stuart Seymour, Project Manager, City of San Diego, 600 "B" Street, Suite 500, San Diego, CA 92101, (619) 533-5214, Fax (619) 533-4267.

[2] Thomas Willoughby, Project Manager, Parsons, Brinkerhoff, Quade & Douglas, 1230 Columbia Street, San Diego, CA 91010, (619) 338-9376, Fax (619) 338-8123.

[3] Richard Trembath, Project Manager, Parsons Engineering Science, Inc., 9404 Genesse Avenue, Suite 140, La Jolla, CA 92037, (619) 453-9650, Fax (619) 453-9652

tunnel system will provide for combined peak flows of 27 cubic meters per second (614 million gallons per day) from the three treatment facilities.

Design of the facility was completed in 1994. Major design elements included tunnel lining system to control infiltration/exfiltration, crossing of three fault zones, surge protection system, and providing for flexibility for multiple modes of operation.

The tunnel system is planned for completion in 1998 with a construction cost of approximately $200 million, and will be designed for an operational life of at least 75 years.

System Overview

The primary function of the PLTO is to convey treated effluent to the Point Loma Outfall Extension (PLOE) for discharge to the Pacific Ocean. Excess treated effluent not used for reclamation purposes will be transported from the North City Water Reclamation Plant (NCWRP), currently under construction, and from other Water Reclamation Plants (WRPs) that may be constructed in the future, to the Point Loma Wastewater Treatment Plant (PLWTP). The WRP flows are combined at the Headworks Structure which is located at the upstream extent of the PLTO. Under normal operating conditions, effluent from the PLWTP and from the WRPs will be combined at a hydraulic tie-in structure at PLWTP and dischared through the offshore tunnel and seabed pipeline, and the recently constructed Point Loma Outfall Extension. The PLTO will also provide the option of discharging combined PLWTP/WRPs effluent through the existing nearshore outfall (old Point Loma outfall), and the option of splitting the effluent between the two outfalls. The PLTO will replace the nearshore outfall that has been operating since 1963. The general arrangement of the system is shown in Fig. 1.

Design flows for the PLTO are shown in Table 1.

Table 1 - Design Flows

	WRP m^3/s (mgd)	PLWTP m^3/s (mgd)	Total m^3/s (mgd)
Ultimate Peak	8.0 (182)	18.9 (432)	26.9 (614)
Ultimate Average	4.8 (109)	10.5 (240)	
Minimum Flow	0 (0)	3.2 (72)	3.2 (72)

The function of the Headworks is to regulate the upstream hydraulic conditions in the pipeline from the North City WRP. The hydraulic grade line (HGL) will be fixed at a location (and elevation) between the Headworks and the North City WRP. This will avoid unfavorable hydraulic conditions, such as excess open channel flow velocities and foaming. The induced headloss will be achieved by a series of sleeve valves. Other functions of the Headworks include surge protection and flow metering.

A schematic of the Tie-In Structure in normal operation mode is shown as Fig. 2.

UNDERGROUND PIPELINE ENGINEERING

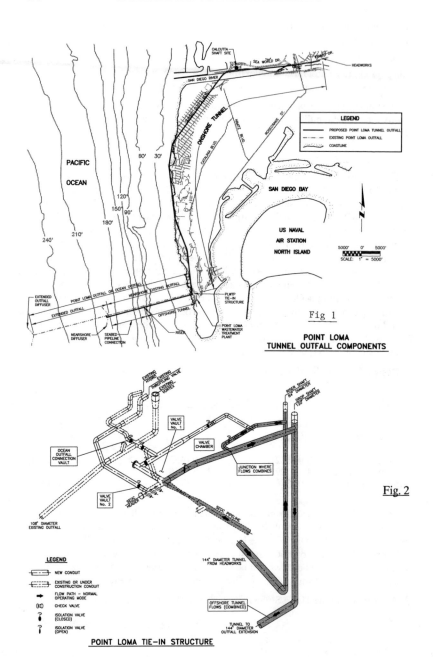

The function of the Pt. Loma Tie-In Structure is to combine flows from the onshore tunnel carrying reclaimed water flows and from the PLWTP. The Tie-In will have the capacity to route flows to the extended outfall diffusers via the proposed tunnel and existing offshore seafloor pipeline, or to the nearshore diffusers via existing nearshore seafloor pipeline.

Two main modes of operation, a normal mode and an emergency mode, have been designed. The normal mode will route WRP and PLWTP flows to the farshore diffuser. In the emergency mode all flows would be routed to the nearshore diffuser.

The Tie-In Structure has to be located in a very congested area at the PLWTP. To provide adequate space the structure will be located partially in a coastal bluff. This will be accomplished by open-cutting of the bluff, constructing the chamber and refilling the bluff similar to the original contours.

Hydraulics

A computer-based steady state hydraulic evaluation was conducted for a number of scenarios. The purpose of the analysis was to confirm conduit diameters, establish maximum steady state hydraulic pressures and calculate maximum pressures at existing facilities to confirm structural capacity. Variables which could be input to the model included flow, conduit diameter, friction factors, headloss coefficients, and ocean elevation. The downstream control is at the ocean surface and HGLs were calculated by considering seawater density head, diffuser losses, sea floor pipeline losses, offshore tunnel friction losses, Tie-In Structure losses, onshore tunnel losses, and Headworks losses.

A summary of the results of the hydraulic analysis is shown on Fig. 3.

A hydraulic transient analysis was conducted for various scenarios such as valve opening and closure at the Headworks or the Tie-In Structure, sudden blockage of the sleeve valves at the Headworks, and increasing or decreasing flows as a result of changing the operational modes. Hydraulic transients will be limited to approximately 6.1 m (20 ft) of water column through the following:

- time limits on valve operation;
- surge tank at the Headworks;
- free water surface connection (through existing vortex) to nearshore seafloor pipeline;
- free water surface at Tie-In Structure drop shaft and riser shaft to protect offshore and nearshore tunnel respectively.

Geologic Setting

The Point Loma Tunnel Outfall alignment is located on the western margin of the Peninsular Ranges geomorphic province, which extends from the Los Angeles Basin into Baja California, Mexico. The alignment is situated along the coastal plain and San Diego Bay. The local terrain is faulted and gently folded, creating fault-bounded

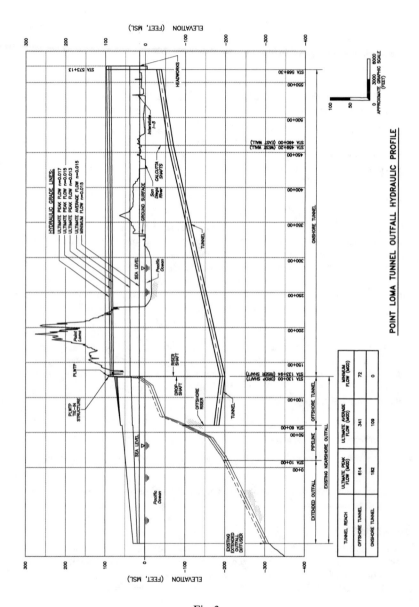

Fig. 3

sedimentary basins, in particular, the San Diego and Mission Bays. The San Diego Bay is flanked on the west by an upwarped sedimentary bedrock ridge, the Point Loma Peninsula. The sedimentary rocks comprising Point Loma Peninsula are composed of upper Cretaceous marine sediments deposited approximately 65 to 95 million years ago upon deeply eroded rocks of the volcanic and granite terrain. These marine sediments represent infilling of a deep marine basin by a series of deep-sea fans.

There are three major fault zones along the project alignment; the Rose Canyon Fault Zone (RCFZ), the Point Loma Fault Zone (PLFT), and an unnamed zone existing approximately 2,440 meters (8,000 feet) offshore just prior to the outfall extension's connection structure. The later two fault zones are potentially active while the RCFZ is classified as active. Based on published information and a correlation between fault length and earthquake magnitude, it is concluded that the Rose Canyon Fault is capable of generating a maximum credible earthquake of magnitude 6.8 to 7.0.

Some of the geologic units the tunnel goes through are:

Fill (Qal):
    Fill soils are widely distributed in the area surrounding Mission Bay and San Diego Bay. These soils generally consist of a wide range of materials from cobbles and gravels to sand and clay.

Alluvium and Offshore Sand (Te):
    Recent alluvial deposits occur within the San Diego River and interfinger with bay deposits in Mission Bay. These materials consist of relatively loose to dense sands and gravels interbeded with soft to firm silts and clays.

Bay Deposits (Qbp):
    Bay Deposits consisting of loose to medium dense sands and soft to firm silts and clays interfinger with the alluvium deposits at the mouth of the San Diego River.

San Diego Formation (Tsd):
    The San Diego Formation consists of poorly indurated layered siltstone, silty sandstone and silty claystone, with occasional cobble conglomerate lenses. These materials are well consolidated but weakly cemented. Upon excavation will typically behave like a soil.

Cabrillo Formation (Kc):
    Within the project area the Cretaceous age Cabrillo Formation consists of cobble conglomerate in a sandstone matrix. The cobbles range in size from fine gravel to boulders as much as 1.5 meters (5-feet) across. The cobbles are composed of rounded granites and metavolcanics and are hard to very hard and fresh.

Point Loma Formation (Kp):
    The Cretaceous age Point Loma Formation outcrops in the sea cliffs of Point Loma and is anticipated to be 300 to 600 meters (1,000 to 2,000 feet) thick in the project area. The majority of the tunnel alignment goes through this formation. The Point Loma Formation consists of interbeded siltstones, silty sandstones, and minor claystones. It contains randomly-occuring, well cemented concretions. At the

proposed depth of the tunnel, the formation is expected to be slightly weathered to fresh with joints generally widely spaced.

Two areas along the alignment uncovered unique challenges for the Project. Our exploration program determined that in these two areas the existing data was misleading. The first area is located 3,660 meters (12,000 feet) from the headworks, south of the San Diego River, beneath an area called "Robb Field". During our investigations we changed tunnel alignment and profiles based on the differing results of each exploratory boring. The second area is located offshore from the PLWTP approximately 2,130 meters (7,000 feet). The existing data indicated that the Point Loma Formation continued 3 to 6 meters (10 to 12 feet) below the seabed. The results of this discovery was to shorten the length of the tunnel and place the riser structure 1,370 meters (4,500 feet) from the outfall extension. From the riser a seabed pipeline connecting the tunnel to outfall is proposed.

Tunneling Equipment and Methods

The PLTO requires excavation equipment that can tunnel in soft ground and soft rock conditions. The choice of tunnel boring machine (TBM) is based on geologic and groundwater conditions and on the length and size of the tunnel. The PLTO project has varying ground conditions and the tunnel length is approximately 15,200 meters (50,000 feet). For descriptive ease the tunnel is broken down into three drives: east, west and offshore. The east and west tunnel drives will originate at the Calcutta construction site, whereas the offshore drive will originate at the PLWTP.

TBMs can be equipped with open-faced, non-pressurized closed-face, or pressurized closed-face shields. Open-faced and non-pressurized closed-face machines require the use of rigorous dewatering plans. A pressurized closed-face shielded TBM (see Fig. 4) is specified for the PLTO for the following reasons:

- the variety of ground conditions anticipated along the alignment - saturated soft ground, to saturated transition zone with cobbles and boulders, to soft rock with randomly occurring concretions;

- the close proximity of the San Diego River, the high hydraulic conductivity of the soils, the presence of the landfill along Sea World Drive, and potential settlement of the highway structures along the tunnel alignment, combine to preclude the use of dewatering;

- the ocean crossings with the ground conditions being the Point Loma Formation consists of interbedded siltstones, silty sandstones, and minor claystones with randomly-occuring faults, fractures, and well-cemented concretions.

The Offshore drive and a portion of the West drive will be under the ocean; consequently, the TBM, machine parts, equipment and materials must be designed to operate in a salt water environment, and the cutterhead seals, articulation joints and tail seals must be adequately designed to resist water pressures of up to 60 meters (200 feet).

The machines are specified to work in open (unpressurized) and closed (pressurized) modes. The closed mode will be applicable in shear zone - where the water column is connected to the ocean, resulting in as much as 67 meters (220 feet) of head; otherwise, the TBM can work in open mode in areas where water inflow is low, to improve TBM productivity. There will be provisions for drilling probe holes ahead of the face, in open mode, to detect adverse conditions where the closed mode would be required. The machines are required to change rapidly from one mode to the other.

In transition zones the TBM cutters will be suitable for effective excavation of gravels and cobbles which may be encountered in the sand, silt and clay matrix of the San Diego and Cabrillo Formations under high hydrostatic pressure. The spoil conveyance system from the face must be designed to collect and transfer all types of material anticipated in both dry and wet conditions. In addition, the ribbon section of the primary screw conveyor should be designed to pass boulders, cobbles and fragments up to at least 4.5 cm (18 in) in size to a stone trap, where they are removed.

In the Point Loma Formation, the clayey nature of the claystones (and to some extent the siltstones), would tend to promote caking inside the excavation chamber if the excavated material is not very well mixed. The TBM cutterheads will be capable of mixing the excavated material well enough to prevent caking inside the excavation chamber; injection of water/polymer/clay/bentonite additives into the excavation chamber with a view to creating a more homogeneous mix. Shield tail seals are specified to withstand the maximum pressures at tunnel inverts plus an adequate safety margin.

<u>Tunnel Lining Design</u>

The functional requirements of the liner design are as follows:

- must be watertight and capable of withstanding 67 meters (220 feet) of external head and 37 meters (120 feet) of internal water pressure;

- capable of carrying all long-term external pressures and operating loads, and all short-term construction loads, particularly jacking forces from the TBM propelling itself from the installed rings;

- meet exfiltration criteria for maximum and average crack width of 0.01cm (0.004 in) and 0.005cm (0.002 in), respectively;

- be durable, providing at least the specified 75-year design life of the tunnel.

After evaluation of available options, a single pass, gasketed, bolted, precast concrete segmental lining was selected as the cost-effective system that meets all of the functional requirements.

The liner ring is 1.2m (4-feet) wide, 28 cm (11-in) thick, and consists of five segments. The liners utilize two bolt pocket types to reduce packet filling. The standard segment will weigh 2300 kg (5 kips).

To prevent water leakage into the tunnel during construction, the precast concrete segments should be bolted together in the tail of the TBM. Bolts, required across the joints in order to erect the lining, are set in pockets formed along the edges of the segments. The

UNDERGROUND PIPELINE ENGINEERING

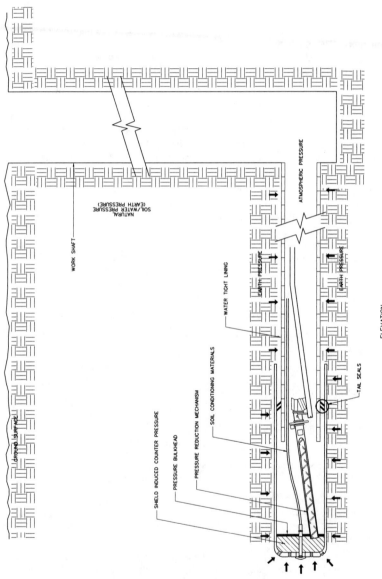

Fig. 4

POINT LOMA TUNNEL OUTFALL 477

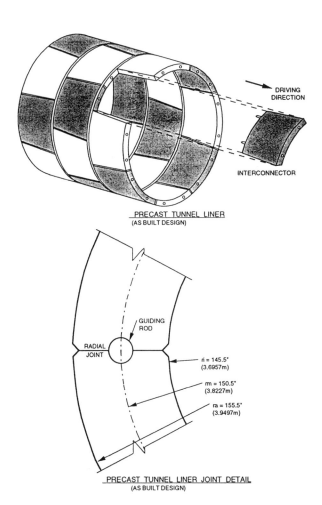

Fig. 5

lining is provided with a longitudinal key segment to permit closure of ring without dislodging the gasket. Dense high quality concrete is specified for the liners.

As an added means of dealing with the 37 meters (120 feet) of internal water pressure, a hydrophilic gasket system, with gaskets located along each segment edge, provides a watertight seal against both inflow into the tunnel during construction, and exfiltration during operation. The gaskets will be thicker than the available groove to ensure full contact, even when not subject to moisture. A schematic of the liner system is shown in Fig 5.

A 28 cm (11-inch) thick liner with a clear concrete cover of 5 cm (2-in) to the tie rods, provides an added level of corrosion protection, and allows for possible reinforcing cage offsets during manufacturing.

The prescribed lining system is similar to European and Japanese practice. Rigorous quality control standards are required in the manufacturing and storage operations, erection and procedures, in order to meet the rigorous tolerance requirements for the tunnel segments.

Project Benefits

The City of San Diego, as part of their master plan to comply with Clean Water Act's secondary treatment requirements, implemented an aggressive water reclamation program. A project within that program was the construction of PLTO. The benefits this project would provide are:

- Provide "fail safe" discharge of wastewater effluent from North City Reclamation Plant and other reclamation plants in the central and northern service areas of the City.
- Offload existing sewer lines from the North and Central systems leading to Pump Station No. 2 which pumps the City's effluent to the PLWTP.
- Bypass the PLWTP with as much as 7 cubic meters per second (162 mgd) (secondary) treated effluent.
- Provide additional sewer capacity and infrastructure to the increasing population of the City of San Diego.
- Parallel the 30-year old seabed pipeline outfall that currently connects to the recently completed outfall extension.

Environmental and Public Acceptance

Public acceptance of the reclamation program is one of the City's principal goals. To that end they have implemented a comprehensive public information program that combines face-to-face meetings, group (community) meetings, implementation of "public advisory groups" and the distribution of pamphlets and newsletters addressing the program's plans and status. The advisory group meets, at least monthly, with the technical staff in order to maintain a constant dialogue of status as it relates to the project. It also provides a forum for public input to the planning and design process.

The environmental process for PLTO initiated with the projects predecessor, the San Diego River Outfall (SDRO) in the fall of 1991. SDRO followed the State's CEQA-process and has completed and filed a Final Environmental Impact Report in February 1993. The PLTO environmental process is underway with the preparation of an amendment document to the SDRO document.

Choice of Pipes for Micro tunnels

Y.G. Diab[1]

Abstract

Actually, the development of microtunneling in France is important because many underground crossings are realized in urban areas and big cities. A national research program was established to develop and improve this technology.

This paper is divided in two parts. In the first part, the method of choosing pipes for microtunneling realization and the categories of applied loads are analysed. In the second part, the first controlled experimentation is presented and recommendations for other realizations are given.

Introduction

The development of microtunneling particularly in urban areas, has been made possible by a combination of new technological advances :

- improved knowledge concerning the geology of sub-soils through the use of more reliable measurement techniques with a degree of complexity adapted to the nature of the project and permitting the identification of structures located in vicinity. This approach is important in the choice of microtunneling machine and the design of pipes for the lining of the micro tunnel.

- the related development of precast pipes for lining micro tunnels, specially designed for this purpose.

- the development of precision-guided tunneling machines, particularly for small diameters in the micro tunnel range, not accessible from the inside and hence fully controlled from the outside.

The microtunneling system has been used in Japan for twenty years. It was used for the first time in France in 1989. Since that time many sites and experimentation were realized, and a national research and development program was

---

[1]LGCH, ESIGEC, Savoy University, Savoie Technolac, 73376 Le Bourget du Lac Cedex. France

establish by the IREX " Institut pour la Recherche appliquée et l'EXpérimentation en génie civil : National Institute for Applied research in Civil Engineering" (Audouin, 1994), (IREX, 1993), (Launay and Notin,1994), (Mermet et al, 1990). However, some questions remain, especially about : the microtunneling system driving according to the soil nature ; and the behaviour of pipes under the actions of various loading.

Jacked pipes

The loads applied on a jacked pipe are the following :

1- Long term loads : in this category are considered all loads applied on a pipe placed in trench or embankment. Diab (1992) analyses these loads by using analytical approaches and numerical models.

- soil and ground water external pressure : the vertical pressure of the earth on the crown of the buried pipe can be evaluated from the weight of the soil corrected for the internal friction in the earth. In the Finite Element modelling usually used (Guilloux, 1987), this load is taken into account via the unit weight of the earth in each of the elements.

In a similar way, the horizontal pressures are distributed, depending on the nature of the earth and the stiffness of the pipe. The problem can be simplified by assuming that these are constant over the height of the pipe and that there is a factor of proportionality (depending on the internal angle of friction of the soil). The evaluation of these loads was a subject of great deal of research by authors such as Terzaghi (Spangler, 1960), Burns and Richards (1964), Marston (1930), Spangler (1956, 1958) who proposed analytical models more or less adapted to project conditions (depth of cover relative to the size of pipe, Young modulus of the soil, etc.) but whose principles and results are comparable. In our opinion the best method able to evaluate earth loading applied on a pipe is the Finite Element because it permits the analysis of various and different soils scheme.

- water internal pressure if any, the value and the frequency of this pressure is defined by the companies managing the network.

- effect of surface live loads transmitted (and diffused through the earth). Many authors developed methods to evaluate stress and strain induced by live loads .

- overall shear and bending forces induced by any movements of the surrounding earth (settling, consolidation).

2- Forces arising from the jacking technique : Two main forces have to be considered.

- longitudinal thrust inducing compressive forces that are to some degree deviated from the centre of the pipe wall. The misalignment and angular deviation create an eccentric aspect of the thrust transmitted to the pipe column and passing through the pipe junctions.

- additional external pressure from the fluid grouted into the annular space generated by the excavation process. This pressure might reach several MPa (PSI).

Design of pipes for jacking

A good design of a pipe has to have :

- a good resistance to axial thrust : utilization of materials with a high compressive strength and thicknesses appropriate to applied loads.
- adequate joints to transmit these forces in the most evenly distributed manner, while maintaining a seal to keep flow inside the pipe and to prevent infiltration from outside. The way in which these junctions respond to angular deviations is a decisive factor in both, the guidance of the pipe and its capability to support the thrust.

Evaluation of thrust forces : the thrust driving the pipe column must overcome the combination of the head force at the front of the tunnel and the embrace of the surrounding soil over the whole external surface of the pipe. During jacking, the thrust balances the sum of :

1- the head force : reaction of the tunnelling machine cutting action on the column, reaction of the cutting front, and penetration of the cutting head. The cutting front reaction is a function of the excavation method and on the type of earth : it is calculated for the characteristics of the soil, whose active and passive thrust phenomena it involves.

2- the friction along the pipe column : this is characterised by a series of static and dynamic coefficients of friction (the angle of friction between the soil and the pipe depends on the roughness of the outer surface of the pipe and lies between two-thirds and three thirds of the internal angle of friction of the soil. The value of the friction depends also of the weight of the pipe, the presence of water in the earth and the quality of the contact between the outer surface of the pipe and the surrounding soil ; this contact depends is a function of the size of the annular clearance that the tunnelling machine leaves around the outer surface of the pipe. the friction depends on the shape of the pipe and its stiffness (a flexible pipe generates additional lateral reaction as it deforms).

Finally, friction is reduced by using lubricant in the annular space between the pipe and the soil, when the nature of the soil so permits. In practice, the lubricant is injected through openings made in the outer skirt of the tunnelling machine. the injection of an appropriate amount of lubricant under controlled pressure also limits the convergence of the earth on the pipe and reduces the friction at the pipe earth interface.

Designing pipes for jacking : the two basic choices concern the length of pipes and the type of system used to join them together.

The length of the jacked pipe should not exceed that of the systems making up the tunnelling machine, and particularly the length of the first follower tube. This precaution is essential in order to limit the number of contacts points and hence the friction between the earth and the pipe. The occurrence of such contact points is also aggravated by the convergence of the earth on the pipe, whatever clearance is left by the machine around the outside of the pipe.

The junctions should be designed to remain leak-tight under moderate deviations and misalignments, while still transmitting the axial forces uniformly between two adjoining pipes. Consequently special attention will be paid in designing the junction to the surfaces in contact, and an appropriate material can be inserted between the two contiguous pipes in order to reduce the undesirable stress concentrations that the geometrical imperfections of the surfaces in contact will generate. This material has to have an equivalent strength to the pipe material. The junction must also be stiff enough to maintain adequate compression at the seal during the driving phase, taking account of the misalignments to which the junction is subjected.

Pipes for microtunneling

In microtunneling, the diameter of the pipes is kept between 200 to 1200 mm. This characteristic leads to a number of limits :

- a higher stress level in the pipe material during driving. In fact the friction opposing the movement of the pipe column through the earth is fairly closely proportional to the external surface area of the pipe, and hence to its diameter. In normal production pipes, regardless of the constituent material, wall thickness is roughly proportional to diameter. For this range therefore, the available thrust is proportional to the square of diameter. In conclusion, at stresses that are acceptable in the material, the driven length will be shorter the smaller the diameter.

- the difficulty (or even the virtual impossibility in the range of the smallest diameters) of effectively transmitting the thrust inside tunnels by means of intermediate stations, a method used in traditional jacking and with which there is no theoretical limit on the length of a pipe section capable of being jacked. Pipes for microtunneling are characterised by the following factors :

Dimensional requirements :
- the inside diameter has to allow the passage of muck removal tubes, cables and laser sighting lines, compatible with the control of the microtunneling machine.
- the outside diameter is ideally some 10 to 30 mm less than the outside diameter of the machine tubes.
- squaring off in order to distribute the stresses uniformly at the junction,
- straightness : in order to limit any additional spurious friction,

In general the preferred manufactured methods are those which can guarantee tight tolerances for these functional dimensions.

High level of mechanical performance : For considerations of resistance to axial thrust, checks will be made to ensure that pipes with a high length/diameter ratio are unlikely to buckle.

Transmission of loads in line with the junction : it is useful to introduce a thrust distribution material with the dimensional requirements in the seal plane between pipes (possibly retained by the outer ring or attached to one of the pipes).

In addition to these special requirements, pipes for micro tunnelling must satisfy the functional requirements of the pipe systems of which they form part : sewage system operating under gravity, pressure pipes, etc.

Current work

Many works are actually realized in France by the FSTT (French Society for Trenchless Technology) to obtain an harmonisation of the products used in microtunneling as regards both dimensions and functional performance.

Since the designers and operators of pipeline systems nevertheless identify pipes in term of inside diameter, a range of nominal diameters have been selected in parallel (Henri,1992), sufficiently wide to cover the majority of products in any material (table 1).

De = 360/ 404/ 535/ 640/ 750/ 850/ 960/ 1090/ 1180/ 1275/ 1400

ND = 200/ 250/ 300/ 350/ 400/450 /500/ 550/ 600/ 650/ 700/ 800/ 900/ 950/ 1000/ 1050/ 1100/ 1150/ 1100/ 1150/ 1200

Table (1) Nominal diameters (ND) and outside diameters D_e

Also, theoretical approach is undertaken to incorporate design of jacked pipes in service into the general methods used for designing pipes laid in trenches. In fact, the micro tunnel installation induces environmental conditions in the final stage which must be specified in the guide of the design of pipelines. This theoretical approach, when available, will be used on planned and instrumented experiments which permits obtaining a better idea of the stress distributions in the wall of the pipe and in the junctions during the jacking phases.

Experimentation

To identify the soil, some boreholes were realised on the last two drives in order to characterise soils with Atterberg limits determination, and triaxial tests.

Results showed that the soil is an homogenous clay. During the experimentation, some measurements of driving have been made. The aim of these measurements was to recognise soils during the machine advancement (Quebaud et al, 1992). A metallic ring with 4 hollow cylinders has been put between the machine and the first pipe. The aim of the experimentation was to answer :

- the joint efficiency when the piping is not in a straight line,
- the knowledge of ground pressure distribution along the piping.

In order to answer these questions, 6 strain gages were fixed on some pipes (4 longitudinal gages, 2 transversal gages). The thrust loading analysis allowed to study the pipe behaviour during the boring. When a pipe is bored, it undergoes a load equal to the thrust value, then the soil reaction acts on the pipe, the thrust pressure must be increased in order to compensate this reaction. When the following pipe is bored, the loading is transmitted to the bored pipes by the joints and the surface of the pipe submitted to the soil reaction increases. In fact, the soil homogeneity means a constant loading in the front of the machine while the load applied on the pipe will change.

Figure (1) shows the average of applied loads (F_m) in function of the bored length. This load (F_m) is calculated with the average strain (ε_m) and the pipe properties.

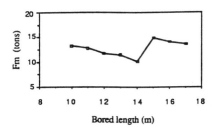

Figure (1) : average loading recorded during realization

We can observe that the curve of figure (1) decreases linearly, then suddenly increases and then decreases again. This phenomena can be explained as follows :

- the curve sudden increase is due to a soil changing : in fact, mechanical tests realised on samples taken off around 18 meters showed the presence of local limestone blocs in this area.
- the decrease is maybe due to the fact that the piping is not in a straight line : the joints dissipate the thrust energy in the ground.

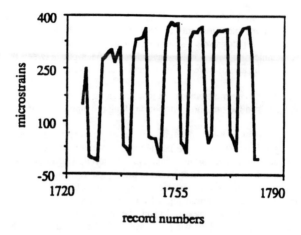

Figure (2) : viscosity phenomena observed during drive realisation.

Figure (2) permits to analyse the viscosity phenomenon. The longitudinal gage response (strain value) increases as the bored length increases, and becomes constant in the end : it means that thrust loading which acts on the pipe becomes more and more important, the compression is directly linked to the thrust loading : it is a typical viscosity phenomenon. Its analysis is interesting because its influence increases with the increasing of the machine diameter.

Analysis of driving parameters

The over thrust phenomenon, when the thrust jacks stop : during the pipe drive, where that was equal to 18-20 tons, these picks could reach 110 tons : this phenomenon must be due to a synchronization problem of starting between thrust jacks and auger. These high values provoke an over design of the pipes. Figure (3) shows the lubricant influence during the tunnelling ; the two graphs can be devised into 3 steps :

- step 1 : the thrust increases, due to the successive pipes addition : the torque (torsion) increase is surely due to an important rotation speed which provoke the filling of the auger and some difficulties to evacuate excavated soil ;
- step 2 : in opposition to the first step, we bore with lubricant ; the thrust becomes constant (homogeneous soil), and the torque decreases, because of a better soil excavation ;
- step 3 : when the lubrication is stopped, the thrust and torque values come back to a linear response : the soil reaction and the filling of the auger acts again.

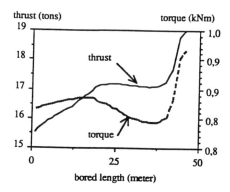

Fig (3) : lubrication influence observed during the drive

Conclusion

This instrumented experimentation was the beginning of many tests and realizations realised by many universities in France (INSA de Lyon, Laboratoire de mécanique de Lille, ENSG Nancy and LCPC) in collaboration with Civil Engineering companies (Fougerolle-Ballot, Devin Lemarchand, GFC, SADE, etc.) and pipes manufactories (Bonna, Eternit,etc.). The co-ordination of these field is insured by the FSTT and the National Project "MICROTUNNELS". Results show the necessity to carry others studies on new sites with different soil categories.

References :

Audouin, M. ; " Notions de base sur les microtunnels " Conférence Pollutec, Lyon / France, 10/1994. 9 pp.
Burns, J. Q. and Richard, R. M., "Attenuation of stresses for buried cylinders". Proceedings of the Symposium on soil-structure interaction, University of Arizona, Tuscon, U.S.A., 1964, p. 376-392.
Diab, Y. G. "Comportement structurel des conduites rigides entérrées" Ph.D thesis from the University Claude Bernard. Lyon / France, 12/1992. 445 pp.
Guilloux, A., "L'utilisation de la méthode des Elements Finis dans les tunnels et ouvrages souterrains". Comptes rendus des Journées d'Etudes sur l'utilisation de la méthode des Elements Finis dans les projets géotechniques, Paris / France, 1987,15 pp.
Henri, Ph., " Pipes for microtunnels : Suitability of use " Proceedings of the conference " No trenches in Town " Paris / France, 1992, Balkema Editions, pp 95-101.
IREX " Le projet national Microtunnels : présentation " Report from the National Institute for Applied Research in Civil Engineering, Paris / France, 01/1993. 48 p.
Launay, J.; Notin, Y. " Les chantiers Microtunnel à Dreux " Conférence Pollutec, Lyon / France, 10/1994. 7 pp.

Marston, A., "The theory of external loads on closed conduits in the light of the latest experiments". Records of the Highway Research Board, v. 9, 1930, p. 138-170.

Mermet, M. ; Crosnier, D. ; Legaz, C., and Audouin, M. " Réalisation d'un collecteur à Bry-sur-Marne par la technique du Microtunnelier " Proceedings of the conference "Underground Crossings for Europe " Lille / France, 1990, Balkema Editions, pp 347-350.

Quebaud, S.; Morel, E. and Henry J.P. " Proposition of a method for the study of the microtunneling machine behaviour and the soil - structure interaction " Proceedings of the confernce " No trenches in Town " Paris, 1992, Balkema Editions, pp 11-14.

Spangler, M. G., "Loads and supporting strengths of underground conduits". Proceedings of the Washington University Conference, St Louis, U.S.A., 1956, p. 193-213.

Spangler, M. G., "A practical application of the imperfect ditch method of construction". Records of the Highway Research Board, v. 37, 1958, p. 271-277.

Spangler, M. G., "Soil Engineering", International Textbook Company, Scranton, U.S.A., 1960, 573 pp.

Construction of the Big No-Dig

John J. Struzziery, P.E.[1], Member
Arthur A. Spruch, P.E.[2], Member

Abstract

This paper will describe construction related considerations for large diameter cured-in-place pipe (CIPP) sewer rehabilitation. A case study example will be presented to describe the construction phase of a $12.2 million, 12 km (7.5 mile), 122 to 152 cm (48 to 60 inch) diameter pipeline, which is the largest sewer rehabilitation project ever to be undertaken in this country. The paper will also describe construction logistics, installation details, testing, problems encountered, and other considerations that have been addressed on this project and which can be applied to other similar projects.

Special considerations such as design for a fully deteriorated pipe in tunnel applications; manhole and structure connection details; on-site liner saturations; liner tube advancement and overlaps; terminations; and repairs will be described. Cleaning and internal inspection considerations as well as quality control elements related to finished product and testing verification will be described.

This project has advanced the technology of the CIPP sewer rehabilitation market by successfully addressing job related logistics such as scheduling, weather conditions, material transportation, flow handling, site access, construction staging, and environmental restrictions. The experience gained on this project, including record installation lengths, has provided greater understanding of the capabilities and constraints of the CIPP systems, thereby improving technological and engineering confidence to specify this technology on larger and more complicated projects in the future.

---

[1]Mr. Struzziery is an Associate at S E A Consultants Inc., 485 Massachusetts Avenue, Cambridge, MA 02139-4018
[2]Mr. Spruch is a Principal at S E A Consultants Inc., 485 Massachusetts Avenue, Cambridge, MA 02139-4018

Introduction

Large diameter sewer rehabilitation technology and capabilities have advanced as a result of the Massachusetts Water Resources Authority (MWRA) Wellesley Extension Relief Sewer (WERS) project. This $12.2 million project is the largest sewer rehabilitation project ever to be undertaken in this country. This paper will describe construction related considerations that have been addressed as part of this project for the cured-in-place pipe (CIPP) and sliplining methods of sewer rehabilitation. The paper will also describe construction logistics, installation details, testing, problems encountered, and other considerations that have been addressed on this project and which can be applied to other similar projects.

The WERS is a 35 year old, 12 km (7.5 mile), 122, 137, and 152 cm (48, 54, and 60 inch) diameter, precast reinforced concrete pipeline that includes three double barrel siphons and a 915 m (3,000 foot), 152 cm (60 inch) diameter tunnel. During the design phase of the project, a condition survey was performed that indicated over 75 percent of the pipeline had moderate to severe deterioration with more than 13 to 25 mm (1/2 to 1 inch) of concrete loss due to hydrogen sulfide corrosion. Based on this condition survey as well as physical verification at access locations, the entire length of pipeline, which included 83 manholes and ten structures was recommended to be rehabilitated by either CIPP or sliplining methods. Performance specifications were prepared to allow competitive bids for each of these methods from several product manufacturers.

The project's environmental impact was one of the driving forces in advancing the rehabilitation technologies utilized. Over 75 percent of the alignment is located in the 100 year floodplain of the Charles River and crosses a number of wetlands and public and private water supply aquifers. As a result, permitting agencies had significant input in establishing conditions of work or work restrictions to minimize environmental impacts in these areas. The alignment also crossed 100 private properties which resulted in the need to minimize surface disruptions in these areas as well.

It should also be noted that prior to the rehabilitation phase, there was open excavation construction for an adjacent 137 and 152 cm (54 and 60 inch) diameter pipeline across the entire alignment. This new sewer was designed to provide increased flow carrying capacity to reduce the occurrence of raw sewage overflows during heavy rain storms. This new sewer and the WERS are designed to work together to meet future peak flows during a 40 year planning period.

The new sewer was also designed to provide for flow diversions during the rehabilitation of the WERS to avoid the need for bypass pumping. As there were numerous bends at each manhole over half the alignment, CIPP construction was preferred because it allowed inversions through multiple bends, eliminating the need for sliplining pits and thereby reducing disruptions in wetland areas. This flow diversion capability as well as the reduced environmental impact of CIPP construction were probably the principle reasons why CIPP technology was more competitive than the sliplining method during the bidding phase. By way of comparison, the new sewer was constructed over a 2-1/2 year period at a cost of about $38 million, while the rehabilitation work was bid at a cost of approximately $12 million with a duration of about two years.

The project is presently (December 1994) under construction with most of it being completed using the CIPP method of rehabilitation by Insituform of New England, Inc. At the time of this writing, consideration is being given by the contractor to slipline the 915 km (3,000 foot) tunnel. As will be pointed out later in this paper, we believe that costs and construction logistics were the primary reasons for changing construction methods for the tunnel.

Design Considerations

As a result of the condition survey, it was determined that the rehabilitated pipe design would be based on a fully deteriorated condition of the existing or host pipe. This was evident from the loss of concrete, the prominence of exposed aggregate, and visible reinforcing steel throughout the pipe length. This design basis provides for the rehabilitated pipe to support all dead loads, live loads, and groundwater load imposed, including the 100 year flood elevation, with the assumption that the existing sewer pipe cannot share any loading, nor contribute to the structural integrity of the completed pipe.

Design Loadings - The design of the rehabilitated pipe also needed to address the trench and tunnel construction methods used for the WERS pipeline. Design calculations were submitted to support each inversion reach to be installed. Calculations were provided for the following parameters: (1) maximum deflection, (2) minimum pipe stiffness, (3) ring bending strain, (4) hydrostatic collapse resistance using short-term modulus of elasticity (for sliplined pipe), and (5) constrained buckling resistance using long-term modulus of elasticity.

Consideration for the short term loading condition which applies while the rehabilitated pipe is being installed and the long term condition once the new pipe is in place needed to be addressed for both the CIPP and sliplined systems. The short term condition for the CIPP process turned out to be negligible since the internal head used to install the liner tube needs to be greater than the groundwater pressure to prevent liner tube collapse. The long term loading condition was found to be the controlling basis of design using unconstrained buckling resistance with the long term modulus of elasticity for design of the CIPP liner thickness. Liner thickness ranged from 19 mm (0.75 inches) for the 122 cm (48 inch) CIPP to 33 mm (1.3 inches) for the 152 cm (60 inch) CIPP.

Conversely, for the sliplining method of pipe rehabilitation, the designed wall thickness of the pipe is generally controlled by the unconstrained collapse resistance of the highest external pressure resulting from groundwater above the pipe or from grout pressure in filling the annular space. The hydrostatic collapse pressure is one of the more important design parameters and is imposed on the sliplined pipe after it has been inserted into the existing host pipe and while it is being grouted in place. This condition occurs since the sliplined pipe has no side support at the springline of the pipe. This is considered a short term loading condition because once the liner pipe is grouted and the grout sets up, a constrained condition exists and the grout now fully supports the liner pipe approaching ideal bedding or support conditions similar to concrete encasement of the liner pipe. The long term design condition is determined by the constrained buckling resistance after the sliplined pipe has been grouted in place and when the sliplined pipe is below the groundwater table. Constrainment, or pipe support from the annulus space grouting of the sliplined pipe, greatly increases its resistance to wall buckling under hydrostatic load and therefore did not control in the design of the pipe.

Tunnel Design - For the 915 km (3,000 foot) tunnel, acknowledgement was made during the design to allow for the soil arching action of the tunnel (soil overburden on the existing pipe varied from 8 to 26 m (25 to 85 feet)). This was considered prudent since the tunnel was constructed using drill and blast techniques in the rock and mixed face portions, and by compressed air in a soft ground portion. Precast concrete pipe sections were used for the finished tunnel lining. Since the overburden was undisturbed, it was reasonable to design for an equivalent 6 m (20 foot) prism earth loading for the CIPP or sliplined pipe.

If the CIPP process was used for this tunnel, a tower 17 m (55 feet) tall would be needed to provide sufficient hydraulic head during installation of the liner tube to overcome the external ground water pressure on installation forces (maximum groundwater above the crown of pipe was 17 m (55 feet)). While such an installation would not be impossible, a number of factors would need to be considered. These factors included forces on the tube and tube strength to prevent it from bursting under such a high head, controlling advancement of the tube during installation (over 445 kN (100 kips) force would be exerted on the liner tube), tower structural design, protection of downtube from overstress, liner tube transportation limitations and the need to splice sections at the job site as compared to more controlled conditions at the factory, inversion limitations on a tube of this diameter and size, the likelihood of an overlap if the work was done half at a time from two directions, man entry safety concerns to prepare an overlap in such a confined space, and the need for ensuring proper wet-out and curing times. For these types of reasons and their associated costs, the contractor opted to rehabilitate the tunnel using the sliplining technique.

Manhole and Structure Rehabilitation - Another factor that was considered during the design was that the manhole and special structures also be rehabilitated based on compatibility with the pipe rehabilitation system as well as providing a structural rehabilitation for a 40 year planning period. There are three types of manholes along the WERS, 122 cm (48 inch) diameter precast concrete, 122 cm (48 inch) diameter brick, and 107 cm by 137 cm (42 inch by 54 inch) oval brick manholes. In general, the manholes were in good structural condition. Since at the time of bidding there were limited manhole rehabilitation technologies with a proven 40 year design life to resist hydrogen sulfide corrosion, the approach taken was to provide a proven PVC structural liner system rather than use a protective coating system. The contractor chose to use the Permaform manhole system which consists of pouring a new 76 mm (3 inch) thick concrete wall on the inside of the manhole and placing interlocking PVC sheets on the inside of the form prior to pouring the concrete. The PVC sheets provide the corrosion resistance required for hydrogen sulfide.

For the existing concrete structures such as flow connection chambers and siphon head houses, varying degrees of deterioration were evident. However, sufficient concrete was in place in most areas to use a protective coating system since there were limited structural liner systems on the market. A 100 percent solids epoxy system was chosen and applied at a thickness of 65 mils for corrosion resistance with thicknesses of up to 250 mils in heavily corroded areas.

The contractor was required to ensure that both the manhole rehabilitation system and the structure coating system were compatible with the pipe rehabilitation system used. In all cases except the tunnel, this meant that the manholes and structures would be rehabilitated after the CIPP work was complete as a means of achieving a quality finished construction. Similarly, the manholes at either end of the tunnel will be completed after the sliplining is complete.

Environmental Concerns

As indicated previously, the environmental significance of this project became one of the driving forces in both setting the design performance specifications as well as in effecting the least environmentally disruptive method of construction. Even though many permit requirements were placed on the recently completed new sewer construction by open cut construction, the permitting agencies wanted to avoid the need to impact the already revegetated wetlands along the pipeline alignment during the rehabilitation phase. This objective set the stage in pushing the limits on the CIPP and sliplining technologies since long pipe reaches of generally over 305 m (1,000 feet), and in some cases over 610 m (2,000 feet), would be needed on a regular basis.

Since the permitting agencies required the designation of which areas would be impacted during the rehabilitation work, the access and staging locations for each method of construction were shown on the plans. In this way, permit conditions were placed specific to any environmentally sensitive ares. For example, reduced easement widths were stipulated to minimize impacts across the wetland areas; tree counts were required at staging locations in the event additional clearing would be needed so as to replace a tree for a tree; time restrictions were placed for work alongside the river bank during the flood season of February to May; sliplining access shafts were designated and selected locations adjacent to the river were required to have steel sheeting for earth support; and any disturbed wetlands were required to be revegetated within two years.

Working with the environmental agencies early in the project allowed them to gain greater understanding of the design and construction issues. We believe that involving them in the design and construction efforts promoted greater cooperation among the parties and contributed to the success of the project.

Construction Considerations

Several of the more important construction considerations are described below:

Access and Staging Areas - Environmental restrictions and permit conditions required the contractor to plan his access and staging area locations to minimize impacts along the environmentally sensitive areas. Private property owners adjacent to a staging area or access location also needed to be recognized by reducing surface disruptions as well as monitoring construction noise during work activities. Since the Insituform process requires 24 hour work periods, this became an essential element of the work.

Long Installation Reaches - As mentioned previously, the contract was set up to provide for long installation reaches to minimize these types of impacts. In the past, this type of work was generally limited to a localized section of pipe needing repair. This project, which required a continuous length of 12 km (7.5 miles), required careful planning of the various work activities. The first inversion installation consisted of 450 m (1,400 l.f.) of 122 cm (48 inch) CIPP, which was a record length for this installer. Over the course of the project, new records have been set and broken, with the current record length at 720 m (2,352 l.f.) of 122 cm (48 inch) CIPP. Lengths originally considered exceptional are now common practice, with over 30 inversions to be completed with inversion lengths ranging from 335 m (1,100 l.f.) to 720 m (2,352 l.f.).

Flow Handling - Flow handling was facilitated by use of the flow splitting chambers along the route of the WERS and the parallel new interceptor sewer. Sewage pumps were also used to handle local flows from intermediate service connections or for pipeline infiltration. Once the flow was bypassed from the pipe to be rehabilitated, cleaning and internal inspection activities were performed.

Cleaning and Internal Inspections - Cleaning was required to remove all debris, sludge, rocks, grease spalled concrete, and other solids. An internal man entry video inspection followed the cleaning operation to verify that the line was clean and that there were no physical obstructions that would prevent installation of the inversion liner such as collapsed or crushed pipe, protruding rebar, protruding service connections, excessive joint leakage or reductions in cross sectional area at offset joints. The contractor was required to furnish inspection reports with video tapes and photographs for pre and post construction internal inspections.

Construction Sequencing and Scheduling - Proper sequencing and scheduling of work activities were important considerations to assure timely completion of each work phase. The first seven liner inversions were completed during snow storms and -20°C (-5°F) temperature conditions. Resin tanker deliveries were delayed because of the poor travel conditions and precautions were needed to keep the resin flowable by heating it when it arrived on-site. In comparison, inversions were also completed during the extended mid-summer heat wave and precautions were needed to prevent the resin from early curing from the heat by applying ice baths to the resin as it was pumped from the tanker to the liner tube. It was readily apparent that these types of issues were important aspects that needed to be factored in the project schedule since they would be prevalent throughout the project.

CIPP Installation - The size and length of each installation reach required the use of on-site wet outs to saturate the liner tube with resin as opposed to a factory wet out for smaller diameters. The wet out machines consisted of a roller bed system and roller frame system that would uniformly distribute the resin throughout the liner tube. A static mixer was used to blend the resin and a catalyst before it was pumped into the liner. Vacuum pumps were used to remove any trapped air while the liner was being saturated with resin. The tube then was advanced through a roller frame which set the thickness of the resin before it was inverted into the manhole or structure.

As the liner left the roller frame, it was inverted down the manhole and into the pipe. The liner tube was then turned inside out at the pipe/manhole opening and cuffed back on a frame at the top of the manhole. Water was then pumped inside the tube to extend the liner to the termination point. The water head provided progressive rounding of the tube to hold the tube tight against the pipe wall as the liner was installed.

After the inversion was completed, boilers would be brought on-site to heat the water within the tube to a temperature of 82°C (180°F) and pump heated water through recirculation hoses throughout the pipe. The heating schedule for each inversion varied according to the length, temperature of inversion water, and the capacity of the boiler. Thermocouples were placed between the existing pipe and the liner tube at the inversion and termination locations to monitor the temperature on the backside of the tube and verify that the cure was effective. Generally, the heating and curing period took about 36 to 48 hours. Upon proper curing, cool water was then added to the recirculating water within the pipe to gradually decrease the temperature of the cured pipe to 38°C (100°F) and allow discharge of the water down the existing sewer system. The cooldown process took 36 to 48 hours.

Special Considerations

Other special construction considerations that were addressed throughout the project are described below.

<u>Liner thickness transitions at manholes</u> - The CIPP was designed based on the required pipe stiffness for each pipe installation reach between the inversion manhole and the termination point. Any liner thickness transitions or changes were required to occur at manhole locations so that the controlling design condition between manhole reaches would be used similar to open cut pipe design.

<u>Liner overlaps</u> - The long inversions on this project required the use of overlaps between contiguous upstream and downstream locations in many areas. All overlaps were preferred to occur in the downstream location so as not to be affected by the flow, however, because of liner installation sequencing this was not always possible and upstream overlaps were used. The ends of both CIPP sections were specified to be bevelled to provide a smooth transition and to limit exfiltration problems. A watertight joint was prepared by placing three hydrotite seals that would swell when water contacted them on the second CIPP section. The contractor was also required to provide a bond between the laminates of the two installed CIPP sections by grinding off a 1.5 m (5 foot), 360 degree strip of the coating at 3 to 4.5 m (10 to 15 feet) from its end and applying a compatible resin to the strip just prior to beginning the overlap inversion. A mechanical type lock was also achieved by cutting a groove in the first CIPP section.

<u>Liner tube advancement</u> - This consideration was important in order to control the speed or advance of the liner tube while it was being installed. Since each inversion location had varying terrain conditions and availability of inversion water, proper planning was essential to avoid any unexpected conditions during the liner installation. Inversion water was generally supplied by a 152 mm (6 inch) pump drawing water from the nearby river. The inversion head water would be increased to advance the liner. Advancement of the liner was normally controlled by the use of hold back ropes tied off to the roller frame. In one case, use of a tractor on the opposite side of the river was needed to control the hold back forces during liner installation.

<u>Liner Terminations</u> - The liner was normally completed before the manholes or structures were rehabilitated. The liner was extended into manholes to allow for a smooth, clean cut to match the configuration of the manhole base and riser sections. The crown section of the liner was used as a base form for the manhole rehabilitation system. At structures, the liner terminated flush with the inside face of the structure wall to provide a smooth, clean cut and continuous transition from the liner pipe to the structure wall.

<u>Material Properties</u> - The materials for the CIPP consisted of a polyester resin

which normally has a modulus of elasticity of 1,725 MPa (250,000 psi). The manufacturer had recently tested and proposed use of a polyester resin having a modulus of elasticity of 2,415 MPa (350,000 psi). This was a 40% increase over what the manufacturer had used in the past. These values were supported by long term testing conducted under ASTM D2990. Test results from field samples are supporting these higher values with numerous results between 2,760 MPa to 3,100 MPa (400,000 psi to 450,000 psi) for the modulus of elasticity.

Material Testing - For each inversion, the contractor was required to provide two flat plate or unrestrained samples which were tested in accordance with ASTM F-1216 and ASTM D-790. The purpose of this testing was to validate that the design basis for the pipe was achieved. In addition, preselected locations were designated where restrained samples would be obtained by using a pipe form of the same diameter of the rehabilitated pipe or by cutting a section of liner at overlap locations. The restrained samples would be more representative of the finished cured-in-place pipe and provided further support to the unrestrained sample testing.

The restrained samples generally provided results for the flexural modulus of elasticity of approximately 70 to 90 percent of the flat plate samples. These results can be attributed to the variability in the preparation of the flat plate samples since they are made by hand from the same material and are cured on both sides in the down tube. Through statistical testing results and comparison of flat plate versus restrained samples, we established a correlation of using 80% of the flat plate results to represent the restrained or actual field installed liner. The same liner thickness as being installed is prepared and then placed in a metal frame for insertion in the downtube where it is placed during the curing and cool down periods to simulate the conditions that the liner is undergoing. This procedure is an acceptable method of verifying the design data and is included in ASTM F-1216.

The contract specifications required that the contractor supplement and confirm the flat plate testing with restrained samples taken from the pipe. It was recognized that this procedure was not practical for every inversion but wherever it was possible, these samples were obtained. This type of testing was considered essential in order to maintain the quality and consistency of each pipe installation. Testing was performed for flexural modulus, flexural strength, liner thickness, and peel strength. Results in the flat plate samples were reduced to acknowledge the more controlled conditions at which those samples were prepared and cured. There were instances where the contractor was required to submit revised design calculations to support actual field data where the test results fell below the original design values. All inversions were verified to meet the liner thickness and strength characteristics required for each pipe reach.

Tunnel Rehabilitation by Sliplining - Since the contract documents provided for the alternative of sliplining, when the contractor indicated he intended to complete the tunnel by this method, it was noted that he would need to meet the qualification and performance specifications already included in the contract. Access was limited to manholes at either end of the 915 m (3,000 foot) tunnel. However, during the design, the sliplining shaft was designated at a location approximately 305 m (1,000 feet) downstream to avoid excavation and surface disruptions to a commercial golf driving range. To further complicate this type of construction, no shaft was allowed at the upstream tunnel manhole to avoid surface disruptions in that residential neighborhood.

The current construction plan provides for a 1,220 m (4,000 foot) sliplining installation for a 137 cm (54 inch) diameter pipe inside a 152 cm (60 inch) diameter pipe. Since this eventuality was considered during the design, the critical issues were incorporated as part of the specifications and included the following items:

> Grout pressure - Grouting of the annular space was required to be completed from internal ports in the pipe since surface activities above the pipe were prohibited. The pipe was specified to be designed to withstand a grout pressure of 70 kPa (10 psi) due to the sensitivity of the tunnel, the limited access, and the unknown and perhaps unconventional methods that could be chosen by a contractor.

> Groundwater pressure - The 70 kPa (10 psi) maximum grout pressure was not a governing factor since the specifications required the pipe design to be based on the highest hydrostatic groundwater pressure along the tunnel which is 17 m (55 feet) or 165 kPa (24 psi).

> Pipe strength - The specifications also called for the highest class or strength of pipe to be based on the design of the pipe between the tunnel manholes. (Pipe stiffnesses of 830 kPa (120 psi) are expected for this sliplined pipe.)

> Annulus space - The minimum annulus space was specified as 51 mm (2 inches) to provide greater clearance within the tunnel during sliplining. The pipe joint was also required to be smooth with the same outside diameter as the liner pipe. These considerations provided for a greater degree of confidence that the sliplining could be successfully completed.

> Fiberglass pipe by Hobas USA was the only sliplining product that could meet these criteria.

Installation of sliplining pipe for a 1,220 m (4,000 foot) reach was envisioned to be performed by training segments of pipe at a time rather than jacking individual pipes as is normally done. This method allows a segment of several pipes to be pulled to the upstream end and successively jointed to the previous train segment. Jacking and winching forces will be important considerations to monitor for the pipe design. Based on this scenario, this method of rehabilitation construction is expected to advance the capabilities of this industry.

Summary and Conclusions

Due to the size and significance of this project, special considerations such as design for a fully deteriorated pipe with high groundwater levels in tunnel and trench applications; cleaning and internal inspection; manhole and structure connection details; on-site liner saturations; liner tube advancement and overlaps; terminations; repairs; quality control elements related to finished product; and testing verification could not be taken for granted nor considered as routine. Job related logistics such as scheduling, weather conditions, material transportation, flow handling, site access, construction staging, and environmental restrictions were important for a successful outcome. These types of issues presented job difficulties which needed to be addressed at various times throughout the project. By addressing these problems, greater understanding of the capabilities and the constraints of the CIPP and sliplining systems were gained and can be applied to future projects.

The experience gained on this project has demonstrated that large diameter sewer rehabilitation can be successfully performed in long installation reaches. Most CIPP inversions were at least 305 m (1,000 feet) with the longest being 720 m (2,352 feet) for the 122 cm (48 inch) diameter; 550 m (1,793 feet) for the 137 cm (54 inch) diameter; and 485 m (1,594 feet) for the 152 cm (60 inch) diameter pipe. This project has shown that, with proper planning and coordination, this technology can break through the limits of the past.

When specifying CIPP technology, readers should be aware of four specific problems encountered on this project which are summarized below:

- During one inversion performed in hot weather in which proper precautions were not taken to cool the resin, the liner was only partly inverted before the resin prematurely hardened and halted the liner advancement. The contractor was required to absorb the time and expense to manually cut and remove the section of liner from the pipe. Additional inspections and testing were performed to verify that the installed liner was acceptable.

- Pipe cleaning became an issue, particularly in areas of limited or restricted access where the vacuum truck could not access each manhole. In some cases, manual cleaning was needed where mechanical or hydraulic cleaning did not work because of this restricted access or because of heavier sediments than originally anticipated. Verification of pipeline sediment depths is difficult to determine prior to bid, and readings taken at manholes did not provide representative conditions along the pipeline.

- Cold weather conditions including freeze-ups during cleaning operations, resin heating needs, and reduced productivity affected the project.

- The manufacturer and the contractor were reluctant to substantiate the design basis of their product even though the contract required them to do so. Due to the significance and magnitude of this project, it was considered essential for the contractor to fully document use of their product.

Other suggestions that should be considered in projects of this type include:

- Understand the capabilities, design basis, and limitations of each system to be used. One should not rely on manufacturer's literature or standard specifications.

- Understand and use ASTM references in specification development and product evaluation, installation, and testing parameters.

- Observe rehabilitation systems first-hand on another project to understand considerations that may apply to your project.

- Be firm and consistent in making the contractor adhere to the contract provisions for submittals, product descriptions, design basis, product installation considerations, and testing verification.

This project has advanced the technology of the CIPP sewer rehabilitation market by successfully addressing these types of issues. Record installation lengths and the experience on this project have provided the technological and engineering confidence to advance this industry on larger and more complicated projects in the future.

This paper represents the opinions and conclusions of the authors and not necessarily those of the Massachusetts Water Resources Authority.

I:\reports\cambridg\ascepipe.rpt

Computer Aided Pipeline Design:
A Step-by-Step Description of the Process

Jorge A. Garcia, Ph.D., P.E.[1], Assoc. Member, ASCE

Abstract

State-of-the-art pipeline engineering utilizes modern computer hardware and software throughout the design process. Survey data collected by computerized instruments or GPS systems allow digital transmission of data to personal computer systems. Data is then manipulated in the computer utilizing Computer Aided Design Software (CAD) to develop terrain surfaces, horizontal and vertical alignments and cut/fill volume computations if required. The complete design process takes place in real time interface between the engineer and CAD. The final design is then transferred to a selected output media.

The process just described involves several intermediate steps and possible problems and options that are not fully described in the literature. The potential for erroneous design is very real when digital data are electronically transmitted between instruments and are blindly manipulated inside the computer environment. A step-by-step description of the process helps in visualizing potential problems and optimizing design efforts.

This paper will address in detail the steps involved in CAD pipeline design, including potential problems and limitations. A design case study of a 457-mm (18-in) water transmission pipeline in the City of Las Cruces, New Mexico is presented as part of the paper.

---

[1]Design Engineer, Utilities Engineering Dept., City of Las Cruces, P.O. Drawer CLC, Las Cruces NM 88004.

Introduction

Pipeline design involves a number of processes from hydraulic analysis to the development of final construction plans. State-of-the-art computer technology allows the entire process to be performed with the use of personal computers. Hydraulic network analysis software is used in the preliminary stages of design, where different layouts, pipe sizing and networking are explored. Next, a detailed analysis of the physical setting is performed during the field survey which also utilizes computerized digital instruments. Survey data in digital form is then brought into a computer aided design environment (CAD) where all of the design process will take place. The outcome is a set of final construction drawings which, coupled with technical specifications, constitutes the complete contract document.

The process of taking field survey data into CAD to generate final construction drawings is the subject of this paper. This process, outlined as a process flowchart in Figure 1, requires the manipulation of data to generate a working terrain model on which to lay the pipeline. A horizontal alignment which includes pipe centerline, stationing, and construction and permanent easements is created. The horizontal alignment is used in conjunction with the terrain model to generate vertical pipeline profiles at desired stationing intervals and a selected vertical scale. Next, vertical tangents and vertical curves are designed to provide finish grades. Vertical layout of the pipeline follows, usually maintaining a uniform depth. Location of combination air valves, gate or butterfly valves and other fittings are designed on the final profile.

Construction drawings include a general vicinity map, plan and profile sheets (P&P) for the pipeline alignment, fittings and other details. P&P sheets are generated by linking together horizontal alignments and vertical profiles. The linkage is performed in a paper environment which takes into account a selected media size and scale for the drawings.

The sections that follow will address in detail the different facets of the process just described. Conclusions will be derived on proper and improper procedures in CAD pipeline design, tips for more effective use of the CAD environment will also be presented. Several P&P sheets generated from the design drawings of a 457-mm (18-in) water transmission pipeline for the City of Las Cruces, New Mexico are presented in the paper.

This paper will specifically describe the CAD design process utilizing AutoCad R.12 and AdCADD Design Modules. The commands and terminology employed are software specific and might not directly apply to other software packages. The intent of the paper, however, is to outline a design methodology and expose potential problems and limitations that can hopefully be extrapolated to other design software.

COMPUTER AIDED PIPELINE DESIGN

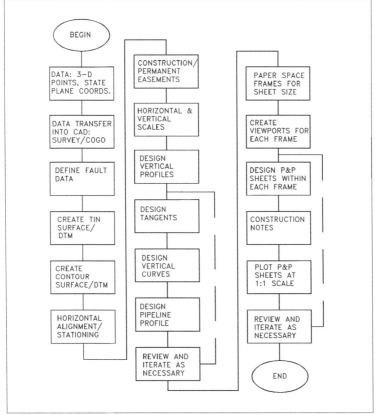

Figure 1. Process flowchart for CAD pipeline design

Field Survey Data and CAD

The basis for field survey information is a three dimensional space with axes North, East and elevation. Points are collected along the proposed pipeline alignment utilizing modern surveying instruments which are electronically transmitted or downloaded to a personal computer for analysis.

Figure 2 shows a series of points surveyed along a pipeline alignment. The points have a State Plane coordinate position and an elevation. A collection of points constitutes the basis for the development of a terrain model on which the pipeline will be designed. Field survey points also cover a wide enough area to provide terrain information for construction and permanent easements.

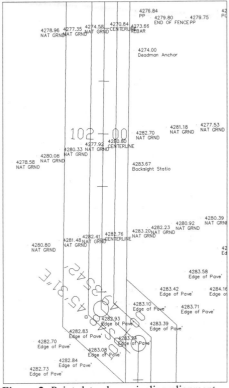

Figure 2. Point data along pipeline alignment

The transfer or import process of three dimensional point data into the CAD system is performed using a Survey module. Data can be in a variety of different formats, with the most common being either space delimited or comma delimited ASCII files. Once the desired format is selected, data is imported into a coordinate geometry (COGO) point file where it can be fully utilized by the CAD system. As part of the field survey, existing utilities and other structures are also entered as point and line data.

Point data shown in Figure 2 also depicts the terrain model surface boundaries. Outside points need to be joined with a three dimensional (3-D) polyline to define fault data.

Point data together with fault data enable the creation of a 3-D triangulated irregular network (TIN) surface where individual points are linked together by a series of contiguous triangles. The sides of these triangles can then be broken into elevation increments, thus creating an interpolation mechanism for development of a contour surface. The TIN is developed using a DTM module and can be edited as necessary to adequately fit project boundaries.

Once a final TIN has been developed, existing ground contours are created within the DTM at selected contour intervals. It is advisable to keep existing ground contours in properly labeled CAD layers in case there is a need to develop finish ground contours.

The section that follows will describe the use of the terrain model to design both horizontal alignment and vertical profiles for the pipe. The section will also introduce the distinction between model space and paper space inside the CAD system, and will address the importance of coordinate reference system consistency in successful CAD pipeline design procedures.

Horizontal Alignment and Vertical Profiles

The pipe centerline is designed on the terrain model surface developed with point and fault data. Construction and permanent easements are also designed, sometimes simply using the OFFSET command on both sides of the pipe centerline. Using appropriate surveying convention, the pipe centerline alignment here referred to as the horizontal alignment is stationed, points of intersection (PI) are labeled and bearings are given for each straight section. This part of the design process is performed with the Design module within the CAD system.

Figure 3 shows the horizontal alignment developed for the South Zone-1 Project in Las Cruces, New Mexico. This design project involves a 457-mm (18-in) water transmission pipeline from the Jornada storage tank south to the Telshor booster station. Although the primary purpose of the water line is to bring water to new developments located within the Zone-1 pressure zone, the line will also feed a lower pressure zone through pressure reducing valves (PRVs).

The 6144.5 m (20,159 ft) alignment of this transmission pipeline involves crossing Federal and State lands with rugged topography. The pipeline also crosses a subdivision in the process of development. Figure 3 also shows location of the pipeline parallel to section lines, portions of existing subdivisions and a tank site. Although the scale of the figure in this paper makes it difficult to see any detail, this sheet illustrates some of the CAD labeling of bearings, horizontal curves, stationing and several general surveying notes.

The horizontal alignment has been constructed in a terrain surface developed from field data referenced to the State Plane coordinate system. The alignment, therefore, is fully consistent with this reference coordinate system. At this point it is important to emphasize that the entire design process is performed utilizing the State Plane coordinate system or any pre-determined coordinate reference. Entities such as lines, points or any object that represents the real world must be kept in its surveyed position. This allows for additions or any modifications that might be needed later. Any additional survey data can be easily imported and added to the present design if this design has preserved real world coordinates. At this point, no scaling of the final construction drawing product has been addressed or is even relevant. The design itself takes place at a one-to-one scale, meaning that the distance between consecutive 30.48 m (100-ft) stations, for example, is 100 units.

Design of the horizontal alignment complete with other related objects such as easements, is followed by the development of vertical profiles. The terrain must now be viewed in profile to address cut and fill areas and to design the finish grade. This process takes place also within the Design module of CAD and involves setting up a vertical scale, horizontal scale and reference frames for the development of plan and profile sheets. In the case of the project referenced in this paper, a vertical scale of 25.4 mm (1-in) equals 1.524 m (5 ft) and horizontal scale of 25.4 mm (1-in) equals 15.24 m (50 ft) was utilized. A total of sixteen stations at this scale fit the traditional D-size drawing of 864x559 mm (34x22 in), including pertinent axis labels. The frames created here are not designed to be used in the development of final P&P sheets, but for reference only. For example, the first frame will contain the profile from station 0+00 to station 16+00, the second frame will contain stations 16+00 to 32+00, and so on. Each pipeline frame is named using the VIEW command for prompt retrieval at a later stage.

The development of vertical profiles involves four phases. The first phase is to create the existing ground profile using the terrain model and the horizontal alignment. The second phase is to design vertical

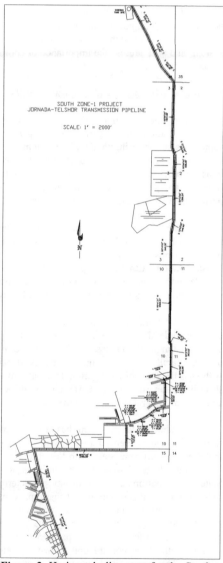

Figure 3. Horizontal alignment for the South Zone-1 water transmission pipeline

tangents and vertical points of intersection (PVI). The thirds phase involves the design of vertical curves using the tangents and a desired length of curve. Phases two and three may involve an iterative process where consideration is given to cut/fill volumes, haul distances, etc. The fourth and last phase is to lay the pipeline at the desired depth. If a uniform depth is utilized throughout the alignment, the OFFSET command can be used to accelerate the process. The design of combination air valves is dependent on the pipeline profile and is shown on corresponding profiles and on the horizontal alignment. Also, isolation gate or butterfly valves are shown on profiles and horizontal alignment.

Development of Plan and Profile Sheets

The design of the horizontal alignment and vertical profiles was performed in real world space, with no consideration given to output media size other than the reference frames utilized for setting up profile stationing. Linking the horizontal alignment to corresponding vertical profile to generate a plan and profile (P&P) sheet, however, requires consideration of the paper environment in which such sheets will be generated. This paper environment is called Paper Space in AutoCad R.12.

The paper environment is entered by switching the CAD system from Tilemode 1 to Tilemode 0, where the coordinate reference system will no longer be State Plane, but rather paper reference with distances in inches. Within Tilemode 0, a series of paper space frames within the appropriate layer are created with dimensions consistent with the desired output media size. For example, a D-size sheet will require a series of frames with dimensions 864x559 mm (34x22 in). Once the first frame is created, it can be copied as many times as required by the number of sheets in the construction plans. Each paper frame is then labeled with the VIEW command for later retrieval. When using the VIEW command, it is recommended to use the ASSIST/INTERSECTION feature so that exact dimensions of the frame are captured.

After the frames have been named, a VIEWPORT layer is created and made the current layer. Taking the first frame, two horizontal viewports are created. Remaining within Tilemode 0, the CAD system is switched to Model Space opening two windows into the real world environment as illustrated by Figure 4. Making the bottom window active, the first vertical profile is retrieved using the command VIEW/RESTORE followed by the first profile name. The scale is then adjusted using the ZOOM/XP command. If the profile requires more space than the default one-half sheet, switch back to paper space and use the STRETCH command to expand the size of the viewport.

Next, making the top window active, the horizontal alignment is retrieved using again the VIEW/RESTORE command. In this case, the complete alignment as it exists in the real world appears in the window as illustrated by the top portion

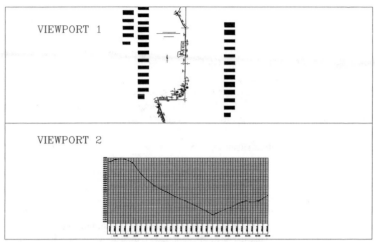

Figure 4. Horizontal viewports created in paper space

of Figure 4. Using the ZOOM command, the approximate portion of the plan view that will match the profile stationing is windowed. The plan view is then aligned horizontally with the paper axes using the DVIEW/TWIST command. The proper scale is set using the ZOOM/XP command as before. It is important to note that the MOVE and ROTATE commands should not be used in this case because this would move the horizontal alignment from its State Plane coordinate position in the real world.

The process just described is repeated for each P&P sheet. Figure 5 shows a completed sheet for stations 96+00 to 112+00. The top portion of the sheet shows the plan view including pipe centerline, easements, contours, stationing and bearings. Note how the DVIEW/TWIST command was utilized to align the view with the horizontal axes of the paper, leaving the North arrow pointing to the right. Note also that the viewport layer has been frozen for reproduction purposes, thus eliminating the horizontal line shown in Figure 4.

A situation may arise where the plan view changes direction within the selected stationing as was the case for stations 48+00 to 64+00 shown in Figure 6. In this case, the top viewport is stretched to the left, leaving room for the creation of a third viewport to fit stations 60+00 to 64+00. This third viewport is made active and the horizontal alignment is restored. The DVIEW/TWIST command is then used to align the plan view with the axes of the paper. The ZOOM/XP command is again used to set the proper scale.

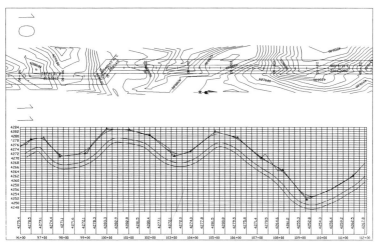
Figure 5. P&P sheet for stations 96+00 to 112+00

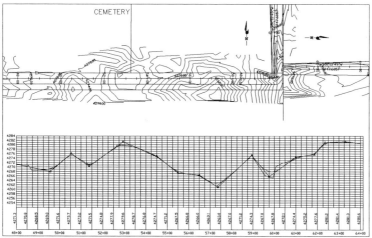
Figure 6. P&P sheet for stations 48+00 to 64+00

The use of the paper environment to generate P&P sheets allows the production of drawings to any desired sheet size and scale. The figures generated for presentation in this paper were developed by creating frames 139.7x85.3 mm

(5.5x3.36 in) in paper space. Horizontal viewports were created and plan and profile views were restored following the same process as the D-size construction sheets.

Some CAD Tips, Problems and Limitations

Computer aided design lends itself to numerous techniques that make the process more efficient. Depending on the software being utilized, the designer and design conventions, several lessons can be learned. Some design tips dealing with the manipulation of the CAD environment are given below:

1. Perform the complete design in real world reference coordinate system. Never MOVE or ROTATE any object or feature from its original survey position. Perform a complete project design on the same file if hardware is not a limitation.

2. Utilize the VIEW command to name the horizontal alignment, individual profiles and final P&P sheets in paper space. This allows easy retrieval when generating or plotting P&P sheets.

3. Always use the ASSIST feature in CAD to define views. Exact dimensions or intersections are obtained in this manner.

4. Use DVIEW/TWIST commands in model space within Tilemode 0 to develop plan views for individual P&P sheets. This process rotates the view rather than the object or feature.

5. Set proper scale in model space within Tilemode 0 using the ZOOM/XP command. This will insure exact dimensions when a drawing is generated on paper.

The design tips outlined above are directly linked to potential problems that commonly arise. For example, any intentional or accidental move of objects can introduce serious problems which might not be obvious until the construction process takes place. Any linkage to other projects or survey of additional points or features are also affected if real world reference position is not maintained.

Survey data is electronically collected and imported into CAD as described earlier. It is good practice to utilize the database query features of the CAD modules to check elevation ranges in points against hard copies of the original data. In this manner, any extreme discrepancies can be detected and resolved.

When field data has been collected in a band along a proposed centerline, changes in alignment will cause the TIN surface to inaccurately extrapolate the data. This limitation can be overcome by carefully editing the TIN outside fault line data, so that the terrain model contours accurately represent field topography.

Final P&P sheets must include construction notes that clearly indicate the stations where combination air valves, isolation valves and fittings are to be installed. Also, appropriate notes must be included pertaining to allowable joint deflections. Stations for horizontal points of intersection (PI) and points of vertical intersection (PVI) must be properly labeled.

Summary

Computer aided pipeline design is an interactive process that involves several steps. First, the engineer creates a representation of field topography utilizing field survey data along a proposed pipeline alignment. Point data and fault line data are used to create a TIN surface which is edited to reflect surveyed areas. Next, a terrain model is created by generating a contour surface at desired intervals.

Once a terrain model is created, a pipeline centerline is designed on this surface. Appropriate construction and permanent easements are also designed. The centerline is stationed and bearings are labeled using standard conventions.

Taking the terrain model and the horizontal alignment, vertical profiles are generated at selected station increments. Next, vertical tangents and vertical curves are designed for the finish grade. The pipeline is then designed at a given depth below finish grade, air and isolation valves are located and properly labeled.

After the design has been performed in a real world reference system, a paper environment is used to create the construction drawings. Frames are created in this paper environment taking into consideration media size and scale. Viewports are opened in these frames to accommodate plan and profile drawings for each sheet. Utilizing previously defined views, individual sheets are plotted.

Appendix

References

1. "AdCADD Civil Survey Modules", Softdesk Inc., 7 Liberty Hill Road, Henniker NH 03242.

2. "AutoCad Release 12 Reference Manual", Autodesk, Inc., publication 100752-01, August 6, 1992.

3. South Zone-1 18" Water Transmission Pipeline CAD design file, City of Las Cruces Utilities Engineering Department, Las Cruces, New Mexico.

TRANSFER MATRIX TECHNIQUE and PIPE STRUCTURAL ANALYSIS
Marcel GERBAULT[1]

Abstract

In the previous paper, *'A Soil-Structure Interactive Model'* (Ref [1]), given in this 2nd Conference on Underground Pipeline Engineering, a soil-structure interactive model has been developped where the soil is simulated by distributed springs placed around the pipe. An accurate analysis of this type of model needs to take account of 2nd. order theory and of the influence of axial forces. Transfer matrix technique gives an appropriate answer to this question. The relevant transfer matrices were established and are given here.

1- Introduction

This paper supposes a previous reading of the here above mentioned paper [1], from which it should be considered as a continuation. The same symbols and definitions are used. In order to introduce the transfer matrix technique, a pipe section has been discretised into n elements and n+1 nodes. Elements are rectilinear, so the section is a polygonal line. All the elements are rigidly fixed in sequential way one to the other.

At node #I a state-vector is defined equal to : $\mathcal{E}_i^T = [u_i, v_i, \omega_i, M_i, N_i, T_i, 1]$ where u_i and v_i are the local coordinates of the displacement of node I, ω_i its rotation, M_i the bending moment oriented as ω_i, N_i the normal force (positive for a compression) and T_i the shear force. The last component 1 allows the introduction of a 7th. column reserved to loads in the transfer matrix. A transfer matrix **T** allows to pass from state-vector $\mathcal{E}_i$ at node I to state-vector $\mathcal{E}_{i+1}$ at node I+1, by the equation :
$$\mathcal{E}_{i+1} = \mathbf{T}_{i,i+1} \mathcal{E}_i \qquad (1)$$

The matrix $\mathbf{T}_{i,i+1}$ is a 7x7 matrix. The transfer matrices of elements for which 2nd order is neglected are well known (see Ref [2] and [3]) ; they are given in relevant text books. For buried flexible pipe design analysis, it is important to take into account the 2nd order effects, in order to deal with the amplification of deformations (pipe void or under vacuum) or with the stiffening due to internal pressure (pressure pipe). These matrices were established, but first we recall here the classical results when 2nd order is neglected.

---
[1] Prof. E.N.P.C. ; SADE, 28 rue de La Baume, F-75008-PARIS, FRANCE

2 - Transfer matrix

2.1 Free element

Let ℓ be the length of the element, EI its flexural rigidity and S its cross section per unit length.

$$\mathbf{T}_{i,i+1} = \begin{bmatrix} 1 & 0 & 0 & 0 & -\dfrac{\ell}{ES} & 0 & -\left(\dfrac{\tau_i}{3} + \dfrac{\tau_{i+1}}{6}\right)\dfrac{\ell^2}{ES} \\ 0 & 1 & \ell & \dfrac{\ell^2}{2EI} & 0 & \dfrac{\ell^3}{6EI} & -\left(\dfrac{4p_i + p_{i+1}}{120}\right)\dfrac{\ell^4}{EI} \\ 0 & 0 & 1 & \dfrac{\ell}{EI} & 0 & \dfrac{\ell^2}{2EI} & -\left(\dfrac{3p_i + p_{i+1}}{24}\right)\dfrac{\ell^3}{EI} \\ 0 & 0 & 0 & 1 & 0 & \ell & -\left(\dfrac{2p_i + p_{i+1}}{6}\right)\ell^2 \\ 0 & 0 & 0 & 0 & 1 & 0 & (\tau_i + \tau_{i+1})\dfrac{\ell}{2} \\ 0 & 0 & 0 & 0 & 0 & 1 & -(p_i + p_{i+1})\dfrac{\ell}{2} \\ 0 & 0 & 0 & 0 & 0 & 0 & 1 \end{bmatrix} \quad (2)$$

The loads applied to the element are varying linearly from p_i to p_{i+1} for transverse load, and from τ_i to τ_{i+1} for longitudinal load. p_i represents the sum of normal soil pressure $\sigma_t(I)$, of hydrostatic pressure, internal as well as external, pipe own weight and possibly the soil reaction pressure Δp_i.

This matricial equation is set from two differential equations of elasticity in the element, and of three differential equations of equilibrium.

They are, for projection on local x-axis :

$$\begin{cases} \dfrac{dN}{dx} = \tau_i\left(1 - \dfrac{x}{\ell}\right) + \tau_i\dfrac{x}{\ell} & \text{equilibrium} \\ \dfrac{du}{dx} = -\dfrac{N}{ES} & \text{elasticity} \end{cases} \quad (3)$$

and for projection on y and z local axes :

$$\begin{cases} \dfrac{dT}{dx} = p_i\left(1 - \dfrac{x}{\ell}\right) + p_{i+1}\dfrac{x}{\ell} & \text{equilibrium} \\ \dfrac{dM}{dx} = T & \text{equilibrium} \\ \dfrac{d^2v}{dx^2} = \dfrac{M}{EI} & \text{elasticity} \end{cases} \quad (4)$$

The solution of this system is :

$$\begin{cases} u_{i+1} = u_i - \dfrac{\ell}{ES} N_i - \left(\dfrac{\tau_i}{3} + \dfrac{\tau_{i+1}}{6}\right)\dfrac{\ell^2}{ES} \\ v_{i+1} = v_i + \ell\omega_i + \dfrac{\ell^2}{2EI} M_i + \dfrac{\ell^3}{6EI} T_i - \left(\dfrac{4p_i + p_{i+1}}{120}\right)\dfrac{\ell^4}{EI} \\ \omega_{i+1} = \omega_i + \dfrac{\ell}{EI} M_i + \dfrac{\ell^2}{2EI} T_i - \left(\dfrac{3p_i + p_{i+1}}{24}\right)\dfrac{\ell^3}{EI} \\ M_{i+1} = M_i + \ell T_i - \left(\dfrac{2p_i + p_{i+1}}{6}\right)\ell^2 \\ N_{i+1} = N_i + \left(\dfrac{\tau_i + \tau_{i+1}}{2}\right)\ell \\ T_{i+1} = T_i - \left(\dfrac{p_i + p_{i+1}}{2}\right)\ell \end{cases} \qquad (5)$$

This system of six equations (5) is condensed in the symbolic matricial equation (1) above.

NOTE : if shear deformations are not negligible (as may happen for rigid pipes), then an additional term may be added in the 2nd. equ : $\quad -\dfrac{\ell}{GS_1} T_i$

with , for homogeneous material $\quad G = \dfrac{E}{2(1+\nu)} \quad$ and $\quad S_1 = \dfrac{5}{6}S$

2.2 Element resting on an elastic soil

The equilibrium equation projected on local y-axis -1st equ. of (4) - is changed to :

$$\dfrac{dT}{dx} = -\beta y(x) + p_i\left(1 - \dfrac{x}{\ell}\right) + p_{i+1}\dfrac{x}{\ell} \qquad (6)$$

leading to the differential equation, using (4) :

$$EI\dfrac{d^4 y}{dx^4} + \beta y = -p_1 + (p_1 - p_2)\dfrac{x}{\ell} \qquad (7)$$

The other equations of (4) are not changed.

Let :

$$\gamma = \left(\dfrac{\beta}{4EI}\right)^{\frac{1}{4}} \quad \text{and} \quad \tilde{u} = \gamma\ell$$

$$\begin{aligned} F_1 &= \cosh\tilde{u}\cos\tilde{u} & F_2 &= \dfrac{1}{2}(\sinh\tilde{u}\cos\tilde{u} + \cosh\tilde{u}\sin\tilde{u}) \\ F_3 &= \dfrac{1}{2}\sinh\tilde{u}\sin\tilde{u} & F_4 &= \dfrac{1}{4}(\cosh\tilde{u}\sin\tilde{u} - \sinh\tilde{u}\cos\tilde{u}) \end{aligned} \qquad (8)$$

PIPE STRUCTURAL ANALYSIS

The solution of this new system of differential equations is :

$$\begin{cases} u_{i+1} = u_i - \dfrac{\ell}{ES} N_i - \left(\dfrac{\tau_i}{3} + \dfrac{\tau_{i+1}}{6}\right)\dfrac{\ell^2}{ES} \\ v_{i+1} = F_1 v_i + \dfrac{F_2}{\gamma}\omega_i + \dfrac{F_3}{E I\gamma^2} M_i + \dfrac{F_4}{E I\gamma^3} T_i + \dfrac{1}{\beta}\left(F_1 p_i - p_{i+1} + F_2 \dfrac{p_{i+1}-p_i}{\tilde u}\right) \\ \omega_{i+1} = -4\gamma F_4 v_i + F_1 \omega_i + \dfrac{F_2}{E I\gamma} M_i + \dfrac{F_3}{E I\gamma^2} T_i + \dfrac{1}{\beta}\left(-4\gamma F_4 p_i + (F_1 - 1)\dfrac{p_{i+1}-p_i}{\ell}\right) \\ M_{i+1} = -4 E I\gamma^2 F_3 v_i - 4 E I\gamma F_4 \omega_i + F_1 M_i + \dfrac{F_2}{\gamma} T_i - \dfrac{1}{\gamma^2}\left(F_3 p_i + F_4 \dfrac{p_{i+1}-p_i}{\tilde u}\right) \\ N_{i+1} = N_i + \left(\dfrac{\tau_i + \tau_{i+1}}{2}\right)\ell \\ T_{i+1} = -4 E I\gamma^3 F_2 v_i - 4 E I\gamma^2 F_3 \omega_i - 4\gamma F_4 M_i + F_1 T_i - \dfrac{1}{\gamma}\left(F_2 p_i + F_3 \dfrac{p_{i+1}-p_i}{\tilde u}\right) \end{cases} \quad (9)$$

which results in the matrix $\mathbf{T}_{i,i+1} =$ equ. (10)

$$\begin{bmatrix} 1 & 0 & 0 & 0 & -\dfrac{\ell}{ES} & 0 & -\left(\dfrac{\tau_i}{3}+\dfrac{\tau_{i+1}}{6}\right)\dfrac{\ell^2}{ES} \\ 0 & F_1 & \dfrac{F_2}{\gamma} & \dfrac{F_3}{E I\gamma^2} & 0 & \dfrac{F_4}{E I\gamma^3} & \dfrac{1}{\beta}\left(F_1 p_i - p_{i+1} + F_2 \dfrac{p_{i+1}-p_i}{\tilde u}\right) \\ 0 & -4\gamma F_4 & F_1 & \dfrac{F_2}{E I\gamma} & 0 & \dfrac{F_3}{E I\gamma^2} & \dfrac{1}{\beta}\left(-4\gamma F_4 p_i + (F_1-1)\dfrac{p_{i+1}-p_i}{\ell}\right) \\ 0 & -4 E I\gamma^2 F_3 & -4 E I\gamma F_4 & F_1 & 0 & \dfrac{F_2}{\gamma} & -\dfrac{1}{\gamma^2}\left(F_3 p_i + F_4 \dfrac{p_{i+1}-p_i}{\tilde u}\right) \\ 0 & 0 & 0 & 0 & 1 & 0 & \left(\dfrac{\tau_i+\tau_{i+1}}{2}\right)\ell \\ 0 & -4 E I\gamma^3 F_2 & -4 E I\gamma^2 F_3 & -4\gamma F_4 & 0 & F_1 & -\dfrac{1}{\gamma}\left(F_2 p_i + F_3 \dfrac{p_{i+1}-p_i}{\tilde u}\right) \\ 0 & 0 & 0 & 0 & 0 & 0 & 1 \end{bmatrix}$$

We shall now examine what are the changes in these matrices when axial effects and 2nd. order theory (i.e. equilibrium equations written after displacement) are taken into account.

2.3 Free element *(2nd order)*

Let 1 be the origin node of the element and 2 the other node. From the two following equations (11), written in local axes,

$$\begin{cases} EI\dfrac{d^2y}{dx^2} = M - Ny \\ \dfrac{d^2M}{dx^2} = -p_1\left(1-\dfrac{x}{\ell}\right) - p_2\dfrac{x}{\ell} \end{cases} \quad (11)$$

we get the differential equation: $EI\dfrac{d^4y}{dx^4} + N\dfrac{d^2y}{dx^2} = -p_1\left(1-\dfrac{x}{\ell}\right) - p_2\dfrac{x}{\ell}$ (12)

Integrating twice this equation and knowing the boundary conditions in node 1:
$y(0) = v_1$, $y'(0) = \omega_1$, $EIy''(0) = M_1$, $EIy'''(0) = T_1$

we get: $\quad EIy'' + Ny = v_1 N + \omega_1 Nx + M_1 + T_1 x - p_1\dfrac{x^2}{2} + \dfrac{p_1 - p_2}{6\ell}x^3$ (13)

2.3.1 First case: *N>0 compressive force*

The solution of the equation (13) can be written, with $k^2 = \dfrac{N}{EI}$:

$$y = A\cos kx + B\sin kx + v_1 + \omega_1 x + \dfrac{M_1}{N} + \dfrac{T_1}{N}x - \dfrac{p_1}{2N}x^2 + \dfrac{p_1}{k^2 N} + \dfrac{p_1 - p_2}{6\ell N}\left(x^3 - \dfrac{6x}{k^2}\right)$$

Constants A and B are determined by: $y(0) = v_1$ and $y'(0) = \omega_1$, so we get:

$$\begin{aligned} y = v_1 + \omega_1 x + M_1\dfrac{1-\cos kx}{N} + T_1\dfrac{kx - \sin kx}{kN}x \\ - \dfrac{p_1}{N}\left(\dfrac{x^2}{2} - \dfrac{1-\cos kx}{k^2}\right) + \dfrac{p_1 - p_2}{6\ell N}\left(x^3 - \dfrac{6x}{k^2} + \dfrac{6\sin kx}{k^3}\right) \end{aligned} \quad (14)$$

We have now to calculate: $v_2 = y(\ell)$, $\omega_2 = y'(\ell)$, $M_2 = EIy''(\ell)$, $T_2 = EIy'''(\ell)$ in order to find the coefficients of the relevant transfer matrix, knowing that the 1st and 5th rows and columns are not changed, as they concern only the longitudinal displacement and the axial force. We give the matrix **T**, reduced only to the modified elements, i.e. ignoring 1st and 5th rows and columns, with $\tilde{u} = k\ell$:

$$\mathbf{T} = \begin{bmatrix} 1 & \ell & \dfrac{1-\cos\tilde{u}}{N} & \ell\dfrac{\tilde{u}-\sin\tilde{u}}{N\tilde{u}} & -p_1\dfrac{\ell^2}{N}\left(\dfrac{1}{2} - \dfrac{1-\cos\tilde{u}}{\tilde{u}^2}\right) + (p_1-p_2)\dfrac{\ell^2}{N}\left(\dfrac{\tilde{u}^3 - 6\tilde{u} + 6\sin\tilde{u}}{6\tilde{u}^3}\right) \\ 0 & 1 & \dfrac{\tilde{u}\sin\tilde{u}}{N\ell} & \dfrac{1-\cos\tilde{u}}{N} & -p_1\dfrac{\ell}{N}\left(\dfrac{\tilde{u}-\sin\tilde{u}}{\tilde{u}}\right) + (p_1-p_2)\dfrac{\ell}{N}\left(\dfrac{\tilde{u}^2 - 2 + 2\cos\tilde{u}}{2\tilde{u}^2}\right) \\ 0 & 0 & \cos\tilde{u} & \dfrac{\ell\sin\tilde{u}}{\tilde{u}} & -p_1\ell^2\left(\dfrac{1-\cos\tilde{u}}{\tilde{u}^2}\right) + (p_1-p_2)\ell^2\left(\dfrac{\tilde{u}-\sin\tilde{u}}{\tilde{u}^3}\right) \\ 0 & 0 & -\dfrac{\tilde{u}\sin\tilde{u}}{\ell} & \cos\tilde{u} & -p_1\ell\dfrac{\sin\tilde{u}}{\tilde{u}} + (p_1-p_2)\ell\left(\dfrac{1-\cos\tilde{u}}{\tilde{u}^2}\right) \\ 0 & 0 & 0 & 0 & 1 \end{bmatrix}$$

Obviously, these relationships permit calculation of the Eulerian critical forces, which depend upon the boundaries conditions. For instance, if the element is biarticulated, i.e. these conditions are $v_1 = v_2 = 0$ and $M_1 = M_2 = 0$, and for sake of simplicity we take $p_1 = p_2 = p$, the first row allows calculation of ω_1 :

$$\omega_1 = -\frac{T_1}{N}\left(\frac{\tilde{u}-\sin\tilde{u}}{\tilde{u}}\right) + p\frac{\ell}{N}\left(\frac{1}{2} - \frac{1-\cos\tilde{u}}{\tilde{u}^2}\right)$$

The 2nd row gives ω_2

$$\omega_2 = \omega_1 + \frac{T_1}{N}(1-\cos\tilde{u}) - p\frac{\ell}{N}\left(\frac{\tilde{u}-\sin\tilde{u}}{\tilde{u}}\right)$$

In this case, due to symmetry, we have $\omega_2 = -\omega_1$; we can calculate for instance T_1.

$$T_1\left(1-\cos\tilde{u} - 2\frac{\tilde{u}-\sin\tilde{u}}{\tilde{u}}\right) = p\ell\left(\frac{\tilde{u}-\sin\tilde{u}}{\tilde{u}} - 1 + 2\frac{1-\cos\tilde{u}}{\tilde{u}^2}\right)$$

The 1st positive value of $\tilde{u}$ for which the factor of T_1 vanishes is $\tilde{u} = \pi$, while, for this figure, the factor of $p\ell$ is $\frac{4}{\pi^2}$.

We see that T_1, thus ω_1 and $y(x)$, increase infinitely when :

$\tilde{u} \to \pi$, i.e. $k\ell \to \pi$ or $N \to N_{cr} = \pi^2 \frac{EI}{\ell^2}$ the Eulerian critical force in this case.

If the Eulerian critical force is reached, displacements are infinite : this instability corresponds to a singularity of the rigidity matrix.

Moreover we notice that if $N \to 0$, i.e. $\tilde{u} \to 0$, we obtain the coefficients of the matrix established at 1st order, in 2.1.

2.3.2 *Second case : N<0 tensile force*

In this case $k^2 = -\frac{N}{EI}$ and we proceed as in 2.3.1, obtaining the solution :

$$y = A\cosh kx + B\sinh kx + v_1 + \omega_1 x + \frac{M_1}{N} + \frac{T_1}{N}x - \frac{p_1}{2N}\left(x^2 + \frac{2}{k^2}\right) + \frac{p_1-p_2}{6\ell N}\left(x^3 + \frac{6x}{k^2}\right)$$

The constants A and B are determined by the conditions $y(0) = v_1$ and $y'(0) = \omega_1$. Then the same calculations are developed as here above. The reduced matrix is, with $\tilde{u} = k\ell$, **T** =

$$\begin{bmatrix} 1 & \ell & -\dfrac{\cosh\tilde{u}-1}{N} & -\ell\dfrac{\sinh\tilde{u}-\tilde{u}}{N\tilde{u}} & p_1\dfrac{\ell^2}{N}\left(\dfrac{\cosh\tilde{u}-1}{\tilde{u}^2} - \dfrac{1}{2}\right) - (p_1-p_2)\dfrac{\ell^2}{N}\left(\dfrac{-\tilde{u}^3-6\tilde{u}+6\sinh\tilde{u}}{6\tilde{u}^3}\right) \\ 0 & 1 & -\dfrac{\tilde{u}\sinh\tilde{u}}{N\ell} & -\dfrac{\cosh\tilde{u}-1}{N} & p_1\dfrac{\ell}{N}\left(\dfrac{\sinh\tilde{u}-\tilde{u}}{\tilde{u}}\right) - (p_1-p_2)\dfrac{\ell}{N}\left(\dfrac{-\tilde{u}^2-2+2\cosh\tilde{u}}{2\tilde{u}^2}\right) \\ 0 & 0 & \cosh\tilde{u} & \dfrac{\ell\sinh\tilde{u}}{\tilde{u}} & -p_1\ell^2\left(\dfrac{\cosh\tilde{u}-1}{\tilde{u}^2}\right) + (p_1-p_2)\ell^2\left(\dfrac{\sinh\tilde{u}-\tilde{u}}{\tilde{u}^3}\right) \\ 0 & 0 & \dfrac{\tilde{u}\sinh\tilde{u}}{\ell} & \cosh\tilde{u} & -p_1\ell\dfrac{\sinh\tilde{u}}{\tilde{u}} + (p_1-p_2)\ell\left(\dfrac{\cosh\tilde{u}-1}{\tilde{u}^2}\right) \\ 0 & 0 & 0 & 0 & 1 \end{bmatrix}$$

In this situation we can see a diminution of the displacements, when compared to 1st order results.

2.4 Element resting on an elastic soil *(2nd order)*

This case leads to the differential equation :

$$EI\frac{d^4y}{dx^4} + N\frac{d^2y}{dx^2} + \beta y = -p_1 + (p_1 - p_2)\frac{x}{\ell} \qquad (15)$$

Dividing by *EI*, and with $s = \frac{\beta}{EI}$, two cases have to be considered.

2.4.1 *First case* : $N>0$ compressive force

The equation (15) can be written, with : $k^2 = \frac{N}{EI}$

$$\frac{d^4y}{dx^4} + k^2\frac{d^2y}{dx^2} + sy = -\frac{p_1}{EI} + \frac{p_1 - p_2}{EI}\frac{x}{\ell} \qquad (16)$$

Looking for solutions of the form $\exp(rx) = e^{rx}$, the associated characteristic equation is : $r^4 + k^2 r^2 + s = 0$
This twice-square equation has a discriminant $\Delta = k^4 - 4s$
Consideration on the sign of Δ shows that $\Delta < 0$. As a matter of fact, $\Delta = 0$ means : $k^2 = 2\sqrt{s}$ or also $N = pR = 2EI\sqrt{s} = 2\sqrt{EI\beta}$ from which we recognize the buckling pressure : $p_{cr} = \frac{2}{R}\sqrt{EI\beta}$ critical pressure of a circular pipe placed in an elastic medium.(Ref [4]).

Obviously, for purposes of stability, we must have $p < p_{cr} \Leftrightarrow \Delta < 0$.

Let now Δ designate the following expression : $\Delta = \sqrt{\frac{4s}{k^4} - 1}$ \qquad (17)

The 1st step of the resolution leads to consider :

$$\begin{cases} r_1^2 = -\dfrac{k^2}{2}(1 + i\Delta) \\ r_2^2 = -\dfrac{k^2}{2}(1 - i\Delta) \end{cases} \qquad (18)$$

In order to find the four solutions of the characteristic equation, we write

$1 + i\Delta = \rho e^{i\theta}$ with : $\rho = \sqrt{1 + \Delta^2} = \dfrac{2\sqrt{s}}{k^2}$; $\theta = \arccos\dfrac{1}{\rho} = \arcsin\dfrac{\Delta}{\rho}$, so :

$$\begin{cases} r_1^2 = -\sqrt{s}\, e^{i\theta} = \sqrt{s}\, e^{i(\theta + \pi)} \\ r_2^2 = -\sqrt{s}\, e^{-i\theta} = \sqrt{s}\, e^{-i(\theta + \pi)} \end{cases} \text{ with } 0 \leq \theta \leq \frac{\pi}{2}$$

Let then again : $\rho_1 = s^{\frac{1}{4}}\sin\dfrac{\theta}{2}$ \quad $\rho_2 = s^{\frac{1}{4}}\cos\dfrac{\theta}{2}$ in order to be able to find the 4 solutions of the characteristic equation,

$$\begin{cases} r_1 = \pm(-\rho_1 + i\rho_2) \\ r_2 = \pm(\rho_1 + i\rho_2) \end{cases} \qquad (19)$$

and to obtain the solution in the form of :

$$y(x) = A\cosh\rho_1 x \cos\rho_2 x + B\sinh\rho_1 x \sin\rho_2 x + C\sinh\rho_1 x \cos\rho_2 x$$
$$+ D\cosh\rho_1 x \sin\rho_2 x - \frac{p_1}{\beta} + \frac{p_1 - p_2}{\beta\ell} x$$

A, B, C, D, are determined by : $v_1 = y(0); \omega_1 = y'(0); M_1 = EIy''(0); T_1 = EIy'''(0)$
We put $x = \ell$ in order to find v_2, ω_2, M_2 and T_2 in function of v_1, ω_1, M_1 and T_1.
The calculations are long. In order to simplify the presentation, we designate by :

$$u_1 = \rho_1 \ell \qquad u_2 = \rho_2 \ell$$
$$F_1 = \cosh u_1 \cos u_2 \qquad F_2 = \sinh u_1 \sin u_2$$
$$F_3 = \sinh u_1 \cos u_2 \qquad F_4 = \cosh u_1 \sin u_2$$

and we give the coefficients of the reduced transfer-matrix by writing :

$$\begin{bmatrix} v_2 \\ \omega_2 \\ M_2 \\ T_2 \\ 1 \end{bmatrix} = \begin{bmatrix} \ddots & & \cdots & \cdots & \\ & \ddots & & \vdots & \\ & & A(i,j) & & \\ & \vdots & & \ddots & \\ & \cdots & & & \ddots \end{bmatrix} \begin{bmatrix} v_1 \\ \omega_1 \\ M_1 \\ T_1 \\ 1 \end{bmatrix}$$

$$A(1,1) = F_1 + \frac{F_2}{\Delta} \qquad A(1,2) = -\frac{(u_2^2 - 3u_1^2)\frac{F_3}{u_1} + (u_1^2 - 3u_2^2)\frac{F_4}{u_2}}{2\ell\sqrt{s}}$$

$$A(1,3) = \frac{\ell^2}{EI} \frac{F_2}{2u_1 u_2} \qquad A(1,4) = \ell \frac{\frac{F_4}{u_2} - \frac{F_3}{u_1}}{2EI\sqrt{s}}$$

$$A(1,5) = \frac{p_1}{\beta}(A(1,1)-1) + \frac{p_1 - p_2}{\beta}\left(1 - \frac{A(1,2)}{\ell}\right)$$

$$A(2,1) = \frac{1}{\ell}\left(\left(u_1 + \frac{u_2}{\Delta}\right)F_3 - \left(u_2 - \frac{u_1}{\Delta}\right)F_4\right) \qquad A(2,2) = F_1 + \frac{F_2}{\Delta}$$

$$A(2,3) = \ell \frac{u_1 F_4 + u_2 F_3}{2EI u_1 u_2} \qquad\qquad A(2,4) = \ell^2 \frac{F_2}{2EI u_1 u_2} = A(1,3)$$

$$A(2,5) = \frac{p_1}{\beta} A(2,1) + \frac{p_1 - p_2}{\beta\ell}(1 - A(2,2))$$

$$A(3,1) = -\frac{2EI\,u_1 u_2}{\ell^2} F_2\left(1 + \frac{1}{\Delta^2}\right)$$

$$A(3,2) = \frac{EI\,u_1 u_2}{\ell(u_1^2 + u_2^2)} \left(\frac{u_2^2 - 3u_1^2}{u_1}\left(\frac{F_3}{\Delta} + F_4\right) + \frac{u_1^2 - 3u_2^2}{u_2}\left(\frac{F_4}{\Delta} - F_3\right)\right)$$

$$A(3,3) = F_1 - \frac{F_2}{\Delta}$$

$$A(3,4) = \ell \frac{u_1 u_2}{(u_1^2 + u_2^2)}\left(\frac{1}{u_1}\left(\frac{F_3}{\Delta} + F_4\right) - \frac{1}{u_2}\left(\frac{F_4}{\Delta} - F_3\right)\right)$$

$$A(3,5) = \frac{p_1}{\beta} A(3,1) - \frac{p_1 - p_2}{\beta \ell} A(3,2)$$

$$A(4,1) = \frac{EI}{\ell^3}\left(u_1(u_1^2 - 3u_2^2)\left(F_3 + \frac{F_4}{\Delta}\right) + u_2(u_2^2 - 3u_1^2)\left(F_4 - \frac{F_3}{\Delta}\right)\right)$$

$$A(4,2) = -\frac{EI}{\ell^2}\frac{(u_1^2 + u_2^2)^2}{2u_1 u_2} F$$

$$A(4,3) = \frac{1}{2\ell}\left(\frac{u_1^2 - 3u_2^2}{u_2} F_4 - \frac{u_2^2 - 3u_1^2}{u_1} F_3\right)$$

$$A(4,4) = F_1 - \frac{F_2}{\Delta}$$

$$A(4,5) = \frac{p_1}{\beta} A(4,1) - \frac{p_1 - p_2}{\beta \ell} A(4,2)$$

At last $A(5,1) = A(5,2) = A(5,3) = A(5,4) = 0 \quad A(5,5) = 1$

It can be easily verified by limited developments that, if $N \to 0$, the coefficients of the matrix, established at 1st order in 2.2, are obtained.

2.4.2 *Second case* : <u>N≤0 tensile force</u>

The equation can be written, with : $k^2 = -\frac{N}{EI}$

$$\frac{d^4 y}{dx^4} - k^2 \frac{d^2 y}{dx^2} + sy = -\frac{p_1}{EI} + \frac{p_1 - p_2}{EI}\frac{x}{\ell} \tag{20}$$

The associated characteristic equation is : $r^4 - k^2 r^2 + s = 0$
The discriminant of the twice-square equation is $\Delta = k^4 - 4s$
In this case, there is no danger of buckling, and full consideration of the sign of Δ is necessary.

a) 1st case : $\Delta < 0$ $(|N|\text{ small})$

We use the same process as in 2.4.1. We designate by :

$$\Delta = \sqrt{\frac{4s}{k^4}-1} \qquad \begin{cases} r_1^2 = \dfrac{k^2}{2}(1-i\Delta) \\ r_2^2 = \dfrac{k^2}{2}(1+i\Delta) \end{cases}$$

$1+i\Delta = \rho\, e^{i\theta}$

$\begin{cases} r_1^2 = \sqrt{s}\, e^{-i(\theta+2n\pi)} \\ r_2^2 = \sqrt{s}\, e^{i(\theta+2n\pi)} \end{cases}$ with $0 \le \theta \le \dfrac{\pi}{2}$

So the 4 solutions are : $r_i = s^{\frac{1}{4}}\left(\pm\cos\dfrac{\theta}{2} \pm i\sin\dfrac{\theta}{2}\right)$

Let :
$$\rho_1 = s^{\frac{1}{4}}\cos\frac{\theta}{2} \qquad \rho_2 = s^{\frac{1}{4}}\sin\frac{\theta}{2}$$
$$u_1 = \rho_1 \ell \qquad u_2 = \rho_2 \ell$$
$$F_1 = \cosh u_1 \cos u_2 \qquad F_2 = \sinh u_1 \sin u_2$$
$$F_3 = \sinh u_1 \cos u_2 \qquad F_4 = \cosh u_1 \sin u_2$$

The coefficients of the transfer-matrix are the same as in 2.4.1.

b) 2nd case : $\Delta = 0 \Leftrightarrow k^4 = 4s$

The characteristic equation is : $\left(r^2 - \dfrac{k^2}{2}\right)^2 = 0$

The solutions are double : $r = \pm\dfrac{k}{\sqrt{2}}$; and with $r = \dfrac{k}{\sqrt{2}} = s^{\frac{1}{4}}$, we can write :

$$y(x) = (Arx+B)\cosh rx + (Crx+D)\sinh rx - \frac{p_1}{\beta} + \frac{p_1-p_2}{\beta\ell}x$$

The 4 constants are determined in the same way, as well as the coefficients of the transfer-matrix. With $\tilde{u} = r\ell$, we obtain :

$A(1,1) = \cosh\tilde{u} - \dfrac{\tilde{u}\sinh\tilde{u}}{2}$ $\quad A(1,2) = \dfrac{3\sinh\tilde{u} - \tilde{u}\cosh\tilde{u}}{2\tilde{u}}\ell \quad$ $A(1,3) = \dfrac{\ell^2}{EI}\dfrac{\sinh\tilde{u}}{2\tilde{u}}$

$A(1,4) = \dfrac{\ell^3}{EI}\dfrac{\tilde{u}\cosh\tilde{u} - \sinh\tilde{u}}{2\tilde{u}^3} \quad A(1,5) = \dfrac{p_1}{\beta}(A(1,1)-1) + \dfrac{p_1-p_2}{\beta}\left(1 - \dfrac{A(1,2)}{\ell}\right)$

$A(2,1) = \dfrac{\tilde{u}}{2\ell}(\sinh\tilde{u} - \tilde{u}\cosh\tilde{u}) \quad A(2,2) = A(1,1) \quad A(2,3) = \dfrac{\ell}{EI}\dfrac{\tilde{u}\cosh\tilde{u} + \sinh\tilde{u}}{2\tilde{u}}$

$A(2,4) = \dfrac{\ell^2}{EI}\dfrac{\sinh\tilde{u}}{2\tilde{u}} \qquad\qquad A(2,5) = \dfrac{p_1}{\beta}A(2,1) + \dfrac{p_1-p_2}{\beta\ell}(1-A(2,2))$

$$A(3,1) = -\frac{EI}{\ell^2}\frac{\tilde{u}^3 \sinh \tilde{u}}{2} \qquad A(3,2) = EI\, A(2,1) \qquad A(3,3) = \cosh \tilde{u} + \frac{\tilde{u} \sinh \tilde{u}}{2}$$

$$A(3,4) = EI\, A(2,3) \qquad A(3,5) = \frac{p_1}{\beta} A(3,1) - \frac{p_1 - p_2}{\beta \ell} A(3,2)$$

$$A(4,1) = -\frac{EI}{\ell^3} \tilde{u}^3 \frac{\tilde{u} \cosh \tilde{u} + \sinh \tilde{u}}{2} \qquad A(4,2) = A(3,1)$$

$$A(4,3) = \frac{\tilde{u}}{2\ell}(\tilde{u} \cosh \tilde{u} + 3 \sinh \tilde{u})$$

$$A(4,4) = A(3,3) \qquad\qquad A(4,5) = \frac{p_1}{\beta} A(4,1) - \frac{p_1 - p_2}{\beta \ell} A(4,2)$$

$$A(5,1) = A(5,2) = A(5,3) = A(5,4) = 0 \qquad A(5,5) = 1$$

c) <u>3rd case</u> : $\Delta > 0$ ($|N|$ large)

Then we get : $\Delta = \sqrt{1 - \dfrac{4s}{k^4}}$ $\qquad \begin{cases} r_1^2 = \dfrac{k^2}{2}(1-\Delta) \\ r_2^2 = \dfrac{k^2}{2}(1+\Delta) \end{cases}$

r_1^2 and r_2^2 are positive, we have 4 real roots $\pm r_1$ and $\pm r_2$.

Let : $r_1 = \dfrac{k}{\sqrt{2}}\sqrt{1-\Delta} \quad r_2 = \dfrac{k}{\sqrt{2}}\sqrt{1+\Delta}$, the solution y(x) can then be written :

$$y(x) = A \cosh r_1 x + B \sinh r_1 x + C \cosh r_2 x + D \sinh r_2 x - \frac{p_1}{\beta} + \frac{p_1 - p_2}{\beta \ell} x$$

We proceed in the same way in order to determine the constants and the coefficients of the transfer-matrix :

$$u_1 = r_1 \ell \qquad u_2 = r_2 \ell$$
$$ch_1 = \cosh u_1 \quad ch_2 = \cosh u_2 \qquad F = u_2^2 - u_1^2$$
$$sh_1 = \sinh u_1 \quad sh_2 = \sinh u_2$$

$$A(1,1) = \frac{u_2^2 ch_1 - u_1^2 ch_2}{F} \qquad A(1,2) = \frac{\ell}{F}\left(\frac{u_2^2}{u_1} sh_1 - \frac{u_1^2}{u_2} sh_2\right)$$

$$A(1,3) = \frac{\ell^2}{EI} \frac{ch_2 - ch_1}{F} \qquad A(1,4) = \frac{\ell^3}{EI\, F}\left(\frac{sh_2}{u_2} - \frac{sh_1}{u_1}\right)$$

$$A(1,5) = \frac{p_1}{\beta}(A(1,1) - 1) + \frac{p_1 - p_2}{\beta}\left(1 - \frac{A(1,2)}{\ell}\right)$$

$$A(2,1) = \frac{u_1 u_2}{\ell} \frac{u_2 sh_1 - u_1 sh_2}{F} \qquad A(2,2) = \frac{u_2^2 ch_1 - u_1^2 ch_2}{F}$$

$$A(2,3) = \frac{\ell}{EI} \frac{u_2 sh_2 - u_1 sh_1}{F} \qquad A(2,4) = \frac{\ell^2}{EI} \frac{ch_2 - ch_1}{F}$$

$$A(2,5) = \frac{p_1}{\beta} A(2,1) + \frac{p_1 - p_2}{\beta \ell}(1 - A(2,2))$$

$$A(3,1) = -\frac{EI}{\ell^2}\frac{u_1^2 u_2^2}{F}(ch_2 - ch_1) \quad A(3,2) = EI\, A(2,1)$$

$$A(3,3) = \frac{u_2^2 ch_2 - u_1^2 ch_1}{F} \qquad A(3,4) = EI\, A(2,3)$$

$$A(3,5) = \frac{p_1}{\beta} A(3,1) - \frac{p_1 - p_2}{\beta \ell} A(3,2)$$

$$A(4,1) = -\frac{EI}{\ell^3}\frac{u_1^2 u_2^2}{F}(u_2 sh_2 + u_1 sh_1) \quad A(4,2) = A(3,1)$$

$$A(4,3) = \frac{u_2^3 sh_2 - u_1^3 sh_1}{\ell F} \qquad A(4,4) = A(3,3)$$

$$A(4,5) = \frac{p_1}{\beta} A(4,1) - \frac{p_1 - p_2}{\beta \ell} A(4,2)$$

$$A(5,1) = A(5,2) = A(5,3) = A(5,4) = 0 \quad A(5,5) = 1$$

3 - Conclusion

The matrices, established here above, are used in a computer program which allows the calculation of buried pipe in interaction with an elasto-plastic soil. The relevant model is presented in Ref [1].

References

[1] M. GERBAULT : 2nd International Conference A.S.C.E. Seattle - 'Advances in underground pipeline engineering' - *A Soil-Structure interactive model* (June 1995).

[2] E.C. PESTEL and F.A. LECKIE : *Matrix Methods in Elastomechanics.* (1963) McGRAW-HILL.

[3] J. COURBON : *Cours de théorie des structures, ch. 1 - Méthode des matrices-transfert.* (1970) Ecole Nationale des Ponts-et-Chaussées.

[4] M. GERBAULT : *Calcul de canalisations circulaires semi-rigides.* Annales de l'I.T.B.T.P. N° 439, Novembre 1985.

PCCP Design Concepts Made Simple

Richard I. Mueller, P.E., Member, ASCE[1]

Abstract

Prestressed Concrete Cylinder Pipe (PCCP) supplies virtually every metropolitan area in the United States and Canada with raw and treated water. In 1992, the design procedure for PCCP expanded from two alternative, one-page procedures to the biggest and most complex American Water Works Association (AWWA) standard, the 109-page AWWA C304. C304 is so big, who has time to read it, much less digest and comprehend the procedure?

This paper segments the AWWA C304 design procedure into short, understandable sections and concepts. Each section or concept is compared to the previous design procedures to highlight the revisions and reasons for them. Design considerations and resulting criteria to be included in consulting engineer's specifications are clearly presented.

Introduction

Prior to the publication of AWWA C304 in 1992, PCCP designs were generally determined from one of two short appendices in AWWA C301. Appendix A of AWWA C301 provided a semi-empirical, "cubic parabola" design method, (see figure 2), and Appendix B of AWWA C301 defined a stress analysis design method, (see figure 3). The adequacy of these two design methods to determine appropriate prestressed pipe designs for given operating conditions has never been seriously challenged. So why has the PCCP design procedure changed? Why has the PCCP industry put forth all this effort to make things more complicated?

There are three basic reasons for the design procedure change. First, the

---

[1]Vice President, Engineering, Gifford-Hill-American, Inc., 1003 Meyers Road, Grand Prairie, Texas 75050

PCCP DESIGN CONCEPTS

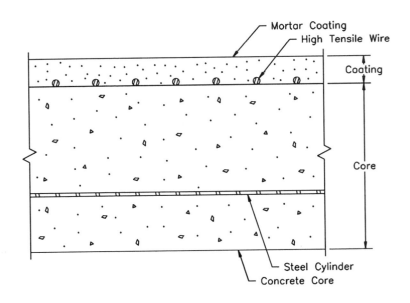

Figure 1 Cross Section of Embedded-Cylinder PCCP

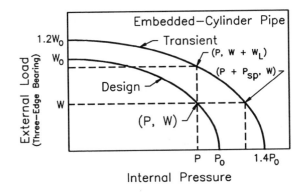

Figure 2 AWWA C301-79 Appendix A "Cubic Parabola" Design Curve

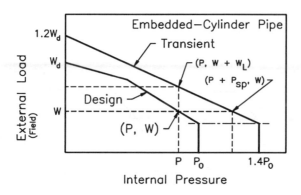

Figure 3 AWWA C301-79 Appendix B "Stress Analysis" Design Curve

two old design procedures gave very similar, but still different, pipe designs, and the differences caused some confusion. With the edition of AWWA C304, there is now a unified design procedure, (UDP), used throughout the country. Also, the old procedures were based only on limiting the concrete core tensile stress. By comparison, UDP provides comprehensive stress analysis of all pipe component materials. And finally, the old procedures used 40-year-old approximations to characterize component material performance, whereas the new procedure uses state-of-the-art modeling of component materials, and accommodates the differing environments at each pipe manufacturing plant.

Development of the New Procedure

The new design procedure was developed very methodically over approximately 8 years. The development started with an international literature search and a review of papers published through the American Concrete Institute and similar technical organizations. This search revealed the state-of-the-art developments applicable to prestressed concrete which would be included in the new design procedure.

Data from hundreds of PCCP tests performed over more than 40 years were then gathered and analyzed. This data included not only typical tests for pipe external load and internal pressure capability, but also specialized strain-gauge measurements used to determine the state of stress of the pipe component materials.

The state-of-the-art concepts were used to develop an interim, step-by-step, computerized integration procedure which was used to evaluate time-related component stress variations. The historical pipe tests were used to verify the step-by-step computer model. From this model were developed the simplified and more practical design equations which were ultimately included in AWWA C304.

As stages of the developmental efforts were completed, the results and the proposed new design procedures and equations were published for peer review. The peer review process was accomplished first by the publishing of papers in well-known, widely circulated structural journals and magazines, and, secondly, through the review by the AWWA Concrete Pipe Committee members and the AWWA Standards Council.

Basic Design Concepts

The basic concept for the old design procedures is still used in AWWA C304. That design concept, (see figure 4), recognizes that tension in the concrete core is the usual limiting parameter for ordinary pressure pipe installations. Consequently, the tension in the pipe wall which will be necessary to contain the internal pressure is calculated and added to the pipe wall tension necessary to support the external loads and weights. Then the amount of prestressing wire necessary to compress the pipe concrete enough to offset those future tensile forces is determined.

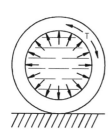 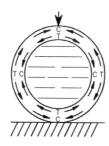 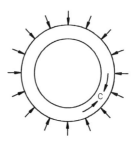

Add Pipe Wall Tension Containing Internal Fluid Pressure...

To Pipe Wall Tension Supporting Loads And Weights...

Then Compress The Pipe Concrete to Offset That Future Tension.

Figure 4 Basic Design Concept for PCCP

However, the new design procedure has been enhanced and made more comprehensive. UDP now limits stresses or strains in <u>all</u> component materials, not just the concrete core, to preclude cracking or yielding. This is accomplished for PCCP in the same manner as is done for the design of beams. That is, stress and strain diagrams, (see figure 5), are used to calculate the working and transient stresses and strains in the pipe component materials.

Updating Component Material Performance Data

Besides comprehensively evaluating all the component materials, the new design program has replaced the 40-year-old design approximations with more accurate, state-of-the-art formulas and information. For example, the modulus of elasticity of pipe steel, whether wire or cylinder, had been set at 193,050 MPa (28,000,000 psi). But UDP more-accurately separates the moduli. The wire modulus now remains at 193,050 MPa (28,000,000 psi) but the cylinder modulus has been changed to 206,850 MPa (30,000,000 psi).

In the old procedures, the initial ratio "n_i" of the modulus of elasticity for vertically-cast concrete to the modulus of elasticity of steel was 7 for young concrete, and the resultant ratio "n_r" was set at 6 for concrete which was fully mature. In UDP, the cast concrete modular ratio is calculated from a formula based upon the strength of the concrete. And since there are now separate moduli for wire and cylinder steel, there are separate formulas needed to represent the ratio between the concrete modulus and either the wire or

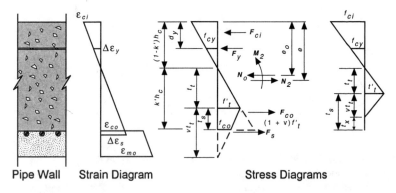

Figure 5 Schematic of Strain and Stress Distributions in Pipe-Wall Cross Section at Pipe Springline (See AWWA C304 for Variable Definitions)

cylinder modulus. The new initial modular ratio for cast concrete with cylinder steel is $n'_i = 117(f'c)^{-0.3}$, and with prestressing wire is $n_i = 109(f'c)^{-0.3}$, where f'c is the design 28-day compressive strength of the core concrete. The new resultant modular ratio for cast concrete with cylinder steel is $n'_r = 99(f'c)^{-0.3}$, and with prestressing wire is $n_r = 93(f'c)^{-0.3}$. As with the old design procedures, different modular ratio values are also determined for pipe made with concrete placed by spinning.

A similar update has been applied to wire relaxation. Wire relaxation was set by the old procedure at 5 percent. But in UDP there are separate formulas used to calculate wire relaxation, depending on whether the concrete is placed by vertical casting or spinning. The wire relaxation loss factor "R" for cast concrete is $R = 0.111 - 3.5(As/Ac)$, and for spun concrete is $R = 0.132 - 3.1(As/Ac)$, where As/Ac is the ratio between the area of wire and the area of core concrete.

The stress/strain curve for mortar or concrete has been updated from the old diagram where the tension increases linearly with strain until fracture, to the more-accurate diagram, (see figure 6), where it is recognized that concrete and mortar have a maximum tensile strength somewhat similar to the yield point in steel. At strains beyond that causing the maximum tensile stress, the concrete or mortar can continue to carry lesser amounts of tension up to the point of visible cracking.

Another set of design factors which have been updated are those for creep and shrinkage. The old procedures lumped creep and shrinkage together into a single factor, which was 2.0 for cast concrete. And that factor was used

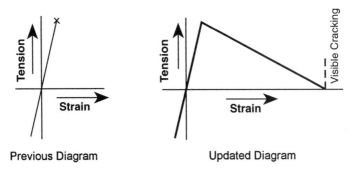

Figure 6 Previous and Updated Stress/Strain Diagrams for Mortar and Concrete

regardless of the thickness of the coating or inner or outer concrete core. However, creep and shrinkage are primarily due to evaporation of water from the concrete or mortar. Pipe made in Phoenix will have more creep and shrinkage than pipe made in Seattle. Likewise, thicker concrete will lose less water to evaporation than thinner concrete, so the creep and shrinkage of thicker concrete will also be less. Creep and shrinkage factors are now calculated separately. The creep factor "ϕ" and shrinkage strain "s" are now

$$\phi = \frac{(h_{co} + h_m) \phi_{com} - h_m \phi_m + h_{ci} \phi_{ci}}{h_{ci} + h_{co}}$$

$$s = \frac{(h_{co} + h_m) s_{com} - h_m s_m + h_{ci} s_{ci}}{h_{ci} + h_{co}}$$

where h_{co}, h_m, and h_{ci} are the thicknesses of the outer core concrete, the mortar coating, and the inner core concrete, respectively, and ϕ_{ci}, ϕ_{com}, ϕ_m, s_{ci}, s_{com}, and s_m are creep factors and shrinkage strains for the inner core, the outer core plus the coating, and for the coating, respectively. The effect of the pipe manufacturing environment and the pipe wall thickness are included in the new formulas. This is an important point to keep in mind as one considers items to include in pipe specifications. The pipe exposure period, the ambient conditions to which the pipe are exposed, and the pipe wall thickness all effect the resultant creep and shrinkage factors.

An area between PCCP design and manufacture which also needed improvement was that all pipe had been designed using generic material design characteristics, but no testing to assure adequate material performance was required. Now, although there are default material design characteristics, the material used in each separate pipe factory must be tested to meet or exceed the default performance. If the available material does not perform as well as the default assumptions, actual material performance characteristics are used to design the pipe.

Derivation of the Core Compression Formulas

The derivation of the old formulas for PCCP concrete core compression is very similar to deriving the hoop stress formula; all are derived by balancing the forces acting on the pipe wall, (see figure 7). The initial PCCP concrete core compressive stress times the transformed area of the core concrete balances the stress in the prestressing wire times the area of the wire. When the representative formula is solved for the initial concrete core compression "fci", the result is

PCCP DESIGN CONCEPTS

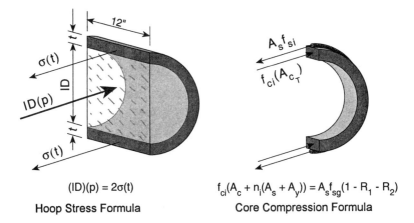

(ID)(p) = 2σ(t) $f_{ci}(A_c + n_i(A_s + A_y)) = A_s f_{sg}(1 - R_1 - R_2)$

Hoop Stress Formula   Core Compression Formula

Figure 7 Balancing Forces Acting On Pipe Wall

$$fci = \frac{As\,(fsg)\,(1-R1-R2)}{(Ac + n_i\,(As + Ay))}$$

where As, Ac, and Ay are the areas of prestressing wire, core concrete, and steel cylinder, respectively, fsg is the gross wrapping stress of the wire, R1 and R2 are wire stress loss factors, and n_i is the initial modular ratio. The formula for resultant core compression "fcr" is similarly calculated to be

$$fcr = \frac{fci\,[Ac + n_r\,(As + Ay)]}{Ac + n_r\,(As + Ay)\,(1 + Cr)}$$

where n_r is the resultant or "final" modular ratio, Cr is the creep factor, and all other variables are as previously defined. These formulas for initial and resultant core compression were used in both the old AWWA C301 Appendix A and Appendix B design procedures.

In UDP, these formulas are updated to include the state-of-the-art material performance calculations. The updated formulas are

$$fci = \frac{As\,(fsg)}{Ac + n_i\,As + n'_i\,Ay}$$

and

$$f_{cr} = \frac{f_{ci}(A_c + n_rA_s + n'_rA_y) - (A_sE_s + A_yE_y)s - A_s(R)f_{sg}}{A_c + (n_rA_s + n'_rA_y)(1 + \phi)}$$

where E_s and E_y are the moduli of the prestressing wire and steel cylinder, respectively, and all other variables are as previously defined.

<u>Putting It All Together</u>

The old AWWA C301 Appendix B design formula is derived by combining the resultant core compression, f_{cr}, with the stress from the pipe internal pressure and the stress from the external loads and weights carried by the pipe, (see figure 8). Plotting the resulting equations gives linear design curves which are very similar to the parabolic design curves of the old AWWA C301 Appendix A, (see figures 2 and 3). UDP uses the same general design approach as the old AWWA C301 Appendix B, but updates the formula variables and expands the design to include similar formulas for not only the core concrete, but the cylinder, wire, and mortar coating as well.

When the resulting new design curves are plotted, the controlling portion of the UDP curves for concrete core and mortar coating tension typically are as shown in figure 9. But since UDP also limits the stresses and strains in the other pipe components, figure 9 also includes these other members of the UDP design curve "family". The outer curves are the performance limits for concrete in compression, and the cylinder and wire in tension.

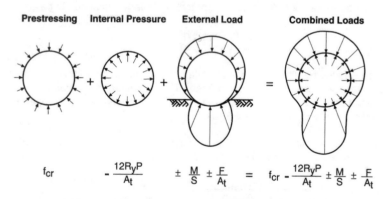

Adding concrete tension allowance ($7.5\sqrt{f'_c}$) and solving for p:

$$P = (f_{cr} + 7.5\sqrt{f'_c} \pm \frac{M}{S} \pm \frac{F}{A_t}) \frac{A_t}{12R_y}$$

Figure 8 Derivation of the AWWA C301-79 Appendix B Design Equation

PCCP DESIGN CONCEPTS 533

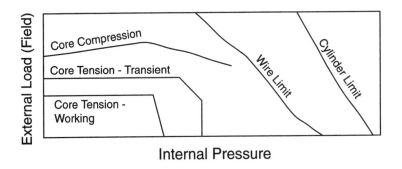

Figure 9 Typical AWWA C304 (or "UDP") Design Curves

Simplifying the AWWA C304 Design Process

There are many steps involved to calculate a PCCP design in accordance with AWWA C304. Taken one-at-a-time, each step is not that complicated. But the whole design procedure is very comprehensive, and it is laborious to attempt to design PCCP by hand. Consequently, two steps have been taken to simplify and streamline PCCP designing. First, tables of pipe design areas of prestressing wire for all diameters (410 - 1520 mm) (16 - 60 inch) of lined cylinder pipe under various bedding and external loading conditions have been included in AWWA C304.

Second, the American Concrete Pressure Pipe Association has compiled a computer program to perform the calculations using the design procedure of AWWA C304. This program is used by all PCCP manufacturers in the United States and Canada, and is available to anyone interested in performing their own PCCP designs.

Comparison of Wire Areas for New and Old Designs

One might wonder if the new design procedure is only an elaborate smoke screen to allow PCCP to be made more cheaply using thinner cores or less prestressing wire. It is not. The review of PCCP performance tests proved the old PCCP design procedures were actually quite accurate and needed only fine-tuning to provide a new, more-flexible design procedure which would better accommodate unusual installations. Table 1 shows a comparison of area ratios of prestressing wire calculated using the old and new design procedures. All comparable wire area ratios are very similar, but on average, the new design procedure requires a few percentage points more wire than the old PCCP designs.

Table 1 Prestressing Wire Area Ratios for AWWA C301-79 Appn. A, AWWA C301-79 Appn. B, and AWWA C304 Design Methods

Working Pressure = 1.03 MPa (150 psi)
Earth Cover, Meters (feet)

Pipe Size (mm, in) & Type	1.8 (6)			2.4 (8)			3.0 (10)		
LCP	"A"	"B"	UDP	"A"	"B"	UDP	"A"	"B"	UDP
610 (24)	.24	.24	.26	.24	.24	.26	.25	.24	.26
910 (36)	.33	.33	.35	.33	.33	.36	.35	.33	.39
1070 (42)	.38	.38	.39	.38	.38	.43	.40	.38	.46
1220 (48)	.46	.42	.46	.46	.42	.50	.48	.47	.53
ECP*									
1370 (54)	.41	.39	.43	.45	.40	.43	.52	.48	.49
1520 (60)	.45	.42	.47	.48	.45	.49	.56	.54	.54
1830 (72)	.54	.49	.58	.57	.56	.60	.65	.66	.64

*ECP wire area ratios are applicable only to specific core thicknesses.

Criteria to be Included in PCCP Specifications

Obviously, most pipe specifications and plans already state or show the internal working pressure and external dead loads for which pipe must be designed. Owners and engineers should know AWWA C304 includes default assumptions for transient pressure or loading conditions and for pipe installation rates which should be considered when PCCP projects are designed. If the default assumptions do not apply for a particular project, the specifications should clearly state what actual conditions will apply. Primary points to be considered include:

- what transient external loads will be on the pipe, if those loads exceed the AASHTO HS20 truck load;

- actual surge pressure, if the pipeline surge pressures will be greater than 40% of the working pressure and greater than 40 psi;

- field test pressure, if the field test pressure is greater than 20% higher than the pipe working pressure;

- actual expected time between pipe manufacture and installation, if it is anticipated the pipe will be manufactured but not installed for more than 9 months, (that is, 270 days),(this will require the creep and shrinkage factors used in the pipe design to be adjusted);

- the expected time between installation of pipe and filling the pipe with water, if the pipe will be buried but remain empty for more than 3 months (90 days) after installation, (again, to require the default creep and shrinkage factors to be adjusted).

Other requirements should also be included in the PCCP specifications to avoid pipe corrosion if there are cyclic exposures to wetting and near-complete drying, exposure to freeze/thaw conditions, or exposures to man-made waste dumps. Recommendations for handling such considerations are given in AWWA's Manual M9, Concrete Pressure Pipe.

References

AWWA C301-79, Standard for Prestressed Concrete Pressure Pipe, Steel Cylinder Type, for Water and Other Liquids. American Water Works Association, Denver, CO (1979).

AWWA C304-92, Standard for Design of Prestressed Concrete Cylinder Pipe. American Water Works Association, Denver, CO (1992).

AWWA Manual M9, Concrete Pressure Pipe. American Water Works Association, Denver, CO (1995).

Structural Performance Criteria for Fitness-for-Service
Evaluations of Underground Natural Gas Pipelines

Wen-Shou Tseng[1] and Chih-Hung Lee[2]

Abstract

This paper describes a methodology for developing a set of strain-based structural performance criteria for application to structural fitness-for-service evaluations of underground steel pipelines and presents an example application of these criteria to the evaluation of a high-pressure underground gas transmission pipeline which crosses a geologically unstable region. The development of these criteria and their applications to structural fitness-for-service evaluations utilize the present state-of-the-art nonlinear methods of pipeline structural analysis.

Introduction

Underground steel pipelines are vulnerable to damages due to ground-settlement-induced deformations. In-line geometry inspection, using a specially developed pipeline inspection gauge (pig), can presently establish the deformed geometry of an underground pipeline with sufficient accuracy for determining the pipeline's deformation demands, e.g., bending curvatures of straight pipeline segments and smooth bends, angular deflections of elbows, and magnitudes and shapes of cross-sectional ovalization. Knowing the demands, a fitness-for-service evaluation of the pipeline based on code-allowed stress

---

[1] President, International Civil Engineering Consultants, Inc., 1995 University Ave., Suite 119, Berkeley, CA 94704

[2] Senior Gas Engineering Specialist, Pacific Gas and Electric Company, 123 Mission Street, San Francisco, CA 94106

criteria for the deformation-controlled loading situation would be overly conservative; a more reasonable approach is to use a set of properly developed strain-based performance criteria which take into account the inherent ductility of the pipeline's material.

This paper describes a recently-developed methodology for establishing a set of strain-based performance criteria for application to a structural fitness-for-service evaluation of a 27 km-long, high-pressured underground steel gas-transmission pipeline which crosses a geologically unstable region in Northern California. In developing the criteria, the concept of limit-state, empirical results derived from analyses of test data, and rigorous state-of-the-art elasto-plastic nonlinear structural analysis methods have been utilized. The application of these criteria to the evaluation of the above-mentioned underground pipeline will now be presented.

Structural Fitness-for-Service Criteria

To ensure fitness-for-service of a pipeline is to ascertain that the pipeline will achieve the following performance goals under normal service loads:

(1) Carry its contents within the design operating pressure range without cracking, leaking, or rupturing, thus preventing unintended loss of content and pressure.

(2) Permit normal operational maintenance inspections by allowing the passage of in-line inspection or maintenance tools without undue impediment.

For an underground pipeline, the normal service loads under which the above basic performance criteria need be satisfied include: (1) internal pressures ranging from zero to the maximum design value, (2) external soil pressures, and (3) ground-settlement-induced pipeline deformations.

In order to assure the above basic criteria are met, appropriate structural performance criteria must be established to ensure that an adequate margin of safety exists against each potential failure mode of the pipeline under any combination of the above-mentioned service loads. The failure modes considered relevant to the pipeline's performance are: (1) tensile failure by fracture and net-section-collapse, (2) compression failure by local buckling leading to cross-sectional collapse causing cracking and leaking, and (3) unacceptably large cross-sectional ovalization. To develop suitable criteria to prevent each of these failure modes, the concept of "limit-state" criteria as used in American Gas Association (AGA) topical report (Stephens, et. al. 1991) has been adopted. In accordance with the concepts and definitions used in this report, three levels of limit-sate, in descending order of conservatism, can be defined and used, namely, "yield limit-state", "damage limit-state", and "ultimate limit-state".

The yield limit-state, which is the basis used in engineering design codes, is suitable as a design limit. It is, however, considered to be too conservative for fitness-for-service evaluation purposes. The ultimate limit-state, on the other hand, is considered unconservative since the current state-of-the-art for predicting it still contains large uncertainties. Thus, the intermediate damage limit-state is considered most appropriate as the basis for developing structural fitness-for-service criteria. Adopting this limit-state for each of the three failure modes mentioned above, appropriate fitness-for-service criteria have been established as described below.

<u>Maximum Flow-Stress Strain Criterion</u> - To guard against tensile failure, a strain limit has been established to ensure a negligible potential for fracture of the pipeline's steel and weld materials. The tensile damage limit-sate is considered to be attained when the maximum longitudinal tensile strain reaches a level corresponding to the "flow-stress" of the material, which is defined to be 10 ksi over the minimum yield strength (defined to be the stress at the strain of 0.5%), with a factor of safety of 1.5. The flow-stress is the uniform stress level at which crack growth and fracture can be initiated resulting in a net-section-collapse (Stephens, et. al. 1991). The determination of the flow-stress strain limit for fitness-of-service evaluation purposes is illustrated in Fig. 1 for a typical ductile pipeline material. As shown in this figure, this damage limit-state is between the yield limit-state and the ultimate flow-stress limit-state.

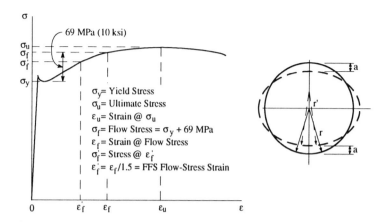

Fig. 1. Maximum Flow-Stress Strain Criterion

Fig. 2. Pipe Cross-Sectional Ovalization

Buckling Damage-State Criterion - To prevent initiation of compression failure by local buckling, the buckling damage-state criterion as recommended in the above-mentioned AGA topical report, which is based on a statistical analysis of available test data on straight pipelines, can be adopted. Thus, compressional damage limit-state for a straight pipeline initiated by local buckling is considered to be reached when the maximum longitudinal compression strain reaches the AGA-recommended empirical buckling damage-state criterion. In accordance with the AGA recommendation, the 95%-confidence-level maximum compression strain at zero pressure, i.e., ε_o, at which local buckling may be initiated is given by

$$\varepsilon_o = 2.42 \left[\frac{t}{D}\right]^{1.59} \quad (1)$$

where t is the pipe wall thickness in inches and D is the pipe diameter in inches. This criterion as illustrated in Fig. 3 does not explicitly take into account the influence of cross-sectional ovalization due to bending of the pipe and internal and/or external pressure. To account for the cross-sectional ovalization due to bending, a modification of the above criterion has been made, based on the theoretical analysis by Gresnigt (1986), by replacing the pipe's cross-section radius $r = D/2$ by the radius of curvature of the ovalized pipe, r', at the location of maximum compression strain where buckling will ultimately occur (see Fig. 2); in which case, Eq. (1) becomes

$$\varepsilon_o = 2.42 \left[\frac{t}{2r}\right]^{1.59} = 0.804 \left[\frac{t}{r}\right]^{1.59} \quad (2)$$

$$r' = \frac{r}{1 - (3a/r)} \quad (3)$$

where a is the amplitude of the ovalization shown in Fig. 2. To account for the effect of internal or external pressure, p, on the criterion, the recommendation of Gresnigt based on test data for pressurized pipes is adopted. Thus, the buckling damage-state criterion with internal pressure ($p > 0$) and/or external pressure ($p < 0$) can be expressed in terms of the maximum compressive strain, ε_c, as follows:

$$\varepsilon_c = \varepsilon_o \pm 3000 \left[\frac{pr}{Et}\right]^2; \quad p \gtrless 0 \quad (4)$$

where ε_o is the strain limit for zero pressure given by Eq. (2) and E is the Young's modulus of elasticity of the pipe material. This modified criterion, also shown in Fig. 3, is based on empirical results for straight pipelines; thus, it does not apply to pipe elbows for which an empirical criterion equivalent to that for straight pipes is not yet available.

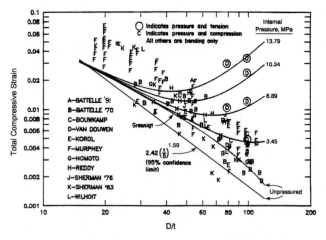

Fig. 3. Buckling Damage-State Criterion (AGA Report 1991)

Maximum Moment-Resistance Criterion - To account for the overall pipe behavior observed in tests of straight pipe specimens under combined bending and internal/external pressure loadings (Gresnigt 1986), in which the point of maximum bending moment-resistance of the pipe is often coincident with the onset of local buckling for thin-wall pipes and the onset of large plastic deformations for thick-wall pipes or thin-wall pipes with significant internal pressure, a damage limit-state deformation criterion to control the overall pipe behavior can be defined to be the bending curvatures at the maximum moment-resistance capacities under combined bending and internal/external pressure loadings. For the load-controlled loading situation, this point corresponds to the start of unrestrained buckling or unbounded plastic deformations in the pipe leading to eventual collapse of the cross-section; therefore, it is an ultimate limit-state. However, for the displacement-controlled loading situation such as that subjected to by a buried pipe, the point of maximum moment-resistance does not lead to unrestrained buckling or unbounded plastic deformations; thus, it is a damage limit-state similar to the buckling damage-state described above. Consequently, it is suitable to serve as a fitness-for-service criterion for a buried pipeline under ground-settlement-induced pipeline deformations. Furthermore, unlike the local buckling mode which generally occurs in a sudden manner and, therefore, is difficult to predict analytically, the point of maximum moment-resistance is generally attained in a gradual and stable fashion; thus, it is more amenable to analytical prediction with sufficient accuracy. The analytical methods that can be utilized to establish this criterion generally fall into two categories: (1) classical mechanics method of elastic-plastic nonlinear analysis and (2) modern finite-element numerical method of nonlinear analysis. For straight pipelines, because of their geometric simplicity, both methods can effectively be utilized for developing the criterion. However, for pipe elbows having a more complex geometric configuration, only the finite-element method can be used effectively. Examples of the maximum moment-resistance criterion obtained from nonlinear analyses for a straight pipeline are shown in Fig. 4.

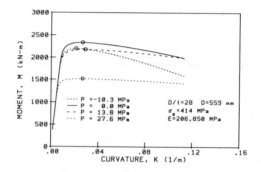

Fig. 4. Maximum Moment-Resistance Criterion for Straight Pipelines

<u>Ovalization Limit-State Criterion</u> - To allow unimpeded passing of normal maintenance tools, such as pipeline cleaning pigs or inspection gauges, pipeline cross-sectional ovalization under combined bending and internal/external pressure loadings should be controlled by a limit-state criterion expressing the maximum allowable ovalization as a percentage of pipe diameter. Following the recommendation by Gresnigt (1986), this criterion can be set reasonably at 15% of the pipe diameter.

Given a straight pipeline with known pipe diameter (D) and wall thickness (t), and its material stress-strain relationship subjected to combined loadings of ground-settlement-induced bending deformations and internal and/or external pressures, the above four limit-state criteria can be expressed as functions of two loading parameters, namely, the bending curvature (K) and resultant pressure (p). So expressed, they can readily be used to determine whether or not the pipe under a particular service loading condition (K and p) satisfies each of the criteria.

Similarly for pipe elbows, except for the buckling damage-state criterion which is not applicable for elbows, the other three criteria can be developed in terms of the same two loading parameters; however, because of the presence of initial curvature, the parameter for measuring bending deformation for the elbows need be expressed in terms of the average curvature change (ΔK), which is defined to be the deformed angle of the elbow (α_j) minus its initial angle

(α_i) averaged over a selected centerline curve length (s), i.e.,

$$\Delta K = \frac{\alpha_j - \alpha_i}{s} = \frac{\Delta \alpha}{s} \qquad (5)$$

in which, if s is selected to be the exact centerline curve length of the elbow, ΔK becomes the average curvature change of the elbow itself. However, in practice, the length over which the elbow angular deflection ($\Delta \alpha$) is obtained from field measurements or in-line geometry inspection data does not correspond precisely to the elbow's curve length; rather, it normally includes a portion of the straight pipe joining either end of the elbow. In this case, ΔK defined by Eq. (5) gives a measure of the "composite" curvature change averaged over the "gauge" length of the elbow joint over which $\Delta \alpha$ is measured.

Application of the Fitness-for-Service Criteria

To illustrate the development and application of the above-described criteria for fitness-for-service evaluations of underground pipelines, an actual example of application to a 27-km (17-mile) long underground gas transmission pipeline is described. This line is located in the Sacramento/San Joaquin Delta region in Northern California where the land is predominantly below sea level and protected by perimeter levee systems. As the line traverses across the Delta, it crosses 4 islands, 4 rivers (or sloughs), and 8 levees. Due to the presence of a relatively thick layer of highly compressible peat soil, the levees have been continuously subsiding requiring periodic fills to maintain their elevations. Because of this, the line, which is buried underground, is continuously subjected to ground-settlement-induced deformations causing concerns about its structural fitness for service, especially those segments crossing the 4 rivers and their associated 8 levees.

The pipeline was constructed of Grade X-60 steel pipes and Grade Y-65 steel elbows having a 559 mm (22 inch) outside diameter with 17 mm (0.660 inch) wall thickness except at the river crossings where the pipe wall is 20 mm (0.792 inch) thick. The mechanical properties of the pipe steel and weld materials including their yield and ultimate strengths, and the stress-strain curves up to their ultimate strain levels were ascertained by tests of coupons taken from spare or replaced pipe segments and elbows.

To define the present state of the pipeline's deformed geometry, an in-line geometry survey was made using a specially designed pipeline inspection gauge (pig), which is capable of monitoring the instantaneous position and altitude of the pig in the pipe and capable of mapping the inside geometry of the pipeline cross-sections at close intervals. This survey yields measured data which can be analyzed to obtain the horizontal and vertical curvatures of the straight pipe sections and smooth bends, and the angular deflection of each elbow over a gauge length which can also be determined from the measured data. For example, a typical curvature profile determined from the survey data for a straight section of the pipeline is shown in Fig. 5 and a typical angular deflection and associated gauge length as determined from the measured data are shown in Fig. 6. Based on results of this type, the present pipeline bending deformation demands, in terms of curvatures (K) for the straight pipe sections and changes of the curvature (ΔK) for bends and elbows can be established quite accurately, which can then be compared with the fitness-for-service criteria developed specifically for the pipeline to evaluate its condition.

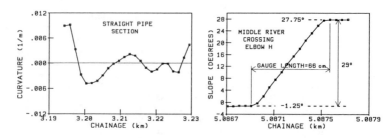

Fig. 5. Curvature Profile for Straight Pipeline Calculated from Survey Data

Fig. 6. Deformed Angle for Elbow Calculated from Survey Data

To develop the pipeline-specific fitness-for-service criteria using the methodology described previously, detailed nonlinear analyses were performed for the straight sections and the elbows of the pipeline based on its specific geometric configurations and material properties. These are separately described below.

<u>Straight Pipe Sections</u> - Detailed elasto-plastic nonlinear analyses for developing the fitness-for-service criteria for straight sections of the pipeline were conducted using both the classical mechanics method such as that developed by Gresnigt (1986) and the finite

element method using three-dimensional shell models. To account for the pipe cross-sectional ovalization under combined bending and pressure, the analyses considered large deformation geometry. In these analyses, pressure loading at different levels was applied first and kept constant in each case while bending deformations were applied in incremental steps. By carefully monitoring the resulting amplitudes of maximum tensile and compressive strains in the pipe wall, bending moment of the pipe cross-section, and cross-sectional ovalization, and applying the four fitness-for-service criteria described previously, the resulting criteria expressed as functions of bending curvature (K) and resultant pressure (p) obtained using the classical mechanics method are shown in Fig. 7. In this figure, the maximum flow-stress strain criterion is not shown because it falls outside the range of the graph. From the results shown in this figure, one can see that for this pipeline having a relatively thick wall (D/t = 28), the fitness-for-service criteria are governed by the maximum moment-resistance criterion. Furthermore, within the operating internal pressure range of 0 to 14.89 MPa (2,160 psi), the zero internal pressure case governs. To illustrate the variations in the criteria, the criteria developed for the same pipe with a thinner wall, e.g., t = 6.7 mm (0.265 inch), D/t = 83, corresponding to those shown in Fig. 7 are shown in Fig. 8, which show that the governing criterion for this case becomes the buckling damage-state criterion, as would be expected. Based on the criteria so established and the pipeline bending curvature demands as determined from the pipeline survey data mentioned previously, the fitness-for-service conditions of the straight portions of the pipeline were then assessed.

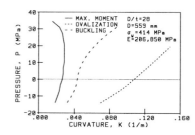

Fig. 7. Structural Fitness-for-Service Criteria for Straight Pipe (D/t = 28)

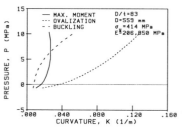

Fig. 8. Structural Fitness-for-Service Criteria for Straight Pipe (D/t = 83)

Pipe Elbows - The elbows of this line are of typical elbows having a bend radius equal to 1.5 D. Given such a tight bend, the stress and strain distributions in the elbows under the loadings considered will be quite different from those for a straight pipe under the same loading conditions. Consequently, for establishing their fitness-for-service criteria, each elbow having a unique initial bend angle (α_i) had to be analyzed separately. Because of the more complex geometry, detailed nonlinear analyses for developing the fitness-for-service criteria were conducted using the finite element method based on a model such as that shown in Fig. 9. Typical results for the maximum moment-resistance criterion for several elbows, expressed as a function of pressure (p) and composite curvature change (ΔK) are shown in Fig. 10. Similar results were obtained for the maximum flow-stress strain and ovalization limit-state criteria. The buckling damage-limit-state criterion, however, was not developed for the elbows for the reason mentioned previously. Using the criteria as established and the composite-curvature-change demands determined from the field survey data, the fitness-for-service conditions of the elbows were determined.

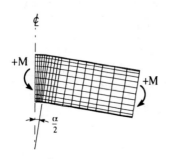

Fig. 9. Typical Finite-Element Model for Elbow

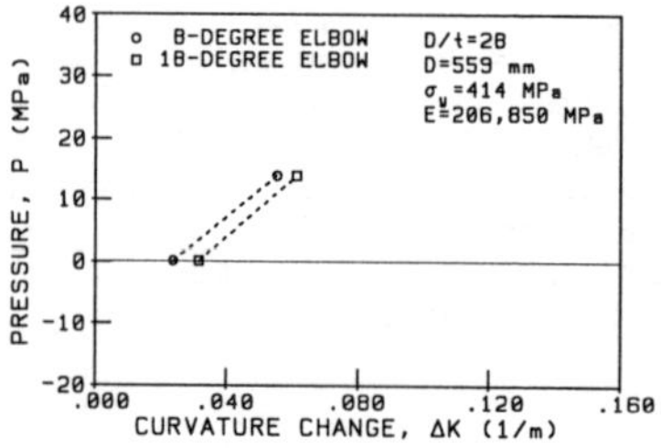

Fig. 10. Typical Maximum Moment-Resistance Criterion for Elbow

Smooth Bends - The field and factory bends of this line are limited in their initial bend curvatures by the small angular deflection rate, i.e., 1.5° per one pipe diameter, specified in their fabrication specifications. This results in a bend with a rather long bend radius of curvature, i.e., of the order of 38 D. Since the fitness-for-service criteria for straight pipes are more conservative than the corresponding criteria for elbows which have much tighter bend radius (1.5 D), it is appropriate and conservative to use the criteria for the straight pipes for fitness-for-service evaluations of smooth bends. This approach was adopted in evaluating the fitness-for-service conditions of the smooth bends in this line.

Based on the fitness-for-service criteria developed and the associated evaluations conducted for this pipeline, none of the straight sections and smooth bends of this line was found to have their demands exceeding those of even the most conservative maximum moment-resistance criterion, even though a preliminary evaluation using a simplified linear analysis method and the code-allowed stress criteria had indicated over-stressed conditions at several locations. The evaluations, however, did identify a few critical elbows located at specific levee crossings having demands exceeding those of the governing maximum moment-resistance criterion; however, they were within the allowable ranges covered by the other two criteria. These critical elbows identified have since either been replaced or been placed under close surveillance.

Concluding Remarks

Application of the strain-based structural performance criteria developed herein to the 27-km long underground gas transmission pipeline described shows these criteria to be effective in assessing the fitness-for-service condition of underground pipelines subjected to ground-settlement-induced deformations. By appropriate modifications, they can also be used in assessing the integrity of underground steel pipelines subjected to transient (short-term) cyclic ground deformations produced by seismic events. The transient (short-term) nature of seismic deformations warrants a relaxation of the structural performance criteria proposed herein; whereas, to minimize the potential of low-cycle fatigue failure due to the cyclic nature of the seismic deformations requires a tightening of the proposed criteria.

References

Gresnigt, A. M. (1986) "Plastic Design of Buried Steel Pipelines in Settlement Areas," Heron, Vol. 31, No. 4, pp. 3-112.

Stephens, D. R., Olson, R. J., and Rosenfeld, M. J. (1991) "Topical Report on Pipeline Monitoring -- Limit State Criteria," Report to Line Pipe Research Supervisory Committee of the Pipeline Research Committee, American Gas Association, NG-18, Report No. 188, SI 4.3-86.

State-of-the-Art Review: Trenchless Pipeline Rehabilitation Systems

M. Najafi, M. ASCE[1] and V. K. Varma, F. ASCE[2]

Abstract

This paper describes the state-of-the-art review for trenchless pipeline rehabilitation conducted under the Construction Productivity Advancement Research (CPAR) program. For the purpose of this study, the existing trenchless rehabilitation methods will be classified into six groups: cured-in-place pipe (CIPP), sliplining, in-line replacement, close-fit pipe, point source repair, and sewer manhole rehabilitation. For each group, specific rehabilitation procedures are provided and some of the advantages and limitations of each method are discussed. Such factors as ability to add to structural integrity of old pipe, maximum and minimum size of pipe, lateral connections, existing flow, ability to adjust bend and curves, and maintaining a grade, are considered. This study provides an overview of these methods and presents guidance on the situations where each method is applicable.

Introduction

The state-of-the-art review was one of main elements of the research project, "Trenchless Construction: Evaluation of Methods and Materials to Install and Rehabilitate Underground Utilities." The research was a Corps and Industry cost-shared project funded under the Construction Productivity Advancement Research (CPAR) program. The laboratory partner was the U.S. Army Engineer Waterways Experiment Station (WES), Geotechnical Laboratory. The industry partner was the Louisiana Tech University's Trenchless Technology Center. Industry participants contributed over half the total cost of this research. The overall objective of the CPAR program is to improve productivity in the U.S. construction industry, thereby enhancing the competitiveness of the U.S. industry in domestic and overseas

---

[1]Assistant Professor, Engineering Technology, Missouri Western State College, St. Joseph, Missouri.

[2]Professor and Chair, Department of Engineering Technology, Missouri Western State College, St. Joseph, Missouri.

markets. The Corps and other federal and state agencies will benefit by realizing reduced costs, enhanced safety and better products and services as the products of this research program are developed, tested, and commercialized.

The basic trenchless pipeline rehabilitation methods can be categorized into the following types: (1) Cured-In-Place Pipe (CIPP), (2) Sliplining, (3) In-Line Replacement, (4) Close-Fit Pipe, (5) Point Source Repair, and (6) Sewer Manhole Rehabilitation. Table 1 provides a summary of these basic methods, with additional information provided in later sections.

Table 1
Comparison of Different Trenchless Pipeline Rehabilitation Methods

Method	Diameter Range (mm)	Maximum Installation (m)	Liner Material	Applications
CIPP:				
Inverted in Place	100-2700 (4-108 in.)	900 (3,000 ft)	Thermoset Resin/Fabric Composite	Gravity and Pressure Pipelines
Winched in Place	100-1400 (4-54 in.)	150 (500 ft)	Thermoset Resin/Fabric Composite	Gravity and Pressure Pipelines
Sliplining:				
Segmental	100-4000 (4-158 in.)	300 (1,000 ft)	PE, PP, PVC, GRP (-EP & -UP)	Gravity and Pressure Pipelines
Continuous	100-1600 (4-63 in.)	300 (1,000 ft)	PE, PP, PE/EPDM, PVC	Gravity and Pressure Pipelines
Spiral Wound	100-2500 (4-100 in.)	300 (1,000 ft)	PE, PVC, PP, PVDF	Gravity Pipelines Only
In-Line Replacement:				
Pipe Displacement	100-600 (4-24 in.)	230 (750 ft)	PE, PP, PVC, GRP	Gravity and Pressure Pipelines
Pipe Removal	up to 900 (36 in.)	100 (300 ft)	PE, PVC, PP, GRP	Gravity and Pressure Pipelines
Close-Fit Pipe:				
Modified Cross Section	100-400 (4-15 in.)	210 (700 ft)	HDPE, PVC	Gravity and Pressure Pipelines
Drawdown	62-600 (3-24 in.)	320 (1,000 ft)	HDPE, MDPE	Gravity and Pressure Pipelines
Rolldown	62-600 (3-24 in.)	320 (1,000 ft)	HDPE, MDPE	Gravity and Pressure Pipelines

Table 1 (Continued)

Method	Diameter Range (mm)	Maximum Installation (m)	Liner Material	Applications
Point Source Repair:				
Robotic Structural	200-760 (8-30 in.)	N/A	Epoxy Resins/Cement Mortar	Gravity
Grouting	N/A	N/A	Chemical Grouting	Any
Link-Seal	100-600 (4-24 in.)	N/A	Special Sleeves	Any
Point CIPP	100-600 (4-24 in.)	15 (50 ft)	Fiberglass/Polyester, etc.	Gravity
Spray-on-Lining	76-4500 (3-180 in.)	150 (500 ft)	Epoxy Resins/Cement Mortar	Gravity and Pressure Pipelines
Manhole Rehabilitation	Any	N/A	Spray-On Lining, PVC, CIPP	Sewer Manholes

Cured-In-Place Pipe (CIPP)

CIPP is a liquid thermoset resin-saturated material that is inserted into the existing pipeline by hydrostatic or air inversion or by mechanically pulling with a winch and cable. The material is heat-cured in place. Insituform introduced CIPP in the United Kingdom in 1971 and entered the U.S. market in 1977. Since 1971, over 22 million feet of CIPP have been installed worldwide with Insituform process. In 1989, the InLiner USA® process was introduced in Houston, Texas. In the 1980's, Ashimori Industries developed a method called hose lining in Japan basically for pressure pipe applications in the Japan gas industries. This product, known as Paltem, was further developed in collaboration with Tokyo Gas. Recently a number of CIPP systems have been introduced into the market. This increased number of available methods provides many technical and economical benefits for utility owners through increased competition.

The primary components of the CIPP are a flexible fabric tube and a thermosetting resin system. For typical CIPP applications, the resin is the primary structural component of the system. These resins generally fall into one of the following generic groups: (1) Unsaturated Polyester, (2) Vinyl Ester and (3) Epoxy. Each resin has distinct chemical resistance and structural properties. All have distinct chemical resistance to domestic sewage.

Unsaturated polyester resins were originally selected for the first CIPP installations due to their chemical resistance to municipal sewage, good physical properties in a CIPP composite, excellent working characteristics for CIPP installation procedures, and economic feasibility. Unsaturated polyester resins have remained the most widely used systems for the CIPP processes for over two decades.

Vinyl ester and epoxy resin systems are mainly used in industrial and pressure pipeline applications where their special corrosion and/or solvent resistance and higher temperature performance are needed and higher cost justified. In drinking water pipelines, epoxy resins are required.

The primary function of the fabric tube is to carry and support the resin until it is in-place in the existing pipe and cured. This requires that the fabric tube withstand installation stresses with a controlled amount of stretch but with enough flexibility to dimple at side connections and expand to fit against existing pipeline irregularities. The fabric tube material can be woven or non-woven with the most common material being a non-woven, needled felt. Impermeable plastic coatings are commonly used on the exterior and/or interior of the fabric tube to protect the resin during installation. The layers of the fabric tube can be seamless, as with some woven materials, or longitudinally joined with stitching or heat bonding.

The primary differences between the various CIPP systems are in the composition and structure of the tube, method of resin impregnation (by hand or by vacuum), installation procedure, and curing process. There are two primary approaches to installing the flexible tube—inverting in place and winching in-place. Specific variations of installation procedures and materials are employed by different manufacturers.

Sliplining

Sliplining is one of the earliest forms of trenchless pipeline rehabilitation. A new pipe of smaller diameter is inserted by pulling or pushing into a deteriorated host pipe and the annulus space between the old pipe and new pipe is grouted. This method has the merit of simplicity and is relatively inexpensive. However, there can be a loss of hydraulic capacity.

This system is used where the host pipe does not have joint settlements, misalignments, deformations, and similar defects. The new pipe can form a continuous, water-tight pipe within the existing pipe after installation. Where the new pipe has to be laid to an even grade, the use of plastic or metal locators/spacers is essential. Spacers also maintain the pipe location during annular grouting to ensure a uniform surrounding. The service connections are then re-connected to the new pipe. Sliplining can be categorized into three types—continuous, segmental and spiral wound. Each method is discussed below.

Continuous. The continuous sliplining method involves accessing the deteriorated pipe at strategic points and inserting HDPE pipe, joined into a continuous tube, through the existing pipe structure. This technique has been used to rehabilitate gravity sewers, sanitary force mains, water mains, outfall lines, gas mains, highway and drainage culverts, and other pipeline structures. The technique has been used to restore pipe as small as 25.4 mm (1 in.), and the maximum pipe diameter is limited by the availability of factory-made pipes. Over 30 years of field experience shows that this is a proven cost-effective means that provides a new structure with minimum disruption of service, surface traffic, or property damage that would be caused otherwise by extensive excavation.

Segmental. This method involves the use of short sections of pipe that incorporate a flush sleeve joint commonly used in microtunneling and pipe jacking processes. A number of plastic pipe products, such as GRP, PVC, PP and PE, which include short-length sections with a variety of proprietary smooth joints (both inside and outside) have been specially developed for sliplining sewers. This method is applicable for diameters more than 900 mm (36 in.).

Segments of the new pipe are assembled at entry points and pushed inside the host pipe. After the new pipe is positioned in place, the annular space is grouted. This is a very simple method and can be carried out by a general pipe contractor. The laterals are usually re-connected by excavation from outside.

Spiral Wound. This technique is based on forming a pipe in-situ by using PVC-ribbed profiles with interlocking edges. This method can be used for either structural or non-structural purposes, depending on the grouting requirements. There are two variations of this method: one fabricates a pipe in the manhole by helically winding a continuous PVC fabric; the second, used for larger diameters (over 762± mm or 30± in.), uses preformed panels inserted in-place in the existing pipeline. Excavation is not required for this process. House connections and laterals can be reconnected by local excavations or by a remote-controlled cutter.

In-Line Replacement

When pipelines are found to be structurally failing or inadequate, then in-line replacement should be considered. This is a relatively expensive method of pipeline rehabilitation since the existing pipe is removed and a completely new pipe is installed. In-line replacement or trenchless methods of replacing of an existing pipeline can be categorized as Pipe Displacement and Pipe Removal.

Pipe Displacement. Pipe displacement, also known as pipe bursting, is a technique for breaking out the old pipe by use of radial forces from inside the old pipe. The fragments are forced outward into the soil and a new pipe is pulled into the bore formed by the bursting device. The Pipeline Insertion Machine (PIM) was developed in the late 1970's in England. Although it originally was originally developed for the gas industry, the method subsequently found application in the replacement of waterlines and gravity sewers.

The pipe displacement equipment usually uses a standard pneumatic mole with a special pointed shield. Twin hydraulically operated breaker arms are attached in the front of the tool. These breaker arms are remotely operated and can exert pressure to overcome difficult joints or surroundings. A standard air compressor is also needed to provide power to the pneumatic mole, and a hydraulic winch is required as well.

The deteriorated host pipe is broken outward by means of an expansion tool and the new pipe is either towed behind the bursting machine or jacked into place using conventional pipe jacking techniques. This method requires the reconstruction of laterals by excavation from the surface.

Pipe Removal. The development of microtunneling machines with the capability of crushing rocks and stones has led to their use in excavating existing pipelines for replacement. Pipe removal equipment used today are slightly modified remote-controlled microtunneling systems with crushing capacities. The various elements include a modified shield, a jacking unit, a control console and a conveying system. A jacking pit of about 3 m (10 ft) in diameter and a smaller receiving pit are required. As the machine is driven forward (excavating the old pipe and surrounding ground), the new pipe sections are pushed up behind the advancing shield. This method can be used to upsize the new replacement pipe.

Close-Fit Pipe

This process of trenchless pipeline rehabilitation has been carried out in the United States since 1988. Compared with CIPP, and assuming other project factors to be the same, closed-fit technology does not require a long curing process therefore require less time to complete a project.

This type of trenchless pipeline rehabilitation uses coiled deformed new pipe before it is installed, then expands it to its original size and shape after placement to provide a close fit with the existing pipe. Most lining pipe first is deformed in the manufacturing plant, shipped to job site, then inserted and finally reformed by heat and pressure or naturally. This method can be used for both structural and non-structural purposes.

There are three versions of this approach: Modified Cross Section including fold and formed PVC (NuPipe, AM-Liner), deformed and reformed (U-Liner) and sectionalized lining (Sureline), Drawdown and Rolldown.

Modified Cross Section. Modified cross section technology originated as the result of a five-year research and development project in France to develop a trenchless method of rehabilitation without reducing flow capacity. The first U.S. installation was in Toms River, New Jersey, around September of 1988, using an 8-inch diameter pipe. In the U.S.A., modified cross section technology originated in the oil industry and has been developed by Pipe Liners, Inc., of New Orleans, Louisiana, for use in sanitary sewers. This development started in 1987.

The method uses a jointless extruded PVC or HDPE pipe which is folded or deformed to reduce the cross-sectional area. The folded pipe is mechanically pulled into the existing pipeline, then formed to the shape of the existing pipeline using heat, pressure, and in the case of NuPipe or AM-Liner, a mechanized rounding device. Service connections are reinstated internally using a remote-controlled robotics cutter.

In most modified cross section technologies, protruding taps should be fully removed or reduced to less than 5% of the host pipe diameter. The taps will normally be reduced by some type of internal cutting device.

Drawdown. This method has evolved from methods developed for British Gas. The Roll-Down System, from Subterra Ltd., Dorset, U.K., involves a hydraulic pusher that forces the pipe through multiple pairs of rollers to reduce its diameter. Swage-Lining, from British Gas draws the new pipe through a die to reduce its diameter. For larger diameters, the pipe is heated. Overall diameter reduction ranges from around 20% for a 100-mm (4-inch) diameter pipe, to 7% for a 610-mm (24-inch) pipe.

This method sliplines a HDPE solid wall pipe into an existing pipeline after joints are butt-fused and the HDPE pipe is swaged down in diameter. Diameter reduction depends on historic memory which is a function of the deforming temperature. Compressing the pipe temporarily crushes the chain structure, allowing the pipe to be reduced in diameter and later reverted to its original size without affecting performance. Pipes of 76 to 610 mm (3 to 24 in.) in diameter can be installed utilizing this method.

After long, continuous lengths of tube are pulled into the existing pipe, pressure is applied to the inside of the new pipe to speed up the reversion process. The pipe in its reverted form usually fits closely to the old pipe wall, and no annular space remains.

Rolldown. The Rolldown system is similar to drawdown except that the new pipe diameter is reduced for insertion by running the new pipe through a cold rolling machine. This rearranges the long chain structure of the plastic pipe to produce a smaller diameter pipe with thicker walls and minimal elongation.

The Rolldown system, from Subterra Ltd., Dorset, U.K., involves a hydraulic pusher that forces the pipe through multiple pairs of rollers to reduce its diameter. Long, continuous lengths of MDPE or HDPE tube are pulled into the existing pipe. Then pressure is applied to the inside of the pipe to speed up the reversion process.

The pipe in its reverted form usually fits closely to the old pipe wall, and no annular space remains.

Point Source Repair

Point source repair of pipeline rehabilitation covers a broad range of techniques such as robotics structural, grouting, CIPP, link-seal, shotcrete, coatings, spray-on linings, etc. as are discussed in this section.

When local defects are found in a structurally sound pipeline, due to cracking or joint failure, point source repairs are considered. Coatings are fixed to the interior wall of the existing pipe by adhesion, or in the case of robotics repairs, the defect is filled with epoxy. Systems are available for remote-controlled resin injection to seal localized defects in the range of 100 to 760 mm (4-30 inches) in diameter.

The new spot repair devices are used to address four basic problems. The first purpose involves maintaining the loose and separated pieces of un-reinforced old pipe aligned to ensure the load bearing equivalent of a masonry arch. The second purpose is to provide added structural capacity or support to assist the damaged pipes to sustain structural loads. The third purpose provides a seal against infiltration and exfiltration. Finally, the fourth purpose is to replace missing pipe sections.

Robotic Structural. Robotic structural repair is one of the newest technologies (over a decade) to trenchless pipeline rehabilitation. Robots are used to structurally repair isolated defect areas in pipelines. First, robots are used to grind the defect area exposing a clean and smooth surface. Then these grooved areas are injected with epoxy based resins, which bond to surrounding host pipe, creating a structural as well as a permanent barrier impervious to interior or exterior chemicals or objects.

Robotic structural point repair is used either as stand alone or as a precursor to other rehabilitation methods. As a stand alone, robotics structural point repair is used to repair radial, longitudinal and spider cracks. The process also lends itself to repairing broken joints, slip joints, open joints, protruding service connections, recessed service connections, roots and other foreign objects that are usually found in collection pipelines systems.

The robotics process uses the epoxy resin as the final structural fix. The epoxy bonds to the pipe medium and permanently seals the wall from further infiltration of outside matter. Also due to the epoxy hardness and structural adhesion, a repair to the pipe wall stops the occurrence of further cracking with respect to the location repaired.

Robotic structural point repairs are carried out by an operator manipulating the robotics functions by remote control with the aid of a closed circuit television (CCTV). As a first step, the robot is positioned at the defect area and is surveyed for the best starting position. Chemical grouting is implemented if any infiltration of water is present. The operator then begins to grind out the crack(s). This accomplishes two goals. One, the crack is cleared of all foreign material and is by physical properties stopped from further cracking due to the grove cut. Second, the grove created gives a larger surface area to inject the epoxy resin.

The second step is to fill the void area with the epoxy. This is carefully accomplished, making sure that the groove is fully filled and flush with the pipe wall. Once the epoxy cures, (1-2 hour, complete 8 days) the pipeline is structurally mended and returned to the previous state without reduction in flow.

Grouting. Grouting, as sliplining, is one of the oldest method of pipeline rehabilitation. In recent years, there have been new advances in products and equipment for a wide variety of grouting applications.

Chemical grouting is normally used for sewer lines and is applied to seal leaking joints and circumferential cracks. Small holes and radial cracks may also be sealed by chemical grouting. Chemical grouts can also be applied to pipeline joints, manhole walls, using special tools and techniques. Chemical grouting is used in concrete, brick, clay sewers and other pipe materials to fill voids in backfill outside the sewer wall. Such backfill voids can reduce lateral support of the wall and allow outward movement resulting in the rapid deterioration of the structural integrity of the pipe. Chemical grouting has no structural properties capable of ensuring an effective seal where joints or circumferential cracking problems are due to ongoing settlement or shifting of the pipelines. Therefore, it is not effective to use chemical grouting to seal longitudinal cracks or to seal joint where the pipe near the joint is longitudinally cracked.

All chemical grouts are applied under pressure after appropriate cleaning and testing of the joint or the location where grout is going to be applied. Methods are available to use robots to clean, cut and grout pipelines. Reaming is the process of scraping the internal walls of the pipe to remove the debris and create a rough surface for application of a chemical grout. However, a rough surface may or may not be required for the application of the chemical grout, depending on the individual manufacturer. The sealant is applied at locations of infiltration/inflow (I/I) or exfiltration to prevent the surrounding soil from being eroded away thereby prolonging the life of the pipeline.

Spray-On Lining. The spraying of a thin mortar lining or a resin coating onto pipes is a well established technique. Spray-on lining systems use either epoxy resins or cement mortar to form a continuous lining within the host pipe. These systems result in improved corrosion resistance and hydraulic characteristics but, except for shotcrete and gunite, have little value to enhance structural integrity of the pipe.

For man-entry pipes, diameters between 900 to 3,600 mm (36 to 142 in.), structural reinforced sprayed mortars (shotcrete or gunite) are effective and widely used for rehabilitating pressure pipes and gravity sewers. Specialty concrete are available in three types: cement mortar, shotcrete, and cast concrete. Acid resistance mortars have been used in industry as linings in tanks or as mortar bricks. Development of mechanical in-line application methods (centrifugal and mandrel) has established mortar lining as a successful and viable rehabilitation technique for sewer lines, manholes and other structures. Specialty concretes containing sulfate resistant additives such as potassium silicate and calcium aluminate have shown greater resistance than typical concrete to acidic attack on sewer pipes and manhole structures. As with any other trenchless pipeline rehabilitation, the pipeline must be thoroughly cleaned and dried before a rehabilitation method can be applied. If the lining is carried by machine or carts and applied manually with a trowel, the application distance is limited by the length of the hose available and the distance between valves, bends, tees, etc.

For non-man entry pipes, diameters between 100 to 900 mm (4 to 36 in.), the lining is sprayed directly onto pipe walls using a remote controlled travelling sprayer. The lining materials include concrete sealers, coal tar epoxy, epoxy, polyester, silicone, urethane, vinyl ester, and polyurethane. These linings are intended to form an acid resistant layer that protects the host pipe from corrosion. These linings have been applied to sewer lines since 1960's, with mixed success. Failures of some applications are largely due to the specification of the material on the basis of manufacturers' claims without actual field testing for a specific project conditions.

Link Seal. This method of pipeline rehabilitation utilizes a sleeve to correct localized structural damage. Spot repairs can be conducted with this method on pipe diameters ranging from 150 to 2,800 mm (6 to 110 in.). For diameters between 150 and 600 mm (6 to 24 in.), a stainless steel sleeve which is wrapped in polyethylene foam is used. This sleeve and an inflatable sewer plug are placed over the damaged area. With the aid of a CCTV, the plug is inflated until the sleeve lock is in place. The plug is then deflated and a visual inspection takes place.

There are different proprietary techniques for this method. Prior to the operation, the pipe must be thoroughly cleaned and inspected by CCTV to identify all the obstructions such as displaced joints, crushed pipes and protruding service laterals. The operation then continues according to instructions from the system manufacturer and type of application.

Point CIPP. The application of point CIPP are for pipelines that are structurally sound but may contain isolated pipe lengths which have structurally failed. The materials used in point CIPP repair are same as regular CIPP methods with over two decades of proven performance.

Sewer Manhole Rehabilitation

Sewer manholes require rehabilitation to prevent surface water inflow and groundwater infiltration, to repair structural damage and to protect surfaces from damage due to corrosive substances. When rehabilitation methods will not solve the problems cost-effectively, manhole replacement should be considered. Selection of a particular rehabilitation method should consider the type of problems, physical characteristics of the structure, location, condition, age and type of original construction. Extent of successful manhole rehabilitation experiences and cost should also be considered. Manhole rehabilitation methods are directed either at the frame and cover or sidewall and base. The sewer manhole rehabilitation for sidewall and base can be divided into the following methods: (1) Spray-on lining, (2) Cast-in-place, (3) Cured-in-place and (4) Profile PVC.

Conclusions

This report described some of the features of relevant rehabilitation methods. New and potentially viable methods were also mentioned. The design engineer and utility owner should be aware of these methods and others as their technologies develop and capabilities and limitations of each. The choice of methods depends on the physical condition of the pipeline system components, such as, pipeline segments, manholes, and service connections; and the nature of the problem or problems to be solved.

Definition of Acronyms

PE - Polyethylene
PP - Polypropylene
EPDM - Ethylene Polypelene Diene Monomer
PVC - Poly-Vinyl-Chloride

PVDF - Poly-Vinylidene Chloride
GRP - Glassfiber Reinforced Polyester
HDPE - High Density Polyethylene
MDPE - Medium Density Polyethylene

Acknowledgment

This study is a part of the state-of-the-art review which is prepared by the Trenchless Technology Center (TTC) for the U.S. Army Corps of Engineers' Construction Productivity Advancement Research (CPAR) program. To obtain a complete report, please contact the Trenchless Technology Center, Louisiana Tech

University, P.O. Box 10348, Ruston, LA 71272, Phone (318)257-3204, Fax (318)257-2562.

References

Najafi, M. (1994). *"Trenchless Pipeline Rehabilitation: State-of-the-Art Review,"* Trenchless Technology Center, Ruston, Louisiana.

U.S. Environmental Protection Agency (EPA) (1991). "*Handbook of Sewer System Infrastructure Analysis and Rehabilitation,*" Office of Research and Development, Cincinnati, Ohio.

CONDITION ASSESSMENT AND REHABILITATION PROGRAM FOR LARGE DIAMETER SANITARY SEWERS IN PHOENIX, ARIZONA

Thomas M. Galeziewski[1], P.E., M. ASCE,
Samuel A. Edmondson[2], P.E., D.E.E., and
Robert Webb[3]

ABSTRACT

A condition assessment program, focusing on large diameter (600 mm (24-inch) and above) unlined concrete sewer pipe, was undertaken by the City of Phoenix in 1992. Following an inventory of the system to identify the locations of these sewers, the pipe segments were lamped (photographed) during manhole inspections to provide an initial assessment of corrosion conditions. Approximately 29.6 km (97,000 ft) of pipe were recommended for CCTV inspection, and a ranking algorithm was used to prioritize the pipe segments based on the level of corrosion and pipe defects observed. Potential rehabilitation methods were evaluated and planning level costs were generated to develop a capital improvements program. A total of 14 projects, composed of multiple pipe segments with corrosion conditions ranging from "severe" to "dirt visible", were recommended for rehabilitation/replacement. Two of the projects involved the Salt River Outfall (SRO), a regional trunk sewer which collects wastewater from four communities for treatment at the 91st Avenue WWTP. The SRO segments included over 550 m (1,800 ft) of 2,050 mm (81-inch) pipe, and approximately 397 m (1,302 ft) of 1,200 mm (48-inch) pipe. The design of these rehabilitation projects included an assessment of structural and material requirements, as well as hydraulic capacity. The selection of rehabilitation techniques for each pipe segment included consideration of construction access, pipe size, traffic, utilities, and environmental concerns.

---

[1] Project Manager, HDR Engineering, Inc., 5353 North 16th St., Suite 205, Phoenix, AZ 85016; PH (602) 248-6622; FAX (602) 265-6472

[2] Regional Vice President, Brown and Caldwell, 3636 North Central Ave., Suite 300, Phoenix, AZ 85012; PH (602) 222-4444; FAX (602) 222-4466

[3] Project Manager, City of Phoenix Water Services Department, 200 West Washington St., 8th Floor, Phoenix, AZ 85003; PH (602) 261-8232; FAX (602) 495-5843

INTRODUCTION

The wastewater collection system in the City of Phoenix (City) includes approximately 5,955 km (3,700 miles) of sewers, ranging in size from 200 mm (8-inch) diameter to 2,300 mm (90-inch) diameter pipe, and over 72,000 manholes. The City estimated that approximately two percent of the system consists of unlined concrete pipe, 600 mm (24-inch) diameter and larger, constructed prior to the mid-1960's. Because of concerns about hydrogen sulfide corrosion damage, the City requested a study to identify both the condition of the unlined concrete pipes and the required corrective action on a prioritized basis (Condition Assessment Program), followed by design projects to correct the problems identified. To accomplish the Condition Assessment Program, the following objectives were established:

1. Identify all unlined concrete sewers within the Phoenix collection system.

2. Assess the condition of the unlined sewers, optimizing the use of limited funds for closed-circuit television (CCTV) inspection of the pipes.

3. Prioritize sewer replacement/rehabilitation requirements over a five-year period, identifying recommended rehabilitation methods and planning level costs.

4. Provide a point of reference to which future system inspections can be compared.

In addition to Phoenix sewers, the study included a regional trunk sewer known as the Salt River Outfall (SRO). The SRO is owned by the Subregional Operating Group (SROG), a multi-city agency which includes Phoenix, Glendale, Mesa, Scottsdale, Tempe, and Youngtown. The SRO collects wastewater from the cities of Phoenix, Mesa, Scottsdale, and Tempe and transports it to the regional 91st Avenue Wastewater Treatment Plant owned by SROG. The SRO generally follows the north bank of the Salt River, conveying flow from east to west through the Phoenix metropolitan area. The City of Phoenix is responsible for maintaining the SRO, which is constructed of unlined concrete pipe for most of its length.

CONDITION ASSESSMENT PROGRAM

The Condition Assessment Program began in 1992, using the structured, step-by-step approach described in the following paragraphs.

System Inventory. City staff compiled a preliminary system inventory which identified the total suspected unlined concrete pipe lengths by diameter in each quarter-section of the City, and served as the starting point for the project. To refine the preliminary inventory, quarter-section maps of the sewer system and project files were reviewed to identify the pipe material used. Inventory data sheets were completed for each segment of pipe either identified as, or suspected of being unlined concrete. This information was entered into a computer database.

Based on the inventory data collected, 35.5 km (116,347 ft) of sewers in the City of Phoenix were identified as unlined concrete pipe. The pipes ranged in size from 600 mm (24-inch) to 1,500 mm (60-inch) diameter, and are connected to 258 manholes. All the pipes are located within a 64.7 km^2 (25 mi^2) area in the central downtown area of the City. In addition, the SRO included 35.2 km (115,631 ft) of unlined concrete pipe, ranging in size from 1,275 mm (51-inch) to 2,300 mm (90-inch) diameter, and connecting to 114 manholes. Thus, the total length of unlined concrete pipe identified was 70.7 m (231,978 ft), together with 372 manholes.

Physical Inspection. Physical inspection of the unlined concrete sewers consisted of manhole inspection and pipeline lamping. Manhole inspection required the entry of field personnel into the manhole to identify general characteristics and to obtain structural and hydraulic information about the manhole. Pipeline lamping involved photographing the first 1.5 to 3.0 m of all pipes connected to the manhole to assess corrosion conditions in the upstream and downstream ends of the pipes. Two crews performed the field work over a six week period. A Manhole Inspection/Pipeline Lamping Form was completed for each manhole, and this information was added to the computer database.

A total of 370 pipe segments were photographed from the upstream and downstream ends. Each end was given a corrosion code based on the conditions observed in the photographs. The corrosion condition codes used are listed below:

Code	Condition	Description
D	Dirt	Dirt is visible
R	Rebar Exposed	Exposed rebar clearly visible
S	Severe	Exposed aggregate, rebar indentations apparent
MS	Moderate Severe	Exposed aggregate over most of the pipe surface
M	Moderate	Flaking/pockmarked pipe surface, with onset of exposed aggregate
ML	Moderate Light	Initial "flaking" of pipe surface, most of original surface still visible
L	Light	Minimal to no corrosion apparent
U	Unknown	Condition unknown
PU	Partially lined	Partially lined pipe - condition unknown

For pipes which had different corrosion codes on each end, the worst case condition was used to establish the corrosion condition of the pipe. The PU condition code applied to the SRO, which in some cases was lined for several pipe sections upstream and downstream of a manhole. Based on this evaluation, pipes in the worst condition were identified for further inspection to optimize the use of available funding for CCTV inspections. Approximately 56 percent (18 km) of the Phoenix unlined sewers were classified with an MS condition code or worse, and were recommended for CCTV inspection. For the SRO, no D or R conditions were observed in the photographs. However, approximately 33 percent of the pipe segments (11.6 km) were classified as either S or MS, and were also recommended for CCTV inspection.

CCTV Inspection. The CCTV inspection work was conducted by one crew over a four-week period. Inspection data were recorded on standard forms using defect condition codes to standardize the entry of pipe condition information into the computer database. A total of 183 pipe segments were inspected, including 144 Phoenix pipes and 39 SRO segments. Pipe corrosion was the most common defect observed, with the severity ranging from N (None) to D (Dirt visible). Corrosion was generally most severe above the springline of the pipe, where the pipe wall is periodically wetted during peak flows, and then exposed to the atmosphere during low flows. Other defects identified included pipe holes, collapsed pipe sections, protruding laterals, infiltration, and debris.

Prioritization Procedure. Based on the condition assessment of the pipe during the CCTV phase, a numerical score was developed for each pipe segment. This score was obtained from points assigned to each pipe defect and condition observed, such that sewers with the most defects and in the worst condition received the highest scores. The individual pipe segments are then prioritized for rehabilitation or replacement in descending order of pipe score.

Scoring Algorithm. Prioritization of individual pipe segments required the development of a scoring algorithm to assess pipe conditions. The score for each pipe segment includes corrosion points, structural points, "other" defect condition points, and impact factors. These four components are combined as shown below to calculate the pipe score:

$$\text{TOTAL PIPE SCORE} = (\text{COR} + \text{STR} + \text{OTH}) \times \text{IF}$$

where COR = Total Corrosion Score OTH = Total Other Defect Score
      STR = Total Structural Score IF = Impact Factor

The total corrosion score includes both corrosion in the pipe and at laterals, and is obtained by adding the points of various corrosion levels over the total length of pipe. The "in pipe" corrosion score is based on a percentage of pipe with a given corrosion level, rather than length, so that a short pipe segment and a long pipe segment with the same corrosion condition receive the same score. Corrosion points ranged from 0 to 5,000 per one percent of pipe length. Thus, a pipe coded "D" for 100 percent of its length received a score of 500,000. For corrosion at laterals, the points ranged from 0 to 1,000.

The total structural score is obtained by adding the structural defect points in the pipe and at the joints. Structural points ranged from 50 to 50,000 for pipeline defects, and from 10 to 200 for joint defects. "Other" defects include laterals, roots, infiltration, and alignment. The total other defect score is obtained by adding all the defect points for four components. Defects points ranged from 10 to 100 for laterals, from 2 to 10 for roots, from 2 to 20 for infiltration, and from 10 to 40 for alignment problems.

The impact factor (IF) is a multiplier, ranging from 1.0 to 2.0, which slightly adjusts the pipe score based on three components: pipe location, number of traffic lanes, and pipe wall penetration depths. The IF provides a secondary method of prioritization for pipes with similar defect scores, providing a higher score adjustment to pipes which would cause greater inconvenience if a failure occurred.

Pipe Ranking. Using the scoring algorithm and the information from the CCTV database, pipe scores were calculated for each segment televised. The pipe segments were then sorted to provide a prioritized list that placed the pipes in the order of descending pipe score. A separate prioritization list was prepared for the SRO, as it is a single conveyance facility composed of numerous pipe segments. Pipe scores for the City of Phoenix pipes ranged from 0 to 304,506. Scores for the individual pipe segments of the SRO ranged from 0 to 161,999.

Condition Categories. Condition categories were developed using the prioritized pipe lists to develop the recommended rehabilitation program. The condition categories are defined as follows:

- Category I: Pipes containing 3 m (10 ft) or more of corrosion level "D", or 2 or more collapsed pipe sections. Pipes in this category require immediate action.

- Category II: Pipe scores greater than 24,000, indicating at a minimum severe corrosion, together with exposed rebar and possibly dirt conditions. Pipes in this category require immediate action.

- Category III: Pipes which have severe corrosion throughout the pipe length, but no worse, with pipe scores ranging from 20,000 to 24,800. These pipes require action within the next 5 years.

- Category IV: Pipe scores ranging from 7,500 to 19,999, in which the predominant corrosion condition is moderate severe. No immediate action is required, and reinspection is recommended in 5 years.

- Category V: Pipes with predominantly moderate corrosion conditions throughout the pipe length, with pipe scores less than 7,500. No immediate action is required, and reinspection is recommended in 10 years.

Rehabilitation Methods. To develop a capital improvements program for sewer repair, a preliminary assessment of various rehabilitation methods was conducted. At the planning level, the selection of an appropriate method for a particular sewer reach depends upon the condition assessment of the pipe, site constraints and cost, together with qualifying assumptions. Several assumptions were generated from discussions with City staff in determining the applicability of rehabilitation methods to a given pipe segment. These assumptions are listed below:

1. Pipe replacement would be required if 3 m (10 ft) or more of pipe are collapsed or coded as "D", or if the pipe reach contains two or more collapsed sections.

2. Bypass pumping is feasible for all flow conditions and in all locations.

3. Pipe capacity reductions are acceptable in the pipe evaluated.

The four rehabilitation methods investigated during the study phase included sliplining, cured-in-place pipe (CIPP), spiral-wound pipe, and pipe replacement. Using the pipe condition data from the CCTV inspection and cost data for each rehabilitation method, costs were prepared for each televised pipe segment, and a preliminary selection of rehabilitation method was made.

Rehabilitation Projects. To develop rehabilitation projects, pipe segments were color coded on quarter-section maps, based on their identified condition category. Individual pipe segments with similar corrosion conditions were then combined with other contiguous pipe segments to form multi-segment projects. All Category I and Category II pipe segments were included in these projects, so that all pipes with the worst conditions would be included in the initial rehabilitation projects. A total of 14 rehabilitation projects were recommended, including 12 Phoenix projects and 2 projects along the SRO. These projects are summarized in Table 1, together with the estimated cost of the recommended rehabilitation option for each project. The projects listed encompass 56 of the 144 televised Phoenix pipe segments, and 4 out of 39 SRO pipe reaches.

REHABILITATION DESIGN

Following completion of the Condition Assessment Program, the City moved immediately into the design phase for the 14 rehabilitation/replacement projects. The design of the two SRO projects, which included the largest pipe diameters, is discussed below. The pipe segments which comprise these projects are as follows:

1. A segment of 2,050 mm (81-inch) diameter reinforced concrete pipe extending 560 m (1,838 ft) westward (upstream) from the intersection of 23rd Avenue and Lower Buckeye Road. The segment was constructed with plastic lining in five locations to serve as corrosion protection in areas of more turbulent flow. The pipe has an earth cover of about 7.9 m (26 ft).

2. A segment of 1,200 mm (48-inch) diameter reinforced concrete pipe extending 397 m (1,302 ft) northward (downstream) from a City of Tempe metering station,

TABLE 1 - SUMMARY OF RECOMMENDED REHABILITATION PROJECTS

Project No.	Location	Existing Pipe Diameter mm	Project Length m	Recommended Rehabilitation Method	Rehabilitated Pipe Diameter mm	Construction Cost $ (millions)	Capital Cost[1] $ (millions)
P[2]-1	23rd Avenue	900	1,525	Sliplining	760	0.99	1.60
P-2A	Jackson St.	750	1,684	Sliplining	610	0.93	1.52
P-2B	12th Street	675	874	Sliplining	610	0.46	0.74
P-3	7th Avenue	600	153	Replacement	610	0.17	0.28
P-4	12th Street	675	166	Replacement	685	0.20	0.32
P-5	McDowell Road	675	218	Replacement	760	0.16[3]	0.16[3]
P-6	19th Ave./Van Buren St.	825	822	Sliplining	685 - 760	0.43	0.69
P-7	Durango St.	1050	107	Replacement	1050	0.14	0.23
P-8	17th Ave./Grand Ave.	675	66	Spiral Wound	585	0.074	0.12
P-9	15th Avenue	1050	489	CIPP	1020	0.45	0.72
P-10	2nd Avenue	900	117	Sliplining	750	0.07	0.12
P-11	McDowell Road	600	197	Sliplining	500	0.08	0.12
SROG[4]-1	Lower Buckeye Rd.	2050	560	Sliplining	1,825	0.78	1.25
SROG-2	Priest Drive	1200	397	CIPP	1,150	0.38	1.86
Totals			7,375			5.27	8.47

(1) Includes 30% for contingencies and 25% for engineering, administrative, and legal costs.
(2) "P" denotes City of Phoenix project.
(3) Actual cost for this project.
(4) "SROG" denotes projects with SRO pipe segments.

located immediately east of Priest Drive adjacent to the south bank of the Salt River. Earth cover over the pipe ranges from 8 m (26 ft) at the Tempe metering station to approximately 1.2 m (4 ft) at the center of the Salt River. During high flow conditions in the Salt River, the pipe is subjected to a hydrostatic head of about 10.7 m (35 feet), resulting in significant infiltration through joints and cracks in the pipe.

The scope of work for the rehabilitation design was divided into two major tasks: 1) a Phase II study which included reviewing background information, conducting field investigations, analyzing data and information, and developing and analyzing various alternative rehabilitation techniques; and, 2) preliminary and detailed design.

Engineering Evaluation. A detailed evaluation was made to assess the condition of the pipes and to develop design criteria for rehabilitation. The following paragraphs briefly summarize the work accomplished.

Closed Circuit Television (CCTV) Inspection. The CCTV tapes from the Condition Assessment Program were reviewed, and each sewer segment was graded into one of five corrosion and structural condition grades. The 2,050 mm (81-inch) diameter segment was assigned a single condition grade consistent with exposed reinforcement bars and corrosion in an advanced stage, with collapse likely in the near future. For the 1,200 mm (48-inch) diameter segment, the southerly 163 m (534 ft) was graded based on the rebar cage being visible, but not exposed, with collapse unlikely in the near future, but with further deterioration likely. Corrosion was found to be in an advanced stage and heavy infiltration was noticeable in the northerly part of the 1,200 mm (48-inch) diameter segment.

Manhole Inspection. The field investigation included three manhole inspections, two on the 2,050 mm (81-inch) diameter pipeline and one on the 1,200 mm (48-inch) diameter sewer line. Severe corrosion conditions were found in the two manholes on the 2,050 mm (81-inch) diameter sewer. Moderate corrosion, with some exposed rebar, was observed in the manhole connected to the 1,200 mm (48-inch) sewer.

Geotechnical Investigation. The geotechnical investigation consisted of four soil borings at the 2,050 mm (81-inch) diameter sewer line site and two at the 1,200 mm (48-inch) diameter pipeline location. The primary purpose of the geotechnical investigation was to determine if the soils surrounding the pipelines create a corrosive environment. The results of the Standard Penetration Test (SPT) were similar for both pipe locations. The soil surrounding the pipes is medium-dense to very dense, and has a permeability of 1.0×10^7 cm/sec, which is indicative of a soil that groundwater can readily travel through. The soil resistivity and slightly alkaline pH values indicated that the existing conditions are not contributing to external deterioration of the sewers. The soil density values indicated that the soil is adequate to provide support for the rehabilitated or replaced pipe in either location.

Concrete Investigation. The concrete investigation consisted of obtaining four concrete corings from the crown of the 2,050 mm (81-inch) diameter sewer and two cores from the crown of the 1,200 mm (48-inch) diameter pipe. The locations of the corings were strategically placed to obtain representative information at points on the pipelines, including both the unlined and the plastic lined portions of the sewers. The results of the concrete corings, summarized in Table 2, indicated that the wall thickness of the 2,050 mm (81-inch) diameter pipe has been corroded to the point that 90 percent of the rebar thickness is exposed in some locations. In other areas the rebar is corroded to at least 50 percent of its original cross-section, and, in some instances has debonded from the concrete. The circumferencial steel has no cover at many locations, and the inner longitudinal steel is also exposed in some areas. No significant corrosion was found on those cores taken from plastic lined sections of the pipe. The concrete investigation confirmed the conditions seen in the CCTV and the manhole inspections.

TABLE 2 - CONCRETE TESTING SUMMARY

Parameter	Pipeline Segment	
	2,050-mm (81-inch)	1,200-mm (48-inch)
Average Unit Weight, kg/m^3 (pcf)	2,467 (154)	2,611 (163)
Average Compression Strength, kPa (psi)	56,668 (8,219)	46,712 (6,775)
pH	11.7	6.5
Sulfate Concentration, mg/l	>1,000	>1,000
Core Thickness[1], mm (inches)	130 to 140 (5.1 to 5.5)	108 to 144 (4.25 to 5.65)
Original Wall Thickness, mm (inches)	216 (8 1/2)	146 (5 3/4)

[1] Unlined Pipe Section Only

For the 1,200 mm (48-inch) diameter pipe, the concrete testing information (shown in Table 2) and the CCTV inspection results indicated that circumferencial rebar is exposed at some locations immediately downstream of the Tempe metering station and at the northerly end of the segment. The remaining sections of the pipe have surface corrosion but with a maximum material loss on the pipe wall of 5.1 cm (2-inches).

Structural Analysis. The pipes were analyzed as rigid conduits using working stress design theory. Earth loads were computed based on the Marston formula for underground conduits. Assumptions included a unit weight of backfill of 178.6 kg/m^3 (120 pcf), a maximum allowable concrete compressive stress of

45 percent of the maximum concrete compressive strength, and a maximum allowable reinforcing steel stress of 1,547 kg/cm^2 (22,000 psi).

The analysis of the 2,050 mm (81-inch) diameter pipe showed that the unlined sections of pipe have experienced loss of section from corrosion, rendering them structurally inadequate for current loadings. It is believed that a complex soil/structure arch is the primary load carrying mechanism. As further degradation occurs or as soil stress is redistributed, this mechanism will ultimately result in failure and should not be relied upon to meet future loads.

The 1,200 mm (48-inch) diameter pipeline has also experienced loss of structural section. This section loss is approximately 5.1 cm (2-inches) at the crown and 2.5 cm (1-inch) at the springline. This section loss is not significant enough to cause structural distress for the existing imposed loadings, however, a corrosion protection lining is required to prevent further deterioration.

Hydraulic Requirements. One consideration in evaluating pipeline rehabilitation alternatives is the effect rehabilitation has on the sewer system flow capacity. Changes in pipe size or shape associated with pipe rehabilitation typically reduce the flow area and the hydraulic radius components of pipe flow calculations. Thus, cost considerations often entail comparing the cost of a less expensive rehabilitated pipe with a lower sewer system capacity, to the higher cost of a replacement pipe which will maintain the existing sewer system capacity.

As noted previously, the SRO is jointly owned by the multi-cities in the SROG. Not all cities have ownership in all segments of the SRO. Contractually, each city has been allocated an ownership capacity in the SRO, which is summarized in Table 3 for the two pipe segments. If the contractual capacities in the SRO are to be maintained, the rehabilitated pipe segments must have a flow capacity equal to or greater than the amounts shown in Table 3.

Rehabilitation Alternatives. Numerous methods exist to repair and rehabilitate wastewater collection systems. Some can only be used where the existing pipe structure is fundamentally sound and/or where adequate or spare hydraulic capacity exists. Some can be used to restore the structural capacity of a sewer while others can be used to reduce infiltration and inflow into the system. Many of the systems also provide protection to corroded sewers. Oftentimes, rehabilitation methods cost less than conventional replacement, and generally minimize open trench excavation, reducing traffic disruption and public inconvenience. The choice of method or combination of methods depends upon the physical attributes of the sewer system and the nature of the problem(s) to be solved. The categories of alternatives for rehabilitating the SRO included: sliplining, using cured-in-place pipe, deformed pipe, and spiral wound pipe; segmental linings; coatings; minimum excavation, such as trenchless replacement; conventional replacement; and, maintenance and repair techniques. In total, 29 alternatives were considered.

TABLE 3 - CAPACITY OWNERSHIP[1]

Allocated Ownership (City of)	2,050 mm[2] (81-inch)		1,200 mm[3] (48-inch)	
	m³/sec	mgd	m³/sec	mgd
MESA	1.01	23.0	0.66	15.0
PHOENIX	1.52	34.7	-	-
SCOTTSDALE	1.01	23.0	-	-
TEMPE	0.60	13.8	0.57	13.0
TOTAL	4.14	94.5	1.23	28.0

[1] Values shown are per Exhibit C-1, Multi-City Purchased Capacity in Jointly-Owned Sewage Transportation Facilities, Revised December, 1984.
[2] Values shown are Average Daily Flows, based on a Mannings "n" = 0.011 and d/D = 0.9.
[3] Values shown are Average Daily Flows, based on a Mannings "n" = 0.012 and d/D = 0.9.

The viable rehabilitation alternatives were selected using a two-step process. During an initial screening, methods were eliminated because of their lack of structural support, excessive structural support, insufficient flow capacity, and failure record. The remaining alternatives were then evaluated with respect to three categories: implementation criteria, project cost, and durability. Implementation criteria included structural, quality control, hydraulics, access, flow bypassing, traffic and public impact, and construction issues. The purpose of identifying implementation criteria was to recognize the impacts or difficulties inherent in the various rehabilitation alternates. The screening process identified several viable alternatives, described in Table 4, that will provide the required level of rehabilitation for each sewer segment.

Selected Alternatives. For the 2,050 mm (81-inch) diameter segment, the selected alternative was replacement, even though the total project cost for rehabilitation (by sliplining) was approximately $800,000 to $900,000 less. The primary consideration was the need to maintain the contractual ownership capacities (see Table 3) in the pipe to accommodate future increases in flow to the 91st Avenue WWTP. Cured-in-place pipe was selected as the rehabilitation alternative for the 1,200 mm (48-inch) diameter pipe. This alternative will enable the contractual ownership capacities (see Table 3) in the pipe to be maintained, and will save approximately $340,000 as compared to the cost of replacement. Although considerably less costly, the sliplining alternative was eliminated due to the capacity reduction which would occur with this method. It is anticipated that the preliminary and detailed design phases of the projects will be completed during the summer of 1995. Construction should be completed and the projects fully operational in late 1995 or early 1996.

TABLE 4 - VIABLE REHABILITATION ALTERNATIVES

Pipeline Segment mm (inches)	Option	Proposed Pipe Material	Proposed Pipe ID mm (inches)	Pipe Capacity SROG Contractual Requirement (ADF)[1] m³/sec (mgd)	Pipe Capacity Proposed Project m³/sec (mgd)	Construction Access Requirements	Limitations	Advantages	Total Project Cost[2] $(millions)
2,050 (81)	Rehabilitation (slipline)	Spirolite	1,700 (66)	4.14 (94.5)	3.38 (77.1)	Two access pits.	Reduced flow capacity.	Can be installed with some flow in the pipe.	1.4
2,050 (81)	Rehabilitation (slipline)	Hobas	1,800 (72)	4.14 (94.5)	4.26 (97.3)	Two access pits.	Reduced flow capacity.	Can be installed with some flow in the pipe.	1.5
2,050 (81)	Replacement	T-Lock Lined RCP[3]	2,130 (84)	4.14 (94.5)	6.72 (153.4)	Open cut in street. Open cut railroad crossings.	Railroad crossing. Existing sewer to be abandoned. Major disruption of street.	No loss in flow capacity.	2.3
1,250 (48)	Rehabilitation (slipline)	HDPE[4]	1,000 (38)	1.23 (28.0)	0.98 (22.3)	Two access pits.	Reduced flow capacity.	Can be installed with some flow in the pipe.	0.35
1,200 (48)	Rehabilitation	CIPP[5]	1,100 (44)	1.23 (28.0)	1.44 (32.9)	Two access pits.	Reduced flow capacity.		0.86
1,200 (48)	Replacement	T-Lock Lined RCP[3]	1,200 (48)	1.23 (28.0)	1.76 (40.1)	Open cut construction.		Little loss in flow capacity.	1.2

(1) See Table 3, ADF = Average Daily Flow
(2) Cost includes construction, contingency (20%), and allowances for engineering, legal and administrative costs (25% for Rehabilitation Project; 30% for Replacement Project)
(3) Reinforced concrete pipe
(4) High density polyethylene pipe
(5) Cured-in-place pipe

Denver's Experience in Trenchless Technology

Joseph Barsoom, P.E & P.L.S. Director of Engineering, Wastewater Management Division, City and County of Denver.

ABSTRACT

In 1982, the City of Denver was faced with a situation in downtown area, where a 12-inch sanitary sewer line located in a 12-foot wide alley, with utilities surrounding both sides of the sanitary sewer. There were utility vaults extended throughout the alley leaving only 4-feet of separation between these structures located above the sewer line. Routine maintenance of the 12-inch sanitary sewer became a problem when it started to develop cracks and off-set joints. In the process of televising the line to evaluate the extent of the damage, the T.V. camera became lodged in the line. The problem area had to be excavated to retrieve the camera. The extent of the problem of trying to dig 15 feet deep between utilities and vaults, through a 4-foot hole, in sand and gravel, to your imagination.

A decision was made to replace this sewer. Due to the proximity of the conflicting utilities the conventional open cut construction methods were ruled out, as did, jacking and slip lining, because of the inability to reconnect service lines. In order to solve this problem, the design team improvised a design to construct a 60-inch tunnel, and suspend the sanitary sewer from the top, which would allow connection of the services from inside of the tunnel. This approach was a successful solution, but very expensive, therefore, rehabilitating all the sewers in downtown Denver using this approach would be cost prohibitive and out of the question. Until a more favorable rehabilitation procedure could be conceived all T.V. inspection and jetting of sewer lines having similar physical constraints were suspended.

INTRODUCTION AND BACKGROUND

In 1984, the cured-in-place process was introduced to the City, and was implemented on a limited basis. The experience with the product showed considerable savings and a practical approach to solving many of the rehabilitation problems. Since 1984, the City of Denver has rehabilitated 140,000 linear feet of sewer lines. These sewers range in sizes from 8-inches to 48-inches in diameter, and various sized egg-shaped sewer lines that extend up to 6-feet by 3-feet 6-inches in dimension.

The purpose of this paper is to present the results of 8-years of experience in sewer rehabilitation, and to compare the cost of rehabilitating sewer lines with the cost of removal and replacement.

Including are two lists of projects as appendices. The first list identifies sewer projects using the Cured-in-place process of rehabilitation. The second list identifies the conventional method of removal and replacement of existing lines. Each of these lists shows the project number, the bid date, the total cost of rehabilitating the line per inch diameter per linear foot, the pipe size, the footage, and contract costs.

In order to compensate for different sizes of lines, different length of inversions, and different size of projects, the attached tables and costs are based on the weighted diameter of sewer lines in each project.

In recent years, many new reliable materials and techniques have became available in this market place, each with a long history of success. Engineers should explore these new methods, consult manufacturers and technical societies, and form their own opinion about these new methods of rehabilitation.

When selecting a particular rehabilitation method the following factors should be carefully considered:

- Functional overlap
- Durability
- Structural and Design Condition
- Cost
- Construction Issues

- System Wide Implication

Functional overlap

It is important to search the reasons for Rehabilitation. The reason could be to add structural capabilities to the existing system, or to enhance the flow characteristics, or decrease the infiltration inflow in the line. Knowing the purpose of the rehabilitation will impact the method needed. In case of flow characteristics, some of the Rehabilitation methods will be eliminated that results in decrease of the line size. Also increase of the structural requirements some of the material will have more structural capabilities than others. During the planning stage all these variables should be explored and the methods recommended should be capable of serving these requirements.

Durability

Many of the Rehabilitation methods use material with long history of use. In some cases, it is simply a new method of installing traditional material. Knowing the history of use and the life expectancy of these material will help the specifier to recommend the appropriate material for the needed purpose.

Structural Design Condition

Two conditions should be considered in designing rehabilitation process, depending on the condition of the existing pipe.
These two conditions are.
    a)   Partially Deteriorated pipe. The existing pipe can support soil and surcharge loads through the design life of the rehabilitated pipe, and the soil adjacent to the existing pipe must provide adequate side support. The rehabilitation process for this condition, should be designed to support only the external hydraulic loads due to ground water (and internal vacuum). Since the soil and surcharge loads can be supported by the existing line.

    b)   Fully Deteriorated Pipe. The existing pipe is not structurally sound and cannot support soil and live loads or expected to reach this condition over the design life of the rehabilitation method used. The rehabilitation process for this condition should be designed to support hydraulic, soil and live loads. The groundwater level, soil type and depth, and surface live loads should be determined prior to design of the proposed rehabilitation method.

Cost

It is evident from Denver experience, that the rehabilitation of existing sewers are more economical than complete removal and replacement. In addition to the savings in direct costs, there are other indirect cost savings. These can include the cost of disruption caused by the construction work (e.g traffic delays and diversions, loss of trade and access restrictions to business, and noise, dirt and health hazards) and the effects on other utility equipment. The cost of these items should be considered when evaluating different alternatives.

System-wide Implications

It is important to consider the system-wide effects of a proposed solution to ensure that other problems are not created. This includes increase or decrease of the hydraulic capacity and its impact on the system. In larger systems, this can be achieved by analyzing the proposed solution in a hydraulic model of the system. The results of the assessment may be the deciding factor in selecting one type of solution over another.

As of 1992 the following rehabilitation process has been approved in Denver:
Cured-in-Place (CIPP) and Folded PVC pipe

a) Cured-inplace Pipe (CIPP)
The City Contracts require that each Bidder must be licensed by INSITUFORM of North America Inc. to perform the Insituform Method of Construction, or must be licensed by ASHIMORI INDUSTRY COMPANY, Ltd. to perform the Paltem Method of Construction, or must be licensee of INLINER, USA Process, or must be licensed of KM INLINER Process, at the time of bid opening to do the specific type of sewer Rehabilitation Work.

The CIPP processes involve the insertion or inversion of a flexible lining into the sewer. The lining inserted or inverted via existing manholes and, depending on the system selected, is installed using one of the following installation methods:

- Water inversion, where the lining is inverted under the pressure of water, and cured by circulating hot water.

- Winched insertion, where the lining is winched into place and inflated against the sewer wall by a calibration hose inverted into the lining under the pressure of water, and cured by circulating hot water.

-Air inversion, where the lining is inverted under the pressure of air, and cured by introducing steam.

The lining may be manufactured to suit many sewer shapes, and can accommodate small deformations and changes in direction of the pipeline.

Sewers from 4- through 114-inch diameter have successfully been lined using these techniques.

Depending on size, the City specifications require that, a maximum of 1500 feet can be installed in a single insertion. Full bypass pumping or diversion of flows is necessary.

In many cases the lining will de designed to be a close fit to the existing sewer and therefore annulus grouting is not required.

Preparation

Preparation of the installation of the CIPP includes the following:

a) Safety Inspection

Prior to entering access area such as manholes, and performing inspection or cleaning operations, an evaluation of the atmosphere to determine the presence of toxic or flammable vapors or lack of oxygen must be undertaken in accordance with local, state, or federal safety regulations.

Cleaning of pipeline

All internal debris should be removed from original pipeline. Gravity pipes should be cleaned with hydraulically powered equipments, high-velocity jet cleaners, or mechanically powered equipment as per NASSCO Recommended Specifications for sewer Collection System Rehabilitation. Pressure pipelines should be cleaned with cable attached devices or fluid propelled devices as per AWWA Manual for Cleaning and Lining Water Mains, M28.

Inspection of Pipelines

Inspection of pipelines should be performed by experienced personnel trained in locating breaks, obstacles, and service connections by closed circuit television or in person for man entry pipelines. The interior of pipelines should be carefully inspected to determine the location of any conditions that may prevent proper installation of the impregnated tube, such as protruding service-taps, collapsed or crushed pipe, and reductions in cross-sectional area. These conditions should be noted so that they can be corrected.

Line obstruction

The original pipeline should be clear of obstructions such as solids, dropped joints, protruding service connections, crushed or collapsed pipe, and reductions in the cross-sectional area of more than 40% per (ASTM F-1216), (Engineer should decide for himself if this value is adequate for any particular situation). If inspection reveals an obstruction that cannot be removed by conventional sewer cleaning, a point repair excavation should be made to uncover and remove the obstruction.

Bypassing

If bypassing of the flow is required around the sections of pipe designated for reconstruction, the bypass should be made by plugging the line at a point upstream of the pipe to be constructed and pumping the flow to a downstream point or adjacent system. The pump and bypass lines should be of adequate capacity and size to handle the flow. Services within this reach will be temporarily out of service.

Public advisory services will be required to notify all parties whose service laterals will be out of commission and to advice against water usage until mainline is back in service, which will vary up to 12 hours.

Resin Impregnation

The tube should be vacuum-impregnated with resin (wet-out) undercontroled conditions. The volume of resin used should be sufficient to fill all voids in the tube material at normal thickness and diameter. The volume should be adjusted by adding 5 to 10% excess resin for the change in resin volume due to polymerization and to allow for any migration of resin into the cracks and joints in the original pipe.

Folded Pipe (NUpipe)

The folded PVC pipe system involves the insertion of polyvinyl chloride (PVC) thermoplastic U-shape pipe. The U-shape pipe then rerounded and expanded in the sewer to form a tight fit with the existing sewer.

The pipe is either extruded in a folded from or extruded in the conventional circular form and subsequently folded. The result is a pipe that is ultimately capable of having a tight fit inside the sewer, and is easily installed because its cross section during installation is less than the final section. The pipe is rerounded by heat and pressure in the form of steam and a rerounding device.

Pipeline preparation

The pipeline preparation require the following:
    a) Safety inspection
    b) Cleaning of pipeline
    c) Inspection of pipeline
    d) Line obstruction removals,and point repair if needed
    e) Bypassing
These processes are common for both CIPP and folded PVC,and they are all described before in this paper.

Installation of Folded PVC

The installation of Folded PVC requires the following process:

    a)   <u>Insertion</u>

        The folded PVC pipe arrives on the site in a steam heated truck. The spool of folded pipe should be positioned at the insertion point and contained in a heating chamber. A minimum temperature of 200F should be maintained in the heating chamber for a one hour period to fully heat the length of folded pipe to be inserted. Shorter insertions may be fully heated over a shorter time period as recommended by the manufacturer.

A containment tube should then be pulled through the existing conduit, secured at both ends, and inflated with air at low pressure.

A cable should be strung through the containment tube (within the existing conduit) and attached to the folded pipe. The folded pipe should be fully contained, heated along the entire length, and pulled with a power winch unit and the cable, directly from the spool, through the insertion point, and through the containment tube within the existing pipe to the termination point. A dynamometer should be provided on the winch or cable to monitor the pulling force. The pulling force should not exceed 1,000 pounds for any pipe size so as not to exceed the allowable strain limits of the folded pipe.

After insertion is complete, the tension from the winch should be relieved and the folded pipe should be cut-off at the insertion point and restrained at the termination point.

b) Expansion

To determine temperature during expansion, suitable monitors to gauge temperature should be placed between the folded pipe and the containment tube at the insertion and termination ends.

Through the use of heat, pressure, and a rounding device, the folded pipe should be fully expanded. Expansion pressure should be sufficient to unfold the PVC pipe, press against the wall of the existing conduit, and form dimples at service connections.

The pressure to expand the pipe is typically in the range of 8 to 10 psi (55 to 60kpa)

The expansion process should eliminate the trapping of any entering or standing water between the formed pipe and the existing pipe by employing a mechanical rounding technique. A rounding device should be propelled at a controlled rate within the folded pipe expanding it in a sequential manner. The rounding device should be flexible and inflated with continual pressure so that it presses the formed pipe against the existing pipe wall while pushing water ahead of the expansion process. The expansion rate (or rounding device speed) should not exceed 5ft. per second (1.52m/s).

Once the rounding device has reached the termination point, the expansion pressure should be stabilized for a minimum period of 2 minutes to ensure the complete expansion of the pipe at local deformities and to allow for complete dimpling at side connection.

Cool Down

The formed pipe should be cooled to a temperature below 120F(49C) before relieving the pressure required to hold the PVC pipe against the existing pipe wall.

Service Connections

After the formed pipe, or the CIPP pipe has cooled down, the terminating ends should be trimmed and the existing active services be opened. This should be done, without excavation, from the interior of the pipeline by the use of a television camera and a remote control cutting device.

Inspection and Acceptance

The inspection of the finished installation may be visual, if appropriate, or by closed-circuit television if visual inspection cannot be accomplished. Variations from true line and grade may be inherited because of the condition of the original piping. No infiltration of groundwater should be observed. All service taps should be accounted for and be unobstructed.

Sampling

a) For CIPP

For each insertion or inversion length, the preparation of two CIPP samples is required, one from each of the following methods:

a) The sample should be cut from a section of cured CIPP at an intermediate manhole or at the termination point that has been winched or inverted through a like diameter pipe which has been held in place by a suitable heat sink, such as sand bags.

b) The sample should be fabricated to the same thickness as the CIPP. The sample should be made from the same tube and resin as the CIPP, and cured in a clamped mold placed in the downtube.

b) For Folded PVC

For each insertion length a sample should be cut from a section of rounded pipe at the insertion and/or termination point that has been inserted through a like nominal diameter pipe acting as a mold.

Testing

Should be in accordance with ASTM F-1216 for CIPP and other applicable ASTM for folded PVC. (An ASTM standards for both material and installation of folded PVC in process, and should be approved in 1994 or about, and should be followed for testing).

It should be noted that the methods used in this comparison are based on personal experience, and this does not imply any product endorsement or lessen the responsibility of an engineer in designing and specifying a rehabilitation system nor does it imply that other systems are not viable.

CONCLUSION

Total average cost to remove & replace existing sewer pipe per inch-diameter per foot.	Total average cost to rehab existing sewer pipe per inch-diameter per foot.
$ 11.62	$ 8.77

Items not included in the above comparison:

1. Manhole Rehabilitation.

2. Relocation of Utilities.

3. Cost of Engineering and Construction Management.

Considering the above factors, the cost of complete removal and replacement could be increased by 7% to 15% and the cost comparison will be as follows:

Total average cost to remove & replace existing sewer pipe per inch-diameter per foot.	Total average cost to rehab existing sewer pipe per inch-diameter per foot.
$ 13.01	$ 8.77

This information is based on work done between 1984 and 1994.

Recently several Trenchless Technology methods were approved for use, that resulted in fair competition and lower prices. The average cost per inch diameter per foot for the work done in 1994 was $5.6.

It is suggested that Engineers working on sewer rehabilitation projects consider evaluating other available alternatives and benefit from the competition among these rehabilitation methods.

As of December,1994. Comparing the cost of Rehabilitation to the cost of complete removal and replacement shows the following:

a. Cost to remove and replace existing sewer pipe per inch-diameter per foot $13.01
b. Cost to rehabilitate existing sewer pipe using Trenchless Technology per inch-diameter per foot $ 7.13
c. The weighted average diameter of sewer rehabilitated 13.52 inch
d. Savings per inch diameter per foot $ 5.88
e. Total length of sewer rehabilitated as of 1994 177,157 Ft
f. Total savings to the city as of December,1994 $14,715,959

It is evident that there are savings in rehabilitation of aged sewers using Trenchless Technology over the conventional method of removal and replacement.

In addition to the economical advantages, using Trenchless Technology will have minimum impact on the public and business in the area. Rehabilitation of existing line will not require more than few pieces of equipments for few hours. Complete removal and replacement of deep lines such as sanitary sewers will require complete closure of the streets for weeks, disrupting the daily lives of citizens. During the construction of several sewer projects, using the conventional open-cut method, contaminated soil were found in the trench. Removal and disposal of these material delayed the completion of these projects, and added substantial expenses to the cost of these projects. Before using the Trenchless Technology, an alley sewer in Downtown Denver, was replaced at a cost of $600.00 per foot. The alley had to closed disrupting business for two months. Using the Trenchless Technology in sewer rehabilitation eliminate these type of problems and added costs to the project.

Using the conventional method of removal and replacement has showed that, no matter what the contractor does to protect the job site, or what area in town the project is located, the results were the same. Angry citizens and business owners.
In many cases the City was threatened to be sued for loss of business, or property damage. Many claims for damages have been filed.

In the last few years, all the sanitary sewers around the City Hall, in the Grand Prix route, the sewers through Auraria Higher education, the lines around Denver General Hospital. Also many alleys in Downtown Denver, Under several viaducts ,and under railroads. Under highways, under heavily travelled streets, under South Platte River, and under Cherry Creek. These projects were rehabilitated without closure of a single street or routing any traffic. The City did not receive one single call or complain from any citizen.

Other issues have been noticed in this experience, the open cut method of complete removal and replacement, change orders due to change of conditions ranges from 10-15% of the total cost of the project. Using the Trenchless Technology to rehabilitate the sewers, the change orders dropped down to less than 0.5%.

As of 1993 the following rehabilitation methods are approved for use in the City of Denver:

      a.   Inliner, USA process.
      b.   Insituform.
      c.   KM Inliner process.
      d.   Nu-Pipe.
      e.   Paltem method.
      f.   Rib Loc.

As a closing statement with many approved well

tested methods of rehabilitation available, Engineers and owners will benefit from the competition, also the industry will compete to improve their products.

Trenchless Technology is no longer an art, it is a science. The Technology of the future is now on hand, and should be considered whenever it is applicable. There is still more work to do to introduce this Technology to Engineers and owners. Some institutes such as Louisiana Tech University, the authors of the Trenchless Technology Magazine, and the City and County of Denver. These agencies are active in the education, training and research of the application and use of the Trenchless Technology, and should be recognized for their efforts. The industry needs to continue working with ASTM and ASCE to get standards written and approved for these products.

APPENDIXES

REMOVAL AND REPLACEMENT OF EXISTING SEWERS

PROJECT NUMBER	BID DATE	PIPE SIZE	TOTAL COST INSITUPIPE INSTALLATION /INCH DIAMETER /LINEAR FOOT	PROJECT FOOTAGE	CONTRACT AMOUNT	REMARKS
W79-075B	2/12/82	8,21,24,27"	$13.08	1,541	$ 197,468.00	
CBD-PH.II W76-003	5/28/82	15"	$48.65	422	$ 307,960.00	Not used in calculations
CBD-20TH ST W79-075D	9/30/82	15,24 & 27"	$15.12	1,378	$ 545,165.00	
CBD-BROADWY W79-075C	3/04/83	10,18,21,24	$ 8.67	3,427	$ 862,976.00	
ALLEY REP. W82-183	7/15/83	8" & 12"	$11.56	7,541	$ 862,976.00	
ALLEY REP. W84-106	9/28/84	8"	$13.92	11,043	$ 1,229,827.00	
SPEER DELG. W83-012A	3/21/88	15,18,30,33	$ 7.53	9,450	$ 2,193,279.00	
1990 IMPROV W89-345B	12/21/90	8" & 10"	$11.45	1,460	$ 148,888.00	
TOTAL INSTALLED AVERAGE COST			$11.62	35,840	$ 5,788,950.00	

TRENCHLESS TECHNOLOGY EXPERIENCE 587

SEWER REHABILITATION

PROJECT NUMBER	BID DATE	PIPE SIZE	TOTAL COST INSITUPIPE INSTALLATION /INCH DIAMETER /LINEAR FOOT	PROJECT FOOTAGE	CONTRACT AMOUNT	REMARKS
W82-602	1/15/88	8",12",15", 18",24"x36" & 32"x48"	$ 8.85	15,654	$ 1,399,108.00	
W87-166A	9/04/87	8"	$10.22	2,718	$ 222,314.00	Deep installation deformed pipe
W87-011	3/11/88	10"	$ 9.31	581	$ 54,103.00	Done as subcontractor
W83-012A	9/09/88	24x36,26x39 28x42,30x45 & 42"x63"	$ 6.83	6,963	$ 1,639,742.00	
W79-078	8/10/89	10" & 18"	$ 7.19	4,987	$ 598,174.00	
W89-345A	11/16/90	8,9,10,12, 15 & 18"	$ 6.67	10,018	$ 594,469.00	
W90-318	1/25/91	8,9,10,12, 15,18 & 21"	$ 6.73	14,070	$ 1,006,802.00	
W90-329A	8/09/91	32"x48"	$11.00	2,037	$ 896,929.00	
W88-083B	5/10/91	2' x 3', 2'4" x 3'6"	$ 9.03	3,219	$ 957,626.00	
	TOTAL INSTALLED AVERAGE COST		$ 8.74	103,777	$11,765,186.00	

Rev. 10/23/91

SEWER REHABILITATION

PROJECT NUMBER	BID DATE	PIPE SIZE	TOTAL COST INSITUPIPE INSTALLATION /INCH DIAMETER /LINEAR FOOT	PROJECT FOOTAGE	CONTRACT AMOUNT	REMARKS
W84-096	8/17/84	8"	$ 9.36	5,905	$ 442,617.00	First Insituform Project
W83-143	10/15/84	8"	$ 7.42	5,646	$ 335,508.00	
W84-177A	4/05/85	8"	$ 8.09	7,490	$ 484,763.00	
W84-177B	7/12/85	8"	$ 8.69	6,188	$ 430,208.00	
W84-177C	3/07/86	8" & 12"	$ 7.79	6,036	$ 408,798.00	
W75-066C	6/27/86	8"	$ 8.11	1,295	$ 82,225.00	Done as subcontractor
W79-075F	9/26/86	8",24",30" 48",21"X31" & 32"X48"	$ 9.67	4,389	$ 1,458,965.00	
W87-010B	5/22/87	9" & 24"	$10.63	3,827	$ 491,035.00	Installed under highway
W84-300A1	7/02/87	8",9" & 10"	$ 9.83	2,003	$ 181,175.00	
W84-300A2	11/20/87	8"	$11.75	751	$ 80,625.00	Done as subcontractor

THE VALUE OF INTERNAL MANHOLE INSPECTIONS

By: James D. McGregor, Member, ASCE[1]

ABSTRACT

This paper summarizes the value of internal manhole inspections. Manholes are convenient access points within a wastewater collection system at which valuable information can be derived. Manholes function like windows to the underground network of pipes. Visual observations into the pipes connected to manholes provide data on structural condition, maintenance problems, extraneous flows and exfiltration sites. Photographs, flow and pipe measurements, and atmospheric monitoring document existing conditions and can be used for assessing sewer capacity, extraneous flow impacts , pipe condition, cross contamination of storm sewers and verifying computer modeling.

INTRODUCTION

Sanitary sewer manholes are distributed throughout wastewater collection systems. The manholes are nodal points in the collection systems where sewer pipes end, can change gradient, direction and size. Manholes serve as access points for sewer cleaning and inspection. The uniform distribution of manholes makes them ideal locations from which to gather information relevant to the condition of the existing collection system.

Surface manhole inspections can be helpful in determining the internal condition of a manhole and flow conditions in the manhole channel. However, a surface inspection provides no clues about the internal condition of sewer pipes connected to the inspected manholes.

Within a sanitary sewer system having an average spacing of approximately 230 linear feet between manholes about 20 percent of the collection system can be visually inspected by internal manholes inspections. This projection

----

[1]Sr. Associate, Killam Associates, 27 Bleeker Street, P.O. Box 1008, Millburn, N.J. 07041

is based on inspecting all manholes in a basin and being able to visually inspect an average distance of about 20 linear feet into each pipe connected to manholes. The internal inspection is accomplished by shining a high intensity light into sewer pipes while making visual inspections. The procedure is sometimes known as lamping.

Documentation of Findings

During an internal manhole inspection it is important to document the observed conditions. There are two (2) basic forms of documentation - a manhole inspection report and a photograph. A written report is necessary to record the following:

- Atmospheric conditions
- Dimensional information
- Flow rates
- Maintenance problems
- Manhole location

A photograph is a permanent visual record of observed conditions. It is an excellent method of recording internal pipe conditions both good and especially bad. A color photograph depicting large root intrusions, grease accumulations, structurally damaged pipe, the handy work of vandals or severe extraneous flow problems is an extremely valuable aid for illustrating sewer conditions.

Photographs and Slides Facilitate Presentations

Sanitary sewers frequently receive low budgetary priorities because the sewers are out of sight and largely out of mind. The subsurface location of sewers prevents the general public from detecting maintenance problems like they can detect on roadways having pot holes. Another unfortunate aspect of sanitary sewers is that the general public is largely uneducated about how a sewer system functions and the importance of preventative maintenance. Among the general public are the administrators, politicians and authority members who control budgeting for sanitary sewer systems. Sometimes a written or verbal description of existing sewer conditions cannot adequately convey the severity of existing sewer problems. This may delay the authorization of funding for sewer maintenance or rehabilitation work.

Sanitary sewer photographs and slides obtained from internal manhole inspections are valuable tools for educating the general public about sanitary sewers. Slides are the best method of depicting sewer conditions at public information meetings or presentations to public officials. Sewer problems which the general public are not aware of can be illustrated and their impact explained. Such presentations are important in establishing justification for preventative maintenance programs and sewer rehabilitation.

DETECTING MAINTENANCE PROBLEMS

Many sanitary maintenance problems can be detected by internal manhole inspections. Frequently, maintenance problems occur in the vicinity of manholes and are associated with the following:

- Changes in pipe direction
- Changes in pipe gradient
- Poorly designed pipe connections

The deposition of sediment, rocks and debris as well as the accumulation of grease and detergent on pipe walls occurs within pipes having low cleansing velocities. Frequently, depositions accumulate at manholes (see Figure 1).

Internal manhole inspections can also detect other types of maintenance problems near manholes such as root intrusions, building pipes which protrude into municipal sewers and sags or vertically misaligned pipe. This information determines the nature of the maintenance problem and permits the inspector to assess whether the problem can be resolved with available cleaning equipment.

Some cleaning procedures only poke holes in sewer obstructions allowing liquids to pass while the major obstruction such as large root intrusions or grease accumulations remain in the sewer. Therefore, a second internal manhole inspection after sewer cleaning is particularly helpful in determining if the sewer cleaning methodology was effective.

Internal manhole inspections can detect submerged objects deposited in manhole channels by vandals. Dropping large rocks and other debris into interceptor sewers seem to be a favorite pastime of vandals. Manholes in wooded easements are favorite targets.

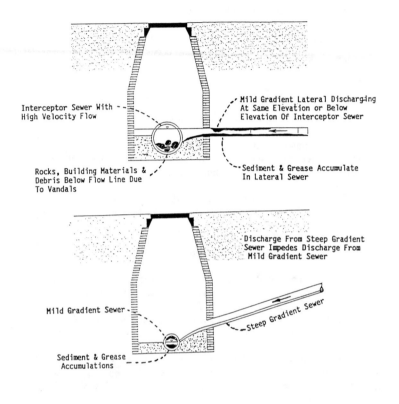
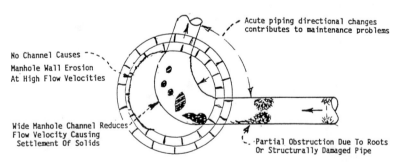

Figure 1

Typical Pipe Maintenance Problems
Detectable By Internal Manhole Inspections

ASSESSING STRUCTURAL PROBLEMS

Internal manhole inspections can determine the structural condition of manhole components and the visible segments of pipes connected to the manholes. It is not uncommon to find structurally damaged pipe in the vicinity of manholes because of the differential settling rates between some manholes and pipes. The structural problems can also be associated with other factors. Listed below are typical manholes piping conditions where structural damage might be detected.

- External drop connections
- Very shallow pipes
- Very deep pipes
- Pipes which cross below the alignment of nearby utilities
- Pipes whose loading conditions differ from the original design
- Pipes conveying industrial wastewater
- Pipes with hydrogen sulfide problems

Structurally damaged pipe detected at manholes provides clues about the overall structural condition of the sewer system and the possible cross contamination of nearby storm sewers from wastewater exfiltration. Sometimes detected structural problems provide clues to potential problems in other segments of the collection system where similar pipe was used or a particular contractor worked. Manhole inspection clues are helpful in planning television inspection work to evaluate the condition of similar sewers (see Figure 2).

Internal manhole inspections also provide information which normal television inspection cannot provide such as the following:

- Pipe diameter measurements
- Pipe wall thickness
- Extent of pipe invert deterioration due to chemical attack
- Extent of pipe wall deterioration due to hydrogen sulfide
- Wastewater samples
- pH of pipe wall material
- Wastewater temperature
- Characteristics of bench and channel deterioration
- Chlorine residual testing

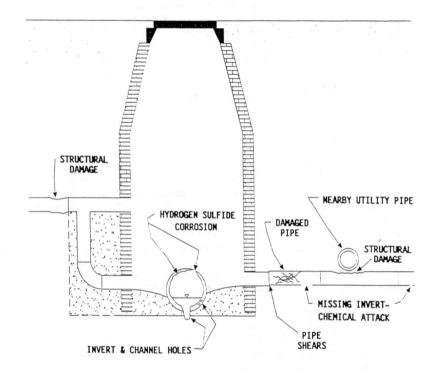

Figure 2
Typical Manhole & Pipe Structural Defects
Detectable By Internal Manhole Inspection

SEWER CAPACITY ASSESSMENT

Instantaneous flow measurements obtained during internal manhole inspections are useful in assessing existing wastewater discharges. Accurate flow measurements should be obtained which may require the following:

- Accurate depth measurements
- Accurate velocity measurements
- Weired flow rates
- Description of weir discharge conditions (free or submerged flow)
- Obstructed pipe area (sediment and grease)

Instantaneous flow measurements can be conducted at a variety of times and locations to determine various information such as the following:

- Nighttime low flows
- Daytime high flows
- Peak wet weather flows

An evaluation of the instantaneous flow data and pipe capacity data projected from record plans can determine the following:

- Available capacity
- Existence of free discharge
- Existence of obstructed flow conditions

When free discharge conditions do not exist, the sewer may be obstructed by some type of maintenance problem. Additional manhole inspections should be conducted to determine the cause or causes of the downstream obstruction.

EXTRANEOUS FLOW ASSESSMENT

Extraneous flows in a sanitary sewer system can cause significant operational and maintenance problems within the collection system and at the wastewater treatment plant. Peak extraneous flow rates which coincide with large prolonged rainfalls have the most adverse impacts on sewerage systems. Peak extraneous flow rates can affect the sewerage system by causing the following:

- Sewage backups into buildings
- Sewage overflows and bypasses
- Increased treatment and pumping costs
- Poor treatment plant effluent quality
- Sewer capacity limitations
- Community growth restrictions
- Imposition of sewer connection ban
- Fines by regulatory agency

Sanitary Sewer Evaluation Survey

In order to assess peak extraneous flows, a sanitary sewer evaluation survey (SSES) must be conducted. The major objective of the investigation should be to identify and quantify peak extraneous flow sources. The ideal time to investigate a sewer system is during rainfalls which activate severe peak extraneous flows. An integral part of the investigation should be internal manhole inspections. The manhole inspections provide data as follows:

- Dry weather base line flow rates
- Peak wet weather flow rates
- Surcharging locations
- Manhole extraneous flow sources
- Pipe line extraneous flow sources
- Flow rates for specific extraneous flow sources

Manhole Inspections Provide Clues To Required SSES Tasks

Sanitary sewer evaluation survey (SSES) work involves a variety of testing and inspection methods used to identify extraneous flow sources. Among the methods are manhole inspections, flow isolation, television inspection, smoke testing, dye flooding and tracing, building plumbing inspections and building lateral sewer testing. These methods can be costly to conduct and can provide misleading information if not conducted properly.

A general guideline for evaluating sanitary sewer systems projects that 80 percent of the extraneous flow problems originate from about 20 percent of the sanitary sewer system. Therefore, it is not necessary to undertake many of the testing and inspection methods throughout an investigated drainage basin. Dry weather and wet weather flow measurements recorded in strategic sewer basin manholes provide the key to determining where SSES tasks should be conducted.

Internal manhole inspections conducted during a peak wet weather period in conjunction with flow isolation (the insertion of temporary sewer plugs) can identify and quantify extraneous flows. Internal manhole inspection data determines the following:

- Segments of collection system without excessive extraneous flows
- Sewers to be television inspected during peak wet weather conditions
- Building laterals with extraneous flow
- Sewer pipes to be cleaned
- Sewers to be smoke tested
- Manhole rehabilitation sites
- Type of required manhole rehabilitation
- Some types of required pipe rehabilitation

Extraneous Flow Sources Identifiable By Internal Manhole Inspections

An internal manhole inspection by a knowledgeable inspector can provide valuable extraneous flow information. The inspector should be aware of potential extraneous flow sources, know how to evaluate wastewater flow characteristics and listen for audible evidence of extraneous flows. He should also be aware that lateral sewers from buildings can account for 30 to 80 percent of the extraneous flow within a drainage basin. Therefore, he should be familiar with typical wastewater discharges from residential buildings and monitor visible lateral pipes for evidence of continuous clear flow or sump pump type discharges.

Typical Manhole Extraneous Flow Sources

Extraneous flows enter manholes from a variety of sources. Figure 3 illustrates some typical sources. The manhole has several typical manhole construction materials (brick, block and precast concrete) to illustrate some extraneous flow sources associated with those construction materials.

Extraneous flows entering at the cover, frame and riser locations can be documented by a surface inspection. However, an internal inspection can identify leaks occurring in the lower portion of a manhole and quantify all internal leaks.

Monitoring Techniques

Within a manhole the inspector can perform visual inspections and listen for audible evidence of extraneous flows. Cascading sounds in sewers where no drop connections exist, sump pump operational noises, and steady loud discharges in sewers with mild slopes are audible evidence of extraneous flows. Visual inspections include the following:

- Inspecting visible portions of manhole for obvious leaks
- Observing channel flow for evidence of invert and channel leaks
- Evaluating the flow characteristics from all tributary pipes

    - discharge clarity
    - discharge variability
    - steady crystal clear flow-evidence of extraneous flow
    - steady crystal clear flow shortly after a rainfall-evidence of rainfall induced infiltration
    - steady crystal clear flow in dry weather may be water main leakage

- Lamping all pipe connections for evidence of extraneous flows

During the pipe lamping, the inspector should also listen in pipes with large rates of suspected extraneous flows for audible evidence. Figure 4 depicts

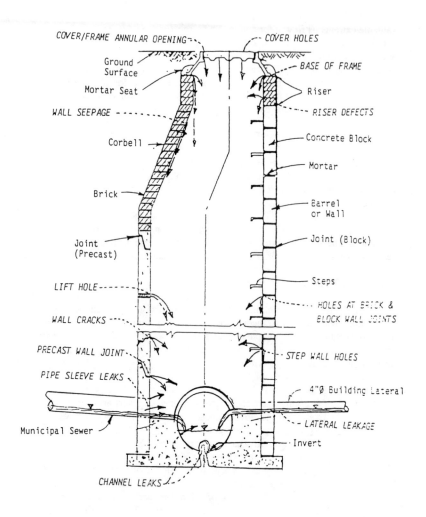

Figure 3

Typical Manhole Extraneous Flow Sources

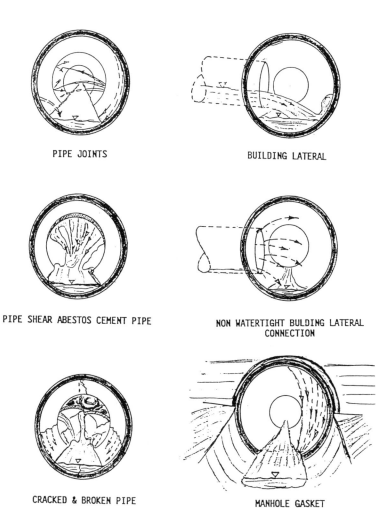

Figure 4
Typical Sewer Pipe Extraneous Flow Sources
Visible From Manholes

typical sources of extraneous flow which can be detected during lamping (internal pipe inspections).

Quantifying Extraneous Flows

Extraneous flows detected by internal manhole inspections can be quantified by using sewer plugs and weirs or containers of known volume. Some manhole leaks can be directly measured by determining the time it takes to fill a container. The total leakage in manholes with numerous leaks or leaks which cannot be measured with a container can be determined by plugging all influent lines and weiring the total flow at the effluent pipe.

The extraneous flow within a sewer pipe between adjacent manholes can be isolated by plugging influent pipes in the upstream manhole and weiring at the downstream manhole. Any significant leakage in the upstream manhole should be deducted to obtain a net flow for the isolated sewer pipe.

Quantifying extraneous flows is important because it provides the following:

- Inventory of identified extraneous flows
- Documents severity of specific extraneous flow sources
- Ability to relate extraneous flow variability to types of rainfall events
- Data base for projecting distribution of extraneous flow between municipal and private sewers
- Basis for prioritizing TV and rehabilitation work

Maximum Measured Extraneous Flow Rates

The extraneous flow rates from specific sources can vary dramatically in accordance with the type of source, rainfall magnitude and intensity, soil saturation, local geology, sewer depth, sewer condition and quality of sewer construction work. Extraneous flow sources fluctuate in response to the hydrostatic head created by groundwater in sewer trenches. Some individual extraneous flow sources can have a significant impact on small sewer systems.

Listed below are some measured extraneous flow rates from individual sources. These values, measured during periods of peak extraneous flow, illustrate the importance of reducing building lateral extraneous flows. The rates are listed in liters per second (l/sec) and in gallons per day (GPD) rates and have an assigned equivalent dwelling (ED) number. The ED number is an allowance of 0.0122 l/sec (280 GPD) for the average wastewater from a single family dwelling. The ED unit is used as a comparison value because politicians, developers and the general public can easily relate to the flow comparison.

Typical of Source	Maximum Measured Extraneous Flow Rate l/sec	Maximum Measured Extraneous Flow Rate (GPD)	Equivalent Dwelling Units
MH steps	0.13	(3,000)	10
Block MH wall joint	0.17	(4,000)	14
Drop connection joint	0.31	(7,000)	25
Pipe joint (8"ϕAC)	0.43	(10,000)	35
Precast MH lift hole	0.66	(15,000)	53
Precast MH wall joint	1.05	(24,000)	85
Precast MH flat slab top	1.66	(38,000)	135
Non watertight building lateral bolt on saddle	1.75	(40,000)	142
Pipe shear (8"ϕAC)	1.97	(45,000)	160
MH pipe sleeve/channel	2.19	(50,000)	178
Broken pipe (8"ϕ)	3.55	(81,000)	289
Building lateral (4"ϕ)	5.69	(130,000)	464

CONFIRM COMPUTERIZED MAPPING AND MODELING

Computer inventorying, managing and modeling sanitary sewer systems is increasing, especially in large wastewater collection systems were large quantities of data must be processed. Existing collection system data can be manipulated and evaluated to predict a variety of future needs. The hydraulic modeling of interceptor sewers can rapidly evaluate different flow scenarios within a collection system and predict where relief lines will be required.

Computer models rely on available sanitary sewer information for input data. This information can include the following:

- Record plans
- TV inspection data
- Sewer stoppage reports
- Manhole inspection records
- Flow metering data

Frequently office personnel trained in data entry and computer applications input sanitary sewer data into computers. The usefulness of computer models developed from available information depends on the accuracy of available data, the accuracy of input data and the interpretation of data by the computer analyst. The computer data base can be adversely affected by a variety of existing field conditions such as the following:

- Pipe diameters different than record plans
- Existing sewers not on record plans
- Maintenance problems
- Emergency spot repairs with smaller diameter pipe
- Structurally damaged pipe

Internal Manhole Inspections Confirm Field Conditions

Internal manhole inspections provide existing sanitary sewer information to collaborate if input data is correct and whether computer modeling assumptions are realistic. The manhole inspections are particularly valuable for confirming modeling within sanitary sewer systems having limited records and having widespread maintenance and extraneous flow problems.

The manhole inspections can pertain to any of the field problems described in this paper. A significant finding of internal manhole inspections is the need for thorough sewer cleaning in some interceptor sewers. The most troublesome interceptor sewers are usually within easements where limited access is available for sewer cleaning and TV inspection equipment. Manhole inspections have revealed that some of these sewers simply need to be cleaned to restore their design flow capacity. In those instances, the need for computer predicted relief sewers may not be necessary.

SUMMARY

Internal manhole inspections provide sanitary sewer operation and maintenance personnel a means of assessing conditions within a sewer system. In order to derive the maximum benefit from the inspections, personnel should be well equipped and trained to achieve inspection objectives. Field inspection data should be catalogued and evaluated in the office. The assessment should include a debriefing with field inspection personnel and formulation of plans to resolve detected sewer problems.

Acknowledgement

The author extends his thanks to the Water Environment Federation for granting permission to reprint this paper which appeared in the proceedings for Collection Systems Operation and Maintenance, June 1993.

Microtunneling Design Considerations

Stephen J. Klein, M. ASCE[1]
Randall J. Essex, M. ASCE[2]

Abstract

   Microtunneling methods for installing pipelines are becoming more common every day. This popularity is due to the need to replace and upgrade our aging and undersized pipeline infrastructure, coupled with the need to minimize construction impacts on the public. Microtunneling techniques can address both of these needs in a cost effective manner. Some of the design considerations for microtunneling projects are similar to conventional open cut pipeline projects but there are several unique aspects to be considered. Successful completion of a microtunneling project depends on recognizing and addressing the unique aspects of these projects.

Introduction

   Microtunneling methods are being called on more and more today because the environmental and social impacts of pipeline construction can be significantly reduced when these methods are used. In addition, microtunneling methods can accomplish difficult pipeline installations in challenging geologic conditions with tighter line and grade control than was previously considered possible.

   The purpose of this paper is to review and discuss some of the key design considerations for microtunneling projects. Important design considerations include: establishing appropriate project design criteria; completing adequate subsurface investigations; determining feasible construction methods; selecting an appropriate alignment with adequate

---

[1] Senior Associate, and [2] Vice President and Principal, Woodward-Clyde Consultants, 500 12th Street, Suite 100, Oakland, CA, 94607.

staging areas; assessing jacking pipe requirements; and developing appropriate specifications and contract provisions.

Project Design Criteria

An important first step involves determining the basic design criteria that must be addressed in design of the project. Some of the more important design criteria include:

- Hydraulic Requirements
- Design Profile
- Alignment Routes
- Manhole Requirements
- Cover Requirements
- Line-and-Grade Tolerances
- Corrosion Considerations

Hydraulic requirements include information regarding design flows, minimum grades, and internal pressures (if any). Minimum grades are particularly important in the design of sanitary sewers where minimum flow velocities have to be maintained for adequate scour. Alignment routes involve identifying criteria for determining the pipeline location. These criteria may include utilizing existing easements and identifying the location of connection points to existing lines. Manhole and cover requirements will usually be established in accordance with the owner's requirements. Typical line-and-grade tolerances that can be achieved using remote-controlled microtunneling methods are about 50 mm (2 in) for line and 25 mm (1 in) for grade when geologic conditions are favorable and proper equipment operation and steering procedures are employed. Corrosion considerations are an important factor in pipe selection. Both internal corrosion resulting from the fluids carried by the pipe and external corrosion due to chemical reactions with the soil and groundwater surrounding the pipe should be addressed.

Geotechnical Considerations

Subsurface Investigations - It is essential that a thorough subsurface investigation be carried out before construction begins to identify the geologic conditions along the pipeline alignment. The anticipated geologic conditions are the most important factor in the selection of appropriate microtunneling equipment for a prospective project. More than one microtunneling project has come to a virtual standstill when unanticipated bedrock or boulders were encountered and the equipment selected for the project could not advance the heading. An accurate knowledge of the geologic conditions at the site is critical to identify the need for special microtunneling equipment features that will minimize loss-of-ground and surface settlement. Important physical properties which need to be determined include the strength, grain size, moisture content, plasticity characteristics, compressibility, and permeability of the deposits. Geotechnical considerations are discussed in greater detail by Klein (1991) and Essex (1993).

Exploratory borings are usually spaced at about 300 m (1,000 ft) intervals for open cut pipeline projects. For a microtunneling project the boring spacing should be reduced to about half of this interval (i.e., 150 m) and in difficult geologic deposits a 60 m (200 ft) spacing might be required. Borings should extend at least two pipe diameters or a minimum of about 3 m (10 ft) below the pipe invert to obtain complete data within the zone of influence of the tunnel. Continuous soil sampling or rock coring from one diameter above the pipe to one diameter below the pipe is highly recommended. This more intensive sampling is necessary because it is important that potentially adverse conditions be identified prior to construction. Conditions such as thin saturated sand lenses, thin clay beds that perch groundwater, or soft interbedded clay layers may not be noticed in a conventional sampling program. Such conditions could have a large impact on the stability of the heading, and on the potential for loss-of-ground and surface settlement.

Encountering boulders in the heading may severely hinder or even halt progress of a microtunneling operation. Boulders may be likely in glacial deposits and also in coarse-grained alluvial and/or talus deposits. The contractor needs to know the maximum size of the boulders so that appropriate microtunneling equipment can be selected. It is very difficult if not impossible to obtain a representative sample and to determine the maximum boulder size with conventional small diameter borings. A more effective investigation technique is to use large diameter bucket auger borings or backhoe test pits to obtain representative samples of the boulder deposits. Bucket auger borings 0.75 to 0.9 m (2-½ to 3 ft) in diameter have been particularly useful in evaluating the size and nature of boulder deposits.

Groundwater conditions will have an important influence on the behavior of the ground. Groundwater levels should be determined in the borings, and pumping tests or other field tests to estimate the hydraulic conductivity should be conducted if dewatering will be necessary and is considered to be feasible. Chemical analyses of groundwater samples should be performed if groundwater contamination is suspected, so that appropriate treatment and disposal methods can be implemented during construction for water removed from excavations and dewatering wells.

The results of chemical analyses can also be used to identify potentially corrosive groundwater conditions.

Ground Behavior - Ground behavior refers to the anticipated performance of the ground at the tunnel heading. Ground behavior for soft ground tunneling is typically described according to six categories (Terzaghi, 1950). Fine grained, cohesive soils (silts and clays) are usually characterized as firm, swelling, or squeezing ground. Coarse grained, cohesionless granular soils (sands and gravels) are characterized as firm, running, or flowing ground depending on the degree of cementation and on the groundwater conditions. Both cohesive or cohesionless soils may be characterized as raveling ground in certain situations. Squeezing, running, or flowing ground behavior indicates potentially unstable ground conditions with low stand up time. In such conditions, significant loss-of-ground will occur at the tunnel heading resulting in surface settlement unless the behavior of the ground is controlled. Loss-of-ground may occur in raveling ground conditions requiring face support in a timely manner to minimize this behavior.

Settlement Due to Microtunneling - Surface settlement is mainly a result of loss-of-ground during tunneling and dewatering operations that cause subsidence. For microtunneling techniques, loss-of-ground may be associated with soil squeezing, running, or flowing into the heading; losses due to the overcutter on the machine; and steering adjustments. The actual magnitudes of these losses are largely dependent on the type and strength of the ground, groundwater conditions, size and depth of the pipe, equipment capabilities, and the skill of the contractor in operating and steering the machine. Sophisticated microtunneling equipment that has the capability to exert a stabilizing pressure at the tunnel face, equal to the in situ soil and groundwater pressures, will minimize loss-of-ground and surface settlement without the need for dewatering. Surface settlement using this type of equipment for a recent project in Pleasant Hill, California was limited to about 3 to 6 mm (0.125 to 0.25 in) for a 1.4 m (54 in) diameter pipeline installed in medium stiff to stiff clay interbedded with medium dense sand layers. It is important to carefully control the pressures applied at the tunnel face with this type of equipment because, if the overburden pressure is exceeded, heave of the ground surface can occur causing damage to nearby utilities and other improvements.

Microtunneling Equipment

One large advantage of microtunneling methods is the special equipment that is available to handle difficult and variable geologic conditions. In unstable ground conditions such as squeezing or flowing ground (soft clays or saturated sand deposits), pressurized face microtunneling equipment can be used to install the pipeline without excessive loss-of-ground and significant surface settlement. Face stability is maintained by applying a stabilizing pressure to the tunnel face using a slurry (consisting of natural clay or bentonite mixed with water), or an earth pressure resulting from the weight of excavated muck contained in a closed auger system. The slurry systems are capable of applying a pressure to the tunnel face that can balance the hydrostatic pressures preventing groundwater inflows and eliminating the need for dewatering.

Special microtunneling machines are also available that can crush boulders and also bore through rock further extending the capabilities of this equipment. Machines are available that can be equipped with either drag bits, button cutters, or disc cutters (Coller, 1993). Recently, an 125 mm (5 in.) mini-disc cutter has been developed at the Colorado School of Mines for efficiently excavating hard rock with a microtunneling machine (Friant et al., 1994).

Alignment Considerations

Similar to all pipeline projects, identifying feasible microtunneling alignments involves evaluating available right-of-way, and easement acquisition issues, and determining the location of existing utilities. In some cases, alignments not considered feasible with open cut methods may be possible if microtunneling methods are used. A feasible pipeline alignment must avoid any existing underground utilities. Microtunneling techniques may have advantages in this regard because in congested areas it may be possible to locate the proposed pipeline deeper below existing utilities avoiding potential conflicts and relocations with only a small increase in construction costs.

Straight horizontal alignments are generally preferred for microtunneling projects. Straight alignments provide for more accurate control of line-and-grade and for a more uniform force distribution at the pipe joints reducing the risk of eccentric loads which could damage the pipe.

In order to be feasible a prospective alignment must have adequate jacking and receiving pit locations available. Prospective jacking and receiving pit sites must be spaced at distances that are compatible with microtunneling techniques. The maximum distance that pipe can be jacked from a pit without the use of intermediate jacking stations is about 120 to 150 m (400 to 500 ft) depending on variables such as; type of pipe, pipe size, structural capacity of the pipe, thrust capacity of the main jacks, soil conditions, and effectiveness of the bentonite lubrication system. Intermediate jacking stations can be used to extend the drive length that can be installed between jacking and receiving pits.

Intermediate jacking stations (IJS) consist of a series of hydraulic jacks spaced evenly around the pipe circumference within the pipe string that can be activated to push a section of the pipe string forward, supplementing the capacity of the main jacking system. The pipe must be large enough to permit access for removal of the jacking stations once the drive is completed. The minimum pipe size for which the use of IJS is considered practical is 0.9 m (36-in) ID pipe, although, some manufacturers are developing remote control retrieval systems to be able to use them in pipe sizes as small as 0.6 m (24-in) ID. IJS have been used to extend jacking distances from 120 m (400 ft) to over 300 m (1000 ft). For the project in Pleasant Hill, California discussed above, an 1,150-foot drive was completed. Two IJS were installed for the drive but only one of them was actually used because jacking loads were much lower than anticipated.

Staging Area Requirements

As opposed to open cut construction, only limited space is required at the ground surface for staging a microtunneling project. The main construction staging area is at the jacking pit where all the pipe is actually installed. Smaller space requirements apply at the receiving pit where the access is needed mainly to retrieve the tunneling machine, although it may be convenient to construct a manhole at the receiving pit location also.

It is important to provide adequate space for staging construction operations so that pipe installation can be completed in an efficient manner. Construction access to the jacking pit must be provided for transporting tunnel muck, pipe sections, and tunneling equipment. A typical jacking pit site needs enough space for the jacking pit, slurry separation and recycling tanks, a crane, the control cabin, pipe storage, and support facilities such as a generator, power pack, and bentonite lubrication unit. The jacking pit should be a sufficient distance from overhead electrical lines to avoid hazards in operating the crane although, in some areas a gantry system can be used instead of a crane for smaller pipe sizes. The size òf the jacking pit excavation depends on the pipe size and pipe lengths to be used, and generally ranges from about 3 m (10 ft) by 4.5 m (15 ft) for 0.6 m diameter pipe to about 5 m (16 ft) by 7.5 m (25 ft) for 2 m diameter pipe. The size of the jacking pits must also be selected to accommodate the jacking equipment and a thrust wall capable of providing the required resistance. The equipment can be arranged around the jacking pit in a space ranging from about 200 to 230 square meters (2000 to 2500 square feet), if the site is ideal, i.e., no physical obstructions are present (Figure 1). The equipment arrangement is quite flexible and space requirements can be minimized to use smaller sites, if necessary. Frequently, jacking pits have been located in the parking strip at the edge of a street with the equipment set up in a linear arrangement as indicated in Figure 2. Similar linear arrangements have also been used to stage microtunneling operations from the median of wider, more heavily travelled street without significantly impacting traffic flow. Staging area requirements can be further minimized by using each jacking pit to install two drives, one in each direction. This approach reduces the number of jacking pit locations and provides for more centralized efficient construction operations.

The space required at receiving pits is much less. The actual shaft size of the receiving shaft is about 3 to 5 m (10 to 15 ft) in diameter (or square) depending on the pipe size and the length of the tunneling machine. The necessary staging area is only that required to construct the shaft and manhole, if needed.

Jacking and Receiving Pits

Jacking and receiving pits will generally be vertical excavations with shoring and bracing systems. These excavations must be supported in accordance with OSHA requirements. Although, the contractor should determine the plan dimensions of the pits, general design criteria for these excavations should be developed such as recommended lateral earth pressures diagrams for active and passive cases. Several shoring systems are commonly used; sheetpiles systems with internal bracing, soldier pile or circular steel rib systems and timber lagging, and liner plate systems with steel rib supports. In Europe caisson sinking methods have been frequently used to construct jacking pits using circular precast concrete sections. Recently these methods have also been used on projects in the U.S. in Houston, New York, Maryland, and Seattle.

Another key factor in the design of jacking and receiving pits is groundwater control. Dewatering systems using deep wells or well points are frequently employed. Alternatively, a groundwater cutoff can be used if relatively impermeable soils are present below water bearing soils. Sheetpiles, for instance, could be driven into the impervious soils to cutoff groundwater inflows. Grouting or some other method of groundwater control is necessary when launching the tunneling machine and advancing out of jacking pits or advancing into receiving pits unless groundwater levels are temporarily lowered in these areas. Where penetration grouting is not feasible, jet grouting techniques have been

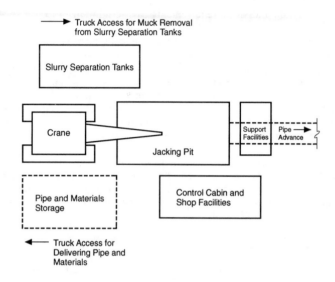

Figure 1. TYPICAL SITE LAYOUT FOR MICROTUNNELING

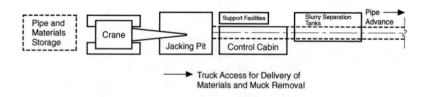

Figure 2. TYPICAL LINEAR SITE LAYOUT FOR MICROTUNNELING

utilized to control groundwater inflows when breaking out of the jacking pit.

Jacking Pipe Design Considerations

The main difference in the design of pipe for jacking applications, as opposed to normal trenching methods, is that the pipe must be capable of withstanding the axial forces applied to the pipe during installation. These axial jacking forces can be substantial, approaching 1000 tons or more for larger pipes, and may be the controlling load case in design of the pipe. As well as jacking forces, the pipes must be designed for external forces due to soil and groundwater pressures and live loads and for internal forces due to internal pressures, if applicable. Analysis procedures for external forces due to soil and groundwater pressures are covered elsewhere (Stein et al., 1989, Hancher et al., 1989, and O'Rourke et al., 1991). Live loads are typically due to highway or railroad traffic. For pipelines beneath streets and or highways, live loads consider the maximum allowable axle load, and for railroad crossings the maximum live load is due to a Cooper E-80 axle load (Stewart and O'Rourke, 1991).

Jacking pipes typically used in the U.S. include:

- Steel Casings
- Reinforced Concrete Pipe (RCP)
- Glass-Fiber Reinforced Thermosetting Resin Pipe (Fiberglass Pipe)
- Vitrified Clay Pipe

Steel casings are used mainly where required by local highway or railroad officials or in situations where the carrier pipe cannot be installed directly by jacking. RCP is the most common pipe material currently used in microtunneling. In most cases it can be installed by direct jacking and functions as both the casing and carrier pipe. Rubber-gasketed joints can be designed for situations where leakage control is important. Fiberglass pipe has been used for a number of projects in the last five years. It has a number of advantages, compared to RCP: lighter weight and easier to handle; smoother exterior with very consistent dimensions resulting in lower pipe friction; higher tensile and compressive strength that is less susceptible to spalling; thinner wall sections allowing smaller bore for the same pipe size; and better corrosion resistance. Another recent trend is the use of vitrified clay pipe. Several recent projects have involved the installation of vitrified clay pipe using microtunneling methods. Clay pipe has some of the same advantages as fiberglass in terms of strength, smoothness, and corrosion resistance. In Europe polymer concrete pipe is now being used (Bloomfield, 1994). This pipe is constructed with polyester resin instead of cement and has several of the same advantages discussed above for fiberglass pipes in comparison to RCP.

Specification Requirements

There are several unique aspects about microtunneling projects that should be addressed in the project specifications. The following list includes subjects that should be evaluated in developing specifications for a microtunneling project.

1. Microtunneling equipment requirements.
2. Installation tolerances.
3. Settlement control requirements.
4. Acceptable jacking pipes, joint details, and structural requirements.
5. Guidance and control systems.
6. Installation records.
7. Muck and slurry disposal.
8. Jacking and receiving pit requirements.
9. Grouting requirements.
10. Instrumentation and monitoring.
11. Submittals. The following should be submitted for review:

    (a) Microtunneling equipment description and literature.
    (b) Jacking system and maximum jacking loads.
    (c) Pipe shop drawings and calculations demonstrating ability to sustain maximum jacking loads.
    (d) Intermediate jacking system details.
    (e) Bentonite injection system details.
    (f) Groundwater control details, launching seals.
    (g) Jacking and receiving pit shoring design and shop drawings.
    (h) Thrust block details.

Suggested Contracting Provisions

The success of a microtunneling project depends on the efforts of an experienced contractor. Therefore, it is strongly recommended that contractors be required to prequalify prior to submitting a bid. Prequalification criteria should be developed based on the technical requirements of the project. Usually, it is more desirable to prequalify contractors prior to issuing bid documents because it is difficult for many agencies to reject a low bid, even if the low bidder does not meet the qualifications requirements established for the project.

Three provisions that should be considered for reducing the potential for and assisting in the resolution of disputes that sometimes arise during construction include: the Geotechnical Design Summary Report (GDSR), Dispute Review Boards, and Escrow Bid Documents. Use of these provisions has been very successful on many tunneling projects. They are addressed in detail in the 1991 ASCE publication, "Avoiding and Resolving Disputes During Construction". Merritt et al., (1991) discuss the benefit of preparing a GDSR for microtunneling projects based on their experience in Houston. These three provisions can be implemented independently of each other but are considered to be more effective if implemented together because each of these provisions will enhance the benefits of the others.

Conclusions

Microtunneling techniques will be essential for the construction of future pipeline projects. There are many significant benefits of these techniques, but the ability to minimize the environmental and social impacts of pipeline construction using these methods are the most important. These benefits will increase the popularity of microtunneling methods in constructing new pipeline projects particularly in the urban areas of this country. Successful completion of these future projects will depend on adequately addressing the design considerations discussed in this paper.

References

ASCE, (1991). <u>Avoiding and Resolving Disputes During Construction, Successful Practices and Guidelines</u>, Prepared by the Underground Technology Research Council, American Society of Civil Engineers (ASCE), New York, NY, 82 p.

Bloomfield, T.D., (1994). "Polymer Concrete Pipes," No Dig International, Vol. 5, No. 9, October, pp. 14-16.

Coller, P.J. (1993). "Introduction of Iseki, Rock Machine and Perimole," North American NO-DIG '93, San Jose, CA

Essex, R.J., (1993). "Subsurface Exploration Considerations for Microtunneling/Pipe Jacking Projects", in <u>Trenchless Technology: An Advanced Technical Seminar</u>, Trenchless Technology Center, Louisiana Tech University, Baton Rouge, LA, pp. 275-288.

Friant, J.E., and Ozdemir, L., (1994). "Development of the High Thrust Mini-Disc Cutter for Microtunneling Applications," NO-DIG Engineering, Vol. 1, No. 1, June, pp. 12-16.

Hancher, D.E., White, T.D., and Iseley, D.T., (1989). "Construction Specifications for Highway Projects Requiring Horizontal Earth Boring and/or Pipe Jacking Techniques," Joint Highway Research Project, Final Report, JHRP-89/8, Purdue University, West Lafayette, IN, July 285 p.

Klein, S.J., (1991). "Geotechnical Aspects of Pipe Jacking Projects," in Pipeline Crossings, ASCE Specialty Conference Proceedings, Pipeline Division, Denver, CO, March, pp. 113-128.

Merritt, B.K., Crisci, A. and Klein, G.H., (1991). "Houston Pipe Jacking - Large and Small," RETC Proceedings, AIME, pp. 391-407.

O'Rourke, T.D., El-Gharbawy, S., and Stewart, H.E., (1991). "Soil Loads at Pipeline Crossings," in Pipeline Crossings, ASCE Specialty Conference Proceedings, Pipeline Division, Denver, CO, March, pp. 235-247.

Stein, D., Mollers, K., and Bielecki, R., (1989). Microtunneling, Ernst & Sohn, Berlin, 352 p.

Stewart, H.E. and O'Rourke, T.D., (1991). "Live Loads for Pipeline Design at Railroads and Highways," in Pipeline Crossings, ASCE Specialty Conference Proceedings, Pipeline Division, Denver, CO, March, pp. 317-329.

Terzaghi, K., (1950). "Geologic Aspects of Soft Ground Tunneling." Chapter 11, in Applied Sedimentation, P.D. Trask, ed., John Wiley and Sons, Inc., New York, NY, pp. 193-209.

From Conception to Completion: Watershed 22 Trunk Sewer Upgrade

Curtis W. Swanson[1]; Glenn E. Hermanson,[2] M. ASCE; Charles W. Joyce,[3] M. ASCE

Abstract

Nobody likes raw sewage running down their street. In the past that is exactly what was happening to the residents in two areas in Watershed 22 of the Central Contra Costa Sanitary District (CCCSD). Now, after the application of various trenchless technologies and innovative construction techniques, the overflow problems have been eliminated. Presented below are the steps taken and lessons learned from conception to completion of the Watershed 22 Trunk Upgrade Project.

Introduction

CCCSD is approximately 112 kilometers east of San Francisco in the East Bay Area as shown in Figure 1. CCCSD collects wastewater from 280,000 people and currently treats wastewater from 390,000 people due to contributions from neighboring areas. The collection system contains approximately 2,080 kilometers of sewers ranging in size from 100 mm to 2600 mm in diameter. Approximately 80 percent of the collection system consists of sewers 150- and 200-mm in diameter. CCCSD operates 18 pumping stations and a 170,300 cubic meters/day (45mgd) activated sludge wastewater treatment plant.

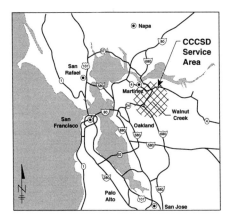

Figure 1
Location of
Central Contra Costa Sanitary District

Watershed 22 is one of the many watersheds within CCCSD. Two areas have had overflow problems, the Los Arabis Drive Area and the Woodland Way Area. Each of the project areas were evaluated in six areas of concern that could be causing or contributing to the overflow problem. The areas of concern are maintenance problems, capacity problems, flat sewers, structural condition, access problems, and slope and geological concerns. The design storm for this project is a 20-year storm at build out conditions. After substandard sewer reaches were identified, project alternatives were evaluated and a solution recommended.

Maintenance Problems. CCCSD maintains a computerized database of each sewer reach and structure (manhole, rodding inlet, etc.) in the collection system. This database includes descriptions of overflows, plugged sewers, routing cleaning and maintenance, and other types of service calls. This information along with television inspection and ferreting (field locating) were used to assess the condition of exiting sewers in Watershed 22.

Preventive maintenance and cleaning activities include hydroflushing and power rodding. For sewers in fair to excellent condition, cleaning frequencies are every 2, 5, or 10 years. For sewers in poor to fair condition, more frequent maintenance and cleaning are required to prevent plugged sewers and overflows. Depending on the condition of the sewer, cleaning may be performed weekly, monthly, or every 3, 6, or 12 months to remove roots, grease, rocks, grit, and other types of solids.

A sewer reach with a maintenance frequency of 12 months or less is considered a candidate for replacement or rehabilitation.

Capacity Problems. Wastewater flows and sewer capacity in Watershed 22 were evaluated using a static computerized hydraulic flow model. Wastewater flows were based on a 20-year wet weather event and ultimate build out conditions for the project area. These flows were compared to the capacity of the existing sewers to determine any capacity deficiencies. The sewers were ranked according to the following criteria:

1. Wastewater flows greater than 130 percent of full pipe capacity.
2. Wastewater flows between 100 and 130 percent of full pipe capacity.
3. Wastewater flows less than 100 percent of full pipe capacity.

Sewers falling in the first and second categories were considered to be capacity deficient.

Flat Sewers. Sewers laid on "flat" slopes allow grease and debris to accumulate. Sewers are considered "flat" when the slope is insufficient for the flow to reach the CCCSD minimum velocity criteria. The CCCSD criteria is a velocity of 1 meter per second for main sewers and 0.67 meter per second for trunk sewers, both when flowing full.

Structural Condition. Television inspection revealed structural defects such as cracks, sags, and offset joints in the existing sewers. The majority of sewers in Watershed 22 are vitrified clay ranging in size from 150 mm to 200 mm. The sewers were installed in the late 1940's and early 1950's. The television inspection also revealed areas of root intrusion, grease accumulation, and debris build-up.

Access Problems. Over half of the sewers in Watershed 22 are located in backyard easements as opposed to public or private streets. Easements, especially older ones, physically restrict access.

Easements in Watershed 22 are mostly narrow, overgrown, landscaped, fenced, and otherwise developed. In some cases, homeowners have unknowingly constructed improvements such as redwood decks, spas, or storage sheds over sewers of manholes within the easements. New construction or rehabilitation may require the time and cost of acquiring a legal easement where one currently does not exist.

Photo 1: Existing Easement Obstructions

Slope and Geological Concerns. Some of the sewers in Watershed 22 are located near creeks or on slopes. The soils in Watershed 22 are generally silt and clay which tend to expand and shrink during the wet and dry weather periods experienced in California. There was evidence of surface cracks and erosion along existing sewer alignments. Some of these surface movements or cracks corresponded with structural defects or sags in the existing sewers observed by television inspection. Geotechnical studies were conducted to evaluate long-term soil movement and the viability of replacing or rehabilitation an exiting sewer in these locations.

Los Arabis Drive Area

The majority of Los Arabis Drive area is residential. The area is served with 150 mm to 250 mm diameter sewers as shown in Figure 2. The sewer reaches within the Los Arabis Drive area were researched, evaluated, and inspected for potential problems and it was determined that the alignment shown in Figure 2 would require upsizing and rehabilitation.

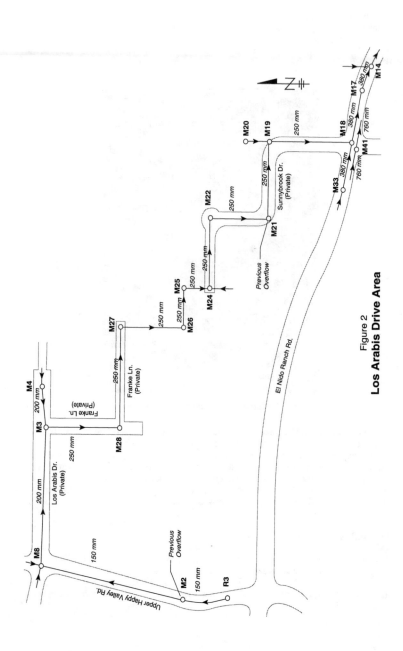

Figure 2
Los Arabis Drive Area

Various alternatives were evaluated. The two most feasible alternatives were the base project alternative and the bypass alternative, both alternatives would use open cut construction methods. The base alternative consisted of open trench construction with upsized sewers either in the existing alignment or in a parallel alignment within the same street. The bypass alternative would divert flow at manhole M8 and reroute the flow south along Upper Happy Valley Road and east along El Nido Ranch Road to an existing trunk sewer. Pipe bursting was added as a third alternative after the District completed a pilot project which indicated this technology was viable for the Los Arabis alignments.

Recommended Alternative. The recommended alternative is shown in Figure 3. Approximately half of the alignment was constructed using open trench method and half was up-sized using pipe-bursting method. Pipe bursting was chosen because it allowed the existing sewer to be up-sized without digging up the entire alignment. The pipe-bursting increased the existing 250 mm vitrified clay pipe (VCP) sewer to a 350 mm High Density Polyethylene (HDPE) sewer.

During final design, the following requirements were specified:

- The insertion head used shall be either a hydraulic or pneumatic mechanism. This requirement minimized the disruption to the above ground terrain and nearby structures, as well as minimizing the potential for the pipe bursting head from getting stuck.

- Inside weld bead on the HDPE pipe joints shall be removed. This requirement kept the pipe flow surface smooth.

- The inner wall of the HDPE pipe shall be white, light green, light red (vitrified clay color), or natural. Yellow and light purple are not acceptable. A non-black pipe interior makes TV inspection of the pipe easier.

- Where possible, the laterals were required to be externally reconnected. Each lateral was dug up and a HDPE saddle was heat fused onto the HDPE sewer. This requirement makes a water tight connection between the lateral and the sewer main.

Woodland Way Area

The Woodland Way area had been a source of problems for many years. Initial investigations of alternate routes for the sewer were conducted in 1974. In February 1990, manhole M35 on Woodland Way overflowed during a two-year storm, most likely due to poor conditions in the pipe.

The majority of the Woodland Way area is residential. The area is served with 150 mm to 250 mm diameter sewers as shown in Figure 4. The sewer reaches within the Woodland Way area were researched, evaluated, and inspected for potential problems. The areas of concern for the sewers in this area included: insufficient capacity, excessive root intrusion, structural cracks and sags in the pipe, and poor access to manholes for maintenance.

The two most feasible alternatives were the pump station alternative and the bypass alternative. The pump station alternative consisted of the following improvements:

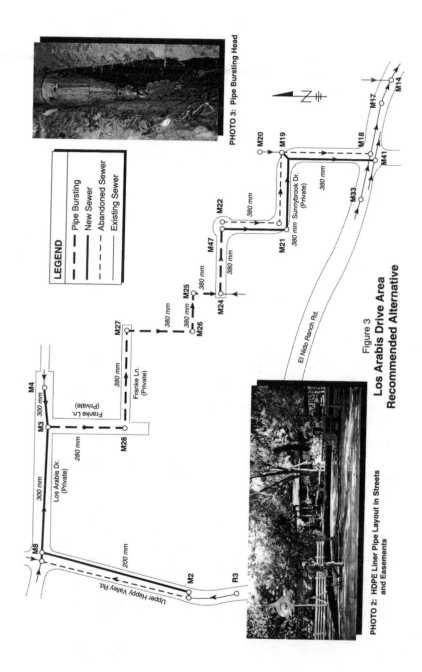

WATERSHED 22 TRUNK SEWER UPGRADE

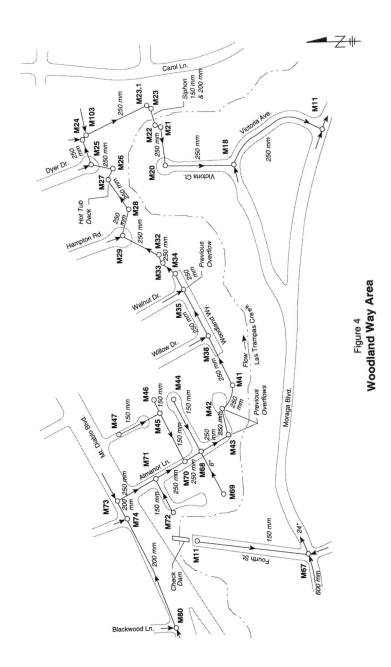

Figure 4
Woodland Way Area

- Replacement of the existing sewer with larger pipe along the same alignment from M70 to M29 using conventional construction methods.
- Use of a directional drilling technique to construct a new alignment from M29 to M24 consisting of straight connections and increased slope.
- Installation of a pump station and force main installed by directional drilling under Las Trampas Creek.

The bypass alternative would divert flow at manhole M71 across Las Trampas Creek to Fourth Street. This would divert 49% of the flow from the Woodland Way area. The bypass alternative would consist of the following improvements:

- Either a pipe bridge or siphon crossing of Las Trampas Creek.
- Rehabilitation of sewers downstream of M43 in poor condition.

The bypass alternative including a siphon is undersirable because of frequent maintenance.

Recommended Alignment. As the design process continued, another creek crossing location was investigated. This location, downstream of the check dam, allowed the use of a pipe bridge and eliminated the need for a new siphon. The recommended alternative is shown in Figure 5. Approximately half of the alignment was constructed using open trench methods and half was lined.

During final design, the following requirements were specified:

- To encourage competition, two different lining methods were allowed in the Contract Documents. These were : cured-in-place pipe and folded plastic pipe liner (U-Liner™ and Nu-Pipe™).
- Where possible, the laterals were required to be externally reconnected. Each lateral was dug up and a rubber-based saddle was strapped to the lined sewer. This arrangement makes a water tight connection between the lateral and the sewer main.

Pipe Bridge. Four different pipe bridge types were evaluated. Each was evaluated for ease of maintenance, constructability, and aesthetics. Las Trampas creek is approximately 35 meters wide in this location. The following pipe bridge types were evaluated:

- Space Truss. The space truss is triangular in cross-section with a major structural member at each of the three vertices. The sewage flows through the top structural member which is a mortar coated steel pipe.
- I-Beam. Two I-beams side-by-side spanning the creek. A ductile iron pipe between the two I-beams would carry the sewage.
- Suspension Bridge. A tower at each abutment would support a suspension cable. A steel pipe would be supported from the cable. Stiffening cables to resist lateral forces would also be used. This type of pipe bridge is applicable to longer spans.

WATERSHED 22 TRUNK SEWER UPGRADE 623

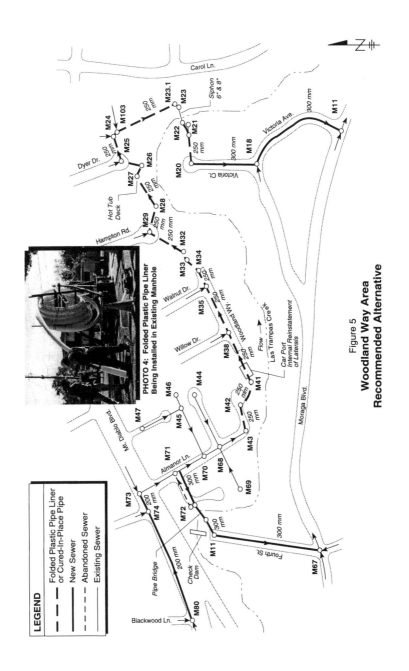

Figure 5
Woodland Way Area
Recommended Alternative

- Pipe-Within-A-Pipe Bridge. The structural system consists of a steel pipe spanning the creek as shown in Figure 6. Inside the steel pipe, a HDPE pipe carries the sewage flow as shown in Figure 7. Due to the differences in thermal expansion, an expansion pipe sleeve must be used as shown in Figure 8.

The Pipe-Within-A-Pipe Bridge was selected. The inner HDPE pipe is impervious to corrosion and would therefore require low maintenance in the years to come. Also, if it ever becomes necessary to replace the inner HDPE pipe, it can be removed without affecting the structural integrity of the bridge. The pipe-within-a-pipe has the smallest silhouette of the alternatives and therefore would be less obtrusive from an aesthetics point of view.

Photo 6: Installed Pipe Bridge

Conclusion - Lessons Learned

As with any new technology, there were successes and surprises (not always positive). This project was no exception.

Pipe Bursting. First, the use of pipe bursting to increase the sewer size from 250 mm inside diameter to 350 mm diameter was accomplished. Upsizing in this pipe size had only been attempted once before in California and in different soil conditions. In this case the decision to use only a pipe bursting system with a hydraulically or pneumatically operated bursting head was prudent.

The pipe bursting segment of the project was accomplished quickly, within one month. Overall, there was much less disruption to the neighborhood and to the public than if conventional open cut construction had occurred.

The reinstatement of laterals with heat-fused saddles provided a sound and long-lasting service connection.

With the two pipe size increase, there was a potential for surface displacement from the bursting operation. However, the surface displacement was greater than expected. In reaches where the sewer was 1.5 to 2.5 meters deep, vertical surface rising of up to 7 cm was observed. A crack, approximately 2 to 5 cm wide, developed along the pipe alignment, as the bursting head progressed. The vertical displacement was limited to the area right above the newly installed pipe and did not extend more than 60 cm from the center line on either side. In most areas of the project, the displacement occurred in landscaped areas and did not have any adverse impact. In one reach, approximately 45 meters long, a badly-deteriorated asphalt driveway was cracked and was replaced. In another area, a brick and mortar sidewalk was removed prior to pipe bursting, and a 15 cm deep by 30 cm wide trench was dug along the pipe alignment, prior to pipe bursting, to prevent damage to an adjacent tennis court. This mitigation was successful. The brick and mortar sidewalk was replaced after pipe bursting.

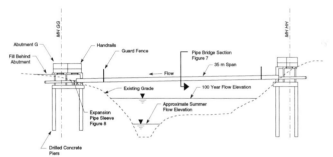

Figure 6
Pipe Bridge Profile

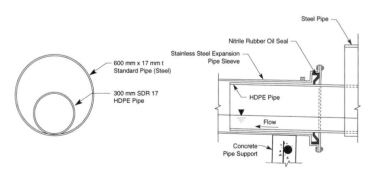

Figure 7
Pipe Bridge Section

Figure 8
Expansion Pipe Sleeve

In areas where the pipe depth was 3 to 4 meters deep, the vertical displacement from pipe bursting was not observed. Surface cracking of the ground or pavement areas did not occur.

Another "learning experience" was the laydown or staging area requirements for the 350 mm inside diameter SDR-17, HDPE pipe prior to installation. CCCSD has pipe bursted extensively with smaller diameter HDPE pipe, both SDR-17 and SDR-35. In these sizes, the HDPE pipes are more flexible than the larger sizes. The larger HDPE pipes require a straight laydown area directly behind the installation pit. Two of the six original laydown areas could not be used. Fortunately, alternate straight laydown areas were available.

U-liner™. Overall, installation of the HDPE liner in the existing 250 mm vitrified clay sewer was successful;. The installation was relatively fast and caused little disruption to the public and neighborhood. The installation did require two more weeks than originally planned due to equipment breakdowns, difficulties in pulling the liner into the existing pipe due to bends and disruption to the planned work sequence due to required point repairs.

One surprise was the amount of point repairs to the old clay sewer that was necessary before the liner could be installed. During design, approximately 17 meters of point repairs were estimated. The actual length of point repairs was 67 meters. Some of these additional repairs were due to unsound pipe encountered during construction. However, approximately half of the repairs were the result of recommendations by the liner subcontractor so that the liner could be installed.

Most of the lateral service reconnections on the liner pipe were accomplished externally with strap-on saddles. However, because of buildings adjacent to, or other surface improvements over the sewer, internal lateral reinstatements were made at ten locations using a television camera and remotely operated grinder. The specifications called for the lateral reinstatements to be circular, and between 95% and 100% of the lateral opening. The subcontractor had difficulty locating the lateral openings and did not achieve specification. The actual lateral openings are significantly greater than the 100 mm diameter laterals. While the laterals are functioning, the quality of the lateral reinstatements is still a point of discussion with the liner subcontractor.

[1] Principal Engineer, Central Contra Costa Sanitary District, 5019 Imhoff Place, Martinez, California, 94553-4392.
Phone: (510) 228-9500. Fax: (510) 228-4624.

[2] Project Engineer, Montgomery Watson, 777 Campus Commons Road, Suite #250, Sacramento, California, 95825-8308.
Phone: (916) 924-8844. Fax: (916) 924-9102.

[3] Project Manager, Montgomery Watson, 355 Lennon Lane, Walnut Creek, California, 94598-2427.
Phone: (510) 933-2250. Fax: (510) 945-1760.

MICROTUNNELING FORCES: The Pipe's Perspective

Robert Lys, Jr., P.E.; Thomas M. Garrett, Ph.D.[*]

Abstract

This paper reports a statistical method for calculating the maximum axial force required in microtunneling. Other forces put on the pipe when steered are discussed. A theoretical method is given for the calculation of the centripetal force.

Introduction

Much work has been done, ranging from the elegant to the practical, in the field of microtunneling forces from the perspective of the MTBM or "mole"(1,2). For the owners of pipelines, however, a perhaps more important matter are the forces exerted on the pipe. It is, after all, a pipeline that they are purchasing. Axial forces exerted on the pipe in microtunneling range from tens to hundreds of tons. If a pipe fails in microtunneling it almost always fails during the installation due to these forces. The negative consequences of axially overloading a pipe can be dramatically increased by steering.

This situation parallels that of open trench installation. In cut and fill the majority of pipe failures are due to vertical forces from soil loads when backfilling the trench. The negative consequences of these forces can be dramatically increased by improper bedding or trench width. For these reasons owners always specify the vertical pipe strength (D-load or 3 edge bearing), bedding, and trench width. They do this for their own protection.

For exactly the same reasons, the specifying agency has a need to know *a priori* what forces are likely to be exerted on the pipe. In this paper we will describe both statistical and theoretical methods to calculate the forces that will be exerted on a microtunneling pipe during its installation.

---

[*]NO-DIG, A Division of MCP Industries, Inc., 826 E. Fourth St., Pittsburg, KS 66762

I. A Statistical Method for Calculating Axial Forces

One approach to the problem of calculating axial forces on the pipe is to calculate the actual force on the mole, and then say that the force on the pipe must be the same as or less than this amount. To this end, voluminous amounts of data have been collected. Head pressure, soil type, depth, frictional forces, weight forces, and many others have been examined (2-5). Stein (1) lists no fewer than five formulas for calculating skin friction alone. The problem is none of them have wide range of application. It is not difficult to understand why. Microtunneling is a human endeavor. Unlike soil, which is controlled only by nature, the forces applied to the main jacks, the steering jacks, the rate of excavation, and whether or not to use bentonite or EZ-Mud are all controlled by people. Since all these factors strongly influence the jacking force, theoretical predictions of the main axial force are problematic.

Luckily, as John Graunt showed in 1662, human behavior is readily analyzed by statistical methods (6). We therefore gathered data from microtunneling jobs concerning pipe diameter, drive length, and jacking force for the maximum force that occurred on the job. A partial list of this data is tabulated in Table 1 (7).

Table 1.
Maximum Jacking Force Occurring on Microtunneling Jobs
Partial List

Maximum Force (Tons)	Drive Length (Ft)	Pipe O.D.	Project (Ft)
34	360	2.042	Oso Creek
55	156	2.042	Middletown
59	240	1.292	Maxey Road
66	300	1.292	Everglades Park
85	385	1.792	Woodway
100	278	1.292	Homestead
68	320	1.792	Tollway
24	208	1.292	Kapaa
70	260	2.042	Kapaa
95	270	1.292	Indian Trails

The form of frictional force was chosen to analyze the data. This form was chosen because it was observed that, in general, the job maximums occurred during times of high frictional forces, e.g. after a work stoppage. The analysis, however, is statistical and does not depend on whether the maximum force originated from penetration resistance or frictional resistance. Frictional resistance is given by (8)

$$R = M \times O.D. \times \pi \times L$$

where $M = \mu \times N$ from classical mechanics (9), and L = drive length.

We are seeking, therefore, a statistical value for M.

A histogram of the force/surface area of pipe/soil contact is shown in Figure 1. The normal distribution was calculated for this histogram and is also plotted in Figure 1 as a bell shaped curve. The mean thus calculated for M was 0.05 +/- 0.02 tons/ft^2. We can therefore predict the maximum jacking force on a given job with the following probabilities:

Table 2.
Probability of Occurrence of Maximum Axial Force

% Probability	Of a Maximum Axial Force ≤
84.1 (1 σ limit)	0.07 tons/ft^2
97.7 (2 σ limit)	0.09 tons/ft^2
99.9 (3 σ limit)	0.11 tons/ft^2

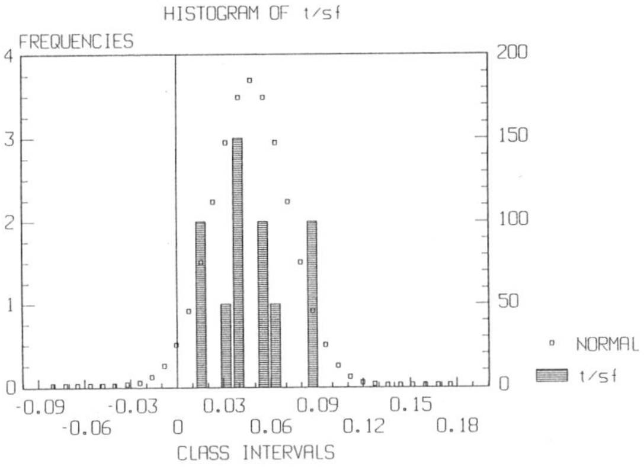

Figure 1. Histogram and normal distribution of force/surface area.

For example, if we want to jack a pipe with a 24.5" O.D. 333 feet we have a 84.1 % chance that the force required will be

$$0.07 \times \frac{24.5}{12} \times \pi \times 333 = 150 \text{ tons or less}$$

We have a 97.7 % chance the axial force required will be

$$0.09 \times \frac{24.5}{12} \times \pi \times 333 = 192 \text{ tons or less}$$

We have a 99.9 % chance the axial force required will be

$$0.11 \times \frac{24.5}{12} \times \pi \times 333 = 235 \text{ tons or less}$$

Current empirical methods to estimate the jacking force can be evaluated by these statistical methods. NO-DIG currently uses a figure of 0.06 tons/ft^2 to estimate the maximum jacking force required (10). This figure is seen to be the mean +1/2 standard deviation. In Europe, 10 kN/m2 is currently used in much the same manner (8). This is a very conservative number at 0.10 tons/ft^2, 2.5 standard deviations from the mean. A major machinery manufacturer (11) in the USA uses a figure of 2.8 short tons per inch of mole O.D. per 100 feet of pipe installed. The formula is even more conservative at 0.11 tons/ft^2, 3 standard deviations from the mean.

It should be noted that this method is a statistically based one. The 0.05 tons/ft^2 is not a friction factor for either the pipe or the soil but a statistically derived average of human behavior on microtunneling jobs. Different types of soils, pipes, machines, and operators have therefore been taken into account in this analysis. Only a paradigm shift in the behavior of microtunneling contractors in the USA would cause this analysis to change.

In specifications, typically a safety factor is added to these figures. For example, if we choose the mean +1 standard deviation as the likely maximum force required, a safety factor of 2.5, and a maximum drive length of 333 feet for a 24.5" O.D. pipe, the pipe would be required to have the following strength:

$$\left(0.07 \times \frac{24.5}{12} \times \pi \times 333 \right) \times 2.5 \text{ S.F.} = 374 \text{ tons before axial failure}$$

Most pipe manufacturers take the safety factor into account and publish tabulations of axial strengths ÷ safety factors.

II. Theoretical Calculations of Forces When Steering

Please note that in this section σ is used to mean stress, while in the prior section this symbol was used for standard deviation.

A. Edge Stress

Because mircotunneling pipes are not solid bodies, but hollow cylinders, the stress on the edge of the pipe is significant. This is especially true when a jacking pipe is steered, causing the joint to open up. This in turn distributes the main axial jacking force over a much smaller surface on the end of the pipe. Hence, much less jacking force can be applied. The calculations for this have been known for some time (12) and have been adopted in several specifications (13,14).

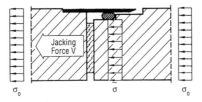

Figure 2. Transfer of force with pressure resistant pipe coupling under axial load (13).

In general the allowable jacking force is reduced by a fractional "stress ratio" $\frac{\max \sigma}{\sigma_o}$ so that

$$allowable\ F = \frac{axial\ pipe\ strength}{\frac{\max \sigma}{\sigma_o}}$$

where the stress ratio is given by the ratio of the closed diameter, Z, to the pipe O.D., d_a, according to the following graph in Figure 3.

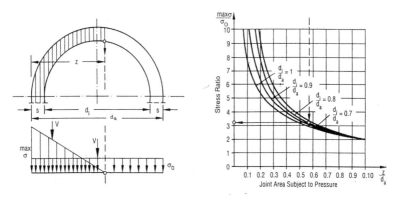

Figure 3. Dependence of stress ratio on ratio Z/d_a (13).

If the pipe joint is not gaping, $\frac{\max \sigma}{\sigma_o} = 2$. For this reason safety factors for microtunneling are never chosen less than 2. If the pipe joint begins to open the stress ratio increases exponentially, decreasing the allowable jacking force. Typical safety factors for microtunneling are 2.5 to 4.0.

B. Centripetal Forces

When steering, a centripetal force must be applied to the mole, usually through steering jacks. From the pipe's perspective this is translated into a force perpendicular to the jacking force. If two pipes are deflected through an angle, θ

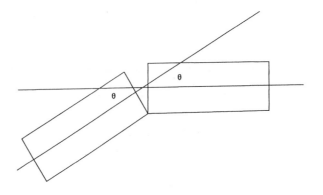

Figure 4. Deflection of two pipe by θ degrees.

we generate the following vector diagram.

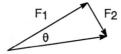

Figure 5. Vector diagram of forces on a steered pipe.

The assumption here is that the centripetal force is proportional to the jacking force, F_1, i.e. that it must overcome the same types of resistance as the axial force. That the centripetal force cannot be due to a change in the velocity can be shown by the calculations of classical mechanics which yield forces orders of magnitude too small[15].

We then have

$$F_2 = F_1 \tan \theta$$

For example, if there is a 5 degree deflection and a 100 ton jacking force

$$F_2 = 100 \tan 5 = 9 \; tons$$

Since microtunneling pipes are in general not designed for such forces it is important to have line and grade tolerances (typically +/- 1") on microtunneling jobs. And if during the installation the pipeline should come off of these tolerances, it is important to come back to line and grade very, very slowly so as not to put edge stresses and centripetal forces on the pipe. It is for this reason installation logs are kept of jacking forces, drive length, and amount of steering. A final consequence of these forces is that it is far better to design curvilinear systems using straight lines than to attempt to steer pipe through an angle.

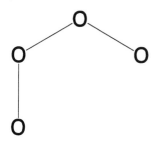

Figure 6. Use of straight pipelines to construct a curvilinear project.

Conclusions

We have demonstrated statistical and theoretical methods for calculating axial and centripetal forces in microtunneling and have also reviewed methods for calculating the edge stress when steering. This information, from the pipe's perspective, provides the engineer with the ability to calculate *a priori* the data needed to specify pipe strengths, safety factors, and line and grade tolerances necessary for a good installation.

References

(1)   Stein, D.; Mollers, K; Bielecki, R. *Microtunneling*; Ernst & Sohn: Berlin, 1989; pp 253-265.

(2) Atalah, A.L.; Iseley, T.; Bennett, D. *Conference Papers*, NO-DIG '94, Dallas, TX; North American Society for Trenchless Technology: Chicago, 1994; D2:1-24.

(3) Scherle, M. *Rohrvortrieb*; Bauverlag: Wiesbaden, 1977; Part 2.

(4) Salomo, K.P. Dissertation, Technical University of Berlin, 1979.

(5) Weber, W. Dissertation, Rhenish-Westphalian Technical College, Aachen, 1981.

(6) Boorstein, D.J. *The Discoverers*; Random House: New York, 1983; pp 667-669.

(7) Competitive pipes to NO-DIG were included in the calculations but have been excluded from the tablulation.

(8) Stein, D.; et al. *ibid.* p 258.

(9) Feynman, R.P; Leighton, R.B.; Sands, M. *The Feynman Lectures on Physics*; Addison-Wesley: Reading, 1963; Vol. 1, Chapter 12, pp 3-5.

(10) Lamb, E.; Lys, R. Jr.; Garrett, T.M. *Proceedings*, Trenchless Technology: An Advanced Technical Seminar, Vicksburg, MS; Trenchless Technology Center, Louisiana Tech University: Ruston, 1993; pp 458-466.

(11) Garrett, T.M., personal communication, 1995.

(12) Stein, D.; et al. *ibid.* pp 266-267.

(13) Hornung, K. *ATV A161*; German Association for Water Pollution Control: **1987**, *10*, 257-264 and *11*, 303-310.

(14) *EN 295-7*; European Committee for Standardisation: **1994**, Annex B.

(15) Halliday, D.; Resnick, R. *Physics*; Wiley: New York, 1978; pp 59-61.

TRENCHLESS REPLACEMENT & CORROSION PROTECTION OF DETERIORATED MANHOLES

William E. Shook[1]

Abstract

Hydrogen sulfide gases commonly generated within sanitary sewer environments fosters corrosion of concrete manholes and similar underground structures and pipe. Left unprotected this corrosive action will completely destroy such structures. When the potential for sulfide generation is anticipated, pipe and manholes are usually installed with an integral locking plasShookl1tic liner. Since many manholes have been installed in some regions without such protection, an effective method for retrofitting is necessary. A proven system which avoids costly and disruptive dig-up and replacement while still providing complete structural replacement is PERMAFORM. This patented and versatile process employs the same plastic T-rib liner to create an impermeable protective liner within existing manholes at the same high quality standards as newly installed and tested.

Introduction

According to figures reported by the U. S. Environmental Protection Agency, there are about 20,000 collection systems in the United States serving seventy-five (75%) percent of the total population. These sewer systems have more than 15 million manholes of which about one-half are over fifty (50) years old. EPA estimates that half of these manholes, about 4 million, are in immediate need of structural replacement due to age, traffic wear, infiltration and deterioration. Manholes affected by sulfide corrosion show effects in much less time. It is not uncommon for

---
[1]President, Action Products Marketing Corp., Box 555, Johnston, IA 50131

unlined precast manholes to lose wall thickness at the rate of 10-20mm/year. Severe deterioration substantially shortens the useful life of these unprotected structures.

According to their 1990 Needs Survey (Report to Congress), the annual capital expenditures required to Replace/Rehabilitate the collapsing sewer infrastructure (Category IIIA) is $4.2 billion dollars. This is in addition to the $3.5 billion dollars annually that is estimated for the correction of Inflow/Infiltration (Category IIIB). With about five (5%) percent of this portion for manholes, expenditures would range from $300 to $500 million dollars annually.

There are some communities that have already launched comprehensive programs to correct their sewer system deficiencies. Houston, for example, has initiated a program to spend $250 million dollars over each of the next five years to upgrade their system and bring it into compliance. There are more than 80,000 manholes in their system. Orlando has spent over $350 million dollars in a recent upgrade of their waste water system and received the Outstanding Achievement Award in Water Pollution Control for 1991 for their efforts. Orlando has an annual budget of $350,000 per year just for inflow/infiltration control.

Evaluating the problem

Even though structural deficiencies arise from a variety of sources, such as materials, traffic loads, ground conditions and age. The most serious deterioration with the greatest potential for immediate repairs results from sulfide corrosion and ground water intrusion.

Ground water intrusion at precast joints has a much greater effect on structural stability than once considered. Furthermore, it is exacerbated by sulfide decay. Each leak is also washing in soil fines that create voids in the fill material immediately surrounding the manhole. The size of the leak, the fill material and the length of time that it has been leaking are critical factors to the seriousness of the structural degradation. Visual inspection of the interior manhole surface alone is not often sufficient to determine the extent of this problem. Probing of the mortar joints and precast section joints in search of hollows outside the wall is a much better test. Unfortunately, ground settlement or collapse of the roadway immediately adjacent to the structure is too often the clearest evidence of this problem.

In the southern region of the United States, structural decay caused by hydrogen sulfide attack on the interior concrete surfaces of manholes is the principal problem. Primary factors contributing to the generation of this problem are: high bacterial growth in the sewerage, warm climates and line turbulence from pumping stations and drop inlets. The sulfuric acid produced by aerobic bacteria growing on the interior concrete surfaces turns Portland cement concrete into gypsum. This soft, spongy material sloughs from the walls and exposes the interior reinforcing steel. According to a May 1991 EPA report to Congress on <u>Hydrogen Sulfide Corrosion in Wastewater Collection and Treatment Systems</u>, corrosion rates of 10mm per year are common. In a period as short as ten years, 100mm of a manhole wall can be lost. This often represents more than one-half of its original wall thickness.

This report goes on to state that plastic linings that are embedded into the new concrete pipe and manholes during their manufacture and before they are buried provide the best protection against this corrosion based on over forty years of field experience.

In addition, the American Concrete Pipe Institute reports that when 25mm or more of the interior of a manhole that is in a concrete roadway is corroded, its designed H-20 load bearing capacity is lost. If the manhole is in an asphalt roadway, its H-20 load bearing capacity is lost if as little as 12mm of the manhole wall is corroded.

Structural rebuilding is imperative regardless of the cause of the serious degradation. For the very same reason that trenchless pipe replacement has become important, namely cost savings and convenience, replacement of manholes without digging and without interrupting sewer flows is likewise highly beneficial to the sewer district.

The PERMAFORM solution

Structural replacement of manholes and lift stations with the PERMAFORM System is a convenient and cost-effective method that has been proven in over 2000 installations over the past nine years without a single failure. This unique system uses patented technology and equipment to internally form and place a 75mm wall of high strength concrete wholly within the existing structure while keeping the sewer flows active. The new interior wall is placed at one time from the bench to the casting and consolidated into place to provide a one-piece, joint-free and structurally independent manhole within a manhole. It does not require a bond to the old substrate; the existing interior acts rather as an outside mold. Voids, cracks and seams are completely sealed; and, where the old wall is most deficient, the new wall is at its thickest. In a standard 1200mm manhole, the new interior would be 1050mm and the new cone is one-piece with the wall. It may be concentric or eccentric according to the existing cone.

For corrosion protection, a white, T-ribbed plastic liner of the type referred to earlier, is positioned like a skin around the exterior of the forms as they are erected within the manhole. Once the plastic lined forms are totally assembled, the concrete is placed. After the concrete is sufficiently set, in about 60 minutes, the forms are removed and any joints in the plastic sheeting (remember, there are none in the new concrete wall) are heat fused with an overlapping weld strip. The new plastic, embedded liner is then tested with a holiday detector at 10,000 volts to ensure that there are no pinholes that would allow sewer gases or wastes to come into contact with the new concrete.

When the lines on which the manholes are situated and lined with high-density polyethylene pipe, the liner in the newly rebuilt manhole is welded directly to the pipe liner. If the lines are made of PVC, then a T-ribbed PVC liner is used. If the pipe is made of some other material or if a cured-in-place pipe liner is used, an adapter boot is employed to ensure a water tight seal at the wall penetration. The sewer district is then totally protected against any corrosive materials or gases leaking out or any ground water leaking in.

Sewer mains that have been lined in areas of high ground water generally do not diminish infiltration unless the manholes are completely sealed at the same time. Sealing of lines and manholes provides a completely rebuilt and protected collection system.

APPENDIX

KEY WORDS

- manhole rehabilitation
- PERMAFORM
- T-rib plastic lining
- hydrogen sulfide corrosion
- sewer collection systems

REFERENCES

U.S. Environmental Protection Agency <u>Needs Survey</u> Report to Congress, 1990

U.S. Environmental Protection Agency <u>Hydrogen Sulfide Corrosion in Wastewater Collection and Treatment Systems.</u> May 1991

CIPP PIPE PERFORMANCE DURING THE 1993 FLOOD

By: Gary T. Moore[1], P.E., M. ASCE &
Charles H. Nance[2], P.E., M. ASCE

Introduction

This paper, along with another paper[1], provide information regarding efforts which the Metropolitan St. Louis Sewer District (MSD) has undertaken towards completion of a major sewer rehabilitation program. The goal of the program was to improve the structural integrity of numerous large diameter combined masonry sewers. A variety of rehabilitation technologies including: sliplining, shotcrete, cured-in-place-pipe (CIPP), and replacement have been successfully installed in combined sewers within the district. The inspection, design and construction processes for the program began in 1987, and are continuing today. The work has progressed steadily with construction having been completed in some watersheds and inspections of conditions and design currently being performed in others. During the Spring and Summer of 1993, the rehabilitated pipes experienced loadings which equalled the maximum loads assumed during design.

The District is approximately 48% complete with the implementation of the Overflow Regulation System (ORS) improvement program in the Bissell Point Service Area. The objectives of the ORS program are: to eliminate sewage overflows during dry weather to the Mississippi River at high River stages, to prevent Mississippi River fluctuations from adversely affecting the sewerage facilities and to provide control and storage within the collection/trunk sewer system to optimize wastewater treatment. Under some of the planned operating conditions, the old brick sewers are to be pressurized. This design condition caused MSD to be very concerned regarding the structural integrity of the system.

---

[1] Engineering Manager, St. Louis Metropolitan Sewer District, 2000 Hampton Avenue, St. Louis, Mo. 63139

[2] Eastern Region Manager, Insituform Mid-America, Inc.
17988 Edison Avenue, Chesterfield, Mo. 63005

The implementation of the ORS program required the District to evaluate the structural impact which the change in operations imposed on the sewers. The evaluation concluded that the existing combined sewers would be subjected to significantly higher external groundwater pressure. A considerable amount of the existing combined sewers required rehabilitation to improve their structural integrity.

This paper reports on the performance of the completed rehabilitated piping during the flooding of 1993. A description of the results of a failed pipe section which had not been rehabilitated is also included for comparison. An emphasis on the construction methods used for rehabilitation along with results of a post-flood inspection of the pipes are included.

Design Criteria

The selection criteria for the pipes to be rehabilitated are covered in the previously referenced paper by Collins and Stude. The general design criteria are presented here so that they may be related to pipe performance during the flooding. In general a decision was made to rehabilitate all masonry sewers above 760mm (30") in diameter including 600mm by 900mm (2'x3') egg shaped. Under anticipated operating conditions, these sewers could become pressurized when the level of groundwater outside the sewer was lower than the internal pressure. Under other conditions, the sewers could be subjected to a full depth (3 to 7 meters) of groundwater while the sewer was nearly empty. These conditions are dictated by the operating plan for the system and the objective to take all dry weather flows to treatment no matter how high the river stage. When the river level is low and localized rains result in high sewer flows, the gates are closed and the sewers back up. This creates a surcharge condition. When the river levels (and consequent groundwater levels) are high, and localized rain is not present, low flows in the sewer present the other design extreme. During 1993, the assumed design conditions actually developed as the flood affected river levels, groundwater levels and flow conditions within the sewer. During the late winter and early spring rainy periods in St. Louis, Mississippi River levels, and groundwater levels were below flood stage and the sewers were internally surcharged due to heavy local rains. In the late summer, when the Mississippi reached record levels, there were periods of time when groundwater levels were at the surface while the sewers were experiencing low levels of flow due to the absence of localized rains.

Methods of Rehabilitation Used

There are nearly 1,126km of sewers in the Bissell Point Watershed with approximately 34km of sewer requiring rehabilitation for the ORS program. The sewers which were rehabilitated ranged in size from 680mm (27") in diameter to 7.3m x 5.2m horseshoe shaped sewers. The sewers range from 74 to 144 years old and were constructed of various materials such as stone and brick masonry, segmented concrete block or tiles and reinforced concrete. The various types, shapes and materials of construction dictated that different methods of rehabilitation be used for the ORS improvements. Approximately 16km of sewers, at a cost of $30 million, have been rehabilitated as of December 1994. Figures 1 & 2 illustrate the amount of usage of the four rehabilitation techniques and costs as related to the overall improvement program.

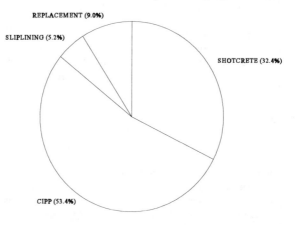

Figure 1   Rehabilitation Type Footage

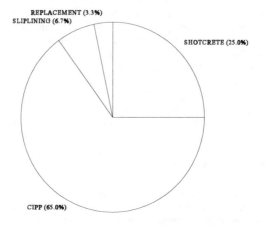

Figure 2   Rehabilitation Type Costs

The methods which were used for the program are: reinforced shotcrete, cured-in-place pipe, sliplining and complete sewer replacement. Generally, sewers greater than 1.5m in diameter were rehabilitated using reinforced shotcrete. Sewers less than 1.5m in diameter were rehabilitated using CIPP, sliplining or replacement by new construction. Although open cut reconstruction has been used, it is not dealt with in detail in this paper because of its wide use and acceptance. A brief description of the work completed by each method follows.

Approximately 1.5km of sewers to be rehabilitated have been replaced with new sewers at a cost of $1 million. Typically, in the City of St. Louis, sewer replacement must be evaluated to determine the feasibility due to being located within street pavement, under railroads, next to and under buildings, limited working room and location of numerous other buried utilities. However, there were three (3) sewers for which sewer replacement was the only option, because of: their large size, they were collapsed or they were severely hydraulically undersized.

Sliplining

Sliplining has accounted for approximately 0.8km of sewers which have been rehabilitated. Sliplining pipe ranged from 450mm (18") to 1,370mm (54") in diameter while materials used were ductile iron pipe, and centrifugally cast fiberglass/polyester composite pipe (HOBAS). The annular space between the sliplined pipe and the existing sewer was filled with a cement grout.

Sliplining has advantages in that many contractors are able to perform the work; the installation can be performed without bypass pumping in combined sewers during dry weather; and minimum to moderate disruption to surface and subsurface improvements occurs. During the rehabilitation method selection, the engineers rejected sliplining as a possible alternative in sewers which have moderate to sharp bends or curves. Sewers which have a high number of laterals (reconnection of laterals can be costly and is dependent on the number, size and location), and non-circular sewers (i.e. egg-shaped, bracing may be required to restrict movement during grouting and hydraulic capacity requirements were carefully analyzed) were not considered as viable sliplining candidates.

For this project, sliplining was used on a limited basis. Only those pipes which were round and those with moderate alignment curvature were considered for this alternative. Also, only those pipes which had adequate capacity for the design flows after the sliplining were considered. The sliplining results in a reduced internal diameter which can, especially for non-circular host pipes, result in flow capacity reductions. The improved "n" value, for the sliplined pipes as compared to the existing masonry sewers, was carefully considered during the design to assure that adequate hydraulic capacity was maintained. Construction by this method requires excavation of at least one access pit at the insertion end of the pipe.

In the diameters under consideration for this project, short segments of HOBAS pipe were used with gasketed joints. The pipe segments were lowered into the host pipe at the access pit and pushed down the line with the joints being pushed home in the final pipe position. Placement in this manner allows for proper blocking of the liner pipe prior to grouting and easier location and reconnection of services. Services were carefully located and reconnected from inside the slipliner using prefabricated inserts. Following installation of the sliplined pipe and connection of the services, the annular space was grouted with cementitious grout to provide proper side support and lock the liner in place relative to the host pipe. Grouting was performed in several lifts in order to prevent the liner from floating and deforming.

Reinforced Shotcrete Lining

Approximately 5.1km of sewers were rehabilitated using shotcrete at a cost of $7.5 million. Shotcrete was used in sewers larger than 1.5m in size due to working room and quality control issues. The physical access requirements for men and equipment for this construction technique are such that a decision was made to limit the lower end of the diameter application range to 1.5m. Room is needed for flume piping, air hoses, product hoses, and free movement of workers into and out of the work zone. Recommended nozzle spacing from the work surface of from two to four feet, dependent upon mix type, also entered into this decision. Typically, two types of structural reinforcement were used, selection was dependent on shape and type of materials of the "host" sewer. Generally, welded wire fabric reinforcement was used in the circular and/or egg-shaped brick or tile sewers, while steel reinforcing bars were used in the large concrete horseshoe-shaped sewers.

Shotcrete has the advantages that it conforms to various irregularities easily; is a well documented and time tested material and can be installed with minimum disruption to the surface and subsurface improvements. Limitations of which the engineer must provide for are: lack of corrosion resistance (a primary reason for failure of unprotected existing concrete sewers), protection from infiltration and flow in the sewer (typically flumes are used which greatly restrict movement of men and materials in smaller sewers); certification of nozzlemen to ensure a quality product; and reduced hydraulic capacity of the resultant smaller diameter sewer. The process is difficult to thoroughly inspect and material properties testing of the end product requires extensive effort and expense.

For shotcrete sections, the sewer was first thoroughly cleaned. Following cleaning, reinforcing wire was anchored to the pipe wall and shotcrete was pneumatically applied to the interior of the pipe. Multiple passes of shotcrete were used to achieve the required design thickness of the material. Design varied with the condition of the host pipe and the shape of the sewer. Microsilica was added to the shotcrete mix to improve performance. The microsilica mix results in several benefits. Greater density was achieved in the finished material. This greater density provides for increased corrosion resistance and decreased permeability of the pipe wall. When compared to conventional non-microsilica mix designs, less rebound occurs during installation thereby reducing installation time and significantly reducing project cost for materials and labor to remove and dispose of the rebound materials. Higher compressive and flexural strengths were achieved due to the reduction of rebound and resultant higher cement content in the finished product which improved structural performance of the liner.

CIPP Rehabilitation

Approximately 8.5km of sewers were rehabilitated using this method at a cost of $19.5 million. The sewers ranged in size from 200mm (8") in diameter to 2,100mm (84") in diameter. Much of the work was bid competitively against sliplining, shotcrete or open cut methods. In many cases, the CIPP was judged to be the only acceptable rehabilitation technique and this work was contracted for following price negotiations.

CIPP has advantages in that it conforms to the "host" pipe easily, provides minimum disruption to surface and subsurface improvements, the minor area reduction of "host" sewer is more than offset by improved hydraulic characteristics, infiltration does not have to be stopped prior to placement, and CIPP is cost effective in confined urban areas. CIPP, like any method, has limitations which the engineer must consider: largest installation to date is an eight foot diameter sewer; existing flows must be bypassed, there is difficulty in scheduling work in combined sewers to avoid rain events; round-the-clock operations during installation may pose some noise problems in residential areas; some wrinkling may occur at size transitions or direction changes; if methods which don't use inversion techniques are used standing water could pose a problem; sewers with "flats" (flat walls or inverts) require special design; and size and slope of "host" sewer can affect access shaft locations (if needed).

Design of the cured in place pipe for the project conditions was performed by engineers at Horner & Shifrin, Inc's., St. Louis, Missouri office. The designs considered the existing condition and shape of the sewers as determined during internal inspection. Designs were performed in accordance with ASTM F1216 for circular pipes and as recommended by Insituform of North America for oval or egg shaped pipes. Loading included only groundwater for structurally sound host pipes. For existing pipes which were structurally deteriorated, loads included soils, surface surcharges and groundwater. The CIPP lines were generally designed to withstand 3m to 7m of external pressure which resulted in a wall thickness based on a standard dimension ratio (SDR) from 50 to 75. A major factor in the design is if the "host" sewer is structurally adequate to carry the soil and live loads. If the "host" sewer was found to be unable to carry the soil and live loads then the CIPP was designed to accommodate all loading conditions or the "host" sewer was structurally rehabilitated with liner plates prior to the CIPP installation.

A unique approach was adopted when dealing with extreme groundwater conditions. Design of CIPP pipes requires that physical properties of the CIPP pipe be known. The properties which are most important include the flexural modulus and allowable flexural stress. For high groundwater conditions in round pipe, the flexural modulus controls the thickness of the pipe. An adjustment is made to the short term flexural modulus to account for long-term performance. This reduction in modulus is an allowance for the long-term effects of sustained loading, or "creep", of the material when subjected to buckling conditions. For the ORS system, the extreme design condition of maximum possible groundwater levels was determined to have a limited time frame of occurrence, approximately 244 days as compared to normal design criteria which are established for fifty years of sustained loading. When considering this design condition, the short-term modulus was reduced by 50% to account for the long term effects of sustained loading, but the safety factor used in design was reduced to 1.5 from the normal 2.0. The engineers determined that the most extreme design conditions would occur infrequently for limited periods of time (weeks or months) during the fifty year design life of the project. The experience provided by the flooding in 1993 confirms that this approach is reasonable. Groundwater during the peak of the flooding did reach the surface while minimal flows were within the sewers during the months of June, July and August. During the Spring, prior to the maximum flooding, the sewers were at times surcharged with lesser groundwater levels. The adjustment in safety factor resulted in a more economical CIPP design. Flawless performance of all of the CIPP pipe which had been installed in the flood zone, during conditions which matched the most extreme design assumptions, indicates that this design approach was appropriate. Long term loading conditions from lower levels of groundwater than occur during maximum flood conditions were analyzed using the standard safety factor of 2.0, but did not control the design thickness.

Although the results of Louisiana Tech's CPAR (Construction Productivity Advancement Research, sponsored by the Corps of Engineers), CIPP testing program[1] were not yet available during the design of this project, they are worthy of mention here. Briefly, the performance of the Insituform CIPP was as predicted by ASTM F1216 design formula. The long-term performance of the standard resin Insitupipes indicate that the 50% reduction in flexural modulus for the long term effects of sustained loading is appropriate. For enhanced resin Insitupipes, the testing indicates that the greater stiffness material warrants only a 40% reduction in elastic modulus to predict long-term performance. The results of the CPAR testing also indicate that it is very important to use the appropriate material properties and creep reduction factors for the actual CIPP materials and processes which are being used. The Insitupipes were the only CIPP pipes for which the recommended properties proved to be conservative and appropriate.

There is another important aspect of the CIPP performance which impacts this project. When the CIPP is placed within the existing sewer, and the combination is subjected to groundwater forces, the loadings to the existing pipe are actually reduced. For this project, the host pipes were either brick or concrete. These pipe materials are permeable. Prior to placement of the CIPP within the existing pipe, external hydrostatic loads create a pressure gradient across the pipe wall. This pressure gradient loads the pipe as the pressure on the outside of the pipe is equal to the full hydrostatic head, and the pressure on the inside is zero if the pipe is not flowing full. Once the CIPP is installed, the pressure gradient across the existing pipe wall changes. The CIPP resists the full external hydrostatic load and the pressure on the inside of the host pipe is equal to the pressure on the outside. This results in a net reduction in load to the host pipe thereby increasing it's ability to continue to carry the soil loading. This net load reduction to the host pipe will only apply where the host pipe remains capable of carrying the external soil loads independently of the CIPP. This loading condition, where the host pipe carries the soil loads and the CIPP carries the hydrostatic loads, is commonly referred to as "partially deteriorated" design condition for the CIPP.

All CIPP construction for this project was performed using the Insituform Process. For the diameters of pipes involved, installation was typically performed from access shafts located at manholes. The manholes were excavated and replaced with precast manholes. Installation of the CIPP was performed using standard water inversion techniques. The non-woven polyester felt tube was saturated with a liquid thermosetting resin prior to installation. For this project, enhanced polyester resin was used. Installation techniques turn the tube inside out as it enters the host pipe and the water pressure holds the resin saturated felt tight against the host pipe. Resin is forced into cracks and gaps in the host pipe wall thereby locking the CIPP tightly into the host pipe. Once the tube is inverted, the water is heated and the liquid resin cures forming a new tight fitting structural pipe inside the host pipe. Flow capacities are increased and storage capacity is maximized due to the relatively thin resultant pipe wall. This was the primary method of rehabilitation for the system. Construction work has been provided by Insituform Mid-America, Inc.

The Flood of 1993

The flooding which occurred throughout the upper Midwest during the spring and summer of 1993 represent a very unusual hydrologic event. Unusually heavy rains began in October of 1992 and continued over a widespread area through August of 1993. Flooding began in February of 1993 in Minnesota and Wisconsin. Iowa was hard hit in May and June. Illinois and Kansas got their dose of flood conditions in June and July. St. Louis, at the confluence of the Missouri and Mississippi rivers, began to experience flood conditions on the Mississippi river in March of 1993. The flood levels on the Mississippi continued throughout the spring with only a short drop being experienced in May. From June through August flood levels rose steadily setting one record after another. Finally in August the record river level of approximately 15.1m was reached. This is a full 5.8m above flood stage and nearly 2m higher than any previously recorded river level. With the extended length of the flooding and the record levels, the design conditions of groundwater at the surface was experienced for many of the installed lines.

There were periods of time during the summer of 1993 when rainfall within the drainage basins for these sewers surcharged the sewers. There were also periods of time where the sewers were not surcharged but were operating at normal dry weather flow levels. However for both conditions, record levels were recorded for the Mississippi River and for the associated groundwater conditions. The observed performance of the CIPP reconstructed sewers shows, at least subjectively, that the design assumptions were correct and the design appropriate and correct for the project. Inspection following the flooding of portions of the CIPP sewers has not found any damage or indications of overstress or failure within the sewers.

Unfortunately, while the CIPP sewers were performing flawlessly, other sewers which had not been rehabilitated were failing. One of the first decisions regarding rehabilitation for this program was to rehabilitate non-reinforced, brick, masonry or concrete sewers. Reinforced concrete pipe (RCP) sewers were not on the program unless the structural condition observed during inspection required improvement. A dramatic failure of a section of RCP sewer occurred during the flooding. At the Salisbury pump station, numerous sewers come together and the flow is diverted at low levels to the interceptor sewer for transport to the treatment plant. At high flow levels, caused by local rainfall, flows discharge to the river. In July of 1993, with Mississippi River levels at record heights, a failure occurred in a section of RCP pipe near the pump station. The pump station is positioned just inside the main river levee upstream from downtown St. Louis. When the failure occurred, the combined sewers leading into the pump station, and the pump station wetwells were filled with sand and silt. It was feared that the levee would collapse at this location and emergency action was implemented to stop the leakage. This emergency action included shutdown of the pump station and dumping of over 181,400kg (200 tons) of rip rap onto the river side of the levee. Once the leak was stabilized and the levee protected, it was observed that ground settlements in the area of the pump station averaged approximately one foot over a football field sized area. This area included the portion under the main line railroad tracks which run parallel to the river behind the levee.

An inventory of piping in the area indicated that there was a 1.4m diameter pipe which had been rehabilitated with Insituform prior to the flooding. Prior to any field investigation, it was suspected that this was the line which had failed and nearly led to catastrophic flooding of a large industrial area being protected by the levee. However, once the flood had subsided and the lines could be cleaned and investigated, it was determined that the Insitupipe was in excellent shape and that the failure had occurred in a section of RCP which had been previously judged to be in sound condition. The inspection found open and separated joints along with severely cracked pipe sections which had not been recorded during pre-flood inspections. Remedial work has been completed to reroute the flows in this area including additional Insituform work and installation of some new lines using hand tunnelling methods. This particular failure has challenged the MSD to reconsider the original assumption that the RCP lines which were observed during the initial investigation to be in good condition would not need rehabilitation. Inspection of these sewers during high groundwater has been performed to make a more detailed analysis of these lines or the risk associated with failure of them. Again, the CIPP rehabilitated pipes were judged to be properly designed and installed, reinforcing the strong confidence the District has with this rehabilitation technique.

Post Flood Investigations

As a result of the severe conditions experienced, critical sewers within the ORS system have been reinspected to observe their condition. To date, no problems have been located within the rehabilitated sewers other than some minor infiltration occurring within a stretch of shotcrete lined sewer. These leaks will be repaired using pressure grouting techniques.

Conclusions

A significant planning, design and construction effort has been undertaken to provide an improved level of sewage treatment in the St. Louis area. This project included use of unique and innovative solutions to eliminate dry weather overflows to the Mississippi River during high river stages. The design approach for the project was carefully thought out and provided for an efficient and economical use of resources. The modified sewer system received its first major test during the flooding which occurred in the spring and summer of 1993. The design assumptions used and the quality achieved in the field resulted in a successful handling of this extreme test. Additional work is being performed in other drainage basins to bring them up to the same level of service as has been achieved in the Bissell Point basin. Due to the experience gained during the flood, the District is confident that the design approach used, and reported by Collins and Stude, is appropriate.

Based upon the performance of the rehabilitated sewers during the flooding of 1993, the District will carry out future rehabilitation improvements without concern of failures, knowing that the engineering and rehabilitation techniques used for the ORS improvements will produce the expected results.

1. Collins, M.A. and Stude, C.T. "Rehabilitation of Masonry Combined Sewers in the City of St. Louis", to be presented at ASCE Advances in Underground Pipeline Engineering Conference, Seattle, Washington, June 1995.

2. Guice, Dr. L.K., Straughan, Dr. T., Norris, C.R. & Bennett, D.R., "Long-Term Structural Behavior of Pipeline Rehabilitation Systems", August 15, 1994

Thermal Performance of Trench Backfills for Buried Water Mains

Caizhao Zhan[1], Laurel Goodrich[2], and Balvant Rajani[3]

Abstract

A field evaluation study was carried out to compare the thermal performance of several different backfill materials, in conjunction with a water mains renewal project in the City of Edmonton. In the autumn of 1993, seven test sections were constructed using different backfill materials. Each section was instrumented to monitor temperatures, thermal conductivity and moisture content within the backfill as well as water temperatures and flow rate in the water mains. This paper describes the thermal histories of each section during the winter of 1993-1994 and assesses the effectiveness of different backfills for frost protection. Two dimensional finite element analyses were carried out to simulate the field test results using measured thermal properties. Furthermore, finite element analyses were performed to evaluate the frost protection effectiveness of different backfill materials.

Introduction

Backfill materials for municipal water lines are chosen based on initial costs, ease of placement, and mechanical performance. In colder regions, frost protection of service lines and, ultimately, of the water mains themselves is an additional consideration, and this is generally addressed by placing the system below the maximum frost depth anticipated in the native soil. But frost penetration is affected by backfill material type and certain choices may exacerbate the problem whilst others offer the possibility of decreasing the required depth of cover and thus significantly reducing new construction or rehabilitation costs.

A strategic study (Goodrich and Sepehr, 1993) by the Infrastructure Laboratory of Institute for Research in Construction (IRC) at National Research Council of Canada (NRCC) indicated that it was possible that the PVC water mains buried in unshrinkable fill or controlled low-strength material (CLSM), one of the

---

[1] Research Associate, National Research Council of Canada, Institute for Research in Construction, Ottawa, Ontario, Canada K1A 0R6
[2] Senior Research Officer, National Research Council of Canada, Institute for Research in Construction, Ottawa, Ontario, Canada K1A 0R6
[3] Research Officer, National Research Council of Canada, Institute for Research in Construction, Ottawa, Ontario, Canada K1A 0R6

backfill materials currently used by the City of Edmonton, could freeze under severe winter conditions. The City also has a good local source of bottom ash and consequently, the City wanted to explore its possible utilization as a suitable backfill material. A consideration of these factors as well as the increasing share of PVC pipes in the total water distribution system, the City expressed an interest in evaluating the thermal performance of different backfill materials and the structural performance of the PVC water mains when subjected to freezing conditions.

In the autumn of 1993, seven test sections were constructed using backfill materials including native clay, sand, with and without insulation layers, CLSM, expanded shale light-weight aggregate (LWA), thermocrete, and bottom ash from a coal-fired generating station, in conjunction with a water mains renewal project in the City. In three sections, strain gauges were installed on the water lines, while each section was instrumented to monitor temperatures, thermal conductivity and moisture content within the backfill as well as water temperatures and flow rate in the active line. The major purpose of this paper is to report the thermal performance of the different backfills used in the field study.

Field instrumentation

The City of Edmonton identified a stretch of water mains on 77 Avenue that required replacement. It was clear from the early planning stages of the project that additional water mains would have to be constructed so that normal water service would not be interrupted in the event that the PVC water main failed. This arrangement provided flexibility in the operation of the water mains to the extent that water flow could be shut off and purposefully freeze the water and attempt to induce failure of the PVC water mains. Concurrently, PVC water mains at a shallow depth of 1 m, which function as a bypass, were installed parallel to the water mains at a normal depth of 2.4 m (Fig. 1). The intent of this arrangement was to induce freezing of water in the bypass water mains and to gain a better understanding of possible failure modes of the PVC water mains.

Construction of the renewal and bypass water lines began on September 27, 1993. Installation of the renewal and bypass water line with instrumentation for the seven test sections was completed on October 15, 1993. Installation of the east and west valve chambers with flow meter, water pressure sensor and emergency shutoff valves was completed during the later part of October and early November 1993.

The site was divided into seven successive 20 m test sections thermally separated by vertical insulation barriers with each test section backfilled with different materials (Fig. 1). The seven test sections are labeled alphabetically from A to G on the basis of their physical positions from west to east. In each test section, 13 to 16 T-type thermocouples were used to record temperatures at various depths on the trench centerline and along the trench wall as well as on the renewal and bypass PVC water pipes (Fig. 1). Time-domain reflectometry (TDR) probes were installed to measure changes in soil moisture at two elevations within the trench. The location of TDR moisture probes are shown in Fig. 1. Twelve thermal conductivity probes were installed at six of the test sections (Fig. 1). In addition, sensors to measure earth pressure in the backfill and strains on the pipes were also installed (Rajani and Kuraoka, 1995).

Data acquisition equipment was used to monitor all sensors with the exception of the soil moisture probes and the thermal conductivity probes.

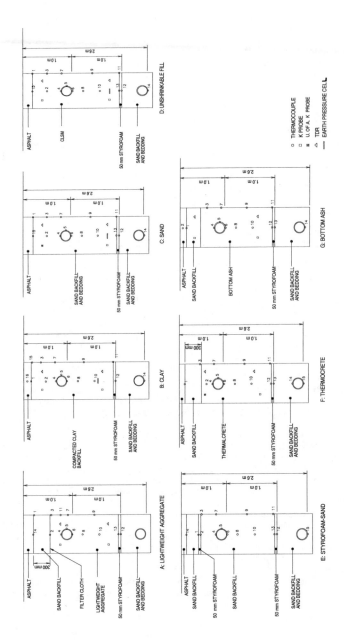

Fig. 1. Seven test sections with sensor locations and backfill materials.

Temperatures were recorded hourly with daily means transmitted via modem to the Infrastructure Laboratory in Ottawa.

Thermal performance of backfill materials

During most of the winter, until early March, 1994, the upper (bypass) water line was active. Unfortunately the flow rate was so large that the presence of the bypass line strongly influenced temperatures in the upper levels of each test section, masking differences due to the material behaviour of the different backfills. In addition, the layout of the experimental site in which the test sections followed in sequence, implies that heat was transferred between the individual sections via the flowing water. Flow rate was typically in the range of 20 m^3/hour, while the temperature drop along the seven test sections was in the order of 2^{0}C (Fig. 2). These numbers are large and imply that, at the depth of the bypass line, the flowing water was supplying heat at an average rate of 380 W for every metre of the pipe. As a result, an evaluation of the thermal performance of the backfills, based on a direct comparison of temperatures at corresponding positions for each test section, cannot be done without ambiguity. Nevertheless, a comparison of temperatures during mid to late winter on the centerline, at points some distance beneath the pipe (thermocouple 8 in Fig. 3) does show a consistent pattern. Ranked in the order from warmest to coldest, the sequence is bottom ash, thermocrete, sand-extruded polystyrene, LWA, sand, CLSM, and clay.

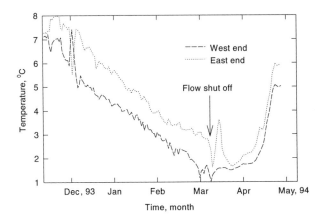

Fig. 2. Variation of water temperature in the pipe with time.

Figure 3 is representative of the temperature behaviour in the backfills between the upper and lower PVC water mains of the trench. It can be seen that the thermocrete (hot, when brought to the job site) was initially much warmer than the other materials and that this significantly biased the comparison. The remaining data strongly suggest a superior performance for the bottom ash section, even after making allowances for the fact that this material, too, was initially warmer than ambient temperature. But, it must also be recalled that flow in the bypass line was from east to west, with the result that the water line temperatures were greatest in the bottom ash, thermocrete, and sand-extruded polystyrene sections and least in the CLSM, sand, clay and LWA sections (in that order). The fact that the deep bottom ash temperature dropped rapidly by 0.5^0C about 10 days after flow was stopped in the bypass line is a measure of the contribution of heat from the active line. The reason why the clay section did not appear to perform better than the sand or CLSM sections may also be, to a large degree, that the bypass water temperature was lowest in the clay section. This is evident in Fig. 3 where it can be seen that about 15 days after no flow in the bypass line, deep temperatures started to rise in the clay section when there was still frost in the overlying zone. Comparing the CLSM and sand sections, it can also be seen that, even though the CLSM temperatures were initially approximately 0.7^0C warmer (Fig. 3), after mid January CLSM temperatures were the second coldest after clay, and this, in spite of the warmer bypass water temperatures in the CLSM section. Following similar reasoning, it is surmised that the deep temperatures in the sand-extruded polystyrene section may have been further biased by heat from the abnormally warm thermocrete located upstream. By the same token, the LWA section had the least amount of additional heat contributed by the warm bypass line, and its relative performance should therefore, in principle, be better than indicated.

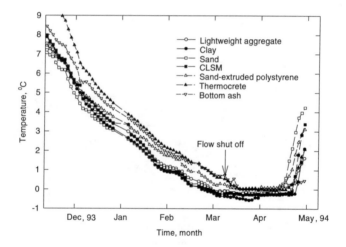

Fig. 3. Variation of temperature with time at thermocouple 8 for all seven sections.

Although subject to the same caveats as the temperature comparison, the relative thermal performance of the backfill test sections can also be assessed by calculating the depth of frost penetration on the centerline below the bypass water line as given in Fig. 4. Ranked in order of increasing frost depth, the results are sand-extruded polystyrene, bottom ash, thermocrete, LWA, sand, CLSM, and clay.

The fact that frost depth is the least in the sand-extruded polystyrene section is understandable in the circumstances. With the heat loss to the surface curtailed by the upper extruded polystyrene insulation layer, heat from the bypass water line would tend to spread uniformly in the strongly conducting sand zone. With this exception, however, the order of increasing frost penetration is consistent with the temperature ranking discussed above. Unfortunately, comparison of the maximum frost depths again does not necessarily indicate which backfill should be preferred because of the large heat contribution from the bypass water line.

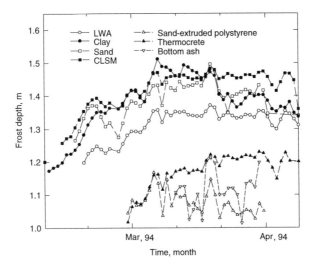

Fig. 4. Variation of frost depth with time at centerline of trench for all seven sections.

In contrast to the temperature data, a very clear pattern emerged from the in-situ thermal conductivity measurements (Table 1). The values listed are averages of measurements made after the initial thermal and moisture contents had stabilized. Ranked in increasing order of their frozen thermal conductivity, the materials are: LWA, bottom ash, thermocrete, clay, sand, and finally CLSM. The highest and lowest conductivity values found for the various materials differed by a factor of nearly 30. The in-situ values obtained for the LWA were approximately 1/3 that measured in previous field studies (Dilger et al., 1994), but in these cases, moisture content were 20% or higher. Values as low as 0.13 W/mK are consistent with the very low moisture contents determined with TDR probes. But since the frost penetration rate is proportional to the ratio of frozen thermal conductivity to latent heat, k_f/L (Table 1), the low thermal conductivity advantage is offset by the low moisture content. Bottom ash had the best ratio of k_f/L, the value being similar to that observed previously for lightweight aggregate. The thermocrete was found to have a thermal conductivity in the range of ordinary concrete, a result which disagrees with earlier laboratory studies. This may have been caused by inadvertent change during processing and handling. The thermal properties obtained for the clay and sand backfills were consistent with the expectations, while the CLSM material had a frozen thermal conductivity approximately 18% greater than frozen sand with a similar moisture content. This corresponds to k_f/L ratio approximately 13% greater than for the sand, or 8.5 times that of bottom ash.

Finite element analyses

As discussed earlier, the flow of water in the bypass line greatly affects the thermal regime in the trench, making it very difficult to assess the frost protection effectiveness of various backfills. In order to provide an alternative assessment, finite element analyses were carried out to determine the effectiveness of frost protection of various backfills used in the field study. As a first step, an analysis was carried out for section C (backfilled with sand) to determine how the finite element results compared with the field measurements. Subsequently, the finite element analyses were performed under identical initial boundary conditions for the trench backfilled with different materials to provide an unbiased evaluation of the frost protection effectiveness of different backfills.

Finite element model

A commercially available finite element program, AFEMS (1990), was used for the two dimensional thermal analyses. The heat transfer within the trench can be idealized as a two dimensional problem. Only half of the problem domain is analyzed because of the symmetry. The finite element mesh close to the trench is given in Fig. 5. Both the upper and lower PVC pipes have a nominal diameter of 200 mm with a thickness of 12.5 mm. The trench is 2.8 m deep and 0.8 m wide. The vertical boundary at the right hand side is located 7 m away from the centerline, and zero horizontal heat flux was assumed. The bottom horizontal boundary also adiabatic was set at 15 m below the surface. The measured temperature at the bottom of the asphalt was used as the input for the top boundary condition. The lower pipe (renewal pipe filled with stagnant air) boundary was assumed to be adiabatic. When the bypass line was active, the measured temperatures of thermocouples 4, 5, and 6 were used for the upper pipe boundary condition. In the case of an inactive bypass, the upper pipe boundary was assumed to be adiabatic.

The initial deep ground temperature was taken as 11 ^{0}C as a rough approximation to the field measurements. The initial ground temperature close to the surface was approximated by running AFEMS for a certain period of time with a linear variation of surface temperature from the initial deep ground temperature to the initial surface temperature, because of the inability of AFEMS to input the initial temperature as a function of depth.

The thermal properties used in the finite element analyses are summarized in Table 1. The volumetric specific heat and latent heat were estimated from the field water content measurements. The unfrozen water content was taken as zero for all the backfill materials except clay, where the unfrozen water content was taken as 40% based on the field measurement on the same type of clay (Rajani and Zhan, 1994). The values for the volumetric specific heat of LWA and CLSM are taken from these used by Goodrich and Sepehr (1993). The thermal conductivities given in the Table 1 were obtained from the field measurements (Rajani et al., 1995). The volumetric moisture content for the native undisturbed clay was estimated as 35%.

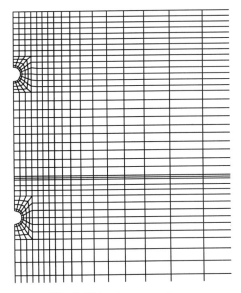

Fig. 5. Finite element mesh.

Table 1 Thermal properties for backfill materials.

Material type	Thermal conductivity W/mK		Estimated volumetric specific heat MJ/m³K		Latent heat, L MJ/m³	Ratio of k_f/L
	Frozen k_f	Unfrozen k_u	Frozen c_f	Unfrozen c_u		
LWA	0.13	0.18	1.413	1.413	4.0	0.03
Clay	2.1	1.52	1.83	2.17	54.1	0.04
Sand	3.02	2.55	1.74	2.16	66.8	0.045
CLSM	3.56	2.52	2.07	2.64	66.8	0.053
Bottom Ash	0.4	0.39	1.74	2.16	66.8	0.006
Undisturbed Clay	2.1	1.52	2.07	2.51	70.14	0.03
PVC	0.26	0.26	1.47	1.47		
Extruded Polystyrene	0.035	0.035	0.075	0.075		

Thermal analyses for Section C backfilled with sand

The first thermal analysis was performed for the section C, backfilled with sand, in order to estimate how the finite element results compared with the field measurements using the field measured thermal properties. Figure 6 presents the comparisons of predicted and measured temperatures along the centerline of the trench. In general, the finite element results predict lower temperatures compared with field measurements. However, the differences between the predicted and measured values are insignificant.

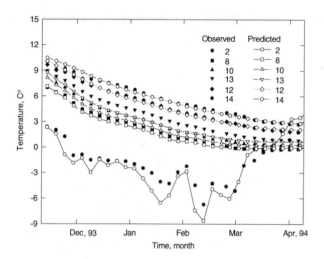

Fig. 6. Comparisons of predicted and observed temperatures along centerline of sand backfill (section C).

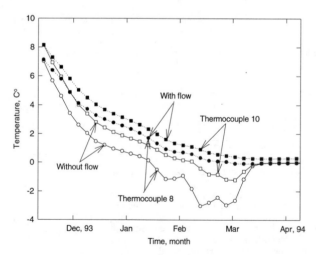

Fig. 7. Comparisons of predicted temperatures in section C at thermocouples 8 and 10 with and without water flow in bypass.

A thermal analysis was also carried out for section C, backfilled with sand, assuming an empty bypass line, in order to assess the effect of water flow in the bypass line on the thermal regime in the trench. The results of this analysis were compared with the results from the former analysis. Figure 7 presents the comparisons of temperatures at thermocouples 8 and 10 below the bypass water main with and without water flow in the bypass. The presence of water flow in the bypass line has a significant impact on the thermal regime under the bypass line, as indicated by the temperature differences at thermocouples 8 and 10 between the cases with and without water flow. The temperature at thermocouple 8 with water flow is about 2 to 4 degrees higher than that with an empty bypass. The lack of fluctuations of temperatures at thermocouples 8 and 10 is another indication that the thermal regime under the bypass is dominated by the water temperature as a result of water flow in the bypass.

The predicted variation of frost penetration depth with time along the centerline is given in Fig. 8 for the cases with and without water flow in the bypass line. The maximum frost penetration depth with an empty bypass is about 1.5 times that with water flow in the bypass. This again indicates the significant impact of the water flow in the bypass on the thermal regime in the trench.

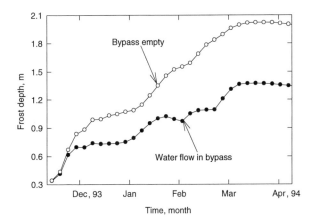

Fig. 8. Comparisons of predicted frost depths along centerline of section C with and without water flow in bypass.

Comparisons of frost protection effectiveness of different backfill materials

As indicated from both the field measurements and finite element analyses, the presence of water flow in the bypass water mains greatly influences the thermal regime in the trench. In order to provide an unambiguous appraisal of the thermal performance of the different backfills, five thermal analyses were carried out for trenches backfilled with LWA, clay, sand, CLSM, and bottom ash, respectively. The initial and boundary conditions were the same for all the five cases, and the bypass was assumed empty. Figure 9 presents the comparisons of frost penetration depth as a function of time along the centerline for the five cases. The superior performance of bottom ash and LWA in terms of frost protection is shown in Fig. 9. The maximum frost penetration depth along the trench centerline is 2.03 m for CLSM, 2.02 m for sand, 1.911 m for clay, 1.632 m for LWA, and 1.29 m for bottom ash. Consistent with the thermal properties measured in the field, clay is better than both sand and CLSM in terms of frost protection of buried utilities. Among all the backfill materials used in the field study, CLSM is the least desirable in terms of its thermal performance.

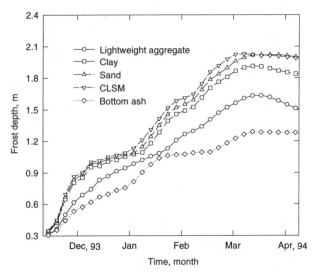

Fig. 9. Comparisons of predicted frost penetration depth along trench centerline for five different backfill materials with empty bypass.

Conclusions

It is disappointing that the backfill temperature data were so dominated by heat flow from the shallow bypass line as to seriously obscure the relative effects of the different backfills themselves. In spite of these difficulties, some patterns do emerge. The ranking of different backfills regarding their frost protection effectiveness is bottom ash, thermocrete, LWA, sand, CLSM, and clay based on temperature as well as frost penetration depth along the centerline of the trench. But the upstream sections received substantially greater heat than available further downstream and this ranking does not unambiguously represent the relative frost protection of the different materials. A more representative ranking can be given based on the measured in-situ frozen thermal conductivity values. In increasing order these were LWA, bottom ash, thermocrete, clay, sand, and CLSM. In spite of its lower thermal conductivity, however, frost penetration within the very dry LWA backfill may be more rapid than within the moist bottom ash backfill, owing to the absence of latent heat released during freezing. Additional data with no flow conditions are being collected for the winter of 1994-1995, in order to provide a clear picture of the thermal performance of the different backfills.

The finite element analysis does produce good predictions of thermal regime in the trench. The evaluation of thermal performance of different backfills using finite element analyses indicates that both bottom ash and LWA are superior in terms of their frost protection effectiveness for buried utilities, while CLSM is the least desirable among the different backfills. Further studies are needed to determine the optimum geometric configurations to fully exploit the advantages of each material combination.

Acknowledgments

The authors extend their appreciation to Mr. John Ward, Director of Engineering Services at The City of Edmonton for his valuable assistance in making this project a success. The authors are grateful to all the technical staff at Infrastructure for their efforts in instrumentation and data acquisition.

References

AFEMS, 1990. "Engineering analysis system." FEM Engineering Corporation, Inglewood, CA.

Dilger, W.H., Goodrich, L.E., Pildysh, M., and Humber, C.A., 1994. "Insulation of buried water lines in cold regions." Proceeding of 7th International Cold Regions Engineering Specialty Conference, pp. 481-497.

Goodrich, L.E. and Sepehr, K. 1993. "Frost protection of buried water mains: Results from a finite element model study." Proceedings of 1993 CSCE Annual Conference.

Rajani, B.B., Goodrich, L.E., Cooke, B., 1995. "Thermal performance of trench backfills and mechanical performance of buried PVC water mains." Report prepared for the City of Edmonton, National Research Council of Canada, Report No. 7005.3, 74p.

Rajani, B.B. and Zhan, C., 1994. "Thermal regime in the vicinity of typical concrete sidewalks." Report prepared for the Cities of Calgary, Camrose, Edmonton, Regina, Saskatoon, and Winnipeg, National Research Council of Canada, Report No. 7013.2, 55p.

Rajani, B.B. and Kuraoka, S., 1995. "Field performance of PVC water mains buried in different backfills." Proceedings of the 2nd International Conference on Advances in Underground Pipeline Engineering, Seattle, U.S.A.

TACOMA'S SECOND SUPPLY PROJECT CHALLENGES AND SOLUTIONS

Roger Beieler[1], Craig Gibson[2], and Tim Larson[3]

Abstract

The Second Supply Project Section of Tacoma Public Utilities (TPU) and a consultant team are currently designing a 33.5 mile long transmission pipeline between the headworks facility on the Green River and the primary service area within the City of Tacoma. The pipeline would also provide service to several water purveyor partners in south King County and possibly to the City of Seattle. The paper discusses the challenges associated with planning, permitting, and designing a large diameter transmission pipeline across sensitive areas including wetlands, rivers, streams, steep slopes, railroads, and highways. The paper includes a discussion of the route selection process, hydraulic analysis, design criteria, computer aided design/drafting techniques, integration of the permitting and design processes, corrosion concerns, and unique construction methods.

Tacoma's Existing Supply System

The Water Division of Tacoma Public Utilities (TPU) provides water service to an area of approximately 150 square miles in both Pierce and south King Counties. The framework of the system as it exists today was essentially developed in 1913 with delivery of water from the Green River. As shown in Figure 1, the Green River originates in the upper Cascade Mountains. The average annual precipitation rate of the Green River watershed is approximately 85 inches. About 75 percent of the flow from the watershed originates above TPU's intake.

---

[1] Design Engineer, CH2M HILL - Bellevue, WA
[2] Project Manager, Tacoma Public Utilities - Tacoma, WA
[3] Project Engineer, Tacoma Public Utilities - Tacoma, WA

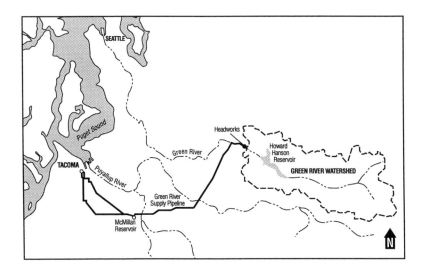

Figure 1. Tacoma's Supply System and Service Area

The watershed and tributary areas above the diversion point consist of about 231 square miles of heavily timbered and mountainous terrain. The source was first developed in 1913 by construction of a diversion dam and gravity transmission pipeline (Pipeline No. 1) which conveyed water into the City. In 1961, the Army Corps of Engineers constructed the Howard A. Hanson Dam three miles above the intake. The dam had two purposes -- to provide flood control to the lower reaches of the Green River and to augment low flows in the river system during the critical late summer and early fall months. The dam has provided significant benefits to the river system by providing storage adequate to maintain 110 cfs in the river during the low flow periods. Water is conveyed from the headworks to McMillin Reservoir, a distance of about 27 miles, by Pipeline No. 1. Two pipelines (No. 2 and No. 4) convey the water from McMillin Reservoir to the primary service area. The primary distribution system is divided into four major pressure zones and includes six major reservoirs.

Need for the Second Supply Project

According to Tacoma's 1987 Water Supply Plan, TPU projected it would need to increase its supply to meet the projected demand in the Tacoma-Pierce County service area in 2007. Similarly, demands exceeding supply in the south King County area were forecast by the year 2010. In 1986, Tacoma was granted the water rights for a second diversion of 65 mgd (100 cfs) from the Green River. The Second Supply Project, consisting of improvements to the headworks and construction of a second supply pipeline (Pipeline No. 5), has been evolving for over 20 years. The flow conveyed by the Second Supply Pipeline (SSP) would meet much of Tacoma's long-term water demands, provide system reliability and flexibility, and form the basis of a regional water system. In addition to Tacoma, the Second Supply Project would provide water to the South King County Regional Water Association, the Lakehaven Utility District, and possibly the City of Seattle via an intertie pipeline between the two supply systems.

Project Challenges

Selection of a route which met cost, environmental, and technical requirements was one of the first challenges. The SSP will traverse 33.5 miles of highly varied terrain posing significant design challenges ranging from steep slopes and multiple wetland passages to river crossings and complicated urban settings. Negotiations and coordination with the Muckleshoot Indian Tribe presented an additional challenge. Complex multi-jurisdictional permitting issues will need to be addressed as preliminary and final design of the project proceed. Successful and cost-effective design will require a carefully planned process which integrates numerous permit agency requirements and deliverables, resolves technical design issues, includes innovative technologies, considers effective construction packaging, and addresses schedule and cost concerns of the water purveyor partners served by the project.

Project Solutions

The progress made prior to 1995 and the anticipated schedule for design and construction of the SSP are shown in Figure 2. Planning and conceptual design of the project began in 1963. Right-of-way for the pipeline was acquired between 1969 and 1975. As of early 1995, predesign activities are complete. Permitting and final design are underway. Construction is anticipated to begin in early 1996.

TACOMA'S SECOND SUPPLY PROJECT 665

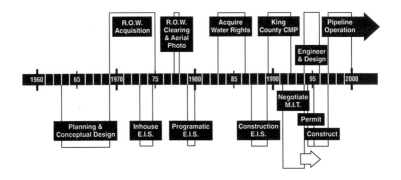

Figure 2. Proposed Schedule

Solutions to many of the challenges have been developed, while work on others is ongoing. Several topics relating to solutions are discussed below, including route selection, the integration of the permitting and design processes, pipe sizing and hydraulic analysis, computer aided design and drafting techniques, and innovative designs for special crossings.

Route Selection

From 1965 to 1970, TPU considered a number of routes for the SSP including routes parallel to Pipeline No. 1. Many of the routes were eliminated due to topographic, property acquisition, and hydraulic concerns. Routes parallel to existing Pipeline No. 1 were eliminated because of concerns regarding simultaneous failure in case of an earthquake or major leak in one of the lines. In addition, it did not meet the project objectives to serve south King County and Seattle. As shown in Figure 3, the current route for the SSP begins at the headworks and terminates at a connection to Pipeline No. 4 near the Portland Avenue Reservoir, a length of approximately 33.5 miles. Approximately 71 percent of the SSP route follows previously encumbered rights-of-way for railroads, roads, pipelines, and BPA power lines. Connecting the SSP to Pipeline No. 4 will allow for backfeeding project partners from Pipeline No. 4 during: 1) periods of extreme low flow in the Green River, 2) when the upper section of Pipeline No. 5 is out of service, and 3) when sufficient quantities of blending water are not available from the North Fork Wells in the controlled Green River Watershed during periods of high turbidity in the Green River.

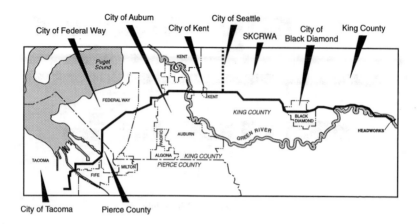

Figure 3. Proposed Route for Second Supply Pipeline

Integration of the Permitting and Design Processes

One of the solutions to meeting the challenges associated with integration of the permitting and design processes was timely development of design criteria with input from the permitting and mitigation teams. The function of the preliminary and final design process is to develop an approach that considers regulatory restrictions, alternative concepts, expectations, and technical requirements. The design must optimize construction scheduling and provide flexibility to address changes and unexpected conditions. The product of the permitting/design process will be a contract document that can be successfully executed by one or more contractors.

The permitting/design team, consisting of TPU staff and several consulting firms, identified the preliminary tasks necessary to begin the permitting/design process. Technical memorandums were prepared on many of the topics to facilitate development of recommendations and provide a record of the decisions. The technical memorandums were incorporated into a project manual to provide a single reference source for use by the design team. The tasks included the following:

- Assisting with negotiations and preparation of agreements
- Assisting with communication issues including the news media, resource agencies, property owners, and the general public
- Developing flow delivery and decision criteria
- Permit coordination, tracking, and reporting
- Developing design criteria and standards
  * Pipe structural design methods
  * Pipe lining, coating, joints, and thrust restraint
  * Line valve location and operation
  * Hydraulic analysis and operational protocols
  * Surge analysis and control techniques
  * Metering and telemetry requirements
  * Corrosion protection
  * Disinfection and testing procedures
- Evaluating alternative construction strategies including prepurchase
- Identifying and evaluating risks including seismic events
- Preparing an erosion and sedimentation control plan
- Developing a restoration plan and cultural resource plan
- Providing assistance with easement and property acquisition
- Identifying off-site wetland mitigation opportunities
- Developing computer-aided design and drafting techniques to effectively integrate permitting and design requirements

Pipe Sizing and Hydraulic Analysis

A hydraulic model of the existing transmission system and the SSP was developed. The design gravity flow to be conveyed by the SSP is 76 mgd. Sizing the pipeline for this flow will allow TPU's first diversion water right to be conveyed from the headworks to the distribution system in case Pipeline No. 1 is out of service. In order to maintain the hydraulic grade line above a high elevation area near mile 7 and provide an allowance for a future filtration plant, the diameter of the pipeline must be 72 inches for the first 7 miles, 60 inches for the next 20 miles, and 48 inches for the last 6 miles. In general, maximum design velocities will be in the 6 to 9 feet per second range.

The Second Supply Pipeline will be a welded steel pipeline. Because the relative roughness of a pipeline varies with diameter, the Darcy-Weisbach equation was used to more accurately compute head loss due to friction. The headloss over the length of the pipeline using the Darcy-Weisbach equation was equivalent to using a Hazen Williams "C" value of 143. Because the "C" value cannot be accurately determined during design, a range of "C" values was used with the hydraulic model to determine how sensitive the flow rate was to the pipe roughness. A "C" value of 138 resulted in a computed flow capacity of 74 mgd, a "C" of 143 resulted in 76 mgd, and a "C" of 148 resulted in 79 mgd.

Computer Aided Design and Drafting Techniques

To the extent possible, design of the pipeline will be accomplished using computer aided design and drafting (CADD) techniques. Mapping and visualization products developed for design purposes will also be used for permitting and project presentations. The three-dimensional topographic model will be used as a tool to compute quantities, determine the extent of cut and fill slopes, determine easement requirements, and prepare profiles.

Aerial photographs were converted to electronic raster files and combined with the planimetric information on the plan-profile drawings. The combination of photographic and planimetric data proved to be very useful as detailed routing choices were being made, both on-site and in the office. Graphic simulations based on photographs and electronic images will be used in project presentations. The simulation shown in Figure 4 below was used to illustrate the impact of trenching and pipe installation in an urban setting.

Figure 4. Graphic Simulation

Innovative Designs for Special Conditions

A profile showing the topography along the SSP route and the hydraulic grade lines for various conditions is included as Figure 5. The hydraulic grade line elevation at the headworks is 894 feet. The hydraulic grade at Pipeline No. 4, the terminus of the SSP, will vary depending on system demands and whether the pumps at the City's wells are operating. The hydraulic grade line will be less than 576 feet, the spill elevation of McMillin Reservoir, which supplies Pipeline No. 4. The pipeline will cross two major valleys with floor elevations near sea level, the Green River Valley near mile 20 and the Puyallup River Valley near mile 30. Static pressures in the pipeline at the valley crossings will be in the range of 360 psi. Design challenges associated with three areas of the alignment are described in this section.

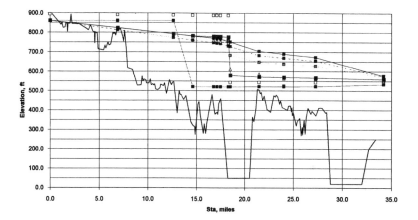

Figure 5. Pipeline Profile

Puyallup River Crossing. Three alternatives were considered for the Puyallup River crossing: a bridge supported pipeline, directional drilling, and microtunneling. A total of 20 sub-alternatives were developed and analyzed using a matrix evaluation technique. A subjective value between 1 and 5 was assigned to each of the categories of the matrix. The categories included cost, constructability, permitting, property acquisition, traffic disruption, business disruption, and maintenance/reliability issues. The alternative selected was microtunneling. The proposed tunnel profile is shown in Figure 6.

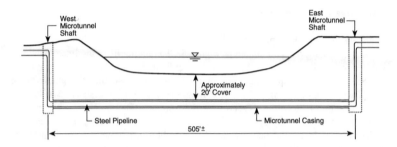

Figure 6. Proposed Profile - Puyallup River Crossing

Upper Green River Crossing. The proposed crossing is located at the site of an abandoned railroad bridge where the river is about 160 feet wide. The bridge and tracks were removed several years ago. The concrete piers and rubble abutments which were used to support the bridge are in relatively good condition. The pipe crossing will use these piers to span the river. As shown in Figure 7, the total length of the crossing will be about 300 feet.

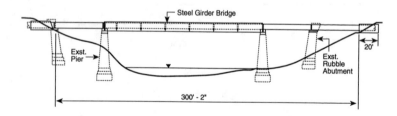

Figure 7. Upper Green River Crossing

The pipeline between the piers would span the main channel of the river and be supported by a steel plate girder bridge. The bridge would consist of two longitudinal girders, diagonal cross bracing, and transverse floor beams. The pipeline would be located between the girders, and ring girders would be used to connect the pipeline to the floor beams. Sleeve couplings would be provided on the upstream ends of each span to allow movement associated with thermal expansion and contraction. A cross section of the proposed bridge is included as Figure 8.

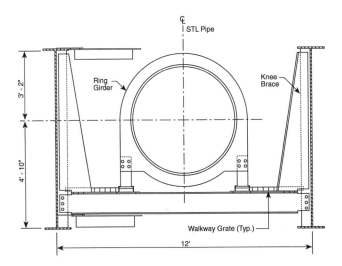

Figure 8. Bridge Cross Section

Paralleling High Voltage Transmission Lines. Approximately seven miles of Pipeline No. 5 will be constructed along Bonneville Power Administration's (BPA) high voltage transmission lines. Induced voltage associated with the lines is a safety concern. Static charges can affect personnel who come into contact with the pipeline or appurtenances. Dangerous voltages can be caused by both electromagnetic or electrostatic induction. Electromagnetic induction primarily affects buried pipe and is a concern for operator safety. Electrostatic induction occurs primarily in aboveground or exposed pipe and is, therefore, a concern for construction workers. In addition, if a transmission tower were to fail during a seismic event and a conductor came in contact with either the ground or the metallic tower frame, the steel pipeline would likely become charged for a brief period of time.

Several homes have recently been constructed adjacent to the high voltage transmission lines. Some of the yards and improvements extend into the BPA right-of-way and into the area reserved for construction of the SSP. The homeowners have encouraged TPU to locate the pipeline closer to the transmission towers than originally planned. However, the potential dangers and concerns associated with paralleling high voltage power lines becomes more severe as the distance between the pipeline and towers is decreased.

Mitigative measures will be required to reduce the risk of shock. The most effective measures involve grounding the pipeline. Grounding can be accomplished by deep vertical electrodes, vertical electrode beds, or horizontal wires placed just below the ground surface. Based upon an evaluation of the options, it appears that locating a zinc or magnesium ribbon anode parallel to the pipeline is the best solution. Counterpoise wires will also be installed at several transmission towers to reduce the risk of inducing high voltage on the pipeline during power line fault conditions. The wires will be bare aluminum and extend approximately 250 to 500 feet parallel to the pipeline depending on the soil resistivity. The longer wires are necessary in locations with high resistivity soils. At the appurtenances, grounding mats will be installed to raise the potential of the earth around the mat to the potential of the pipeline, thereby reducing the risk of shock.

Conclusion

The Second Supply Project was proposed in 1963 as a two phase, eleven year project scheduled to begin in 1970. Numerous challenges related to technical, environmental, political, and financial aspects of the project have arisen and solutions have been developed. The regional partnerships developed in recent years have expanded the viability of the project. Additional technical challenges lie ahead as final design of the project proceeds. Completion of Tacoma's Second Supply Project will be a major accomplishment for the region's future water supply needs.

Sewer Trench Subsidence Due to Severe Flooding

Mohammed S. Islam,[1] Assoc. Member, Quazi S. E. Hashmi,[2] and Steven C. Helfrich,[3] Members ASCE

Abstract

This paper presents results of an investigation performed to evaluate the possible cause(s) of severe ground distress along a completed sewer pipeline trench. Ground along portions of the trench caved in to depths ranging from about 6.0 inches (15.2 cm) to about 42.0 inches (106.7 cm). The investigation proved that a number of factors, including poorly compacted trench backfill, use of pipe bedding material which did not meet generally accepted filter criteria, and above all, severe flooding, contributed to the observed ground failure. The main factor responsible for the subsidence proved to be the migration of fines from the base soils into the coarse aggregate used as pipe bedding.

Introduction

Ground surface along about 3,700 linear feet (1,128 m) of a 16,000-linear-foot (4,878 m) pipe trench subsided within a year after construction. Depth of the compacted trench backfill ranged from 15 to 40 feet (4.6 to 12.2 m). Trench failure, predominant around the manholes, was observed after heavy rainfall flooded the project area. Such subsidence is generally attributed to poor quality and inadequate compaction of the trench backfill soils. In this case, however, failure occurred after severe flooding. A post-failure investigation to identify

---

[1]Senior Staff Engineer, Converse Consultants Inland Empire, 10391 Corporate Drive, Redlands, CA 92374

[2]Principal Engineer/Branch Manager, Converse Consultants Inland Empire, Redlands, CA

[3]Principal Engineer/Senior Vice President, CCIE, Redlands, CA

reasons for the trench backfill failure comprised review of existing information, field exploration and laboratory testing.

Project Background

The 36-inch-diameter (91.4 cm) pipeline was constructed from May to October of 1992. The trench backfill was placed under the full-time observation and testing of a soils consultant. In order to meet project specifications, a minimum of 90 percent relative compaction as per ASTM Standard D1557-91 was required for the backfill. Required minimum compaction for the upper 12 inches (30.5 cm) of backfill in paved areas was 95 percent. After construction, the soils consultant certified the trench backfill as being properly compacted.

From late December of 1992 to March of 1993, heavy rainfall inundated the area, remaining for over two weeks. Following, two approximately 2,800- (654 m) and 900-foot-long (275 m) isolated segments of the sewer trench subsided. Ground loss and the resulting gradual cave-in occurred primarily around and/or adjacent to the manholes. The observed cave-in depths ranged from 6.0 inches (15.2 cm) to about 42.0 inches (106.7 cm) in depth. Exhibits 1 and 2 show the damage.

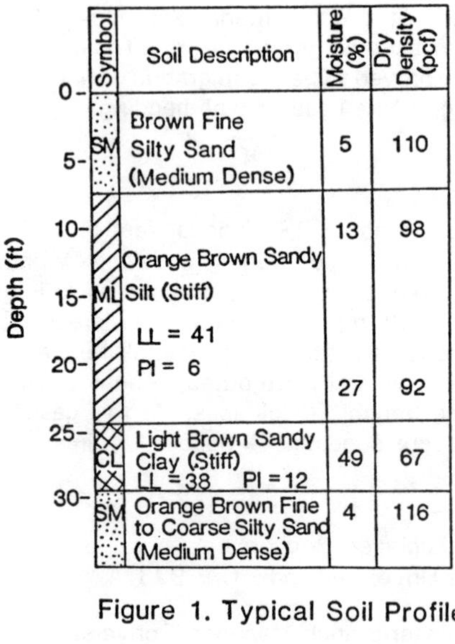

Depth (ft)	Symbol	Soil Description	Moisture (%)	Dry Density (pcf)
0–5	SM	Brown Fine Silty Sand (Medium Dense)	5	110
10	ML	Orange Brown Sandy Silt (Stiff) LL = 41 PI = 6	13	98
15–20	ML			
20	ML		27	92
25	CL	Light Brown Sandy Clay (Stiff) LL = 38 PI = 12	49	67
30	SM	Orange Brown Fine to Coarse Silty Sand (Medium Dense)	4	116

Figure 1. Typical Soil Profile

Review of Existing Information

Prior to construction, the soils consultant conducted a preliminary geotechnical investigation report for the pipeline project. The subject alignment is situated within a valley filled with Pleistocene-age alluvium. Exploratory borings show soils along the alignment to vary significantly, from fine-grained silty clay (CL) to fine to coarse-grained sand (SP). A typical soil profile is shown in Figure 1. Grain-size distributions of the native soils fall within the range shown in Figure 2. The finer fractions of the site soils have liquid limits ranging from 38 to 41 and plastic limits ranging from 6 to 12. The maximum dry density and the optimum moisture content ranged from 102 to 139 pcf (16.3 to 22.2 kN/m^3), and 7.0 to 23.0 percent, respectively. Based on consolidation tests performed on undisturbed 2.42-inch-diameter (6.15 cm), 1.0-inch-thick (2.54 cm) ring samples, the coefficient of compressibility for the site soil at *in situ* densities ranged from 0.025 to 0.25. Collapse tests indicated that native site soils are slightly susceptible to hydroconsolidation (< 2 percent) at 1.0 to 2.0 psf (0.05 to 0.10 kPa) overburden pressures.

Depth to groundwater in 1992 was about 65 feet (19.8 m) (elevation 416 m) below existing ground surface. National Flood Insurance Rate Maps (FIRM) show that about 90 percent of the alignment was situated within Flood Hazard Zone A, depicted as "areas of 100-year flood; base flood elevation and flood hazard factor not determined."

Exhibit 1. Ground Subsidence Along Sewer Trench

Exhibit 2. Ground Subsidence Around Manhole (Typical)

The soils consultant's daily reports on construction observation and the final compaction report were reviewed. Records indicate no established guidelines regarding the frequency of field density tests. Construction was initiated without any approved shoring plan. This prevented the field technician from performing a sufficient number of tests at deeper depths for some portions of the trench. At the demand of a representative of the county's Public Works Department, field density tests were performed at 2.0-foot (61 cm) depth intervals around the manholes. As a result, the test locations around the manholes were concentrated. To keep pace with construction, few tests and limited construction observation were performed for the portion of the trench between the manholes.

Construction records indicate instances when the geotechnical consultant's field representative suspected that the contractor had placed uncompacted fills in the trench in his absence. However, no corrective actions were taken. Based on the project specifications for the pipeline backfill, uncompacted lift thicknesses of up to 5.0 feet (1.52 m) were allowed for this project. Typically, however, fills were placed in 2.0 to 3.0-foot- (60 cm to 90 m) thick lifts and compacted with sheepsfoot rollers. Field density tests were usually performed within the upper 12 inches (30.5 cm) of a given lift.

Contract documents for the project specified that the pipe bedding zone,

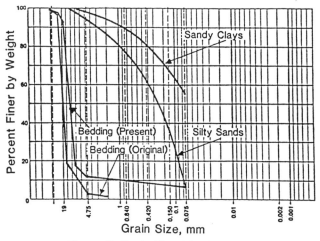

Figure 2. Grain Size Distribution

from about 6.0 inches (15.2 cm) below the pipe invert to about 12 inches (30.5 cm) above the pipe crown, be backfilled with coarse aggregate. Pipe bedding was not required to be either compacted or tested for density. Based on the compaction report, in order to expedite trench backfilling, the contractor placed coarse aggregate in excess of that required by the project specifications. It is estimated that, on average, bulk volume of the coarse aggregate placed per linear foot (0.3 m) of the trench was about 35 cubic feet (1.0 m^3). Additional uncompacted coarse aggregate and imported clean sands were placed around the manholes. The gradation of the coarse aggregate used as pipe bedding is shown in Figure 2.

Field Exploration

Preliminary field reconnaissance was conducted to map the horizontal and vertical extent of the distress. Based on the observed distress, 4 exploratory trenches were excavated across the distressed trench into the bedding materials. *In situ* density tests of the backfill were performed using the Sand Cone Method (ASTM Standard D1556-86). Bulk samples from both the trench zone and pipe zone material were collected for laboratory testing. Thirty-seven cone penetration tests (CPT) for sounding were performed at intervals of 200 feet (61 m) along the alignment, including distressed as well as non-distressed areas, to determine the present conditions of the trench backfill and in particular, the existence, if any, of loose unstable zones or cavities underground.

Laboratory Testing

Compaction tests were performed to determine maximum dry density and optimum water content according to ASTM Standard D1557-91. Consolidation tests were performed on remolded soils to determine compression characteristics of the trench soils, and laboratory tests included three pinhole tests (ASTM Standard D 4647-87) to determine dispersive characteristics of site soils. Specific gravity and unit weight tests were performed on pipe bedding material to determine its in-place volume and density. To aid in understanding the subsidence mechanism, a laboratory model was fabricated to simulate the field conditions. Details of the model test and test results follow.

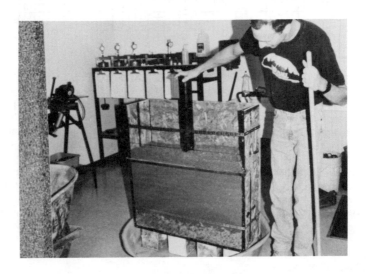

Exhibit 3. Laboratory Model

The model, shown as Exhibit 3, was designed at a scale of about 1 inch (2.54 cm) equals 1 foot (30.5 cm) (in the field). It is a 3'x3'x1'(91.4cmx91.4cmx25.4cm) plexiglass box with a 2.0-inch-diameter (5.1 cm) plastic pipe placed vertically at the center of the box to simulate a manhole. The bottom 6 inches (15.2 cm) of the box was filled with coarser pipe zone materials of the type used in the field. Next, the box was filled with 2.5 feet (76.2 cm) of trench zone soil compacted to 90 percent of the laboratory maximum dry density (ASTM Standard D1557-91). Clear water was then carefully poured into the box to avoid any surface scour and was allowed to rise to about 6.0 inches (15.2 cm) above the surface of the soil. As water seeped into the soil, the rate of water supply at the surface was regulated to maintain a constant level of about 6.0 inches (15.2 cm) above soil surface. Water seeping through the soil layer into the bedding materials was allowed to drain out around the bottom joints of the plexiglass box and collected in a large container. This water flow condition was maintained for 12 days.

Exhibit 4. Ground Subsidence Around Model Manhole

During model testing, the quantity and color of the seeping water and the condition of the backfill soils were observed, and photographs of the model were taken. Progressive migration of fine particles from the backfill soil to the pipe zone material was indicated as the seeping water became increasingly darker in color over time. Some fine particles could also be seen to have filled the voids within the pipe zone materials and deposited at the bottom of the collecting container. By the end of the 10th day, a small piping failure zone was observed. This zone expanded quickly, and after the 12th day, a large caved-in area had developed around the model manhole, as shown in Exhibit 4.

Analysis and Interpretation

The gradations in Figure 2 show that the trench materials contain about 10 to 55 percent fines (passing U.S. Sieve No. 200). The backfill zone materials in distressed areas comprised mainly silty sands (SM) and sandy silts (ML). Laboratory tests on soil samples remolded at 80, 85, and 90 percent relative compaction indicated that the compression index of trench backfill varied from 0.03 to 0.1.

For the type of site soils in question, uncompacted lift thickness should have been less than 8.0 inches (20.3 cm). As mentioned previously, trench fills were placed in 2.0 to 3.0-foot thick lifts. Leflaive (1980), cited by Hausmann (1990), reports that field density within a 16-inch-thick (40.6 cm) layer can vary from 5 to 10 percent. The maximum possible trench settlement may be estimated with this variance in mind. This study assumes that the trench could have been backfilled with soils compacted to relative compaction ranging from 80 to 90 percent.

Based on a compression curve for the sample remolded at 80 percent relative compaction, the maximum subsidence that can occur due to compression under self weight, assuming an average depth of backfill of 24 feet (7.3 m), is about 6.5 inches (16.5 cm). According to the NAVFAC (1986) design manual, a maximum settlement of about 0.2 percent of fill height can occur in fine-grained soils in 3 to 4 years due to secondary compression. Insignificant settlement from secondary compression of the subject trench surface should then be expected at the end of the time period under consideration. However, to estimate the maximum possible settlement, it was assumed that 0.2 percent, i.e., about 0.60 inches (1.5 cm) settlement, had occurred in this case. Consolidation results indicated that very minor settlement (on the order of about 0.15 percent) may have occurred due to wetting, if the backfill had been compacted to about 80 percent relative compaction. Therefore, a maximum of 0.40 inches (1.0 cm) of settlement can be attributed to wetting.

Thus, maximum settlement of about 7.5 inches (19.0 cm) of trench surface could have been expected due to placement of the trench backfill at about 80 percent relative compaction. Based on similar analysis, settlement of about 2.6 inches (6.6 cm) should have been expected if the entire backfill was compacted to 90 percent relative compaction.

The above analysis indicates that the observed settlement cannot occur solely as a result of poorly compacted trench backfill. Reason(s) other than loosely compacted trench backfill must have caused the observed ground loss, especially near the manholes.

Another possible reason for the ground loss is the migration of fine particles from the trench zone to the pipe zone due to seepage and/or piping around manholes. Water seepage through a finer material to a coarser material, such as that which occurs through earth dams (Vaughan and Soars, 1982), can result in loss of fines from the finer material if the voids within the coarser material are too large, compared to the size of the finer material. The rate of migration of fines increases with time, resulting in significant ground loss, subsidence, and subsequent failure of the structure. The proper design of such fine-coarse material interface when seepage is likely requires that certain filter criteria be satisfied to prevent migration of finer particles.

Sherard et al (1984) suggested that to prevent migration of fines from base soils into the sand and/or gravel filters with D_{15} larger than 1.0 mm, the ratio $D_{15}/d_{85} \leq 5$ is the main criteria that need to be satisfied. Here, D_{15} is the particle size in filter for which 15 percent by weight of particles are smaller, and d_{85} is the particle size in base soil for which 85 percent by weight of particles are smaller. Based on the range of gradation of the site soils and the pipe bedding shown in Figure 2, this ratio ranges from 6 to 15. Clearly, the main filter criteria is not satisfied and the possibility of migration of fines from the surrounding soils into the pipe bedding exists.

Following the heavy rain, the ponded water was observed to be seeping into the ground around the manholes. This indicates that the area around the manholes provided the least amount of resistance to the downward flow of water. The cone-shaped sands deposited around the manholes provided a shorter path for the seeping water to reach the pipe zone bedding materials. Upon flooding, water from the surface began to flow through around the manholes at a relatively high velocity (under a head of about 7.6 m), causing erosion. Pinhole tests indicated that the site soils are moderately to highly dispersive, especially at low densities. Thus, the potential of resistance to piping was also low around the manholes. Acting as a conduit, the pipe bedding zone carried these finer particles away from the manholes, providing enough space for the observed ground loss to occur around the manholes.

As observed in the field, initial ground subsidence occurred around the manholes, followed by gradual ground subsidence along the rest of the trench. Ground subsidence occurred predominantly along the pipe center line. Laterally, subsidence was limited to the zone immediately above the pipe bedding. This observed pattern of subsidence also shows that migration of fines was the main reason for the ground loss. Had the subsidence been due to compression alone, distress would have been more randomly distributed and would have occurred simultaneously within the manhole and throughout the trench.

The results of the field density tests performed within the exploration trenches indicated that the relative compaction at test locations varied from 75 percent to 90 percent. Typical results are included as Figure 3. As seen, the measured relative compaction was about 84 percent near surface, increased to 90 at a depth of 10 feet (3.05 m), and then decreased to 76 percent near the pipe top. This observation can be explained as follows. This exploration trench was located adjacent to cave-in areas. The cave-ins had resulted in loss of lateral confinement of the backfill soils at shallow depths. Also, adjacent to pipe bedding zones, migration of fines resulted in a loose unstable structure, rendering trench fill densities much lower than those which existed immediately after construction.

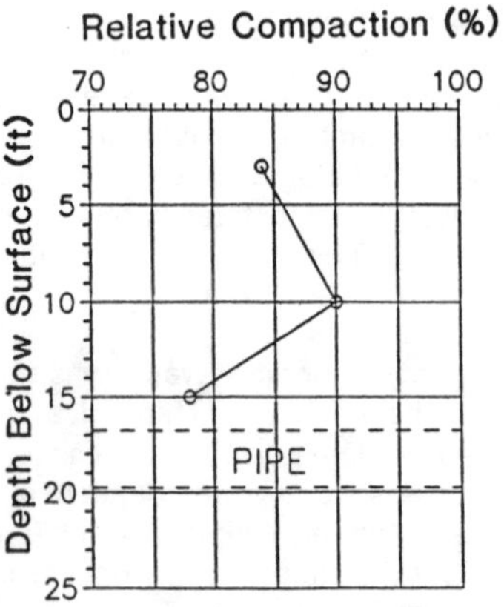

Figure 3. Relative Compaction vs. Depth

A typical gradation of a sample of the bedding materials retrieved from the distressed zones during the present investigation is included in Figure 2. It is seen that the pipe zone materials have been contaminated by about 10 percent of fines by weight. This sample was obtained from above the pipe. It is reasonable to assume that pipe bedding on the sides and underneath the pipe contains more fines.

This analysis confirms that migration of fines and/or piping resulted in the observed ground loss and subsequent trench surface subsidence.

If it is assumed that the fine content of the pipe zone materials is uniformly distributed throughout the entire trench alignment, it follows that about 350 pounds (158.8 kg) of fines per linear foot (0.305 m) of trench migrated into the pipe bedding. A 10-foot-wide (3.05 m) trench would settle about 5.0 inches (12.7 cm) uniformly due to the loss of this amount of fines. About 42 percent of the total volume of the pipe bedding is void (porosity), enough space to hold about 34 lbs. (15.4 kg) of fines per cubic foot (0.0283 m^3) of total volume. Thus, the pipe bedding has enough void to hold about 1,200 lbs. (544 kg) of fines per linear foot (0.305 m) of trench. Migration of this amount of fines would result in about 18 inches (45.7 cm) of settlement at the trench surface. Subsidence greater than 18 inches (45.7 cm) occurred, since, as explained above, fines from the bedding around the manholes were carried away by seeping water.

Based on this analysis, use of inappropriate pipe bedding material appeared to be the main cause of ground distress. Poor backfill compaction appears to have contributed to the observed failure.

Repair Recommendations

Field observation confirmed that toward the end of the summer of 1993, ground subsidence had ceased, despite the fact that some of the cave-in areas were still full of water. This indicated that no further migration was occurring. The fine content of the pipe bedding has now increased, as shown in Figure 2. As a result, the D_{15}/d_{85} of the pipe bedding has now decreased to 5 or less for site soils with $d_{85} \geq 1.0$ mm. Consequently, additional migration should not occur when site soils around the pipe bedding have $d_{85} \geq 1.0$ mm. Furthermore, for the finer soils, a natural filter could have been developed around the pipe bedding. Results of CPT tests confirmed that no cavities exist within the pipe trench backfill. Results of a typical CPT sounding test are shown in Figure 4.

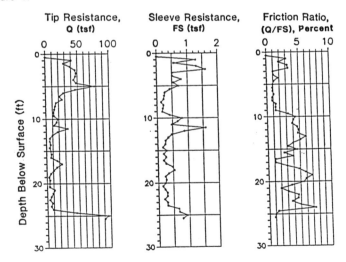

Figure 4. Typical CPT Test Results

The above observations lead to the conclusion that no further migration of fines is likely in the event of future inundation and that no cavities existed within the trench backfill. Thus, the recommendations to repair the distressed trench included removal and replacement with compacted fills of the upper 5.0 feet of the trench backfill soils. To reduce potential of any fine migration from the new compacted fill, a geotextile fabric with an apparent opening size of ≤0.212 mm was placed over the five-foot-deep excavation before backfilling. Approximately 1 year has passed since the completion of the repair work. No evidence of further ground instability can be found along the sewer trench.

Conclusions

The failure investigation has demonstrated that ground distress above pipeline trenches can occur due to a number of factors. In the case of the subject alignment, the contributing factors included the following: severe flooding, migration of fines from backfill soil to pipe bedding aggregate, and poorly compacted trench backfill. In combination, these factors resulted in the observed distress. Although some settlement could have occurred due to poor compaction, the major factor was found to be the use of inappropriate bedding aggregate which did not meet generally accepted filter criteria. Poor trench backfill compaction, however, worsened the situation by contributing to the process of fine migration. Of course, trench subsidence would have been limited to that which occurred because of poor compaction had there been no flooding.

Acknowledgments

Duane Dodson provided valuable editorial suggestions and assistance during the preparation of this manuscript. The work of Theresa Pacheco is also appreciated.

References

Annual Book of Annual Standards, Volume 04.08, Soil and Rock; Dimension Stones; Geosynthetics, 1993

Hausmann, M. R., *Engineering Principles of Ground Modifications*, McGraw- Hill, 1990.

Sherard J. L., Dunnigan L. P., and Talbot, J. R, "*Basic Properties of Sand and Gravel Filters*," Journal of the Geotechnical Engineering Division., ASCE, Vol. 110, No. GT6, June, 1984.

Naval Facilities Engineering Command (NAVFAC) Design Manual 7.02, *Foundations & Earth Structures*, September, 1986.4

Vaughan, P. R., and Soars, H. F., "Design of Filters for Clay Cores of Dams," Journal of the Geotechnical Engineering Division., ASCE, Vol. 108, No. GT1, January, 1982.

Repair Study for the Lafayette Aqueduct No.1

Christopher Dodge [1], A.M. ASCE, Christina Hartinger [2]

The East Bay Municipal Utility District (EBMUD), which provides drinking water to approximately 1.2 million people in San Francisco's East Bay, has experienced leakage and found cracks in its Lafayette Aqueduct No.1. The leakage problem in the aqueduct started almost immediately after the aqueduct was put into service in 1929. EBMUD staff have performed various interior and exterior repairs over the life of the aqueduct, but have not been able to permanently prevent the leakage.

In order to determine the best method to eliminate leakage, EBMUD initiated a repair study. This study presented many unique challenges, including compliance with stringent operational and water quality constraints. Additional challenges were posed by the size of the aqueduct and its location. In addition, high summer demands dictate that all repairs be conducted during a four-month period in the winter.

The goal of the repair study was to develop the most cost-effective, public-sensitive, and environmentally responsible long-term repair strategy. To accomplish this, the study team established a selection criteria process which evaluated and ranked repair alternatives based on economic factors such as capital and maintenance costs, and on non-economic factors such as water quality and construction impacts. These criteria were weighted according to their relative importance.

The study concluded that the most cost-effective, public-sensitive, and environmentally responsible long-term repair strategy consists of sliplining the existing pipe with a steel liner. EBMUD then developed a recommended capital improvement project based upon the study. The estimated project cost is $13.6 million.

This paper discusses the challenges faced by the study team and describes the comparison method used to determine the prevailing leakage repair alternative.

---

[1] Associate Civil Engineer, East Bay Municipal Utility District, 375 Eleventh Street, Oakland, California, 94607-4240, U.S.A.

[2] Engineer, John Carollo Engineers, 2700 Ygnacio Valley Road #300, Walnut Creek, California 94598, U.S.A.

BACKGROUND

Physical Characteristics and Type of Construction

Lafayette Aqueduct No. 1, typically 2.7 meter (9 foot) in diameter, was originally constructed in 1929 for the conveyance of raw water to treatment facilities in the San Francisco East Bay. The length of the aqueduct section for this study is approximately 5.8 kilometers (19,000 feet). The aqueduct runs within a 30.5 meter (100 foot) wide right-of-way.

This section of the aqueduct is mostly buried. It consists of cast-in-place concrete pipe installed by a cut-and-cover trench method and two tunnels: Walnut Creek Tunnel and Pleasant Hill Tunnel. A typical cross section of the cast-in-place portion of the aqueduct is presented as Figure 1.

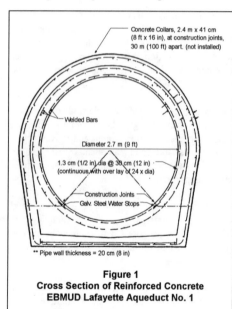

Figure 1
Cross Section of Reinforced Concrete EBMUD Lafayette Aqueduct No. 1

The original design of Lafayette Aqueduct No. 1 did not appear to have provisions for movement joints. EBMUD Operation and Maintenance (O&M) staff have indicated that there are not any visible joints inside the pipe. They believe concrete was placed continuously with 24-hour shifts during the original construction.

Due to the lack of movement joints, any concrete expansion or contraction in response to differential ground settlements, creep, shrinkage, or temperature variation could cause the pipe to crack. The pipe is typically buried by 2.7 meters (9 feet) of earth cover, measured to the springline of the pipe. The soil is thought to provide some insulation from atmospheric thermal variations. The O&M staff have observed that the leakage problems at the buried portions of the pipe are typically aggravated by water temperature drops. However, cracks have appeared on the partially buried portions of the pipe and it is believed that these cracks are more directly influenced by air temperature variations.

Aqueduct Operation

Under normal operating conditions, the maximum gravity flow is 3,900 liters per second (90 million gallons per day) at a pressure of 207 kPa (30 psi). If greater flow is needed, pumping plants can be used to increase the flow. If the local reservoir storage can meet the demand, Lafayette Aqueduct No. 1 can be taken out of service.

O&M staff indicated the leakage problem started almost immediately after the aqueduct was put into service. In 1932, leakage repairs were made by various methods at 1,316 locations. Additional repairs were performed in 1951 and 1968. Between these major repair activities, emergency aqueduct repair is routinely provided by O&M staff as required each year. The repair work continuously imposes a capital burden on staff and budget. The most recent data shows that the water loss is estimated at 410 million liters per year (110 million gallons per year). The leakage through the aqueduct reduces annual revenues by about $180,000.

The majority of the defects observed consist of circumferential and longitudinal pipe cracks and failures of prior repair attempts such as leaking cracks and spalls to gunite rings and mortar patches.

PREVIOUS REPAIR WORK

In dealing with the recurring leakage problem, EBMUD exhausted almost all repair technologies available. The repair methods used can be categorized into two main groups: exterior and interior repair.

Exterior Repair

Exterior repairs are methods applied to the outside of the aqueduct. Although they do not require shutdown, excavation to expose the leaking cracks is necessary.

Concrete collars were installed as a means of covering the leaking cracks. Although the repair was successful in providing temporary relief of the problem, new cracks quickly developed on both sides of the collar. This side effect is caused by discontinuity in the pipe structural stiffness. Lead wool packing with gunite cover was once used to seal the crack. After the lead content in drinking water became regulated, the use of lead wool packing was abandoned. Oakum and hydraulic cement are two commonly used crack sealing compounds. Thin layers of cement, or grout, are generally prone to spalling and cracking. EBMUD has tried several different coating products, including coal tar/bitumastic, which were applied over the cracks. The results were disappointing. EBMUD no longer considers these coating products as viable options.

Currently, EBMUD uses a water activated epoxy for repair of leaking cracks. The epoxy is injected into the crack from outside the aqueduct. When the epoxy comes into contact with the leaking water, it is activated and sets up. Most leaks can be stopped, or at least slowed to a weep, using this method.

Interior Repair

Interior repairs are methods used to seal leaks inside the aqueduct. The majority of these methods require aqueduct shutdown and dewatering.

Similar to the exterior crack sealing methods, polyurethane sealant with cement mortar cover was applied along the crack. The disadvantage of this method is that the cement mortar is susceptible to spalling. Similar to the exterior concrete collar, gunite rings were installed on the inside of the pipe for covering the cracks. During the aqueduct inspections spalling, cracking, and leakage were found at the gunite rings.

Sealing cracks with sawdust was the only interior repair which did not require a shutdown. Sawdust of varying grades was added to the water upstream of the leakage. The sawdust effectively sealed the cracks for the remaining season. EBMUD stopped this practice because of the potential negative impact on water quality and filtration processes.

A proprietary repair method consisting of rubber rings to seal areas was utilized in 1989. The areas sealed varied up to 51 centimeters (20 inches) in width. This repair method appears to be working, but is only effective on existing leaks.

GEOTECHNICAL INVESTIGATION

The repair study includes a geotechnical investigation to identify potential geotechnical hazards along the alignment of Lafayette Aqueduct No. 1. One of the most important geotechnical concerns is that the aqueduct is located in a seismically active area. Several active faults are located in the vicinity of the alignment, but no known active fault crosses the alignment. The study predicted that the aqueduct would remain functional. In support of EBMUD's effort in improving the seismic reliability of their water supply system, the effects of the repair alternatives on seismic performance were considered in the repair study.

CONSTRAINTS

Various constraints to the repair work were identified. These constraints can be categorized into three groups: (1) physical and site constraints; (2) operational constraints; and (3) environmental and permitting constraints.

Physical and Site Constraints

The physical and site constraints of Lafayette Aqueduct No. 1 limit the type of repair methods that may be effectively implemented. The study assumes that all work will be performed within the confines of the existing right-of-way.

In locations where adequate room for installation is available, the repair study considered installing a new pipeline. In locations where adequate room for installation is not available, other types of repair options, such as lining or remove-and-replace were considered.

The aqueduct has very few existing internal access points. The only "walk-in" internal access is at the Walnut Creek Tunnel East Portal. Remaining internal access points are manholes spaced at intervals varying from 210 meters (700 feet) to over 760 meters (2,500 feet). These manholes are mostly 36-centimeter by 41-centimeter (14-inch by 16-inch) ovals.

Operational Constraints

It is essential that the repair work be completed with minimal interruptions to the operation of the aqueduct system. The study identified two main operational constraints: shutdown/start-up requirements and shutdown window.

For the 1990 inspection, shutdown tasks required an O&M crew of six and took about one week. Start-up requirements were approximately the same.

The shutdown window substantially limits the repair options and increases the repair costs because one or more of the following measures would have to be taken:

- Use external repair methods - Based on EBMUD experience, there are few effective external repair methods which provide long-term reliability.

- Increase production by working multiple shifts - Other than the increased costs due to overtime labor rates, there would be limited repair options because of the noise and traffic impacts.

- Schedule work over a number of years - The increased costs are primarily due to the multiple mobilization and demobilization costs.

In order to develop the most cost-effective repair strategy, project components for expanding the shutdown window were explored and discussed during a project workshop. The costs and schedule for implementing each of the components were estimated and included in the recommended implementation plan.

Environmental and Permitting Constraints

It has been EBMUD's practice that water quality shall not be compromised by any repair work. All materials used in repair work which might come into contact with the water must be tested and approved by EBMUD's laboratory. EBMUD's laboratory testing program consists of two tests. The first test is designed to measure the organic leaching potential of the material. The second test is designed to determine if the material supports bacterial growth.

The Regional Water Quality Control Board (RWQCB) is the state government agency that regulates discharges to natural drainage areas. The RWQCB has indicated that chlorinated aqueduct water should not be discharged to drainage basins. EBMUD could obtain a permit to discharge water to sanitary sewers owned by the local sanitary district.

Based on a preliminary investigation, additional permits would be needed from the Department of Fish and Game, Corps of Engineers, and City Encroachment and Tree Removal Permits.

REPAIR ALTERNATIVES

Based on a review of published literature conducted for this study and a telephone survey among manufacturers and contractors, 22 repair methods were identified. All of the manufacturers were contacted to determine which repair methods were feasible for a 2.7-meter (9-foot) pipe. Those manufacturers that cannot accommodate a pipe as large as the Lafayette Aqueduct indicated the market for this size water line is not very large, and they would not be cost competitive in this market. Out of the 22 repair methods, only 12 can accommodate a 2.7-meter (9-foot) pipe. The available repair methods can be classified into three major categories: lining, replacement and point repairs.

Lining

Within the pipeline rehabilitation industry, lining means trenchless (no-dig) construction methods. Since the methods do not require open trenching along the entire pipe alignment, the major advantages are less risk of damaging surface improvements (such as roads, utilities and landscaping); less inconvenience to the public with public acceptance for the project easier to gain; more adaptable to areas where construction clearance and access are limited. The four primary types of lining are: sliplining, cure-in-place pipe (CIPP), fold-and-formed lining, and spiral wound lining. Due to size limitation, fold-and-formed lining is not a feasible repair method for Lafayette Aqueduct No. 1. A spiral wound lining system is not capable of withstanding pressures within the Lafayette Aqueduct No. 1. Therefore, it is likewise not a feasible repair alternative. The two feasible alternatives are sliplining and CIPP.

Sliplining - Conventional sliplining involves insertion of a slightly smaller diameter pipe inside an existing pipe. The new pipe can be in the form of solid high density polyethylene (HDPE) wall with butt fused joints or segmental pipe sections with compression fittings. In addition to HDPE, pipes made of fiberglass, steel and concrete can also be used as the liner. The liner results in a void between the existing and new pipe that can vary in size from five to ten percent of the existing diameter. This void, or annulus, is usually grouted to provide structural stability for the new pipe.

Sliplining requires excavation of access pits. A portion of the crown of the existing pipe at an access pit is typically removed for the new pipe insertion. The new pipe is then pushed and/or pulled through the existing pipe. The sliplining method typically causes greater reduction in pipe diameter and consequent reduction in hydraulic capacity than CIPP. Sliplining pipes have limited ability to negotiate bends. Horizontal and vertical curves and offsets need to be checked to determine the suitability and feasibility of sliplining.

CIPP - CIPP, also known as inversion lining, involves inserting a synthetic resin-impregnated (thermosetting type) flexible liner into the existing pipe. The liner is custom fabricated to match the inside of the pipe being lined. The liner is then expanded against the pipe wall by introducing hot water, steam, or other fluids which provide heat for curing of the thermosetting resin.

There are over a dozen companies that currently have the capability to install CIPP. However, only one manufacturer is capable of installing a CIPP as large as Lafayette Aqueduct No. 1.

Advantages of CIPP include limiting disruption to the pipeline corridor and less reduction in hydraulic capacity. The liner reduces the hydraulic capacity of the pipe as a function of the thickness of the liner. However, one manufacturer claims the liner pipe will increase the hydraulic capacity because it is jointless and has a very smooth interior surface. Disadvantages include a limited service record, a proprietary method and disposal of the hot water.

In summary, Lafayette Aqueduct No. 1 can be lined by either sliplining or CIPP. Four possible choices for the slipliner are steel pipe, reinforced concrete cylinder pipe, fiberglass pipe and HDPE pipe.

Replacement

A new pipe could be installed as a repair method to Lafayette Aqueduct No. 1. Pipe considered for Lafayette Aqueduct No. 1 includes reinforced concrete cylinder pipe, reinforced concrete pipe for pressure applications and steel pipe.

The new pipe can either be installed by conventional trenching or bore-and-jack/ tunneling. Due to the high cost of tunneling, this construction method is used only for special circumstances where the pipe is far below the ground surface and must be placed underneath surface improvements such as structures, utilities, and roads.

Installation of a new pipeline by trenching requires a clear corridor for the new pipeline in both the vertical and horizontal alignment. The repair study established that a clear horizontal corridor of 9.1 meters (30 feet) is necessary to install a new 2.7-meter (9-foot) pipeline. The clearance can be from the edge of the existing right-of-way to the outside face of one of the existing aqueducts. When horizontal clearance is available, the new pipe can be installed parallel to the existing pipe. Where clearance is not available, the new pipe can be installed along the existing pipe alignment by remove-and-replace.

Using the conventional remove-and-replace method, the existing pipe is first removed such that the new pipe can be installed in the same location. The amount of production that can be achieved during the allowable shutdown period using the remove and replace method may be very limited.

Point Repairs

Numerous point repair methods are available. Most of the aqueduct repairs previously performed by EBMUD were point repairs. The primary disadvantage of point repairs is that no restoration, renewal, or improvement will be provided to the overall pipe structure. In other words, there will be no extension of the pipe's original design life following point repairs. In some cases, the point repair may actually weaken the pipe. The repair study identified four major point repair methods as follows:

Internal Rubber Ring Seals - This method utilizes an internal rubber ring held in place by expanded stainless steel bands. The rubber rings allow movement of the pipe and can be effective if the cracks are circular. However, they are not effective for longitudinal cracks and are only useful after the cracks have occurred. Therefore, this method will not solve the recurring leakage problem in Lafayette Aqueduct No. 1 because new cracks are expected to develop in the pipe in the future. The cost of the rubber ring is about $3,000 to $5,000 per location.

Grouting - Grouting the leaking cracks with cement mortar or epoxy can be an effective temporary repair as demonstrated by EBMUD's repair experience. Since the repairs do not address the formation of new cracks, they will not solve the recurring leakage problem in Lafayette Aqueduct No. 1.

Spray-on Membrane - The spray-on membrane typically consists of a polyurethane type coating sprayed on to a thickness of approximately 50 mils. The membrane is used as a protective or waterproofing coating for structures or conduits. Another spray-on membrane consists of shotcreting. This is one of the first methods used by EBMUD for repairs of Lafayette Aqueduct No. 1. Although the shotcrete temporarily stopped the leaks, cracks eventually developed through the shotcrete. Similar to other point repairs, spray-on membrane will not solve the leakage problem in Lafayette Aqueduct No. 1 unless the cause of crack formation is addressed.

Point Repair with Expansion Joints - Since the point repairs are only effective on existing cracks and will not solve the recurring problem of having new cracks forming every winter when the water temperature drops, a scheme developed specifically to address the problem caused by a lack of movement joints is to use point repairs for fixing the existing cracks and to add expansion joints for preventing new cracks in the future. The disadvantages of this scheme are that the repair will not provide any restoration or strengthening of the pipe structure and the repair will be relatively expensive. The estimated total repair cost of using expansion joints with rubber ring seals is $14.6 million.

SELECTION CRITERIA

For evaluation of the identified repair alternatives, the following selection criteria were established.

Capital Cost - Capital cost includes only the cost for implementing the repair/improvement project. Total repairs such as construction of a new pipe or addition of lining will have higher capital costs than point repairs.

Life Cycle Cost - Life cycle cost includes capital cost, annual O&M cost, and loss of revenue due to leakage. Point repairs which neither prevent formation of new cracks nor remove the need of recurring repair efforts would likely have low capital costs but high O&M costs. On the other hand, total repairs would have high capital costs but low O&M costs. Life cycle cost analysis takes into account the potential cost factors and will properly reflect the cost-effectiveness of repair alternatives.

Proven Performance - Proven performance reflects qualifications of repair methods in three areas: local experience, application to water pipes, and application to large diameter pipes.

Long-Term Reliability - Lafayette Aqueduct No. 1 is a critical component in the EBMUD water supply system. Therefore, long-term reliability of the repairs is one of the more important evaluation criteria. In addition to the ability of sealing existing cracks, the ability to prevent formation of new cracks will also be considered under this criteria.

Life Expectancy - Another very important criteria is life expectancy of the repairs. Lafayette Aqueduct No. 1 was constructed more than half a century ago, and may be close to the end of its useful design life. Construction of a new replacement pipeline outside of existing EBMUD right-of-way would not only be cost prohibitive, it would also be environmentally controversial. Therefore, repair methods which will revitalize and renew the structural conditions of the pipe (and thus will extend the useful life of the aqueduct) will be more favorable than the repairs which only address the existing cracks.

Ease of Construction - The construction requirements of each repair alternative were rated based on shutdown requirements; potential for service interruptions; local pit excavation versus trenching; easement requirements; production rate; requirements on curing; cooling and drying of repair materials; ability to accommodate bends, tunnels, possible obstructions and diameter changes due to previous repairs; risk of damage to adjacent utilities and structures; vulnerability to damage during construction; requirements for temporary interior supports; labor intensiveness; requirements for specialty skilled labor; and quality assurance and quality control and inspection requirements.

Pipe Hydraulics / Capacity - Some repair alternatives such as sliplining and CIPP may alter the interior surface frictional characteristics of the aqueduct and also reduce the cross sectional area. These changes could increase the head loss within this section of Lafayette Aqueduct No. 1. On the other hand, construction of a new replacement pipe will provide an opportunity for EBMUD to increase the system capacity.

Water Quality - It is an EBMUD requirement that all repair materials which may come in contact with water carried by the aqueduct be tested and approved by EBMUD. Repairs previously approved by EBMUD and certified by the National Sanitation Foundation for use with potable water will receive more favorable ranking.

Seismic/Structural Favorability - These criteria address structural issues of the repair work including improvement to the structural integrity, ability to accommodate future pipe settlement, offsets or displacements, seismic resistance, and ease of repair for emergency restoration following a major seismic event.

Proprietary Technology - Repairs requiring proprietary technologies will hinder the competitive bidding process for the construction work.

Disruption to Neighbors - Construction associated with the repair work will cause inconveniences to the residential neighborhoods along the Lafayette Aqueduct No. 1 alignment. Repairs requiring trenching would have more severe impacts than point repairs and repairs requiring only insertion pits (headings).

Environmental Impact - These criteria consider the impacts of repairs to the natural environment, including the creek crossings. In general, repairs causing more disruptions to neighbors along the aqueduct alignment will have a greater potential of producing negative environmental impacts. These criteria will not directly affect quality of repair work.

Schedule - Although EBMUD O&M staff have devised creative operational measures to expand the shutdown window for accommodating the aqueduct repairs, it would be advantageous to EBMUD to complete the work as soon as possible.

CAPITAL IMPROVEMENT PROJECT

Based upon the findings established by the repair study, EBMUD developed the following capital improvement project and its implementation plan.

Recommended Repair Strategy

Based on the selection criteria, the evaluation concludes that the most cost-effective, public-sensitive, and environmentally responsible long-term repair strategy consists of sliplining the existing pipe with steel liner along the cut-and-fill sections, and point repairs with internal rubber ring seals along the tunnel sections.

Project Elements

The recommended Lafayette Aqueduct No. 1 repair project includes four major elements:

1. Work required to expand the allowable shutdown window to 4 months
2. Dewatering of Lafayette Aqueduct No. 1
3. Install 15 internal rubber ring seals along tunnel sections
4. Install steel pipe liners along cut-and-fill sections

Cost Estimate

The estimated costs of repairs include the costs of constructing insertion pits, furnishing and installing the steel pipe liners, grouting the annular space between the liner and Lafayette Aqueduct No. 1, and sealing the ends to furnish a smooth hydraulic transition.

To verify the validity and cost effectiveness of the recommended improvement project, the cost of the "do-nothing" alternative was estimated for comparison purposes. The costs of the "do-nothing" alternative are the costs which EBMUD estimates it will incur if the recommended repairs are not implemented. They include the annual revenue loss due to leakage, the costs of annual emergency repairs, costs of periodic major repairs, and the replacement cost at the end of useful life.

The estimated total project cost of the recommended repair strategy is $13.6 million. The cost of "do-nothing" alternative, based on the assumption of 35 years remaining useful life, is $23.4 million (i.e., over 70 percent greater than the cost of repair). The cost breakdowns are presented in Tables 1 and 2.

Table 1 Estimated Cost of the Recommended Project

Item	Cost (x $1,000)
1. Work required to expand shutdown window to 4 months	86
2. Dewatering of Lafayette Aqueduct No. 1	43
3. Install 15 internal rubber ring seals along tunnel sections	75
4. Install steel pipe liners along cut-and-fill sections	8,193
5. Contingency, engineering, legal and administration	5206
Estimated Total Project Cost	$13,603

Table 2 Estimated Cost for "Do-Nothing" Alternative

Item	Cost ($1,000)	Present Worth ($1,000)[a]
1. Annual revenue loss and annual repair [b]	224	4,480
2. Major repair [c]		
Cost at year 10	200	142
Cost at year 20	200	101
Cost at year 30	200	71
5. Replacement by remove and replace (d) Cost at year 35 after aqueduct has been in service for 100 years	62,000	18,599
Estimated Total Cost (Do-Nothing)		$23,393

Notes:
a. Assumed remaining useful life of aqueduct = 35 years ; discount rate = 3.5%
b. Assumed 4 locations at $11,000 per location.
c. Assumed major repair required every 10 years. Including dewatering cost.
d. Including contingencies, engineering, legal & administration.
e. If the aqueduct requires replacement in 25 years, the estimated total cost would be about $30.2 million.

Additional benefits of the Recommended Repair Project

Implementing the project will provide the following additional benefits:
- ☑ Recover 414,300 cubic meters per year (336 acre-feet per year) of EBMUD water for meeting the continually increasing demand
- ☑ Improve seismic reliability of the EBMUD water supply system
- ☑ Facilitate the O&M operations by eliminating the need for unplanned emergency repairs
- ☑ Reduce the risk of the potential catastrophic structural failure of the aqueduct

Based on the foregoing, the Lafayette Aqueduct No. 1 repair project was recommended for implementation.

New Steel Pipe Joining System for Trenchless
Installation and Rehabilitation

Michael E. Argent[1], David A. Pecknold[2] M.ASCE, and Rami M. HajAli[2]

Abstract

The Permalok Steel Pipe Joining System (U.S. Patent No. 5360242) is a mechanical joining system for steel pipe that eliminates field welding, and allows more rapid installation particularly in trenchless construction applications. This paper presents the results of an analytical study, using finite element numerical models, that was carried out to evaluate the strength of the Permalok connector under several idealized loading conditions that are representative of conditions that can be expected to occur during installation and operation of the pipe.

Introduction

Although competitively priced with respect to other products, and despite its attractive strength and ductility, steel pipe is generally used as a casing in trenchless installations only when necessary because it is expensive to install due to down time associated with field welding, and because of corrosion problems. Technologies for cathodic protection and new coatings can now provide the desired longevity. This paper describes the characteristics of a mechanical joining system, the Permalok Steel Pipe Joining System (U.S. Patent No. 5360242) that eliminates the need for field welding. It has been used successfully on numerous microtunneling projects and several boring, jacking and ramming installations. The strength of the Permalok connector is evaluated herein, using finite element numerical models, under several idealized loading conditions that are representative of conditions that can be expected to occur during installation and operation of the pipe.

Description of Joining System

The system consists of a female and a male connector that are shop welded to the pipe sections that are to be connected. The connector design incorporates a series of three circumferential ridges or teeth on the matching external surface of the male and internal surface of the female connectors. See Fig. 4. The connection

[1] Permalok Corporation, P.O. Box 230038, St. Louis MO 63123
[2] Dept. of Civil Engineering, University of Illinois at Urbana–Champaign, Urbana IL 61801

is then made in the field by pushing the pipe segment with the male connector into the already–installed pipe segment which has a female connector. As they are pushed together with a hydraulic jack, the connectors mechanically lock by passing over the series of teeth, which have increasingly larger radial interferences. A silicone adhesive sealant is applied to the connector surfaces, to both reduce frictional resistance during closure of the joint and to seal the closed joint. The completed joint is approximately 12 inches long, depending somewhat on pipe diameter. The characteristic strengths of connectors for three different pipe sizes: 10–inch ID, 0.375–inch wall thickness (WT); 36–inch OD, 0.5–inch WT; and 60–inch OD, 0.75–inch WT, are investigated using finite element numerical modeling. The three connectors that were chosen for detailed investigation also represent different design prototypes, with slight but significant differences in their local geometry.

Loading Conditions

Five loading conditions are considered, and are shown schematically in Fig.1. They are: (1) Closing (mating) of the joint; (2) Compression – advancing (pushing) the joined pipe into the tunnel; (3) Tension – retracting (pulling) the joined pipe from the tunnel; (4) Internal pressure; and (5) Bending – advancing the joined pipe through a misaligned or curved tunnel.

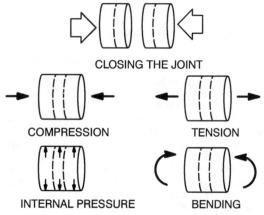

Fig. 1 Loading Conditions

Structural Modeling

For the joint closure, compression, tension, and internal pressure loadings, the same conditions of stress and strain exist on any diametral section of the pipe. As a result, a two–dimensional axisymmetric finite element structural model (Fig. 2a) can be used for these loadings. However, for the bending load condition, the stresses vary across the pipe (vertical) diameter. In this latter case, a three–dimensional finite element structural model, consisting of a full 180°

segment of the pipe cross−section (Fig. 2b), is needed. Finite element models of the Permalok connectors, connected to short segments of the parent pipe were analyzed, using the ABAQUS (Version 5.2) finite element software package. In the analysis, it was essential to accurately model the local geometric details of the connectors; to account for the contact, separation, and frictional slip that can occur between mating surfaces of the female and male connectors as the joint is closed and then subjected to the various load conditions; and to account for the effects of plastic yielding of the steel.

 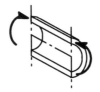

a. Axisymmetric Model for Loading Conditions 1−4

b. 3−D Model consisting of 180° segment of pipe for Bending

Fig. 2 Stress Analysis Models

ABAQUS provides two different solution strategies. Both are used in this study: the joint closure simulations are performed with the "implicit" static option; the other load cases are analyzed using "explicit" dynamic analysis, in which (slowly) time−varying loadings are applied. The selection of the explicit dynamic option for a particular case involves a trade−off between its superior contact simulation capability versus the increased computation time that it requires.

Material properties. The material is specified as A36 steel which has a yield stress of 36 ksi, and a specified ultimate tensile strength of 58−80 ksi. In this study, the ultimate tensile strength is taken as 60 ksi. The behavior of the connector is not expected to be sensitive to this ultimate tensile strength value, because it is reached only at very large strains. The modulus of elasticity E = 29,000 ksi and Poisson's ratio = 0.3. The plastic yielding of steel at high stress levels, exceeding the yield stress, is accounted for by using a von−Mises plasticity model. Strain hardening of the steel, which initiates at 2 percent elongation in a tension test, is modeled in ABAQUS using an isotropic hardening model. Large strains occur locally in the teeth, and are accounted for in the analysis. The pourable silicone adhesive sealant reduces the frictional resistance between contacting surfaces of the connector as the joint is closed; for this loading condition the coefficient of friction is taken as 0.10. After the sealant has hardened, the coefficient of friction is assumed to be equal to 0.20 for loadings acting on the completed joint.

Finite element modeling. The geometric modeling of the connectors is accomplished using simple 4−node quadrilateral elements for the axisymmetric models, and 8−node bricks for the three−dimensional models. The axisymmetric models contained approximately 470 elements and 1100 displacement degrees−of−freedom. The three−dimensional models that are used for the bending load condition contain over 23,000 elements and nearly 90,000 displacement degrees−of−freedom. The male and female connectors are modeled as two distinct bodies that can undergo significant displacements relative to each other; this is necessary for simulating the closing of the joint. In the joint closure analysis, the male connector is advanced in specified incremental stages, with the female connector held fixed at its remote end, until the joint is closed. In all other loading cases, the loads or displacements are applied with the joint assumed to be initially in a closed, *stress−free* configuration. That is, any small stresses remaining in the joint after closure, and before any additional loads act, are assumed to be negligible. This simplifying assumption is supported by the results of the closure analyses.In the tension and compression load conditions, the remote end of the female connector is again held fixed, while the remote end of the male connector is displaced at a specified speed, until a limiting strength is achieved. In the three−dimensional bending analysis, a short segment of the pipe is held fixed at the remote end of the female connector, while the remote end of the male connector is rotated rigidly at a specified speed, about a pipe diameter (Fig. 3). This can be visualized by imagining a thick steel plate closure welded to

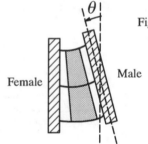

Fig. 3 Bending Loading Condition − a rotation θ is imposed at the remote end of the male connector

the remote end of the male connector; this plate is then rotated at a specified speed, about the horizontal diameter of the pipe. During this rotation the pipe is not allowed to lengthen or shorten along its centerline perpendicular to the axis of rotation. Note that in a homogeneous pipe, there is no tendency for such lengthening or shortening, and therefore only a pure bending moment is generated. In the Permalok connector, the behaviors in tension and compression are different, and an axial compression force is generated in addition to the bending moment. In all cases where a dynamic loading is applied (i.e. all cases except joint closure), the loading speed is reasonably slow; comparison of results at different loading speeds were made (but not shown here) to verify that dynamic effects are not significant.

Results

The results are presented in the following sections, organized by loading condition. Detailed results from the analysis of the 60−inch diameter connector are presented; results for the 36−inch and 10−inch diameter connectors are summarized in Table 1.

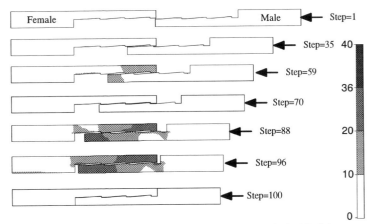

Fig. 4 Von−Mises Stress (ksi) during Closure of 60−inch OD Joint

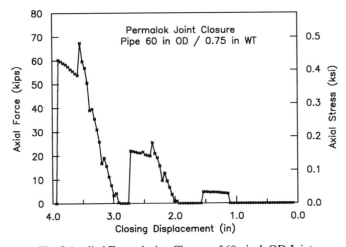

Fig. 5 Applied Force during Closure of 60−inch OD Joint

<u>Joint closure</u>. As the male connector advances, the axial force required to push the joint together changes with the relative position of the male and female connectors (Fig. 5). The closing displacement shown as the horizontal axis in Fig. 5 is the distance that the male connector has advanced into the female connector, so that at the far left of the figure, the joint is closed. Fig. 4 shows the deformed

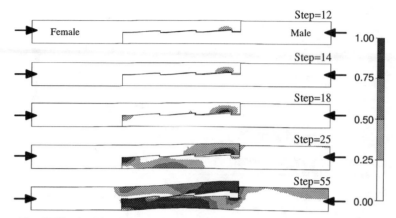

Fig. 6 Effective Plastic Strain (percent) − Compression of 60−inch OD Joint

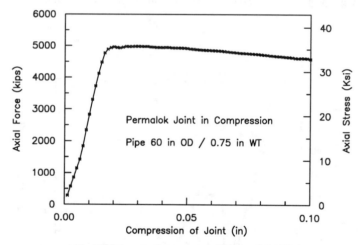

Fig. 7 Compression of 60−inch OD Joint

shape of the connectors as the male connector is pushed into the female connector. Also shown in Fig. 4 are contours of von−Mises stress; when the von−Mises stress meets or exceeds 36 ksi, plastic deformation occurs. Examination of the figure reveals that the resistance to closure consists of a series of three "bumps", or force maxima, of increasing intensity. The first occurs when the first male tooth reaches the first female tooth. At this stage the pushing force is 5 kips in the 60−inch joint (Fig. 5). The von−Mises stress at this stage is less than 10 ksi (Step=35 in Fig. 4), indicating that there is no plastic yielding. When the male connector advances far enough for the teeth to slip past each other, the pushing force and the stresses in the connectors drop to near−zero. As the male connector is pushed still further, the pushing force builds to a second peak (26 kips, Step=59 in Fig. 4), as two pairs of teeth are now in opposing positions. As the teeth slip past each other, the pushing force and stresses again drop to near−zero values (Step=70 in Fig. 4). The third force build−up reaches a maximum (68 kips in the 60−inch joint, Step=96 in Fig. 4) when all three teeth of the male connector are opposite the corresponding teeth of the female connector. This force is the maximum pushing force experienced during the joint closure. When the male connector is pushed further, the teeth once again slip past each other, and the joint finally "snaps" into place, with the stresses again dropping to near−zero values (Step=100 in Fig. 4). The reason that the pushing force increases successively in the three positions is that the number of opposing teeth that are in direct contact increases from one to two to three, and the radial interference (radius of male tooth greater than the radius of female tooth past which it is being forced) between opposing pairs of teeth also successively increases. There is very little plastic deformation in the 60−inch joint; only a small region at the tip of the outer female tooth, not visible in Fig. 4, yields. This behavior depends strongly on the local geometry of the joint design, as well as the pipe diameter and thickness. For comparison, an elastic analysis of the 60−inch joint showed only a slight increase in the maximum pushing force required to close the joint (about 3 percent greater). This suggests that a higher strength steel, or perhaps a steel without a flat−top yield plateau could possibly be used to improve the strength of the connector for other load conditions without incurring much of a penalty in increased closing force.

Compression. Compression failure occurs as the female connector yields and begins to slide radially outward, producing a raised "blister" that extends circumferentially around the joint at the visible seam on the outer surface. Plastic strain first starts to accumulate at the contact surface between male and female connectors in this vicinity, but eventually spreads along the length of the contact surface of the joint (Fig. 6). The axial compression force developed in the joint is shown in Fig. 7. The yield stress (36 ksi) of the material is developed before compression failure of the joint. The maximum compression force that can be sustained is 5,025 kips in the 60−inch joint; see Table 1 for the other cases.

Internal pressure. Under internal pressure, the circumferential (hoop) stress distribution is quite uniform throughout the joint. Failure occurs as general yielding takes place throughout the wall of the pipe, and excessive radial displacements occur. There is no reduction in the strength of the pipe under internal pressure due to the presence of the joint (Fig. 8). See Table 1 for the other cases.

Tension. Tension failure occurs by a prying mechanism in which the female connector displaces radially outward, and the male connector displaces radially inward until the opposing teeth can slip by each other, and the joint then separates (Fig. 9). A large compressive hoop stress appears at the tip of the male connector where it must compress circumferentially as it moves radially inward; a large tensile hoop stress appears at the tip of the female connector where it must

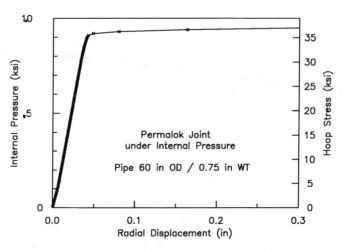

Fig. 8 Internal Pressure in 60–inch OD Joint

elongate circumferentially as it moves radially outward. The plot of plastic strain (Fig. 9) shows the plastic deformation that develops in the 60–inch joint as it unlocks in tension. The axial force that can be sustained by the joint is shown in Fig. 10. The 60–inch joint carries a maximum axial stress of 12 ksi (axial force 1,675 Kips). This axial tension strength, expressed as a fraction of the material yield strength, is 12/36 = 0.33 for the 60–inch joint. Refer to Table 1 for the tension strengths of the other two joints. The tension capacity is even more sensitive to the local joint geometry as well as the pipe diamater and thickness. The yield stress of the flat top steel is the important material parameter determining tension strength because failure in tension occurs when sufficient deformation can take place to allow unlocking of the joint. The 60–inch joint fails in tension at an extension across the joint of about 0.03 to 0.04 inches. A significant aspect of the behavior of the joint may be observed in Fig. 10 – a small extension, about 0.002 inches, is required before the joint develops any resistance. That is, there is a small amount of play in the joint when pulled in tension. In addition, the stiffness (the slope of the axial force vs axial extension across the joint) is greater in compression than in tension (compare Fig. 7 with Fig. 10). These differences between the response of the joints in tension and compression affect the response of the joint in bending.

Bending. The response of the joint in bending can be understood in relation to its behavior in axial tension and axial compression. Fig. 12 shows the bending moment that is developed in the pipe as the rotation is imposed. The lower horizontal axis shows the rotation θ, in degrees; the upper horizontal axis shows the axial extension that occurs across the joint at the bottom (extreme fiber) of the

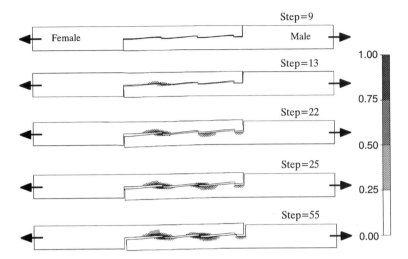

Fig. 9 Effective Plastic Strain (percent) − Tension of 60−inch OD Joint

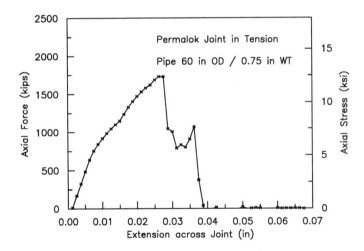

Fig. 10 Tension in 60−inch OD Joint

pipe; the same value of axial shortening occurs at the top of the pipe. The vertical axis at the left of the figures shows the bending moment developed in the pipe. The vertical axis on the right shows this same bending moment as a fraction of the fully plastic bending moment M_p in a solid steel pipe of the same diameter and wall thickness. The 60−inch joint ultimately develops about 84 percent of the fully plastic moment; the other two joints develop similar fractions of M_p (Table 1). Fig. 11 shows the deformed shape at the top and bottom of the joint at the limiting moment; the joint is approaching a condition at the extreme tension fiber that resembles pure tension failure (compare Fig. 9 with Fig. 11), but is at a less advanced stage in compression (compare Fig. 6 with Fig. 11). It is expected that the joint would remain water−tight up to this point. Interestingly, the axial extension that is developed across the extreme fibers of the joint at the limiting state in bending is greater than that developed in pure tension. As already observed, because of the different axial stiffness of the joint in tension and compression, an axial compression force is induced in the joint by the rotation, if the pipe centerline is not allowed to elongate. If, on the other hand, the pipe centerline *is* allowed to elongate, then the axis of rotation shifts upward toward the compression face of the pipe. The compression force and bending moment combination is equivalent to an axial compression force acting at an eccentricity with respect to the pipe centerline.

Conclusions

The numerical finite element modeling has provided some useful preliminary information on strength characteristics for engineering design. Based on the three joints analyzed, which cover a reasonably wide range of pipe diameters and thicknesses, the joint appears to develop virtually the full yield strength of the steel under compression and internal pressure loadings. Under tension loading, only a fraction (ranging from about 1/3 to 3/5) of the yield strength is developed. Because of the behavior of the joint in tension, its bending performance is impaired somewhat, compared to that of a homogeneous section of pipe. The computed limiting moments for the three joints that were studied ranged from 82 to 84 percent of the fully plastic moment, at which point the joints probably still retain their seal.

The insight into the behavior of the joint that has been provided by these results has already led to some design changes in later prototypes to improve performance in tension and bending. Additional, more comprehensive, studies are needed to fully understand the way in which joint behavior depends on its geometry.

A series of laboratory tests is planned, which will be used to validate the numerical modeling results obtained in this study. This is particularly important because a joint with ideal geometry was analyzed, and the sensitivity of the calculated results to fabrication tolerances, to out−of−roundness of the pipe, and to the frictional properties of the sealant is not yet known.

Acknowledgement

Computing support was provided by the National Center for Supercomputing Applications (NCSA) at the University of Illinois at Urbana−Champaign.

References

1. ABAQUS/Explicit, User's Manual, Version 5.2 (1992). Hibbitt, Karlsson & Sorensen, Inc., Pawtucket RI 02860.

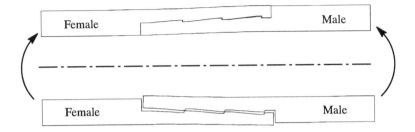

Fig. 11 Deformed Shape at Top and Bottom Fibers of 60–inch OD Joint at Limiting Bending Moment

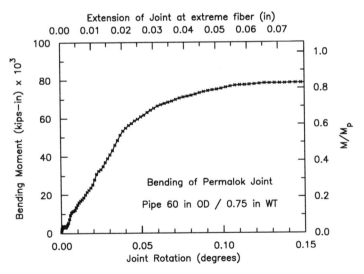

Fig. 12 Bending of 60–inch OD Joint

	10-inch ID 0.375-inch WT			36-inch OD 0.5-inch WT			60-inch OD 0.75-inch WT		
Loading Cond.	Axial Force	Axial Stress	σ_a/σ_y	Axial Force	Axial Stress	σ_a/σ_y	Axial Force	Axial Stress	σ_a/σ_y
	(kips)	(ksi)		(kips)	(ksi)		(kips)	(ksi)	
Closure	32	2.62	0.073	21	0.38	0.011	68	0.49	0.014
Comp.	440	36	1.0	1967	35.3	0.98	5025	36	1.0
Tension	257	21	0.58	956	17	0.48	1675	12	0.33

Loading Cond.	Int. Press.	Hoop Stress[1]	σ_h/σ_y	Int. Press.	Hoop Stress[1]	σ_h/σ_y	Int. Press.	Hoop Stress[1]	σ_h/σ_y
	(psi)	(ksi)		(psi)	(ksi)		(psi)	(ksi)	
Int. Press.	2650	35.3	0.98	1012	35.4	0.98	910	35.5	0.99

Loading Cond.	Bend. Mom.[2]	Axial Force	M/M_p [3]	Bend. Mom.[2]	Axial Force	M/M_p [3]	Bend. Mom.[2]	Axial Force	M/M_p [3]
	(kip in)	(kips)		(kip in)	(kips)		(kip in)	(kips)	
Bending	1194	86	0.82	18,590	424	0.82	79,880	1156	0.84

Table 1. Strengths of Permalok Connectors

[1] Hoop stress $\sigma_h = p D_i/2t$, where p = internal pressure, D_i = inside diameter, and t = wall thickness

[2] Limiting (maximum) bending moment

[3] The plastic moment M_p is calculated from $M_p = \frac{1}{6}[D_o^3 - D_i^3]\sigma_y$ where D_i is the pipe ID, D_o is the pipe OD, and σ_y is the yield stress (36 ksi).

Note: 1 in = 25.40 mm
        1 kip = 4.448 kN
        1 ksi = 6.895 MPa
        1 psi = 6.895 kPa
        1 kip in = 0.1130 kNm

REHABILITATION OF MASONRY COMBINED SEWERS IN THE CITY OF ST. LOUIS

Marie A. Collins, P.E. [1], and Carl Ted Stude, P.E. [2]

Introduction

In order to comply with Clean Water Regulations, the Metropolitan St. Louis Sewer District (the District) is currently implementing an overflow regulation system (ORS) improvement program in its Bissell Point Combined Sewer Service Area. The purpose of this improvement program is to eliminate dry weather sewage overflows during periods of high river stages on the Mississippi River by modifying current pump station and gate closure operations.

As a result of these modifications, gates on the outfall sewers will be closed at high river stages and backwater from the Mississippi River will no longer be allowed into the outfall sewers. Since groundwater tends to seek the same level as the River, these outfall sewers will be subject to external groundwater hydrostatic pressure that will not be equalized by the presence of backwater in the sewer. There is also concern that this condition will create accelerated infiltration into the combined sewers, resulting in the development of voids around them.

To minimize the potential for structural failure of these sewers as a result of this hydrostatic pressure and increased infiltration, a rehabilitation program is being

---

[1] Assistant Manager of Plan Review, Metropolitan St. Louis Sewer District, 2000 Hampton Avenue, St. Louis, MO 63139
[2] Senior Engineer, Horner & Shifrin, Inc., 5200 Oakland Avenue, St. Louis, MO, 63110

conducted to improve the structural integrity of the sewers. This paper addresses the rehabilitation of sewers made of brick and other types of masonry

History of Sewer Construction in the City of St. Louis

The combined sewers in the City of St. Louis are some of the oldest in the United States. Construction of the first combined sewer, the 3.7 meter (12 foot) circular brick sewer in Biddle Street, began in 1850. The first materials used in the construction of the sewers consisted of stone and/or brick masonry with lime mortar joints for all sewers greater than 0.9 meter (3 ft.) in diameter. Smaller pipes were made of either brick, clay or cement. In 1907, reinforced concrete was introduced with the construction of the Harlem Creek Sewer. By 1914, all of the sewers constructed of brick had been built and reinforced concrete was the material of choice for the construction of large sewers. During the 1920's a few masonry sewers were built of curved concrete blocks and clay tiles.

About 480 of the 1,800 kilometers of combined sewers in the City of St. Louis are of brick and/or stone masonry construction and range in size from 0.5 to 4.9 meters (18 inches to 16 feet) in diameter. Sewers built in areas of earth cut or fill were usually constructed entirely of brick. The larger sewers (2.4 meters (8 feet) in diameter and above) were generally circular in cross-section with three or four rings (layers) of brick, whereas the smaller sewers were built with an egg-shaped cross-section with two rings of brick.

Sewers constructed in bedrock or mixed-face strata were generally horseshoe-shaped (inverted "U") with the bedrock serving as the invert. The walls of these sewers were composed of limestone blocks set in mortar (or less frequently brick masonry) from the invert to the springline, with the crown of the sewer made of brick. Occasionally, wood planks were used to form the invert and foundation of these horseshoe-shaped sewers.

The most common bedrock material encountered was limestone or dolomite. These sedimentary rocks typically occur in horizontal layers separated by bedding planes (the geological equivalent of construction joints). In excavating this material, the rock had a tendency to break off in slabs along these bedding planes. As a result, many of the horseshoe-shaped sewers have inverts that change elevation in abrupt "steps", while changes in the elevation of the crown of the sewer occur more gradually. Most "steps" are no higher than 0.3 meter (one foot), but in some cases they are as much as 1.3 meter (four feet) high and take on the appearance of waterfalls.

Numerous repairs to these masonry sewers have been made over the years. Many of the timber bottoms have been replaced with concrete or brick; the sides and crowns have been tuckpointed by hand or coated with a thin layer [less than one centimeter (0.4 inch)] of shotcrete; or entire sections of the sewer have been replaced.

As in many major cities with older sewer systems, the failure and subsequent collapse of these old brick sewers is not an uncommon occurrence in St. Louis. Since the majority of these sewers were constructed under pavement or in street rights-of-way, the failure of a sewer frequently results in a cave-in or collapse of major proportion each year. The District, on the average, experiences one sewer collapse on major proportion each year, with repair costs ranging from about $300,000 to $1,500,000.

Criteria for Identifying Sewers to Be Rehabilitated

Due to funding limitations, it was not feasible for the District to immediately rehabilitate every sewer which would be impacted by the modifications to the pump station operations. It was therefore necessary to establish criteria for identifying the sewers most at risk of failure due to these modifications and determine the reaches of the sewers to be rehabilitated. In establishing these criteria, consideration was given to the total project cost of rehabilitating the sewers, relative to the potential risk of failure and the cost of repair if a collapse should occur. The following criteria were established to determine the scope of the rehabilitation project:

1) Sewers of primary concern would be those located along old stream beds or in areas known to have poor soil conditions.

2) Only sewers greater than 0.6 by 0.9 meter (2 by 3 feet) in size (or greater than 0.8 meter (33 inches) in diameter) would be rehabilitated, unless structural problems where noted on smaller sewers during the field inspections.

3) All sewers of masonry construction would be rehabilitated. Large sewers of reinforced concrete construction which were built in the early 1900's would be rehabilitated only if structural problems were noted during inspection.

4) All of the sewers identified would be rehabilitated from their outfall up to Stage 52 on the Mississippi River at the Market Street gage. This stage corresponds to the level of protection provided by the flood protection system along the River.

Smaller sewers of concrete construction were not considered for rehabilitation unless they had a history of structural problems. These smaller concrete sewers were generally constructed in the late 1960's or early 1970's and were considered capable of handling the additional hydrostatic pressure that would be placed upon them. (During the Mississippi River flood of 1993, the rehabilitated sewers experienced no problems, but a major structural failure occurred in a concrete pipe sewer. This confirmed the validity of the rehabilitation concept, but indicated the need to extend it to the concrete pipe sewers.)

Based on these criteria, 44 different sewers were identified for rehabilitation with a total footage of about 26.2 kilometers (16.3 miles). These sewers were divided into five rehabilitation projects that are being implemented in sequence.

Each of the sewers to be rehabilitated was inspected to obtain accurate field measurements of the cross-section and length, to evaluate the condition of the sewer, and to note features which could have a bearing on the selection of the method of rehabilitation. Utilizing the field information as well as original drawings of the sewers, an evaluation was performed to determine which method of rehabilitation would be the most cost-effective. Since all of the rehabilitation methods are considered structurally adequate, this evaluation was based on considerations of hydraulic adequacy; constructability without excessive disruption of adjoining facilities; and cost.

Once the method of rehabilitation had been selected, construction contract documents were prepared. Where several sewers were to be rehabilitated by the same method, they were included in a single construction contract.

Typical Structural Deficiencies

The structural integrity of masonry sewers is extremely dependent on the support of the surrounding soil. Whether a masonry sewer is circular, egg-shaped, or horseshoe-shaped, its crown is a semi-circular arch. This arch functions as a compression ring reacting to the vertical loading from above and the lateral support of the soil at the sides of the sewer. If the lateral support is lost, the sides of the arch move outward, causing the crown to flatten and cracks to develop.

Minor deformation and cracking of the crown is found in many brick sewers. If the sewer otherwise appears sound, this deformation is generally attributed to the difficulty in achieving densely compacted backfill with the equipment that was available at the time these sewers were built. Sewers in which the deformation has resulted in at least a 10% reduction in the sewer height are considered to be at risk of structural failure. In the sewers inspected as part of this project, this type of deficiency was most commonly found under active or abandoned railroad tracks.

The most common cause of the loss in lateral support around the sewer is the displacement of the surrounding soil and the development of voids outside the sewer. Soil is generally displaced by groundwater seepage into the sewer through deteriorated mortar joints and lateral connections or through weep holes that were built in certain sewers in areas of high groundwater.

Sewers may also experience structural failure as a result of the erosion of rock inverts or the deterioration of timber bottom inverts in horseshoe-shaped sewers, and the loss of bricks in inverts in circular or egg-shaped sewers. Flow in the sewer can then undercut the perimeter or bottom of the sewer and threaten its stability. In this project, this condition was found to occur most often in sewers with slopes of 1% or more.

In brick inverts, the mortar between the bricks usually erodes more rapidly than the bricks themselves. In sewers in this project where bricks in the invert had been stripped away by turbulent flow, the damage had not yet spread to the point where collapse was imminent. In some of the sewers with bedrock inverts, erosion was noted at the "steps", where scour basins had formed. This erosion is of most immediate concern where it extends under the sidewalls of a sewer.

Also of concern is the loss of mortar joining the bricks, which commonly occurs over time as the result of erosion and dissolution by acids in the wastewater or sewer atmosphere. The loss of this mortar allows for the movement and possible loss of the bricks, especially in the crown of the sewer, resulting in the loss of the compressive strength in the arch crown.

A phenomenon that was noted to some extent in all of the sewers in this project was the formation of calcite beneath points of groundwater infiltration. At particularly large infiltration points, and below the connections of "dead" lateral sewers that carry infiltration, ridges of calcite several centimeters thick had formed on the sewer walls. In two of the 0.76 by 1.07 meter (2.5 by 3.5 foot) sewers inspected, the sewers were obstructed with calcite deposits and debris to such an extent that inspection was not possible. In other places, there was widespread infiltration through the brickwork that had left the interior of the sewer coated with a sheet of calcite. Although the calcite, in extreme cases, can reduce the hydraulic capacity of the sewer, it also appears to combine with mortar in a process known as "autogenous healing" to seal leaks.

Design Features of Rehabilitation Techniques

Except for the replacement of an existing sewer, all of the rehabilitation techniques in this project provide a new impermeable lining around the interior circumference of the sewer. These linings were designed with sufficient strength and stiffness to essentially "stand alone" with the maximum possible external hydrostatic pressure acting on their circumference. (In the design of a cured-in-place pipe lining, credit was given for the resistance to buckling added to the support of the original sewer.) This design approach was based on the conservative assumption that groundwater will be able to seep between the liner and the original sewer and exert hydrostatic pressure around the circumference of the liner.

It was generally assumed that the original sewer had sufficient support from the soil to resist the soil load on it. In the few places where the original sewer appeared to have lost its structural integrity, the liner was designed to withstand soil loading as well as hydrostatic pressure.

Resistance to corrosion was not a major factor in the evaluation of rehabilitation alternatives because septicity with its associated generation of hydrogen sulfide seldom occurs in these sewers.

As a result of the design deliberations and observations of the construction work, certain generalizations can be made as to the applicability of the different rehabilitation techniques under various conditions.

Cured-In-Place-Pipe

Cured-In-Place-Pipe (CIPP) is a type of lining for sewers that consists of a polyester fiber felt material of up to several centimeters in thickness, with an impermeable membrane on the side that is in contact with the wastewater. The liner is manufactured as a flexible tube of the length and interior circumference of the reach of sewer to be lined. Just before insertion into the sewer, the felt is saturated with a liquid thermosetting resin. The liner is positioned inside of the sewer and filled with cold water that presses it against the interior of the sewer; the water is then heated to accelerate the chemical reaction that causes the resin to harden. Once the resin has cured, the ends of the tube are cut away to release the water and open the sewer.

The CIPP technology was developed in England and licensed under the trade name Insituform. Although some aspects of the installation procedure are proprietary, CIPP is a generic product for which ASTM Designation F1216 provides a recognized standard. CIPP for man-entry sized sewers is now provided by at least two manufacturers.

A CIPP liner is adaptable to almost any masonry sewer as large as 2.4 meters (8 ft.) in diameter. During insertion, the liner will follow gradual bends, and conform to modest variations in cross-section. It can be fabricated to match transitions in the size of the sewer, and its thickness can be varied to accommodate variations in design loading conditions.

A key parameter in the design of a CIPP liner is the type of resin used, since the resin controls the structural strength and stiffness, and the chemical resistance. The standard resin used on this project has been polyester. In a sewer where strong acids, caustics, or oxidants may be present, a more expensive vinyl ester resin should be utilized.

The required thickness of a CIPP liner is determined by the equations for resistance to deflection and flexural stress that are conventionally applied to the design of buried flexible pipe. For sewers in this project, which were typically designed for a short-term hydrostatic pressure head of 3 to 6 meters (10 to 20 feet), the liner wall thickness typically ranged from one to two percent of the largest dimension of the sewer. (This is equivalent to standard dimension ratios in the range of 100 to 50.)

Because a CIPP liner is relatively thin and quite smooth, it will maintain or slightly improve the hydraulic capacity of most sewers. One of the manufacturers has performed field tests that indicate that its product has a Mannings "n" value equal to 0.009 in a sewer pipe that is straight and uniform between manholes. However, the effective "n" value will be greater in a typical masonry sewer because of hydraulic losses from lateral sewer connections and irregularities in shape, grade, and alignment.

CIPP is not inherently well-suited to installation in sewers with flat surfaces, such as horseshoe-shaped sewers, because of the flexural stress and deflection that hydrostatic pressure will produce in a thin, flat section. This limitation can be overcome by "rounding out" the sewer cross section with mortar or shotcrete before the CIPP is installed.

Installation of a CIPP liner requires a vertical access shaft from the ground surface to the sewer that is essentially as large in diameter as the sewer. Many of the sewers in this project were too large for the liner to be inserted through existing manholes; consequently, access shafts, shored with steel tunnel liner plates, were constructed at points of insertion. These access shafts can be installed around other buried utilities by careful excavation, but constitute a substantial part of the cost of CIPP installation; consequently, the cost of CIPP per unit length of sewer is reduced a the length of sewer lined from each assess shaft increases. Where a sewer gradient is relatively flat, it may be practical to line more than 300 meters (1,000 ft.) of sewer in each direction from an access shaft. (Installing the liner in two directions from the shaft requires two separate operations that can not be done simultaneously.)

After a sewer has been cleaned of debris and protruding objects have been removed, insertion and curing of a CIPP liner typically requires 24 to 36 hours. During this time, all flow must be bypassed; however, it is not necessary to stop whatever infiltration is occurring in the reach of the sewer being lined. After the curing is completed and the sewer is re-opened, lateral sewer connections are easily reinstated by a worker entering the main sewer and cutting the liner away from the connections, which remain visible as outward bulges in the liner.

Cured-in-place-pipe is so versatile it could be specified for almost any sewer without consideration of other methods. CIPP can, however, be relatively expensive in comparison to other rehabilitation methods. For the sewers in this project, the total installed cost has been about $380 per square meter ($35 per sq. ft.) of interior surface lined. If replacement with new pipe did not require disruption of surface improvements and interference with underground utilities, it would be less expensive. However, these improvements and interferences do exist in urban areas, often making cured-in-place pipe the most cost-effective solution.

Reinforced Shotcrete

The reinforced shotcrete method of sewer rehabilitation involves lining the interior of the sewer with a layer of mortar that is conveyed through a hose and "shot" through a nozzle operated by a worker in the sewer. The liner should have reinforcement to limit shrinkage cracking that would permit leakage through it. Normally, welded wire fabric is used for this reinforcement in circular or egg-shaped sewers, while steel reinforcing bars are required in the larger horseshoe-shaped sewers. The reinforcement is set into the sewer with expansion anchors that hold it a specified distance from the interior surface, and the mortar is sprayed over it. Lateral sewers are protected by sleeves or plugs that are removed after the mortar has set.

An alternate type of reinforcement for control of shrinkage cracking consists of steel or synthetic fibers that are mixed into the mortar before it is applied to the interior of the sewer. A major cost associated with conventional reinforcement is the cost of handling and placing the reinforcement. Usually, a shaft or pit must be excavated and a portion of the sewer crown cut away in order to get the reinforcement into the sewer. With fiber reinforced shotcrete, the access provided by existing manholes is usually sufficient.

A limitation of fiber reinforced shotcrete is that it is not a great deal stronger in tension than conventional concrete. When the design loadings on the sewer will only produce compressive stress in the liner, such as with circular or egg-shaped sewers, then fiber reinforced shotcrete can be used. However, where there is a likelihood that loading on the sewer will create bending moments with associated tensile stress, as is the case in horseshoe-shaped sewers, it is advisable to provide conventional reinforcement designed to resist the tensile force. Conventional reinforcement also provides greater resistance to undefinable forces that might be generated by ground movement.

Shotcrete conforms more readily than other types of linings to the various irregularities that can occur in old masonry sewers. For example, where "steps" occur in the invert, they can be levelled by applying a greater thickness of shotcrete to the invert. Nevertheless, the irregularities in the sewer, combined with the confined working conditions, make it difficult to achieve precise control over the structural section that is being created, with respect to its shape, thickness, and placement of reinforcement. To compensate, the District requires a shotcrete liner to be a minimum of 10 cm (4 in.) thick, and shotcrete is generally used only in sewers of 1.8 meter (6 ft.) in diameter or larger.

The shotcrete must be protected from both infiltration and flow in the sewer until it sets. In larger sewers, it is often possible to bypass the dry weather flow through a small pipe that is suspended from the crown of the sewer while the work is done. Infiltration should be controlled before the shotcrete is placed, either by injecting grout through the wall of the sewer or by tapping a tube into the sewer to carry the infiltration while the shotcrete is placed around the tube. The tube is then cut flush with the shotcrete and plugged with hydraulic cement after the shotcrete sets.

Another requirement that the District has adopted for its shotcrete work is for the mortar to contain microsilica as an admixture, amounting to about ten percent of the cement content by weight. This material, which consists of microscopic particles of silicon dioxide, reduces the loss of mortar by rebound during application, and produces a stronger, denser end product that is more resistant to corrosion. Compressive strength greater than 48 MPa (7,000 psi) is achievable.

A structural lining of reinforced shotcrete will reduce the cross-sectional area of a sewer enough that a substantial reduction in capacity will often result, despite an improvement in the smoothness of the sewer interior. For example, a 10 cm (4 in.) thick lining of shotcrete in a 1.8 meter (6 ft.) diameter brick sewer would reduce its hydraulic capacity by ten percent, even if the Manning's "n" factor were improved from 0.016 to 0.013. If the capacity of a sewer were only marginally adequate, this might be cause to reject shotcrete as a rehabilitation technique.

Like cured-in-place-pipe, shotcrete is very versatile as a method of rehabilitation. In addition, it is relatively inexpensive, costing about $130 per square meter ($12 per sq. ft.) of interior surface area. For sewers greater than 1.8 meter (6 ft.) diameter, with adequate hydraulic capacity, shotcrete is generally the preferred method of rehabilitation.

Sliplining

Sliplining involves the insertion of a new pipe into the existing sewer. The space between the liner pipe and the sewer is then filled with grout. Lateral sewers must generally be excavated and reconnected to the new liner from the outside.

Sliplining is most practical for circular sewers, since placing of a circular liner inside a non-circular sewer causes a major reduction in the cross-sectional area and therefore, a major reduction in its capacity. Sliplining was used for a few of the non-circular sewers in this project where the existing sewer had substantial excessive capacity.

To gain access to the sewer for sliplining, it is necessary to excavate an access shaft or pit and to break out the top of the sewer. It is possible for lengths of sewer in excess of 300 meters (1,000 ft.) to be sliplined from a single point, provided the sewer is straight. Flow does not need to be bypassed during installation.

Sliplining is most commonly done with plastic pipe because it is available with low profile joints that minimize the loss of cross-sectional area. In this project, ductile iron pipe was also used for sliplining, in spite of its bell joints, where the liner pipe could be substantially smaller than the existing sewer.

Because of its structurally efficient circular shape and the support of the grout around it, a liner pipe does not require extraordinary strength to withstand external pressure from groundwater. The condition that is critical to the structural design of the liner is the grouting operation. If the liner pipe is buoyed upwards by the grout, there will be a concentrated load along the top of the liner pipe that tends to distort it out of round. In addition, the hydrostatic pressure exerted by the grouting process is likely to be greater than the design hydrostatic pressure from the groundwater, and will be acting on the liner pipe before the pipe has the benefit of the complete circumferential support that the grout will provide once it sets.

To protect the liner pipe from distortion during grouting, the District requires that it have a minimum stiffness of 32 N/cm/cm (46 psi) as determined by the deflection test for plastic pipe in ASTM D2412. Another requirement is that the grout be introduced by gravity rather than pumping, to guard against excessive pressure. The grout that is specified is a type developed for oil drilling operations. It has a low viscosity and no aggregate, so that it will flow easily into the annular space between the liner pipe and the sewer, and has a density greater than water so that it will displace whatever water is trapped in the annular space. The 28-day compressive strength of the grout is required to be 3.4 MPa (500 psi), which is substantially less than the normal strength of mortar or concrete, but is adequate to support the liner pipe.

Sliplining can under the most favorable conditions, be expected to cost approximately $280 per square meter ($26 per sq. ft.) of sewer interior surface area. However, this cost can be increased dramatically where there are a large number of lateral sewers to be reconnected or irregularities in the alignment of the sewer that increase the number of excavations required for access.

<u>Overview</u>

A broad generalization of rehabilitation alternatives is that reinforced shotcrete is likely to be the most cost-effective alternative, provided that the sewer is large enough for proper quality control, some reduction in capacity is permissible, corrosion by acids is not a problem, and infiltration to the sewer during placement is controllable.

Cured-in-place pipe is adaptable to almost any sewer and will maintain or improve its capacity. This method, however, is expensive. Replacement should be considered as an alternative if there are no significant surface improvements, utilities, or other interferences which would make replacement prohibitively expensive.

Sliplining may maintain or improve the capacity of circular sewers, and may be less expensive than cured-in-place pipe if there are not too many lateral sewers to be reconnected. The applicability of sliplining to non-circular sewers is limited to exceptional cases where a substantial reduction in capacity is permissible.

The salient features of these three rehabilitation techniques, as they apply to man-entry sized sewers, are summarized in Table 1.

TABLE 1

Summary of Rehabilitation Techniques for
Sewers Larger than 1 Meter Diameter

Method	Typical Cost	Advantages	Limitations
Cured-in-place pipe	$380/m² ($35/ft²)	Maintains or improves capacity. Adapts to irregularities in shape and alignment. Durability resists fracture from ground movement. Resistant to corrosion. Minimal time spent in sewer by workers. Reinstatement of laterals is easy. Infiltration control not required.	Requires total flow bypass for 24-36 hours. Requires access shaft(s). Maximum sewer size approx. 2.4 m (8 ft.). Requires special treatment of flat surfaces inside a sewer.
Reinforced Shotcrete	$130/m² ($12/ft²)	No limit on maximum sewer size. Adaptable to irregularities in shape and alignment. Reinstatement of laterals is easy.	Requires access shaft(s), unless fiber-reinforced shotcrete is used. Requires extended entry by workers. Requires infiltration to be stopped. Subject to corrosion by acids. Quality control problems and capacity reduction in sewers smaller than 1.8 m (6 ft.)
Sliplining with segmented plastic pipe	$280/m² ($26/ft²)	May be done without bypassing. Infiltration control generally not required. Resistant to corrosion. (Best suited to rehabilitation of circular concrete sewers damaged by corrosion).	Typically reduces capacity. Requires access shaft(s). Reinstatement of laterals is difficult. Requires straight alignment and uniform circular cross section. Maximum sewer size approx. 2.1 m (7 ft.)

Managed Operation and Repair of a Deteriorating Large Diameter Pipeline

Kenneth L. Cramblitt[1]
Thomas J. Lawson[2]
Adrian T. Ciolko[3]

Abstract

Since 1986, Brandon Shores Generating Station has successfully managed a recurring maintenance problem with the buried prestressed concrete cylinder pipe (PCCP) of its circulating water system. It is the intent of this paper to provide a history of the plant's approach to safely operating and maintaining the deteriorating pipe of this system. This paper will briefly discuss the results of failure analysis performed on the ruptured pipe. Findings of engineering studies addressing the structural integrity of the remaining pipeline will be discussed. The inspection methodology used at Brandon Shores to periodically assess the pipe will be described. Several methods of pipe repair, both external and internal will be detailed. Particular focus will be placed on an internal repair design successfully implemented on two pipe sections during a 1993 plant outage. This internal repair resulted in no loss of hydraulic section, and was erected in-place using conventional materials which were entirely transported through 0.9 m (36 in.) manholes. The pipeline was placed back in service and has performed satisfactorily to date.

Introduction

Brandon Shores Station is located on the west shore of the Patapsco River 16 km (10 mi) southeast of Baltimore. The facility is owned and operated by the Baltimore Gas and Electric Company (BGE), a Maryland based, investor-owned utility. Brandon Shores Station consists of two 650 MW steam driven turbine generators, designated Unit 1 and Unit 2. Brandon Shores Unit 1 began commercial operation in May 1984. Brandon Shores Unit 2 began commercial operation in May 1991. Within a steam driven turbine, thermal energy is converted to the mechanical energy used to

---

[1] Senior Engineer Predictive Maintenance, Baltimore Gas & Electric Co., Fort Smallwood Road Complex, 1000 Brandon Shores Road, Baltimore, MD 21226.

[2] Engineer, Structural Development Group, Construction Technology Laboratories, Inc., 5420 Old Orchard Road, Skokie, IL 60077.

[3] Principal Engineer, Structural Development Group, Construction Technology Laboratories, Inc., 5420 Old Orchard Road, Skokie, IL 60077.

drive the generator. After this energy conversion, the steam exits into a main condenser where the latent heat of vaporization is removed as the steam condenses. Brandon Shores' tubular condensers are cooled by brackish circulating water.

After picking up heat in the main condenser, the circulating water is itself cooled by large mechanical draft wet cooling towers. Within each Brandon Shores cooling tower, heat is transferred from the circulating cooling water to the atmosphere at a rate of approximately 938 MW (3.2 billion Btu/hr). Circulating water piping transports the brackish water between the cooling towers and the condensers at a rate of approximately 17.1 kL/s (270,000 gpm). The motive force for fluid transport is supplied by 2 circulating water pumps per unit, each rated at approximately 8.5 kL/s (135,000 gpm) and 3.0 MW (4,000 hp). The distance separating the cooling towers and condensers is approximately 0.8 km (0.5 mi), therefore, the total length of circulating water piping for both units is approximately 3.2 km (2 mi). Figure 1 graphically depicts the route of the Brandon Shores Units 1 and 2 cooling water system.

FIGURE 1. BRANDON SHORES COOLING WATER PIPELINE

Description of Circulating Water Pipeline

Prestressed concrete cylinder pipe is used for the circulating water pipeline at Brandon Shores. In PCCP, a continuous helically-wound high-strength steel wire prestresses a cylindrical concrete core. The compressed concrete resists tensile forces arising from internal pressure and external loads. A thin steel cylinder embedded in the concrete core at the time of casting provides watertightness. A cement mortar coating applied over the prestressing wire provides protection from physical damage and steel wire corrosion. Figure 2 illustrates a section though a PCCP pipe taken at a standard pipe joint.

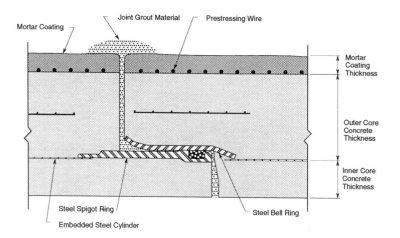

FIGURE 2. PCCP JOINT SECTION

Interpace Corporation fabricated approximately 3,660 m (12,000 ft) of 2,590 mm (102-in.) diameter, approximately 490 m (1,600 ft) of 1,980 mm (78-in.), diameter and 275 m (900 ft) of 1,370 mm (54-in.) diameter PCCP pipe for use in Brandon Shores Units 1 and 2 in 1974. The pipe was designed and manufactured to conform with AWWA Standard C 301-72. The circulating water pipeline design specified a normal operating pressure of 255 kPa (37 psi) and a pipe design pressure of 448 kPa (65 psi). Actual operating pressure varied with grade from a maximum of 255 kPa (37 psi) at the pump discharge to a minimum of 69 kPa (10 psi) at the condenser outlet. In addition, the design specifications for all pipe included an in-place hydrostatic test pressure 1.5 times design pressure 672 kPa (97.5 psi). Static external loading included the Cooper E80 railroad locomotive and AASHO-H20 truck loads.

The in-situ soil at the pipeline site varied from very stiff to stiff silty clay and medium dense sand. The pipe bedding minimum thickness of 300 mm (12 in.) was to be hand placed and normally tamped granular material. Granular backfill material was deposited in 150 mm (6 in.) layers and compacted to a density of 95% modified proctor maximum density.

Initial 1986 Catastrophic Failure and Failure Investigation

In October 1986, and approximately 10 years after construction, one 6.1 m (20 ft) length of 2,590 mm (102-in.) diameter circulating water pipe catastrophically failed (refer to Figures 3-5). The failure occurred during a high pressure event which was created by a control system malfunction and resulting deadheading of the circulating water pumps. The estimated peak transient pressure immediately prior to rupture was 717 kPa (104 psi). The circulating water pipe was to have been designed for a transient pressure of 834 kPa (121 psi).

FIGURE 3. RUPTURED CIRCULATING WATER PIPELINE

FIGURE 4. RUPTURED PIPE

FIGURE 5. RUPTURED PCCP WIRE FRACTURES

A failure investigation was conducted on the failed pipe section. Fractured prestressing wires with open longitudinal splitting defects were observed. Wire fracture surfaces were corroded, indicating that the wires fractured prior to the pressure surge which precipitated pipe rupture. Tensile tests conducted on the remaining intact, but corroded, prestressing wires resulted in brittle failures at stresses substantially below the manufacturer's minimum specified tensile strength of 1.94 GPa (282 ksi). A significant reduction in ductility was noted, suggesting delayed hydrogen embrittlement. The mortar coating of the failed pipe was observed to be significantly thinner than specified and was substantially carbonated (depleted of calcium hydroxide).

The failed pipe section was repaired using a replacement closure assembly, and the pipeline was returned to service in approximately one week (refer to Figure 6). Several steps were immediately taken to prevent possible future pressure surges of the pipeline including defeating the auto control system, chain locking valve actuators, and slowing down valve closure rate. No significant pressure surges have been experienced since this program was initiated in 1986.

FIGURE 6. PIPE CLOSURE ASSEMBLY REPAIR

Following the 1986 failure, several attempts were made to assess the condition of the remaining circulating water pipe and soil conditions in and around the pipe. Included in these studies were above ground potential surveys, geotechnical surveys and soil resistivity surveys. These studies indicated that the soil conditions were similar throughout the pipeline alignment. Random excavations of several locations revealed that the mortar cover thickness and quality varied greatly between pipe segments, indicating a process control problem during manufacture. To assist in the refinement of a pipe maintenance strategy, Construction Technology Laboratories (CTL) was contracted to conduct a study of the structural performance and remaining serviceability of the PCCP circulating water pipelines.

Summary of Pipe Design and Material Deficiencies

CTL's review of the design for the 2,590 mm (102-in.) diameter PCCP Class A and B pipe revealed that these sections were not adequately designed. When evaluated on the basis of AWWA C 301-72 design provisions, it was noted that Pipe Classes A and B were designed for operating pressures significantly less than 65 psi at maximum soil covers of 8 and 10 ft, respectively. This under-design resulted from the manufacturer's overestimation of the three-edge bearing strength of the pipe. Though not the direct cause of the pipeline failure, this design discrepancy likely contributed to the premature deterioration of the pipeline, since the in-situ hydrostatic pressure test to 672 kPa (97.5 psi) conducted following installation may have resulted in mortar coating cracking and disruption of its ability to prevent corrosion.

Deficiencies in the materials and process control used in fabrication of the PCCP were found to be factors which directly impaired the serviceability of the pipeline. The material shortcomings included the use of a Interpace Class IV prestressing wire which possessed unusually high susceptibility to accelerated corrosion phenomena, and the presence of a defective mortar coating. The mortar coating was carbonated, thin, and did not provide a dense, durable protective encasement of the prestressing wire. These design deficiencies and material defects led to premature corrosion of the prestressing wire. Eventual fracture of the distressed wire resulted in a loss of prestress in the affected pipe.

When prestress is lost, the initially compressed embedded steel cylinder may rebound, effectively delaminating the inner and outer concrete cores. To support the external overburden soil and live loads the two cores behave as an unreinforced concrete shell. For a cracked concrete core to support external loads through arch action, adequate passive lateral soil resistance must exist at the pipe springlines. When soil resistance is inadequate, the pipe may collapse. If the concrete core is able to support the external loads, the exposed steel cylinder will resist internal pressure through hoop tension until corrosion reduces its tensile load capacity to the point of leakage or rupture. Using this mechanism for pipeline distress and failure as a framework for the investigation, analytical studies were conducted to evaluate the structural performance of the distressed PCCP under operating conditions.

Results of Structural Performance Evaluation

The ability of the pipe components to maintain the internal pressure and external loads was evaluated assuming a complete loss of prestress. A 1.4 surge pressure factor applied to the 193 kPa (28 psi) proof stress test conducted by the manufacturer on the steel cylinder established a maximum internal operating pressure of 138 kPa (20 psi) for the cylinder at the time of initial prestress loss. An assumed corrosion rate of 0.13 mm (0.005-in.) cylinder thickness per year was used to project operating pressure limits into time following the initiation of cylinder corrosion. The results of this analysis indicated that the pipe would theoretically be expected to maintain leak tightness for several years following a loss of prestress and initiation of cylinder corrosion if operated at an internal pressure range of 70-140 kPa (10-20 psi).

Assuming the internal pressure is resisted by the cylinder, the concrete core must support the external soil and live loads. A series of analytical investigations using nonlinear soil-structure interaction, structural stability and cracked concrete section analyses was conducted to evaluate the capacity of the concrete core to resist the external loads through arch action. Soil properties obtained from soil investigations conducted at pipe excavation sites were used in the finite element analysis. The 51 mm (2-in.) inner core and the 114 mm (4.5-in.) outer core of the 2,590 mm (102-in.) pipe were analyzed independently to evaluate their individual capacities. External loads considered in the evaluation included the design earth covers and the design vehicle loading. Results indicated that the inner core was susceptible to crushing under all load conditions. Soil-structure interaction analysis of the outer core indicated that passive soil resistance would be mobilized along the pipe springlines, and that a minimum factor of safety of 2.4 existed for concrete crushing under external design loads. Calculated factors of safety for buckling of the concrete cores under external loading were large for all load cases evaluated.

In anticipation of continued pipeline deterioration, a study was conducted in 1992 to determine pipe replacement and re-routing options. The results of this study compared the cost associated with replacing the pipeline using different pipe materials, routes, and buried and above ground options. The most feasible options for replacing the pipeline were projected to cost approximately 35 million dollars. This cost did not include lost revenues and purchased fuel and energy replacement costs associated with two extended outages necessary to complete tie-ins to plant equipment.

Pipeline Inspection Strategy

Inspections play a key role in minimizing risks of pipeline failure. Experience has shown that in areas where the pipe loses significant prestress, the pipe cylinder is sufficiently strained to cause longitudinal cracking of the inner core or localized debonding of the steel cylinder from the inner core. Both of these defects can be detected by rigorous internal inspection. During each major unit maintenance outage (approximately every 18 months), the pipeline has been drained and internally inspected using both visual examination for signs of distress and sounding to detect delaminations of the inner core. Detailed inspection records have been maintained which permit identification of changes in individual pipe sections. These records have proved valuable for assessing the condition of the circulating water pipeline. This inspection strategy has been successfully employed at Brandon Shores since 1986 without any incidence of catastrophic in-service failure.

External Repair Methodologies

Prior to 1993, distressed PCCP identified through the ongoing monitoring program have been successfully repaired using three (3) methodologies:

(1) Removal of the damaged section and replacement with an engineered closure assembly.

(2) Repair of damaged pipe section by epoxy injection of cracked inner core, excavation and encapsulation by reinforced concrete (Figure 7).

(3) Application of an external band of rolled steel plate (for a limited number of sections accessibly located in valve pits.

FIGURE 7. REINFORCED CONCRETE PIPE ENCAPSULATION REPAIR

The encapsulation in reinforced concrete repair method offers specific advantages over the use of closure assemblies in that it does not disturb adjacent pipe segments and joint rings. Each of the above repair methodologies requires access to the pipe exterior to accomplish the repair. This is normally accomplished by excavation. However, approximately 40 sections of pipe are located beneath the building foundation near critical facilities inaccessible by external excavation (Figure 8).

Internal Repair Methodologies

Fortunately, pipe sections under building foundations and close to critical plant equipment operate at low internal pressure of approximately 70-140 kPa (10 - 20 psig). Results of engineering studies indicate that, even with a loss of prestress, there was no inordinate safety risk with operation in this pressure range. However, to provide long term serviceability and to further minimize the risk associated with a failure of PCCP piping in these critical areas, preliminary studies were conducted to evaluate internal repair options. Detailed engineering design of one viable procedure for this internal repair of distressed PCCP at Brandon Shores was developed to provide an additional tool for the pipeline maintenance program.

The minimal structural requirements for an internal repair were to provide watertightness and restore the pipe's capacity to resist internal pressure. The repair was not required to contribute to the resistance of soil pressure or overburden loads; it

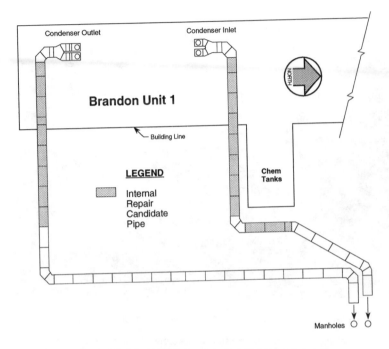

FIGURE 8. PIPE LOCATED NEAR CRITICAL FACILITIES

was assumed that external load would be supported through arch action of the existing concrete core wall. Prior analysis indicated that if the concrete pipe was assumed to have cracked and formed hinges at the crown, springlines and invert, the PCCP would behave much as a flexible pipe in supporting the soil and overburden pressures. This behavior presupposed the lack of bending stiffness of the existing pipe wall section, and thus assumed the total loss of steel wire prestress.

Aside from the functional requirements, a repair method was sought which could be simultaneously implemented on either individual or multiple adjacent or isolated pipe segments. The ability to complete repairs within the two to three week time-frame of a typical scheduled plant outage was necessary to meet plant requirements.

The two broad classifications of internal pipe repair methods evaluated were slip-lining with manufactured liners, and re-lining with liners fabricated or assembled within the pipe. Either method was expected to reduce the internal diameter of the 2,590 mm (102-in.) diameter PCCP, reducing flowrate and negatively affecting condenser performance/generation capability and plant heat rate. BGE engineering determined that the pressure drop and reduced flowrate associated with a reduction of 305 mm (12 in.) in pipe diameter for 20 pipe sections per unit was tolerable, though by no means desirable.

Slip-lining materials evaluated included metal, plastic, concrete and resin-impregnated hose liners. Slip-lining the entire span of 20 critical sections per unit was determined to be feasible and cost effective if implemented during an extended maintenance outage. However, this option could not be easily completed in a typical 2-3 week plant outage duration. Brandon Shores is planning to rehabilitate these critical sections using slip liner technology during the next extended turbine outage currently scheduled in 1998. Therefore, the focus of the investigation shifted to internal repair strategies which could be implemented on individual pipe sections through access provided by existing pipeline openings. Two 0.9 mm (36-in.) manways were available for access of materials, equipment and labor. The additional requirement was imposed that any reduction in diameter, associated with internal repair of individual pipe sections, not restrict the passage of slip liners during the planned comprehensive rehabilitation of the circulating water pipeline.

Methods to internally re-line pipe include formed-in-place liners and erected-in-place liners. Among the former category are pneumatically placed materials such as shotcrete, and hand lay-up materials such as glass-fiber reinforced resins. The lack of water-tightness of these technologies when installed without a secondary liner prohibited their use as primary repair options. Erected-in-place liners for this application required custom fabrication of components, and hand assembly to close tolerances within the confines of the pipe interior. Given the need to withstand internal pressures, and BGE's in-house capabilities, a welded steel plate cylinder with a cementitious liner was selected for detailed design.

Internal Pipe Repair Design

The results of the structural performance evaluation of the pipeline established that the 51 mm (2-in.) thick inner core of the PCCP was not essential for the structural stability of the pipe. It was therefore possible to remove the inner core from a distressed pipe, making the space available for the internal repair components. If the thickness of the reline repair was held to 51 mm (2-in.) there would be no reduction of the inside pipe diameter, no consequent loss of hydraulic section, no reduction in flowrate, and no plant performance penalties. In addition, this design would allow for future installation of the largest diameter slip-liner possible.

Detailed construction specifications were prepared for all aspects of the internal repair procedure. Removal of the inner core by hydrodemolition was to be performed following the erection of bracing within the pipe interior, and then only after any inner core compressive stresses were relieved through a controlled longitudinal cutting procedure. Weld specifications were developed for the assembly of the steel repair liner. Detailed application and material specifications were written for the microsilica shotcrete mortar applied as a protective coating over the steel repair liner.

The main structural component of the internal repair was a 6 mm (0.25 in.) cylindrical steel pipe liner, assembled within the pipe from rolled steel plate segments. The rolled plate segments were dimensioned to fit through the access manways of the pipeline, and could be stacked within the pipe. With the inner core removed, large diameter steel positioning rings were centered within the pipe with shims to provide a 12 mm (0.5 in.) gap between the existing cylinder and the new liner to be assembled. Four rolled plate segments butt welded end-to-end were required to complete a 673 mm (26.5-in.) wide cylindrical section of the repair liner. Each rolled segment was fitted with a steel bar along one edge which served as a backing bar for the longitudinal butt weld. Nine of these cylindrical liner sections were assembled using the positioning rings which served as backing bars for circumferential welds. Longitudinal welds between cylindrical sections were staggered during the assembly. One end of the completed liner was welded to the spigot ring of the repaired pipe and the opposite end of the repair liner was welded to the spigot ring of the adjacent pipe.

The annular space between the existing cylinder and the repair liner was grouted with non-shrink grout through ports installed in the steel liner segments. The grouting operation was controlled to prevent buckling of the assembled liner by hydrostatic grout pressure. Reinforcing steel mesh was attached to the completed repair liner surface, and the repair liner coated with approximately 32 mm (1-1/4-in.) of microsilica fume shotcrete. The ends of the pipe were sealed, the surface of the shotcrete maintained moist, and the shotcrete allowed to cure for 7 days.

1993 Unit 1 Outage and Repair Implementation

During a planned outage inspection of the main circulating pipeline at Brandon Shores Unit 1 in January 1993, six 6 m (20 ft) lengths of 2,590 mm (102-in.) diameter PCCP pipe and one length of 1,980 mm (78-in.) diameter PCCP pipe were identified as being structurally compromised and requiring repair. Two sections of PCCP in difficult to access areas were selected for the trial of the internal repair; the remaining sections were repaired by excavation and encapsulation in reinforced concrete. The internal repair project successfully demonstrated the constructability and economic feasibility of the internal repair design. Work was completed safely (without incident) and within 10 percent of forecast costs for the two internal repairs. The project duration to install these internal repairs was 5 weeks (including 1 week cure) working a single, six day, 10 hour shift. The project was not critical path for the outage therefore an accelerated schedule was not used. Acceleration of the schedule to two or three shifts, and working the remaining seventh day should reduce the duration of construction to 2-1/2 weeks. It may also be possible to reduce the wet cure of the shotcrete to 3 days, bringing the total project duration to 3 weeks.

Logistically, the number of internal re-line repairs that could be completed simultaneously is limited at the site by manway access available to move materials and personnel. It is envisioned that the maximum number may be about four. By comparison, excavation and encapsulation in reinforced concrete can be accomplished in as short a duration as 10 days, weather permitting. During previous outages, as many as four sites and eight pipe sections have been externally repaired concurrently. The internal repair is cost competitive with other repair alternatives for individual pipe sections. Costs for the two internal re-line repairs were $114,000 each versus encapsulation costs which ranged from a low of $90,000 for the 1,980 mm, (78-in.) diameter and $98,000 for the 2,590 mm, (102-in.) diameter pipe to a high of $154,000 for the 2,590 mm, (102-in.) diameter pipe depending on site difficulties encountered. Since the 1993 repairs, the circulating water pipeline has performed satisfactorily.

Conclusions

Changes in operating procedures, implementation of a detailed inspection program, and development of standards for pipe repairs have reduced the risk associated with operation of the distressed PCCP at Brandon Shores Station since 1986. An understanding of the mechanism of the pipe deterioration and the evaluation of the remaining structural serviceability of the pipeline, were key to the development of maintenance strategies. A rigorous internal inspection program has consistently identified significantly distressed pipe sections. As the pipeline ages, it is likely that future inspections will reveal increasing numbers of distressed pipes requiring repair. A major relining effort is planned for the near future on sections of the pipe considered more critical, due to their location to the plant.

An essential component of the maintenance strategy was the development of an internal repair design which could be applied to individual pipe sections not readily accessible by excavation. BGE views the internal relining and external encapsulation methods as complementary repair strategies and plans to continue using the internal relining method primarily on pipe sections which provide limited, difficult access by excavation. The internal repair procedure was successfully implemented on two pipe segments during a scheduled 1993 plant outage. Brandon Shores currently stocks material for two complete internal relining assemblies for possible use during future outages.

Quality Enhancements of Cement-Mortar Coatings

Henry Bardakjian [1]

Abstract

Cement mortar is widely used as a protective coating on pressure pipe incorporating a steel cylinder. Methods of applying cement-mortar coatings are discussed. Three types of mortar coated steel cylinder pressure pipe, designed and manufactured in conformance with American Water Works (AWWA) Standards, are described. Recent revisions incorporated in the applicable AWWA standards include new and revised provisions to increase the density and lower the permeability of mortar coatings. These provisions are discussed and summarized.

The desirable mortar coating quality is discussed and is shown to be dependent on many factors, including the moisture content of the mortar mix and the cement-to-fine aggregate ratio. Mortar coating quality tests are described. Examples of recent mortar coating quality test results are presented.

Introduction

One of the world's first mortar-lined and coated steel pipelines was installed in the City of St. John, New Brunswick, Canada, in 1855. A section of the original 12-inch diameter pipeline removed over a century later was found to be free of steel corrosion (Figure 1). The mortar

---
[1] Chief Engineer, Concrete and Steel Pipe Group, Ameron, Inc., 10681 Foothill Blvd., Ste. 450, Rancho Cucamonga, CA 91730.

coating in this case was mixed in the field and cast in situ around the steel cylinder.

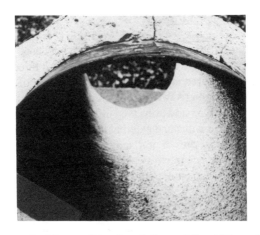

Figure 1. Early Mortar-Lined and Coated Steel Pipe. Installed in the City of St. John, New Brunswick, Canada, this 12-inch pressure pipe has a riveted 16-gage steel cylinder and 3/4-inch cement- mortar lining. The mortar coating was cast in the field. The pipe was removed during relocation work in 1963.

Improved methods of applying cement-mortar coatings and greater understanding of their performance have led to their widespread use on steel cylinder water transmission pipelines.

Cement-mortar coatings for modern steel cylinder pipe (since 1942) consist, by weight, of one part portland cement to not more than three parts of fine aggregate. Water equal to at least 6% of the dry mix weight is added to attain adequate compaction and cement hydration. Application is generally accomplished by high-velocity impaction. The completed coating is cured by steaming or sprinkling.

This paper considers: (1) the principal types of mortar-coated pressure pipe manufactured today; (2) the desirable cement-mortar coating quality; (3) mortar coating quality tests; (4) factors affecting the quality of cement-mortar coating; and (5) overview of the mortar coating quality enhancements in the new AWWA standards.

Types of Mortar-Coated Pipe

Cross sections through walls of three commonly used types of mortar coated pressure pipe are shown in Figure 2. All three types are designed and manufactured in accordance with standards of the American Water Works Association (AWWA). All three pipes are intended to provide the same service; however, they differ greatly in design concepts used and in capacity to withstand external loads. In common, however, all three utilize the properties of portland cement mortar which contribute to pipe stiffness, resistance to physical damage, and prevention of steel corrosion. Of these the most important is corrosion prevention.

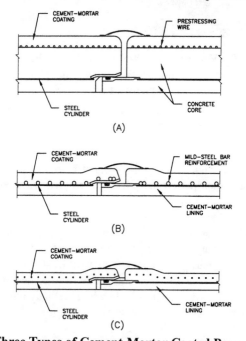

Figure 2. Three Types of Cement-Mortar Coated Pressure Pipe.
(A) Prestressed concrete pipe, embedded-cylinder type, AWWA C301; (B) Concrete cylinder pipe, AWWA C303; and (C) Welded steel pipe, AWWA C200 and C205.

Desirable Mortar Coating Quality

The corrosion-inhibiting properties of portland cement are well known (Scott, 1965) and are attributed to the high alkalinity of hydrated portland cement (pH at least 12.5). Cement mortar coatings effectively prevent steel corrosion of steel cylinder pressure pipe as long as they retain their high alkalinity and remain structurally sound and in intimate contact with the steel. Therefore, the desirable characteristics of the mortar coatings are high compaction and density, low permeability, durability, and structural integrity.

Mortar Coating Quality Tests

Several methods of evaluating the quality of mortar coatings can be used. The four commonly accepted test methods are:

1. Water Absorption Test. This test is performed in conformance with ASTM C497, Method A. The water absorption test is a measure for determining the total porosity of the mortar. A low absorption will indicate low porosity of the mortar. There are two types of porosity that will affect the water absorption:

- Capillary porosity in the hydrated cement paste. This is influenced by the volume of gel formed by the hydraulic reaction between the cement and the water;

- Porosity caused by entrapped air. This is mainly a result of the degree of compaction.

2. Mortar Compression Strength Test. This test is usually used for the purpose of qualifying the mortar-coating application machine and the mortar mix design. The test method procedure is described in Section 3.9.5 of AWWA C301-92 Standard. The compressive strength test is a measure of the degree of compaction of the mortar.

3. Mortar Coating Bulk Specific Gravity(SSD) Test. This test is a measure of the degree of compaction of the mortar. This test is not practical to be conducted on a routine basis.

4. Mortar Permeability Test. The permeability test determines the pattern of voids in the mortar. There are no permeability test methods for mortar or concrete specimens in the ASTM standards. However, different devices and methods are sometimes used. One such test was developed by Ameron in which the mortar specimen is subjected to 20 psi external pressure and permeability is measured in milli-liters per square feet of surface area per inch of thickness per 48 hours. This test is not practical to be conducted on a routine basis.

The most effective single test for mortar coating quality is the absorption test, since a coating with high density and compressive strength and low permeability has a low absorption value.

Factors Affecting the Quality of Cement-Mortar Coatings

The following is a listing of the major factors which influence the characteristics of the mortar coatings:

1. Design, manufacture (including the mortar coating application method and optimization of all the application variables) and installation of the cement-mortar coated steel cylinder pipe.

2. Gradation and properties of fine aggregate used in the mortar coating.

3. Cement-to-fine aggregate ratio.

4. Moisture content of the mortar mix.

The first two items are not in the scope of this paper and thus will not be discussed.

Optimum Cement-to-Fine Aggregate Ratio

Mortar mix proportions by weight of one part cement to not more than four parts of fine aggregate have been specified by some agencies in the early 1940's and 1950's. In the late 1950's, Ameron obtained absorption test results of mortar coatings made with one part cement and four parts fine aggregate and compared them to absorption test results of standard coatings made with one part cement and three parts fine aggregate. The comparative analysis concluded (American Pipe & Construction Co., 1967) that the absorption of the standard coating was lower than the absorption of coating made with a 1:4 ratio mix. It was also believed at that time (no tests were conducted) that a mortar mix richer than a 1:3 ratio would produce a coating with a lower absorption value.

In 1990 through 1992 a Danish consulting firm, G. M. Idorn Consult, conducted a comprehensive test program for Ameron to optimize the cement-mortar coating quality. The test program included optimizing the cement-to-fine aggregate ratio. The absorption for mortar coatings of different mix designs as a function of the water content is given in Figure 3. The results show that the absorption of mortar coatings increases for mix designs richer than a 1:3 mix ratio. The explanation for the higher absorption values with the richer cement content is that the lower air porosity in the mix is offset by an increase of capillary porosity of the hydrated cement paste.

In conclusion, the current standard one part cement to three parts fine aggregate ratio is the optimum mix design for the mortar coatings.

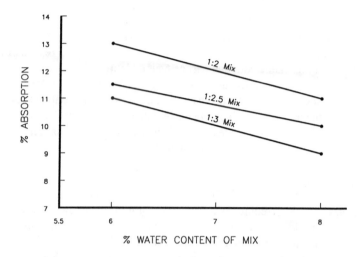

Figure 3. Water Absorption as a Function of Water Content for Different Cement-To-Fine Aggregate Ratios. Tests were done at Ameron's Engineering Experimental Center. The absorption values are not absolute since the experimental mortar applicator is different from that of the actual manufacturing facility applicators. The results show that the absorption values increase with the richer mix designs (richer than 1:3).

Optimum Moisture Content of the Mortar

A minimum water content equal to 6% of the dry mix weight of the mortar has been a requirement in all mortar coating specifications for steel cylinder pressure pipe. The use of this minimum moisture content is ideal for the mortar coating application process; however, the use of a higher moisture content up to an optimum value, based on available raw materials and application method, will result in a higher degree of compaction of the mortar coating. Figure 4 verifies that the absorption of the mortar coating is inversely proportional to the moisture content of the mortar coating.

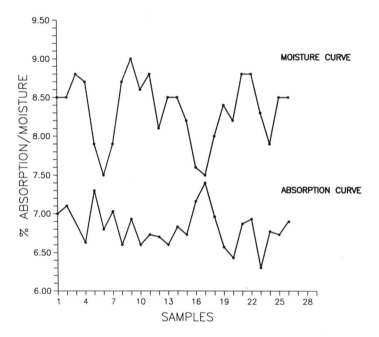

Figure 4. Mortar Coating Absorption of 48-Inch Diameter Prestressed Concrete Pipe as a Function of the Water Content. The pipe was manufactured at Ameron's Phoenix plant. The general trend of the results indicates that higher moisture content will result in lower absorption.

Ameron recently conducted absorption, compressive strength and permeability tests on mortar coating specimens from its Arrow plant. The results are given in Table 1.

Table 1.   Permeability and Absorption Test Results of Mortar Coating Specimens from Ameron's Arrow Plant.

Specimen No.	Moisture Content, %	Absorption %	Compressive* Strength, psi	Permeability**
1	9.36	7.68	6000	0
2	8.16	7.33	6500	1200
3	8.11	7.8	6700	0

*   7-day strength
**   ml/ft^2/in./48 hr. at 20 psi pressure

The results in Table 1 verify that increasing the moisture content of the mortar coating will result in low absorption and permeability, and high compressive strength.

Mortar Coating Quality Enhancements in the New AWWA Standards for Steel Cylinder Pipe

The minimum moisture content of the mortar coating mix has been increased from 6% to 7%. The significance of this change has been already discussed in the above section.

Until recently, the applicable AWWA standards for mortar coatings did not require water absorption or compressive strength tests to verify its quality. The new standards include provision for water absorption tests. AWWA C301-92 Standard for prestressed pipe also includes a provision for mortar compressive strength tests, and a minimum limit on specific gravity for fine aggregate.

All of these new provisions in the applicable AWWA standards will ensure a dense and sound mortar coating, thus enhancing the structural and corrosion protective properties of the cement-mortar coating on steel cylinder pressure pipe. Recent mortar coating absorption and compressive strength test results are given in Figures 5 and 6, respectively.

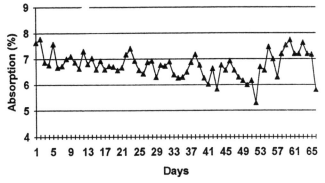

Figure 5. Mortar Coating 7-Day Absorption of 48" Diameter PCCP. The pipe was manufactured at Ameron's Phoenix plant. The daily absorption results are based on the average of 3 samples. The average absorptions are below the 9% requirement by AWWA C301-92.

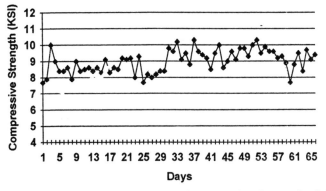

Figure 6. Mortar Coating 28-Day Compressive Strength of 48-Inch PCCP. The pipe was manufactured at Ameron's Phoenix plant. The 28-day compressive strengths are based on the average of 3 samples. The compressive strengths are well above the 5500 psi required by AWWA C301-92.

Conclusions

1. There are several factors which influence the structural and corrosion protective properties of cement-mortar coating on steel cylinder pressure pipe, including pipe design, manufacture, and installation.

2. An increased cement-to-fine aggregate ratio decreases the compaction of the mortar coating. The optimum cement-to-fine aggregate ratio is 1:3 by weight as presently specified in the applicable AWWA standards.

3. The single most critical factor which affects the degree of compaction and quality of the mortar coating is the water content of the mortar mix. The higher the water content, up to an optimum value, the higher the compaction and density of the mortar coating will be.

4. The revised mortar coating quality provisions in the new AWWA standards, applicable to the mortar coatings on steel cylinder pressure pipe, will enhance the structural and corrosion protective properties of the mortar coating.

Appendix

Scott, G. N., "Corrosion Protection Properties of Portland Cement Mortar", J. Amer. Water Works Assoc., Vol. 52, No. 8 (1965), pp 1033-1052.

American Pipe and Construction Co., "Factors Affecting the Quality of Cement-Mortar Coatings", Engineering Topics 10, Monterey Park, CA, Feb. 1967.

SPECIAL PROTECTIVE COATINGS AND LININGS FOR DUCTILE IRON PIPE

A. M. Horton, P.E.[1]

ABSTRACT

Cement-mortar lining has been successfully used to protect the interior of iron pipe and fittings since 1922. Polyethylene encasement has been successfully used to protect the exterior of iron pipe and fittings since 1958. In some cases, either due to installation factors or to the environment to which the pipeline will be exposed, special bonded coatings and linings may be required or specified in lieu of these two standard methods of corrosion protection. This paper discusses the application, advantages, and disadvantages of bonded coating and lining systems which are sometimes specified for iron pipe and fittings.

INTRODUCTION

Since its incorporation in 1899, U. S. Pipe and Foundry Company has evaluated hundreds of coating and lining systems for the protection of iron pipe and fittings. Many test programs, some of which include pipe buried in corrosive test sites in excess of 40 years, are on-going. At the present time, cement mortar linings with an established service record of more than 70 years have furnished the most successful solution to the problem of protecting the pipe from internal corrosion and providing continued high carrying capacity. Polyethylene encasement, with a successful track record of more than 35 years, has furnished the most successful and economical solution to the problem of protecting the pipe from external corrosion in most environments.

In some cases, either due to installation factors or to the environment to which the pipeline will be exposed, special bonded coatings and/or linings may be required or specified. For example, a polymeric, acid resistant lining should be used in lieu of cement mortar linings for conveying septic sewage where the pH of the sewage can become extremely acidic due to the formation of sulfuric acid. A polymeric exterior coating with good weathering resistance may be used in lieu of polyethylene encasement for above ground applications such as pipe supported on piers or installed beneath a bridge. A tough, abrasion resistant polyurethane exterior coating may be specified in lieu of polyethylene encasement for pipe installed using trenchless technology.

[1]Process Engineering Manager, United States Pipe and Foundry Company, Technical Services, 3300 1st Ave. North, Birmingham, AL 35222

This paper discusses the application, advantages, and disadvantages of several bonded coating and lining systems which have been specified or used on iron pipe and/or fittings. These comments are based on U. S. Pipe and Foundry Company's first-hand experience with bonded coatings and linings applied to cast iron pipe and fittings and may not agree with experience of others on steel pipe and other structures. Coating and lining specifications for protective coatings on steel pipe and structures are occasionally cited or specified for ductile iron pipe and fittings. Although ductile iron and carbon steel are both ferrous metals, there are inherent manufacturing/processing differences between the two materials that preclude certain parts of specifications written for carbon steel surfaces from being applied directly to ductile iron pipe. Some of the problems with these standards being specified for ductile iron pipe are also discussed.

HISTORY

The use of a barrier between a corrosive environment and the material to be protected is a fundamental method of corrosion control which is widely specified and which has been used since earliest recorded history. The Egyptians used varnishes as early as 4000 B.C. Polychrome Greek statues became common by 300 B.C. and evidence exists that the Romans used coatings for both decorative and protective purposes.*(NACE, 1984)*

Ferrous metal pipe has been used in the conveyance of water and other liquids for over 300 years. The oldest known cast-iron main still in service was installed at Versailles, France, in 1664. Cast iron pipe was introduced to the United States around 1816. This early iron pipe was not coated nor lined. In certain types of waters, internal corrosion attack (referred to as tuberculation), was causing a build-up of an incrustation which reduced the flow through the pipe by decreasing the effective diameter and increasing the friction in the line. Around 1860 this problem was of sufficient concern that it became general practice to dip pipe in a bituminous molten tar to provide protection to both the interior and the exterior surfaces.*(Wagner, 1969)*

The use of thin film tar dip coatings and linings continued for many years, but the problem of tuberculation in aggressive waters was not solved until a thin layer of cement mortar began to be used for internal corrosion protection. The first recorded installation of cement mortar linings in cast iron pipe was in 1922 at Charleston, S.C. The pipe was lined by means of a projectile drawn through the pipe. This process soon gave way to the centrifugal process. Since 1922 many improvements have been made in the processing of cement lined iron pipe and today, cement mortar linings are normally supplied on ductile iron pipe and fittings for water service unless otherwise specified. This lining is described in American Water Works Association Standard ANSI/AWWA C104/A21.4-90 (*American National Standard for Cement-Mortar Lining for Ductile-Iron Pipe and Fittings for Water*).

In the 1940's, hot tar external coatings began to be replaced with sprayed asphalt cutback coatings. While this material offered some protection in corrosive environments, a better method of external corrosion protection was needed. Polyethylene film became commercially available in 1950, and in 1951-52 testing of polyethylene encased iron pipe exposed to various types of corrosive soils was initiated.*(Wagner, 1964)* The first commercial application of polyethylene encased cast iron pipe was in 1958 in an extremely corrosive, swampy area in Lafourche Parish, LA.

Since 1958 polyethylene encasement has been used to successfully protect millions of feet of gray and ductile iron pipe in thousands of installations across the United States and the entire world. The success of polyethylene encasement has led to the development of an international standard (ISO 8180) and numerous national standards (United States - ANSI/AWWA C105/A21.5 and ASTM A674, Great Britain - BS 6076, Australia - AS 3680 & 3681, Germany - DIN 30674, Part 5, and Japan - JDPA Z 2005). The excellent success of polyethylene encasement to protect iron pipe has been extensively documented in tests, reports, and publications over the past 35 years *(Wagner, 1964 : Collins, 1982: Horton, 1988: Stroud/NACE, 1989: Stroud/AWWA, 1989: DIPRA, 1991)* and it is beyond the scope of this paper to provide additional discussions on this standard method of protection.

Prior to the 1950's, development of coatings and linings in the iron pipe industry was related to the areas of cement mortar linings and bitumastic coating materials. With post war development of new polymeric coatings, an extensive test program was undertaken by U. S. Pipe and Foundry Company to evaluate some of these materials on iron pipe and fittings. Testing and evaluation included not only laboratory testing, but also application trials and field installations at test sites and in operating water and sewer systems. Because new coating materials enter the marketplace almost daily, the test program initiated in the 1950's to evaluate polymeric, bonded coatings and linings is still underway. It is the results of these evaluations, and also experience gained in supplying special coatings and linings over the past 40 years that provide the basis for the following comments regarding coating and lining systems for ductile iron pipe and fittings.

SURFACE PREPARATION

Surface preparation is a critical part in the life of any coating or lining system. Proper surface preparation can depend on many variables such as type of surface being prepared, type of cleaning required (i.e. solvent cleaning, power tool cleaning, abrasive blast cleaning, water blasting, etc.), ambient conditions, degree of cleanliness required, surface profile required, type of cleaning equipment, type of blast media, blast pressure, nozzle diameter, dry and clean air, etc. Following proper safety precautions is also an important part of cleaning operations. All of these considerations are discussed in other publications *(NACE, 1972: NACE, 1973: SSPC, 1989)*, and are beyond the scope of this paper.

Several publications and standards exist for surface preparation on steel surfaces. They include, but are not limited to Steel Structures Painting Council *(SSPC, 1991)*, National Association of Corrosion Engineers Visual Standards, and Swedish Surface Preparation Standards. NOTE: All of these standards are for the surface preparation of steel surfaces and are not directly applicable for ductile iron pipe. Since both gray and ductile iron fittings have a sand cast surface that is more similar to a steel surface, the written SSPC surface preparation standards typically apply for both of these products.

Although ductile iron and carbon steel are both ferrous metals, there are inherent manufacturing and processing differences between the two materials that preclude certain parts of the aforementioned standards from being applied to ductile iron pipe. The Ductile Iron Pipe Research Association addressed these concerns in a 1993 article titled "Surface Preparation for Ductile Iron Pipe Receiving Special Coatings"*(DIPRA, 1993)*. Because the interior and exterior surfaces of ductile iron pipe vary depending on the casting process, the pipe manufacturer should be consulted concerning surface preparation of ductile iron pipe to receive special coatings or linings.

Special Precautions for Surface Preparation of Iron Pipe and Fittings

- Unlike steel pipe, it is possible to "overblast" the external surface of ductile iron pipe. High nozzle velocities and/or excessive blast times can cause "blistering" and "slivering" of some DeLavaud pipe surfaces. The layer that over blasting can cause to "blister" is the epidermal layer on ductile iron pipe which is formed as a result of the DeLavaud centrifugal casting process. This comment does not apply to sand cast fittings.
- Standard asphalt cutback coating is normally supplied on the interior and exterior of ductile iron pipe and fittings. Due to the incompatibility of virtually all polymeric coatings with asphalt, application of special coatings on top of an asphalt is not recommended. Also, because asphaltic cutback is extremely difficult to completely remove from a ductile iron surface, uncoated pipe and fittings should be specified on orders which are to receive a special coating and/or lining.
- As discussed previously, a standard cement-mortar lining is normally applied to ductile iron pipe and fittings unless otherwise specified. Asphalt sealcoat may be used as a topcoat for the cement lining. Because proper surface preparation of a cement mortar lining is extremely difficult (i.e. it must be properly cured for 28 days, stored in an atmosphere where it will not absorb moisture, and then properly blast cleaned without damaging the lining), special polymeric lining materials are not recommended to be applied over cement mortar linings.

THICKNESS OF COATING AND LININGS

The recommended thickness of a coating or lining will depend on the coating or lining material, the product being protected, the desired design life, and the intended usage of the product. The coating material manufacturer will normally publish a "recommended thickness" as part of the technical data sheet for the product: however, this thickness recommendation is normally based on smooth steel surfaces. Due to the rough texture of ductile iron pipe surfaces which is inherent to the manufacturing process, a greater coating or lining thickness is required on ductile iron pipe than is required for steel surfaces. Normally, ductile iron pipe, as cast, will have a higher surface profile than can be achieved with steel through blasting.

As a general rule, a polymeric internal lining material for ductile iron pipe is recommended to be a 40 mil (0.040") nominal thickness and an external polymeric coating is recommended to be a 24 mil (0.024") nominal thickness. Both the coating manufacturer and the pipe manufacturer should be consulted to obtain recommended thickness for specific coatings and linings depending on the products to be protected and the type of service anticipated.

APPLICATION

In addition to proper surface preparation, proper application of the coating or lining material is also essential in achieving the full design life of the material specified. The recommended application procedure for a specific coating or lining material is normally published as part of the product's technical data sheet. These recommended procedures normally include, but are not limited to, the following areas:

- Recommended shelf life and storage conditions for the coating material
- Ambient condition restrictions (i.e. humidity, temperature, etc.)
- Max/min temperature of liquid coating and product to be coated
- Mixing procedure (for plural component materials)
- Maximum pot life and "sweat time" (i.e. time before using)
- Dry film thickness per coat
- Recommended thinner and amount (if allowed)
- Max/min time between coats
- Proper equipment and technique(s) of application
- Proper curing or drying cycles

Detailed discussions of proper application procedures of polymeric coatings are discussed elsewhere in several excellent publications *(NACE, 1972: NACE, 1973: SSPC, 1989)* and are beyond the scope of this paper. In the event of conflicts between general recommendations concerning application procedures, and the coating manufacturer's recommendation regarding a specific coating for a specific product, the coating manufacturer's recommendation should be followed.

INSPECTION

Inspection of ductile iron pipe with special coatings or linings should be carried out during surface preparation, application, final acceptance, and during installation.

- Inspection during surface preparation should assure that the surface to be coated has not been contaminated with any foreign substances, that it has received the specified degree of cleaning using the specified procedures, and that it meets the coating manufacturer's recommendations. Inspection procedure normally consists of visual inspection but can also include surface profile measurement and surface contamination analysis.

- Inspection during application should assure that the coating is applied in accordance with the coating manufacturer's recommendations. Inspection procedure can include temperature & humidity measurements, paint viscosity, wet film thicknesses, and visual observation of the application process.

- Inspection during final acceptance should assure the special coating and/or lining meets specifications. Inspection procedure can include dry film thickness measurements, holiday testing (a test for pinholes or discontinuities in the coating), and adhesion testing. In the event holiday test methods and voltages specified are different than the recommendations of the coating manufacturer, the coating manufacturer's recommendations should prevail. Excessive test voltages can damage an otherwise good coating or lining and this voltage can vary depending on the type and thickness of the material specified.

- Visual inspection during installation should assure that no damage occurred during transit, unloading, and installation at the jobsite.

EXTERNAL COATINGS

Excluding polyethylene encasement and standard asphalt coating, external coatings which have been specified for the exterior of ductile iron pipe include, but are not limited to: polyurethane, coal tar epoxy, coal tar enamel, tapewrap, extruded polyethylene, metallic zinc, zinc/epoxy/polyurethane, and fusion bonded epoxy. Because holidays and voids in bonded organic coatings have proven to be more detrimental than small holidays in polyethylene encasement, many of these bonded exterior coatings are specified to be used in conjunction with cathodic protection or other methods of corrosion control. Brief comments about each of the bonded exterior coating systems listed above are given in the following pages:

Polyurethane

The high build polyurethane coating which has been used extensively on ductile iron pipe since 1988 is an ASTM D-16 Type V system which consists of a polyisocyanate resin and polyol resin which are mixed in 1:1 ratio at the time of application. It is a high solids, high build two component polyurethane coating which cures quickly to form a hard, yet flexible film which is resistant to chipping, cracking, and impact damage *(U. S. Pipe, 1993, POLYTHANE®)*.

Due to its quick setting properties, excellent resistance to handling and impact damage during shipment and installation, and excellent resistance to a wide variety of chemicals, septic sewage, and soils, this material has replaced coal tar epoxy as one of the most popular special bonded exterior coatings and interior linings for ductile iron pipe. One case history with this material on the exterior of 31,000 feet of 12" ductile iron pipe which was shipped over 2,000 miles, reports this exterior coating had little or no shipping and handling damage, and had an installed coating efficiency of 99.66% when tested as part of a cathodic protection system *(Madison Chemical, 1994)*.

Coal Tar Epoxy

Coal tar epoxy has been used to protect the interior and exterior of iron pipe in excess of 40 years. It has proven to be an excellent protective coating if properly applied at sufficient thicknesses. One disadvantage is that most coal tar epoxies are recommended to be applied at approximately 8 mils (0.008") dry film thickness per coat, and the epoxy must be allowed to cure properly (approximately 12 to 24 hours depending on temperature) between additional coats. Thus, for a 24 mil coating, which is the recommended nominal thickness for coal tar epoxy on the exterior of ductile iron pipe, coating application on a pipe can take several days to complete.

Coal tar epoxies are routinely specified to be applied to steel pipe in accordance with American Water Works Association Standard ANSI/AWWA C210-92 *(AWWA Standard for Liquid Epoxy Coating Systems for the Interior and Exterior of Steel Water Pipelines)*. Although coal tar epoxy linings and coatings have been applied for a number of years to ductile iron pipe, the AWWA C210 standard is not directly applicable. This standard references SSPC and NACE surface preparation standards and, as discussed previously, these standards are not directly applicable to ductile iron pipe. Also, the 16 mil (0.016") minimum thickness specified in this standard is insufficient for ductile iron pipe for reasons previously discussed.

Coal Tar Enamel

Coal tar base thermoplastic coatings have been used for many years to successfully protect the exterior of iron pipe and fittings. Some of the earliest iron pipe produced were reported to be dipped in a heated solution of coal tar pitch. Since 1940, coal tar enamel coatings have been routinely specified to be applied to steel pipe in accordance with American Water Works Association Standard ANSI/AWWA C203-91 *(AWWA Standard for Coal-Tar Protective Coatings and Linings for Steel Water Pipelines - Enamel and Tape - Hot Applied)*. This standard for external coating requires a primer, followed by a hot coat of coal tar enamel on which a single layer of specified outerwrap is applied. The coating is then finished with either one coat of water-resistant whitewash or water-emulsion latex paint, or a single wrap of kraft paper.

Although Coal tar enamel and tape has been applied on a limited basis to ductile iron pipe, the AWWA C203 standard is not directly applicable. This standard references SSPC and NACE surface preparation standards and, as discussed previously, these standards are not directly applicable to ductile iron pipe. Also, if the ductile iron pipe to be coated with hot enamel are cement mortar lined, the hot molten coal tar applied to the pipe can potentially damage the cement mortar lining.

Disadvantages of this coating system include health and safety problems associated with asbestos in the felt wrap (can be replaced with glass fiber wrap), fumes from the hot molten coal tar, and handling the hot molten coal tar which is normally several hundred degrees F. Also, the bells on push-on and mechanical joint ductile iron pipe normally present problems on production lines which apply this type of coating system.

Tapewrap

A tapecoating system is normally a prefabricated, cold applied tape system which consists of a primer, a corrosion preventive inner layer tape, and a mechanical protective outer layer tape. Tapewrap coatings are routinely specified to be applied to steel pipe in accordance with American Water Works Association Standard ANSI/AWWA C214-89 *(AWWA Standard for Tape coating Systems for the Exterior of Steel Water Pipelines)*. Although tapewrap coatings have been applied on a limited basis to ductile iron pipe, the AWWA C214 standard is not directly applicable. This standard references SSPC and NACE surface preparation standards and, as discussed previously, these standards are not directly applicable to ductile iron pipe.

The primary advantages of this system are 1) it is capable of very fast production rates on pipe, 2) no cure time is required and the pipe can be shipped almost immediately after coating, and 3) depending on the extent of damage, it can be repaired relatively easily in the field. The primary disadvantages of this coating system are: 1) on large diameter (i.e. greater than 24") heavy ductile iron pipe, the tapewrap is easily damaged during handling, shipping, and storage, 2) the bells on push-on and mechanical joint ductile iron pipe normally present problems on production lines which apply this type of coating system, 3) irregular shaped products such as fittings must be taped by hand which is expensive, time consuming, and dependent upon the skill of the person doing the application, and 4) protection applied to joints after assembly in the field can be time consuming and expensive. Problems associated with applying tapewrap coatings to ductile iron pipe are addressed by the Ductile Iron Pipe Research Association in a product advisory *(DIPRA, 1992)*

Extruded Polyethylene

Extruded polyolefin coatings are routinely specified to be applied to steel pipe in accordance with American Water Works Association Standard ANSI/AWWA C215-94 (*AWWA Standard for Extruded Polyolefin Coatings for the Exterior of Steel Water Pipelines*). Although extruded polyolefin coatings are routinely supplied on ductile iron pipe by one manufacturer in Europe, this coating system is normally not specified for ductile iron pipe domestically. Because AWWA C215 standard references SSPC and NACE surface preparation standards, this coating standard is not directly applicable to ductile iron pipe.

Metallic Zinc/Asphalt Topcoat

Metallic zinc exterior coatings for ductile iron pipe are primarily supplied on ductile iron pipe for the export market. European experience has shown this to be an effective protective, sacrificial coating in moderately aggressive soil, but in very aggressive soils it is recommended only if used in combination with polyethylene encasement.

Zinc/Epoxy/Polyurethane

This system consists of 1) a zinc rich primer for corrosion protection and undercutting resistance, 2) an epoxy "middle" coat for corrosion protection, and 3) a thin (i.e. less than 5 mils) polyurethane topcoat for gloss retention and resistance to weathering. Due to the excellent gloss retention, UV resistance, and availability of colors of the PUR topcoat, this system is specified for above ground applications such as pipe on piers, chemical and treatment plant piping requiring color coding, and under bridges.

Disadvantages of this coating system are: 1) it is normally expensive compared to other systems, 2) it can be easily damaged during transit and installation, 3) it can take several days to apply (or repair) the three layer system due to the cure time required for each layer prior to application of the successive layer.

Fusion Bonded Epoxy

Fusion bonded epoxy coatings and linings are routinely specified to be applied to steel pipe in accordance with American Water Works Association Standard ANSI/AWWA C213-91 (*AWWA Standard for Fusion bonded Epoxy Coating for the Interior and Exterior of Steel Water Pipelines*). Although a few fusion bonded linings and coatings have been applied to small diameter, relatively smooth ductile iron pipe on a limited basis, the AWWA C213 standard is not directly applicable because SSPC and NACE surface preparation standards are referenced.

Also, the 16 mil (0.016") maximum/ 15 mil (0.015") minimum thickness specified for internal linings and the 16 mil (0.016") maximum/ 12 mil (0.012") minimum thickness specified for external coatings in this standard is insufficient for ductile iron pipe for reasons previously discussed. Fusion bonded epoxy coatings and linings applied at a sufficient thickness to ductile iron pipe to obtain a holiday free coating can result in improper cure of the thick epoxy film unless a post-cure procedure is used.

At present, due to difficulties in obtaining holiday free coatings and linings using thin film fusion bonded epoxies on rough internal and external surfaces inherent to ductile iron pipe, this coating system is not recommended for ductile iron pipe. Since both gray and ductile iron fittings and castings have a sand-cast surface that is more similar to a steel surface, AWWA C213 can be applied to these products.

INTERNAL LININGS

Excluding standard cement mortar lining and asphalt sealcoat, linings which have been specified for the interior of ductile iron pipe include, but are not limited to: ASTM C150 Type V cement mortar, Calcium Aluminite cement mortar, polyethylene, polyurethane, ceramic epoxy, coal tar epoxy, drinking water epoxy, fusion bonded epoxy, polyester/vinylester, and glass lining. Brief comments about each of these systems are given below:

Type V Cement Mortar Lining

ASTM C150 Type V sulfate resisting cement mortar is sometimes specified in lieu of Type I or Type II cement mortar linings where increased resistance to sulfates is needed (i.e. as in conveying sea water). In general, Type II cement linings are recommended for waters containing up to 1,000 ppm sulfates, and Type V cement linings are recommended for waters containing up to 2,000 ppm sulfates. Type V cement mortar linings will perform slightly better in sewage environments, but neither of these materials are recommended for septic sewage service where the pH of the sewage can become extremely low due to sulfuric acid formation.

Calcium Aluminate Cement Mortar Lining

This special type of cement mortar lining has been promoted by some manufacturers for use in septic sewage service. It is applied in accordance with American Water Works Association Standard ANSI/AWWA C104/A21.4-90 (*American National Standard for Cement-Mortar Lining for Ductile-Iron Pipe and Fittings for Water*) with the exception that the mortar is composed of fused calcium aluminate binders and fused calcium aluminate aggregates. This lining is recommended down to a pH of 2. Where septic conditions are suspected which could result in pH values lower than 2, an acid resistant polymeric lining material should be considered.

Polyethylene

Fusion bonded polyethylene linings have been applied to ductile iron pipe and fittings and used successfully in corrosive sewer and wastewater service for over twenty years. A 40 mil (0.040") nominal thickness with a 2,500 volt holiday test is recommended for sewer linings. Advantages of this lining material are 1) it is chemically resistant to a wide variety of chemicals, 2) it is extremely abrasion resistant, and 3) it requires no cure time after the pipe has cooled. The main disadvantage to this lining material is that it is susceptible to undercutting at unrepaired cut ends on pipe which are cut to length in the field. Proper repair of damage and field cut ends are strongly recommended for this lining material. Repair materials and instructions are available from the pipe manufacturer.

Polyurethane

Refer to comments under External Coatings-Polyurethane. A 40 mil (0.040") nominal thickness with a 2,500 volt holiday test is recommended for sewer linings. In addition to the polyurethane which has been used successfully as an internal lining for sewer service and as an external coating, a special grade of polyurethane which is approved by the National Sanitation Foundation (NSF) is available for contact with drinking water and foodstuff.

Ceramic Epoxy

The ceramic epoxy used successfully as a lining material on ductile iron pipe for the past ten years is a high build, multi-component, amine cured, and ceramic filled Novalac epoxy. Novalac epoxy, which is one of the most chemical resistant epoxy resins made, combined with a patented blend of ceramic filler for impact resistance, chemical resistance, and undercutting resistance, was specially formulated for protection of ductile iron pipe for sanitary sewer service. It has been used successfully in hundreds of sanitary sewer applications, and has been proven with both laboratory testing and years of actual sewer service on all sizes of ductile iron pipe and fittings *(U. S. Pipe, 1993, PROTECTO 401)*. A 40 mil (0.040") nominal thickness with a 2,500 volt holiday test is recommended for sewer linings.

Advantages of this lining material are 1) excellent resistance to undercutting, even at unrepaired areas of damage, 2) 40 mils (0.040") can be applied in a multipass, one coat process, 3) indefinite recoatability, and 4) ease of field repair. The disadvantages are 1) the Novalac resin may take two to three days (depending on temperature) to cure hard enough for testing, and 2) this material is not recommended for an external coating.

Coal Tar Epoxy

Refer to comments under External Coatings-Coal Tar Epoxy. A 40 mil (0.040") nominal thickness with a 2,500 volt holiday test is normally recommended for sewer linings.

Drinking Water Epoxy

Due to the excellent success and low cost of cement mortar linings, drinking water epoxies are normally not specified for ductile iron pipe. This material has been used, however, to coat the interior of bell sockets and the exterior of spigot ends where the metal components of the joint were to be in contact with a very aggressive water. Only drinking water epoxies tested and approved in accordance with National Sanitation Foundation (NSF) Standard 61 are recommended for contact with drinking water.

Fusion Bonded Epoxy

Refer to comments under External Coatings - Fusion Bonded Epoxy. As discussed, fusion bonded epoxy is not recommended for ductile iron pipe but can be used on ductile iron fittings and castings. If the fusion bonded epoxy is to be in contact with drinking water, it is recommended that the material be tested and approved in accordance with National Sanitation Foundation (NSF) Standard 61.

Polyester/Vinylester

Polyester and vinylester linings normally have excellent chemical resistance to a wide range of chemicals and are also much more resistant to high temperature applications than many of the other lining materials discussed. They are normally very expensive when compared to the other types of linings and, as a result, are specified for severe service applications where no other lining material is suitable. Examples of where this group of lining materials have been used is chemical plant and textile plant waste lines.

Glass Lining

Glass linings have been applied to ductile iron pipe for many years for use in sludge lines. In general, the glass lining process is capable of only building the film thickness to approximately 10 mils (0.010") maximum. On the rough internal surface inherent to most ductile iron pipe, a holiday free lining is extremely difficult to achieve at this thickness. As a result, many ductile iron pipe manufacturers will not supply a glass lined product, but will furnish pipe to customers unlined for application of glass lining by qualified applicators.

SELECTION OF COATING AND LININGS

Tabulated below are some of the criteria which should be taken into consideration when selecting a special coating or lining for application to ductile iron pipe and fittings:

- Applied and Installed Cost
- Track Record/Case Histories
- Adhesion
- Ease of application
- Cure/dry time/recoat time
- VOC regulations
- Health & Safety regulations
- Ease of field repair
- Appearance
- Gloss Retention
- Flexural properties
- Antistick properties

- Chemical resistance
- Abrasion resistance
- Resistance to a given environment
- Impact resistance
- Temperature resistance
- Resistance to sunlight
- Undercutting resistance at damage
- Compatibility with other coatings
- Max/Min thickness requirements
- Production/Delivery time
- Cathodic Protection disbondment
- Dielectric Strength

As discussed previously, polyethylene encasement for external protection and cement mortar lining for internal protection have furnished the most successful and economical solution for protection of iron pipe in most environments. When special conditions preclude the use of these methods of protection, a special coating or lining should be selected after taking into consideration the criteria listed above. Pipe manufactures, coating manufacturers, and the Ductile Iron Pipe Research Association stand ready to offer assistance in selecting the most economical protection system for a specific installation of ductile iron pipe and fittings.

REFERENCES

COLLINS, H.H. 1973, "The Use of Polyethylene Sleeving for the Protection of Buried Spun Iron Pipes", British Cast Iron Research Association, Alvechurch - Birmingham, England (1973)

DIPRA 1991, "Polyethylene Encasement - Effective, Economical Protection for Ductile Iron Pipe in Corrosive Environments", Technical Advisory of the Ductile Iron Pipe Research Association, Birmingham, AL (December 1991)

DIPRA 1992, "Product Advisory: Tapecoat", *Ductile Iron Pipe News*, Fall/Winter 1992, Ductile Iron Pipe Research Association, Birmingham, AL (1992)

DIPRA 1993, "Surface Preparation of Ductile Iron Pipe to Receive Special Coatings", *Ductile Iron Pipe News*, Fall/Winnter 1993, Ductile Iron Pipe Research Association, Birmingham, AL (1993)

HORTON, A. M. 1988, "Protecting Pipe With Polyethylene Encasement, 1951-1988", *AWWA Waterworld News*, Vol. 4, No.3 May/June 1988

MADISON CHEMICAL 1994, "Madison Case History 10.94.13 - 111 Years Projected Design Life", Madison Chemical Industries, Inc., Milton, Ontario, Canada (1994)

NACE 1972, *Coatings and Linings for Immersion Service, TPC Publication No. 2*, National Association of Corrosion Engineers, Houston,TX, p.245 (1972)

NACE 1973, *Industrial Maintenance Painting*, National Association of Corrosion Engineers, Houston,TX, p.245 (1973)

NACE 1984, *Corrosion Basics-An Introduction*, National Association of Corrosion Engineers, Houston,TX, p.245 (1984)

SSPC 1989, *Good Painting Practice - Steel Structures Painting Manual, Volume 1*, third edition, Steel Sturctures Painting Council, Pittsburgh,PA (1989)

SSPC 1991, *Systems and Specifications - Steel Structures Painting Manual, Volume 2*, sixth edition, Steel Structures Painting Council, Pittsburgh,PA (1991)

STROUD/NACE 1989, "Corrosion Control Measures for Ductile Iron Pipe", Paper No. 585, CORROSION 89, National Association of Corrosion Engineers, Houston, TX (April 17-21, 1989)

STROUD/AWWA 1989, "Corrosion Control Methods for Ductile Iron Pipe", *AWWA Waterworld News*, Vol.5, No.4, American Water Works Association, Denver, Co (July/August 1989)

U. S. Pipe 1993, "POLYTHANE® Lined Ductile Iron Pipe and Fittings for Force Mains and Gravity Sewer Lines, 1993 Edition", Brochure of United States Pipe and Foundry Company, Birmingham, AL (1993)

U. S. Pipe 1993, "PROTECTO 401 Ceramic Epoxy Lined Ductile Iron Pipe and Fittings for Force Mains and Gravity Sewer Lines, 1993 Edition", Brochure of United States Pipe and Foundry Company, Birmingham, AL (1993)

WAGNER, E. F. 1969, "Corrosion Control with Cement Mortar Lined Pipe", Proceedings 25th Conference, National Association of Corrosion Engineers, pp.123-129 (March 1969)

WAGNER, E. F. 1964, "Loose Plastic Film Wrap as Cast-Iron Pipe Protection", *Journal of AWWA*, 56, No.3, 361 (March 1964)

Cost Effective Corrosion Mitigation

Gerald A. Craft[1]

ABSTRACT

Agencies who own and operate underground utilities are aware of the tremendous investment required to install such systems, and of how disruptive failure to those systems can be. In order to protect that investment, it is prudent to consider the risk of corrosion and how best to mitigate it.

There are several methods in use in the battle against corrosion with varying opinions on which is best. The recommendations of the ductile iron pipe industry are founded in research, field trials, and over four decades of service history. Yet, these recommendations are often widely divergent from those offered by many corrosion consultants.

Ultimately, of course, it is the design engineer's responsibility to specify which system will be installed. One factor which should not be overlooked is overall economics, and not just the initial acquisition/installation costs.

This paper examines an actual project for the installation of approximately one mile of 30" ductile iron transmission main. Cost figures were obtained from a ductile iron pipe manufacturer and a corrosion protection supplier. Installation estimates were obtained from an underground contractor. Sometimes called Value Engineering, or Life Cycle Cost Analysis, the total cost to purchase, install, and maintain two different corrosion protection systems for the estimated service life of the pipeline are compared. The results are sufficiently dramatic to warrant any design engineer taking a closer look at all of the options.

INTRODUCTION

Corrosion is the deterioration of a material, usually a metal, due to a reaction with its environment. That definition is provided by the National Association of Corrosion Engineers.[2] Everyone recognizes corrosion as a natural phenomena that

---

[1] Regional Sales Engineer, United States Pipe and Foundry Company, 1295 Whipple Road, Union City, California 94587

cannot be eliminated. There are various methods available that are intended to achieve sufficient mitigation of corrosion to allow the products being protected to achieve their designed service life. It is outside the intended scope of this paper to discuss the mechanics of corrosion, or to debate the technical merits of one system of corrosion mitigation over another.

One system that may be specified for the corrosion protection of underground ductile iron pipelines is cathodic protection. Cathodic protection generally requires some type of bonded coating be placed on the pipe prior to installation. This method also requires joint bonding for electrical continuity and some means of applying an electrical current to the pipeline to make it a cathode. The current is supplied in one of two ways: either by impressed current from an outside power source, or through the use of buried sacrificial anodes. Cathodic protection also requires test stations that must be monitored periodically to ensure the system is maintained at optimum efficiency.

Each of the components of a cathodic protection system are project specific, and must be designed after careful consideration of several factors.

The cost of installing a cathodic protection system, while significant in itself, is not the only cost to be considered. Cathodic protection is an active system. The only way to keep it active is to maintain it. The only way to maintain it is to inspect it regularly, test it regularly, maintain and analyze test records, adjust the system as conditions change, replace anode beds as they wear out, and keep the current flowing for the life of the project, all at substantial cost to the owner. These after-installation costs can become substantially larger than the initial acquisition costs.

The ductile iron pipe industry recognizes that a properly designed, installed, and maintained cathodic protection system can be an effective means of mitigating corrosion of underground pipelines. However, the method most often recommended by this industry for the protection of its pipelines when the need for protection is indicated is loose-film polyethylene encasement, manufactured and installed in accordance with ANSI/AWWA C105/A21.5.[3] Within the industry, this method is often referred to as Polywrap, or Baggies.

Polyethylene encasement is available in either 4 mil thick cross-laminated high density polyethylene or 8 mil thick low density polyethylene. It is usually supplied in tubular form in rolls. The 8 mil form is generally perforated at twenty-foot long intervals for use on nominal eighteen foot long pipe, or twenty-two foot long intervals for use on nominal twenty foot long pipe. The tube is slipped on over each length of pipe as the joints are made-up, and secured in place with plastic adhesive tape or plastic strapping.

COST EFFECTIVE CORROSION MITIGATION   759

no maintenance, no power consumption, and no interference with other utilities. Polywrap is a simple system that is not labor intensive to install.

To keep things in perspective, there is no perfect system for corrosion protection. Not one of them is the universal panacea to mitigate corrosion under every conceivable condition, and there is much disagreement among the experts regarding the merits of one system over another. It is still a matter of choice for the design engineer, but one factor which should not be overlooked is economics.

The focus of this paper is to examine the economics of corrosion protection for underground ductile iron pipelines. In order to avoid becoming mired in a technical debate that cannot be won in this forum, it is necessary to proceed on the premise that both cathodic protection and Polywrap are effective methods of corrosion control.

The ductile iron pipe industry generally considers a useful service life for its products of one hundred years. In order to achieve this service life in a corrosive environment, whichever method of corrosion control is specified must remain effective throughout that period of time.

Following are the results of a study done in 1994 to compare the costs of Polywrap vs. the costs of cathodic protection as applied to the same project.

A recent project in Honolulu, Hawaii was selected for this study because there was adequate information available to enable a fair cost comparison. The McCully Street, Waikiki job bid in 1992 with an engineer's estimate of $8 million. This study focused on the approximately one mile of 30" ductile iron pipe. The plans called for 30" Tyton Joint pipe, thickness Class 52, with cathodic protection. The pipe manufacturer had recommended polyethylene encasement for corrosion protection, but the Honolulu Board of Water Supply had hired a corrosion consultant who recommended a cathodic protection system.

If Honolulu had accepted the pipe manufacturer's recommendation, the cost of polyethylene encasement is calculated as follows:
        5,280 ft. of pipe X 1.11 = 5,860.8 ft. of Polywrap needed
            (1.11 is a factor that allows 1 ft. of overlap at each end of the pipe.)
        5,860.8 ft. ÷ 440 ft./roll = 13.32 rolls, or 14 full rolls
        440 ft./roll X 14 rolls = 6,160 ft. of Polywrap
        For 30" pipe, use 67" flat tube width at $1.01/ft. (includes tape)
        6,160 ft. X $1.01/ft. = $6221.60

An underground contractor in the San Francisco Bay Area was contacted for information on the added costs to install the polywrap on this project. After reviewing the plans, he determined that he would not add an additional cost for installing polywrap. By way of explanation, he stated that on a low production job

(30" pipe & larger, city street, many cross services, etc.) there is a lot of crew waiting time. He doesn't have to add a man for polywrap or welding jumper cables, so he doesn't add cost above the basic installation crew. On a high production job (24" & down, open field construction, etc.) he would have to add a man to the installation crew to keep up with the polywrap or welding jumpers, so he adds 50 cents per foot. That means the City would get a mile of protection for just the cost of the materials: $6221.60.

The cathodic protection system that was selected has several components that were examined separately, including the tape coating, the cathodic protection hardware, installation, and inspection.

This project required a coal tar tape coating consisting of a primer layer, an inner tape layer of 19 mils minimum thickness for corrosion protection, and an outer tape layer of 27 mils minimum thickness for mechanical protection, all in conformance to AWWA C-214.

For 30" pipe, this project required double joint bonding using #2 AWG wire thermite welded in place. The design of the cathodic protection system required a test box, one zinc reference electrode and ten 32 lb. magnesium anodes per anode bed. There were to be thirteen anode beds spaced 1000 ft. or less apart. The cathodic protection system was to have a theoretical design life of twenty years.

A coating contractor quoted $10.88 per foot to tapewrap the mile of pipe with a 50 mil system complying with AWWA C-214.
    5,280 ft. X $10.88/ft. = $57,446.40
That is approximately nine times the cost of the polywrap.

Joint bonding required two jumper cables of #2 AWG stranded copper wire with HMWPE insulation, four Cadweld cartridges, four plastic weldcaps, and some mastic per joint. The one mile of pipe contains approximately 295 joints. A corrosion protection firm in the San Francisco Bay area quoted $10.00 per joint for bonding wires and accessories. As mentioned earlier, the underground contractor would not add for installing jumpers on 30" pipe, so installation is "free."

    295 joints X $10.00 per joint = $2950.00

That is a subtotal of $60,396.40 for tapewrap and joint bonds (or almost ten times the cost of the polywrap). These are essential parts of the system for cathodic protection to be effective.

For the anode beds, the project requires one zinc reference electrode, ten 32 lb. magnesium anodes, a test box, two thermite welds, and wiring per bed. The same Bay Area company quoted a cost of $1000.00 per anode bed. The underground contractor looked at the plans and estimated installation cost at $3132.00 per anode bed.

$1000.00/bed X 13 beds = $13,000.00 for materials
$3132.00/bed X 13 beds = $40,716.00 for installation
Total = $53,716.00 for anode beds

That's a subtotal of $114,112.40 to have this cathodic protection system installed, or more than 18 times the cost of the polywrap.

A cathodic protection system is an active system. It requires periodic inspection to ensure it is continuing to function. This project specified a design life for the cathodic protection system of twenty years. A Bay Area corrosion consultant quoted a cost of $1800.00 for annual inspection, based on one day in the field reading voltage potentials and 1/2 to 1 day in the office preparing the report. Since the system is designed for a twenty-year service life, 4% per year was added to the initial cost to account for inflation. The total for nineteen inspections, beginning the year following installation was calculated to be $49,808.21. This brings the total cost of the sacrificial anode cathodic protection system for this project to $163,920.61, or more than 26 times the cost of the polywrap.

As previously mentioned, the cathodic protection system for this project was required to have a theoretical design life of twenty years. The ductile iron pipe industry considers their products to have a service life of one hundred years. That means that, in order to continue to protect the initial investment in the pipeline, the cathodic protection system will have to be replaced when it wears out, as many times as necessary to meet the service life of the pipeline.

This brings us to a term that is being used increasingly by responsible engineers: Life Cycle Cost Analysis. For those not familiar with the concept, it means doing a realistic analysis of all the costs associated with achieving the expected service life of the major investment. Life Cycle Cost Analysis enables a comparison between alternatives to determine which is truly the most cost effective.

The Life Cycle Costs for polyethylene encasement was examined first. It is known that Polywrap does not deteriorate or wear out in service, because every time pipe is exhumed at a test site, a sample of Polywrap is tested and found to meet the requirements for new material. So, there is no need for annual inspection, no maintenance, no periodic replacement. After the initial purchase and installation, polywrap does not cost another cent.

The Life Cycle Costs for the cathodic protection system specified for the McCully Street job was examined next. It has been shown that this sacrificial anode system costs over $160,000.00 to install and maintain for the first twenty years. It has a design life of twenty years, which means it will have to be replaced every twenty years for the life of the pipeline.

Earlier we saw that the materials cost for the anode beds at the time of installation was $13,000.00. Adding a modest 4% per year to account for inflation, the materials cost for identical anode beds in the year 2009 will be over $27,000.00. That cost will increase to almost $58,000 in 2029, about $120,000 in 2049, and over a quarter of a million dollars in 2069.

Our total projected cost to purchase new anodes every twenty years throughout the life cycle of the pipeline is $462,801.03.

The cost to install the anodes initially was $40,716.00. Following a consistent premise of adding 4% annually, the cost to install new anode beds will be over $85,000 in the year 2009, over $180,000 in 2029, over $380,000 in 2049, and over $800,000 in 2069. The total projected cost to install new anode beds every twenty years throughout the life cycle of the pipeline is $1,450,000.

At this point in the Life Cycle Cost Analysis, the sub-total for hardware and installation is $1,912,801.03.

It should be remembered that cathodic protection is an active system that must be inspected periodically. For the sacrificial anode system installed on McCully Street, **annual** inspection was recommended. An impressed current system requires much more frequent monitoring. Someone has to check the power rectifier at least monthly to make sure it is still supplying continuous DC current to the ground anode bed, plus do potential readings, an analysis, and report at least every six months. If the owner agency staff is not equipped to maintain the system in this manner, a corrosion consulting firm can usually provide this service for a fee.

Following the same trend of adding 4% annually for inflation, it was shown that the total cost of inspection for the twenty year life of the initial anode system was $49,808.21. Continuing that trend, the total cost of inspection for each twenty year increment is over $100,000 for the period ending in 2029, over $200,000 in 2049, over $415,000 in 2069, and over $841,000 in 2089. The total cost of inspection of this cathodic protection system to maintain protection for the life of the pipeline is $1,611,849.11.

The total costs to provide cathodic protection for the anticipated service life of the 30" ductile iron pipeline follows:

Initial installation	$114,112.40
Purchase and install anode beds	1,912,801.03
Inspection	1,611,849.11
Grand Total	$3,638,762.54

The one mile of 30" ductile iron pipe sold for just under $375,000.00. Comparing the cost of corrosion protection to the initial cost of the pipe provides some interesting figures:

Cost of Protection /Cost of Pipe = X%

$$\frac{\$6,221.60 \text{ (Polywrap)}}{\$374,880.00 \text{ (Pipe)}} = 2\%$$

$$\frac{\$163,920.61 \text{ (Cathodic, 20 years)}}{\$374,880.00 \text{ (Pipe)}} = 44\%$$

$$\frac{\$3,638,762.54 \text{ (Cathodic, 100 years)}}{\$374,880.00 \text{ (Pipe)}} = 971\%$$

The Life Cycle Cost of the cathodic protection system amounts to approximately 585 times the Life Cycle Cost of the loose-film polyethylene encasement. An important point to consider is that after the project is installed and accepted, the annual maintenance costs would likely have to come from an agency's operating budget.

Present Worth is the initial sum of money that must be compounded for the life span of the project to accumulate the total cost of the project. In terms of Present Worth, in order to earn the more than $3.6 million for cathodic protection that the McCully job is going to cost Honolulu over the 100 year life of the pipeline, the Board of Water Supply would have to deposit:

almost $190,000 at 3% compound interest
$72,000 at 4% compound interest
almost $28,000 at 5% compound interest.

CONCLUSION:

The cost to install a water transmission main in an urban environment is unquestionably substantial. The impact of the premature failure of such a pipeline certainly warrants the consideration of means to protect that pipeline from corrosion. An actual project was used as the basis to compare two different methods of affording corrosion protection: sacrificial anode cathodic protection as specified in the plans and loose-film polyethylene encasement as recommended by the pipe manufacturer.

This study showed the initial acquisition and installation costs of the cathodic protection system were approximately eighteen times the costs to purchase and install loose-film polyethylene encasement.

In performing Life Cycle Cost Analysis for the anticipated 100 year life of the pipeline, a comparison of the two methods showed that the cathodic protection system would cost approximately 585 times the cost of the polyethylene encasement. Such a dramatic difference in the costs of two systems should be worthy of note to any pipeline designer.

REFERENCES:

2. "NACE Standard RPO169-92" - Standard Recommended Practice, Control of External Corrosion on Underground or Submerged Metallic Piping Systems, National Association of Corrosion Engineers, Houston, Texas, Copyright 1992.

3. "ANSI/AWWA C105/A21.5-93", - American National Standard for Polyethylene Encasement for Ductile-Iron Pipe Systems, American Water Works Association, Denver Colorado, Copyright 1994.

New Bonded Tape Coating Systems and Cathodic Protection applied to Non-Steel Water Pipelines: Quality Through Proper Design Specifications

James R. Noonan[1]
Bryan M. Bradish P.E.[2]

Abstract

Anticorrosion considerations are becoming increasingly important in the design of metallic water pipeline structures such as steel, ductile iron and prestressed concrete cylinder pipe (PCCP). While steel water pipelines have traditionally used a variety of anticorrosion protection methods, protection for ductile iron and prestressed concrete cylinder pipe has seldom been considered because those piping systems have served adequately for decades. It was once thought they had innate characteristics that limited or prevented corrosion. However, some municipalities are reviewing corrosion control alternatives for key pipelines installed in "hot" (corrosive) soils (Denn, Charles, 1994).

During the last half decade some service interruptions of ductile iron and prestressed concrete cylinder pipe have been linked to corrosion-related issues, which may play a more important role in service life limits than originally thought. Thus, anticorrosion protection considerations are branching out beyond steel water pipe to embrace all water pipelines that use and/or incorporate metallic materials in their design (Bianchetti, R.L., 1993; Robinson, William, C., 1993).

This paper will discuss the critical performance issues involved in using a combination of new tape coating systems and cathodic protection as employed on steel water pipe. It will review critical performance issues that must be

---

[1] Anticorrosion Consultant, 618, South 11th St., Lafayette, IN. 47905
[2] Chief Engineer, Engineering Div., Newport News Waterworks, 2600 Washington Ave., Newport News, VA. 23607

considered in adapting their use on the other popular water pipeline systems (ductile iron and prestressed concrete cylinder pipe). Are the anticorrosion systems, which have worked well for steel water pipe, directly transferable to non-steel water pipe systems? Should the different construction methods used in the installation of steel water pipe materials be considered in the engineering design specifications for the tape coating and cathodic protection system? What performance and construction issues are most important in adapting the tape coating and cathodic protection systems to water pipelines that are not steel water pipelines? This paper will highlight important engineering design specification issues and give design engineers the information they need to make quality oriented design decisions, optimize adaptation and insure effective performance of bonded tape coating systems and cathodic protection on non-steel water pipelines.

Introduction

The water pipeline industry has experienced extensive and excellent anticorrosion control performance with bonded polyethylene (PE) coating systems on steel water pipe (Steel Plate Fabricators Association, 1992; Thompson Pipenews, 1994). Due largely to the quality track record of the bonded PE system, it is being widely adapted for application to ductile iron and prestressed concrete cylinder pipe when right-of-way anticorrosion control is deemed necessary. Currently, the most widely used vehicle for the adaptation of bonded PE tape coating systems is the American Water Works Association (AWWA) anticorrosion standard as referenced in *AWWA C-214 Tape Coating Systems for the Exterior of Steel Water Pipelines.* The standard stipulates that a 50- or 80-mil-thick coating system be applied to the pipe depending on the pipe diameter. However, the de facto industry standard is to specify a multilayer system at a minimum of 80 mils of bonded PE. Pipe sizes above 144 inches in diameter require multilayer systems greater than 80 mils thick.

The single most important element dictating the quality and success of a pipeline construction project is the quality and thoroughness of the design specifications. This applies to all aspects of the project and especially to the anticorrosion protection coating specification. All too often, a project's total anticorrosion protection coating specification consists of the words "anticorrosion coating to be applied in accordance with AWWA standard C-214." The problem with acceding the entire project's anticorrosion specification to an industry association standard such as AWWA C-214, is that association standards are "minimum" guidelines to be used in writing a full and proper specification. They are not meant to stand alone as a specification for a total project.

Thus, design engineers involved with ductile iron pipe and/or PCCP can avoid common anticorrosion coating specification pitfalls and achieve successful projects of the highest quality through proper design specifications.

Reasons for adapting AWWA C-214 bonded PE coatings to ductile and PCCP

The steel water pipe industry has had a long, extensive, and positive history with the use of bonded PE tape coatings for anticorrosion control. The AWWA C-214 standard for bonded PE on steel pipe was developed over a decade ago and has served as a highly recognized reference standard.

Literally hundreds and hundreds of miles of steel water pipeline projects, from 6- to 144-inch diameter, have been installed in every area of the United States, from coast to coast as well as throughout the world. Operational data from many of the country's top water utility companies as well as bell-hole inspections have shown that bonded PE coatings conforming to the AWWA C-214 standard provide excellent long-term anticorrosion control. Numerous non-destructive, in-ground coating performance tests--such as pipe-to-soil electric potential data, which gives corrosion engineers the ability to gauge an installed coating's in-ground performance quality--have demonstrated current requirements as low as 0.1 microamps per square foot (Polyken Technologies, Pipeline Performance Survey, 1980-94). Current requirements of such low values indicate that an installed pipe and its coating is operating extremely well. Bonded PE systems are not only totally compatible with cathodic protection (CP), they are also most efficient for optimizing the effectiveness of the CP system, thereby extending CP performance life and offering utilities reduced CP operating expense.

Thus, because of bonded PE's strong, long-term quality record for steel water pipe, scores of design engineers of ductile iron pipe and PCCP projects have simply transferred the PE anticorrosion technology to these pipe materials when site right-of-way conditions require anticorrosion control measures.

Why are bonded anticorrosion coatings being used on ductile iron pipe and/or PCCP

It was once thought that ductile iron pipe and PCCP had innate characteristics which limited or prevented corrosion. However, over the last half decade, some water utilities have begun to investigate pipeline site right-of-way conditions for soil resistivity as well as for the presence of stray electrical currents. The use of an anticorrosion coating, such as bonded PE coatings, where right-of-way soil conditions are "hot" (low resistivity/high corrosivity) or in areas where stray currents are a problem is intended to extend the service life of the subject pipelines (Denn, 1994).

We all know that the world has become more complicated, with the rate of complication increasing at an ever-accelerating rate. As we build more pipelines (oil and gas as well as water pipelines) and more mass-transit systems, right-of-way corridors become more congested, and stray currents become a more frequent problem, threatening the integrity of pipelines as never before. The potential of future commercial expansion and urban development in areas that are now rural requires water utility engineers to look many years into the future. Pipelines and their rights-of-way, which may currently be far removed from the hustle and bustle of six-lane traffic, malls, and mass transit systems, could one day be key corridors where stray currents abound and where future service repairs could be extremely inconvenient and costly. Many water utility engineers now believe that the best way to build long-term operating flexibility into a pipeline project, especially into important transmission lines, is to design the pipeline for site conditions as projected twenty years into the future and for service lives far exceeding one hundred years. These engineers believe the best way to gain operational flexibility is to anticipate the potential of congested corridors and stray currents, and prepare for them years in advance by coating important transmission lines with bonded coatings and equipping them with cathodic protection (CP).

Bonded PE coatings and CP also offer the owner/operator a means of continually monitoring the pipe's service condition. If corrosion issues arise, they can be detected early in the scheduled, routine maintenance of the coating/CP system. Thus, operators can detect corrosion problems in advance of a critical failure. Early detection means that the operator can take a proactive, planned approach to pipeline maintenance rather than reacting with crisis management to an urgent service failure. The ability to continually monitor the service quality of the pipeline and to schedule proactive maintenance could save municipalities thousands of dollars and prevent potentially severe service failure, which could result in serious customer dissatisfaction.

Different pipe material and different construction techniques affect anticorrosion coating decisions

Bell profile and transition to pipe barrel

The tape width of bonded PE anticorrosion systems should be of primary importance to the design engineer. All bonded PE coatings should be applied to the pipe at a certified coating application plant. The key concern of many coating applicators is the speed of the through-put; the faster the pipe can be coated, the more profitable the project is to the applicator. Thus, the applicator will often choose an anticorrosion tape with the widest possible width (12 or 18 inches), since the wider the width, the faster the pipe can be coated. Generally, wider widths also mean the applicator will purchase less material and, thus, further reduce production costs. But the wider the tape is, the less likelihood of the coating's wrinkle-free conformance to the bell/barrel profile transition. Thus, the margin for application error increases with wider tape widths, and some coated pipe may have wrinkles near the bell region of the pipe. The coater may blame the trucker, saying the wrinkles occurred during transport; the trucker may blame the contractor, saying the wrinkles occurred during the off-loading; and the contractor may blame the trucker, the applicator, the coating supplier, or all three. The result is that the pipe owner is stuck with a non-optimal coating that must be repaired, possibly at an extra cost.

The entire situation can be avoided if the design engineer specifies that the tape width shall be no more than 9 inches for pipes of 12 through 24 inches in diameter and no more than 12 inches wide for pipe diameters exceeding 24 inches. Choosing the correct tape width should insure that the bell of the pipe is coated with a tough, machine-applied, high-density outer-layer coating, giving the pipe's prone area, the bell, the best possible protection against handling damage. Non-optimum tape widths often mean the bell must be coated by hand, which requires the application of an inherently less robust outer coat that is more susceptible to handling damage. An added benefit of specifying a narrower-width coating system is a tighter more uniformly applied coating that is more resistant to transportation and handling damage. All bonded PE plant-applied coatings should have a minimum of a 1-inch overlap (3 to 4 inches for PCCP). Specifying the proper coating width on the design specification will inhibit the applicator's profit motive or the coating manufacturer's inventory from dictating the width; hence, the project's quality will be enhanced.

Application of pre-heated bonded PE anticorrosion coatings

Polyethylene is a thermoplastic material, which means it becomes more flexible and expands when it is heated. Most quality manufacturers of bonded PE systems recommend that the inner and outer anticorrosion coats be heated to a minimum roll body temperature of 90°F. Pre-heating the coating prior to application dramatically improves the application and conformability of the entire anticorrosion system to the pipe. The pre-heated coating will conform to the bell/barrel transition far more effectively than cold-applied coatings. Furthermore, when the coated pipe cools to ambient temperatures, the cooled polyethylene of the bonded anticorrosion coating shrinks slightly, inducing added tension and conformability, which makes the coating even more resistant to transportation and handling damage. The best way to insure that the bonded PE coating is applied under the correct pre-heat conditions is, first, to specify the minimum pre-heat temperatures as noted above, and second, to require that the coating specifications be strictly followed and substantiated with quality certification that is documented by an in-line, parametric release program.

Indigenous / non-indigenous right-of-way backfill

Ductile iron pipe and PCCP are vastly different materials than steel pipe. Naturally, the differences between the three pipe materials requires vastly different construction practices. Flexible steel water pipe typically uses non-indigenous select backfill with high compaction specifications as a means to maintain the constant circular dimension of the pipe's diameter and limit pipe deflection. The traditional use of select backfill for steel water pipe results in a very uniform and stress-free environment for the pipe and its bonded PE coating. Ductile iron pipe and PCCP, on the other hand, are rigid pipe structures and thus do not require the uniformity of select backfill to maintain their circular dimensions. Therefore, the backfill used in ductile iron pipe and PCCP projects is usually indigenous to the original project site. If the pipe is coated with an anticorrosion system, it (the coating) could experience soil stress characteristics far different from steel water pipe conditions. When site conditions warrant the use of anticorrosion coatings on ductile iron pipe or PCCP, it is important that design engineers consider the backfill characteristics and the potential effect on the pipe's coating.

Pipe diameter and the clay content within the soil generally work together and have a direct stress effect on anticorrosion coatings. That is to say that given a constant backfill type, the greater the pipe's diameter, the greater the subsequent soil stresses on the coating. The stresses are generally manifested as shearing vector forces tangent to the buried pipe's circumference, starting at the twelve and six o'clock positions of the pipe and abating at the three and nine o'clock positions of the pipe (Kellner et al., 1986). Generally, soil stress on pipe diameters of 24 inches and less is of no concern for the bonded PE systems used in the water industry, regardless of soil type. However, if pipe diameters are in the range of 42 inches and greater, then the design engineer should consult the bonded PE coating manufacturer for recommendations regarding non-indigenous backfill specifications or the use of high soil-stress-resistant, bonded PE coatings. Nearly all major manufacturers of bonded PE anticorrosion coatings have high soil-stress-resistant coatings available. If soil stress is an issue, then the design engineer should also consult the coating manufacturer for recommendations on the proper specifications for high soil-stress-resistant joint coatings.

Handling, off-loading, and lowering-in

Steel water pipe has always been anticorrosion coated so traditional steel pipe installation contractors are well versed in I-beam and double sling off-loading and lowering-in. Traditional ductile iron pipe and PCCP contractors are not accustomed to handling their pipe with I-beams and double slings. The historical use of the rigid, uncoated pipe has often allowed the use of wraparound chains for lowering the pipe into the ditch. Large diameter pipe (greater than 30 inches) coated with bonded PE systems should be hoisted utilizing 12- to 18-inch wide double sling webs hung from an I-beam. Hoisting by chains should not be allowed.

Specifications for coating irregular specials and mega-lugs

Irregular specials and mega-lugs are difficult structures to coat with bonded PE systems. Few applicators have workers skilled enough to assure wrinkle-free applications to such convoluted structures. Generally, special structures such as mega-lugs should have separate coating specifications and should utilize epoxies or other corrosion-resistant paint coatings.

Specifications for coating joints, specials, elbows, and patch repair

All major bonded PE manufacturers offer a full array of high-quality hand-applied coating systems for field application to jointed bell and spigot sections as well as to typical T's, stems, and valves. Hand-applied field patch and repair systems are also available. The design engineer should specify that the widths of these systems must never exceed 6 inches and that the pipe surface should be prepared according to the coating manufacturer's recommendations, i.e., clean, primer, etc. Potential voids at the "step-down"" between the bell and spigot should be filled with moldable mastic sealant as recommended by the manufacturer of the bonded PE coating. Once the step-down void areas are filled, the joint should receive a complete, wrinkle-free cigarette wrap of a 2- or 4-inch-wide high-tack joint wrap coating that is hand applied under tight tension. Once the bell and spigot step-down is coated, the entire joint section should be hand coated using a similar high-tack, hand-applied coating of 4 or 6 inches in width. The joint wrap coating should take place starting 2 to 4 inches onto the pipe barrel's mill-applied coating, continuing across the bell and spigot joint to the adjacent mill-coated pipe, either in a spiral or cigarette wrap fashion with a 1-inch overlap as a minimum specification. The specifications of the joint wrap applications should stipulate at least two full layers of a 35-, 40-, or 50-mil-thick joint wrap. If backfill soil is to be indigenous to the right-of-way, then the design engineer should specify that the entire joint wrap coating receive at least one complete overcoat with a perforated plastic rock shield (3/8 to 5/8 of an inch thick).

Cathodic Protection Compatibility

Cathodic protection (CP) was developed nearly a half century ago; its goal is to prevent corrosion of buried metallic structures. There are two terms that are used frequently in the CP industry, anode and cathode, which must be defined here. The "anode" is the location on the metal where corrosion takes place and metal ions are lost. The "cathode" is the location on the metal to which metal ions migrate. There is no metal loss at the cathode.

Corrosion is the destructive attack of metal by chemical or electro-chemical reactions with its environment. Corrosion cannot take place at the anode without the respective presence of the cathode. CP prevents corrosion of a specific metal structure by sacrificially corroding an alternative metal, usually magnesium. A corrosion engineer will use a galvanic anode, such as magnesium, and connects it via a wire to the metal structure, e.g., a pipeline, that needs protection. The magnesium will serve as the anode for the pipeline, corroding while the pipe becomes cathodic and resistant to corrosion. This occurs according to the laws the galvanic series where metals high on the galvanic scale (magnesium) will be anodic and corrode relative to metals low on the scale (steel), which will be cathodic. In this way, a potential corrosion cell is artificially set up by a corrosion engineer by employing an anode, such as magnesium, and connecting it to a pipeline. The driving voltage between the magnesium anode and that of the pipe is greater than the driving voltage found between the indigenous anodes and cathodes on the pipeline. The points on the pipeline that would have been naturally anodic (corroded) are forced to become cathodic by virtue of the higher driving voltage of the sacrificial magnesium anode. Thus, when an electrical circuit is completed, any bare steel that is exposed to the soil environment will resist corrosion. It is saved by the magnesium anode, which corrodes instead. The more bare metal, or holidays, a coating incurs, the greater the rate of the corrosion of the magnesium anodes. Thus, a high-quality coating on metal pipe will require less sacrificial magnesium anodes to protect it.

Direct current rectifiers and their associated sacrificial anode beds are used to supply the CP current for the underground structure. The rectifier/anode bed is a constant voltage source, which is connected to the pipeline so that any holidays that are exposed to the soil become cathodic. In this way, any metal exposed by holidays, which may have been incurred during installation, will resist corrosion.

The voltage potential along the coated pipeline drops from the initial value at the voltage source until it comes under the influence of the next rectifier/anode. This is due to the natural electrical resistance of the entire system. Overapplication of CP can occur when the CP system is designed with fewer anode beds than necessary. Voltage potentials between the anode beds can drop to levels that are too low to make the distant locations of the pipeline cathodic. In order to compensate for this and still continue to protect the distant pipeline areas, CP voltages are set higher, which also often results in an increase in current. The increased current "throws" the cathodic inducing current further down the pipeline, which protects the distant sections. However, while the "distant" section is now protected, the "near" section is now exposed to too much voltage and current. This increase in cathodic current can cause coating disbondment. This phenomena happens to all coatings, but it only happens to holidays, or damaged areas of a coating. It cannot occur on coatings where no holiday exists because there would be no complete circuit and, thus, no cathodic current demanded. Overprotection by CP can be prevented by installing an optimum number of anode beds along a pipeline's right-of-way.

Bonded PE coatings are completely compatible with CP. Bonded PE systems optimize CP economics because PE is an excellent insulator. This means that the CP current and anodes are utilized only at holiday areas and are not wasted or dissipated across the entire pipeline structure as with other more conductive coatings. Thus, bonded PE coatings optimize and reduce CP installation expenses as well as long-term operating costs (Steel Pipe Fabricators Association, Technical Bulletin, 1994).

Conclusion

Right-of-way conditions, such as low resistivity soils and stray currents, or future operational conditions, such as projected urban development, may dictate that bonded PE coatings be used on non-steel metallic pipelines such as ductile iron pipe and/or PCCP. Design engineers should enhance and augment the adoption of AWWA Standard C-214 Tape Coatings for the Exterior of Steel Water Pipelines with specific notice to the criteria outlined in this paper. Enhancing the AWWA C-214 standard for adaptation of bonded PE systems to ductile iron pipe and PCCP will dramatically increase the quality of a pipeline's design specifications and can offer the owner/operator enhanced service life, operational flexibility, and reliable long-term performance.

References

Bianchetti, R. L. "Corrosion & Corrosion Control of Prestressed Concrete Cylinder Pipelines--A Review," *Materials Performance,* NACE, August 1994: 62-66.

Denn, C. "Looking into the Future," *Construction Magazine* 61, no. 18 (Sept. 26, 1994): 10-12.

Kellner, J. D., et al., "Interaction of Soil and Coatings on Buried Pipelines," *Proceedings,* Conference 26, Australasian Corrosion Association, Adelaide, Australia, November 1986.

Polyken Technologies, Mansfield, Mass., "Pipeline Performance Survey," Tape Coated Steel Water Pipe Project Performance Data, 1980-94.

Robinson, C. William. "Lessons from Stray Current Analysis of a Ductile Iron Water Main Failure," *Materials Performance,* NACE, July 1993: 10-13.

Steel Plate Fabricators Association, "Vital Steel Water Pipeline Performing as Promised," *Steel Water Pipe Report* 2, no. 3, 1992.

Steel Plate Fabricators Association, *Technical Bulletin* 1, 1994.

Thompson Pipenews, "15 years, 3.1 Million Feet for Tape Coated Steel Pipe," Winter 1995

CDF - A NEW BEDDING SYSTEM FOR CLAY PIPE
Edward J. Sikora[1] - National Clay Pipe Institute
John Butler - Pacific Clay Products
Robert Lys - Mission Clay Products

Abstract

NCPI, in cooperation with Pacific Clay Products and Mission Clay Products installed two field test lines to measure the bedding factor and installation procedure when Controlled Density Fill (CDF) materials are used in the embedment zone.

Field Test No. 1 - NCPI and Pacific Clay Products - Riverside, CA

Field test line No. 1 consisted of three pipe installed in a crushed stone envelope system with an established 2.2 bedding factor and three pipe installed in a Controlled Density Fill. The pipe were instrumented with strain gages to measure maximum tensile strains for comparison between the two systems for the purpose of developing a bedding factor for the CDF technique. The material for the CDF section contained 100 lbs. cement, 250 lbs. flyash, 2700 lbs. sand and 60 gals. of water per cu. yd. with 1.25% of entrapped air. The pipe in the CDF section were first laid on a flat bed of the crushed stone material. The CDF was then poured from the bottom of the test pipe to the top of the pipe barrel in two equal lifts. There was no indication of flotation in the test area.

The fill was poured by means of a vertical chute directly into the pipe zone from the Ready-Mix truck. The installation was completed without difficulty and the contractor was especially eager to use the system as a replacement for concrete arch or cradle. The strains in the two test sections are reported.

---

[1]Vice President - Director of Technical Services, National Clay Pipe Institute, P.O. Box 759, Lake Geneva, WI 53147

A ratio of the average strains in the two test areas yields a computed bedding factor of 3.0 when CDF material is used. A conservative value of 2.8 is recommended.

Field Test No. 2 - NCPI and Mission Clay Products - Pittsburg, KS

NCPI, in cooperation with Mission Clay Products, conducted additional tests to determine if buoyancy would be a problem with clay pipe and to refine the installation procedure.

The tests were all successful and it was concluded that clay pipe can be bedded with CDF material without special restraints to prevent flotation.

Introduction

Controlled Density Fill (CDF), also called flowable fill and controlled low strength materials (CLSM), is gaining in popularity as a pipe bedding material for installing clay pipe. CDF consists of a sand, cement, flyash and water mixture which has been increasingly used in utility and telephone crossings, backfill in confined areas and stabilization of underground structures. The material flows freely into narrow and irregular spaces and, when solidified, is capable of being paved without further delay. CDF has several advantages over compacted backfill in that it is not subject to time-dependent consolidation, compaction variables or migration. It is also capable of being excavated by conventional equipment at a future date, a feature which adds to its versatility. With compressive strength in the 100 to 300 psi range, CDF is a very good alternative to conventional pipe bedding and backfill materials and, in many instances, is particularly suitable as a replacement for concrete arch and concrete cradle installations.

There have been many authoritative articles written on the merits of CDF and it is not the intent of this paper to present extensive information on the physical characteristics of this material. Additional information may be obtained from the National Ready Mixed Association[1]. Suffice to say that the material, as routinely supplied, possesses qualities that make it very suitable as a bedding material for clay pipe.

Objectives

One of the major reasons for investigating CDF is its potential for replacing concrete reinforcement in very deep and very shallow installations. Contractors are facing increased difficulty when using concrete arch and concrete cradle because of the extended curing time required before backfilling during cut and fill operations. Trench safety requirements require trenches to be backfilled as quickly as possible which places the concrete arch or cradle in structural jeopardy due to an insufficient curing time in many cases. In contrast, the CDF material functions almost entirely in compression and is therefore capable of backfilling as soon as the material solidifies.

To determine an acceptable bedding factor and installation procedure for clay pipe when CDF materials are used, the National Clay Pipe Institute (NCPI) conducted two field tests.

Field Test No. 1

NCPI, in cooperation with the Santa Ana Water Authority, Boyle Engineering Corporation, California Construction and Pacific Clay Products, Inc. installed a field test line in Riverside, CA. The test installation was part of a normal sewer installation. The test portion consisted of three pipe installed in a crushed stone encasement system with a 2.2 bedding factor (Figure 1) and three pipe installed in a controlled density fill (CDF) (Figure 2). The pipe were instrumented to measure maximum tensile strains for comparison between the two installation systems for the purpose of developing a bedding factor for the CDF material. The clay pipe were 39 inches (975 mm) in diameter and 6 feet (1800 mm) in

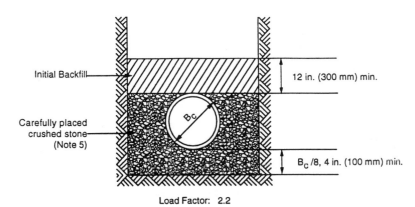

FIG. 1 CRUSHED STONE ENCASEMENT

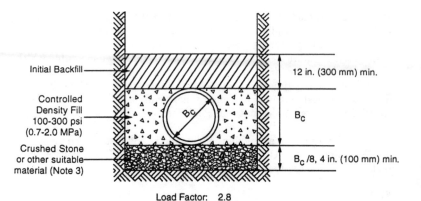

FIG. 2 CONTROLLED DENSITY FILL

length. All pipe were calibrated in a 3-edge bearing machine for load vs stress/strain characteristics with the strain gages positioned to measure circumferential tension in the invert. This is normally the direction and location of highest stress in an installed clay pipe. The two test sections were installed on either side of a manhole starting about 11 feet (3.3 m) back from the manhole face. (Figure 3) The cover depth was 16 feet (4.9 m) and the trench width in the test area was 6 feet (1.8 m) at the level of the top of the pipe. The bedding in the crushed stone section consisted of a well-graded material with a maximum size of about 1 inch (25 mm). The CDF material for the flowable fill section contained 100 lbs. (45.2 kg) cement, 250 lbs. (113 kg) flyash, 2700 lbs. 1222 kg) sand and 60 gals. (227 l) of water per cu. yd. (.91 cu. m) with 1.25% of entrapped air. The estimated flowable fill density was 132 lbs./cu. ft. (2100 kg/cu.m). The pipe in the flowable fill section were laid on a flat bed of the crushed stone material. The CDF was poured from the bottom of the test pipe to the top of the barrel in two equal lifts to prevent buoyancy. The first lift was brought midway up the pipe and the second lift about 20 minutes later was taken to the top of the pipe. There was concern about flotation. The enclosed Point of Buoyancy chart (Figure 4) is a composite value for all clay pipe diameters showing the flowable fill depth as a percent of the diameter of the pipe at the point of potential buoyancy. The chart does not take into account the restraining influence of the pipe previously installed, the

FIG. 3 CDF TEST SECTION

weight of any sand bags or the restraining influence of the internal friction between the CDF material and the pipe surface. There was no indication of flotation in the test area.

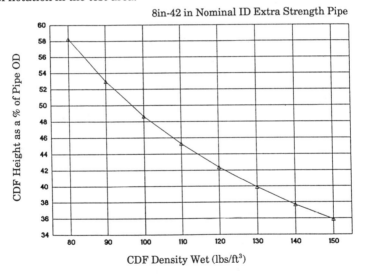

FIG. 4 - POINT OF BUOYANCY

The fill was poured by means of a vertical chute directly into the pipe zone from a concrete delivery truck. The pour was made in the trench box with the bottom of the box positioned 3 feet (1 m) above the top of the pipe (Figure 5). It is expected that a pour could also be made behind the box if

FIG. 5 - POURING THE CDF MATERIAL

necessary. The installation proceeded without difficulty and the contractor was especially eager to use the system as a replacement for concrete arch or cradle. The strains in the two test sections are shown in Table 1. A ratio of the average strains in the two test areas yields a computed bedding factor of 3.0.

	DAYS			
	5	11	19	31
Crushed Stone	88	91	90	84
CDF	65	72	71	51
Bedding Factor	3.0	2.8	2.8	3.6

Average Bedding Factor over 31 days = 3.0

TABLE 1 - COMPUTED CDF BEDDING FACTOR
(Based upon Measured Tensile Strain)

Field Test No. 2

The NCPI Technical Services Committee was most anxious to pursue the development of this promising bedding system and recommended that a second field test be undertaken to make additional buoyancy measurements and to finalize the details of the installation procedure.

This test consist of 6 shallow lines with 3 pipe in each. Two sections each of 8 inch (200 mm), 18 inch (450 mm) and 33 inch (825 mm) diameter pipe were installed. All pipe ends were sealed with 3/4 inch (19 mm) plywood bulkheads with internal restraining cables (Figure 6). Each test section was fitted with two gages to measure flotation (Figure 7). One gage

FIG. 6 - TEST LINE WITH BULKHEADS

was placed at the spigot end and one gage was placed in the center of the middle pipe in each section. The pipe in all six sections were laid upon a flat crushed stone base without haunching. Test Sections 1, 3 & 5 were filled to the top of the pipe in a single continuous pour (Figure 8). Test Sections 2, 4 & 6 were filled in two pours with the first pour to 50% of the pipe diameter. The two-pour series of tests paralleled the Riverside, CA field test where the CDF fill was placed in two equal lifts. The CDF material was allowed to flow around the ends of the test line to maximize the potential for flotation. Additionally, extra water was added to the 18 inch (450 mm) and 33 inch (825 mm) diameter tests to further increase the potential for flotation.

FIG. 7 - FLOTATION GAGE

FIG. 8 - COMPLETED CDF POUR

Test Results

None of the pipe in either the single or double pour tests indicated any potential for flotation despite the extreme liquidity of the CDF mix. A review of the specific gravity of clay pipe indicates that an empty pipe sealed at the ends would be buoyant in water and probably in concrete (See Figure 4) but, for reasons which may be unique to the CDF material, the pipe did not float. Dr. Roy Leonard, President of Alpha Omega Geotech Inc., consultant to the NCPI project, explained the phenomena this way *"as soon as the CLSM (CDF) stops moving, it quickly becomes a plastic mass with density and low shear strength properties. As consolidation takes place, which occurs within a very few minutes, the density and shear strength of the mass increase. In order for the pipe to float, it would have to overcome the shear strength of the CLSM (CDF). The transition from fluid to plastic solid takes a few seconds, and from plastic solid to semi solid a very few minutes. From semi solid to solid requires hydration of the portland cement, and this depends on the cement content. I estimate that about a half hour or perhaps a little more is required for significant hydration of the CLSM (CDF) at Pittsburg. For these increasing strength reasons, the pipe sections did not float."*[2]

Time to Backfill

It is desirable from a safety and construction point of view that the backfilling proceed as soon as practical after the CDF material is poured. To determine the approximate delay time for initial backfilling, penetration tests were made in the CDF material at various intervals. The resistance is shown in Table 2 for the first 3 hours after pour. It should be noted that

	TIME - MINUTES AFTER POUR							
	10	20	30	40	50	60	120	180
Tons/sq. ft.	.000	.031	.047	.063	.078	.078	.125	.156
Lbs./sq. ft.	0	62.5	93.8	125	156	156	250	313

TABLE 2 - PENETROMETER READINGS

in the two pour tests the second pour was successfully made in as little as ten minutes despite the fact that the first pour was not stiff enough to measure penetration resistance. Twenty minutes is probably a minimum delay time. Backfilling could proceed whenever the backfill material does not mix with the CDF material. CDF mix proportions, ground moisture and weather conditions will have some influence on the time to backfill although it is not considered critical. A minimum penetrometer resistance of about 62.5 lbs./sq ft (2976 kP) was adequate. These readings were obtained at 20 minutes following the pour.

Precautions

1. Although no flotation of the clay pipe occurred, one of the 18 inch (450 mm) lines was pushed towards the side of the trench when the flow of the CDF material was directed horizontally towards one side of the pipe. This would be unlikely to occur in deep trenches but because this material may also be used in shallow cuts under roadways it is recommended that the flow of the material should always be directed at the top of the pipe and allowed to flow somewhat equally on both sides. This will prevent horizontal pipe displacement.

2. The CDF material segregates so rapidly it was necessary to continuously rotate the drum and particularly just before pouring.

3. Although untested, users are cautioned to consider the influence of this bedding system when installed in expansive clays.

Additional Advantages of CDF Bedding

1. Trench widths can be reduced since there is no need to provide minimum working space beside the pipe for the purpose of haunching. The trench width can be reduced to the outside diameter of the pipe barrel plus 12 inches (300 mm) or even less on smaller diameter pipe. The narrowest width of trench consistent with maintaining proper horizontal alignment is acceptable. This minimizes the trench backfill load, reduces the amount of backfill material that must be removed and will also minimize the quantity of CDF material required.

2. The CDF material provides uniform and complete filling of the pipe haunches thereby developing a consistent and well distributed support mechanism. Consolidation variations are eliminated. It is likely that a lower factor of safety could be used with this installation.

3. The CDF material will not be subject to material migration or French drain activity as are combinations of crushed stone and sand for example. Geofabric would not be required with CDF bedding.

4. Due to its low compressive strength, CDF materials may be excavated with conventional equipment at a later time if installing other utilities or making connections.

ASTM

The findings of these studies were reported to ASTM Committee C4. The addition of CDF as a new bedding system was balloted at the Society level in December, 1994 and was recently issued as part of ASTM C 12-95 Standard Practice of Installing Vitrified Clay Pipe Lines[3].

Conclusion

Based upon the two field tests reported and subsequent adoption by ASTM, the National Clay Pipe Institute recommends that a bedding factor of 2.8 be used in design of clay pipe systems and, whenever possible, CDF bedding should be selected over concrete arch or concrete cradle because of increased trench safety and improved economic and structural design considerations.

Appendix

1. National Ready Mixed Concrete Association, 900 Spring St., Silver Spring, MD 20910

2. Alpha-Omega Geotech, Inc., 1701 State Avenue, P.O. Box 2670, Kansas City, KS 66110

3. ASTM C 12-95 Standard Practice for Installing Vitrified Clay Pipe Lines

DEFORMATION OF HDPE PIPES DUE TO GROUND SATURATION

Tohda J.[1], Li L.[2], Hamada T.[3],
Hinobayashi J.[4], and Inuki M.[5]

Abstract

HDPE (high density polyethylene) corrugated pipes at construction sites were damaged when a fill reached 2 m high. A series of centrifuge model tests revealed that this accident was caused by collapse of a filling soil due to saturation after raining. Another series of centrifuge model tests clarified the effects of the following factors on deflection of HDPE pipes caused by the ground saturation: type of pipe installation, type and density of filling and backfilling soils, area of backfilling, and thickness of sand bedding.

Introduction

HDPE (high density polyethylene) corrugated pipes having 1.1 m in external diameter were planned to be buried for drainage under a 15 m fill at three construction sites of a housing development area in Japan, the length of each site being 50 m. When the fill reached 2.2 m high over the pipes, it was found that every pipe at all sites was damaged as shown in Fig. 1. The vertical deflection of the pipes, $\delta = \Delta D/2R$, was around 30 %, in which ΔD is the change in the vertical diameter of the pipe and R is the neutral radius of the pipe. Fig. 2a shows a condition of the pipe installation adopted at this construction step. The construction procedure was as follows. After the fill was built with the mixture of clay and sand to the level of the pipe top (1.36 m high

[1]Associate Professor, [2]Doctor Course Student, [3]Master Course Student, Osaka City University, 3-3-138, Sugimoto, Sumiyoshi-ku, Osaka, 558, Japan
[4]Head Engineer, [5]Engineer, Takiron Co. Ltd., 3-7-1, Higashi-okino, Yokaichi, Siga Prefecture, 527, Japan

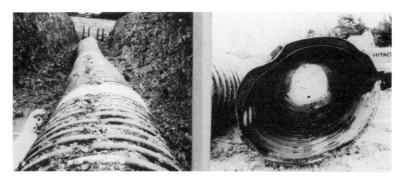

Figure 1. Damage of HDPE Corrugated Pipes

excavated and the pipes were placed on a decomposed from the subgrade), ditches with slopes of 63 degrees were granite bedding of 0.25 m thick; the excavated soil was backfilled inside the ditches and the same soil was placed over the ditches (the terminology of the subgrade, fill, backfill and bedding is shown in Fig. 2b). It was suspected that passing of heavy construction vehicles caused the damage at this time.

At the second construction step, new pipes having the same dimensions were buried in ditches constructed by slope cutting to remove the damaged pipes, in which a countermeasure as shown in Fig. 2b was applied; the bedding was replaced by crushed stones having maximum size of 40 mm and decomposed granite was placed around the pipes. The widths of the ditches at the pipe springline were 2D (D: external diameter of the pipe). Nevertheless, δ was recorded as 7-8 % when the filling height, H, reached 5 m, and they were expected to increase to be more than 20 % when H=15 m. Since the passing of the construction vehicles was prohibited during this construction step, the softening of the filling soil due to saturation

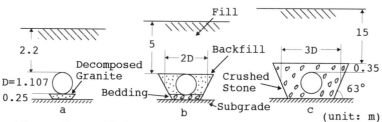

Figure 2. Conditions of Pipe Installation at the Sites

after raining (that is, collapse) emerged as a cause leading to both the pipe damage at the first construction step and the over-deflection of the pipes at the second construction step.

Finally, new pipes were buried again at the same sites by applying another countermeasure shown in Fig. 2c. The crushed stones were placed around the pipes inside the ditches from 0.25 m below the pipe bottom to 0.35 m over the pipe top; the widths of the ditches at the springline were widened as 3D. δ was recorded as 1 % at H=5 m, 2.6 % at H=10 m, and 5 % at H=15 m; these values were located within administrative values of HDPE pipes.

A series of centrifuge model tests was carried out at a 26th scale to understand the cause of the pipe damage at only 2 m high filling. In the tests, the model ground was constructed with a silty sand, and the water table in the model ground was raised to the ground surface and then dropped to the bottom. The tests yielded results that simulated well the pipe damage at the sites. There are many construction sites in Japan, where HDPE pipes are buried with collapsible soils that are easily softened by the saturation (Fukuda M. 1977, Mochizuki A. 1985). Thus, another series of centrifuge model tests was carried out at a 49th scale under various conditions both to investigate effects of the ground saturation on the pipe deflection and to develop effective countermeasures for the construction of flexible pipes when the ground is expected to be softened by the saturation. In this paper, these two series of centrifuge model tests are detailed, and effective countermeasures are recommended on the basis of the test results.

Model, Equipment and Test Procedure

Fig. 3 shows 2-D models and test setup. The 1/26-scaled model using a model HDPE pipe of D=42.6 mm was used in Test series A to simulate the accident at the sites; the cover height was 8.4 cm (2.2 m in the prototype

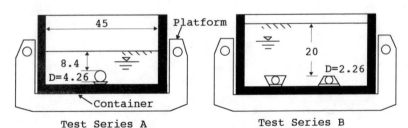

Test Series A   Test Series B
Figure 3. Models and Test Setup (unit: cm)

scale). The 1/49-scaled model using two model HDPE pipes of D=22.6 mm was used in Test series B to investigate parametrically the effects of various factors on the pipe deflection due to the saturation. The cover height in Series B was 20 cm (9.8 m in the prototype scale). The test container having an inside height of 30 cm, a width of 45 cm, and a thickness of 15 cm was installed on a platform; this platform contained a water tank for supplying water to the model ground by air pressure. The platform was attached at one edge of an arm of Mark-5 centrifuge at Osaka City University having a nominal radius of 256 cm.

The walls of the model HDPE pipes were plane and their surfaces were finished to be smooth. Table 1 shows dimensions and flexural stiffness, S_p, of the model pipes, together with those of the prototype HDPE corrugated pipe. S_p is defined as EI/R^3 (kgf/cm^2), where E: elastic modulus of HDPE and I: inertia moment of the pipe wall. The S_p value of the prototype pipe was obtained through concentrated-line loading tests; the wall thicknesses of the model pipes were determined so that their S_p values are coincident with that of the prototype pipe. A deflection gauge was mounted inside each model pipe to measure change in its vertical diameter.

Three types of soils (denoted as F, G, and S) were used in the tests; subscripts, L and D, are used to denote the loose and dense conditions, respectively. Tables 2 and 3 show the primary and secondary properties of the

Table 1. Dimensions of Model and Prototype Pipes

Pipe	D mm	t mm	$S_p = E_p I/R^3$ kgf/cm^2
Model	42.6	1.8	0.56
Model	22.6	0.9	0.56
Prototype	1107	88.5	0.56

Table 2. Primary Properties of Soils

Soil*	G_s	Grain Size Distribution					ρ_{dmax} g/cm^3	ρ_{dmin} g/cm^3	W_{opt} %
		Max. mm	Sand %	Silt %	Clay %	U_c			
F	2.67	2.0	70	17	13	115	1.86	1.18	13.5
G	2.71	2.0	84	9	7	70	1.92	1.37	11.4
S	2.65	1.4	100	0	0	1.75	1.58	1.32	-

* F: Silty Sand, G: Decomposed Granite, and S: Silica Sand.

Table 3. Secondary Properties of Soils

Soil	Density	ρ_d g/cm³	w %	D_r [1] %	D_c [2] %	S_r %	c_d [3] tf/m²	ϕ_d [3] degree
F	Loose	1.50	12	58	81	41	3.0	32
F	Dense	1.70	12	84	91	56	4.6	32
G	Loose	1.50	10	30	78	34	0.9	38
G	Dense	1.70	10	68	89	46	2.3	38
S	Loose	1.43	0	47	91	0	0	37
S	Dense	1.55	0	83	97	0	0	43

[1] D_r: relative density, [2] $D_c = \rho_d/\rho_{dmax}$, and [3] shear strength parameters under drain condition.

soils. Soils F and G are a silty sand (w=12 %) containing fine fractions (under 0.075 mm) of 30 %, and a decomposed granite (w=10 %) with 16 % fine, respectively; they were used as either a filling soil or a backfilling soil. Soil S is an air-dried silica sand without fine; it was used as either a bedding soil or a backfilling soil.

The testing procedure was as follows. The models were subjected to a centrifugal acceleration field of 26 G in Test series A and 49 G in Test series B (G: gravitational acceleration). The ground water table was raised gradually to the ground surface by applying air pressure in the water tank of the platform and then was dropped to the ground bottom; this procedure was repeated twice in Test series B. Pipe deflection was recorded by a digital strain-meter.

Test Conditions and Model Preparation

In Test series A, two pipe installation types, a and b, as shown in Fig. 4, were employed to simulate good and poor construction works in the sites. The models were prepared in accordance with the following procedure. A spacer 5.26 cm high, whose shape is equivalent to the

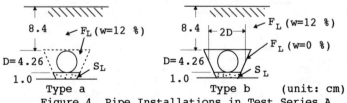

Figure 4. Pipe Installations in Test Series A

ditch constructed at the sites, was set on the container bottom. After the silty sand, F, was filled outside the spacer by light compaction, the spacer was removed and the silica sand, S, 1 cm thick was placed loosely inside the ditch as a bedding, on which the model pipe was placed. The inside area of the ditch was backfilled with Soil F of w=12 % in Type a and with Soil F of w=0 % in Type b. Then, the ditch was filled loosely with Soil F of w=12 % over the pipe 8.4 cm thick.

Fig. 5 shows four types of pipe installations, c-f, and their test conditions employed in Test series B; the cover height of 20 cm (H/D=8.8) was common in all installation types. The influencing factors under studies were: 1) type of pipe installation, 2) type and density of both filling and backfilling soils, 3) width of the backfilling area at the pipe springline, B_s, and 4) thickness of sand bedding, H_b. Actual construction procedures corresponding to the four types of pipe installations and the model preparation in the tests were as follows:

Type c: Pipes are installed on a sand bedding and covered by filling soils without countermeasures. Construction procedure in this type is classified as embankment type and ditch type, as shown in the figure. In the tests, the models were constructed by using the ditch type in the similar way to that in Test series A.
Type d: After fill height reaches the level of the pipe top, a ditch having slope angles of 63 degrees is constructed by excavating the fill. After pipes are installed on a sand bedding, the ditch is backfilled with sand, and covered with filling soils. In the tests, the models were constructed in the similar way to that in Test series A.
Type e: After pipes are installed on soil beddings, directly constructed on the subgrade, backfilling soils are placed around the pipes to form a trapezoid shape with slope angles of 45 degrees. Then, the backfilled area is covered with filling soils. The models in the tests were constructed by using the embankment type shown in the figure.
Type f: Construction procedure is similar to that in Type d, except that a sheet of steel net was placed around either a ditch or a pipe to simulate three types of wire cylinder methods as shown in the figure.

<u>Simulation of the Damage in the HDPE Corrugated Pipes</u>

Fig. 6 shows changes in the pipe deflections, δ, measured for Types a and b during the tests of Series A. δ increased linearly with the centrifugal acceleration; δ at 26 G was recorded as 3.4 % in Type a and 6.6 % in Type b. When the ground water table was raised, δ in both types increased to peak values of 13.3 % and 21.3 %, and

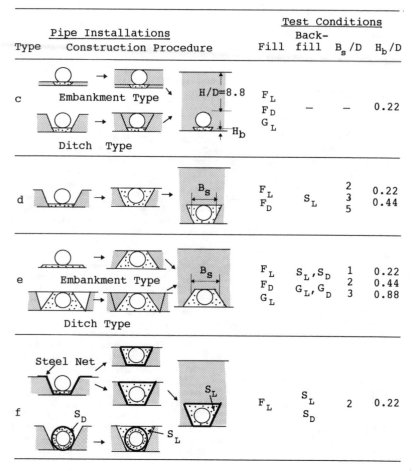

Figure 5. Types of Pipe Installations and Test Conditions in Test Series B

decreased after that owing to the reduction of effective soil weights. The drop of the ground water table slightly increased δ. The measured maximum δ, δ_{max}, were smaller than δ observed in the sites, probably due to the difference in the soils used in the tests and sites. Nevertheless, they were sufficient to cause either the creep or buckling failure in the HDPE corrugated pipes. Thus, it was concluded that shallow fills at two meters high could damage the HDPE pipes when the partially

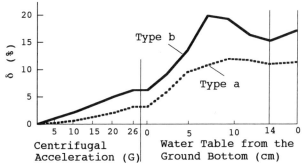

Figure 6. Change in δ during the Tests of Series A

saturated soil, used for the initial filling, was softened later by the saturation.

Effects of Factors on Pipe Deflection Caused by Saturation

Fig. 7 shows the pipe deflections, δ, of Type c, d, and e construction in Test series B. These are the data for the cases when: 1) the filling soil: F_L, 2) the backfilling soil: S_L, 3) $H_b = 0.22D$, and 4) different B_s in Types d and e. In any case, δ increased linearly with the centrifugal acceleration to the values δ_1 at 49 G. The first rise of the ground water table increased δ to a great extent; the second rise slightly changed δ. Type c without countermeasures generated the greatest values of both $\delta_1 = 10.3$ % before the ground saturation and $\delta_{max} = 22.8$ % during the tests. In Types d and e, δ_{max} decreased their values drastically with an increase of B_s.

Fig. 8 summarizes the values of δ_{max} measured in all

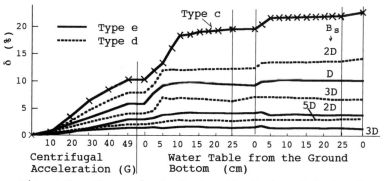

Figure 7. Change in δ during the Tests of Series B

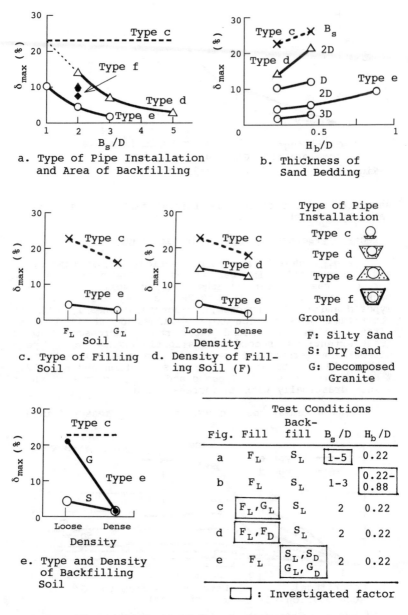

Figure 8. Measured δ_{max} in Test Series B

the tests of Series B. The broken line in each figure denotes the data in Type c without countermeasures. The figure shows the effects of the investigated factors on δ as follows:

Type of pipe installation and Backfilling area (cf. Fig. 8a): Type e generates considerably smaller δ than Type d. In both Types d and e, the greater B_s, the smaller δ it generates. Type f is effective when the pipes must be buried in excavated ditches.
Thickness of sand bedding (cf. Fig. 8b): The smaller H_b, the smaller δ it generates, in particular, in Type d.
Type of filling soil (cf. Fig. 8c): The decomposed granite, G, which contains less amount of the fine fractions, generates smaller δ than the silty sand, F, in particular, in Type c.
Density of filling soil (cf. Fig. 8d): The dense silty sand generates smaller δ than the loose one in any types.
Type and density of backfilling soil (cf. Fig. 8e): The denser backfilling soil, the smaller δ it generates. The dense decomposed granite, G_D, generates almost the same δ as the dense silica sand, S_D.

Discussions on Effects of the Investigated Factors

The results of Test series B showed that Type e is considerably advantageous to minimize δ than Type d. Fig. 9 illustrates the ground deformations for these two types, which were obtained by sketching markers before and after the tests. The figure shows the reason why these two types generated the difference in δ, as follows:

Type d: The filling ground under the backfilling area subsided due to the saturation, and the backfilling area deformed downwards, owing to the loss of support from the filling ground. This must have caused the reduction of the lateral earth pressures acting on the pipe, resulting in the larger pipe deflections.

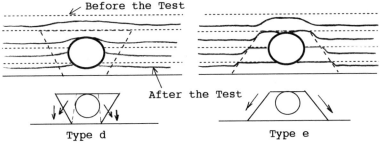

Figure 9. Ground Deformation due to Saturation

Type e: Although the filling soil slipped along the slopes of the backfilling area, the backfilling area around the pipe did not deform. As a result, the backfilling area could support the pipe, generating the pipe deflections smaller than Type d.

The results of Test series B also showed that the pipe deflection caused by the ground saturation is small, when both backfilling and filling soils contain less fine fractions. Fig. 10 shows how the three soils under the loose conditions increase their compressibility and decrease their deformation moduli, E_s, due to the saturation. These data were obtained through K_0-compression tests using a rectangular box; the area of the soil specimen is 12 cm x 12 cm and its height is 10 cm (Tohda et al. 1991). In the tests, axial strain ε_1 and lateral stress σ_3 were measured under K_0-condition (=null lateral strain condition) when axial stress σ_1 was applied stepwise. E_s and ν of the soils were obtained through the following equations derived from Hook's law:

$$E_s = (1-\nu-2\nu^2)/(1-\nu) \cdot \sigma_1/\varepsilon_1, \quad \nu = \sigma_3/(\sigma_1+\sigma_3)$$

Both increase in ε_1 and reduction in E_s due to the saturation were seen in the figure; their differences between

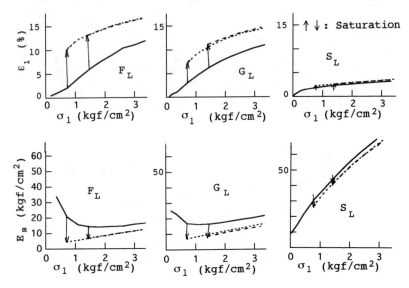

Figure 10. Change in ε_1 and E_s of Soils due to Saturation

each soil explained well the tendency of the test results.

Finally, the tests yielded the results that the small thicknesses of sand beddings produce the smaller deflection. This contradicts with the conventional design standards for flexible pipes (e.g. JMAFF design standard 1988), which specify to select thicker sand beddings under worse ground conditions.

Conclusions

Centrifuge model tests simulated well the damage of HDPE corrugated pipes due to the ground saturation, and quantified the effects of various factors on the deflection of HDPE pipes. Test results showed that the following methods should be adopted in the actual design and construction for flexible pipes such as HDPE pipes to minimize the pipe deflection due to the ground saturation:

1. Type e having a trapezoid backfilling area, shown in Fig. 5, should be selected first. When construction conditions force engineers to select Type d, the width of backfilling area should be widened on the basis of the data presented in this paper, or any other countermeasures such as Type f using wire cylinder methods should be adopted.
2. Soils used for both filling and backfilling should contain small amount of fine fractions, and they should be compacted heavily. For example, Japanese design standard for earth works of road construction (1986) recommends the wet-side compaction of soils to be D_c>90 %, together with S_r>85 % and n_a<10 % (n_a: air porosity) in cohesive soils, and n_a<15 % in sandy soils, to avoid the ground settlement due to the saturation.
3. Thicknesses of sand beddings should be minimized on the contrary to specifications in conventional design methods for flexible pipes.

References

Fukuda M. and Nakazawa J. 1977. Subsidence of Embankment due to Submergence and its Estimation. Soils and Foundation (Journal of JSSMFE). Vol.17 No.2. pp.65-73.
JMAFF (Japanese Ministry of Agriculture, Forestry and Fisheries) 1988. Design Standard of Pipelines.
Mochizuki M. et al. 1985. Investigation of Settlement of Clay Fill at a Housing Development Site. Tuchi to Kiso (Journal of JSSMFE). Vol.33 No.4. pp.26-32.
Tohda J. et al. 1991. Deformation Property of Soils Used in Construction of Buried Pipes. JSCE 46th Annual Meeting. pp.620-621.

Open-cut Sewer Crossing of Railroad Completed in Just 48 Hours

Jeff Garvey, PE [1]
Afshin Oskoui, PE [2]

Abstract

In mid-February of 1993, a major landslide closed the Santa Fe railroad line about 25 miles north of San Diego, California, suspending regular railroad traffic while the tracks were cleared. During this same period of time, construction was already underway on a four-barrel jacked crossing of the rails for the City of San Diego's Morena Boulevard Interceptor Sewer project, and the contractor was encountering severe problems with his jacking operation. On short notice, a combination of public agencies and private firms mounted an effort to complete the pipeline construction during the rail shutdown period, *by open cutting across the rails*. Four 68-inch steel casings for sewer pipe and two 24-inch steel casings for air-jumper lines were constructed on an around-the-clock basis, with on-site design consultation. Construction was completed and the rails returned to service in less than the 48-hour railroad time limit. This incident illustrates how rapid and high-level interagency cooperation led to the successful exploitation of this unique opportunity to trench pipelines across a major railroad.

Overall Project Description

For many years, the City of San Diego had experienced the problem of potential sewer overflows from the old East Mission Bay Trunk sewer. Since these spills impacted Mission Bay, one of the City's prime recreation and tourist attraction areas, it was imperative to correct the problem as soon as possible. The **Morena Boulevard Interceptor Sewer** was designed and constructed to permanently eliminate this problem by intercepting a large portion of these sewer flows before they reached the bayfront area.

---

[1] Principal Engineer, Boyle Engineering Corporation, San Diego, CA
[2] Assistant Deputy Director, City of San Diego Water Utilities Department

The **Morena** project featured 6,700 meters (22,000 feet) of 1800 millimeter (72 inch) reinforced concrete pipe. The alignment of the new pipe could not simply parallel the older sewer, since it occupied some of the highest-traffic-volume streets and most-used recreational areas in the City, which could not tolerate the disruption of prolonged construction activities. Fortunately, an alternative route was available, although it posed significant design problems. Because the project had fixed end points where existing sewers were joined upstream and downstream of the bay, severe grade constraints were imposed on the new design. This resulted in having to build essentially a "straight shot" grade, which resulted in numerous design and construction challenges. In some areas, for example, the grade was 27 meters (90 feet) below the ground surface, so sub-surface construction methods were used. In other areas, the grade was actually above ground level, so the pipe was placed off-road behind protective retaining walls. And in one location, it was necessary to cross under a working railroad track with very little clearance.

Railroad Crossing Design

At the time of project design, the railroad line in question was part of the AT&SF system, and the design of the pipeline crossing was coordinated with their engineering personnel. Several issues were of particular concern to them, including that the rails be crossed perpendicularly, that there be at least seven feet of clearance from the top of the pipe to the rail elevation, and that there be no transition structures in the railroad right-of-way. For the first of these conditions they were willing to compromise on a 37 degree offset, and the clearance requirement could have been met with a few centimeters (inches) to spare, but the restriction on structures was a significant problem.

Just downstream of the tracks, there was a large storm drain, a gas transmission main, and a high-pressure fuel pipeline, all of which it was impossible to dive below due to the grade constraints for the sewer. However, by transitioning from a single large sewer to a smaller diameter multiple barrel configuration, it was possible to avoid both the storm drain and the fuel pipeline. (Unfortunately, the gas main had to be relocated in any case.) Because these interfering utilities were immediately adjacent to the railroad, and because no transition structures were allowed in the railroad property, the multiple barrel configuration had to be extended upstream across the rails. The design configuration approved by the railroad therefore featured four 900 millimeter (36 inch) reinforced concrete pipes in parallel, with about 0.6 meters (2 feet) clearance horizontally between the outside diameters of the pipes. This design is shown in plan view and in profile on Figures 1 and 2.

One other feature of the design bears mentioning. When a sewer is run into a "squash box" type of structure, like the transition structures used here, it is necessary to also provide an "air jumper" pipe, so that continuity of airflow can be maintained. This connects the airspace above the soffit of the upstream structure with the airspace in the downstream structure, which prevents odor problems which could arise if the upstream air was merely vented. For this design, two 400 millimeter (16 inch) pressure-class PVC pipes were provided as air jumpers. To avoid having two more crossings in the same

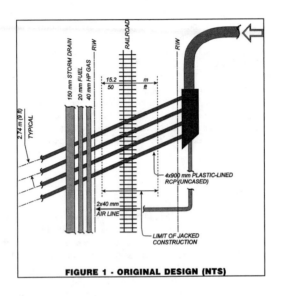

FIGURE 1 - ORIGINAL DESIGN (NTS)

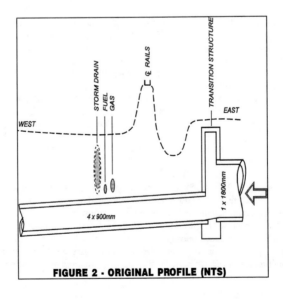

FIGURE 2 - ORIGINAL PROFILE (NTS)

place as the other four, the crossings for the air lines were moved about 15 meters (50 feet) down the tracks. A total of six pipes, then, were to cross the tracks at this project location.

The last design factor to be resolved was the specification of the method of construction of the crossings. For the two smaller pipes, conventional jacked crossings with steel casings were specified. For the four-barrel crossing of the concrete pipes, however, it was believed that modern guided boring machine techniques could install concrete pipes of this size with sufficient accuracy and reliability that steel casings would not be required. The designers determined that sufficient accuracy could be achieved after observing several other sub-surface construction sites, and the space needed for the casing cylinders would greatly increase the clearance problems. In consideration of the fact that this crossing was unpressurized, concrete, and carrying a non-volatile fluid, the railroad company engineers were likewise willing to consider an uncased crossing at this site. Furthermore, they had recently experienced several very successful uncased crossings of their tracks using concrete pipe. The plans were therefore approved by all concerned parties, and the project moved into the construction phase.

Construction of the Original Design

Construction began with the installation of fully-shored jacking and receiving pits on either side of the railroad. Three sides of the pits were shored with solid steel sheeting, but the side facing the tracks was shored with wood lagging marked with ovals indicating the bore insertion points. The boring head, which was fabricated by the contractor, was articulated with four steering jacks, but was not a closed face machine. Excavation at the face was accomplished by a hand-operated "backhoe" arrangement, with a conveyor carrying spoils back to the open pit. Construction of the first 900 millimeter (36-inch) pipe was completed without incident, and the annular space around the pipe was pressure-grouted through periodic holes in the pipe, to seal any void area.

The contractor next chose to install pipe number three, separated from the previously installed pipe by about 2.5 meters (8 feet). During construction of this pipe, problems were noted by the on-site inspector for the railroad company. In fact, the area of the tracks between the two pipe bores had settled over an inch, which was a serious concern for rail operations. A "slow-down" order was issued for trains passing the site, repair activities were instituted by the railroad, and further construction of pipe was put on hold until the situation could be evaluated and resolved. This was a matter with serious financial implications as well, because the rail company served 18 Amtrak trains a day as well as its own freight operation, and there was a $2,000 fee associated with each "track caused" late arrival or disruption of Amtrak's schedule. Fortunately, train operators seemed able to compensate for the slowdown order during the rest of the ride.

Unfortunately, despite restoration of the railroad grade by their repair crews that day, the track was again displaced on the following day, even though no further construction activities occurred. This was an alarming situation, and measurements and observations were immediately taken of the two existing pipe crossings. Much to everyone's dismay, the installed pipes were found to be unstable, with both lateral and vertical movements continuing to occur. Spontaneous movements of up to 2.5 centimeters (1 inch) per day were documented, daily track repair was required, the rail traffic was placed on permanent slow-down status, and the railroad revoked the City's construction permit for this crossing operation and demanded that a plan be prepared immediately to restore the damage. At this point it was unclear whether a permit would ever be granted to complete the construction. Every day the condition of the existing pipe crossings deteriorated, with actual joint separations beginning to occur, so that even these two existing crossings would never be suitable to use. Fortunately, the rate of displacement stabilized to some degree after a few days, but a novel means for correcting this situation still needed to be developed.

Opportunity

It was at this point, while various repair schemes were being contemplated, that a nearly miraculous opportunity arose. In San Clemente, a city about 40 kilometers (25 miles) north of the construction site, a major landslide buried a section of the rail line. Furthermore, because it was feared that a larger slide affecting numerous homes might be set off, the local authorities at that site ordered the railroad not to attempt repairs until the matter could be further studied. As a result, rail traffic on this entire section of track was completely canceled, for an indefinite length of time (although traffic could be restored in as little as 48 hours if repair work was allowed to proceed.) This interruption of rail traffic presented a unique opportunity to correct the problem at the Morena site by open cutting across the now-unused tracks.

Intensive negotiations began between the railroad, the City of San Diego, the contractor, and the design consultant as to how this could be accomplished. Finally, at about seven PM on Friday, February 26, 1993, permission was granted by the railroad for an open-cut crossing operation to begin at 8:00 AM the next morning. Among other special conditions, it was necessary to complete the work in 48 hours and to have the track back in service by 8:00 AM on Monday. If the other track section had been repaired first, and delay in restoring this track resulted in continued rail schedule shutdowns, the penalty for each of Amtrak's canceled trains would have resulted in significant potential total liability for each day of delay. The contractor would canvas local steel suppliers to obtain suitable casing pipes, the railroad maintenance contractor would be on hand to cut out and replace a section of the tracks, and the consultant would provide on-site geotechnical and design services, all under the inspection and supervision of City engineering personnel.

Construction of the Open-cut Crossing

At 7:00 AM on Saturday, the railroad contractor arrived on-site and began to remove a 30-foot long section of track. Modern rail track is installed in continuous welded lengths up to 3,200 meters (two miles) long, so the removal process consists of cutting out the appropriate section and lifting out it entire, with ties still attached. This process was completed and the track section laid out of the way down the rails before 8:00. While they were working, however, someone noticed for the first time that a railroad signal pole was located within the excavation area as well. The railroad signal man arrived at 9:05 to deal with this issue, and the wires were disconnected and the pole removed by 9:55. Finally, a clear field of work was left for the pipeline contractor to begin.

Initial removal of the railroad embankment was accomplished with a "D4" bulldozer, clearing the work area down to ground level and creating a smooth area for construction equipment to move across the track line. Once this was done, the main part of excavating was started with a "245" excavator, while the shoring at the front side of the existing pits was being removed with a crane. The goal of this initial phase of excavation was to expose and remove the two unstable pipes which had been previously installed, and as soon as the lagging and beams had been removed from the face of the jacking pit, excavation began in earnest on that face. Initial excavating had to work around one of the old shoring piles, which was stuck in place and couldn't be pulled out.

As the excavating progressed, the casing pipes were being delivered to the site. The contractor had located four pieces of 1700 millimeter (68 inch) steel pipe, with a wall thickness of approximately 19 millimeters (3/4 inch). The pipe had originally been manufactured as high pressure water pipe for a different project, but was available as surplus from the manufacturer. The first of these casing pipes was on-site when the work began, and the remaining three were delivered during the day, with the last arriving at about 4:00 PM Saturday afternoon. During the morning, as the excavation progressed, the on-site casing pipes were examined by City engineering personnel, and all parties met together to determine a new design using the available materials. The resulting design is shown in Figure 3.

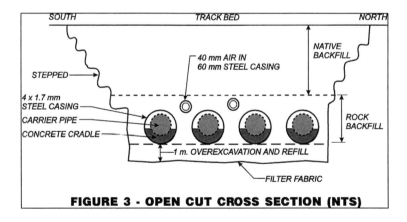

FIGURE 3 - OPEN CUT CROSS SECTION (NTS)

The new design preserved the original 1800-millimeter (6-foot) distance from center to center of the 900 millimeter (36-inch) carrying pipes, and also maintained their original elevations. Preservation of these features was necessary to mate with the transition structure already under construction immediately upstream of the railroad. The bottom of the casing pipe was to be set 25 centimeters (10 inches) below the bottom of the carrier pipe, to allow for the construction of a concrete cradle for the carrier pipe to rest on. The annular space around the carrier pipe would be filled with lightweight cellular concrete to minimize loading on the subgrade. The subgrade would be excavated to a depth one meter (3 feet) below the casing pipes, filter fabric placed on the native material, and the subgrade backfilled with 19 millimeter (3/4 inch) rock. The geotechnical engineer would observe the subgrade as it was exposed, to confirm the adequacy of these measures. There was some disagreement within the group as to how to backfill under the haunches of the casing pipes, especially with only about 10 centimeters (4 inches) of clearance between them in the new design configuration. Finally, it was decided to observe the placement of small rock backfill before deciding whether to require concrete slurry to springline of the casings. As a last design feature revision, it was decided to place the two air jumper pipes inside casing pipes in the same cross section, rather than attempting to jack them into place later in their original design locations. After filling above the air jumper casings with 19 millimeter (3/4 inch) rock, the rest of the embankment would be reconstructed from the material which had been removed. Finally. the upper 30 centimeters (1 foot) of track bedding material, and the rail replacement itself, would be done by the railroad contractor.

By 1:00 PM Saturday, the jacking-pit end of the first of the existing pipes was exposed, and the first section of existing pipe was removed. The removed sections were stockpiled alongside the site for cleaning and eventual reuse. During this phase of the construction there was one crane working with the old pipe pieces, a second crane dealing with the shoring materials, a third crane working with the new casing materials , one excavator working at each end of the cut, trucks to haul off excess excavated material (some of which was being stockpiled for reuse), and the D4 just generally moving things around. By 3:00 PM the first piece of subgrade soil was exposed for examination. It was decided to go ahead as planned, and the placement of filter fabric began. Slowly, the excavation proceeded from the jacking pit end towards the receiving pit end, with successive waves of old pipe removal, excavation to subgrade, filter fabric placement, and rock refill placement. The last of the subgrade was exposed at 11:45 PM on Saturday, and the fabric and rock base all in place by 1:30 AM Sunday, vibrated and compacted with a sheepsfoot roller. At this point a problem developed in that a larger crane would be needed to lift the casing sections out into the excavation for placement on the prepared subgrade. Unfortunately, a concrete pumping truck showed up instead. There was therefore an unscheduled break in the work for a few hours at this point while the equipment ordering problem was resolved.

The large crane arrived about sunrise, and by 9:00 AM three of the casing pipes were in the trench and being welded. (The casings had been shipped in lengths only half as long as needed, so the two halves of each length had to be welded together in-place.) It was also necessary to spot weld repairs to seal up small flaws in the casing pipes at various locations. In fact, the limitation at this point in the construction was a shortage of welders, so this part of the work went on for some time. The last casing was set into the trench at 11:30 AM and the welding continued. As this was being finished, a test piece of the in-place casing was backfilled with small rock and closely inspected. It was decided to use this method for backfill instead of slurry, provided that it was placed in lifts and well vibrated. Backfilling proceeded on that basis throughout the afternoon, and by 5:00 PM the rock fill had been completed and was being compacted with the roller. Placement of the 600 millimeter (24 inch) casings for the air jumper pipes began, which were likewise to be covered with densified rock.

While this was continuing, the stockpile of removed material was examined to select the most appropriate materials for the embankment reconstruction. The removed material was generally quite poor, with a large component of unsuitable clay lumps, so some other material had to be found. The railroad inspector insisted that the embankment would need to be soil and not imported rock materials, so another source of soil was needed. This problem was overcome by having the contractor perform the excavation at a nearby project segment in order to get the soil material for embankment reconstruction at the railroad. Using this as a source, and with a truck shuttle to transport the material, the reconstruction effort continued throughout the night. Finally, the reconstructed area was turned back over to the rail contractor at about 7:00 AM on Monday, a full hour ahead of the 48-hour time limit.

Lessons Learned

The lesson to be learned from the initial problem at this site is that you can never have too much geotechnical information. Even a small pocket of poor material can have very severe negative consequences to a subsurface construction project, and a prudent approach to the unavoidable uncertainties of this kind of work is required. Conservative design measures, such as a continuous steel casing, may provide a useful safety margin even if they at first appear unnecessary.

Even more important than this, however, is the importance of having a well-functioning and cohesive project team in place before the emergency arises. For nearly a year prior to this incident, the City project manager, the consultant, and the contractor had met on a regular weekly basis to discuss problems and progress issues on the job. Because of this, not only were all parties fully up to speed on the day-to-day status of the work, but a highly functional problem-solving team spirit was already thoroughly tested and experienced. It was this team spirit and capability that allowed us all to successfully manage this unique challenge.

Restrained Joint Systems
George F. Rhodes[1]

Abstract

When ductile iron pipe is used in water and sewage transmission lines, there are many occasions where pipe joints must be restrained. This presentation discusses, generally, the areas where joint restraint is required and specifically the methods and products manufactured by U.S. Pipe to effect this restraint. The application and performance of U.S. Pipe's **TR FLEX®** restrained joint pipe and fittings and **FIELD LOK®** gaskets are discussed.

Ductile iron pipe, used primarily for water and sewage service in the United States, is manufactured in sizes 3-inch through 64-inch in nominal lengths of 18 feet and 20 feet. Fittings of many configurations are also cast in sizes 3-inch through 64-inch. U.S. Pipe produces both pipe and fittings in sizes 4-inch through 64-inch. Some 3-inch fittings are also provided.

While several different types of pipe joints are used throughout the industry, the most common is the push-on joint. All push-on joints are proprietary and are only covered by a performance standard (C111) but not a dimensional standard as is Mechanical Joint (C110).

U.S. Pipe's proprietary push-on joint is marketed as **TYTON JOINT®**. The **TYTON JOINT** is a U.S. Pipe patented joint used by U.S. Pipe and other licensees in this county and much of the rest of the world.

Push-on joint pipe and fittings are characterized by ease and speed of assembly using no gland and/or bolts to facilitate the joint seal. When laid in relatively stable soil where no substantial change of direction is encountered, and no thrust forces are developed, unrestrained push-on joint pipe may be successfully used. However, some type of joint restraint is required when thrust forces are developed due to the use of fittings, dead ends, valves, etc.

Concrete thrust blocks have been used for many years to resist this thrust force and prevent joint separation. With the use of thrust blocks, there are several inherent concerns: 1. The bearing capability of the soil, 2. Adequate space, and 3. Future excavations.

A common method for resisting thrust forces in ductile iron pipe lines is the use of restrained joints. Based on certain parameters, the number of required joints or feet of restrained joint pipe on either side of a fitting can be calculated. The method used to calculate the length of restrained pipe requirements was derived by the Ductile Iron Pipe Research Association (DIPRA) in conjunction with the member ductile iron pipe and fittings manufacturers. This method is based on accepted principles of soil mechanics and provides formulas for determining thrust forces and the necessary restraint.

Two sources of the restraining forces are evaluated in this method. One is the static friction between the pipe unit and the soil. The other is the restraint produced by

[1] **Product Development Manager, United States Pipe and Foundry Company, Birmingham, AL**

the pipe as it bears against the side fill soil along each leg of the bend.

Figure 1 depicts a free body of a typical pipe bend where L is the length of the restrained pipe on each side of the bend. Fs represents the static frictional force resisting movement of the pipe in the axial direction and Rs represents the unit bearing resistance.

When polyethylene encasement is used on Ductile Iron pipe for corrosion protection, the value of F_s is reduced by 30%.

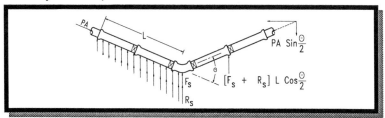

Figure 1

SOIL DESIGNATION	SOIL DESCRIPTION	φ (DEG.)	f_ϕ	C_s (PSF)	f_c	γ (PCF)	CONSTANT K_n A21.51 LAYING CONDITIONS			
							2*	3	4	5
Clay 1	Clay of Medium to Low Plasticity LL < 50 <25% Coarse Particles [ML & ML-CL]	0	0	300	.50* .80	90	.20	.40	.60	.85
Silt 1	Silts of Medium to Low Plasticity, LL <50, <25% Coarse Particles [ML & ML-CL]	29	.50* .75	0	0	90	.20	.40	.60	.85
Clay 2	Clay of Medium to Low Plasticity With Sand Or Gravel. LL<50, 25-50% Coarse Particles [CL]+	0	0	300	.50* .80	90	.40	.60	.85	1.0
Silt 2	Silt of Medium to Low Plasticity w/ Sand or Gravel, LL<50, 25-50% Coarse Particles [ML]	29	.50* .75	0	0	90	.40	.60	.85	1.0
Coh-gran	Cohesive Granular Soils >50% Coarse Particles [GC & SM]	20	.40* .65	200	.40	90	.40	.60	.85	1.0
Sand Silt	Sand or Gravel w/ Silt >50% Coarse Particles [GC & SM]	30	.50* .75	0	0	90	.40	.60	.85	1.0
Good Sand	Clean Sand >95% Coarse Particles [SW & SP]	36	.75* .80	0	0	100	.40	.60	.85	1.0

Table 1: Suggested Values for Soil Parameters and Reduction Constant K_n
*These values to be used for laying condition 2. Definition "Coarse Particles":Held on #200 Sieve.

Whereas both friction and bearing resistance are present for bends and the branch outlet of tees, only frictional resistance is available to oppose thrust on reducers and dead ends.

The two main factors that contribute to the magnitude of the restraining force are the value of certain soil parameters and the laying condition. Table 1 lists the seven basic soil types with values of soil parameters associated with each type. The parameters shown in this table are:

ϕ ..... Internal friction angle of the soil that is a measure of ability of the soil to provide shearing resistance.

f_ϕ .... Ratio of pipe friction angle to soil friction angle.

C_s (P_{SF}) Soil cohesion.

f_c ..... Ratio of pipe cohesion to soil cohesion

γ (PCF) Backfill soil density

K_n .... Trench condition modifier that assures excessive soil movement does not occur. This value depends on the compaction achieved in the trench, the backfill materials and the undisturbed earth.

Laying Conditions	Description
Type 1	Flat-bottom trench. Loose backfill.
Type 2	Flat-bottomed trench. Backfill lightly consolidated to centerline of pipe.
Type 3	Pipe bedded in 4-in. minimum loose soil. Backfill lightly consolidated to top of pipe.
Type 4	Pipe bedded in sand, gravel or crushed stone to depth of ⅛ pipe diameter, 4-in. minimum. Backfill compacted to top of pipe. (Approximately 80% Standard Proctor AASHTO T-99.)
Type 5	Pipe bedded in compacted granular material to centerline of pipe. Compacted granular, or select material to top of pipe. (Approximately 90% Standard Proctor, AASHTO T-99.)

Table 2

The values for restraining lengths of pipe are very conservative and have proven to be very reliable through the years.

U.S. Pipe has several restrained joint systems that can be used when joint restraint is required. One such system is U.S. Pipe's patented **TR FLEX®** pipe and fittings.

The joints of the **TR FLEX** pipe and fittings use the standard **TYTON JOINT®** Gaskets in sizes 4-inch through 42-inch and a special **TYTON JOINT**

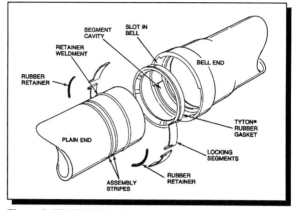

Figure 2: TR FLEX Pipe Components

Gasket in sizes 48-inch through 64-inch. The locking mechanism of the joint consists of a spherical lug (segment) cavity located in front of the gasket seat. A retainer weldment is located on the spigot end of each **TR FLEX** pipe. Once the pipe spigot has been inserted into the **TR FLEX** pipe or fitting bell, ductile iron locking segments are inserted through slots (or a slot) in the bell face. These segments fit behind the lug cavity of the bell and in front of the spigot weldment providing a positive lock. Figure 2 illustrates the components of a **TR FLEX** joint typical of the 4-inch through 36-inch sizes, and Figure 3 shows an assembled joint.

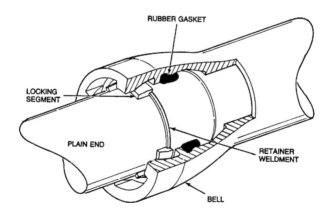

Figure 3: Assembled TR FLEX Pipe Joint

The **TR FLEX®** joint is unique in that, due to the wedging action of the segment between the bell and pipe, a frictional force is created between the segment and pipe (Figure 4). This also causes the weld to be loaded at its strongest point, adjacent to the pipe. The weldments have shear strengths of more than 20,000 pounds-force per linear inch of weld. Taking into account the effect of the design's frictional loading, the weldment is subjected to much lower stresses than in conventional weldment-restrained pipe joints.

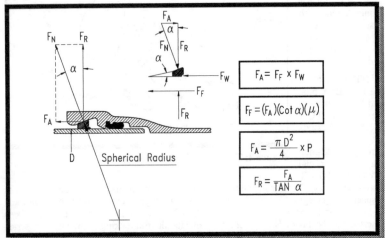

Figure 4: Force Diagram for TR FLEX Pipe Joint

TR FLEX joints in sizes 4-inch through 36-inch have a deflection capability ranging from 5° to 1.5° while the deflection for sizes 42-inch through 64-inch is approximately 0.5°.

The **TR FLEX** joint in sizes 4-inch through 36-inch use from (2) two to (8) eight locking segments inserted through two to four slots. The 42-inch through 64-inch sizes use (11) eleven segments inserted through a single slot. (Figure 5)

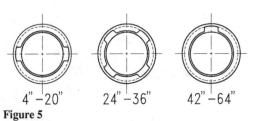

Figure 5

All **TR FLEX** pipe, regardless of length, are normally supplied with the factory applied weldment on the spigot end. However, in many cases pipe must be cut to length in the field. When this is required, some alternate method must be used to facilitate the locking mechanism. One way in which this is accomplished is by using U.S. Pipe's **TR**

FLEX GRIPPER® Ring. The Gripper Ring takes the place of both the weldment on the spigot and the locking segments. The **TR FLEX GRIPPER®** Ring is a segmented ring that is inserted into the segment cavity of the bell normally filled by the locking segments. Stainless steel toothed inserts are located on the inner surface of the Gripper Ring segments. The segmented Gripper Rings consist of two half sections for sizes 4-inch through 20-inch and four quarter sections for 24, 30 and 36-inch sizes. Segments are connected with torque control bolts.

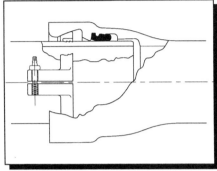

After the gasket has been installed, the Gripper Ring is inserted into the **TR FLEX®** lug cavity and the spigot of the adjoining pipe inserted. The Gripper Ring is secured to the pipe by tightening the torque control bolts on the ring segments. This initial tightening brings the stainless steel teeth into firm contact with the pipe.

As the line is pressurized and thrust forces are created the ramp action between the Gripper Ring and lug cavity forces the teeth into the pipe facilitating a secure lock between the two pipe. (Figure 6)

Figure 6

Gripper Rings are currently produced in sizes 4-inch through 36-inch.

An alternate method of facilitating the joint restraint for field cut **TR FLEX** pipe is by using a steel bar welded to the cut spigot end of the pipe that replaces the normal factory applied weldment.

U.S. Pipe can furnish field weld bars, rolled to the proper diameter, and field welding kits that include fixtures to position the bars in the proper location and welding instructions. Field weldments can be used on all pipe sizes, 4-inch through 64-inch.

An alternate restraining system is one that employs U.S. Pipe's **FIELD LOK®** Gaskets used with the standard **TYTON JOINT®** pipe and fitting. The **FIELD LOK** Gasket is a modification of the standard **TYTON JOINT** Gasket in that a number of stainless steel toothed inserts are embedded into the inner surface of the gasket.

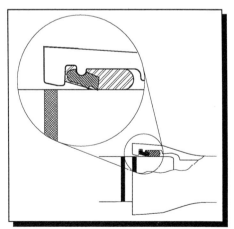

Figure 7

Assembly of the **FIELD LOK** Gasket is identical to the normal assembly of a **TYTON JOINT**. However, upon insertion of the pipe into the socket the stainless steel teeth make firm contact with the pipe surface. Once the line is pressurized, the thrust forces force the teeth into deeper contact with the pipe thus constituting a secure restraint.

U.S. Pipe currently produces the **FIELD LOK®** Gasket in sizes 4-inch through 24-inch.

Both U.S. Pipe's **TR FLEX®** and **FIELD LOK** Gaskets have proven to be an extremely successful, trouble free means of joint restraint for hundreds of thousands of ductile iron pipe and fittings joint assemblies across North America.

REFERENCES

Thrust Restraint For Ductile Iron Pipe. (Third Edition 1992). Ductile Iron Pipe Research Association, Birmingham, AL.

The Use and Application of Restrained Joints for Ductile Iron Pipelines. (1989). United States Pipe and Foundry Company, Birmingham, AL.

Subsurface Utility Engineering:
Upgrading the Quality of Utility Information

James H. Anspach, P.G.[1]

Abstract

  Subsurface utilities are becoming more costly to move, more deadly when damaged, and more prevalent as time passes. Existing records of utilities are incomplete, nonexistent, and hard to reference to existing topologic features. Traditional engineering practice calls for the architect-engineers, designers and constructors to perform a thorough search for, and analysis of, existing records and design accordingly. Until recently, there was no way of knowing the actual location of underground facilities until they were uncovered during construction. Consequently, A/E's and constructors usually disclaim responsibility for anything different from that shown by the records and handle the extra work through change orders for additional compensation. Subsurface utility engineering (SUE) was developed in the last decade as a means to deal with the identification, depiction, and characterization of utilities in a more thorough and professional manner. Quality levels for utility data are now defined and in regular usage by federal, state, and local agencies, and the SUE profession.

  The FHWA, FAA, many state DOTs, utility owners and others studied the benefits of SUE over the past decade. These studies conclude that savings in excess of $10 for every $1 spent in SUE services is a realistic figure (USDOT, October 1994). Subsurface utility engineering is a proven new technology that will continue to grow as an accepted engineering practice.

---

[1] Senior Geophysicist, So-Deep, Inc., 8397 Euclid Avenue, Manassas Park, VA 22111

Introduction

Highways, airports, water systems, sewer systems, power systems, and communications systems are being added or replaced with an increasing pace due to new regulations, technologies, increasing populations and deterioration through use. This is especially true in urban areas. Existing underground utilities create significant project risks during new construction and during retrofits. These risks manifest themselves in many forms.

The most attention grabbing events occur when a high pressure petroleum product pipeline erupts, such as in Allentown, PA, or in Edison, New Jersey (*Pipeline*, 1994). The risks of death, injury, property loss, and environmental releases are obvious in these types of cases. Nationwide, these incidents are more common than we think. Bernold (1994) at the Construction Automation and Robotics Laboratory at North Carolina State University reports that excavating equipment hitting buried utility lines cause an average of one death per day. Other project risks are not so obvious.

Project delays occur when existing utilities are discovered to conflict with new facilities or with excavation during construction. These conflicts create change orders for design and construction. They create claims by the constructor for delays. The delays create loss of revenue for income-producing projects, such as toll roads and commercial developments.

The reason for these project risks due to existing utilities is simple. Existing utility owners' records are not sufficiently comprehensive for design purposes (Anspach, 1992).

Utility Owner Records

Existing records of underground site conditions are usually incorrect, incomplete, or otherwise inadequate because:
- They were not accurate in the first place--design drawings are often not "as-built," or installations were "field run" and no record was ever made of actual locations;
- On old sites, there have usually been several utility owners, A/E's and contractors installing facilities and burying objects for decades in the area. The records seldom get put in a single file and are often lost -- there is almost never a composite;
- References are frequently lost -- the records show something 28'

from a building that is no longer there, or from the edge of a two-lane road that is now four-lane, or part of a parking lot; or
- Lines, pipes and tanks are abandoned, but do not get taken off the drawings.

Even so-called "As-builts" frequently lack the detail and veracity needed for design purposes in a utility congested environment. Furthermore, references on depth are rarely referenced to elevation datum. The amount of cover over a utility can change without obvious visual indications due to interim construction activity, erosion, etc., creating errors on records where "depth of cover" is the sole reference to vertical position.

Traditional Engineering Practices

Engineers realize the problems created by these incomplete records and protect themselves by placing a prominent note on construction plans. (Anspach, 1994) This note varies from engineer to engineer, but the theme is always as follows:

> Utilities depicted on these plans are from utility owners' records. The actual locations of utilities may be different. Utilities may exist that are not shown on these plans. It is the responsibility of the contractor at time of construction to identify, verify, and safely expose the utilities on this project.

Efforts made to field-identify utilities during design are often less than comprehensive. Occasionally, a civil engineer will attempt to characterize the site with limited surface geophysics by leasing ground penetrating radar or metal detection equipment. Without extensive training in this equipment and techniques, judgements are bound to be suspect, and bad data is usually worse than no data at all. Sometimes the engineer will hire a company with this equipment. A vast majority of these companies use this equipment for determining depth to water table, depth to bedrock, extent and existence of contaminant plumes, etc. They rarely have the expertise to characterize small-scale and small surface area utility structures. When utility owners mark their facilities during design due to One-Call statutes, problems arising during construction beg the question of whether the mark on the ground was correct or the survey of the mark was correct.

Sometimes engineers will dig test pits with backhoes and other earth moving equipment to detect the absence, presence, location, and/or condition of existing utilities. The selected locations for these test pits are either from utility records, visual features, or in areas of ambiguity or concern. Pits that contain no utilities ("dry holes") occur regularly. Risk of damage to existing utilities is high, due to the tremendous forces generated by these types of excavating equipment.

Project owners provide little incentive to their design engineers to do a better job of identifying and avoiding existing utilities on projects. Tradition implies that there are always problems with utilities and it is an accepted reason for changes and delays. Therefore, even when engineers are aware of better technologies, these technologies are rarely recommended to the project owner.

Subsurface Utility Engineering

A new engineering technology is emerging that combines the traditional disciplines of geophysics, surveying, and civil engineering in a practice called <u>Subsurface Utility Engineering</u> (SUE). SUE provides data on existing utilities and other cultural subsurface structures at appropriate times in the design process. The subsurface utility engineer takes responsibility for the accuracy and completeness of the utility information on the plans and has specialized professional liability insurance to support his services (Anspach 1992).

To understand SUE, Stutzman and Anspach (1993) define the quality levels of utility information that are available to the design engineer, constructor, and project owner as follows.

Quality Level "D" - Existing Records: Results from review of available records. Gives overall "feel" for congestion of utilities, but is highly limited in terms of comprehensiveness and accuracy. For projects where route selection is an option, this Quality Level is useful when combined with cost estimates for utility relocations following applicable "clear zone" and other accommodation policies.

Quality Level "C" - Surface Visible Feature Survey: QL "D" information for existing records is augmented using surface visible feature survey and digitizing data into CADD. The danger here is that much of the data is "digitized fiction." It is not unusual to find a 15-30% error and omission rate in QL "C" information.

Quality Level "B" - Designating: Two-dimensional horizontal mapping. This information is obtained through surface geophysical methods. It is highly useful for design basis information for conceptual design, and for proceeding prudently to QL "A". It should not be used for design basis vertical information, or where exacting horizontal tolerances are expected.

Quality Level "A" - Locating: Three-dimensional horizontal and vertical mapping. This information is obtained through vacuum excavation of test holes at points of conflict. This is the highest level of accuracy of subsurface utility engineering data. It provides horizontal and vertical design basis information for engineering, construction, maintenance, remediation, condition assessment, and related efforts.

Subsurface utility engineering departs from traditional engineering practice in the Designating (QL "B") component. This component consists of applying surface geophysical methods to the project area, interpreting the results in the field, marking these designations on the ground surface, and surveying the designations to permanent project control. The final work product undergoes a rigorous professional review both in the field and in the office. Existing utility owner information is correlated to the work product and discrepancies are either resolved or forwarded to the client for further recommendations. Deliverables are "sealed" by an appropriately registered professional.

Unlike utility owners (or their contractors) marking their facilities at time of construction, the SUE practitioner has available many surface geophysical methods and equipment (See Table 1) (Anspach, 1994). Method selection is the first step that is crucial to a cost-effective accomplishment of the mission. Utilities and other subsurface structures are composed of differing materials, sizes, and methods of enjoinment. They are emplaced at varying depths. Their surrounding environments can change drastically from point to point. Surface features such as water, buildings, scrap metal, etc., vary from site to site. Utility corridors are often highly congested with many different types of utilities. All these factors combine to defeat the application of any one surface geophysical method from identifying underground structures. Therefore, an integrated approach of different methods is necessary.

Radiofrequency Electromagnetics - ELF, VLF, LF ranges	Inexpensive and highly useful for metallic utilities, or utilities that can be accessed and a conductor or transmitter inserted into them.
Magnetics - Flux gate	Inexpensive and highly useful for utilities or their appurtenances that exhibit a strong magnetic field at the ground surface.
Elastic wave introduction into a non-compressible fluid.	Inexpensive and moderately useful for water lines with sufficient access points (typically fire hydrants) and low ambient noise.
Terrain Conductivity	Moderately inexpensive and useful in non-utility congested areas, or areas of high ambient conductivity. Most useful for tank and drum detection.
Impulse radar (Ground Penetrating Radar)	Moderately expensive and highly interpretative. Useless in areas of high conductivity such as marine clays, or for small utility targets.
Seismic Reflection and Refraction	Expensive and highly interpretative. Usefulness under field conditions extremely limited due to signal/noise ratio problems.
Thermal Imagery	Moderately expensive and interpretative. Sometimes useful for poorly insulated steam systems or other high heat-flux systems.
Radioisotope tracing	Moderately inexpensive to highly expensive. Useful for utilities already impregnated with radioactive isotopes.
Microgravitational	Expensive. Limited to identifying utilities of great mass differential from their surrounding environment.

Table 1. Available surface geophysical methods for subsurface utility characterization.

If a project had an unlimited budget, perhaps all available methods would be used by the SUE practitioner. However, this is a luxury rarely allowed or prudent. Through a combination of existing utility information, visual site investigations, experience, and project owner parameters, the SUE practitioner selects appropriate methods.

Occasionally, utilities exist in the subsurface environment for which no reasonable combination of surface geophysical methods will provide interpretable results. Additionally, the amount of extra cost to identify some utilities through these methods can be counterproductive to the project budget. Therefore, typical cost-effective scopes of work for most projects limit techniques to ELF, VLF, LF, magnetics, and elastic wave propagation through water lines, with additional surface geophysical techniques recommended only on a case specific basis.

This Quality Level "B" data is usually sufficient to accomplish preliminary engineering goals. Decisions can be made on where to place storm drainage systems, traffic management systems, etc., to avoid conflicts with existing utilities. Slight adjustments in design "footprints" can produce significant cost savings by eliminating wide-scale utility relocations. However, potential conflicts will still occur in the complex underground setting.

Contrary to many manufacturers' claims, depth determinations from the surface are not reliable. Neither is the other available data that the design engineer typically uses to produce a reliable design. As design proceeds to more advanced stages of refinement, the engineer needs data about precise width, location and horizontal extent of the utility system, elevation, configuration of non-encased multiple ducts, utility size, utility condition and material type, surrounding environmental conditions, etc. Such data cannot be obtained by simply applying the technologies outlined in Table 1 (QL "B"). Therefore, as a further refinement to the previous designating process, a physical exposure of the utility system at the appropriate location must be made. This is necessary to resolve ambiguities and to obtain more precise data on utilities. This process is termed *Locating* and represents Quality Level "A" data.

Traditional excavation methods using backhoes or other heavy equipment, and even hand shovels or "post-hole" diggers present the real possibility of damage to the utility being exposed. Utility systems such as fiber optic cables, terra-cotta or tile ducts, and small gauge command and control cables can be easily cut. Corroded metallic systems, spalled concrete pipe, and asbestos cement pipes can be quite fragile. Even steel systems in good condition can be unknowingly compromised when their protective coatings and wrappings are nicked or gouged, creating localized corrosion cells.

Air/vacuum excavation systems eliminate the above problems. Additionally, the work area/surface cut is quite small, often measuring no more than eight inches square as compared to a typical three feet by five feet backhoe pit. Dump truck support vehicles for both dirt hauling and backfill material are unnecessary, reducing the imposition on existing traffic flow. The small excavation that exposes the utility system at the precise spot where data is necessary does not require sheeting or shoring. Dewatering of high water tables is easily accomplished when necessary. Backfill and proper compaction of the excavation and paving repair is a simple and inexpensive task. Traffic control and worker safety is better in this small confined work area. The air/vacuum system is a better way to dig a hole, but the real value comes in the data collected from the exposed utility.

By knowing precisely where a utility is positioned in three dimensions at the beginning of the final design process, the designer can make prudent decisions. Small adjustments in design elevations or horizontal locations of new structures might eliminate a utility adjustment. Cut and fill areas might be altered to accommodate the existing utilities while still accomplishing the design mission. Sometimes utility relocations are necessary. By knowing the size, material, and location of the existing utility, the designer can produce realistic cost estimates for moving the utility out of the way of construction. By using the previous *Designating* data, empty corridors for the utility relocations can be quickly identified (Anspach, 1994).

The physical location of the utility is not the only useful data supplied to the designer during the *Locating* stage. Soil conditions, groundwater elevations, possible soil contamination, paving thicknesses and type, condition of the utility, and the depth to rock under the utility trench are all factors that may affect design decisions. Anspach and Wilson (1994) report that significant reductions in bid prices are realized when this comprehensive data, along with construction safeguards such as permanent field markers, is made available to the constructors at the pre-bid meeting,.

The data as discussed above is collected by field engineers, surveyed to permanent survey control, and formatted for easy reference for both the designer and the constructor. As in the *Designating* process, rigorous quality control processes are employed. The data is "sealed" by an appropriately registered professional and insured against errors and/or omissions.

While the utility is exposed, an assessment of its condition can be made. Nondestructive testing techniques to meet the needs of the project owner, project designer, and utility owner are discussed before excavation. Typical techniques used are ultrasonic pipe wall thickness measurements, pipe-to-soil potential measurements, current flow measurements, acoustic emission measurements, temperature gradient measurements, and visual examination by camera insertion. This information is useful for the utility owner, who may decide to replace a system before broadscale failure. The contractor also benefits from a record of the utility condition before construction begins, as he can make the proper excavation and utility protection choices.

Project Cost Savings

The advantages of SUE go beyond that of decreasing the risks of damage to utilities during construction. Tremendous cost savings to the public taxpayer and ratepayer also accrue on projects using SUE. Stevens (1993) approximates the basic areas of project expenditures as follows.

Administrative Costs	20%
Engineering Costs	10%
Construction Costs	45%
Cost "Overruns"	15%
Utility Relocation Costs	10%

Cost savings come in many forms. The total savings on a typical project may range from 10% to 15% (compared with costs from a project not supported by professional SUE). Stevens (1994) approximates the following cost savings.

Administrative (1/10th of 20%) 2%. Projects completed up to 20% faster (Virginia Department of Transportation study) allows financing to be paid quicker. Insurance and bonding costs may be less. Administration of change orders is lessened.

Engineering (1/20th of 10%) 1/2%. SUE techniques save time, therefore expense, by employing direct digital incorporation of utility data from survey into CADD files.

Utility Relocation (1/2 of 10%) 5%. Designers take comprehensive accurate utility information into consideration during the design of a project. Minor changes of design "footprint" or elevation data on paper eliminates wholescale utility relocations before construction.

Construction (1/20 of 45%) 2.25%. Construction bids are lowered because of fewer utility conflicts and an increased assurance of correct data. Liability for identification of utilities is borne by the SUE firm, not the constructor.

Cost Overruns (1/3 of 15%) 5%. Contractor delay claims are reduced (Florida State Department of Transportation). Engineering rework is reduced. Utility damages are reduced.

Anatomy of an Infrastructure Project utilizing Subsurface Utility Engineering

The following flow chart summarizes how utility Quality Levels are used within a typical project. This example is based upon a highway transportation project, but it has application for other types of projects as well.

QL "D" (Unnecessary unless options of route or site selection exist): A Utility Inventory and top-side survey identify utilities that are visible from surface features, but not of record. Utility Relocation Cost Estimates are made, based upon Federal, State, Local clear zone and accomodation policies. The route is selected, in part based upon utility data. The project limits are identified.

QL "C" (Unnecessary when QL "B" data are obtained)

QL "B" : Data are obtained within project limits using surface geophysical methods requiring professional judgements. Known, unknown, and abandoned utilities are identified, surveyed and referenced to project control. A professional review is conducted. Utility data is reviewed with the drainage designer and selected utilities are upgraded to QL "A" (optional) to assist in preliminary drainage or structure design. The utility data is reviewed with utility owners so they can update their existing records and provide input. The designer completes a preliminary design around existing utilities to avoid or to minimize relocations and construction activity near critical utilities. The SUE firm calls "one-call" and other utility owner nonparticipants to advise of excavations for test holes.

QL "A": Data is obtained on all utilities in potential conflict with designed structures, cut and fill, drainage ditches, etc. using appropriate excavation methods. Construction safeguards, such as permanent above-ground markers, color-coded message ribbon in backfill from ground to utility, and constructor-friendly data form, are generated. QL "A" data is reviewed by staff professionals and provided to utility owners. The designer finalizes design, adjusts elevations, construction notes, etc. using QL "A" data. Utility owners design necessary relocations or betterments, based upon design adjustments. Design may include need for additional QL "A" data. The designer reviews all utility data and conflicts, and level of effort to obtain such, with all bidders at preconstruction meeting. Constructor knows abandoned from active utilities. Constructor knows conditions of utilities. Constructor knows depths and elevations of utilities. Constructor knows width of utilities. Constructor knows materials of utilities. Constructor knows exact horizontal locations of utilities. Constructor knows conditions of soils surrounding utilities.

Utility relocations are performed before, or in conjunction with, project construction. QL "A" and "B" data is available to assist utility owner in marking facilities. The project construction begins. All utility relocations, betterment, or any other changes are as-built to project control and combined with the existing database for future retrieval by utility owners and other approved parties.

Summary

An increasing number of projects are proceeding based on better information on the existing utilities within the project limits. Quality levels of utility information are now defined, and are increasingly referenced by project owners in their requests for proposals and in their contracts. With an increasing utility congestion within our urban areas, industrial sites, and military bases, the necessity of accurate comprehensive data is in the public interest. Good engineering practice dictates that everyone involved in a project realize the detrimental effects that poor utility information can create, and recommend an appropriate course of action.

Appendix

Anspach, J. H. (1992). "Subsurface Utility Engineering." *Proceedings from the Twenty-Ninth Paving and Transportation Conference*, University of New Mexico.

Anspach, J. H. (1994). "Locating and Evaluating the Condition of Underground Utilities." *Proceedings from RETROFIT '94*, Stanford University.

Anspach, J. H., and Wilson, S .E. (1994). "A Case Study of an Underground 138kV Transmission Line Design Utilizing Subsurface Utility Engineering." *Proceedings from the American Power Conference*, Chicago, IL.

Bernold, Leonard (1994). In-house study for A Consortium For Safe Excavation, North Carolina State University.

Methfessel, H. A. J. (1993). "Managing Risk Through New Engineering Technologies Such As Subsurface Utility Engineering." *Proceedings of the National Conference on Tort Liability and Risk Management for Surface Transportation*, Pennsylvania State University.

"Pipeline Explosions Create Infernos" (1994). The Building Official and Code Administrator Magazine, p.25. July/August.

Stevens, R. E., and Anspach, J .H. (1993). "New Technology Overcomes the Problems of Underground System Interferences on Power Projects." *Proceedings from the American Power Conference*, Chicago, IL.

Stevens, R. E. (1993). "Subsurface Utility Engineering." *Proceedings of the Society of American Value Engineers*, p. 46.

Stevens, R. E. (1994). "SUE Seminar". Presented at the National Highway/Utility Conference, FHWA, Louisville, KY.

Stutzman, H. G., and Anspach, J. H. (1993). "Site Investigation and Detection." *Proceedings from Research Needs in Automated Excavation and Material Handling*, National Science Foundation.

U.S. Department of Transportation, Federal Highway Division, Office of Engineering. " Subsurface Utility Engineering." October 1994.

PIPE REHABILITATION USING A SLIPLINE PIPE

Ernest R. Hanna, P.E.[1] and Allan J. Scarpine, P.E.[2]

ABSTRACT

Landfilling at a site in western New York was done during the 1950's through 1993. During this time period waste fill was placed over an existing 1.2m (48-inch) reinforced concrete stormwater drain pipe (RCP). The landfill was closed in 1993. Stormwater drains from one side of the landfill to the other through the 1.2m RCP, which is under up to about 46m (150 feet) of waste fill. The landfill operator was concerned with the potential for leachate to escape the landfill via the 1.2m RCP. GZA GeoEnvironmental of New York was retained by the landfill operator to suggest alternatives for stormwater drainage. Following this work, it was decided to slipline the 1.2m RCP. GZA provided final engineering design and prepared contract documents. The landfill operator selected a contractor, administered the project and did its own field construction monitoring.

This paper outlines design considerations for pipe rehabilitation using a slipline pipe with an emphasis on key components related to this project. Items which made design and construction difficult, such as limited access/space and deflections in the pipe, are discussed. Also presented is a summary of the construction and a comparison of key design considerations with the construction procedure followed.

INTRODUCTION

Prior to landfilling at this site, a small stream bisected the area allowing surface water drainage to flow from west to east (see Figure 1). A 1.2m

---

[1] M.ASCE, Associate Principal, GZA GeoEnvironmental of New York, Buffalo, New York

[2] AM.ASCE, Project Manager, Browning-Ferris Industries, Niagara Falls, New York

diameter reinforced concrete pipe (RCP) was constructed along the general alignment of the stream during waste filling operations to allow surface water drainage to be maintained. Existing surface soils at the site are fine grain with low to medium plasticity. Based on a limited number of photographs taken during construction, the 1.2m RCP was bedded in a manufactured crusher run stone. Pipe joints were reportedly pressure grouted. A manway (junction box) was reportedly constructed about midway along the pipe alignment at a pipe deflection.

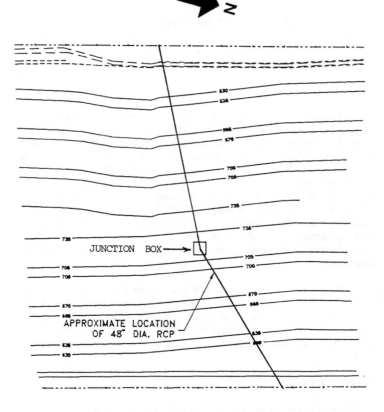

FIGURE 1: PLAN VIEW OF RCP AT LANDFILL SITE.

The environmental monitoring at the site indicated leachate from the landfill may be entering the 1.2m RCP and leaving the site in the surface water drainage. Additionally, because of the relatively flat pipe slope (0.2 percent), sediment build-up was causing surface water back-up within the pipe and at the pipe inlet. The landfill operator decided an engineered solution was necessary to improve surface water drainage and to stop possible leachate outbreaks from leaving the facility.

Consideration was given to either plug the 1.2m RCP and reroute surface water flow or reconstruct/slipline the 1.2m RCP which could stop leachate inflow. The following scenarios were evaluated:

- Plug the 1.2m RCP
    - Route surface water flow to the north
    - Route surface water flow to the south
    - Route surface water to the east over the landfill
- Stop leachate inflow into the 48-inch RCP
    - Regrout the pipe joints
    - Reconstruct the pipe
    - Rehabilitate the pipe

A feasibility study and engineering cost estimates were done to approximate which methods seemed more reasonable. Considering the space constraints and flat topography around the site, the options related to plugging the 1.2m RCP were not considered suitable. These options required pumping (long term operation costs), land acquisition, and tunneling under a state road. Also, reconstruction of the existing 1.2m RCP considering the volume of waste fill over the pipe was considered too costly. Therefore, a tentative decision to rehabilitate the pipe was made.

Pipe rehabilitation can be done using a variety of methods including sliplining, modified sliplining, pipe bursting, cored-in-place pipe, joint grouting, or sprayed on coatings. The next step in the engineered solution involved video or personnel inspection of the pipe to determine existing conditions and to identify problem areas. Before this could be done; however, the pipeline had to be cleaned.

The 1.2m RCP was cleaned on two separate occasions. It was cleaned first in the Spring 1993 using a horizontal auger. A auger machine was set-up at the pipe inlet and about 180m (590 lineal feet) of pipeline was cleaned from this end. Auger spoil was placed as cover material in the active landfill area. Following cleaning operations at the pipe inlet, the auguring equipment was moved to the was moved to the pipe outlet side. An additional 149m (490 feet) of pipe was cleaned from this end also using the auger system to remove pipe sediment. Again, auger spoil was placed within the active landfill area.

Following pipe cleaning the 1.2m RCP was flooded and an underwater physical inspection was done. This inspection reported the following.

- the eastern (outlet side) half of the pipeline had an average sediment depth of about 2.5cm (1 inch).

- the western (inlet side) half of the pipeline had an average sediment depth of about 2 to 8cm (1 to 3 inches).

- the pipeline was apparently not fully cleaned because access was restricted about 180m (590 feet) from the inlet and about 149m (490 feet) from the outlet end of the pipe.

- pipe joint separation was noted to be about 8cm (3 inches) in some sections near the center sections of the pipeline.

- concrete spalling and exposed reinforcement was also noted at the crow of the pipe near the central sections.

- a concrete manway was confirmed to exist at about the center of the pipe (see Figure 1 and 2).

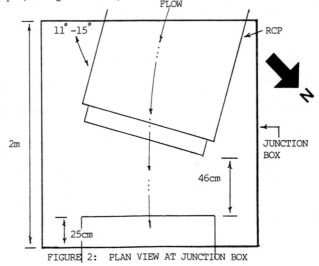

FIGURE 2: PLAN VIEW AT JUNCTION BOX

No construction work was done concerning the pipe rehabilitation following the Spring of 1993 and due to concerns of the landfill operator with selecting a qualified contractor to do the work. A project manual was assembled containing bid documents and specifications. The project was awarded to Skanex Pipe Services, Inc. (Skanex) in the Spring of 1994. Due to the delay from the Spring 1993 pipe cleaning the pipe was recleaned in the summer 1994.

PIPE REHABILITATION

The information obtained following pipe cleaning and inspection indicated that the pipe rehabilitation would require more work than just grouting at the pipe joints. Deterioration at the crown of the pipe indicated structural repair would be necessary. It would also be necessary to provide some protection such that additional concrete deterioration would be limited. The pipeline rehabilitation methods further considered included pipe bursting, sprayed-on-coating, sliplining, and a soft-lining/cured-in-place pipe.

Pipe Bursting

Pipe bursting used radial forces to break out and enlarge the existing pipe allowing a new pipe to be simultaneously installed. Pipe bursting is not typically done for pipe diameters greater than about 0.3m (12 inches) nor for pipe sections greater than about 91m (300 feet). Also, reinforced concrete is difficult to burst. Therefore, pipe bursting was not considered further.

Sprayed-on-Coating

A thin mortar lining or resin coating can be sprayed onto the pipe inner wall to improve hydraulic characteristics and corrosion protection. The sprayed on coating does not improve the existing structural capacities of the pipe. Therefore, a sprayed-on-coating was not considered further.

Sliplining

Sliplining is done by simply pulling or pushing a new pipe into an old pipe. The annular space that remains between the new pipe and the old pipe is filled with grout. Sliplining can improve existing structural capacity and corrosion protection.

Soft Lining System/Cured-in-Place System

A number of systems are available which are based on a soft liner that is put in place and hardened by various methods. The Insituform method is one of the more well known methods. It involves placing a geotextile tube impregnated with a special uncured resin into the pipe, then heating the tube by recirculating hot water through it, which cures the resin and hardens the geotextile forming a pipe lining.

Other similar methods are Nu-Pipe and U-Liner, where a polyvinyl chloride (PVC) or polyethylene (PE) thermoplastic pipe in a collapsed or folded shape is warmed (to make flexible), pulled into the pipe, formed to the pipe shape using hot air, water or steam, and then allowed to harden.

ENGINEERING CONSIDERATIONS

The possibility of the 1.2m RCP bell and spigot joints leaking was a concern because the pipe was under municipal waste fill. To improve the long term stability of the 1.2m RCP and limit the landfills impact on surface water, consideration was given to the more cost effective method to seal the pipe joints to infiltration from the waste fill overburden. Additionally, there were three major engineering concerns related to the 1.2m RCP rehabilitation. The three concerns are

- Pipe Structural Stability

- Pipe Corrosion Resistance

- Pipe Hydraulic Characteristics

Structural Stability

External loading on the pipe due to the landfill waste fill above the pipe must be considered. The existing 1.2m RCP did not visually appear to have collapsed anywhere during the underwater inspection. However, signs of deterioration were noted. although it is assumed that the deterioration was due to sulfide attack on the concrete, this was not confirmed. Therefore, it was assumed that the 1.2m RCP may not be structurally sound. Based on a final landfill elevation of about elevation (El.) 737 and a pipe invert of about El. 586, the pipe maximum external loading is calculated to be about 70,000kg/m (49,000 lbs/ft). The waste fill load above the pipe was estimated to be 470KPa (68 psi). The use of a cured-in place or slipline pipe to remediate the 1.2m RCP may improve its structural stability depending on the type selected.

Corrosion Resistance

Landfill gases were measured within the pipe prior to pipe cleaning activities. The presence of these gases required air monitoring and venting be done when working around or in the pipes. Because hydrogen sulfide is often measured in the landfill gas, it was assumed that concrete deterioration could be attributed to the presence of sulfide. Therefore any pipe rehabilitation had to address resistance to sulfide attack.

Hydraulic Characteristics

The existing 1.2m RCP carries surface water runoff from the adjacent property west of the site and from the western half of the landfill. The contribution of surface water runoff from the landfill is about 1270 liters/sec (45 cfs). The off-site contribution was estimated by the adjoining property owner to be about 113 liters/sec (4 cfs). The approximate discharge capacity of the existing 1.2m RCP was calculated to be about 1800 liters/sec (64 cfs) based on a 1.2m inside diameter, a pipe slope of 0.2 percent, and a Manning roughness coefficient of 0.013. However, it is important to note that the pipe, in its existing state was about one-third to three-quarters full of sediment and the adjoining property to the east at the pipe outlet was also silted in to a higher ground surface elevation than the pipe outlet. To effectively limit sediment build-up in the existing pipe, the downstream channel (on adjoining properties) will have to be recut. This issue has not been resolved.

The use of a slipline pipe would reduce the discharge capacity depending on the pipe diameter. Pipes with an outside diameter of 1067mm (42 inches) and 1000mm (39 inches) were considered. The full flow discharge capacity of these pipes (considering a pipe slope of 0.2 percent and a Manning's roughness coefficient of 0.011) is calculated to be 1416 (50 cfs) and 1190 liters/sec (42 cfs), respectively.

A summary of the applicability of each design consideration mentioned above for each respective method is shown in the Table on the next page.

Method	Design Considerations			Practical
	Structural Capacity	Corrosion Protection	Hydraulic Characteristics	
Pipe Bursting	No	Maybe	Yes	No
Spray Coat	No	Yes	Yes	No
Cured-in-Place	Maybe	Yes	Yes	Yes
Soft Lined System	No	Maybe	Yes	No
Slipline Pipe	Yes	Yes	Maybe	Yes
Joint Grouting	No	No	Yes	No

CONSTRUCTION ISSUES

There were several issues that needed to be addressed during construction which were covered in the project specifications. The issues were the ability to pull the largest diameter slipline pipe possible through the chamber at the 1.2m RCP pipe deflection, limited access, health and safety concerns, grouting the annulus between the slipline pipe and the existing 1.2m RCP, and a quality control check on the pipe joints and pipe grouting. Prior to pipeline rehabilitation, the contractor was required to:

-- Clean the interior of the 1.2m RCP and maintaining it in a clean dewatered condition;

-- Seal any leaks in the 1.2m RCP and junction box;

-- Videotape the interior of the 1.2m RCP following pipe cleaning; and

-- Physically inspect the pipe interior to confirm results of the video inspection and to measure the features within the junction box.

Slipline Pipe Diameter Selection

To address the uncertainty regarding how large a pipe diameter could be pulled/placed in the existing 1.2m RCP, two alternatives were presented for bidding.

Alternative A - High density polyethylene (HDPE Welded) Slipline Pipe, which includes a test pull with a 1000mm (39-inch) pipe through the pipe deflection at the junction box followed by sliplining with a 1000mm pipe. Should the 1000mm test pull be successful, it is the intent of the Company to have a 1000mm continuously welded HDPE pipe used to slipline the existing 1.2m RCP. If the test pull is not successful, it is the intent of the company to slipline the 1.2m RCP with a 1067mm (42-inch) continuously welded HDPE slipline pipe with the exception of a welded joint in the junction box.

Alternative B - Jointed Slipline Pipe which requires slipline with a 1067mm (42-inch) pipe. Pipe joints are bell and spigot type or similar. The pipes considered were polyvinyl chloride (PVC), centrifugally cast fiberglass reinforced polyester, or HDPE.

Limited Access

The Contractor was not allowed to begin rehabilitation work nor was it allowed to enter or work on the adjacent properties to the site unless the Contractor received written permission from the respective property owner(s). The Contractor was responsible to negotiate with the adjacent landowners. Prior to working on adjacent properties, the Contractor was required to provide a copy of all agreements to the Company.

Health and Safety

The existing 1.2m RCP extended underneath a sanitary landfill. Prior to the Contractor working on the site, explosive gases had been measured at the inlet and outlet of the pipe. Additionally, entry into the 1.2m RCP is considered confined space entry. Therefore, concerns regarding health and safety issue were warranted. The construction documents required that the Contractor be solely responsible for performing the construction in a safe manner. The Contractor was required to prepare a project specific health and safety plan which includes considerations for:

- The Company's site policies;
- Municipal and state regulations; and
- Occupational Safety and Health Administration requirements.

A ventilation blower was placed at the upstream end of the 1.2m RCP and air was forced through the pipe. This blower ran for two hours prior to any entry. Air quality monitoring was done continuously. Prior to an entry, it was required that explosimeter readings be 0% and oxygen levels be between 19.5% and 23.5%. Skanex personnel entered the pipe with self contained breathing apparatus, air quality monitoring equipment, two-way radios, and connected to a lifeline. Spotters were placed at both ends of the pipe and radio communication was maintained at all times.

Grouting

The backfill grout was required to be a low shrink (less than .02 percent at 28 days) unreinforced mortar to fill the annulus space between the slipline pipe and the 1.2m RCP. It was required to have a minimum 28 day compressive strength of 5200 KPa (750 psi). Also the heat of hydration was limited to a maximum of 60 C (140 F).

Grout placement procedures were specified to include provisions to stabilize the pipe position during grouting, to limit grout placement in lifts not exceeding two feet, to completely fill the void between the slipline pipe and the existing RCP, and to provide a minimum grout cover of 0.3m (1 foot) over the top of the slipline pipe within the junction box.

Quality control testing was specified for grout compressive strength testing [1 test for every 77 cubic meters (100 cubic yards) of grout] and to check that the annulus was completely grouted. The pipe annulus was required to be sounded at a 15m (50 foot) or less interval spacing. At each sounding location, the Contractor was required to sound at a minimum the pipe crown, springline (2 spots) and invert. In addition to sounding, the Contractor cored the slipline pipe or used nondestructive geophysical methods at a minimum of 12 locations

Air Pressure Testing

At the completion of the slipline pipe installation and grouting, the Contractor was required to do a hydrostatic pressure test to verify that there are no leaks in the slipline. A minimum hydrostatic pressure of 172 KPa (25 psi) was required to be applied and maintained for two hours. A leak was defined as a pressure loss of greater than 35 KPa (5 psi) over the two hour test interval.

Other Considerations

The project was considered straight forward. However, due to regulatory concerns and that the landfill facility was in the closure process, it was important that the project be completed. Therefore it was stated in the project specifications that payment for the work would not be made until the project was completed (i.e. pipe in place, grouted and pressure tested).

CONSTRUCTION

The contractor selected, Skanex, had several concerns prior to starting construction which included health and safety issues, access, dewatering, cleaning the pipe, and pressure testing. Issues related to health and safety and access were worked out with the Company and the adjacent site owners. The work was done primarily in the summer when rainfall is usually lower. The precipitation during the Summer of 1994 was below average which helped dewatering concerns. Sediment removed in the process of pipe cleaning was placed within the landfill boundaries as appropriate. The pipe hydrostatic test pressure required of 172 KPa (25 psi) seemed excessive to Skanex. GZA and the Company considered the hydrostatic pressure testing requirements and agreed that the pressure could be reduced from 172 KPa to 110 KPa (16 psi).

Skanex found, through inspecting the pipe, that the junction box measurements and pipe deflection angle estimated and presented in the bid specifications (see Figure 2) were generally accurate. However, the first thirty feet of pipe downflow from the junction box was a 1372mm (54-inch) diameter RCP instead of 1219mm (1.2m or 48-inch) as originally anticipated.

Skanex performed a test pull to determine which size pipe could be installed as required in the project specifications. A 66m (200 ft.) section of 1000mm diameter pipe was placed at the upstream end of the 1.2m RCP. A nose cone was fabricated on the leading edge of the test section and it was connected to a diesel powered winch, at the downstream end, by a 13mm (1/2 in.) steel cable. This test section was successfully pulled through the pipe. Based on this test pull, it was decided to slipline the

1.2m RCP with a 1000mm HDPE pipe, fusion welded together at the pipe joints.

The 1.2m RCP was maintained in a dewatered condition during the installation of the slipline pipe. The contractor constructed a small earth berm upstream of the pipe to contain surface water runoff. The water collected was diverted through the 1.2m RCP during construction by using a 150mm (6 in.) pump and 150mm diameter HDPE pipe, as needed.

The 1000mm HDPE pipe was delivered to the site in 6.6m (20 ft.) sections. Skanex subcontracted the work related to fusion welding the pipe joints because the large pipe diameter required specialized equipment. As the welding progressed, the slipline pipe was pulled inside the 1.2m RCP. Due to time constraints on the pipe welding subcontractor, the entire length of pipe needed for the project (about 400m) was welded together in one time span.

Skanex fabricated the lead section of slipline pipe into a cone shape to make it easier to pull the slipline pipe through the 1.2m RCP and through the junction box. The only major problem that occurred happened during the pulling of the slipline pipe through the 1.2m RCP. The wire rope used to pull the slipline pipe kept breaking. Skanex, through trial and error, ended up using a larger wire rope to pull the 1000mm diameter slipline pipe. The drag due to the pipe load and the bend at the junction box provided a greater resistance than Skanex anticipated which resulted in increasing the steel cable diameter from 13mm to 19mm (3/4 in.). Other than this issue, Skanex did a professional job and the Company and State regulatory personnel were satisfied with the work. It is also important to note that the pipe manufacturer (Plexco/Spirolite) and the pipe distributer (Vari-Tech Associates, Inc.) stayed involved in the project throughout construction providing assistance to Skanex when requested. This involvement is one reason that the project was completed to the satisfaction of the Company and regulatory personnel.

WAILUPE RECONSTRUCTED TRUNK SEWER
ASCE HAWAIIAN CHAPTER
PROJECT OF THE YEAR
1993

JOHN F. JURGENS[1] M ASCE

Abstract

Common problems of sanitary sewer deterioration and rehabilitation include hydrogen sulfide attack of concrete pipe and related structures, and the restrictions of utilizing earlier construction procedures for making repairs or corrections. Within the 2,000 miles of sewer system operated by the City and County of Honolulu, 10% are critical interceptors. The Wailupe Sewer Trunk is one of these.

Classic conditions conducive to the generation of hydrogen sulfide exist within portions of this system. Flow levels are controlled by a downstream lift station, which had allowed sluggish conditions to occur. This project's unique problems called for a multi-faceted approach to rehabilitation of both hydrogen sulfide-damaged pipes and manholes plus infiltration issues. The result of the project was to restore the structure of the system and eliminate infiltration. Additional operational benefits have resulted from the pipe cleaning.

The Setting

This trunk sewer is 10 to 15 feet deep along the beach east of Honolulu. Part of the system runs through Wailiae Country Club - home of the Hawaiian Open, under the Kahala Hilton Hotel and under numerous properties where homes valued over $3,000,000 are common.

---

[1]Mr. Jurgens is a Technical Representative for Gelco Services, 20606 84th Avenue South, Kent, WA 98032

Project Description and Location

The project consisted of approximately 2,000 lineal feet of existing 30-inch and approximately 4,400 lineal feet of existing 36-inch gravity trunk sewer mains. The sewer main is located along the shore of Wailupe Beach in East Honolulu and extends from the Kahala Wastewater Pump Station to Wailupe Beach Park. It was installed within a sewer easement and crosses through numerous private properties throughout its alignment. The private properties crossed by the sewer include the Kahala Hilton Hotel, Kahala Beach Apartments, Waialae Country Club, and numerous residential lots.

The project was initiated through the Division of Wastewater Management, Department of Public Works for the City and County of Honolulu, which has municipal jurisdiction and responsibility for the line, and the Department's consulting engineering firm, Stanley Yim and Associates.

Tributary Area

The existing trunk sewer main carries wastewater generated by the Waialae-Kahala tributary area which contains approximately 1365 acres. The surface elevations in the area ranges from sea level to about 1,200 feet above mean sea level. The topography of the area is characterized as relatively flat at the low-lying areas and moderately steep at the higher elevations. The existing sewer system in the area consists mostly of gravity sewer lines ranging in size from 6-inches to 36-inches. In addition to gravity lines, force mains and siphons are part of the system.

The tributary area consists mostly of residential developments including some business areas, schools, parks, a golf course and resort areas. These developments were constructed between the early 1950's and mid-1980's.

The existing Wailupe Trunk Sewer main is the major sewer line serving the tributary area. If this main became unable to serve its function, the impact on the entire tributary area would be detrimental. There were no alternate routes available for proper discharging of the wastewater generated by the tributary area should the existing trunk sewer main fall.

INVESTIGATIONS AND EVALUATIONS

Condition of Existing Sewer Trunk Main

According to the city record drawings, construction of the existing trunk sewer main was completed in 1955. It was constructed of reinforced concrete pipe.

Topographic surveying and mapping of the project area was completed in February 1989. The survey indicated that the inverts of the existing sewer manholes were at elevations ranging from (-)11.35 to (-)6.13 feet below mean sea level.

Closed circuit television inspection and videotaping of the existing trunk sewer main was completed in January 1989. Throughout the television inspection, the sewer main was flowing at approximately one-third to one-half full. In order to keep the flow depths as low as possible, the wet well level at Kahala Sewer Pump Station (downstream) was kept below elevation (-)11 feet. Two upstream pump stations (Niu Valley and Paiko Drive) were also shut down during most of the television inspection operations, however, flow depths in the sewer main did not lower significantly when the upstream pump stations where shut down. Only the top one-half to two-thirds of the pipe were visible during the television inspection.

Results of the television inspection indicated the interior of the existing reinforced concrete pipe had suffered medium to heavy deterioration throughout its alignment. At some sections of the existing pipe where severe deterioration occurred, the pattern of the reinforcing material was visible. This deterioration of the pipe appeared to be a result of hydrogen sulfide attacks. Accumulations of sewer grit had been deposited throughout the system. Severity varied, but due to the amount of grit, it significantly affected the hydraulics, increasing the anaerobic condition. Leaking pipe joints were also observed at various locations along the existing sewer main.

Preliminary investigations by the City and County of Honolulu identified settlement of existing sewer manholes. The manhole invert elevations were obtained from the topographic survey, and those elevations obtained by the City's "As-Built" drawings of the existing trunk sewer main differed in every case.

Inspection of the existing twenty-three manholes indicated they were generally in fair to good condition. Infiltration through the walls was found at some manholes. Corrosion of existing manhole rungs was found at all of the manholes. For safety reasons, replacement of all manhole rungs was performed.

The existing trunk sewer main appeared to have the hydraulic capacity to carry the design peak flows throughout most of its alignment, however, the existing sewer lines between manholes #3 and #4 and between manholes #14 and #15 had a "reversed" slope condition. Consequently, upstream portions of the existing trunk sewer surcharged at these sections.

Discussion

Investigations made on the existing trunk sewer main indicate the following:

* It was 34 years old
* It had been installed within a sewer easement crossing through numerous private properties
* Access to the trunk sewer main was very difficult
* It had been installed below sea level
* Some of the sewer manholes may have settled
* As the interior of the pipe had suffered medium to severe deterioration, its remaining service life was questionable
* Except for those sections where "reversed" slope conditions exist, the trunk main has the hydraulic capacity to carry existing design peak flow
* The existing sewer manholes were in generally fair to good condition, however, infiltration through the walls was found at some of the manholes

Determination

The WAILUPE RECONSTRUCTED TRUNK SEWER project was necessary to ensure that the existing trunk sewer main continues to remain functional. Unrepaired, the trunk sewer main was functional, but its remaining service life was questionable. Rehabilitation or reconstruction of the sewer lines was warranted due to the heavy deterioration which the pipe had suffered.

The existing Wailupe Trunk Sewer is the major sewer main serving the Waialae-Kahala tributary area. Therefore, any rehabilitation or reconstruction of the existing sewer main required concepts that would enable continuous flow of sewage throughout the construction period.

Alternatives

The City/County had three possible alternatives to address the condition of the existing trunk sewer main. Those included: taking no action, rehabilitation, and replacement and partial or total reconstruction. All were considered as possible solutions.

As previously noted, the existing trunk sewer was installed within an easement crossing through numerous private properties throughout its alignment. Additional sewer easements would be necessary should corrective measures required realignment, or if installation of new facilities were located outside the limits of the existing easement. Also, temporary construction easements and right-of-entry permits would be required to gain access to the existing trunk main.

Alternative I - No Action

Alternative I is a "No-Action" at all. This was not a viable option, since the existing trunk line was an important and critical link in the sewer system which serves a large tributary area in East Honolulu.

Alternative II - Rehabilitation

Results of the television inspection indicated the interior of the existing pipe had suffered medium to heavy deterioration as a result of hydrogen sulfide attack. Infiltration through pipe joints was also found at various locations along the pipe.

Rehabilitation of the existing pipe required sealing of existing leaks in the pipe interior. Since the interior of the pipe had suffered severe deterioration throughout most of its alignment, its structural integrity was questionable. Additionally, patterns of the pipe reinforcing material were visible at various areas. Therefore, the process selected for this project must have its own structural strength.

The only known process available with the required structural strength was the Insituform® process, a cured-in-place-pipe (CIPP), as identified by ASTM F 1216. This process uses polyester felt materials and special thermoset resins to create a "new" pipe within the existing pipe. Since the Insitupipe® is a continuous new pipe from manhole to manhole, sealing of leaking pipe joints is accomplished at the same time the Insitupipe® is installed, making separate joint sealing operations unnecessary.

The existing sewer manholes were generally in fair to good condition, however, there was some infiltration and moderate corrosion. The existing manholes had to be thoroughly cleaned and any leaks found would be sealed. Additionally, all existing manhole rungs needed to be replaced for safety reasons.

Alternative III   -   Reconstruction

Reconstruction work using standard methods for the existing trunk line would require major excavation work; the removal and reconstruction of existing walls, fences, and walks; the removal and replacement of existing trees, plants, and shrubbery; the tearing up and replanting for a portion of the Waialae Golf Course; the digging up and repaving of part of an existing parking lot; trench shoring for all excavations; dewatering (with the potential of subsidence to surrounding homes and swimming pools); and pipe laying. The work would also be in cramped and crowded spaces due to the homes and swimming pools in the surrounding areas. There were also many areas without adequate space available to permit reconstruction.

The time needed to reconstruct the trunk line using standard construction methods would also be considerable. The alignment could possibly be maintained, but the lack of adequate work space would severely impact heavy equipment movement. Also, this alternative was not cost effective.

In the January of 1991 the City/County of Honolulu awarded a contract to Gelco Services of Salem, Oregon to reconstruct the Wailupe Interceptor, using CIPP technologies.

Summary and Conclusions

This project was challenging, not only due to the time and site constraints in the project, but also because it was necessary to coordinate all activities around the needs and priorities of different agencies, home-owners, businesses and recreational facilities associated with, or in the vicinity of the project.

The schedule for work utilized for this project was prepared in a computerized format utilizing state of the art critical path method (CPM) scheduling software. Although this was a complex schedule to build, the software allowed for continuous review, easy updating, tracing of cost estimates and material needs, and most importantly, quick and easy to understand printouts for Gelco's crews and the City/County personnel.

During the preliminary planning stages, it was assumed power would be supplied to generators by drops from the Hawaiian Electric Company as a noise-control measure. When it became apparent that the drops could not be provided, Gelco secured several specialized silent generators. There were no complaints from the surrounding residents. In preparation for Hurricane Iniki, all potential environmental hazards due to high winds, tide or wet weather were removed from the project site and secured, and no damages occurred.

Specifications required there be a back-up set of by-pass pumps ensuring that if one set of pumps failed the other set would be able to handle all flows without a sewage spill. The by-pass system was created for peak flow capacity minimums of 6.5 mgd or 4,500 gpm. The by-pass conduit used was HDPE 18" SDR 21. This type of conduit insured no leakage of joints and allowed for ease in movement as the project progressed.

At one location, the challenge was to stop the water from flowing through a large hole in the crown of the pipe. This missing segment, approximately 1' by 1.5' in area, was due to significant sulfide attack. The coral backfill allowed water in excess of 20 gpm to enter the pipe. A two-step chemical grout solution was used to seal the crack. The first material used was AV 220, a hydrophobic resin which reacts in the presence of water, expanding in place. This material, if left to react in an unconfined space would expand 10 times it's initial volume.

The high viscosity, of 400 cps for this grout extended the reaction time from 4 to 8 minutes depending on the water temperatures, required a methodology of using a sacrificial carrier, such as a dry jute to be 'wet-out' then hand applied into the void area. This was done in stages until the greatest amount of the void was filled, minimizing the water entering the pipe, then AV 100, and acrylimide chmemical grout, at 3 cps viscosity with a set time of ten seconds was used to finish the grouting process.

The results of the structural rehabilitation can only be described in subjective terms. The condition of the pipes which were repaired was such that a collapse or failure was imminent. The techniques used in making the repairs were chosen so that the problem could be corrected quickly, cleanly, and with the least inconvenience to the public, while still satisfying the technical requirements. The Insituform® process was well received by the residents, media, and technical community of Oahu.

For this project Gelco received the American Public Works Association's CONTRACTOR OF THE YEAR and the City/County of Honolulu was awarded with the American Society of Civil Engineers, Hawaii Chapter PROJECT OF THE YEAR Award.

"Pipeline Integrity Assessment and Rehabilitation"

Peter Brooks and Mark Smith[1]

Abstract

Public Law 102-508 (October 24, 1992), otherwise known as the "Pipeline Safety Act of 1992" established the criterion for the future pipeline safety regulations contained in Title 49 Part 195 of the Code of Federal Regulations (49 CFR 195). Recent and pending regulations are a result of public concern for safety and increased concerns about environmental impact of operating a pipeline. A leak from a pipeline can cost millions of dollars to assess and remediate. Prevention of leaks is regulated and makes fiscal sense as spending thousands of dollars today on prevention is much better (and more politically correct) than spending millions of dollars in a few years to clean up the environment following a spill.

The key to prevention of pipeline leaks is through an integrity assessment of existing facilities utilizing the current state of the art in-line Metal Loss Detection (MLD) tools and alternative methods of integrity assessment using non-destructive techniques. These integrity assessment methods may seem disruptive to operations, however, those operators who are taking the lead in pipeline integrity assessment and rehabilitation of existing facilities will find far less interference (shutdowns, environmental assessments and remediations) in their operations which equates to saving dollars. The paper demonstrates the process of conducting a pipeline integrity assessment and rehabilitation program, and will reflect on the authors' direct experience on this subject.

---

[1]President & Engineer, WGP Engineering, Inc., 1230 Columbia Street, Suite 1050, San Diego, California 92101

Introduction

Pipeline integrity assessment is a program management approach developed by WGP Engineering, Inc. by which pipeline owners and operators ensure regulatory compliance while conducting long term, safe pipeline operations with minimum service disturbance and maximum cost effectiveness. The intent of a pipeline integrity assessment program is to assess the existing system configuration and to make recommendations for inspection, testing, and repair/rehabilitation methods to prevent potential integrity problems and to minimize the severity of uncontrollable pipeline accidents.

Background

The design life of a typical oil or gas pipeline and related terminal facility can realistically vary from, say, 15 to 50 years. For most systems, the original design engineer can establish and predict the service conditions for the first ten years of operation. However, it is not uncommon that as operations progress with time, any number of the following can occur:

- Changes in product type and throughput
- Changes in environmental considerations
- Deterioration of protective systems
- Inadequate maintenance
- Non-use of equipment and appurtenances
- Tightening of regulatory or license conditions
- Increased public concern

Such considerations are of concern to pipeline operators and can lead to rehabilitation requirements for pipelines and related facilities.

Regulatory Review and Compliance

From the "Pipeline Safety Reauthorization Act of 1988" to the "Pipeline Safety Act of 1992", Congress has indicated a broad range of concern for gas and hazardous liquid pipeline safety. The Legislative Regulation process is shown below:

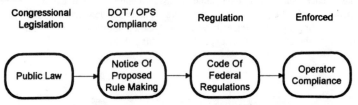

It is normal for there to be 2 - 3 years between each phase.

The areas where there has been emphasis in recent legislation are:

- Pipe Inventory - Operators will be required to keep maps and records to show location and other characteristics of pipelines to be provided to governmental agencies on request.

- Hydrostatic Testing - The requirements for testing will be extended to all pipeline facilities regardless of operating pressure.

- Operations and Maintenance - The minimum requirements for operations, maintenance, and personnel training will be established for operators of all types of product pipelines.

- Emergency Flow Restricting Devices - Will specify the requirements for use of devices and procedures to detect and locate pipeline ruptures and to minimize product release.

- Environmentally Sensitive Areas and High-Density Population Areas - Will require operators to identify these critical areas.

- Increased Inspection Requirements - Will require operators to periodically inspect all pipelines in sensitive areas.

- One Call System Participation - Will require operators to participate in one-call systems and provides funding for establishing these systems even in rural areas.

- Cathodic Protection and Coatings - Will establish the minimum survey requirements.

- Emergency Procedures and Equipment - Will establish the minimum requirements for developing contingency plans.

- Costs and Liability of Spill Cleanup/Remediation - Will establish the minimal spill cleanup requirements and liability.

- Personnel Qualifications - Will define the minimal training and experience requirements to operate pipeline facilities.

Site Assessment

The first step in the pipeline integrity assessment process is to visit the site to obtain pipeline data. Collection of site operational records, drawings, documents, and procedures is necessary to determine the current condition and configuration of the pipeline facility. Direction observation of the facilities is vital for data verification purposes. Facility configuration is recorded on photographs and video tape as well as being recorded by the field engineer on field sketches. Interviews with operations and maintenance personnel are vital. These people are able to fill in the gaps in the pipeline data and are key to identifying problem areas. Picking the brain of an operator who has been at the site for 30 years is far more valuable than any snapshot.

Data Compilation

Once the data has been obtained during the site assessment, it must be organized and maintained in a user friendly format. Geographical Information Systems (GIS) are being developed which can store all site collected information in an electronic media format which can be addressed and updated (as required) both in the field and at the desk of the decision maker. These GIS databases are an important management tool. The data obtained in the field is generally compiled as follows:

- General Information - The basic who, what, and where's of a pipeline facility.

- Physical Data - Specifications, Coatings, Bends, Valves, Branches, Fittings, Route, Terrain, Crossings, Cover, Endpoints, and Cultural Features.

- Operational Data - Pressures, Temperatures, Flowrates, Pumps, Operations and Maintenance, and Contingency Plans.

- System Control - Normal Control, Emergency Shutdown, Communications, and SCADA.

- Pig Traps - Size, Configuration, and Access.

- Pipeline History - Age, Construction Specification, Construction Records, Leaks, and Modifications.

- Inspections and Testing - Pressure Tests, In-line Inspections, Visual Inspections, Cathodic Protection Inspections, and Other Inspections/Testing.

- Geotechnical/Geological - Seismicity, Slope, Erosion, Soil Type, Groundwater, Freezing, Floods, Hurricanes, and Tornadoes.

- Environmental - Leaks, Remediations, Wetlands, Ecologically Sensitive Area, and Endangered Species.

- Other Observation - Items of Interest.

Integrity Assessment Planning

In the integrity assessment planning phase, it must be determined where money should be spent and how it should be spent. A Risk Assessment should be conducted to determine the urgency of the integrity assessment. There are many risk assessment methods available, but all methods generally follow the same flowpath as below:

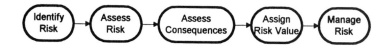

In the risk management block, a cost benefit analysis must be conducted taking into consideration the benefits of pipeline replacement versus pipeline repair and rehabilitation. Careful analysis must be made in this area since new construction pipelines may have different requirements than existing pipeline facilities. A 50 year old rehabilitated pipeline is still a 50 year old pipeline, so it is necessary to determine at what incremental cost increase is it beneficial to replace an existing pipeline system, taking into consideration the appropriate decommissioning costs of the existing pipeline.

Inspection Methods

The state of the art technology for pipeline inspection is incredible. There are electro-magnetic and ultrasonic technology tools which are available today. The inspection tools can be propelled by product flow or pulled through short segments by steel cables. Inspection sensitivities and capabilities can vary, but each basic type listed below (ultrasonic, conventional, and high resolution) has the following advantages and disadvantages:

TYPE	ADVANTAGES	DISADVANTAGES
ULTRASONIC	- Moderate cost - Internal & external evaluation - Accurate defect evaluation - Unlimited thickness - Insensitive to debris - Sensitivity near weld zones - High defect sensitivity	- Couplant required (liquid) - Limited endurance - Insensitive in bends - Long pig - Pipe must be clean - Can miss cracks - Narrow temp. band
CONVENTIONAL	- Inexpensive - Any fluid pipeline - Extensive track record - Good endurance - Moderate length - Moderate cleaning required - Insensitive to non-metal debris	- No defect evaluation - Limited thickness - Poor sensitivity near weld zones - Sensitive to metal debris - Insensitive in bends - Moderate sensitivity to shallow defects
HIGH-RESOLUTION	- Any fluid pipeline - Defect evaluation - Good endurance - Moderate length - Moderate cleaning required - Insensitive to non-metal debris - Sensitive in bends - Good track record - Sensitive near weld zones	- Expensive to run - Limited thickness - Sensitive to metal debris - Narrow temp. band - Moderate sensitivity to shallow defects

In those situations where it is inappropriate to conduct an in-line inspection, it is possible to obtain a representative assessment of the pipeline integrity by alternative methods, including:

- Visual Inspection - Internal (by camera) and external (through excavation and coating removal) inspection at typical and critical locations.

- Radiography - Inspection at typical and critical locations.

- Ultrasonic - Pipewall thickness measurement at typical and critical locations.

- Mechanical - Geometric and caliper measurements of the pipeline at typical and critical locations.

- Pressure Tests - Provides evaluation of current pressure holding capability and required for regulatory compliance.

- Cathodic Protection/Coating Survey - Provides an indication of coating condition and its ability to prevent corrosion.

Pipeline Inspection

Once the integrity assessment method has been selected, the pipeline must be configured prior to and during the pigging sequence to support the passage of the in-line inspection tool. These retrofits include replacement of restricted bore fittings and tight radius/miter bends.

The pigging sequence is generally as follows:

- Cleaning - Multiple runs may be necessary using increasingly rigid tools starting with sponge pigs.

- Scraping - These tools remove cleaning and other types of debris.

- Gauging - This provides an indication (but no location) of possible bore restrictions.

- Geometry - Measures the pipeline ovality, internal geometry, and bend radii of the entire pipeline alignment.

- Dummy Log - has the same dimensions as the live tool without the instrumentation an is used to ensure passage of the live tool.

- Live Tool - Once all retrofits and the preliminary pig runs are completed, there is a reasonable assurance that the instrumented tool will pass the entire length without becoming stuck.

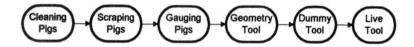

Pig tracking equipment and techniques should be utilized during the pigging sequence listed above. These methods assist in locating stuck equipment. Reasonable tracking capability can be obtained by leap frogging two detectors during the pigging run. Pig tracking and locating can be obtained by using one or more of the following methods:

- Magnetic Tracking - Magnetic Flux Leakage tools transmit a significant magnetic field which can then be detected/tracked by complex magnetic flux detectors or items as simple as a magnetic compass.

- Acoustic Tracking - As a mechanical tool passes through a pipeline it makes a reasonable amount of noise, again there are complex acoustic detectors and simple items such as using a stethoscope.

- Pig Signals - These intrusive and non-intrusive pig signal mechanisms give a positive indication of pig passage.

- Transit Time - The length of unobstructed pig run is approximately equal to the flow velocity times the duration of flow. The length may be adjusted by a small amount to account for bypass flow.

- Pressure Monitoring - Pressure gauges are installed at locations throughout the pipeline alignment. The pressure differential required to drive a heavy inspection tool through the pipeline is monitored at various points as the pig travels through the pipeline.

Integrity Assessment

Once the minimum pipewall thickness has been determined for the pipeline alignment, the remaining strength can be calculated using ASME B31G, "Manual for Determining the Remaining Strength of Corroded Pipelines". This allows possible continued operations at reduced pressure until repairs and rehabilitation are complete.

It is important to again consider repair versus replacement costs in those situations where there is extensive corrosion to the existing pipeline.

The risk assessment conducted as part of the planning process should be updated to reflect the urgency of completing pipeline repairs and rehabilitation.

Rehabilitation

The required method of rehabilitation/repair is determined by the extent of corrosion and the assessment of the risk of failure. In a sensitive area, a temporary repair can be completed to minimize risk until a permanent repair can be completed. These temporary repair methods include repair clamps, patches, and composite material reinforcement. The permanent repair methods are as follows:

- Pup replacement - This involves removing an entire section of pipe and replacing it with new material which is welded into place.

- Patch repair - This involves reinforcing a thin walled section of pipe with a steel patch welded in place which is larger than the corroded section.

- Weld repair - This involves adding fill material to small but deeply corroded sections (pits).

- Composites - This involves reinforcing thin walled sections with composite materials.

- Coating repairs - In the repairs listed above, the coating must be replaced over the repaired section to prevent future corrosion damage. For shallow corrosion defects, simple replacement of the coating material my be sufficient to stop the corrosion process.

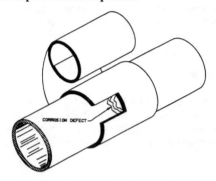

Conclusion

Congress has passed legislation to require that the federal pipeline safety regulations become stiffer in the future. The costs of an integrity assessment may be considered high, but the cost pales in comparison to the costs of a pipeline spill or a pipeline accident. The penalties for non-compliance with federal regulations include large fines and criminal penalties (including retention in federal prisons). The old saying, "An ounce of prevention is worth a pound of cure", may be a gross understatement in the case of pipeline safety. The pipeline integrity assessment process discussed satisfies the objectives of:

- Regulatory Compliance

- Pipeline Inventory Storage Requirements (GIS Database)

- Risk Assessment

- Cost Benefit Analysis

- Budgetary Planning

A well managed integrity assessment program minimizes and controls costly down time and can prevent costly cleanups from a pipeline spill or accident.

Subject Index
Page number refers to first page of paper

Accidents, 786
Acoustic detection, 126
Algorithms, 560
Alignment, 445
Aqueducts, 685
Arizona, 560
Axial forces, 627

Backfills, 13, 138, 650, 673
Bonding, 765
Boring, 456, 468
Buoyancy, 775
Buried pipes, 13, 25, 42, 65, 77, 102, 114, 126, 138, 194, 248, 260, 310, 334, 345, 375, 394, 445, 480, 512, 536, 650, 721

Calculators, 413
Case reports, 183, 501
Cast iron, 285, 298, 310, 322
Cathodic protection, 168, 183, 757, 765
Cements, 375, 734, 745
Centrifuge model, 102, 786
Chemical additives, 383
Chlorides, 168
Clays, 775
Combined sewers, 640, 709
Compaction, 673, 734
Computation, 413, 627
Computer aided drafting (CAD), 501
Computer models, 589
Computer software, 501
Computerized design, 501
Concrete pipes, 1, 150, 156, 168, 183, 272, 524, 560, 721, 837
Concrete, precast, 456
Concrete, prestressed, 524, 721, 765
Concrete, reinforced, 825
Connectors, mechanical, 697
Constraints, 685
Construction, 13, 217, 260, 489, 603, 798
Construction methods, 615, 697
Corrosion, 1, 89, 126, 560, 635
Corrosion control, 168, 183, 322, 745, 757, 765
Cost effectiveness, 603, 757
Cost savings, 813
Costs, 370, 845
Cracking, 1
Cracks, 572, 685
Creep, 345
Crossings, 65, 77, 194, 480, 798
Curing, 548, 640

Damage, 813, 837
Damage assessment, 1
Data collection, 501
Deflection, 126, 238, 260, 272, 445, 786
Deformation, 89, 114, 394, 536, 786
Desalination plants, 217
Design, 603
Design criteria, 260, 524, 825
Deterioration, 837
Displacement, 126, 205
Displacements, 345, 359
Drainage, 114, 825
Ductility, 285, 298, 383, 745, 765, 806

Earth pressure, 102
Earthquake damage, 205, 334, 359
Economic analysis, 370
Economic factors, 375
Effluents, 437
Elastomer, 150
Elastoplasticity, 42, 89, 512
Emergency services, 205
Encasements, 322, 757
Energy dissipation, 437
Equations of motion, 394
Erosion, 65
Erosion control, 77
Estimation, 194, 205

Failures, 65, 77, 89, 425, 721
Fiberglass, 272
Field tests, 138, 775
Fills, 775
Filters, 673
Finite element method, 650

Flexible pipes, 13, 248, 272, 359, 445
Flooding, 456
Floods, 640, 673
Friction factor, 413
Frost penetration, 650
Frozen soils, 345

Gas pipelines, 54, 89, 536
Geotechnical engineering, 217, 813
Germany, 1

Heat treatment, 383
High strength steel, 406
Hydraulic transients, 425, 468
Hydraulics, 217
Hydrogen sulfide, 635, 837

Information, 813
Infrastructure, 548, 603
Inspection, 25, 126, 589, 721, 845
Installation, 42, 260, 310, 322, 489, 697, 775
Intercepting sewers, 798, 837
Iron, 359, 745, 806

Joints, 359, 572

Landfills, 114, 825
Landslides, 226, 798
Laws, 845
Leachates, 114
Leakage, 183, 685, 845
Liners, 150
Liquefaction, 226, 359
Live loads, 194
Loads, 54, 114, 345, 480, 627, 697
Losses, 205, 413

Maintenance, 572, 721
Manholes, 589, 635, 837
Masonry, 709
Materials, properties, 383
Matrices, mathematics, 512
Mechanical properties, 383
Microtunneling, 25, 310, 480, 603, 627
Mississippi River, 709
Mixing, 437
Models, 42, 205

Moisture content, 734
Monitoring, 589
Mortars, 375, 734, 745

Natural gas, 536

Ocean disposal, 437, 468
Oil pipelines, 89
Outfall sewers, 437, 468
Overflow, 709

Performance, 138, 248, 640, 650
Permits, 662
Pipe bedding, 25, 775
Pipe design, 25, 194, 248, 298, 394, 406, 413, 489, 501, 524, 548
Pipe flow, 425
Pipe joints, 150, 697, 806
Pipe lining, 375, 635, 745
Pipeline design, 226, 662
Pipelines, 25, 65, 77, 102, 205, 334, 345, 359, 375, 413, 489, 548, 603, 627, 673, 798, 845
Pipes, 238, 285, 310, 322, 589, 627, 640, 709, 825
Piping systems, 54, 322
Planning, 205, 662
Plastic pipes, 126, 138, 248, 260, 548, 786
Plastics, 635, 745
Polyethylene, 114, 238, 260, 322, 757, 765, 786
Polyvinyl chloride, 138
Potable water, 685
Predictions, 89, 248
Pressure pipes, 168, 183, 298, 310, 734
Pressures, 425
Prestressing, 156, 168
Probabilistic methods, 334
Programming, 413
Projects, 615, 662
Protective coatings, 375, 734, 745, 765
Pull-out resistance, 102
Pumping stations, 709

Quality assurance, 734, 813

Railroads, 798

SUBJECT INDEX

Rainfall, 786
Regulations, 845
Rehabilitation, 25, 489, 548, 560, 615, 635, 640, 697, 709, 825, 837, 845
Renovation, 1, 156, 572, 635
Repairing, 334, 685, 721
Research, 480
Restraint systems, 285, 806
Retrofitting, 183
Rings, 150, 238, 285, 445
Risk analysis, 334
River crossings, 226
Roads, 194
Roughness, 194
Rubber, 150

Safety, 102, 775
Safety analysis, 845
San Francisco, 615
Sanitary sewers, 370, 456, 560, 572, 615, 837
Saudi Arabia, 217
Scale models, 114
Scheduling, 156
Scour, 65, 77
Sea water, 217
Sealing, 150, 285
Seismic effects, 685
Seismic hazard, 334, 359
Selection, 370
Service life, 156
Sewage disposal, 709
Sewer pipes, 260, 370, 489, 560, 806
Sewers, 1, 126, 589, 615, 635, 673
Shotcrete, 456, 640, 709
Slip, 825
Soil mechanics, 272
Soil pressure, 42
Soil saturation zones, 786
Soil-pipe interaction, 25, 248, 272, 345, 445
Soils, 445
Soil-structure interaction, 42, 512
Standards, 298, 322, 524
Statistical models, 627
Steel pipes, 272, 383, 394, 406, 697, 734
Stiffness, 13, 238
Storage, 456

Stormwater, 825
Strain, 394, 536
Stress, 54, 89, 394
Stress relaxation, 238
Structural analysis, 42, 512, 536
Structural behavior, 536
Structural reliability, 640
Structural response, 1
Structural stability, 548
Subsidence, 673
Subsurface investigations, 813
Supports, 54
Sweden, 238

Technology, 480
Tests, 102
Thermal analysis, 650
Thickness, 298, 406
Thin wall structures, 13
Thrust, 806
Transfer functions, 512
Trenches, 445, 650, 673, 798
Tunnel linings, 456
Tunnels, 456, 468, 489

Underground conduits, 480, 536, 757, 813
Underground structures, 548, 572
United Kingdom, 248
Uplift, 345
Utilities, 757, 813
Utility corridors, 572

Valves, 437, 468
Vehicle impact forces, 194
Vibration, 54
Virginia, 156

Wastewater, 168
Wastewater disposal, 589
Wastewater management, 589
Wastewater treatment, 437, 468
Water pipelines, 138, 156, 168, 217, 226, 383, 406, 425, 501, 524, 662, 721, 765, 806
Water quality, 685
Water supply systems, 138, 310, 650, 662

Water transportation, 425
Watersheds, 226, 615
Waterways, 65, 77
Welds, 285
Width, 445

Yield strength, 406

Zinc, 375

Author Index
Page number refers to the first page of paper

Al-Amry, Mohammed, 217
Al-Najdi, Saleh, 217
Al-Shaikh, Abdullah, 25, 217
Anderson, Kenneth W., 150
Anspach, James H., 813
Argent, Michael E., 697

Balasubramaniam, Bala K., 25
Ballantyne, Donald, 205
Bardakjian, Henry, 734
Barsoom, Joseph, 572
Beieler, Roger, 226
Beiler, Roger, 662
Bonds, Richard W., 322
Bornmann, Andreas, 1
Bradish, Bryan M., 156, 765
Bradley, Jeffrey B., 65, 77
Brooks, Peter, 845
Butler, John, 775

Card, Robert J., 406
Carpenter, Ralph R., 310
Carreon, Samuel, 65, 77
Chase, Donald V., 425
Chua, Ken, 456
Ciolko, Adrian T., 721
Collins, Marie A., 709
Conner, Randall C., 285
Craft, Gerald A., 757
Cramblitt, Kenneth L., 721
Cronin, Roger J., 156
Croxton, Richard, 310

Dawson, Karen, 226
Dechant, Dennis A., 406
Diab, Y. G., 480
Dillingham, James H., 425
Dodge, Christopher, 685
Doeing, Brian J., 65

Edmondson, Samuel A., 560
Essex, Randall J., 603

Faragher, E., 248
Fleming, P. R., 248

Fu, Bin, 89

Galeziewski, Thomas M., 560
Garcia, Jorge A., 413, 501
Garrett, Thomas M., 627
Gartung, Erwin, 114
Garvey, Jeff, 798
Gemperline, Mark C., 194
Gerbault, Marcel, 42, 512
Gibson, Craig, 662
Goodrich, Laurel, 650
Guastella, David E., 425

HajAli, Rami M., 697
Hall, Sylvia C., 168
Hamada, T., 786
Hanna, Ernest R., 825
Hartinger, Christina, 685
Hashmi, Quazi S. E., 673
Helfrich, Steven C., 673
Hermanson, Glenn E., 615
Hinobayashi, J., 786
Honegger, Douglas G., 334
Horn, L. Gregg, 298
Horton, A. M., 745
Howard, Amster, 272
Hu, J., 345

Inuki, M., 786
Islam, Mohammed S., 673

Janson, Lars-Eric, 238
Jeyapalan, Jey K., 25
Joyce, Charles W., 615
Jurgens, John, 25
Jurgens, John F., 837

Kawabata, T., 13
Kelly, John, 456
Kentgens, Susanne, 1
Kirkwood, Mike G., 89
Klein, Stephen J., 603
Kormann, Phil, 54
Kuraoka, Senro, 138

Lamb, E. C., 370
Larson, Tim, 662
Lawson, Thomas J., 721
Lee, Chih-Hung, 536
Lewis, Richard O., 156
Li, L., 102, 786
Loeppky, M. W. J., 248
Lys, Robert, 775
Lys, Robert, Jr., 627

Mathew, Ivan, 168
McGregor, James D., 589
Mohri, Y., 13
Moore, Gary T., 640
Mueller, Richard I., 183, 524
Murphy, Stephanie, 226

Najafi, M., 548
Nance, Charles H., 640
Noonan, James R., 765

Oskoui, Afshin, 798

Pecknold, David A., 697
Petroff, Larry J., 260
Prevost, R. C., 394
Price, Ted, 126

Qaqish, Awni, 425

Rajani, Balvant, 138, 650
Rhodes, George F., 806
Robinson, William C., 375
Rogers, C. D. F., 248

Saleira, Wesley E., 25
Sandvik, Arne, 437
Scarpine, Allan J., 825
Selvadurai, A. P. S., 345
Seymour, F. Stuart, 468
Shook, William E., 635
Sikora, Edward J., 775
Siller, Thomas J., 194
Smith, Mark, 845
Spruch, Arthur A., 489
Stein, Dietrich, 1
Struzziery, John J., 489
Stude, Carl Ted, 709
Swanson, Curtis W., 615

Teal, Martin J., 77
Timmermann, David, 437
Tohda, J., 102, 786
Torabi, Hans, 437
Trembath, Richard, 468
Tseng, Wen-Shou, 536
Tucker, Michael S., 359
Tupac, George J., 383
Turnipseed, Stephen P., 183

Varma, V. K., 548

Watkins, Reynold King, 445
Webb, Robert, 560
Willoughby, Thomas, 468

Zanzinger, Helmut, 114
Zhan, Caizhao, 650
Zhou, Z. Joe, 54